# Ultra High Energy Cosmic Rays (UHECR 2022)



EPJ Web of Conferences Volume 283 (2023)

L'Aquila, Italy
3 - 7 October 2022

**Editors:**

**I. De Mitri**
**F.C.T. Barbato**
**D. Boncioli**
**C. Evoli**
**G. Pagliaroli**
**F. Salamida**

ISBN: 978-1-7138-7337-2

**Printed from e-media with permission by:**

Curran Associates, Inc.
57 Morehouse Lane
Red Hook, NY 12571

**Some format issues inherent in the e-media version may also appear in this print version.**



For additional information, please contact EDP Sciences – Web of Conferences at the address below.

EDP Sciences – Web of Conferences
17, Avenue du Hoggar
Parc d'Activité de Courtabœuf
BP 112
F-91944 Les Ulis Cedex A
France

Phone: +33 (0) 1 69 18 75 75
Fax: +33 (0) 1 69 28 84 91

contact-edps@webofconferences.org

**Additional copies of this publication are available from:**

Curran Associates, Inc.
57 Morehouse Lane
Red Hook, NY 12571 USA
Phone: 845-758-0400
Fax: 845-758-2633
Email: curran@proceedings.com
Web: www.proceedings.com

# TABLE OF CONTENTS

# Statement of Peer review

In submitting conference proceedings to *Web of Conferences*, the editors of the proceedings certify to the Publisher that

1. They adhere to its **Policy on Publishing Integrity** in order to safeguard good scientific practice in publishing.
2. All articles have been subjected to peer review administered by the proceedings editors.
3. Reviews have been conducted by expert referees, who have been requested to provide unbiased and constructive comments aimed, whenever possible, at improving the work.
4. Proceedings editors have taken all reasonable steps to ensure the quality of the materials they publish and their decision to accept or reject a paper for publication has been based only on the merits of the work and the relevance to the journal.

Title, date and place of the conference

Ultra High Energy Cosmic Rays 2022 (UHECR 2022)
L'Aquila, Italy, October 3-7, 2022

Proceedings editor(s):

I. De Mitri, F.C.T. Barbato, D. Boncioli, C. Evoli, G. Pagliaroli, F. Salamida

Date and editor's signature

The 6th international symposium on Ultra High Energy Cosmic Rays (UHECR) was held in L'Aquila, Italy, from October 3rd to October 7th, 2022 (www.uhecr2022.gssi.it) hosted by the Gran Sasso Science Institute (GSSI).

The symposium was jointly organized by GSSI, INFN - Laboratori Nazionali del Gran Sasso (LNGS) and the Department of Physical and Chemical Sciences of the University of L'Aquila.

It followed the previous five editions of the series that started in Nagoya in 2010 and continued in CERN (2012), Springdale, Utah (2014), Kyoto (2016) and Paris (2018).

More than 140 participants attended and successfully contributed to the symposium, discussing recent results from UHECR observations, theoretical developments, and future plans. Invited review talks were given on the following topics:

- UHECR sources
- Astrophysics of CR accelerators
- CR acceleration processes
- Magnetic fields and UHECR propagation
- Transition from galactic to extragalactic CRs
- Results from high energy direct measurements and future prospects
- Multi-messenger connections of UHECRs
- Hadronic interactions studies at LHC
- Beyond Standard Model physics with UHECRs

In addition, contributed plenary talks covered all the main aspects of the field and a poster session completed the program.

The symposium was attended by key scientists in the field as well as by PhD and graduate students, in a friendly atmosphere that also encouraged lively and informal discussions among all the participants.

The session of Wednesday was held in the auditorium of INFN LNGS and a visit to the underground facility was organized on the same day.

A dedicated committee selected the two best posters presented at the symposium. The awards have been assigned to Teresa Bister and Hiroko Miyamoto (from the Pierre Auger and the JEM-EUSO collaborations, respectively).

During the final editing phase of this volume we received the sad news that Veniamin Sergeyevich Berezinsky, one of the members of the International Advisory Committee of the UHECR conference series, passed away.

V.S. Berezinsky was one of the pioneer investigators of the origin of UHECRs, and a leading figure in the field of theoretical astroparticle physics at large. His seminal article written with Zatsepin in 1969 on the so-called cosmogenic neutrinos, inspired generations of scientists working in this field. His interests ranged on a wide spectrum,

https://doi.org/10.1051/epjconf/202328300001

from supersymmetry to topological defects, from neutrinos and gamma rays to cosmic rays in a variety of situations (supernovae, clusters of galaxies, and many others).

He spent the initial part of his scientific career in the former Soviet Union, from his graduation in 1961 to the beginning of his Italian adventure in 1991. He carried out his research work at the Lebedev Institute of Physics and the Institute of Nuclear Research. In 1991 he moved to Italy and was the founder of the theoretical astroparticle physics group at the Gran Sasso National Laboratory of INFN and later a professor at the Gran Sasso Science Institute, since its very inception.

Many of us have learned a lot from working, interacting and even just talking with Venya. He will be sorely missed.

L'Aquila, April 2023

Ivan De Mitri

for the Local Organizing Committee

# The Snowmass UHECR White Paper on Ultra-High-Energy Cosmic Rays

*Frank G.* Schroeder[1,2,*], *Alan* Coleman[1,3], *Johannes* Eser[4], *Eric* Mayotte[5], *Fred* Sarazin[5], *Dennis* Soldin[1,6], and *Tonia M.* Venters[7] for the co-conveners and co-authors of the Snowmass 2021 UHECR Whitepaper[8]

[1]Bartol Research Institute, Department of Physics and Astronomy, University of Delaware, Newark, DE 19716, USA
[2]Institute for Astroparticle Physics (IAP), Karlsruhe Institute of Technology (KIT), 76021 Karlsruhe, Germany
[3]Department of Physics and Astronomy, Uppsala University, Uppsala, SE-752 37, Sweden
[4]Department of Astronomy and Astrophysics, University of Chicago, Chicago, IL, USA
[5]Department of Physics, Colorado School of Mines, Golden, CO, USA
[6]Institute of Experimental Particle Physics (ETP), Karlsruhe Institute of Technology (KIT), 76021 Karlsruhe, Germany
[7]Astroparticle Physics Laboratory, NASA Goddard Space Flight Center, Greenbelt, MD, USA
[8]See reference [1] for the complete list of authors and affiliations.

**Abstract.** This proceeding summarizes the talk given at the opening of the UHECR 2022 conference in L'Aquila on the whitepaper 'Ultra-High-Energy Cosmic Rays: The Intersection of the Cosmic and Energy Frontiers' [Astroparticle Physics 149 (2023) 102819 - arXiv:2205.05845] that has been prepared for the Snowmass survey in the USA. The whitepaper provides an overview of recent progress and open questions regarding the particle physics and astrophysics related to ultra-high-energy cosmic rays (UHECR) and outlines the connections between the particle and astrophysics aspects of cosmic rays. It also discusses what instrumentation is needed to address the major scientific questions in ultra-high-energy cosmic-ray physics. While the upgraded Pierre Auger Observatory and Telescope Array will remain the workhorses at the highest energies in the current decade, new experiments with significantly higher exposure are needed in the coming decade. Ground arrays featuring simultaneous detection of the position of the shower maximum and the size of the muonic component will enable particle astronomy by measuring the rigidity of individual events. They should be complemented by other detectors maximizing the total exposure. This can be achieved by a few next-generation experiments using the latest developments in detection and analysis techniques: GRAND as a ground-based radio array, and POEMMA as a space-borne stereo fluorescence telescope will feature complementary approaches to provide maximum exposure; IceCube-Gen2 with its surface array, and GCOS aim at increased statistics with high accuracy for particle physics and rigidity-based galactic and extra-galactic astrophysics. While designed to discover the astrophysical cosmic-ray sources at the highest energies, the same experiments also contribute to particle physics, e.g., by studying the muon puzzle in cosmic-ray air showers, and by their discovery potential for exciting new physics, such as certain Dark Matter candidates. With the full whitepaper available as a reference, this proceeding will briefly present the science cases of the experiments, highlighting their individual strengths and outlining how they complement each other.

## 1 Introduction

Snowmass is a 'particle physics community planning exercise' which is run every decade in the United States to discuss the scientific questions and experiments of the coming decades in order to inform U.S. funding agencies [2]. Cosmic-ray physics is part of the Snowmass 'Cosmic Frontier' [3] topical group 7 'CF7 - Cosmic Probes of Fundamental Physics' [4], whose conveners solicited a whitepaper about the status and future of ultra-high-energy cosmic rays (UHECR). By eliciting the help of the international UHECR community, we have prepared the requested whitepaper, which meanwhile has been published in Astroparticle Physics [1].

The whitepaper presents the recent progress and open questions on both the particle physics and astrophysics side of UHECR physics, and additionally summarizes relations to other fields of science. Earlier, for the Astro2020 survey in the U.S., two shorter whitepapers were published with a more specific focus on Galactic and extragalactic cosmic rays, respectively [5, 6]. While these are still worth reading, the Snowmass whitepaper is more comprehensive and is the result of a more extensive community process, including input from the major UHECR collaborations and individual scientists through several small online workshops. Finally, scientists who did not contribute directly were given the option to comment on and endorse the whitepaper. As the whitepaper reflects a wide community effort, it is highly recommended to read at least its executive summary [1].

Since that executive summary of the whitepaper stands on its own, this proceeding does not provide a redundant full summary, but instead provides an overview on

*e-mail: fgs@udel.edu

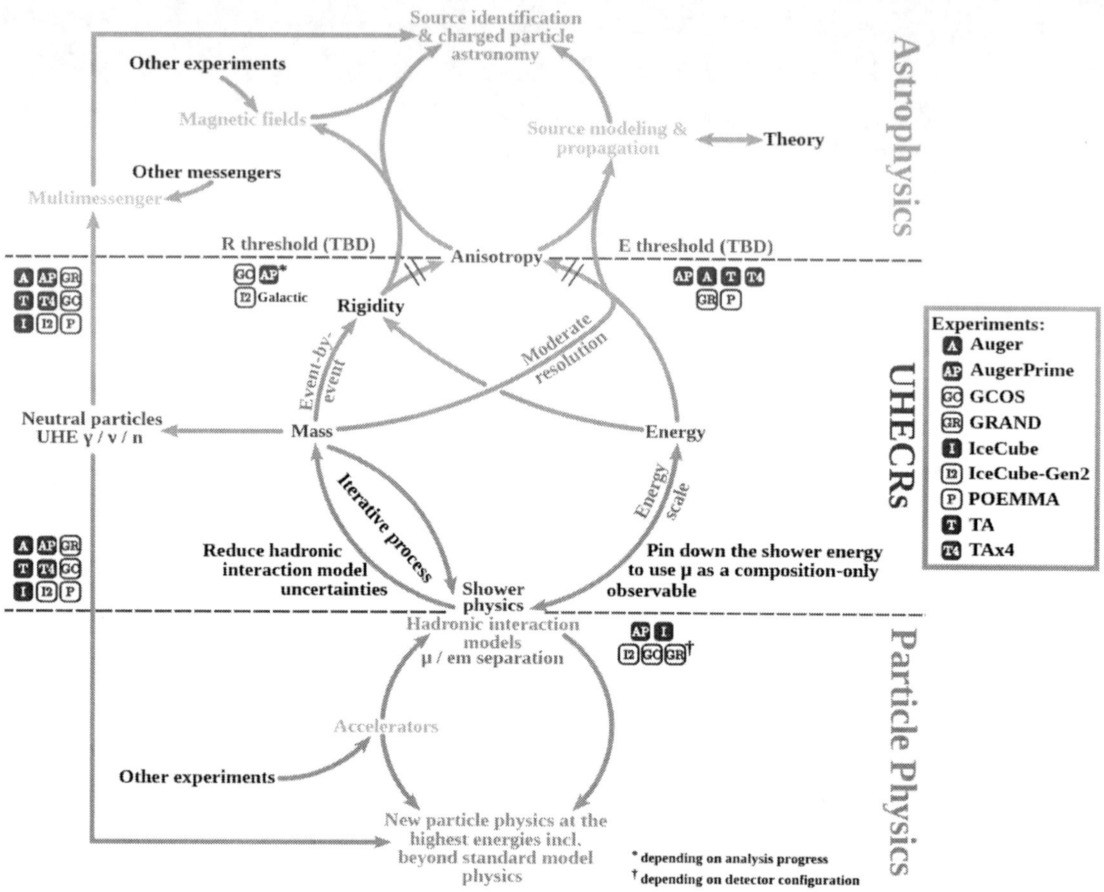


**Figure 1.** Sketch of how ultra-high-energy cosmic rays (UHECRs) are related to astrophysics and particle physics, and how current and future experiments contribute to each of these areas of physics. The mass and energy of the primary UHECRs are derived from air-shower measurements. These air-shower measurements are also used to study hadronic interactions and complement accelerator experiments, particularly in the forward region. Improvements of hadronic interaction models, in turn, will increase the accuracy of the mass derived from the measured air showers. Depending on the mass resolution, the astrophysics question of the UHECR sources is targeted either by rigidity-enhanced anisotropy measurements, which potentially enables particle astronomy by back-tracking through the magnetic fields in space, or by statistically comparing source models to measured anisotropy and mass composition. Finally, the search for neutral EeV particles, such as photons and neutrinos, enables a multi-messenger access to the UHECR sources and offers discovery potential for new particle physics beyond the standard model. (Figure from Ref. [1])

two particular aspects discussed in the whitepaper. These are connections between the particle and astrophysics aspects of UHECR and the major experiments planned to address open questions regarding UHECR physics. Nonetheless, also for these topics, the whitepaper itself contains a wealth of valuable information that are not presented here in the same detail.

*Definition of ultra-high-energy cosmic rays (UHECR)*: UHECR are typically defined as cosmic rays above an arbitrary energy threshold. Often, 1 EeV is chosen as the threshold because of a prefix step when using the unit eV for energy, although no apparent physics change is happening at this energy. For the whitepaper, we have instead chosen approximately $10^{17}$ eV, which is close to the heavy/second knee(s) in the energy spectrum [7–10], where presumably the transition to extragalactic cosmic ray starts.

## 2 UHECR: Particle Physics and Astrophysics at the Highest Energies

Ultra-high-energy cosmic rays are the most energetic particles known in the Universe. In the past decades, we have learned that UHECR are atomic nuclei of a mixed mass composition [11, 12]. Neutral particles, such as photons or neutrinos have yet to be discovered at EeV energies, setting strong limits on their flux [13–15]. Together with the UHECR anisotropy discovered above 8 EeV [16], these experimental results provide evidence that (at least the majority of) UHECR are not the decay products of some un-

 https://doi.org/10.1051/epjconf/202328301001

| Experiment | Feature | Cosmic Ray Science* | Timeline |
|---|---|---|---|
| Pierre Auger Observatory | Hybrid array: fluorescence, surface $e/\mu$ + radio, 3000 km$^2$ | Hadronic interactions, search for BSM, UHECR source populations, $\sigma_{p\text{-}Air}$ | AugerPrime upgrade |
| Telescope Array (TA) | Hybrid array: fluorescence, surface scintillators, up to 3000 km$^2$ | UHECR source populations proton-air cross section ($\sigma_{p\text{-}Air}$) | TAx4 upgrade |
| IceCube / IceCube-Gen2 | Hybrid array: surface + deep, up to 6 km$^2$ | Hadronic interactions, prompt decays, Galactic to extragalactic transition | Upgrade + surface enhancement; IceCube-Gen2 deployment; IceCube-Gen2 operation |
| GRAND | Radio array for inclined events, up to 200,000 km$^2$ | UHECR sources via huge exposure, search for ZeV particles, $\sigma_{p\text{-}Air}$ | GRANDProto 300; GRAND 10k; GRAND 200k multiple sites, step by step |
| POEMMA | Space fluorescence and Cherenkov detector | UHECR sources via huge exposure, search for ZeV particles, $\sigma_{p\text{-}Air}$ | JEM-EUSO program; POEMMA |
| GCOS | Hybrid array with $X_{max}$ + $e/\mu$ over 40,000 km$^2$ | UHECR sources via event-by-event rigidity, forward particle physics, search for BSM, $\sigma_{p\text{-}Air}$ | GCOS R&D + first site; GCOS further sites |
| | | | 2025 2030 2035 2040 |

*All experiments contribute to multi-messenger astrophysics also by searches for UHE neutrinos and photons; several experiments (IceCube, GRAND, POEMMA) have astrophysical neutrinos as primary science case.

**Figure 2.** Current and future experiments leading the field of ultra-high-energy cosmic rays. To some extent, all experiments contribute to both astrophysics and particle physics related to UHECR. Nonetheless, they have different strengths and complement each other: the Pierre Auger Observatory currently provides the highest measurement accuracy and largest exposure for UHECR and is in the southern hemisphere; the upgraded Telescope Array will provide a similar exposure at the highest energies in the northern hemisphere; IceCube(-Gen2) with its surface and in-ice arrays is a unique laboratory to measure air showers up to a few EeV, i.e., up to the energy range of the Galactic-to-extragalactic transition; GRAND and POEMMA will offer the highest statistics for UHECR with very complementary approaches and will additionally target UHE neutrinos; GCOS will be optimized for high statistics and high accuracy at the same time to provide an event-by-event rigidity measurement of UHECR. (Figure from Ref. [1])

known superheavy particles, but originate from natural accelerators, and are, at the highest energies, of extragalactic origin.

The composition of UHECR is lightest in the energy range of the ankle of the cosmic-ray energy spectrum at about 5 EeV [12], which implies that the fraction of protons is high at the ankle energy and decreases towards higher energies. Protons are not excluded making up a small fraction of the UHECR flux at the highest energies, but as of yet, no strong evidence exists for protons at energies significantly above $10^{19}$ eV. Therefore, although several air showers beyond $10^{20}$ eV have been measured, the highest energy per nucleon might be limited to around $10^{19}$ eV, and the highest rigidity of UHECR reaching Earth might be around 10 EV.

For astrophysics, this situation makes reaching the goal of particle astronomy with UHECR more challenging, since the magnetic deflection at 10 EV is still significant. Consequently, if really no particles would exist with rigidities significantly about 10 EV, then particle astronomy cannot be achieved by simply collecting higher statistics at the highest energies, but will additionally require tracing individual UHECR deflections back through the Galactic and extragalactic magnetic fields. On the one hand, this means the magnetic fields will need to be known well enough, which will possibly be achieved during the coming decade. On the other hand, particle astronomy will be difficult without an approximate measurement of the rigidity of the individual cosmic rays and, thus, per-event mass sensitivity at least for the energy range from $10^{19}$ eV to several $10^{19}$ eV.

Possibly, the composition will be nearly pure at extremely high energies around and above $10^{20}$ eV, i.e., dominated by nuclei of one mass group. In that scenario, once the mass composition around $10^{20}$ eV is measured statistically, the rigidity can easily be determined from the energy alone, and per-event mass sensitivity will be unnecessary at the highest energies. At lower energies, higher statistics will even help without knowing the mass on per-event level, as the average mass composition and anisotropy can be tested against theoretical models. Therefore, while it is mandatory to have at least one UHECR observatory featuring event-by-event rigidity resolution below $10^{20}$ eV, the need for event-by-event rigidity resolution above $10^{20}$ eV is not yet clear and depends on whether or not UHECR have a mixed or pure mass composition at the highest energies.

Given the unknown mass composition above $10^{20}$ eV, it is advisable to have at least one UHECR observatory of each type, i.e., at least one observatory that maximizes the exposure at the highest energies, and one observatory that features per-event mass sensitivity enabling rigidity-based particle astronomy. In any case, full sky coverage for UHECR is important, which requires a space-borne observatory or at least two sites for ground arrays.

For particle physics, even $10^{19}$ eV is far beyond the reach of accelerator experiments. In addition to guaranteed progress in particle physics, such as the measurement of the proton-air cross-section [17, 18], UHECR also provide a probe for new physics, such as the search for Lorentz invariance violation [19], or super-heavy Dark Matter [20, 21]. Moreover, cosmic-ray air showers will continue to be a unique probe of hadronic interactions in the forward region, a phase space which is difficult to cover in accelerator experiments.

Of particular interest in this respect is the study of the muon puzzle. Several UHECR air-shower experiments have measured with high significance that Monte Carlo simulations of air showers using state-of-the-art hadronic interaction models predict fewer muons than measured by these experiments. The discrepancy increases with energy [22], and there are tensions between hadronic interaction models and measurements also regarding other muon-related observables, such as the dependence on zenith angles [23] or the muon production depth [24]. More accurate air-shower measurements are needed to test the predictions of improved hadronic interaction models. Air-shower arrays also offer many synergies with accelerator experiments, including, but not limited to, the investigation of the muon puzzle [25].

Finally, a better understanding of the particle physics in air showers also has a positive impact on UHECR astrophysics. This is because the hadronic interaction models are used to interpret air-shower measurements, and better models will improve the accuracy of the air-shower reconstruction, in particular, regarding the mass of the primary particle. In a virtuous circle, air-shower detectors can iteratively contribute to a better understanding of the particle physics in these showers. Then, this better understanding improves the quality of the cosmic-ray mass reconstructed from air-shower measurements, which in turn can be used to test for finer discrepancies between the hadronic interaction models and the measurements. Consequently, also the astrophysics science related to UHECR benefits from progress on the particle physics side.

## 3 Experiments for the next Decades

This section provides an overview of current and future UHECR experiments that are needed to address the particle and astrophysics questions. In addition to the running experiments Auger, TA, and IceCube with their ongoing/planned enhancements AugerPrime, TAx4, and IceCube-Gen2, three future experiments have been identified as being crucial for UHECR science in the coming decade. These are the Giant Radio Array for Neutrino Detection (GRAND), the Probe of Extreme Multi-Messenger Astrophysics (POEMMA), and the Global Cosmic Ray Observatory (GCOS), which complement each other in detection techniques and science goals. Fig. 2 contains those experiments that have been identified in the whitepaper as the major efforts needed in the field. This list is not complete. It does not contain experiments for lower cosmic-ray energies, and there can be additional experiments making a unique contribution to a specific aspect of UHECR physics. Except for IceCube-Gen2 and GRAND, all of the major experiments were represented by dedicated contributions at the conference, which is why this section only provides a short overview, instead focusing on the complementary approaches to the particle and astrophysics sides of UHECR science.

With their upgrades close to completion, the **Pierre Auger Observatory** in Malargüe, Argentina [26], and the **Telescope Array (TA)** in Utah, USA [27] will remain the workhorses of UHECR physics until at least the end of the decade. The upgrade of TA, called **TAx4**, primarily increases the size of TA: at the highest energies, TA will feature an aperture in the northern hemisphere similar to the aperture of Auger in the southern hemisphere. The upgrade of Auger, called **AugerPrime**, does not change the size of the surface array, but rather increases the measurement accuracy for individual showers, which enables a number of new analyses, such as mass-sensitive anisotropy studies or more stringent tests of hadronic interaction models. For recent results and the status of the upgrades of Auger and TA, please check the dedicated contributions

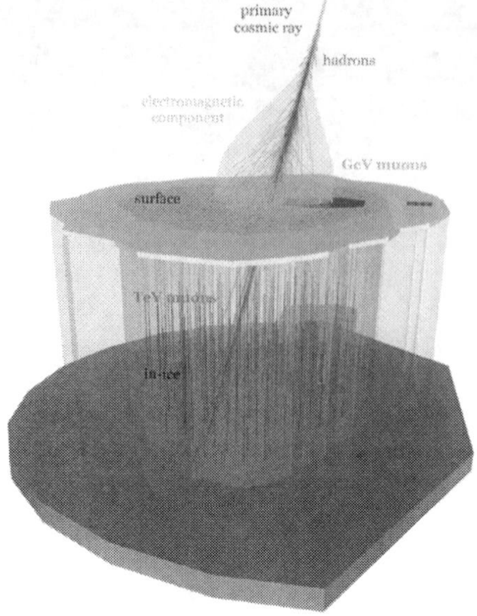


**Figure 3.** IceCube-Gen2, though optimized for neutrino astronomy, will also be a unique lab to study particle physics in air-showers because a surface measurement of the electromagnetic component and low-energy muons is combined with a high-energy muon measurement in the ice. Compared to the current IceCube detector with its surface array IceTop (in red), the larger IceCube-Gen2 detector (light blue) will measure surface-deep coincidences for a wider range of zenith angles. Together with the larger surface area, this will result in an overall 30x increase of aperture for such surface-deep coincident events. [Figure courtesy of the IceCube-Gen2 collaboration [28]]

presented at this conference. Important in the context of next-generation experiments is that AugerPrime will pioneer the concept of a detector dedicated to an event-by-event mass sensitivity, realized by adding scintillators (for vertical and mildly inclined events) and radio antennas (for very inclined events) to the water-Cherenkov detectors of the surface array.

Built primarily for neutrino astronomy, **IceCube** also is a unique lab for cosmic-ray air showers due to the combination of a surface array and a deep optical detector [29], and makes essential contributions to the physics of the most energetic Galactic cosmic rays in the energy range up to around one EeV. With **IceCube-Gen2** [30], both arrays will be extended, and a new design for the surface array will enhance the accuracy. Mitigating the problem of snow accumulation above the existing ice-Cherenkov tanks of the IceTop surface array, the IceCube-Gen2 surface array will consist of elevated scintillators and elevated radio antennas [31, 32]. The latter provides a measurement of energy of the electromagnetic shower component [33–35] and of $X_{\max}$ [36–39], which enhances the total accuracy for studies of the cosmic-ray mass composition. In combination with the high-energy muons from the in-ice optical detector, IceCube-Gen2 will feature per-event mass sensitivity for the energy range of the Galactic-to-extragalactic transition. Through its higher aperture, it will extend IceCube's anisotropy measurements from currently a few PeV [40] to around 100 PeV.

In addition to the astrophysics of the most energetic Galactic cosmic rays, IceCube-Gen2 provides distinct capabilities to study particle physics in air showers. This is because the deep optical array in the ice can measure the TeV and PeV muons produced in the same shower detected by the surface array (see Fig. 3). Through its larger area and larger range of zenith angles for surface-deep coincidences, IceCube-Gen2 will have a thirty times larger aperture for such cosmic-ray events measured simultaneously in the surface and deep arrays. These can be used to study atmospheric leptons in general, such as the muon puzzle, and specifically also prompt decays, such as muons originating from charm decays. Studying hadronic interactions is a science goal by itself. Nonetheless, improved knowledge on the hadronic interactions and particularly on the flux of prompt leptons will also lower uncertainties on atmospheric backgrounds for neutrino astronomy, and it will help in general to interpret air-shower measurements more accurately.

Other experiments reaching into the energy range above 100 PeV are **LHAASO**, potentially **SWGO**, and the **SKA**. While SWGO [41] and LHAASO [42] have gamma astronomy as their primary target, they will also make important contributions to cosmic-ray physics with their ultra-dense arrays and, in the case of LHAASO, through the simultaneous measurements of air showers by several detection techniques. The low-frequency array of the Square Kilometer Array (SKA-low) will feature several 10,000's of radio antennas and enable unprecedented precision on the radio signal of air showers [43]. While radio detection of air showers has become a standard technique [44, 45], e.g., for energy and $X_{\max}$ measurements, SKA can measure finer details of the shower profile beyond $X_{\max}$ [46]. On the one hand, this increases the accuracy of the proton and Helium fractions among the cosmic rays around the second knee energy; on the other hand, the ultra-high precision enables new ways to test hadronic interaction models, as these differ not only in the number and distribution of muons, but also predict slightly different longitudinal profiles of the electromagnetic component [46]. Therefore, even without the capability to measure high-energy muons, the SKA will make an important complementary contribution to cosmic-ray astrophysics and particle physics below 1 EeV.

For the highest energies, GCOS will provide high measurement accuracy for air showers in combination with a larger full-sky aperture. It will be complemented by two experiments aiming at huge full-sky exposure of UHECR that exceeds the currently leading Pierre Auger Observatory by more than an order of magnitude: POEMMA and GRAND.

**POEMMA** [47] will consist of two space-borne fluorescence telescopes for stereo observation of air showers. The goals of POEMMA are searching for UHE neutrinos and measuring UHECR with unmatched statistics. The fluorescence technique is mature and provides high accuracy for energy and $X_{\max}$, which translates into high precision for the average mass composition. Building on the experience of the JEM-EUSO program [48, 49], POEMMA will bring key advances to multi-messenger astronomy at the highest energies.

**GRAND** [50] will be a giant radio array aiming at UHE neutrino as primary science case. Nonetheless, if built to its full size of about 200,000 antennas distributed among several sites totalling 200,000 km$^2$, GRAND will provide a similar exposure for UHECR as POEMMA. During the next years, further R&D is required to apply radio detection as a stand-alone technique on such a large scale, and the radio emission of nearly horizontal showers needs to be understood in further detail. For this purpose, various prototypes of GRAND are planned, and dedicated simulation studies are performed. Although from the UHECR perspective the science goals are similar to POEMMA, GRAND is very complementary in the approach (huge ground array vs. space-borne stereo telescope, and promising radio vs. established fluorescence technique).

Finally, **GCOS** [51] will be the UHECR ground array that will be designed for the maximum per-event measurement accuracy feasible, enabling rigidity-enhanced particle astronomy. The exact design of GCOS is currently under discussion in the community, with the plan to conduct necessary R&D and tests during this decade. At least two sites of several 10,000 km$^2$ total are needed, better a few sites for smooth full-sky coverage. Several sites will also facilitate the participation of several countries in the project, and with both GRAND and GCOS featuring several sites, at least one of the sites should be in common for the purpose of cross-calibration and to benefit from shared infrastructure.

The baseline design of a GCOS site is a ground array of particle detectors, possibly segmented or nested water-Cherenkov detectors enhanced by radio antennas, to provide a 24/7 measurement of the electromagnetic (for energy and $X_{\max}$) and muonic shower component over a wide range of zenith angles. Due to the complementary mass sensitivity of muons and $X_{\max}$, a simultaneous measurement of both observables for each individual event is essential to estimate the rigidity of the primary cosmic rays. Also, some of the sites can feature additional detectors, such as next-generation fluorescence telescopes which would further enhance the measurement accuracy of the electromagnetic component during clear nights [52, 53]. Nonetheless, decent $X_{\max}$ measurements may already be possible with the particle detectors alone, e.g., by using neural networks for the analysis of the traces of water-Cherenkov detectors [54]. The high accuracy for both, the muon number and $X_{\max}$, will also help other science cases, such as the selection of protons or the search for neutral particles.

## 4 Conclusion

Thanks to the increasing quality of UHECR measurements in the past decade, UHECR science has entered an interesting era with the resolution of its most exciting puzzles possibly in reach. The discovery of anisotropy and mixed mass composition in the past decade provided evidence for UHECR originating from astrophysical accelerators, and the first UHECR sources may be discovered by the upgraded observatories during this decade. Nonetheless, it is unknown to what extent this answers the question of the origin of UHECR, and it is likely that a comprehensive picture of the sources can only be obtained by the next generation of UHECR observatories.

Also, in regard to the particle physics of air showers, in particular, for discrepancies between hadronic interaction models and measured muons, progress is expected from the upgraded UHECR observatories, as well as new accelerator experiments. While the muon puzzle may be solved during the next decade, it often happens that a more accurate generation of experiments discovers new puzzles and discrepancies between models and measurements. Furthermore, searching for BSM physics and for neutral EeV particles certainly requires a new generation of air-shower observatories with higher exposure.

In parallel to increasing the exposure, the mixed mass composition mandates an increase in accuracy for the reconstruction of the mass of the primary cosmic-ray particles. It is important to realize that the event-by-event precision for the mass is intrinsically limited by shower-to-shower fluctuations. Perfect resolution of either the muon number or $X_{\max}$ alone is insufficient, and therefore, a simultaneous measurement of both shower parameters is necessary to improve the accuracy of the rigidity of individual events.

Current observatories, such as AugerPrime, provide ideal testbeds for next generation instrumentation. Especially R&D of optimized detectors designs, such as surface detectors for simultaneous $X_{\max}$ and muon measurements, is needed during the next years to prepare for the construction of future ground arrays. Featuring an event-by-event rigidity estimate, GCOS will be the ground array leading the field of rigidity-enhanced particle astronomy. GCOS' UHECR physics will be complemented by the observation of other messengers up to EeV energies, and by the huge UHECR exposure of GRAND and POEMMA.

## References


[1] A. Coleman et al., Astropart. Phys. **149**, 102819 (2023), 2205.05845

[2] J.N. Butler et al., *Report of the 2021 U.S. Community Study on the Future of Particle Physics (Snowmass 2021) Summary Chapter*, in *2022 Snowmass Summer Study* (2023), 2301.06581

[3] A.S. Chou et al., *Snowmass Cosmic Frontier Report*, in *2022 Snowmass Summer Study* (2022), 2211.09978

[4] R.X. Adhikari et al. (2022), 2209.11726

[5] F.G. Schröder et al., Bull. Am. Astron. Soc. **51**, 131 (2019), 1903.07713

[6] F. Sarazin et al., Bull. Am. Astron. Soc. **51**, 93 (2019), 1903.04063

[7] W.D. Apel et al. (KASCADE Grande), Phys. Rev. Lett. **107**, 171104 (2011), 1107.5885

[8] R.U. Abbasi et al. (Telescope Array), Astrophys. J. **865**, 74 (2018), 1803.01288

[9] M.G. Aartsen et al. (IceCube), Phys. Rev. D **100**, 082002 (2019), 1906.04317

[10] P. Abreu et al. (Pierre Auger), Eur. Phys. J. C **81**, 966 (2021), 2109.13400

[11] A. Aab et al. (Pierre Auger), Phys. Lett. B **762**, 288 (2016), 1609.08567

[12] H.P. Dembinski, R. Engel, A. Fedynitch, T. Gaisser, F. Riehn, T. Stanev, PoS **ICRC2017**, 533 (2018), 1711.11432

[13] P. Abreu et al. (Pierre Auger), Astrophys. J. **933**, 125 (2022), 2205.14864

[14] A. Aab et al. (Pierre Auger), JCAP **10**, 022 (2019), 1906.07422

[15] R.U. Abbasi et al. (Telescope Array), Astropart. Phys. **110**, 8 (2019), 1811.03920

[16] A. Aab et al. (Pierre Auger), Astrophys. J. **891**, 142 (2020), 2002.06172

[17] P. Abreu et al. (Pierre Auger), Phys. Rev. Lett. **109**, 062002 (2012), 1208.1520

[18] R.U. Abbasi et al. (Telescope Array), Phys. Rev. D **92**, 032007 (2015), 1505.01860

[19] P. Abreu et al. (Pierre Auger), JCAP **01**, 023 (2022), 2112.06773

[20] P. Abreu et al. (Pierre Auger) (2022), 2203.08854

[21] P. Abreu et al. (Pierre Auger) (2022), 2208.02353

[22] D. Soldin (EAS-MSU, IceCube, KASCADE-Grande, NEVOD-DECOR, Pierre Auger, SUGAR, Telescope Array, Yakutsk EAS Array), PoS **ICRC2021**, 349 (2021), 2108.08341

[23] W.D. Apel et al. (KASCADE-Grande), Astropart. Phys. **95**, 25 (2017), 1801.05513

[24] A. Aab et al. (Pierre Auger), Phys. Rev. D **90**, 012012 (2014), [Addendum: Phys.Rev.D 90, 039904 (2014), Erratum: Phys.Rev.D 92, 019903 (2015)], 1407.5919

[25] J. Albrecht et al., Astrophys. Space Sci. **367**, 27 (2022), 2105.06148

[26] A. Aab et al. (Pierre Auger), Nucl. Instrum. Meth. A **798**, 172 (2015), 1502.01323

[27] T. Abu-Zayyad et al. (Telescope Array), Nucl. Instrum. Meth. A **689**, 87 (2013), 1201.4964

[28] IceCube-Gen2 Collaboration, e.g., IceCube-Gen2 Technical Design Report (in preapration); and proceedings of the 6th ARENA 2022 (submitted)

[29] R. Abbasi et al. (IceCube), Nucl. Instrum. Meth. A **700**, 188 (2013), 1207.6326

[30] M.G. Aartsen et al. (IceCube-Gen2), J. Phys. G **48**, 060501 (2021), 2008.04323

[31] F. G. Schroeder for the IceCube-Gen2 Collaboration (IceCube-Gen2), PoS **ICRC2021**, 407 (2021)

[32] A. Leszczynska for the IceCube-Gen2 Collaboration (IceCube-Gen2), PoS **ICRC2021**, 411 (2021)

[33] A. Aab et al. (Pierre Auger), Phys. Rev. Lett. **116**, 241101 (2016), 1605.02564
[34] P. A. Bezyazeekov et al. - Tunka-Rex Collaboration (Tunka-Rex), JCAP **01**, 052 (2016), 1509.05652
[35] W.D. Apel et al. (LOPES), Eur. Phys. J. C **81**, 176 (2021), 2102.03928
[36] W. D. Apel et al. - LOPES Collaboration (LOPES), Phys. Rev. D **85**, 071101 (2012), 1203.3971
[37] S. Buitink et al. (LOFAR), Phys. Rev. D **90**, 082003 (2014), 1408.7001
[38] W. D. Apel et al. - LOPES Collaboration, JCAP **09**, 025 (2014), 1404.3283
[39] P. A. Bezyazeekov et al. - Tunka-Rex Collaboration, Phys. Rev. D **97**, 122004 (2018), 1803.06862
[40] M.G. Aartsen et al. (IceCube), Astrophys. J. **826**, 220 (2016), 1603.01227
[41] P. Abreu et al. (2019), 1907.07737
[42] C. Zhen et al. (LHAASO), Chin. Astron. Astrophys. **43**, 457 (2019)
[43] T. Huege et al., EPJ Web Conf. **135**, 02003 (2017), 1608.08869
[44] T. Huege, Phys. Rept. **620**, 1 (2016), 1601.07426
[45] F.G. Schröder, Prog. Part. Nucl. Phys. **93**, 1 (2017), 1607.08781
[46] S. Buitink et al., PoS **ICRC2021**, 415 (2021)
[47] A.V. Olinto et al. (POEMMA), JCAP **06**, 007 (2021), 2012.07945
[48] A. Santangelo et al. (JEM-EUSO), Exper. Astron. **40**, 3 (2015)
[49] E. Parizot et al. (JEM-EUSO), *The JEM-EUSO program*, in *These Proceedings* (2023)
[50] J. Álvarez-Muñiz et al. (GRAND), Sci. China Phys. Mech. Astron. **63**, 219501 (2020), 1810.09994
[51] J.R. Hörandel (GCOS), PoS **ICRC2021**, 027 (2021), 2203.01127
[52] T. Fujii et al., Astropart. Phys. **74**, 64 (2016), 1504.00692
[53] M. Malacari et al., Astropart. Phys. **119**, 102430 (2020), 1911.05285
[54] A. Aab et al. (Pierre Auger), JINST **16**, P07019 (2021), 2101.02946

# A Brief History of the Study of High Energy Cosmic Rays using Arrays of Surface Detectors

A A Watson[1]

[1] School of Physics and Astronomy, University of Leeds, Leeds, UK

**Abstract.** A brief history of the development of surface detectors for the study of the high-energy cosmic rays is presented. The paper is based on an invited talk given at UHECR2022 held in L'Aquila, 3 – 7 October 2022. In a complementary talk, P Sokolsky discussed the development of the fluorescence technique for air-shower detection.

## 1 The discovery of extensive air showers

Cosmic-ray physicists have always studied the highest-energy particles that exist in nature, even when, in the earliest days, they thought that the ionisation they were detecting at ground-level originated from primary photons. In this article, I will focus mainly, but not exclusively, on the efforts developed to study cosmic rays with energies above ~0.1 EeV, and I will introduce these in a historical manner. We will find that the techniques now widely-used are, often, simply scaled-up versions of what has gone before: those of us active in the field, should be very grateful to the work of a small number of pioneers.

Above an energy of ~100 TeV, the flux of cosmic rays is so low that direct detection with equipment flown on a balloon or a spacecraft is difficult. Instead, one must make use of the secondary particles that the higher-energy cosmic rays produce, which reach ground level and are spread out so that they can be observed, more or less simultaneously, by arrays of detectors. This phenomenon is known as an *extensive air shower*. The separation of the detectors usually grows as cosmic rays of higher energy are studied, largely for financial reasons, and deployment at mountain altitudes is useful for studies at the lower energies.

B Rossi made the first detection of extensive air showers in 1934 while searching for the East-West effect in Eritrea, [1]. He observed the correlated arrival of particles, in widely-separated Geiger counters, at a rate well above that expected by chance, after taking into account the known resolving time of his coincidence circuitry. Rossi named this phenomenon '*sciami*', '*swarms*' in English. Earlier [2], he had found a rapid increase in the rate of coincidences between a triangular arrangement of Geiger counters, as the thickness of lead above the detectors increased. All three counters could not have been triggered by a single particle travelling in a straight line. The results were so surprising that the paper describing them was at first rejected, and only published after W Heisenberg vouched for Rossi's reliability. These results, now known as the *Rossi transition curves*, led K Schmeiser and W Bothe to conclude that showers would also develop in air. They made the first systematic study of this possibility [3] using Geiger counters separated by up to 50 cm, introducing the term *'luftschauer'* to describe what they observed. At around the same time, W Kolhöster, I Matthes and E Weber [4] made similar investigations with counter separations up to 75 m.

In addition to this systematic work, two other earlier observations pointed to showering in air. Firstly, in 1932, P M S Blackett and G Occhialini [5] noted that some of the cascades seen in a cloud chamber triggered by coincidences between a set of Geiger counters, appeared to originate in the air surrounding their instruments. Secondly, in 1935, E Regener and G Pfotzer [6][1] observed a maximum in the rate of coincidences between three vertically-separated Geiger counters as they were carried upwards by a balloon. The peak intensity, at ~14 km above sea-level, was interpreted by Regener in 1938 [7] as being due to the multiplication of electrons in the atmosphere, which he called *'Schauer'*, his thinking being guided by the theoretical work of H Bhabha and W Heitler [8]. Neither Regener nor Rossi appear to have recognised that the same physical mechanisms lay behind their respective observations. Rossi was unable to follow up his work in part because of his central role in creating a new physics department at the University of Padua and, tragically, because the anti-Semitic laws of the Fascist regime led to him losing his position[2].

Despite the work of Rossi and the two German groups, credit for the discovery of extensive air showers has usually been given to P Auger and his collaborators

[1] This maximum is commonly known as the Pfotzer maximum, though this naming is in fact inappropriate: see P Carlson and A A Watson, Hist Geo Space Sci **5** 1 2014

[2] In 1973, a law was introduced by the Italian Government offering a university chair to those Italians who, like Rossi, had been exiled in this manner. Between 1974 and 1980, Rossi held a professorial position in the University of Palermo.

[9] for what was actually another serendipitous discovery arising from the use of the classical method to measure the resolving time of the new coincidence circuitry developed by R Maze [10][3]. Auger appears to have been unaware of the earlier work of Rossi, or of that of the two German groups, and for a while the commonly used abbreviation for extensive air showers, EAS, was interpreted as '*Extensive Auger Showers*' [11]. On the Jungfraujoch (3463 m), Auger and his team were able to increase the counter separations to over 300 m, and later, using cloud chambers and Geiger counters shielded by lead, to make the discovery that penetrating particles (muons) were associated with the showers. Using the new techniques of quantum electrodynamics, Auger et al. argued [12] that the primary particles causing the largest showers must have energies of ~1 PeV: at this time the highest-energy cosmic rays, as inferred from geomagnetic studies, were only ~10 GeV, so this was an extraordinary extension of Nature's energy scale. Auger thought that the primary particles were electrons.

Auger left France in 1939, spending time in Blackett's laboratory in Manchester, *en route* to Chicago before moving to Canada in 1943. There, his team at Chalk River, that led the British-Canadian-French efforts in support of the Manhattan project, included B Pontecorvo, a talented experimentalist as well as an exceptional theorist. Auger's time in Manchester coincided with the presence there of Rossi who was also *en route* to the United States, where he worked first in Chicago, with A H Compton, and later in Cornell, where K Greisen was his first American graduate student. Both Rossi and Greisen joined the Manhattan project at Los Alamos. The discussions that took place while Auger and Rossi were in Manchester surely enhanced Blackett's long-standing interest in the shower phenomenon [13].

## 2 Activities during World War II

Neither Bothe nor Kolhörster were able to carry on their cosmic-ray work after the outbreak of WWII, although Bothe did attend the International Conference on Cosmic Rays held in Chicago in June 1939[4]. While in Chicago, Auger was able to continue his shower studies and, in collaboration with A Rogozinski and M Schein, flew ten Geiger counters, arranged along a 10 m line, to an altitude of ~15 km (~120 g cm$^{-2}$) in January 1943. Four of the counters were close together at the centre of the array [14]. From the observation that coincidences between these central counters were unaccompanied by hits in the more distant ones, Auger deduced that, at this altitude, the shower consisted of a narrow bunch of particles, an observation now easy to understand from what we know about high-energy interactions.

In the Soviet Union, soon after the Battle of Stalingrad had ended in February 1943, D V Skobeltzyn initiated studies of showers in the Pamirs at an altitude of 3860 m. Results obtained in 1946 [15] extended the work of Auger at a similar altitude and were discussed in the context of the models of cascades then available[5].

A novel way of detecting air showers was discussed in the UK during World War II. On 4 September 1939, just after the outbreak of the War, A C B Lovell and J G Wilson (who was later to lead the Haverah Park project) were visiting a radar station in Yorkshire. Watching over the shoulder of a radar operator, they saw echoes on the screen, which the young woman ignored telling them that these were not from aircraft, but were reflections from the ionosphere. This led Lovell to conceive of using radar to detect the ionisation trail left by shower particles. A paper was written, with Blackett, proposing that the *'radio echo technique'* might be used to detect air showers [16]. A letter from T L Eckersley to Blackett [17] soon followed, in which it was pointed out that the recombination time of the electrons adopted in their calculation was too large by a factor of $10^6$ and consequently that the technique was less promising than claimed. This criticism was forgotten – war-work had precedence - and, soon after the War ended, Blackett encouraged Lovell to look for showers using this technique with surplus war equipment that was readily available. Lovell soon discovered that the trails that he and Wilson had spotted in 1939 were in fact due to meteors. In his writings, Lovell makes clear that had the Eckersley letter been recalled after the War, the Jodrell Bank radio astronomy facilities might never have developed [17].

[3] During the third European Cosmic Ray Symposium held in Paris in 1972, I was invited one evening, along with others including Arnold Wolfendale, Cormac O'Ceallaigh, and George Wdowczyk, to the home of Roland Maze. During the evening, Wdowczyk asked Maze to recount the story of the discovery of extensive air showers. Speaking very slowly, in French, he made it clear that his task had been to improve the resolving time of the coincidence circuitry then available. This he did and, to measure the resolving time, he used the random coincidence method. There was nothing in what Maze said to suggest that a search for showers had been planned. Rather it seems very likely that this, like Rossi's, was a serendipitous discovery.

[4] Heisenberg organised two symposia on cosmic rays in Berlin in 1941 and 1942 but neither Bothe nor Kolhörster are listed as contributors (*Cosmic Radiation*, edited by W. Heisenberg, translated by T.H. Johnson, Dover Publications 1946)

[5] It is remarkable, that at this crucial time in Soviet history, research in cosmic rays was supported. During a meeting in Pylos in 2004, G T Zatsepin, a student of Skobeltzyn, commented that Skobeltzyn saw work on cosmic rays as a reconnaissance for the study of high-energy interactions. At the time, the possibility that understanding these interactions might yield a source of energy may have looked appealing to political leaders. A H Compton, when seeking support for his latitude studies in the 1930s, had advanced a similar argument.

## 3 Activities in the immediate post-War period: 1945 - 1965

### 3.1 Results from arrays of Geiger counters used at Moscow State University

It will be apparent from the discussions in sections 1 and 2 that the Geiger counter was the major tool used in the discovery of extensive air showers, and this device remained an important technology well into the 1950s, before being superseded by scintillation counters and water-Cherenkov detectors. A key result from the use of Geiger counters was that obtained by G V Kulikov and G B Khristiansen who, in 1958 [18] reported a steepening of the number spectrum of showers at a size of $\sim 10^6$ particles. At this time, the primary energy corresponding to such a size was known only very approximately and was thought to be ~10 PeV. The steepening, what we now call *'the knee'*, was interpreted by Kulikov and Khristiansen as evidence for an astrophysical feature, specifically a demonstration in favour of an extragalactic origin of cosmic rays above 10 PeV.

### 3.2 Impact of the work of the MIT Group

After finishing work on the Manhattan project, Rossi established a powerful cosmic-ray group at MIT. Perhaps unsurprisingly, one of his interests was the study of extensive air showers. Rossi targeted the determination of the direction and energy of the incoming primaries as being of the highest priority. In his autobiography [19], he states that the program which he initiated *"because of the originality of its conception and the significance of it results, ranks amongst the foremost accomplishment of the MIT group."* Many of the instrumental and analysis techniques developed by his team form the bedrock of what we use today.

Rossi was aware of the invention of the scintillation counter in 1944 by S C Curran who, in support of the Manhattan project, had coupled a photomultiplier to a scintillating crystal to measure radioactivity. Rossi saw the possibility of introducing scintillation counters for the study of air showers. However, larger areas were needed and the first MIT detectors were liquid scintillators made of benzene, with the addition of a small amount of para-terphenol to act as a wavelength shifter. Using three liquid-scintillation counters, each of 600 cm$^2$, P Bassi, G Clark and Rossi [20], in a seminal paper, established that the spread of arrival times of the shower particles was sufficiently small to make determination of shower directions by fast-timing feasible. They also showed that, close to the centre of the shower, the electrons preceded the muons, and that an appreciable fraction of the electrons observed at sea-level arose from nucleons having energies above 20 GeV. This paper remains fascinating reading for its original insights, and as a demonstration of how clear physical reasoning, coupled with observations from a small number of detectors, can lead to profound understandings.

The next step was to lay out an array of 11 scintillators on a triangular grid covering 0.17 km$^2$ at the Agassiz station of Harvard University. A lightning strike led to some of the detectors catching fire and these were then replaced with large-area plastic detectors designed and manufactured by the MIT group [21]. The largest event recorded at the Agassiz array had a primary energy of ~5 EeV.

The work at Agassiz was the forerunner for two important shower projects. One was the construction of arrays in Bolivia, using much of the Agassiz instrumentation, first at a site on the Alto Plano (El Alto) at 4200 m (620 g cm$^{-2}$), and then at an altitude of 5200 m (500 g cm$^{-2}$) on Mt Chacaltaya. At the El Alto site, important measurements of shower development were made using what is now known as the *Constant Intensity Cut method* [22]. The work at El Alto led to the first estimate of the maximum of shower development, ~600 g cm$^{-2}$ at ~0.1 EeV, in reasonable accord with later direct measurements made using the fluorescence technique. In addition to measuring the energy spectrum of cosmic rays to beyond 0.1 EeV, the work at Mt Chacaltaya, carried out by a collaboration of physicists from Bolivia, Japan and the USA, made use of a 60 m$^2$ shielded detector to search for showers deficient in muons, the signature of photon primaries. None was observed which we now know to be reasonable as the photon flux at the energies in question are heavily attenuated by the CMB, which had not been observed when this activity was initiated, and also the photon flux is much smaller than had been anticipated.

The second project, the initiative of J Linsley, was the construction of an array of 8 km$^2$ at Volcano Ranch, New Mexico (1770 m, 834 g cm$^{-2}$), a desert site near Albuquerque. This was the first device to cover an area of more than 1 km$^2$. In its first manifestation, the array comprised 19 plastic scintillation counters of 3.26 m$^2$ laid out on a triangular grid[6] with a spacing of 442 m. This array operated during 1959-1960. For an enlarged version, which ran from 1960 to 1963, the same detectors were spread out 884 m apart. A 20$^{th}$ detector was placed at various locations, sometimes unshielded, and sometimes covered with 10 cm of lead for muon studies.

Linsley was the first person in the shower community to face the many challenges of building such a large instrument. While for the first phase he had the full-time

[6] There is currently discussion about the best choice of grid geometry. It is unclear why the triangular layout, which has certain advantages (Q Luce et al., Proc. 27$^{th}$ European Symposium on Cosmic Rays), was chosen for so many arrays. In a letter to Rossi written in 1986, as part of a discussion about the evolution of the arrays stemming from the MIT group, Linsley comments "On the question of rings vs. squares or some other form of lattice, after all, concentric rings make a bullseye, the classical symbol of a target". Linsley liked targets and visual allusions.

https://doi.org/10.1051/epjconf/202328301002

assistance of L Scarsi (who was visiting Rossi's group from Italy for three years on a Fulbright Fellowship), later, for the expansion, and for operation of the larger configuration, Linsley worked virtually single-handed, with only limited support to help with some of the heavier work and with some of the data-reduction. The analogue signals from the detectors were recorded by photographing oscilloscope traces. The recording equipment contained ~500 thermionic valves and was operated in the trying conditions associated with a desert climate, including dealing with rattlesnakes. Operations had to be suspended for extensive periods during the summer because of the frequent lightning storms: strikes to ground, picked-up on the earthing of the cables, damaged the electronics. The single-handed operation of the larger array was a remarkable achievement with an exposure of 25.5 $km^2$ sr year being achieved.

Linsley was the first to report the observation of a flattening in the number spectrum of showers at a shower size of $\sim 10^9$, now known as *'the ankle'*. This was interpreted as corresponding to a similar property of the energy spectrum at ~1 EeV [23], a feature which Linsley suggested might mark a cross-over between galactic and extragalactic cosmic rays. In [23] the arrival distribution of 1034 events with an average energy of over 1 EeV was presented, with the limit to any anisotropy set at <10%. The estimates of shower sizes and energies were subsequently revised following a more accurate measurement of the lateral distribution function [24] with a denser configuration of detectors laid out over 1 $km^2$. The iconic event of 100 EeV [25] was assigned a revised energy of 139 EeV [26].

In addition to developing his own group, Rossi invited scientists from Australia, Bolivia, China, France, Italy, India, Japan and elsewhere for extended stays[7]. The visits of M. Oda were of particular importance for the development of cosmic ray research in Japan.

Greisen took advantage of the scintillator factory developed at MIT, to make extensive studies over a range that extended to both lower and higher energies than achieved at the Agassiz array [27]. He used essentially the same analysis techniques as those pioneered by the MIT group. Studies of the momentum spectrum of muons were also carried out at the Cornell array.

### 3.3 Impact of work at the Geiger counter array at Harwell

The work at MIT led, indirectly, to another important development in instrumentation. At Harwell, where the UK carried out post-war atomic work, an array of Geiger counters, covering 0.61 $km^2$, was constructed *'outside the Wire'*. One of the initiators of this project was Pontecorvo who had moved to the UK following his time with Auger at Chalk River. Pontecorvo was aware of the MIT work with liquid scintillators, but, as para-terphenyl was very expensive (even now, it is more costly than printer ink), he asked J V Jelley to investigate the efficiency of the light output as a function of the quantity used [28]. Jelley found that even with no additive he could detect light – it was Cherenkov radiation – from benzene and from other liquids, including distilled water, as a muon passed through [29]. This discovery led N A Porter to develop a large water-Cherenkov detector [30] subsequently adopted for shower studies at Haverah Park and at the Pierre Auger Observatory. Porter's detector is compared (figure 1) with one of those used at the latter site. Little has changed and Pontecorvo might perhaps be regarded as the father of the water-Cherenkov detector.

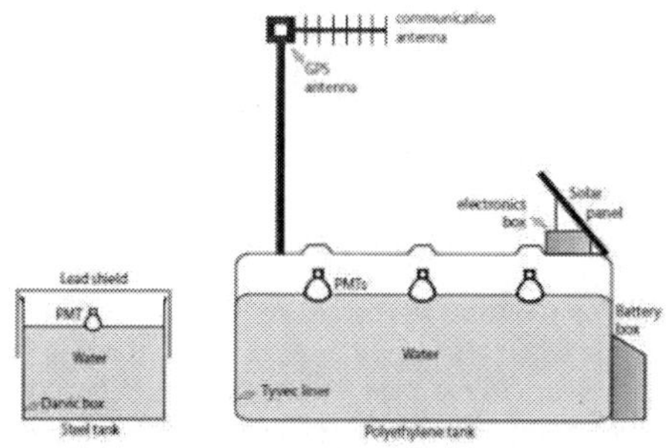


**Fig. 1.** Scale comparison of Porter's water-Cherenkov detectors with a detector at the Auger Observatory [from 31].

The array at Harwell was also used to explore the possibility of detecting Cherenkov light produced as showers traverse the atmosphere. This arose directly from a study by Blackett [32] of the essential features of Cherenkov radiation in air. He calculated that the energy thresholds for electrons and muons to produce Cherenkov radiation were 20 MeV and 4 GeV respectively, and showed that the Cherenkov radiation from cosmic rays was $\sim 10^{-4}$ of the night-sky background. According to Lovell [33], Blackett was at this time interested in the quantum efficiency of the eye and had concluded that extensive air showers would produce a flash of light that he should be able to see by lying down and looking upwards under suitable dark-sky conditions, an effort which Blackett undertook himself. There is no record of the outcome of Blackett's efforts, but the idea inspired Galbraith and Jelley to a successful detection of light from air showers [34]. To establish that the light detected was indeed Cherenkov radiation, further work was carried out at Pic du Midi.

The detection of air-Cherenkov radiation became a common activity at several shower arrays, notably in the Soviet Union where the method blossomed under the leadership of A E Chudakov and G T Zatsepin [35]. Important observations made there with this technique led to model-independent estimates of the primary

[7] Rossi's initial efforts to have Blackett visit were thwarted, as during the McCarthy era it proved impossible for him to get a visa. Blackett did visit later when Eisenhower was President.

https://doi.org/10.1051/epjconf/202328301002

energy of cosmic rays that produced showers of size $10^5$ to $10^6$, by Greisen [36] and S I Nikolsky [37].

### 3.4 Shower studies in Japan

Cosmic-research was initiated in Japan in the 1930s by Y Nishina in his laboratory at RIKEN and flourished particularly in the period after the end of World War II when the study of nuclear physics was forbidden, an edict reinforced by the destruction of the cyclotrons at RIKEN, Kyoto and Osaka. S Tomanaga led studies of extensive air showers at Mt Norikura (2770 m) and later played a key role in establishing the Institute of Nuclear Studies (INS) in Tokyo. He was also influential in encouraging J Nishimura and K Kamata to develop three-dimensional calculations of electromagnetic cascades, work which they began after reading the Rossi and Greisen article in Reviews of Modern Physics [38], a copy of which had reached Japan just before the attack on Pearl Harbour [39].

Experimental work on showers at sea-level at INS was activated by Oda on his return from MIT in 1956. He led the construction of a series of ever-more complex arrays, culminating in one that contained 14 1 $m^2$ scintillators for measuring signal sizes and a further 5 smaller scintillators for direction measurement [40]. In addition, 8 $m^2$ of scintillator was deployed in a tunnel, 15 m deep, to measure the muon signal above 4.5 GeV. The most important insights came from combining the data from the 14 unshielded scintillators and the muon detectors. The INS group was the first to point out the information that could be derived from a study of plots of muon versus electron number, $N_\mu$ vs. $N_e$ plots in today's notation. Some data are shown in figure 2.

Large fluctuations in muon number are seen at fixed electron size, with the sharp upper boundary suggesting that the first interaction dominates the fluctuations found in showers. This type of diagram, made with greatly enhanced statistics, when combined with the detailed shower simulations that became available from the late 1960s, proved to be an important tool for extracting information on primary mass.

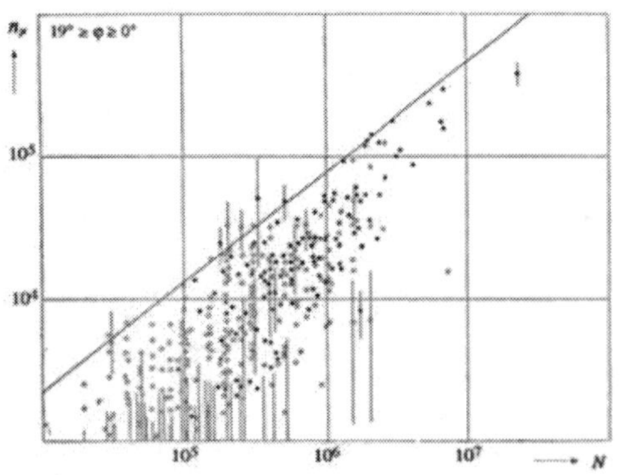


**Fig. 2.** Reconstructed muon number $N_\mu$, vs. shower size N for vertical showers [40].

These simulations also led to the recognition that, at a fixed energy, the fluctuations in electron number were much greater than those in the muon signal, so that muon number was an effective surrogate for primary energy, although of course the energy deduced was dependent upon the hadronic model adopted. Plots such as those of figure 2 led to the idea of identifying photon-initiated showers through a paucity of muons and to the efforts at Chacaltaya.

### 3.5 The Haverah Park Array

The array at Harwell was decommissioned in 1958 as the land was required for the Culham Nuclear Fusion Laboratory. A decision was taken to continue shower studies in the UK, but in the universities, and a new array was constructed on a moorland site at Haverah Park, about 25 km from Leeds. This initiative, strongly supported by Blackett, was led by Wilson who had worked with Blackett at Birkbeck in the late 1930s and moved with him to Manchester in 1937. Wilson had also interacted with both Rossi and Auger during their brief spells in Manchester [13].

Scientists from Durham University, Imperial College London and Leeds University formed the initial collaboration. Following extensive studies by the Imperial College group, led by H R Allan, the decision was taken to use water-Cherenkov detectors as the basic detecting elements. The construction and operation of the 12 $km^2$ array was the responsibility of the Leeds group. The Durham team built a magnetic spectrograph to study the momenta of muons, with the Imperial College group (later joined by researchers from Nottingham University) developing shielded detectors to measure muon signals.

In the 1970s, prompted by Greisen, the Durham group began measurements of the air-Cherenkov signal using 5 inch photomultipliers pointing upwards. The potential of using Cherenkov radiation to cross-calibrate estimates of shower energies at different arrays was explored by the Durham team, but attempts to do this for energy estimates from the Volcano Ranch and Haverah Park arrays floundered for a variety of reasons.

The problem of obtaining the energy of showers detected using an array of water-Cherenkov detectors was challenging, as it was impossible to infer the number of electrons in the events from the measurements. When the project was started, determining the shower size was still commonly seen as the first step to estimating the energy. However, Hillas [41, 42] introduced the concept of classifying the size of showers through the use of the signal at an appropriate distance from the shower axis (now called $S(r_{opt})$ and widely adopted across the community). Also, through a series of simulations [42], Hillas and colleagues established a conversion between this value and the primary energy, which was used to determine the energy spectrum [43].

Other important results obtained at the array included identifying experimentally the magnitude of the photon, electron and muon components [44], and the first detection of shower-to-shower fluctuations in showers of energy ~1EeV, deduced from a study of the risetimes of the signals from the large area water-Cherenkov detectors. This work also gave hints of a change in the Elongation Rate of showers at ~5 EeV [45, 46].

Extensive studies of radio emission in showers were also carried out under the direction of Allan (see below).

### 3.6 The Yakutsk Air Shower Array

By far the most complex, and most northerly, of the early giant arrays, was that operated by the Institute of Cosmophysical Research and Aeronomy at Yakutsk, Siberia (latitude 62° N, 105 m). This array, originally proposed by D D Krasilnikov in 1959, began taking data with 13 stations in 1967 and by 1974 had been developed to cover 18 km$^2$. A detailed description of the array, which used 58 scintillation counters as the workhorse detectors, and included 192 m$^2$ of muon detectors with a threshold of 0.5 GeV, can be found in [47]. A particularly important feature was the use of 35 photomultipliers of various sizes to measure the air-Cherenkov radiation associated with the showers. The exceptionally clear Siberian skies make the data from air-Cherenkov detectors particularly valuable and allow a unique approach to the measurement of the energy spectrum and to parameters that relate to shower development. During winter, the temperature falls to -40 C, with Cherenkov light absorption in the atmosphere negligible: only molecular and aerosol scattering are considered.

The Cherenkov-light measurements can be used to find the energy lost by electrons through ionisation, essentially using the well-known track-length integral method. At 10 EeV, this accounts for about 77% of the energy of the incoming cosmic ray and is estimated from the total Cherenkov light reaching the ground by integrating over all distances using empirically-measured lateral distribution functions. These measurements also allow access to indirect information about the longitudinal development of showers. The results on the energy spectrum [48] and the mass composition [49] of cosmic rays above 0.1 EeV are in good agreement with those from other devices using different techniques.

### 3.7 The SUGAR Array

The team from the University of Sydney, who designed 'The Sydney University Giant Air Shower Recorder (SUGAR)', introduced a totally novel method of the recording of extensive air-showers with an array of ground detectors. Before this innovation, the practice was to link detectors with cables to some common point where coincident triggers between them could be made, and the signals recorded. This method becomes impractical for areas much larger than ~10 km$^2$ as it is rarely possible to have the relatively unrestricted land access enjoyed by Linsley at Volcano Ranch. In addition, the cost of cable, their susceptibility to accidental damage, lightning strikes, and the problems of generating fast signals over many kilometres, are further handicaps. The new idea was the inspiration of M M Winn and was first described in [50]. The Sydney group proposed the construction of an array of detectors that ran autonomously, with the time of a trigger above a certain number of particles recorded with respect to a timing signal transmitted across the array.

The concept was realised in the Pilliga State Forest near Narribri (250 m), New South Wales, where 47 stations were deployed over an area of ~70 km$^2$. Most of the detectors were on a grid of 1 mile (1.6 km), with 9 on a smaller spacing to enable showers of lower energy to be studied. Time and amplitude data were recorded locally on magnetic tapes and coincidences between different stations were later found off-line at the University of Sydney.

A difficulty was that the rate of triggers of a local station, above a level low enough to be useful, is very high and the rate could not be handled with technologies then available. The solution adopted was to require coincidences between pairs of detectors, 50 m apart, buried under 2 m of earth. The concept was brilliant but it was ahead of its time in terms of the technology available. Calor gas had to be used to supply the power at each station and the reel-to-reel tape recorders proved difficult to maintain in the dusty environment. The problem of handling many magnetic tapes at a single computing site also proved to be a challenge. A further problem was that the PMTs used were 7 inches in diameter and suffered from severe after-pulsing, complicating the measurement of the signals as logarithmic time-to-height converters were used to find the amplitudes. There was also a serious difficulty in estimating the energy of events as only muons were detected and therefore there was thus total reliance on shower models with no possibility to test which model was the best because of a lack of other types of detector in the array. Attempts to overcome this issue by using a fluorescence-light detector and a small number of unshielded scintillators were unsuccessful. An energy spectrum was reported [51].

The measurement of the shower directions with a precision of a few degrees was a demonstration that the timing stamp method was effective and the most valuable data from the SUGAR array were undoubtedly from the measurements of directions, the first to be made in the Southern Hemisphere at energies above 1 EeV [52]. No anisotropies were detected.

The concept of autonomous detection was tested at Haverah Park [53] in an early attempt to devise methods to make construction of an array of ~1000 km$^2$ economical, but the innovation of autonomous detectors has had its most effective realisation first at the Pierre Auger Observatory and subsequently at the Telescope Array. The innovation of Winn lives on.

Catalogues containing details of the largest showers recorded at Volcano Ranch, Haverah Park, SUGAR and Yakutsk are available [54].

## 4 The Situation in the 1980s

It is probably not widely realised how few people were working on ultra-high energy cosmic rays over forty years ago. In the mid-1970s, the number of faculty members involved with Volcano Ranch, Haverah Park, SUGAR and Yakutsk, and the efforts focussed on the Fly's Eye detector, was around 25, so that assuming that there was 1 post doc and 1 PhD student per staff member (perhaps generous), the number involved was certainly below 80. An unfortunate consequence was that publications were almost entirely in the Proceedings of the Biennial International Cosmic Ray Conferences and consequently the work attracted little attention beyond a very limited audience. The field was not highly regarded. For example, a referee of a seminal paper submitted by Linsley and Scarsi [55] to Physical Review, wrote that it was *"on a subject mainly just of technical concern to the few specialists working on air-showers, rather than being of general and fundamental significance, i.e., much of this could equally well be a technical report privately circulated"* [56]. Fortunately, the editors saw sense and this key paper was published.

Although Linsley remained active and influential in the cosmic field until his death in 2002, data-taking at Volcano Ranch with the 8 km$^2$ array was terminated in 1963, although important work was carried out there in the 1970s with a contracted array [24]). The SUGAR effort ended in 1979. The Haverah Park work with the 12 km$^2$ array ceased in 1987, and only Yakutsk continued to operate, but with a contracted area.

Efforts by the Haverah Park group to develop an array of 100 km$^2$, seen as a pilot for an area of 1000 km$^2$, were not supported, in part because the timing signal, planned for the autonomous system was derived from TV transmissions, which were then available for only 16 hours per day. Also, the promise of the Utah Group [57] to build a fluorescence detector capable of recording $10^6$ events per year above 10 PeV, and $10^4$ per year beyond 1 EeV, imposed a planning blight on some ambitions. It is interesting to note that fewer than $10^4$ events above 1 EeV have so far resulted from the totality of efforts in Utah.

In Japan, the construction of a complex and well-provisioned array of ~1 km$^2$ started in 1983, but with the upper energy targeted at 1 EeV. Construction of the 100 km$^2$ AGASA instrument got underway in 1989, with the full array operational in 1991, the year that the first discussions of what became the Auger Observatory started.

One of the strong drivers for building the Haverah Park array was the expectation that anisotropies associated with the Galaxy would be observed. Nothing significant was found for, as we now know, the instruments of that era were over two orders of magnitude too small to observe a significant signal. What undoubtedly sustained the field was the search for the steepening predicted by Greisen and by Zatsepin and Kuzmin (the GZK effect). Searching for this feature was seen as a significant target and shows the importance of having theoretical predictions when one is competing for money. Without the GZK goal, the possibility of gaining continuing support for any large projects to study air showers would have been small.

The absence of other predictions made it difficult for J W Cronin and me to argue for construction of an array of the size of the Auger Observatory. The hardest question that we had to face, and we were asked it many times, was *'why do you want to make the array so large?'* However, had we not built 3000 km$^2$ we would still be waiting to observe significant anisotropies. Phenomenologists must recognise that they have a key role to play in predicting what might be seen, and therefore of the scale of the instruments that are needed.

## 5 Work in Europe in the 1990s

Although, in the 1990s, activities in the field of ultra-high cosmic rays slowed internationally, work at lower energies blossomed in Europe, in both Germany and Italy. Additionally, a claim from the Kiel group in 1983 [58] of the observation of PeV gamma rays from Cygnus X-3, although now regarded as likely to have been incorrect, proved to have enormous impact on the development of both the TeV $\gamma$-ray and UHECR scenes.

In Italy, a group under the leadership of G Navarra constructed and operated an array, EAS-TOP, at Campo Imperatore (2005 m) as part of the National Gran Sasso Laboratory. The facility operated from 1989 to 2002 and, in its final configuration, consisted of 35 10 m$^2$ scintillation counters spread over ~$10^5$ m$^2$, together with a 144 m$^2$ calorimeter, designed for measurements of muons and hadrons, with the target of getting information on primary mass in the region of the knee [59]. Of particular importance are the data on the anisotropy of cosmic rays in the 0.1 to 1.0 PeV range [60]. In addition to a highly-significant detection of the long-sought Compton-Getting effect, the anisotropic feature that is expected due to the motion of the Earth around the Sun, a first harmonic amplitude in sidereal time, based on an analysis of 2.1 x $10^9$ events at ~0.1 PeV, was observed with a significance of over 10 sigma. The amplitude of the sidereal signal was 2.6 x $10^{-4}$ at a phase of ~6°. At higher energies (around 0.4 PeV) a morphology change in the anisotropy was observed for the first time, with the sidereal signal having a larger amplitude ($6.4 \times 10^{-4}$ ) and a different phase (~200°).

Contemporaneous with the work at Campo Imperatore, a team led by G Schatz, constructed and operated an array at the Institute für Kernphysik in Karlsruhe (110 m) in Germany. The instrumentation was spread over 200 x 200 m$^2$, with 250 shielded and unshielded detectors, a muon tracking device and a large calorimeter [61]. The main aim of this project was to determine features of the cosmic ray spectrum, and the variation of the primary mass, across the knee region. Detailed results were reported, with the knee position and the mass composition extracted being dependent upon the hadronic physics model adopted. A major result from the group was the demonstration, *without recourse to models*, that the number spectrum of muons steepened around the knee, but only in showers that were rich in electrons (i.e. showers produced by light nuclei) [62]. This work finally established that the knee feature in the energy spectrum, first demonstrated by Kulikov and Khristiansen [18] in 1959, was due to astrophysics and not to a change in hadronic interactions, thus putting to rest a debate that had lasted for over 40 years. I find such *model-independent conclusions* very compelling: trends are demonstrated that can be interpreted readily and unambiguously with an elementary understanding of the principles of shower development.[8] Another important product from the work in Karlsruhe was the creation of the CORSIKA code for shower simulations [63], now widely used across the cosmic ray community. The report has been cited over 1000 times.

Towards the end of the periods of running of the EAS-TOP and KASCADE arrays, the groups merged to build a larger device, KASCADE-Grande, at the KASCADE site [64]. This array, spread over 700 x 700 $m^2$, contained an additional 37 10 $m^2$ scintillation detectors. A major result was the discovery, *again without resorting to assumptions about hadronic models*, of a second knee at around 0.1 EeV [65]. This spectral break is suggestive of iron nuclei having attained a rigidity such that they can no longer be contained within the Galaxy.

It is worth noting that around 60 people signed papers from the KASCADE-Grande effort, not so different from the total number involved in UHECR activities in the 1970s. With this project, the era of the large collaborations needed for effective research in the field had arrived.

### 5.1 The impact of Cygnus X-3

From the mid-1960s, there was considerable interest in studying the central regions of air showers. This was possible, of course, only at relatively low energies (~1 PeV), but the observation of showers with single and multiple cores was pushed particularly strongly by the Sydney group [66] who used a chessboard configuration of 64 scintillators at the centre of an array of 300 Geiger counters. In addition to a claim for a flux of deuterium above 1 PeV, they presented evidence for anomalously large transverse momenta of pions in the showers. Similar effects were reported from the INS array were a neon hodoscope system was used to investigate the core region [67]. It was eventually recognised that many of the sub-cores seen in the detectors were the products of showering of particles from denser regions of the roofs covering the detectors. At this time, the power of shower simulations to investigate the impact of such configurations lay some way off.

In Kiel, West Germany, a group under the leadership of J Trümper made a particularly careful study of cores of showers. A neon hodoscope of 34 $m^2$, under a very light roof, was used along with an array of 16 1 $m^2$ scintillation counters. No anomalous effects were detected and the claims of large transverse momenta, based on core structure, slowly evaporated over the next decade. After Trümper moved to Tübingen and a career in X-ray astronomy, operation of the Kiel array continued, and in 1983 Samorski and Stamm [58] reported evidence of a signal, attributed to γ-rays of ~1 PeV from Cygnus X-3. In addition to a 4.4 sigma excess in the region around the object, a variation with the $4.8^h$ modulation characteristic of the source was noted. Similar time variations had previously been seen at lower energies with TeV telescopes, and subsequently there was an apparent confirmation of the effect by the Haverah Park group at ~ 1 PeV [68].

The evidence never strengthened but the excitement generated by the claim led a number of eminent particle physics to join the field. While several eventually moved to TeV astronomy, the most distinguished of the defectors remained in cosmic-ray physics. This was Cronin, who led the construction of the CASA array. Located at the Fly's Eye site in Utah , it covered 0.23 $km^2$ with 1089 scintillation detectors [69]. The array, which was built with the goal of detecting γ-rays from Cygnus X-3 several times each day, also contained a muon detector of 2560 $m^2$, which was used as a veto against charged cosmic rays. Although no γ-rays were ever detected, Cronin's appetite for working in cosmic rays was whetted and, in my view, without his singular input and dedication, it is unlikely that the Auger Observatory would ever have been developed. Life would have been very different for many of us without Cygnus X-3.

## 6 Subsequent developments

### 6.1 Hardware

It may be surprising, and perhaps even a little depressing, to realise that the major detection techniques, the scintillation counter and the water-Cherenkov detectors, first used for cosmic ray work in the 1950s, still retain such a central role in our work. Of course, there have been developments: in particular, photomultipliers and scintillator material are hugely cheaper than in the early days, optical fibres now aid

[8] An identical method had been introduced some years earlier by the CASA-MIA Collaboration to reach the same conclusion, although from a smaller number of events (M A K Glasmacher et al. Astroparticle Physics 12 1 (1999)).

light collection from large areas, and photomultipliers have much higher photocathode efficiencies.

The development of autonomous detector arrays, following the innovation of the Sydney group in the 1960s, has been of immense importance. Fluorescence detection (discussed in the talk by P Sokolsky) and measurements of radio emission have come to the fore, but it is worth recalling that the use of these were first discussed in the 1960s. In the case of the fluorescence method, it took until the late 1980s before a measurement of the energy spectrum was reported from it, and successful exploitation of the radio technique took even longer.

In 1965, at the ICRC in London, Greisen gave an invited paper, still worth reading, that was insightful and influential [70]. Amongst the topics he discussed was the detection of air showers using the radio method, which had just been demonstrated [71]. Greisen commented: '*The technique is barely in its infancy.... We feel confident it is a significant breakthrough and further study will reveal ways of obtaining information about showers not available by other means*'. His prediction would eventually turn out be correct but it was to be an infancy lasting nearly 40 years as the method fell out of favour around 1975, largely because of the inability to monitor the geo-electric field adequately at that time. The approach was revived early in the 21$^{st}$ century and is now a key tool, being exploited at LOFAR to study the depth of shower maximum and at the Auger Observatory for the same purpose, but also to improve estimates of the primary energy. The technique is also being applied in several projects targeted at the detection of very high-energy neutrinos.

Despite several efforts over the years, the radar echo method has yet to be successful.

### 6.2 Developments from growth in computing power

It is hard to imagine how our field could have advanced without the availability of the Monte Carlo technique, but it took around 20 years from the invention of the method by Ulam and von Neumann for sufficient computer power to be available for calculations relevant to air showers to become practical. Results from the first such work, using computers, were presented at the London ICRC in 1965. Those reported by Hillas were carried out at Leeds University on a KDF9 mainframe machine with 64 k of memory weighing about 5 tonnes, housed in a disused church and costing over €35M in today's money. Technology has advanced and the extensive use of the CORSIKA code had already been mentioned.

Adoption of the ideas of Artificial Intelligence (AI) and Neural Networks, now gaining wide use across our field, have also had a long gestation with early claims for the power of these concepts often over-hyped. Historians speak of two 'AI winters', 1974-1980 and 1987-1993, during which funding for AI research largely dried up. In 1991, for example, massive Japanese investment was curtailed when objectives, such as using computers to have a casual conversation, had not been met, a target only achieved in 2010. During 1991, around 300 commercial AI companies were shut down or went bankrupt.

Interestingly, during the second winter, AI was being used to address problems of importance to the astroparticle physics community. Halzen, Vazquez and Zas [72] and Reynolds [73], developed Neural Network techniques to analyse images from TeV gamma-ray telescopes. Perrett and van Stekelenborg, part of the team working in the Bartol Institute on analysis of data from the SPASE experiment at the South Pole, used the Neural Net approach to improve the speed of shower reconstruction [74].

### 6.3 Hadronic Physics

A big surprise, at least to me, has been that it has proved possible to obtain interesting results relating to hadronic physics at energies well-beyond those reached at the LHC from air-shower studies. Data from the Auger Observatory have been used to measure the p-air cross-section at a centre-of-mass energy of 57 TeV [75]. An excess in the number of muons, over that expected from hadronic models, which describe LHC data has also been uncovered [76].

Such work was not foreseen when the project was planned. However, if it does prove possible to identify the mass of individual events, as is envisaged with the extension of the Auger Observatory, this line of research might well develop strongly.

## 7 The Pierre Auger Observatory and the Telescope Array

The idea of building a truly giant shower array was developed during the 1990s and evolved into the Pierre Auger Observatory [77]. The goal of the project was to measure the properties of the highest energy cosmic rays with unprecedented precision. Data taking began on 1 January 2004 with a partial deployment of the instrumentation. The full Observatory of 1600 12 m$^2$ water-Cherenkov detectors on a 1500 m triangular grid spread over 3000 km$^2$ and overlooked by four fluorescence detectors, was completed in June 2008. Subsequently, 30 m$^2$ of muon detector was added, along with a number of radio antenna. Phase I of the effort ended on 31 December 2020. For phase II, named AugerPrime, 4 m$^2$ scintillator are being placed above each water-Cherenkov detector, with radio antennas added at every station. A major target of phase II is the determination of the mass of the primary particles individually rather than on an average basis.

The only giant array operating during the 1990s was the AGASA instrument, which covered ~100 km$^2$. Striking results were reported using data from it including a claim that there were events with energies well-beyond the steepening expected from the GZK effect [78] and a report of clustering of events above 10 EeV [79]. To check these results with higher statistics, the decision was taken to build the Telescope Array, a hybrid instrument with 507 scintillators spread across 700 $km^2$ on a 1200 m square grid [80], and overlooked by three fluorescence detectors [81]. Data-taking at the Telescope Array started in 2007.

These two observatories are on-going projects and, in a paper devoted to history, no further discussion is offered.

**7 Concluding remarks**

Study of cosmic rays with air-showers across the energy spectrum from 0.1 PeV to the very highest energies is in a lively state. What I have tried to show is the evolution that has led to the present vibrancy of our field. Our present position owes much to the input of a small number of iconic figures, most notably, in my view, Auger, Blackett, Chudakov, Greisen, Hillas, Jelley, Linsley, Rossi and Zatsepin. We continue to exploit the ideas and techniques which they developed but with much better financial support. We should never forget that we are standing on the shoulders of giants.

I am very grateful to the organisers of UHECR2022 for inviting me to indulge in one of my fascinations – the history of our field. Parts of the presentation above draw on the paper by Karl-Heinz Kampert and myself [31]. I recall how enjoyable it was to work with him on that project. I would also like to thank Valerie Higgins (archivist at FNAL) for her help during a visit to the Linsley archives in July 2019.

## References

I have used the format *PICRC date place* for work appearing in the Proceedings of the Biennial International Cosmic Ray Conferences. Proceedings can be accessed at
https://ads.harvard.edu/pubs/icrc_proceedings.html

1. B. Rossi, Supplemento a la Ricerca Scientifica **1** 579 (1934)
2. B. Rossi, *Z. Phys.* **82**: 151 (1933)
3. K. Schmeiser and W. Bothe, Ann. Phys. **424** 161(1938)
4. W. Kolhörster, I. Matthes and E. Weber, Naturwissenschaften **26** 576 (1938)
5. P.M.S. Blackett and G. Occhialini, Nature **130** 363 (1932)
6. E. Regener and G. Pfotzer, Nature **136** 718 (1935)
7. E. Regener and A. Ehmert, Z.Phys. **111** 501(1938)
8. H. Bhabha and W. Heitler, Proc Roy Soc A **159** 432 (1937)
9. P. Auger, R. Maze and Robley, Comptes Rendus **208** 1641 (1939a)
10. R. Maze, J. Physique et le Radium **9** 162–168 (1938)
11. P. Auger, in *Early History of Cosmic Ray Studies*, pp 213-218, editors Y. Sekido and H. Elliot, D. Reidel, Dordrecht (1985)
12. P. Auger et al., Rev Mod Phys **11** 288 (1939b)
13. A.C.B. Lovell, private communications (2003)
14. P. Auger, A. Rogozinski and M. Schein, Phys Rev **67** 62 (1945)
15. D. V. Skobeltzyn, G. T. Zatsepin and U. V. Miller, Phys Rev **71** 315 (1947)
16. P.M.S. Blackett and A.C.B. Lovell, Proc Roy Soc (London) **A177** 183 (1941)
17. For details, see A C B Lovell, Notes and Records of the Royal Society **74** 119 (1993)
18. G.V. Kulikov and G.B. Khristiansen, JETP **35** 441 (1959)
19. B. Rossi, *Moments in the Life of a Scientist*, Cambridge University Press (1990)
20. P. Bassi, G.W.Clark and B. Rossi, Phys Rev **92** 441 (1953)
21. G.W. Clark, F. Scherb and W.B.Smith, Rev Sci Inst **28** 433 (1957)
22. J. Hersil et al., Phys Rev Lett **6** 22 (1961)
23. J Linsley, PICRC 1963 Jaipur **4** 77
24. J. Linsley, PICRC 1977 Plovdiv **12** 56
25. J. Linsley, Phys Rev Lett **10** 146 (1963)
26. J Linsley, in *Catalogue of Highest Energy Cosmic Rays: Giant Extensive Air Showers: No. 1 Volcano Ranch and Haverah Park*, M Wada (editor)1980 Itabashi, Tokyo Japan by World Data Center C2 for Cosmic Rays The Institute of Physical and Chemical Research Report number: CHECR-1-1980 https://inspirehep.net/literature/1293873
27. J. Delvaille, F. Kendziorski and K. Greisen, PICRC 1961 Kyoto, J Phys Soc Japan **17** Suppl A-III 76
28. J.V. Jelley, in *Very High Energy Gamma Ray Astronomy* pp 27 – 37, Editor K.E. Turver, D. Reidel Dordrecht (1987)
29. J. V. Jelley, Proc Phys Soc A **64** 82 (1951)
30. N. A. Porter et al., Phil Mag **3** 826 (1958)
31. K-H. Kampert and A. A. Watson, European Physical Journal H **37** 359 (2012)
32. P. M. S. Blackett, Physical Society of London, Gassiot Committee Report p34 (1948)
33. A. C. B. Lovell, Biographical Memoirs of Fellows of the Royal Society **21** 1(1975)
34. W. Galbraith and J.V. Jelley 1953 Nature **171** 349 (1953)
35. A. E. Chudakov et al., PICRC 1959 Moscow **2** 50
36. K. Greisen, *Progress in Cosmic Ray Physics*, North Holland Vol III p 3, Editor J G Wilson (1956)
37. S. I. Nikolsky, Proc 5$^{th}$ InterAmerican Seminar on Cosmic Rays, La Paz, Bolivia **2** XLVIII (1962)
38. B. Rossi and K. Greisen, Rev Mod Phys **13** 240 (1941)
39. J Nishimura, private communication (2015)
40. S. Fukui et al. Prog Theor Phys Suppl **16** 1 (1960)
41. A. M. Hillas, PICRR 1969 Budapest, Acta Physica

Acad Sci Hungaricae **29** 355 (1970)
42. A. M. Hillas et al., PICRC 1991 Hobart **3** 1001
43. M. A. Lawrence, R. J. O. Reid and A. A. Watson, J Phys G **17** 733 (1991)
44. E.W. Kellermann and L. Towers, J Phys A **3** 284 (1970)
45. A.A. Watson and J.G. Wilson, J Phys A **7** 1199 (1974)
46. R. Walker and A.A. Watson, J Phys G **7** 1297 (1981)
47. B.N. Afanasiev et al., in *Proc. Tokyo Workshop on techniques for studying the Extremely High Energy Cosmic Rays,* edited by M. Nagano p 35 (1993)
48. A. A. Ivanov, A. P. Knurenko and I. Ye. Sleptsov, New Journal of Physics **11** 065008 (2009)
49. S. Knurenko and I. Petrov, Adv. Space Res. **64**, 2570 (2019)
50. C. B. A. McCusker, H.D. Rathgeber and M. M. Winn PICRC 1963 Jaipur **4** 306
C. B. A. McCusker and M.M. Winn, Il Nuovo Cimento, **28** 175 (1963)
51. M. M. Winn et al. ,J Phys G **12** 653 (1986a)
52. M. M. Winn et al., J Phys G **12** 675 (1986b)
53. G. Brooke et al., PICRC 1983 Bangalore **11** 185
54. *Catalogues of Highest Energy Cosmic Rays: Giant Extensive Air Showers:* No. 1: Volcano Ranch and Haverah Park (1980), No 2: SUGAR (1986), No 3: Yakutsk (1988), Editor M. Wada, Itabashi, Tokyo Japan by World Data Center C2 for Cosmic Rays The Institute of Physical and Chemical Research Report number: CHECR-1-1 https://inspirehep.net/literature/1293873
55. J. Linsley and L. Scarsi, Phys Rev **128** 2834 (1963)
56. *John Linsley Archive*, Fermi National Accelerator Laboratory, Box 49.9
57. J. Keffel et al., University of Utah NSF proposal GP 2445 p6 (1974)
58. M. Samorksi and W. Stamm, ApJ **268** L17 (1983)
59. M. Aglietta et al., NIM A **336** 410 (1993)
60. M. Aglietta et al., ApJL **692** L130 (2009)
61. T. Antoni et al., NIMA **513**, 490 (2003)]
62. T. Antoni et al., Astropart Phys **16** 373 (2002)
63. D. Heck et al., FZKA 6019 (1998)
64. W.D.Apel et al., NIMA **620** 202 (2010)
65. KASCADE Grande Collaboration, Phys Rev Letters **107** 171104 (2011)
66. A.D. Bray et al., PICRC 1965 London **2** 668
67. S. Shibata et al., PICRC 1965 London **2** 672
68. J. Lloyd-Evans et al., Nature **305** 784 (1983)
69. A. Borione et al., NIMA **346** 329 (1994)
70. K. Greisen PICRC 1965 London **2** 609
71. J. V. Jelley et al. Nature **205** 327 (1965)
72. R.Vazquez, F. Halzen and E Zas, Phys Rev D **45** 356 (1992)
73. P. T. Reynolds, PICRC 1991 Dublin **4** 496
74. J.C. Perrett and J T P M Stekelenborg, J Phys G **15** 1291 (1991)
75. P. Abreu et al., Phys Rev Lett **109** 062002 1208.1520 (2012)
76. A. Aab et al., Phys Rev Lett. **117** 192001 (2016)
77. A. Aab et al., NIMA **798** 172 (2015)
78. M. Takeda et al. Astroparticle Physics **19** 447 (2003)
79. M. Chikawa et al., PICRC 2001 Hamburg p 341
80. T. Abu-Zayyad et al., NIMA **689** 87 (2012)
81. H. Tokuno et al., NIMA **676** 54 (2012)

https://doi.org/10.1051/epjconf/202328301003

# The History of Air Fluorescence

Pierre Sokolsky[1,*]
[1]University of Utah, Salt Lake City, Utah

**Abstract.** I describe the history of the use of the air-fluorescence technique in UHECR physics.

## 1 Introduction

This article is a brief review of the history of the use of atmospheric air-fluorescence in the study of Ultra-high Energy Cosmic Rays. This review is very much centered on the work of US groups, both because of the historical importance of the Utah detectors and because of my own involvement. This is not in any way to diminish the very significant early work in the Soviet Union and Japan and the many essential later improvements to the technique by the Pierre Auger collaboration. Since many essential elements of air-fluorescence detectors were developed in the Fly's Eye and HiRes era, I treat these in some detail. The Auger and TA upgrades are the state of the art whose details can be found elsewhere in many contemporary reviews.

## 2 Pre-history: First Ideas

It is difficult to assign a single person or group to the original idea of using atmospheric air fluorescence to detect and study extensive air showers (EAS) produced by cosmic rays in the UHE range. Between the mid-1950's and the early 1960's, three groups in the Soviet Union, Japan and the US were considering the possibility. Publications by K. Greisen[1], G. Tanahashi, K. Suga, Oda et al.[2]and A. E. Chudakov[3] date to between 1960 and 1962, but there is anecdotal evidence of discussions at earlier cosmic ray conferences. Chudakov was mainly concerned with the development of the Cherenkov technique for gamma-ray astronomy. Here air-fluorescence would be a background which needed to be estimated. The first attempts at building detectors specifically to observe air-fluorescence from EAS were in the US and Japan.

## 3 Snakes in the Haystack: Early Attempts at Detection

The US effort centered on the K. Greisens Cornell University group, which built a prototype detector using Fresnel lenses to focus the EAS light on an array of photomultipliers [4]. The instrument was housed in a plywood "Bucky-dome". Greisen's group worked out all the basic elements of air fluorescence detectors - optics, light collection, phototubes, and fast electronics. They also developed the basic ideas of geometrical shower reconstruction using relative phototube signal timing. Another lasting contribution from this group was a compilation of information on the air-fluorescence efficiency in A. Bunner's Ph.D. Thesis [5]. While they failed to observe any air-fluorescence events, primarily due to the infelicitous Ithaca atmosphere and poor signal to noise, the conceptual design they developed has stood the test of time. The subsequent 50 years of development has rather been a matter of refinement of technique.

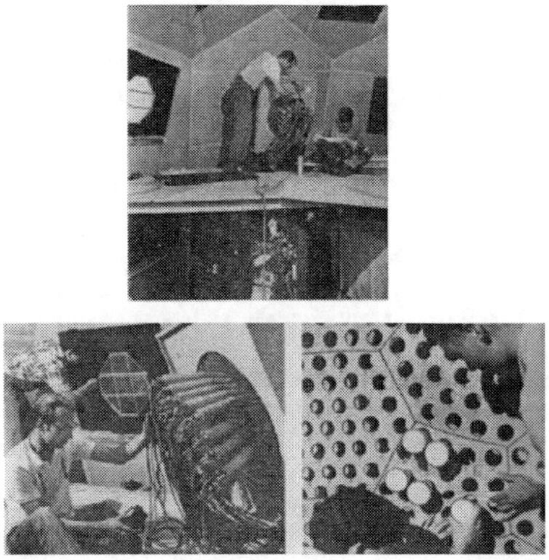

**Figure 1.** Greisen group prototype air-fluorescence detector.

The Japanese group independently developed a similar design also using Fresnel lenses. The detector was set up at the University of Tokyo's Dodaira Observatory. While most of the events that triggered their detector were attributed to Cherenkov light production, one event with energy near $10^{19}$ eV was very likely seen in air-fluorescence[6].

*e-mail: ps.protopop@gmail.com

**Figure 2.** Japanese air-fluorescence prototype building. On the left is Prof. Nagano.

The next attempt at detecting cosmic rays using air fluorescence was of a different nature. By the early to mid 1970's a number of ground arrays were collecting data. These used either plastic scintillation (Volcano Ranch) or water Cherenkov detectors(Haverah Park) and covered areas of 1 to 10 sq km. It was clear that a flux of UHECR up to and beyond $10^{19}$ eV existed. Could such arrays be used to trigger an air-fluorescence detector?

A group at the University of Utah became interested in pursuing this idea. The group grew out of a long established research program in underground cosmic ray physics established by J. Keuffel. Keuffel together with H. Bergeson and others had built an underground muon detector in the Silver King Mine in Park City, Utah to study neutrino interactions. In 1970, they observed an anomaly in the atmospheric muon flux which some interpreted as due to W boson production. This was dubbed " the Utah effect" and generated much attention. As it turned out, the effect was due to an error in the analysis, but Keuffel's complete honesty in disclosing the problems earned him great respect in the community.

In the aftermath, Keuffel became interested in Greissen's work at Cornell and brought several Cornell physicists to Utah. George Cassiday and Gene Loh became instrumental in the design of a new prototype air fluorescence detector. After Keuffel's untimely death in 1974, Cassiday, Bergeson and Loh established a collaboration with John Linsley of the University of New Mexico, whose Volcano Ranch plastic scintillation array had been operating for some years in the New Mexico desert.

It was decided to build a small three-mirror prototype and move it to the Volcano Ranch site[7]. The Utah prototype differed from Greissen's or the Japanese prototypes in several respects. Coated spherical mirrors were used, which produced sharper spot sizes than Fresnel mirrors. The front end of the sample and hold electronics were triggered using discriminator circuits equipped with filters which were optimized to reduce sky noise while passing thru the expected pulse width from an EAS travelling with nearly the speed of light. And, of course, the ground array could confirm that an actual EAS had passed through the atmosphere.

Conditions at Volcano Ranch were primitive. Linsley used hay to cover the plastic scintillation counters so that they would be at least partially thermally insulated. A constant danger to the experimenters was the presence of rattle-snakes who would curl up in this insulation and needed to be poked out when any maintenance was performed. The three-mirror Utah prototype was hauled out to the remote location notwithstanding muddy dirt roads. In the end, the run produced 15 well-reconstructed coincident events.

After comparison with expectations, based on air-fluorescence efficiency calculations from Bunner's thesis, it was clear that these were the first events clearly detected using air-fluorescence light. Of particular importance in proving this was the ability to show that the air-fluorescence contribution to the phototube signals could be separated from the air-Cherenkov signal. The signal's angular dependence with respect to the shower axis showed a rapid drop from zero to 30 degrees, followed by a flat tail independent of emission angle. This was a clear signature of fluorescence. A crude estimate of the EAS energy also tallied reasonably with the more traditional surface detector energy estimates.

Much of the analysis software of this and the subsequent full Fly's Eye experiment was based on the detailed work of J. Elbert, a research professor with the Utah group. The September 1977 issue of PRL featured a paper entitled "Measurement of Light Emission from Remote Cosmic Ray Showers"[8]. The era of air-fluorescence as a practical methodology for the study of UHECR had begun.

## 4 Desert dreams: the Utah Fly's Eye

Not long after the PRL publication, the Utah group, now led by George Cassiday and Eugene Loh, made a proposal to the NSF to build a full scale, 2pi solid angle air-fluorescence detector dubbed the "Fly's Eye". The detector site was selected to be on Five Mile Hill at Dugway Proving Ground, a remote Army testing facility several hours drive from Salt Lake City. The site was chosen because of its remoteness from light pollution, clear skies and dry climate, and the safety and security provided by the surrounding base.

Funding was approved and construction was essentially complete by 1981[9]. The original Fly's Eye con-

https://doi.org/10.1051/epjconf/202328301003

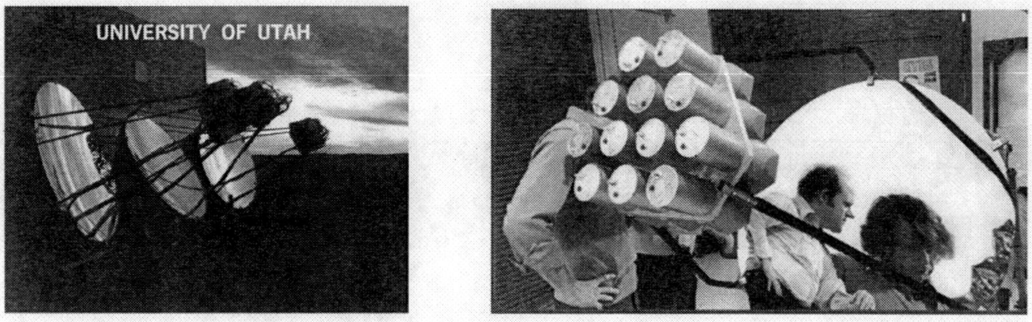


**Figure 3.** Utah air-fluorescence prototype detector. H. Bergeson (L) and G. Cassiday (R) reflected in mirror.

sisted of 67 1.6 m diameter mirrors placed in ingeniously designed galvanized metal cans that rotated down during the day. Each mirror had a camera consisting of 14 3.5 inch diameter round PMT's coupled to individual hexagonal Winston light cones. The PMT signal was fed into a sample and hold integration circuit which was triggered by any one of four different discriminator channel settings. Each discriminator channel had a different filter chosen to maximize signal to noise for a wide range of signal pulse widths corresponding to the expected range of cosmic ray energy and shower distances.

After some initial turn-on difficulties, the Fly's Eye took data, with essentially the same configuration, for more than a decade. Of great assistance to this effort was the work of an impressive series of Australian postdocs from the University of Adelaide: Peter Gerhardy, Bruce Dawson, Dave Liebing, and Dave Bird as well as Brian Fick who joined from Virginia Tech. Academic staff additionally included R. Baltrusaitis, M. Salamon and P. Sokolsky. Software for the on-line PDP-11 computer was developped and maintained by David Steck who worked closely with Cassiday.

Considerable effort was spent in the first few years developing event reconstruction and analysis programs. The lead in this effort was J. Elbert who adapted and further developed ideas from the Cornell group's work. The geometry of each event was determined by a plane fit to the direction of the triggered phototubes followed by a nonlinear fit to the arrival times in the event-detector plane. While this monocular reconstruction worked well for long tracks that exhibited significant time-angle curvature, it generated biases for shorter and more linear tracks. Reconstruction of well surveyed Xenon flasher tracks showed that geometrical reconstruction was not well enough understood. To fully understand reconstruction biases an additional detector was required.

Beginning in 1983, a second partial Fly's Eye was constructed at a site 3.5 km from the first. It had identical mirrors and electronics and events that triggered both detectors could be reconstructed using a stereo technique. The geometry of such an event was simply determined from the intersection of the two shower-detector planes. This method was largely free from the monocular timing biases and was used to fully understand FE I monocular reconstruction. By 1985 a sufficient number of new mirrors were added to FE2 to be able to do real physics with the increased aperture. The development of FE2 hardware and analysis was led by Pierre Sokolsky, who had joined the group in 1982, with the assistance of postdoc Y. Mizumoto.

The Fly's Eye produced the first real physics results that were based on the use of air-fluorescence. Earliest publications included the first limit on cosmic electron-neutrino fluxes (using the LPM effect in the Earth)(1983)[10], a measurement of the p-Air inelastic cross section(1984)[11], and a first attempt at measuring the UHECR spectrum, titled "Evidence for an Ultra-high Energy Cosmic Ray Spectrum with no sign of"(1985)[12]. The low-bias stereo data allowed the Utah group to produce the first fluorescence based EAS elongation rate measurements as well as studies of the EAS Xmax distribution. Comparison with the relatively crude hadronic interaction models that then existed already showed that the mass composition in the $10^{17}$-$10^{18}$ energy range was changing from heavy to light in apparent correspondence with the appearance of an ankle in the stereo cosmic ray spectrum in this energy region[13].

In 1991 FE 1 detected an event which generated a great deal of attention. First reported by H. Dai and carefully analyzed by him and P. Sommers, it was a very well reconstructed EAS with an energy of $3.6 \times 10^{20}$ eV. This was far beyond the expected GZK with no sign of in the spectrum[14]. Dubbed the "Oh My God Particle" by the news media, it added a mystery to our understanding of cosmic rays since the FE1 monocular spectrum was already showing evidence for a with no sign of at or near the predicted GZK energy of $6 \times 10^{19}$ eV.

Meanwhile, spectra measured by ground arrays such as Haverah Park, Akeno and AGASA seemed to show a continuing spectrum with no hint of a with no sign of. Now the "Oh My God" particle seemed to indicate possible new physics as well. The time was ripe for proposing larger and better detectors.

 https://doi.org/10.1051/epjconf/202328301003

**Figure 4.** The Utah Fly's Eye detector (L). G. Cassiday at the control terminal. The display behind him was a unique planetarium dome with LED's representing each tube's pointing direction. Events were displayed as they came in with the relative timing slowed down so the human eye could see the track development.

**Figure 5.** P. Gerhardy checking a phototube base.(L); B. Fick, B.Dawson, D. Liebing on 5 Mile Hill, the FE 1 site. One of the many 4w drive vehicles used up over the course of the experiment to access 5 Mile Hill, the F.E. 1 site.

 https://doi.org/10.1051/epjconf/202328301003

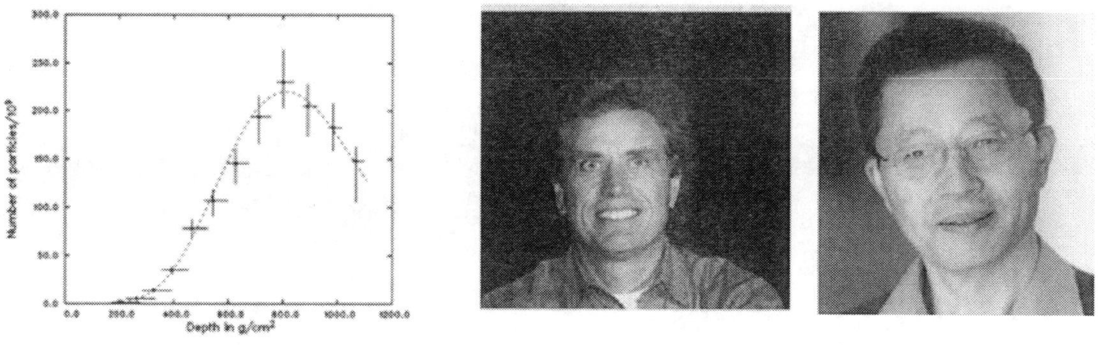


**Figure 6.** "Oh My God Particle" profile (L), P. Sommers and H. Dai

## 5 Funding Follies: The Battle For High Resolution

It was clear from the Fly's Eye experience that the next generation detector should have smaller pixel size. The 3.5" diameter phototubes used in the Fly's Eye subtended a 5 deg by 5 deg solid angle on the sky. This limited the number of track samples for distant EAS tracks making reconstruction difficult. Additionally, since fluctuations in sky noise determine the signal to noise, reducing the tube aperture would allow triggering on more distant tracks. After a number of studies, a 1 degree by 1 degree tube aperture was found to be optimal. Newer hexagonal phototubes from EMI and Hammamatsu made it possible to close-pack tubes without having to use Winston cones. The basic design of the new detector, to be called HiRes (for High Resolution Fly's Eye) was settled. It included larger 2 m spherical mirrors to be manufactured in Utah. Gene Loh and engineer Stan Thomas had developed an ingenious vacuum slumping technique to form plate glass into a spherical shape. This was followed by surface aluminization and the application of a novel hard coating to maintain reflectivity in the dusty desert environment.

Several initial proposals for funding to the NSF were unsuccessful. Review committees were moderately encouraging but strongly suggested increasing the size of the group and enlarging the collaboration. The Utah group which at this point was led by Gene Loh and Pierre Sokolsky and included Mike Solomon and Dave Kieda was then joined by Tom O'Halloran from the University of Illinois and Wonyong Lee and Bruce Knapp from Columbia University. The NSF then provided sufficient funds for the construction of the HiRes Prototype detector, consisting of 14 mirrors, each with 256 phototubes with 1 degree by 1 degree solid angle. The Prototype was built at the location of FE 1 overlooking the University of Chicago CASA gamma-ray astronomy array 3.5 km distant.

The CASA detector[15] was a 1 km by 1 km array of closely spaced plastic scintillation detectors whose purpose was to study gamma rays from Cyg X-3 and other astrophysical objects. It included an underground muon array built by the University of Michigan (MIA). Jim Cronin, who led this group, was beginning to be interested in UHE cosmic rays and the idea of combining CASA-MIA and HiRes prototype data to enlarge the scope of CASA-MIA became attractive. While each detector was independently triggered, a light signal from HiRes prototype flagged CASA-MIA data in time coincidence for future joint analysis. This became the first hybrid cosmic ray detector[16], combining shower Xmax determination with a measure of muon multiplicity from the underground detectors (the surface detector was saturated for most of the coincident data).

While the small area of the CASA array prevented the gathering of statistics much beyond $10^{18}$eV, results on the spectrum and composition from below $10^{17}$ to just above $10^{18}$ eV were published, showing evidence for a second knee in the spectrum. The muon elongation rate measurement using this data was probably the first indication that EAS models could not reproduce the muon multiplicity, even for a very heavy composition. The Xmax data confirmed the stereo Fly's Eye result: the composition was changing from heavy to light in this energy region[17]. Analysis of this data was largely done by B. Fick and Z. Cao who were postdocs at the time.

## 6 The High Resolution Fly's Eye

After successful demonstrations of the HiRes prototype detector, the group resubmitted the HiRes proposal. A difficulty arose because of the project cost, which would require approval by the National Science Board to move forward. In a stroke of luck, the minimum threshold for NSB approval was moved upwards and the NSF program manager W. Chinowsky was able to authorize funding for the full HiRes. However, only two of the three proposed stations were approved[18].

The final HiRes proposal envisaged stations spaced 12 km appart, so that a substantial fraction of triggered events could be reconstructed in stereo. The funded stations included HiRes I at Five Mile Hill ( site of the original FE and the HiRes prototype) and HiRes II at Camels

**Figure 7.** HiRes Prototype detector on 5 Mile Hill [L], CASA/MIA array with FE2 at the center [R]

Back Mountain. The detector had a total of 66 mirrors, each with 256 phototubes covering a 1 deg by 1 deg solid angle. The data acquisition electronics at HiRes I were sample and hold while HiRes II used a FADC system designed by Bill Sippach and Bruce Knapp at Columbia University's Nevis Labs . The expanded HiRes collaboration grew to include groups from Rutgers University (led by Gordon Thomson), University of New Mexico (John A.J. Matthews), Los Alamos National Laboratory (G. Sinnis) and Montana State (J. Belz). The University of Tokyo built several CTA-type steerable telescopes for gamma ray astronomy at a nearby site at Dugway. This group was led by M. Teshima. The spokesman for HiRes was Pierre Sokolsky. The scale of the project required a project manager. John Matthews was hired and continued on to manage all the Utah-based projects, including the Telescope Array (which he is currently co-spokesman of).

A substantial effort, led by Lawrence Wiencke, was devoted to understanding atmospheric transmission. A linear array of vertical Xenon flashers spaced one km apart was installed between HiResI and HiRes II. In addition, a steerable YAG laser system at HiRes II swept over the fiducial volume of the experiment on a regular basis. A nightly database of atmospheric transmission and cloud obscuration in the fiducial volume was maintained. Episodic manned hot air balloon flights carrying a Xenon flasher were performed to check the optical response of the detector to a point light source at various distances in the atmosphere. This was the first systematic attempt at understanding the impact of atmospheric variations on air fluorescence data[19]. The HiRes detector took data for eleven years, shutting down in 2006. It even survived the closure of Dugway Proving Grounds to non-military personnel after 9-11. The experiment was subsequently run by collaborators and volunteers from Los Alamos for nearly a year. They had national security clearances and were acceptable to the Army.

The most important physics results from HiRes included: the first > 5 sigma observation of a cut-off in the UHE spectrum at $6 \times 10^{19}$ eV, consistent with the GZK prediction[20] (soon after confirmed by Auger[21]) and the study of the Xmax distribution using well reconstructed stereo events[22]. The resultant elongation rate measured from $10^{18}$ to above $10^{19}$ eV was entirely consistent with a light composition (as interpreted by early versions of hadronic model simulations QGSjet and Sybill).

Under the direction of G. Thomson, the HiRes group also pioneered the detailed simulation of the response of the detector to EAS, enabling a comparison between Xmax data and predictions from hadronic models for varying compositions. The methods used were based on the succesful simulation of detector response in High Energy Physics experiments. The shape of the Xmax distribution in the $10^{18}$ to $10^{19}$ eV region was found to be consistent with predictions for a primary proton flux by several hadronic interaction models.

## 7 Snake Array, Northern Auger

After the publication of the HiRes spectrum with no sign of result, a major controversy arose with the Japanese AGASA experiment. AGASA surface detector results pointed to a continuing flux beyond $10^{20}$ eV with no sign of a cutoff. A series of debates at various physics meetings were held but there was no sign of a resolution. In the event, members of the AGASA group began to work with HiRes on plans for a future, larger aperture and possibly hybrid detector. This grew out of the realization that

**Figure 8.** HiRes Camel's Back detectors (L); HiRes pmt array and optical filter panel (R).

had HiRes and AGASA been co-sited, important checks on systematics and energy scale uncertainties could have been performed.

However, the first idea considered was an HiRes-like purely air-fluorescence detector that was to extend in a long chain resulting in a total stereo aperture 10x larger than HiRes. Dubbed the "Snake Array", significant work was done in finding 10 to 12 sites in NW Utah along the remote Snake Valley bordering Nevada. The site surveys were done by Lawrence Wiencke and Shigeru Yoshida from the University of Tokyo.

In a significant development, Jim Cronin convened the six month giant air shower array study at Fermilab in 1995. Cronin was focused on resolving the issue of the possible violation of the GZK cutoff (if the AGASA results were correct and the "Oh My God Particle" was not a fluke). The initial plans that emerged from the study were for a surface array at least 10x larger than AGASA. Cronin was persuaded by Paul Sommers and others to add an air-fluorescence detector making the energy scale estimation much less dependent on hadronic model simulation. Planning for the new Pierre Auger hybrid observatory began. It was to have complete sky coverage with a southern and northern site.

The HiRes group and the AGASA group were working together now on new ideas and a collaboration with Auger was proposed. The Northern site was to be chosen either at a new Millard County, Utah site favored by the HiRes/AGASA group, or at a site in Colorado. After considerable study of both sites, the Auger collaboration, meeting at Fermilab in 2000, chose the Colorado site for Northern Auger. This eventually led to a split with the HiRes/AGASA group which had already done preparatory work for the site in Millard County. Northern Auger was never funded by the NSF. Instead J. Cronin and A. Watson were able to put together a large international collaboration to build the southern site. Construction began on the PAO site in Malargue, Argentina in 2002 [23]. The PAO detector would have water Cherenkov detectors covering a 3000 km$^2$ surface area and four HiRes-like air fluorescence stations overlooking the array. The fluorescence stations had important detailed differences and improvements.

At the same time a Japanese group, now led by M. Fukushima from the University of Tokyo, was successful in getting funding for what was to be called the Telescope Array. Japanese funding provided plastic surface detectors covering 700 km$^2$ and two air-fluorescence stations with new mechanical design by Mitsui Engineering including segmented spherical mirrors. US groups, led by Utah were funded by the NSF to relocate refurbished parts of the HiRes detector to the new TA site thus adding a third air-fluorescence detector for full hybrid coverage. Construction on the new site began in earnest in 2003 and data taking began in 2006 [24]. The TA collaboration eventually included institutions from Japan (led by the ICRR of the University of Tokyo), Korea, Belgium (The Free University of Brussels), Russia (INS) and the United States. The first spokesmen for the experiment were M. Fukushima and P.Sokolsky.

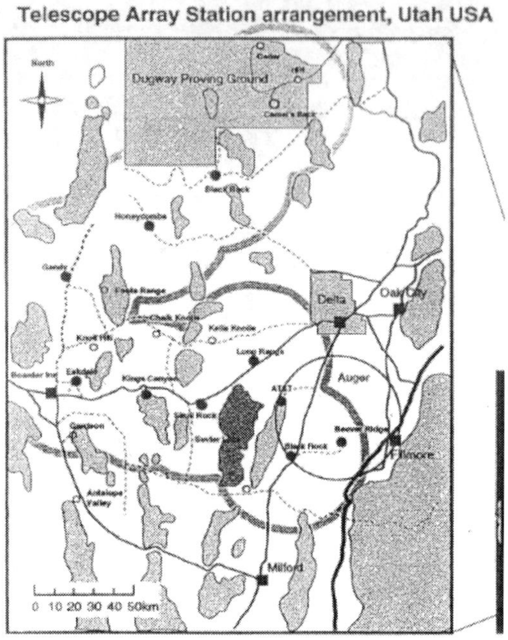


**Figure 9.** Proposed Snake Array cites in NW Utah.

## 8 TA and Auger

While the details of experimental design are different between Auger and TA, the basic hybrid detector triggering and and event reconstruction are similar, so that together, these two detectors conform to the original idea of a global N-S coverage envisaged by Cronin. The original TA had less than a third of Auger's surface aperture and began taking data at a later date, so statistics in the North were smaller than in the South. Regardless, TA surface data showed, among other results, an interesting possible medium scale anisotropy dubbed the "Hot Spot"[25]. More statistics would be necessary to actually claim a discovery, however. To facilitate this, the TAx4 proposal was developed and funded. A significant fraction of the final 3000 $km^2$ surface array was added in 2021. Two additional HiRes-like air fluorescence stations overlooking the new parts of the array were completed by the Utah group. Above $10^{19}$ eV Auger and TA now have roughly similar apertures and the search for anisotropies, global or local is proceeding apace. Of particular interest is the discovery of a large scale dipole anisotropy by the Auger collaboration[26].

## 9 The Future of the Past

Where does one go from here? While the obvious way forward is increased detector area and sensitivity to the highest UHE cosmic rays, looking to lower energies is also growing in importance. The HiRes prototype-CASA experiment showed how to move to exploring lower energy cosmic rays. Since the flurescence light is proportional to deposited energy in the atmosphere, signal to noise becomes problematic below $10^{17}$ eV. However, the Cherenkov component, which is emitted at angles up to 20 degrees from the EAS axis, is much stronger and can be used both to trigger the detector and to reconstruct the shower profile. Untill fairly recently, events with a strong Cherenkov component to the signal were discarded from analysis. In this instance, the detailed and well-vetted Cherenkov light simulation programs developed by gamma ray astronomers have made it possible to understand the EAS shower light production as a mixture of Cherenkov and air-fluorescence light, down to almost $10^{15}$ eV. The whole range of the cosmic ray spectrum from the first knee to the GZK cutoff can now be measured with a consistent energy scale.

The first steps in this direction were made by the Utah group, led by C. Jui. They built the TALE detector [27] adjacent to the Utah TA site. Expanding on the HiRes prototype idea, they increased the azimuthal and elevation angle acceptance of the HiRes-like detector. TALE was able to extend the spectrum measurement to near $10^{16}$ eV using events with a mixture of Cherenkov and air-fluorescence light. Further refinements in the analysis of events completely dominated by Cherenkov light extend the measurement to near $10^{15}$ eV. Much of the success of this analysis is due to the meticulous work of T. Abu-Zayyad, a research professor at Utah. An infill surface array near the detector was deployed to allow hybrid measurements of these events as well.

A similar set of detectors at the Auger site called HEAT and AMIGA [28] also extended the range of the Auger spectrum and composition measurement to below $10^{17}$ eV. Results from this foray into low energies have been very interesting. The spectrum exhibits three features: the first knee, a broad dip, a second knee, an ankle structure and finally a flattening, followed by the GZK like feature. Many of these individual features have been previously seen in surface detector experiments, but never in one experiment with over five decades of energy range[29].

For the highest energies, there are several paths for the air-fluorescence technique: more of the same but on a larger scale (Snake Array), reducing the cost and increasing reliability and stability of mirrors/phototubes/electronics so that they can cover huge areas and run semi-autonomously (FAST) [30], or placing a downward looking detector into space, either on the ISS or as one or more free-flyers (EUSO, POEMMA) [31]. Radio detection is increasingly competitive, in terms of ability to reconstruct shower profile and energy but requires a much higher detector density on the ground. It remains to be seen what the international cosmic ray community will decide is the appropriate mix of detectors to increase apertures 10 to 100x beyond Auger and TA. It is very likely that air-fluroescence will continue to play an important role because of its nearly model indepent energy scale determination and well-understood systematics.

**Figure 10.** [Top L] BR air-fluorescence detector station for TA, [Top R] Middle Drum detector station for TA; Bottom [L] Auger mirror and pmt array [R] Auger detector housing.

## References

[1] K. Greisen, Ann. Rev. Nucl. Sci.**10**, 63 (1960)

[2] K. Suga, *Proceedings of the 5th Interamerican Seminar on Cosmic Rays* (La Paz, Bolivia, 1962), Vol. 2, XLIX

[3] A. E. Chudakov, *Proceedings of the 5th Interamerican Seminar on Cosmic Rays* (La Paz, Bolivia, 1962), Vol. 2, XLIX

[4] A.N. Bunner, K. Greisen and P.B. Landecker, Canadian Journal of Physics **46**, S 266 (1968).

[5] A. Bunner *Ph.D Thesis, Cornell University* Report number: NYO-3038-5 (1967)

[6] T. Hara et al.,*Proc. Of 11th ICRC* (Budapest, 1969) EAS-4/2

[7] H. E. Bergeson et al., PRL**29**, No.13 847 (1977)

[8] H. E. Bergeson et al., op. cit.

[9] R. M. Baltrusaitis et al., NIM A**240**, 410 (1985)

[10] R. M. Baltrusaitis et al., Ap J Letters**281**, 9 (1984)

[11] R.M. Baltrusaitis et al., PRL**52**, 1380 (1984)

[12] R. M. Baltrusaitis et al., PRL**54**, 1875 (1985)

[13] D. Bird et al., PRL**71**, 4301 (1993)

[14] D. Bird et al., ApJ**441**, 144 (1995)

[15] K. G. Gibbs et al., NIM A**264**, 67 (1988)

[16] T. Abu-Zayyad et al., Astropart. Phys.**16**, 1(2001)

[17] T. Abu-Zayyad et al. Ap. J. **557**, 686 (2001)

[18] G. Thomson et al., Nucl. Phys. B**136** , 28 (2004)

[19] L. Wiencke et al. NIM A**428**, 593 (1999); R. U. Abbasi et al., Astropart. Phys.**25**, 74 (2006)

[20] R. U. Abbasi et al.,PRL **100**, 10111 (2008) ;R. U. Abassi et al., Ap J **622**, 910 (2005)

[21] J. Abraham et al., PRL**101**,061101 (2008)

[22] R. U. Abbasi et al. PRL**104**, 161101 (2010)

[23] P. Neesal et al., J.Phys. Conf. Ser.**293**, 012036 (2011)

[24] K. Martens et al., Nucl. Phys. B Proc. Suppl. **165**, (2007)

[25] R.U. Abbasi et al., ApJ Letters **790(2)**, L21 (2014)

[26] J. Abraham et al., Science**357**, 1266 (2017)

[27] R.U. Abbasi et al., ApJ **865**, 74 (2018)

[28] C. Meurer et al.,Astrophysics and Space Sciences Transactions **7**, 2 (2011)

[29] G. I. Rubtsov et al. *Proceedings of 2021 ICRC*, PoS(ICRC2021) 012

[30] T. Fujii et al., *Proceedings of 2019 ICRC*, PoS (ICRC2019) 259

[31] J.F. Krizmanic, *Large Area Networked Detectors for Particle Astrophysics* (World Scientific, 2023) Ch. 4, 93

 https://doi.org/10.1051/epjconf/202328302001

# Results from high energy direct measurements and future prospects

*Oscar* Adriani[1,2,*] and *Lorenzo* Pacini[2,**]

[1]Department of Physics and Astronomy, University of Florence, I-50019 Sesto Fiorentino, Florence, Italy
[2]INFN Sezione di Firenze, I-50019 Sesto Fiorentino, Florence, Italy

**Abstract.** In this contribution a review of the recent results from high energy cosmic ray measurements, in the 'above TeV' energy regions, will be presented. The future experiments that will be realised to significantly improve the current measurements, aiming to explore the PeV region with direct measurements, will also be described.

## 1 Introduction

Thanks to the experimental results obtained before and during the nineties, a 'standard' picture of cosmic ray (CR) physics has been at that time established. According to this picture, the all-particle spectrum was well approximated with a single power law from few *GeV* (above the solar modulation effect dominated region) to the *PeV* region, where a first spectral break (knee) is present: this could be connected to the effect of lack of magnetic containment of the CR in our Galaxy and/or to the limit of the galactic accelerator mechanism [1]. The value of the spectral index (about −2.7) is explained by well known theoretical models related to the acceleration of particle inside CR sources [2], e.g. SuperNova Remnants (SNRs) and diffusive process [3] inside the Galactic magnetic field. To better understand the CR physics, it is useful to separate different species in two categories: primaries, i.e. CRs originated and accelerated by astrophysical sources and secondaries, i.e. particles produced by the interaction of other particles (mostly primaries) with the InterStellar Matter. Several acceleration-diffusion models predict that all the primary CR fluxes have universal (species independent) spectral indices, while the spectral index of secondaries is expected to be different with respect to the one of primaries. Furthermore, another well accepted hypothesis was that the anti-matter components of CR are purely of secondary origin (i.e. no sources of CR anti-matter is present). Unfortunately, no accurate direct measurements of individual spectra were available during the nineties to confirm these hypothesis.
Thanks to the recent astonishing and huge progresses of balloon and space detectors, very detailed measurements of many CR species (both primaries and secondaries) have been carried out in the last decades. These results, presented in this paper, created serious challenges to the 'conventional scenario' of CR.
A brief description of the main running CR space experiments is presented in Section 2, while several relevant results concerning CR direct measurements, by exploiting different and complementary probes, will be discussed in Sections 3, 4, 5, 6 and 7. Finally, a brief discussion on future CR space experiments is also presented in Section 8.

*e-mail: oscar.adriani@fi.infn.it
**e-mail: lorenzo.pacini@fi.infn.it

## 2 Space experiments devoted to CR measurements

In the late 1990s - early 2000s, new complex and refined balloon experiments (e.g. CREAM [4]) have opened a new era of accurate measurements of the individual CR spectra. Then, the high precision and complete space experiments PAMELA [5] and AMS-02 [6] started a decade of important measurements with magnetic spectrometer, followed by space calorimeters (CALET [7] and DAMPE [8]) launched in orbit around 2015.
Generally speaking, the spectrometer-based experiments are able to separate matter and anti-matter particles, thanks to the high magnetic field and accurate tracking capabilities. Silicon strip detectors are used to identify the sign of the particle charge by precisely reconstructing the track curvature, which also allow the reconstruction of the particle rigidity (i.e. the momentum divided by the charge). Additional sub-detectors are needed to provide complementary and redundant information, e.g. a calorimeter allows to to properly separate electromagnetic and hadronic showers. As an example, the scheme of the AMS-02 instrument is shown in Figure 1. Due to the weight, volume and power consumption constraints of space instruments, the main limitations of current on-orbit spectrometers are the acceptance and maximum detectable rigidity (MDR); as a result of these limitations, these detectors can not measure particle above few TeV.
The calorimetric-based experiments are not able to separate anti-matter from matter, but with a weight similar to the one of a typical spectrometer, these instruments feature significantly larger acceptances. Moreover, the

 https://doi.org/10.1051/epjconf/202328302001

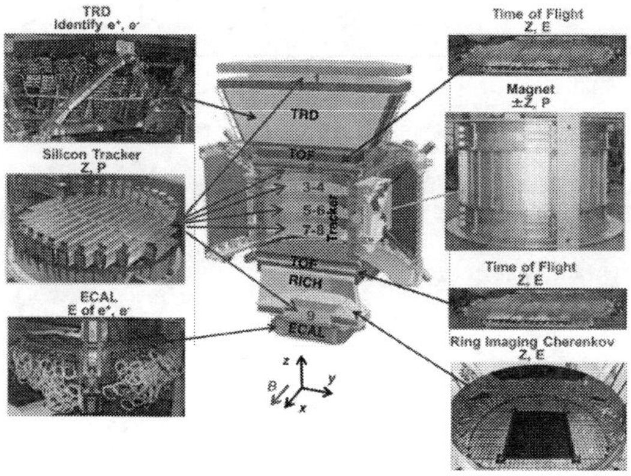


**Figure 1.** A scheme of the AMS-02 detector, showing the spectrometer and the main sub-detectors [6].

calorimeter energy resolution does not strongly decrease with the particle energy; space calorimeters are hence able to measure particles up to hundreds of TeV. Ancillary subdetectors are anyway necessary to measure the particle direction and charge, in order to well reconstruct individual species. The schemes of the two space calorimeters, CALET [9] and DAMPE [8], are shown in Figure 2.

Beside the spectrometric and the calorimetric techniques

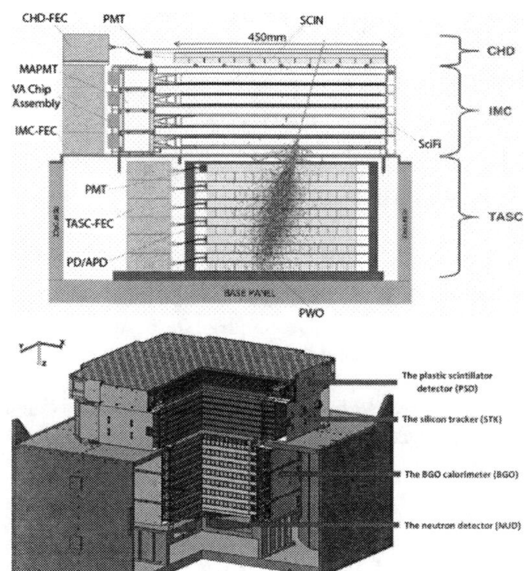


**Figure 2.** Schemes of the CALET [9] (top image) and DAMPE [8] (bottom image) detectors.

here briefly described, several additional different detection techniques can be used to achieve specific goals in the field of CR observation. A brief reference of few peculiar detection methods which will be employed for future detectors is presented in Section 8.

## 3 First unexpected results: the hardening and softening of proton and helium.

The first unexpected feature discovered during this new era of CR detectors is the proton and Helium (He) spectral hardening, i.e. the increase of the spectral index at a total energy around 200 $GeV$. This hypothesis has been firstly suggested by CREAM [4], and later accurately measured by PAMELA [10], as shown in Figure 3. Afterwards, the

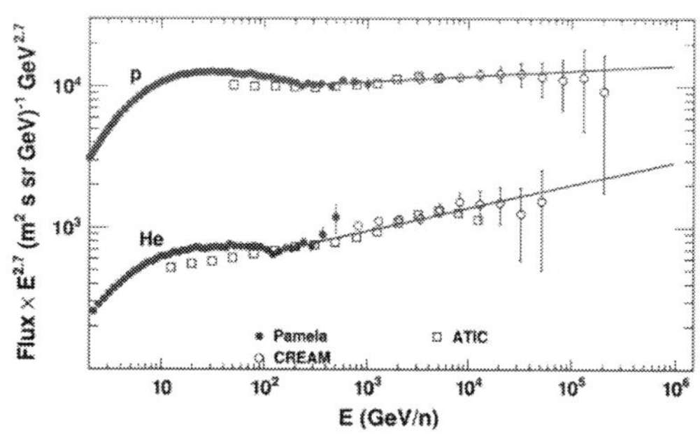


**Figure 3.** Proton and Helium spectra measured by PAMELA, CREAM and ATIC [10]. The hardening of the spectra at about 200 $GeV$ total energy is clearly shown.

hardening has been confirmed by several different experiments, even if the value of the difference in the spectral index and the turning energy (or rigidity) are not completely consistent among different results: for instance, the value of the spectral index increase ($\Delta\gamma$) measured by AMS-02 is $\sim$ 0.22 [6] while the one measured by CALET is $\sim$ 0.28 [11]. Furthermore, the He spectrum hardening is slightly smaller than the one of protons, with $\Delta\gamma \sim 0.12$ according to AMS-02 observation [12]. The hardening can be related either to new physics in the CR sources or to propagation phenomena. Several theoretical model predict that the hardening could be related to a change of the diffusive regime, e.g. [13].

Beside the hardening, DAMPE recently observed also a softening of the proton [14] and He [15] spectra above 10 $TeV$ per nucleon: the proton softening has also been confirmed by CALET [11]. The spectra of proton and He measured up to tens $TeV$ per nucleon are shown in fig. 4. At the time of the publication of this article, there are not well accepted models predicting the observed softening.

Another important observation is related to the proton/He ratio. Being these particle primaries, according to the assumptions explained in Section 1, the spectral indexes of both species should be the same. On the contrary, experimental results show that even if proton and He spectral features (hardening and softening) are similar, the overall spectral index is different. As a result, the proton/He ratio is not flat, as shown in Figure 5. This is another challenge for the CR models, and currently there is not a well accepted theory which is able to properly predict the proton/He ratio.

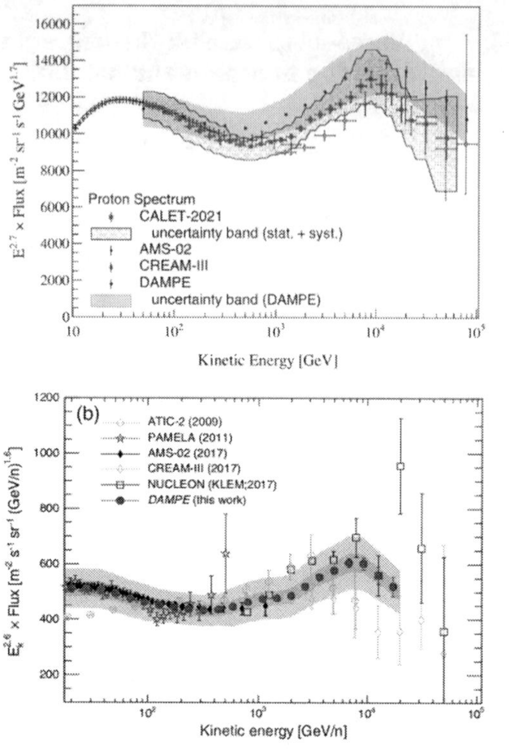


**Figure 4.** Proton and Helium fluxes measured by different experiments [11] [15].

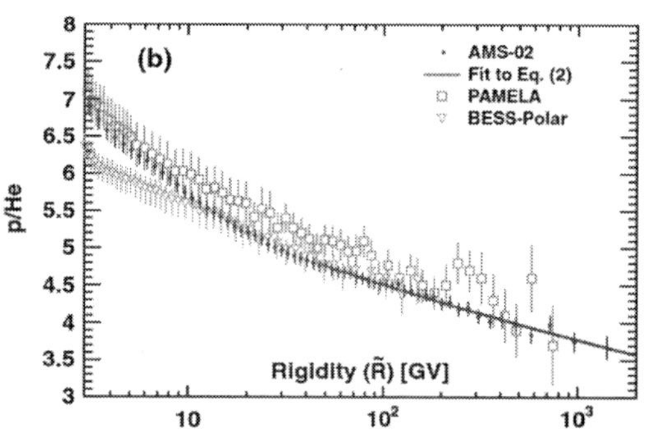


**Figure 5.** Proton over Helium ratio measured by different experiments [12].

## 4 Primary heavier nuclei spectra: a clear confirmation of the hardening.

Measurements of spectra of nuclei heavier than He are necessary to understand if the same hardening is present in every primary spectrum.

The first measurements showing an hardening consistent with the one observed for He are the Carbon (C) and Oxygen (O) fluxes, measured by the AMS-02 [16] and CALET [17] collaborations. As an example, the $\Delta\gamma$ measured by CALET for C and O is $\sim 0.15$. C and O spectra are shown in Figure 6. The figure also shows that

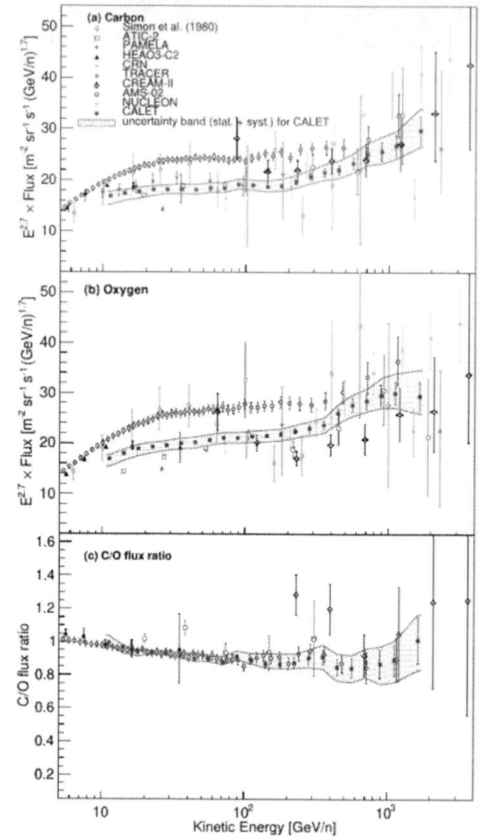


**Figure 6.** Carbon (top panel) Oxygen (central panel) and ratio C/O (bottom panel) measured by different experiments [17].

the spectral shape of C and O measured by AMS-02 and CALET are similar, but the absolute normalizations are significantly different; the flux measured by AMS-02 is systematically larger than the one measured by CALET and previous experiments (e.g. PAMELA). Understanding the reasons of this discrepancy is an important challenge for the CR experimental community.

The measured spectra of Neon, Magnesium and Silicon are compatible with the hardening at a rigidity of about $200\,GV$. Even if the hardening is present, AMS-02 data [6] shows that the rigidity dependence of these three nuclei is different with respect to C, O and He, as shown in Figure 7. Even if Neon, Magnesium and Silicon are mainly primaries, there are models that could explain these different spectral shape with respect to lighter nuclei, by taking into account parameters like the different spallation rate and the non-homogeneous diffusion of heavier CR, e.g. [18].

Regarding even heavier nuclei, the Iron spectrum has been recently measured by CALET [19] and AMS-02 [20] (see

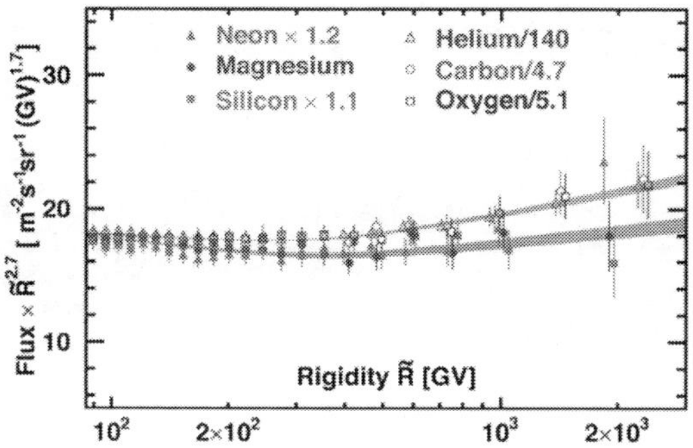


**Figure 7.** Neon, Magnesium and Silicon spectra compared with Carbon, Oxygen and Helium spectra measured by AMS-02 [6].

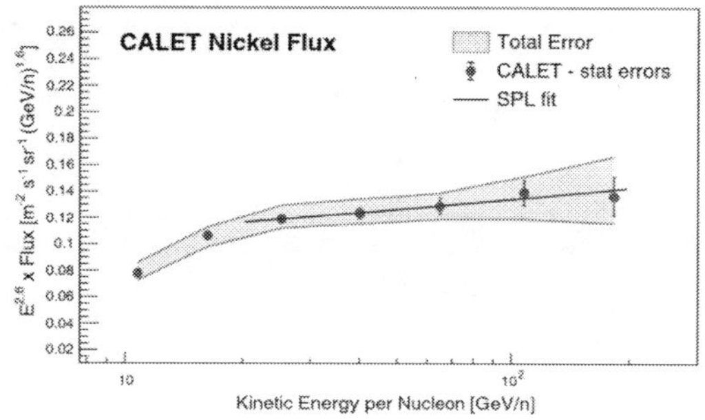


**Figure 9.** Nickel spectrum measured by CALET [21].

Figure 8 ), while a new result regarding the Nickel spectrum has been published by the CALET [21] collaboration (see Figure 9). AMS-02 Iron result suggests that the Iron spectrum is similar to the spectra of light nuclei (He, C, O). The Iron spectral index measured by CALET is consistent with the one measured by AMS-02, but the normalization is considerably different. This confirms that a normalization issue for nuclei is really present, as has already been pointed out in the case of C and O fluxes. Furthermore, CALET data shows that Nickel/Iron ratio is flat above 10 GeV. Unfortunately, by taking into account the CALET and AMS-02 published data, the limited statistic of heavy ions does not allow to clearly confirm the presence of the hardening in the Iron and Nickel spectra.

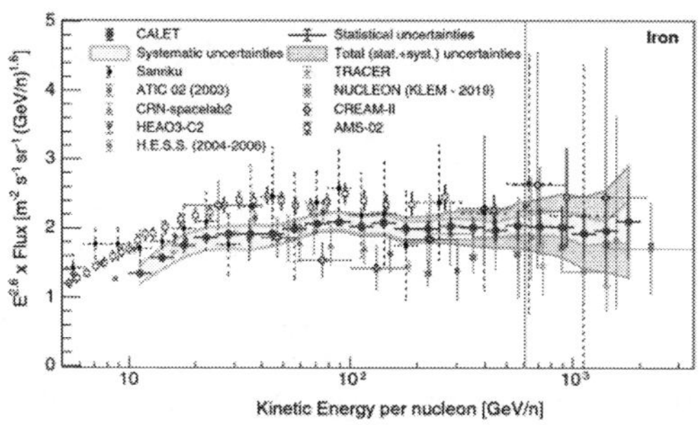


**Figure 8.** Iron spectrum measured by several experiments [19].

## 5 A unique probe for CR propagation: secondary nuclei.

Secondary nuclei, e.g. Lithium (Li), Beryllium (Be) and Boron (B), are mainly produced by the spallation of primary nuclei with the ISM. Being mainly produced after the acceleration and propagation of primaries inside the Galaxy, the spectral index of secondary nuclei is more affected by diffusion with respect to the one of primaries. Thus, the observations of secondary nuclei provide complementary information with respect to primary fluxes. Particularly, secondary/primary ratio rigidity dependence does not strongly depends on CR source physics, while it is strongly affected by CR propagation and mean path length inside the galactic magnetic field. Furthermore, secondary spectra could provide additional information about the physics of the hardening. A simplified speculation regarding this issue could be formulated as follow:

- if the hardening of secondaries and primaries is similar, this feature should be related to CR acceleration, i.e. to the physics of the sources;
- if the secondaries harden more than primaries, with $\Delta\gamma$ which is about twice the one of primaries, this feature should be related to propagation effects.

Few experiments have measured several secondary spectra and secondary/primary ratios. AMS-02 [6] data shows that Li, Be and B spectral shape are very similar each other: the hardening of these spectra is more evident with respect to He, C and O. This result supports the hypothesis which links the hardening to propagation effects. Recently, the CALET [22] collaboration published the observation of B spectrum and B/C ratio, while the DAMPE [23] collaboration published the B/C and B/O ratios. These results are compatible with AMS-02 ones and they support the presence of the hardening in secondary/primary ratio. Several measurements of B/C and B/O ratios are shown in Figure 10.

Other confirmations of the relation between the hardening and CR propagation is provided by the measurement of heavier nuclei spectra, made by AMS-02 [24]. A summary of the nuclei measurements by AMS-02 is shown in Figure 11. These results are consistent with the scenario already discussed in this section. Being Sodium, Aluminum, and Nitrogen a mixture of primaries and secondaries, the value of the spectral index of these particles is smaller than the one of He, C and O and larger than the one of B, Li and Be. Finally, Fluorine is a heavy secondary and it features slightly harder spectra with respect to light secondaries: currently it is not clear if this result is consistent with the CR standard scenario.

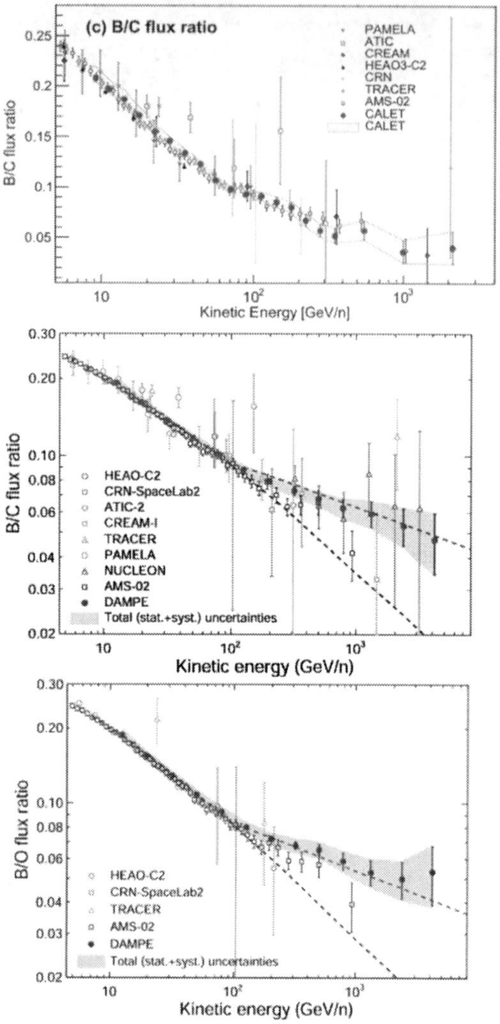


**Figure 10.** B/C and B/O measured by different experiments [22] [23].

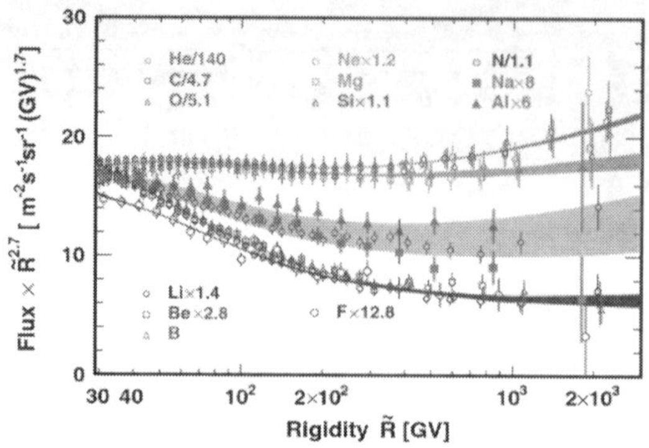


**Figure 11.** Nuclei spectra measured by the AMS-02 collaboration [24].

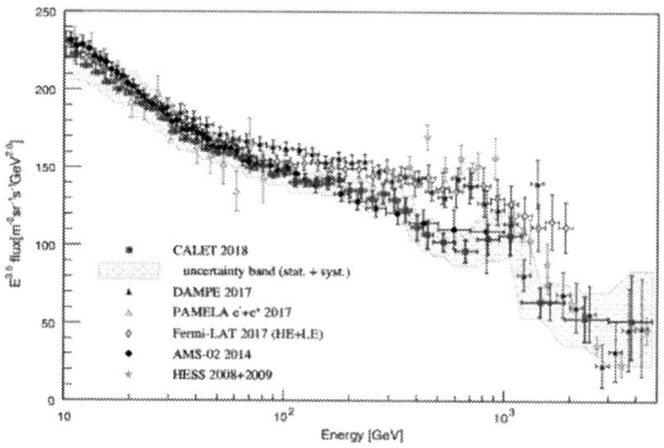


**Figure 12.** Electron flux measured by several experiments [25].

## 6 Investigating nearby CR sources with the electrons plus positron spectrum.

Differently form protons and nuclei, electrons (and positrons) loose a sizeable amount of their energy during the diffusion inside the Galaxy, due to synchrotron radiation and inverse Compton scattering. Thus, electrons with energies larger than few $TeV$ could be observed at Earth only if they are originated, roughly speaking, in sources within $\sim 1$ $kpc$ distance from the Solar System. Therefore, the high energy electron flux is strongly affected by nearby CR sources and provide complementary information with respect to heavier high energy particles, which are accelerated by sources distributed inside the entire Galaxy. Several measurement of the electron+positron spectrum are shown in Figure 12. The DAMPE experiment (blue points in figure) reported the presence of a cut-off of the spectrum at around 1 $TeV$ [26], which is consistent with the absence of CR nearby sources which significantly contribute to the TeV energy region. The data of CALET [25] (red points in the figure) are consistent with the ones of DAMPE above 1 $TeV$. During the ICRC2021 conference [27] the CALET collaboration have confirmed the presence of the electron cut-off observed by DAMPE, thanks to the improved statistics with respect to the already published data. Regarding the lower energies, around hundreds GeV, two group of experiments are clearly presents in the data reported in Figure 12; DAMPE and Fermi-LAT measured fluxes that are significantly higher with respect to the other experiments, in particular CALET and AMS-02. Understanding this difference is an important challenge for the experimental communities, since it could be related to unknown systematic errors; for instance, the work in [28] investigated the impact of scintillator non-linearity on the absolute energy scale of space calorimeters.

## 7 The puzzle of positron and anti-proton high energy spectra.

As described in Section 1, the standard scenario foresee that the anti-matter particles in CR are only secondaries. This hypothesis has been first questioned by the

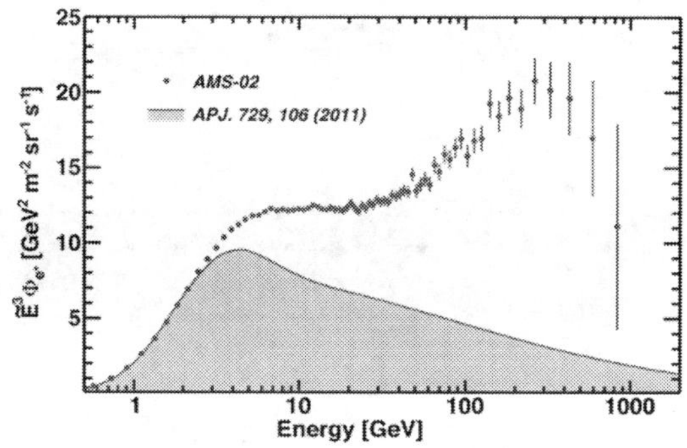


**Figure 13.** Positron spectrum measured by AMS-02 (red) compared with the prediction of the Galprop model (green) for the purely secondary positrons [6].

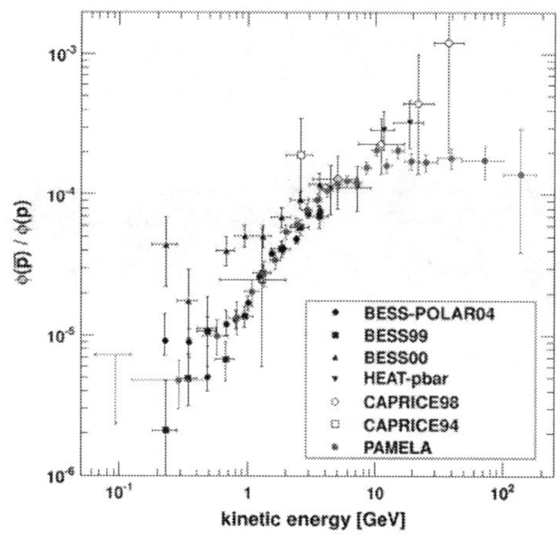


**Figure 14.** Anti-proton/proton ratio measured by several experiments [10].

positron/electron ratio published by PAMELA [10], that observed a clear positron excess above tens $GeV$, which is not consistent with the theoretical models about secondary production of anti-matter. Later, AMS-02 [6] confirmed this result and extended the positron spectrum up to 1 $TeV$: the measured flux features a peak around 300 $GeV$, as shown in Figure 13. The positron excess could be related either to nearby CR standard sources, like pulsars, not properly modelling the production and acceleration of anti-matter component, either to dark matter (DM) annihilation in our Galaxy [6]. An accurate understanding of positron excess also require accurate measurement of electron+positron spectrum above the TeV region, as discussed in previous section.

Another unexpected feature is present in the anti-proton/proton ratio, shown in Figure 14. Typical models of purely secondary production predict a decrease of this ratio above tens $GeV$, while PAMELA [10] and AMS-02 [6] experiments observed an almost flat ratio at high energies. Unfortunately, the theoretical models predicting anti-proton secondary production are strongly affected by large errors due to the uncertainties in the knowledge of relevant cross sections. Thus, to accurately evaluate the difference between the measured spectrum and the model predictions, more accurate measurements of cross sections are needed. The necessity of new measurements is well described in [29].

## 8 Future experiments for CR direct measurements

In this section, few, not exhaustive, examples of future experiments, aiming to continue this prolific era of CR direct measurements, are described. Starting from the near future, a brief description of some experiments with specific relevant goals is presented.

- GAPS [30]. This balloon experiment will measure low energy anti-nuclei by exploiting exotic atoms formation in the anti-matter annihilation in the detector. Being based on a completely innovative design, GAPS will have a sensitivity significantly higher with respect to previous experiments. The presence of anti-nuclei in CR is strongly connected to DM physics, thus this experiment will be a unique probe for DM search. The experiment should for the first time be launched in 2023/2024.
- HELIX [31]. This is an experiment based on a superconductive magnetic spectrometer, instrumented with drift chambers. It will be installed onboard of a long-duration balloon and it will measure the composition of low energy light CR, providing new information about the isotopic abundance ratios (not describer in this paper).
- TIGERISS [32]. This experiment will measure the abundance of CR nuclei from $^5$B to $^{82}$Pb. It is based on the original design of the SuperTIGER balloon instruments and it will improve it's capability in the identification of heavier nuclei. It will be installed onboard the International Space Station, greatly profiting of a very long term exposure, much larger with respect to balloon experiments.

A future general purpose next generation CR experiment is HERD [33]. This is a large acceptance calorimetric experiment. Thanks to a fully 3D design of the detector, it will be able to measure particles coming from every face of the detector (excluding the bottom one). This translates in an acceptance at least ten times larger than the current experiments, with a payload weight of about 4 Tons. The instrument will be installed on-board the Chinese Space Station around 2027. A preliminary design of the detector is shown in Figure 15. The main goal of HERD is the first direct measurement of the CR knee. Differently from the ground based experiments providing up to now information of CR knee, HERD will be able to identify the charge of every detected particle, providing unique information on the composition of CR up to the knee region. Other relevant goals of the mission are the measurement of electron+positron flux up to tens TeV and the observation of gamma ray sky. Furthermore, the

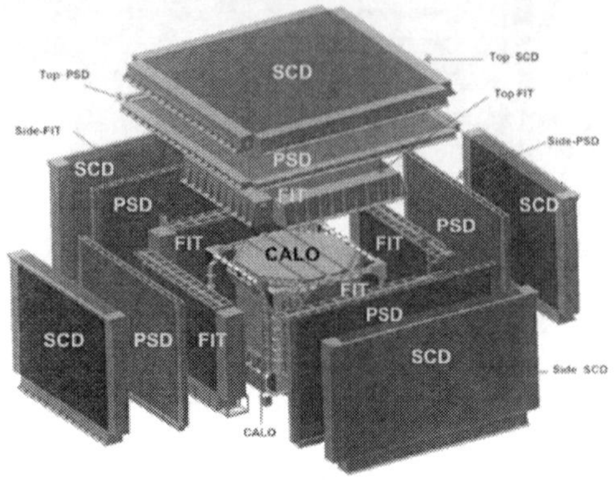


**Figure 15.** Preliminary schematic design of the HERD experiment. The calorimeter (CALO), the scintillating FIber Tracker (FIT), the Plastic Scintillator Detector (PSD) and the Silicon Charge Detector (SCD) are shown [33].

calorimeter will be equipped with two independent readout [34] systems and a TRD will also be employed to validate the energy scale of TeV protons. This innovative design will significantly reduce the energy scale systematic error thanks to unique cross-calibration capabilities, greatly helping to understand the normalization issue of nuclei and electron+positron measurements (described in Section 4 and in Section 6).

The two long term projects for the development of new generation space spectrometer with superconductive magnets currently planned are AMS-100 [35] and ALADInO [36]. Both instruments are designed to operate in the Sun–Earth L2 Lagrangian point, for several years. By following the same base approach of HERD, these detectors will accept particles coming from every directions, to increase the acceptance. Graphical renderings of the two instruments are shown in Figure 16. Thanks to the huge acceptance and to the high Maximum Detectable Rigidity, these projects aim to provide revolutionary observations of electrons, positrons, and antiprotons up to tens $TeV$. They will also be able to measure nuclear CR up to PeV energies, and investigate the presence of low-energy antideuterium and anti-helium components in CR.

## 9 Conclusion

Starting from the 2000s, a new golden era of accurate direct cosmic ray observations has been opened. The observational improvements occurred during the past decades allowed to identify in the CR spectra many new unexpected features below the knee, revealing new physics phenomena, that should still properly be incorporated in a coherent model for cosmic ray origin and propagation.

Many new open questions have hence appeared and still need to be clarified, e.g.:

- do measurements of secondary and primary nuclei confirm that the hardening is due to propagation effect?

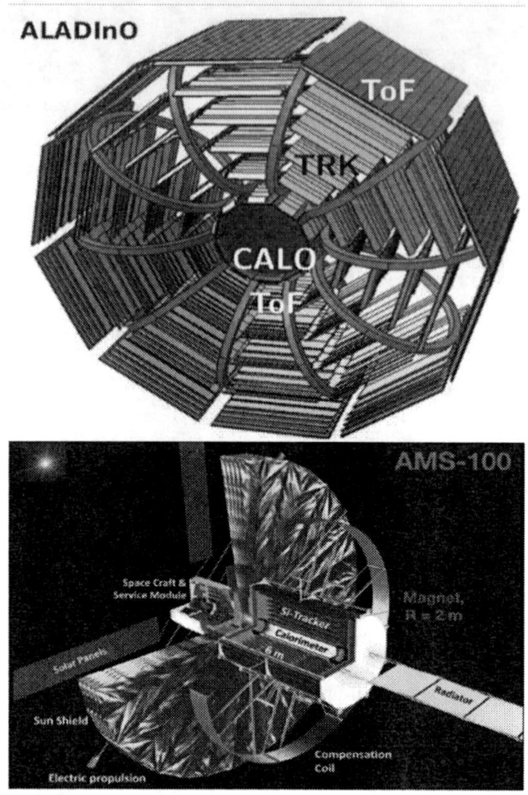


**Figure 16.** Graphical rendering of the baseline design for the ALADInO [36] (top) and AMS-100 [35] (bottom) detectors

- why is the slope of cosmic rays proton and helium spectra different?
- what is the origin of the cutoff observed at $\sim 1\ TeV$ in the electron spectrum?
- why do the proton, positron, and antiproton spectra have roughly the same slope for particle energies larger than $10\ GeV$?
- is the origin of the positron fraction rise above $10\ GeV$ related to the presence of nearby CR sources (e.g. pulsars)?
- why the antiproton/proton ratio is constant above $60\ GeV$?

The increase in statistics of the existing experiment and the new future experiments (GAPS, HERD, ALADINO, AMS-100, etc...) will allow to significantly extend the current measurements, and will also improve our still quite limited knowledge on cosmic rays. A common effort between experimentalist (in different fields) and theorists is currently nicely under way and it is absolutely necessary to shade light on the many still open intriguing questions connected to the exciting field of Cosmic Rays.

## References


[1] A. De Angelis, M. Pimenta, *Introduction to Particle and Astroparticle Physics* (Springer Cham, 2018)

[2] A. Bell, Astroparticle Physics **43**, 56 (2013)

[3] J. Skilling, Astrophysical Journal **170**, 265 (1971)

[4] Y.S. Yoon, H.S. Ahn, P.S. Allison, M.G. Bagliesi, J.J. Beatty, G. Bigongiari, P.J. Boyle, J.T. Childers, N.B. Conklin, S. Coutu et al., The Astrophysical Journal **728**, 122 (2011)

[5] O. Adriani, G.C. Barbarino, G.A. Bazilevskaya, R. Bellotti, M. Boezio, E.A. Bogomolov, L. Bonechi, M. Bongi, V. Bonvicini, S. Borisov et al., Science **332**, 69 (2011)

[6] M. Aguilar, L. Ali Cavasonza, G. Ambrosi, L. Arruda, N. Attig, F. Barao, L. Barrin, A. Bartoloni, S. Başeğmez-du Pree, J. Bates et al., Physics Reports **894**, 1 (2021)

[7] Y. Asaoka, Y. Akaike, Y. Komiya, R. Miyata, S. Torii, O. Adriani, K. Asano, M. Bagliesi, G. Bigongiari, W. Binns et al., Astroparticle Physics **91**, 1 (2017)

[8] J. Chang, G. Ambrosi, Q. An, R. Asfandiyarov, P. Azzarello, P. Bernardini, B. Bertucci, M. Cai, M. Caragiulo, D. Chen et al., Astroparticle Physics **95**, 6 (2017)

[9] O. Adriani, Y. Akaike, K. Asano, Y. Asaoka, M.G. Bagliesi, G. Bigongiari, W.R. Binns, S. Bonechi, M. Bongi, P. Brogi et al. (CALET Collaboration), Phys. Rev. Lett. **119**, 181101 (2017)

[10] O. Adriani, G. Barbarino, G. Bazilevskaya, R. Bellotti, M. Boezio, E. Bogomolov, M. Bongi, V. Bonvicini, S. Bottai, A. Bruno et al., Physics Reports **544**, 323 (2014)

[11] O. Adriani, Y. Akaike, K. Asano, Y. Asaoka, M.G. Bagliesi, E. Berti, G. Bigongiari, W.R. Binns, S. Bonechi, M. Bongi et al. (CALET Collaboration), Phys. Rev. Lett. **122**, 181102 (2019)

[12] M. Aguilar, D. Aisa, B. Alpat, A. Alvino, G. Ambrosi, K. Andeen, L. Arruda, N. Attig, P. Azzarello, A. Bachlechner et al. (AMS Collaboration), Phys. Rev. Lett. **115**, 211101 (2015)

[13] P. Blasi, E. Amato, P.D. Serpico, Phys. Rev. Lett. **109**, 061101 (2012)

[14] Q. An, R. Asfandiyarov, P. Azzarello, P. Bernardini, X.J. Bi, M.S. Cai, J. Chang, D.Y. Chen, H.F. Chen, J.L. Chen et al., Science Advances **5**, 3793 (2019)

[15] F. Alemanno, Q. An, P. Azzarello, F.C.T. Barbato, P. Bernardini, X.J. Bi, M.S. Cai, E. Catanzani, J. Chang, D.Y. Chen et al. (DAMPE Collaboration), Phys. Rev. Lett. **126**, 201102 (2021)

[16] M. Aguilar, L. Ali Cavasonza, B. Alpat, G. Ambrosi, L. Arruda, N. Attig, S. Aupetit, P. Azzarello, A. Bachlechner, F. Barao et al. (AMS Collaboration), Phys. Rev. Lett. **119**, 251101 (2017)

[17] O. Adriani, Y. Akaike, K. Asano, Y. Asaoka, M.G. Bagliesi, E. Berti, G. Bigongiari, W.R. Binns, M. Bongi, P. Brogi et al. (CALET Collaboration), Phys. Rev. Lett. **125**, 251102 (2020)

[18] B. Schroer, C. Evoli, P. Blasi, Phys. Rev. D **103**, 123010 (2021)

[19] O. Adriani, Y. Akaike, K. Asano, Y. Asaoka, E. Berti, G. Bigongiari, W.R. Binns, M. Bongi, P. Brogi, A. Bruno et al. (CALET Collaboration), Phys. Rev. Lett. **126**, 241101 (2021)

[20] M. Aguilar, L.A. Cavasonza, M.S. Allen, B. Alpat, G. Ambrosi, L. Arruda, N. Attig, F. Barao, L. Barrin, A. Bartoloni et al. (AMS Collaboration), Phys. Rev. Lett. **126**, 041104 (2021)

[21] O. Adriani, Y. Akaike, K. Asano, Y. Asaoka, E. Berti, G. Bigongiari, W.R. Binns, M. Bongi, P. Brogi, A. Bruno et al. (CALET Collaboration), Phys. Rev. Lett. **128**, 131103 (2022)

[22] O. Adriani, Y. Akaike, K. Asano, Y. Asaoka, E. Berti, G. Bigongiari, W.R. Binns, M. Bongi, P. Brogi, A. Bruno et al. (CALET Collaboration), Phys. Rev. Lett. **129**, 251103 (2022)

[23] DAMPE Collaboration, Science Bulletin **67**, 2162 (2022)

[24] M. Aguilar, L.A. Cavasonza, B. Alpat, G. Ambrosi, L. Arruda, N. Attig, F. Barao, L. Barrin, A. Bartoloni, S. Başeğmez-du Pree et al. (AMS Collaboration), Phys. Rev. Lett. **127**, 021101 (2021)

[25] O. Adriani, Y. Akaike, K. Asano, Y. Asaoka, M.G. Bagliesi, E. Berti, G. Bigongiari, W.R. Binns, S. Bonechi, M. Bongi et al. (CALET Collaboration), Phys. Rev. Lett. **120**, 261102 (2018)

[26] G. Ambrosi, Q. An, R. Asfandiyarov, P. Azzarello, P. Bernardini, B. Bertucci, M.S. Cai, J. Chang, D.Y. Chen, H.F. Chen et al., Nature **552**, 63 (2017)

[27] S. Torii, Y. Akaike, PoS **ICRC2021**, 105 (2021)

[28] O. Adriani, E. Berti, P. Betti, G. Bigongiari, L. Bonechi, M. Bongi, S. Bottai, P. Brogi, G. Castellini, C. Checchia et al., Journal of Instrumentation **17**, P08014 (2022)

[29] M. Korsmeier, F. Donato, M. Di Mauro, Phys. Rev. D **97**, 103019 (2018)

[30] F. Rogers, T. Aramaki, M. Boezio, S. Boggs, V. Bonvicini, G. Bridges, D. Campana, W. Craig, P. von Doetinchem, E. Everson et al., Astroparticle Physics **145**, 102791 (2023)

[31] N. Park, P. Allison, J. Beatty, L. Beaufore, Y. Chen, S. Coutu, M. Gebhard, N. Green, D. Hanna, H. Jeon et al., PoS **ICRC2021**, 091 (2021)

[32] B.F. Rauch, N.E. Walsh, W.V. Zober, PoS **ICRC2021**, 087 (2021)

[33] F. Gargano, *The High Energy cosmic-Radiation Detection (HERD) facility on board the Chinese Space Station: hunting for high-energy cosmic rays*, in *Proceedings of 37th ICRC — PoS(ICRC2021)* (2021), Vol. 395, p. 026

[34] O. Adriani, M. Antonelli, A. Basti, E. Berti, P. Betti, G. Bigongiari, L. Bonechi, M. Bongi, V. Bonvicini, S. Bottai et al., Journal of Instrumentation **17**, P09002 (2022)

[35] S. Schael, A. Atanasyan, J. Berdugo, T. Bretz, M. Czupalla, B. Dachwald, P. von Doetinchem,

https://doi.org/10.1051/epjconf/202328302001

M. Duranti, H. Gast, W. Karpinski et al., Nuclear Instruments and Methods in Physics Research Section A: Accelerators, Spectrometers, Detectors and Associated Equipment **944**, 162561 (2019)

[36] O. Adriani, C. Altomare, G. Ambrosi, P. Azzarello, F.C.T. Barbato, R. Battiston, B. Baudouy, B. Bergmann, E. Berti, B. Bertucci et al., Instruments **6** (2022)

# LHAASO on Cosmic Ray Knees

*Zhen* Cao[1,2,3,*] *on behalf of* LHAASO Collaboration

[1]Key Laboratory of Particle Astrophyics, Institute of High Energy Physics, Chinese Academy of Sciences, 100049 Beijing, China
[2]University of Chinese Academy of Sciences, 100049 Beijing, China
[3]Tianfu Cosmic Ray Research Center, Chengdu, Sichuan, China

**Abstract.** LHAASO construction was complete in July 2021. The full array is operating very stably since then. All arrays, KM2A, WCDA and WFCTA are calibrated, including the absolute energy scale at 21 TeV, which was measured by using WCDA and propagated to WFCTA, with the uncertainty will be reduced down <10% in 4 years. The knee of pure proton spectrum will be measured in the first phase of the hybrid measurements of showers using four types of EAS detection techniques started in 2019 winter. Sufficient data have been collected and the analysis is in progress. Since the last run, the second phase were started in 2021 winter. The knee of the iron spectrum is the goal which will take at least 3 years of data collection. CR all-particle spectrum, composition and anisotropy are under analysis.

## 1 Introduction

For 110 years, Cosmic Rays (CR) have been measured for their spectrum and composition with tremendous progresses. Particularly in the last two decades, space borne spectrometers have measured spectra of individual CR species up to few hundreds of TeV with high precision based on large statistics[1][2][3][4]. New features of the spectra of CR nuclei reveal the origins and propagation in our galaxy of CRs are much more complicated than what was learned in the era in which the detection of CR composition and spectrum was limited by the precision. However, the situation has not been significantly improved in the energy range around the knee, which is still too high to the current space borne detectors. The dominant instruments are still small scale Extensive Air Shower (EAS) detector arrays with single detection techniques. Few complex EAS detection instruments with hybrid techniques are either having very high threshold energy, such as TA and PAO experiments, or at very low altitude sites, such as KASCADE, where showers at energies near the knee decayed very much, thus the measurements of EAS suffer from the shower-to-shower fluctuation. In such a way, the uncertainty in the description of the EAS development are magnified. This leads a deep entanglement between the CR composition, EAS modeling and EAS energy reconstruction. In the 30 years of development of EAS technique as a $\gamma$-ray detection technique, the high altitudes of the EAS array sites that are closer to the shower maximum depth have been found very effective in reducing the shower-to-shower fluctuation effect[5]. Multi-parameter measurements of EAS by using hybrid detection techniques eventually bring the EAS measurement in a new 'high precision' era. More conventional way of particle detection, namely identifying the primary particles before trying to reconstruct the shower energy, or multiple iteration of shower energy and composition reconstruction, eventually becomes available in EAS detection at energies around the knee. LHAASO, by taking many cutting edge techniques in the EAS detection[6], has been built and operated for this purpose. In different phases, the spectra of pure proton, pure iron nuclei, mixture of proton and helium nuclei will be measured at energies from 100 TeV to 100 PeV in years. Meanwhile, spectrum and composition of all particles will be investigated in a way that suffers less fluctuation and has huge statistics thanks to the size of the LHAASO array. In this paper, we describe the status of LHAASO as a CR detector, the multi-parameter detection, particle identification, energy reconstruction, calibration and their performances. Observational scheme for covering different energy ranges is also discussed. Some preliminary results, such as the anisotropy of CR arrival directions, and expectations are shown in the end of this paper.

## 2 The LHAASO Experiment

The Large High Altitude Air Shower Observatory (LHAASO) had been completely built on top of Mt. Haizi (4410 m above sea level) in the Sichuan province, China by July 2021[7]. The vertical atmospheric depth is ~600 g/cm$^2$. Since then, the kilometer-square array of 5216 scintillator counters and 1188 muon-detectors (KM2A) and water Cherenkov detector array (WCDA) gaplessly covering 78,000 m$^2$ with 3120 detector cells are fully operated with a duty cycle more than 98%. Daily rate is $2\times10^8$ events above 10 TeV and $3\times10^9$ events above 300 GeV, respectively. The wide field of view (FoV) Cherenkov telescope array (WFCTA) composing of 18 telescopes with a

*e-mail: caozh@ihep.ac.cn

5 $m^2$ light collector for each is also operated in dry seasons for more than 1400 hours per year. Up to now, 70 million CR events accumulated and matched with either WCDA or KM2A, thus forming a hybrid data set of CRs above 10 TeV. The shower geometry is reconstructed using the EAS arrays with very high resolutions, i.e. the shower arrival direction and the core location are measured with errors smaller than 0.2° and 3 m, respectively.

In KM2A, scintillator counters are deployed in a triangle grid with a space of 15 m between counters which is covered by a 5 mm thick lead plate, which converts secondary $\gamma$-rays into $e^+e^-$ pairs in the counter, and measures the total deposited energy by all charged particles passing through the scintillator plates. The number of particles is measured according to the charge of single particle measured during the data taking. Muon-detectors are also in the triangle grid with a distance of 30 m between detectors, and the total active area is 40,000 $m^2$. With the overburden of 2.5 m soil, the muon detection has nearly zero backgrounds of $e^+$s, $e^-$s and $\gamma$-rays in a shower, except for one or two detectors very closed to the shower core. The $\mu$-content with the threshold energy of muons ~0.9 GeV is measured for each shower. The lateral distributions of both electromagnetic particles and muons in a shower are well measured using KM2A.

In WCDA, all energy carried by $e^+$s, $e^-$s and $\gamma$-rays in a shower is deposited in water inside the detector. Cherenkov light generated in water is proportion to the deposited energy and sampled by the PMTs at the center of each cell and 4 m beneath the water surface. For showers with energy higher than 10 TeV, the lateral distribution of the energy flux is well measured as a rather smooth function. Muons in the shower could generate larger signals than electromagnetic particles since they have longer tracks in water. They could generate exceptionally large signals in cells that the muons pass through, particularly for those cells far from shower cores.

Cherenkov telescopes are operated at higher threshold energy than those of WCDA and KM2A, so events recorded by the telescopes can find their counterparts either in WCDA or/and in KM2A. With the well reconstructed shower geometry by the arrays, the images recorded by the telescopes can be used to extract a couple of shower parameters. The shower energy can be reconstructed with the total number of photons in the image by knowing exact distance between the telescope and the shower axis. Simultaneously, the angular distance between the centroid of the image and the arrival direction measures the atmospheric depth of the shower maximum as well. The shape of image has a correlation with the shower composition at certain level.

## 3 Detector Calibrations

The pointing calibration of the ground array KM2A and water Cherenkov detector WCDA are calibrated by measuring the standard candle, the Crab, and is found better than 0.1°[8][9]. The accurate pointing direction is essential in the hybrid measurements of showers using the EAS arrays and WFCTA.

### 3.1 Pointing Direction and Photometric Calibration of WFCTA

WFCTs record bright stars in their FoVs as trajectories cross over the cameras by using the integrated counting rate in 10 second of each pixel. The known directions of the stars are used to calibrate the main axes of the telescopes for their zenith angle which is measured by an inclinometers mounted on the telescope frames. The calibration accuracy is 0.023°. Within 23 days, the daily monitoring results show that the pointing errors (RMS) in zenith and azimuth angles are 0.022° and 0.029°, respectively. The minimum to maximum is less than 0.2°. Such a pointing accuracy is essential in combining the FoVs of adjacent telescopes for events that have the images cross the boundaries of individual telescopes. By including those crossing-boundary events, an enhancement of event accumulation of 30% is achieved. The images of the starts also provide perfect uniform parallel light for the calibration of the spot size for each telescope. The measured spot size with an accuracy of 3% are hard-coded in the detector simulation and the image reconstruction programms.

The photometrical calibration of telescopes has been done at two levels. The cameras are calibrated and monitored using LED sources mounted 2570 mm from faces of cameras. LEDs generate uniform surface light illuminating the cameras. The light intensity distribution across the surfaces of the cameras is measured by a photo sensor mounted on a portable 2-dimensional platform that touches to the camera surface. Absolute luminosity at each pixel is derived by the photon counts measured by a portable calibrator placed besides the cameras. The calibrator is periodically sent to National Institute of Metrology, China for calibration. Comparing the LED signals measured by pixels with nominal values, every pixel is calibrated with an overall uncertainty less than 2.6%[10]. The other level of calibration is carried out by using lasers that shoot pulses crossing the FoV of telescopes at a distance ~1 km. The light reaches the telescopes through Rayleigh scattering is calculated with a uncertainty of 7% as the known light sources. They are used to calibrate each entire telescope, including entrances, reflectors and cameras. The calibration results will be available soon.

### 3.2 Absolute Energy Calibration

The shadow of the Moon in galactic CRs is observed shifting westwards due to the geo-magnetic field(GMF)[11]. Given the daily updated measurements of GMF, we calculated the deflection of a mixed sample of protons and Helium nuclei according to the composition ratio measured by DAMPE[2][3] and the corresponding shift of the shadow, $\Delta$, as a function of energies, $E$ in TeV, of CR particles at the LHAASO site, i.e. $\Delta = 2.1°/E$. Showers detected by WCDA are grouped according to the energy estimator, namely the total charge, $Q$ recorded by the registered WCD cells in a shower event, to measure the shift of the Moon shadow, thus we have the WCDA calibrated as a shower energy detector, i.e. $E = 1.33Q^{0.95}$ GeV, where $Q$ is in unit of number of photo-electrons. For the minimum

$\Delta = 0.06° \pm 0.03°$ that can be measured by WCDA which has an angular resolution of ~0.15°, the median energy is found to be 35 TeV. This is the highest calibration energy in CR experiments.

WFCTs can not be operated directly towards the Moon. The energy scale measured by WCDA is propagated to WFCTA by using commonly measured showers by the both detectors. Those showers arrive in a certain zenith window matches the FoV of the telescopes and central range of WCDA so that they are well reconstructed. The requirement of having well measured images of showers by WFCTA sets a rather high threshold above 15 TeV according to the reconstruction by WCDA which is calibrated as described above. The median energy of the group of events is (23.4±0.1) TeV. The median energy of those events reconstructed using the images measured by WFCTA is (21.9±0.1) TeV. This verifies the energy reconstruction by WFCTA with a discrepancy of 5.7%. To estimate the uncertainty of the scale itself, the event selection criteria are applied to the data set for Moon shadow measurement and the shift of the Moon shadow is measured again. It is found a shift of $0.10° \pm 0.03°$ of the shadow, translating to (21.0±6.5) TeV. The uncertainty of 30% is statistics dominant in the measurement of the shift. It dominates over the propagation error of 5.7% as well. The results are shown in Figure 1[12]. With time being, the statistic error of the shift measurement could reduced down below 10% in 4 years. The systematic difference between the two techniques will not longer be negligible.

## 4 Hybrid Measurements of Spectra of Individual CR Species

Identification of primary particle of EAS is a very challenging task. According to the experience in particle physics, measuring as many independent parameters of showers as possible is a promising way to approach. In LHAASO, four types of detectors are deployed to measure showers from different aspects, as an analog of spectrometers in collider experiments. They are the lateral distribution of shower particles (or energy flux) both at very closed region of the core within 2-3 meters and over a very extended area up to hundreds meters from the core, the content of muons and their lateral distribution over a very wide area, and the Cherenkov image of the shower development through the atmosphere. Depending on where the shower core is, at least two type of hybrid measurements of EAS are carried out in the LHAASO experiment.

One of the goals is measuring the spectra of pure proton samples and mixed proton plus helium samples (denoted as H+He) at energies around the knee, and extend the energy range down below 100 TeV where the spectra have been measured in space borne detectors such as DAMPE[2] and [3]. The knee of the proton spectrum will be unveiled with a clear transition from the spectral index from 2.7 to greater than 3.0, so that the energy range should at least start from 100 TeV and cover all energies up to 10 PeV. Showers with cores in one of the pools of WCDA of 22,500 $m^2$ (denoted as WCDA-1) collected in 2019-2022 observational seasons are sufficient for this purpose. This data set is also sufficient for the spectrum measurement of the H+He samples, by which a fine structure that indicates the knee of the Helium spectrum is expected to be uncovered. The coverage of higher energies might be necessary, thus larger statistics may be required. It could be achieved by having higher selection efficiency than the proton samples discussed below in detail.

The other goal is measuring the spectrum of iron around its knee at an energy that might be above 10 PeV and below the second knee around 200 PeV. This requires a much larger accumulating effective aperture for sufficient statistics at such higher energies than that for the lighter nuclei. All events with cores falling into the entire KM2A should be collected. With the spacing of 15 m between the scintillator counters, information from the shower core will be missing. For shower at such high energies, more inclined showers are selected to avoid those not having developed sufficiently so that the showers touch down the ground before reached the maxima.

For the two goals, WFCTA are configured differently and naturally divided into two phases of the observation. The first phase started with 6 telescopes deployed at the southwest corner of WCDA-1 from 2019-2020 observational season. All telescopes have the elevation angle of 60°, i.e. the zenith angle range from 22° to 38° is covered, so that there is at least 647 $g/cm^2$ for showers to develop in the air. Before April 2021, the hybrid operation accumulated more than 70 million events with the cores inside WCDA-1 for 720 hours during which 1/2 of KM2A were operated. The configuration of all four types of detectors is shown in Figure 2. Showers have the cores in WCDA-1, the green square area in the figure, have the most complete measurements of all parameters. The pure proton samples will be selected among them. A larger hybrid event sample can be selected with cores falling in the KM2A, e.g. the region enclosed by the red line in Figure 2. Without the parameters measured by using WCDA-1, the species identification is more difficult but it could be compensated by measuring the $\mu$-content better because more muon detectors are registered in those showers. A mixed H+He samples could be selected among them. The large area of KM2A and higher selection efficiency than that of pure protons allow a precise measurement of the spectrum with large statistics. It is essential for searching for the fine structure of the H+He spectrum around the possible Helium knee.

In phase II starting from September 2021, all 18 telescopes are deployed to the sites near the southeast corner of WCDA-1, and form a ring-shaped FoV covering all azimuthal direction between zenith angle range from 37° to 53°. Such a setup is aiming at the detection of CRs at higher energies around the possible knee of iron spectrum, i.e. >10 PeV, so that more atmospheric depth, >751 $g/cm^2$, allows better developments of showers. To collect sufficient statistics,the entire KM2A needs to be used to achieve the hybrid measurements. The layout the detector arrays is illustrated in Figure 3. Since the detector configuration for the hybrid measurement of showers that have cores in WCDA-1 is very similar to Phase I, except for more inclined, those events are also used for the pure proton analysis. This offers an important opportunity of cross checking on the proton spectrum measurements. It is also found that the events for proton showers at lower energies as those in Phase I could be also well measured at deeper atmospheric depth. It is due to the large fluctuations in shower development for protons. Showers in WCDA-1 are found measured better as well with the help of the rest part of entire WCDA in phase-II.

### 4.1 Multi-Parameter Analysis for Separation between Species

A typical hybrid shower event with the core inside WCDA-1 recorded in Phase-I is shown in Figure 4. In the right-upper panel, WCDA-1 measures the shower core with a huge dynamic range up to 0.2 million photoelectrons recorded in a cell as indicated by the color scale in the figure. This allows a measurement of the fine structure of the lateral distribution of the energy flux in the core. Here, shower particles have not experienced much scattering in the air so that they represent the fresh population generated not far above the surface, in other words, the hadronic components of the shower before them hit the surface. The energy flux weighted average lateral distance $r_i$ of detector cells $< ER >= \Sigma Q_i r_i / \Sigma Q_i$, where $Q_i$ is the charge measured by the $i$-th cell, is chosen as a parameter sensitive to the primary species, denoted as $p_F$ after normalized by the preliminarily determined shower energy. The right-lower panel in Figure 4, the detection of the muon lateral distribution is shown as the number of muons in each MD. It is noticed that only handful muons recorded by each MD at distance more than 100 m from the core. Summing over all MDs between 30m and 380 m from the core for the total measured number of muons. It is a rather traditional parameter for species selection in EAS observation due to the relation $ln\mathrm{N}_\mu \sim 0.1 ln\mathrm{A}$, where $\mathrm{N}_\mu$ is the muon content of the shower induced by a primary particle with the atomic number A. Normalized by using the total muon content in the shower, the parameter $P_\mu$ is selected as a sensitive variable for primary species identification. In the left-upper panel of Figure 4, a typical image of the shower is recorded in the camera of WFCTA. It is well known that the shower elongates longer for proton initiated shower than that by heavier nuclei. However, the length of the image depends the distance of the the shower to the telescope, $R_p$, due to a pure geometrical effect. Corrected by $R_p$ and the preliminarily reconstructed shower energy, the parameter defined as the ratio of the length and the width of the image, denoted as $P_C$, is found sensitive to the composition of the primary species.

Before applying the multi-variate analysis to those parameters, it is important to exam the capability of the parameters individually in the primary particle identification with a simulation toolkit that takes all detector effects into account for all four type of detectors in LHAASO. In Figure 5, the distributions of the three parameters, $P_F, P_C$ and $P_\mu$, of proton initiated showers are plotted against the same distributions of showers initiated by Helium and all heavier nuclei. It is clear that the parameters are well selected to represent the characteristics of shower initiated by different nuclei. Here, the composition model by Gaisser *et al.*[16] is assumed. It looks a very challenging if a high purity of the proton sample, e.g. 90%, is the demand, the event selection efficiency is only ~20%. On the other hand, correlations between those parameters are found not negligible. Further efforts to find more independent parameters is still on going. Moreover, for cores falling in KM2A, $P_\mu$ and $P_C$ are available. Selection efficiency using $P_\mu$ is even improved due to better muon content is measured than those in WCDA. Large statistics provided by the huge KM2A, makes this measurement very useful, particularly serves as cross checking.

Same parameters can be used in H+He sample selection. It is found the situation is much better than that for protons. The purity is easily to reach 95% with selection efficiency higher than 40%. Similarly, groups of independent data could cross check between the different approaches.

Finally, multivariate analysis will be applied to those data sets particularly for those have more than 3 parameters.

### 4.2 Energy Reconstruction

For pure proton or iron samples, the shower energy reconstruction is rather straightforward by using the total charge measured in the Cherenkov image of a shower at the impact parameter $R_p$ from the telescope. Given the well determined shower geometry by using the EAS arrays, the shower energy uncertainty is dominated by the fluctuation of shower longitudinal development. In case of proton showers, quite symmetric resolution functions are found to be fitted with a Gaussian form with the $\sigma$ to be 20% around 0.1 PeV and 15% above 1 PeV. The systematic shift of the Gaussian central expectations are less than 2% over the energy range from 0.1 to 10 PeV. The energy response function for showers at impact parameters of 120-130 m from the telescope, resolution and systematic shift as functions of shower energy and a resolution distribution at 1 PeV are shown in Figure 6. Some non-Gaussian tails are found beyond $2\sigma$ region due to the deeply developed showers were favored in the proton selection procedure. Further modification of the energy reconstruction by taking into account the EAS parameters is underdevelopment to improve the resolution.

For the mix samples with proton and Helium nuclei, the reconstruction is found more difficult due to the systematic difference as large as 7% between the two types of showers. For the hybrid samples measured by KM2A and WFCTA, the $\mu$-content of a shower is found very helpful to compensate the difference, if a term proportional to the $\mu$-content is included in the shower energy estimator, which was selected as the total charge of the Cherenkov image as that used for the pure proton samples. The latest progress[17] in this direction shows that the systematic difference has been reduced less than 2%.

## 5 All Particle Spectrum, Composition and Anisotropy Measurements

A long standing problem is that the composition, $ln$A, and shower energy, $E$, are entangled each other in CR detection using the EAS technique, if a sole shower detection technique is used in an experiment. Using the multiple EAS parameters detected by LHAASO, we are developing ways to measure the energies and the composition of the primary particles. One way is using KM2A array which measures both $\mu$-content and the lateral distribution of shower charged particles and $\gamma$-rays, defined as EM-particles, simultaneously. For showers in the knee region, i.e. $0.1 < E < 10$ PeV, the 1 km$^2$ array collects sufficient statistics even only quasi-vertical events being used for the measurements. Two orthogonal linear combinations of the two densities of the EM particles and muons are found quite independent measurements of shower energy and composition[13, 14], respectively. The energy resolutions are ~20% at 0.1 PeV and ~10% above 10 PeV, respectively, maintaining a rather small reconstruction bias at the level below 2%. Independently, a difference of $ln$A at a level of ~1 can be resolved, using the relation between average $\mu$-content of showers induced by nuclei with atomic number $A$ and $A$, i.e. $N_\mu^A = A(E/A\epsilon_c)^\beta$[15], where $\beta$ is a constant index of ~0.9 around the knee, and $\epsilon_c$ is the critical energy also a constant. The average composition is indicated by $< ln\mathrm{A} >= (lnN_\mu^A - lnN_\mu^p)/(1-\beta)$, where the average $\mu$-content of proton induced showers can be measured by using the pure proton samples discussed above.

The other way is using the hybrid measurements of events that detected also by WFCTA. The shower-maximum depth $X_{max}$ is measured by using the angular difference between the direction in which the Cherenkov image has the maximum and the shower arrival direction. For showers arriving between $27° < \theta < 34°$ ($\theta$ is the zenith angle), the resolution is found to be ~45 g/cm$^2$ for proton-like shower and ~34 g/cm$^2$ for iron-like shower around 1 PeV, respectively. $< ln\mathrm{A} >= (X_{max}^p - X_{max}^A)/k$ indicates the average composition varies with shower energy. Here, $X_{max}^p$ and $X_{max}^A$ are the average depths of maxima of showers induced by protons and nuclei with A nucleons. $k$ is the *elongationrate* ~85 g/cm$^2$ per decade around the knee. $X_{max}^p$ can be measured by using the pure proton samples discussed above.

Here we see a good opportunity of cross checking between different techniques for the CR composition measurements. It is a very important investigation particularly for any possible change over the knee. Moreover, it also opens a window for a precise investigation in the PeV regime about the long standing $\mu$-content puzzle.

The preliminary measurement of the anisotropy of CR arrival directions is done by using the KM2A data above ~10 TeV. A clear structure with a deficit around a tenth of percent of the flux is found consistent with the 'loss cone' structure. The 'tail-in' enhancement also clearly measured at energies below 160 TeV. However, an evolution with energy of the anisotropical pattern has been revealed[18]. The corresponding systematic uncertainty has been estimated by measuring the flux variation over the period of anti-Sidereal time. It is found at least one order of magnitude lower than the signal level. Giving the huge statistics and capability of primary particle identification by KM2A, the further investigations for different species and over an even wider energy range will reveal more interesting results, particular attentions need to be on the energy dependent transition as shown in Figure 7[18].

## 6 Expectations and Summary

In summary, LHAASO has collected sufficient data in the hybrid measurements of charged CRs in phase-I. Both combinations by using WCDA-1 plus WFCTA and KM2A plus WFCTA are under analyzing towards producing pure proton spectrum from 100 TeV to few PeV and H+He spectrum up to 10 PeV. Sufficient number of events allows to resolve the knee of proton spectrum and fine structure around knee of Helium spectrum. The KM2A alone has also collected huge data set for all particle spectrum and composition measurements using the measurements of lateral distributions of both electromagnetic particles and $\mu$s. The hybrid data setKM2A plus WFCTA also used in the all particle spectrum and composition analyses. Many cross-checking opportunities become available for the fundamental measurements in the knee energy region. At the first time, the energy absolute calibration by measuring the 'negative beam', i.e. the Moon shadow, has been pushed to as high as 21 TeV. The operation of hybrid CR detection using the entire KM2A and WFCTA is stable with high duty cycle in phase-II, aiming at the higher energy range covering the knee of iron spectrum above 10 PeV.

**Acknowledgments** The author would like to thank all staff members who work at the LHAASO site above 4400 meter above the sea level year round to maintain the detector and keep the water recycling system, electricity power supply and other components of the experiment operating smoothly. We are grateful to Chengdu Management Committee of Tianfu New Area for the constant financial supports to the researches with LHAASO data. This research work is also supported by the National Key R&D program of China under the grant 2018YFA0404201.

## References

[1] M. Aguilar *et al.* (AMS Coll.), Phys. Rep. **894**, 1-116, (2021).

[2] An,Q.,*et al.* (DAMPE Coll.), Science Advances, **5**, eaax3793 (2019).

[3] Alemanno, F.*et al.* (DAMPE Coll.), Phys. Rev. Lett. **126**. 201102 (2021).

[4] F. Alemanno *et al.* (DAMPE Coll.), Science Bulletin **67**, 2162, (2022).

[5] Zhen Cao, Universe **7**, 9, 339, (2021).

[6] H.H. He, Radiation Detection Technology and Methods **2**, 7 (2018).

[7] Z. Cao, Nature Astronomy **5**, 849-849 (2021).

[8] F.Aharonian, *et al.* (LHAASO Coll.), Chin.Phys. **C 45**, 025002 (2021).

[9] F.Aharonian, *et al.* (LHAASO Coll.), Chin.Phys. **C 45**, 085002 (2021).

[10] F.Aharonian, *et al.* (LHAASO Coll.), NIM, **A 1021**, 165824 (2022).

[11] A.Bartoli,*et al.* (ARGO-YBJ Coll.), Phys.Rev. **D 84**, 022003 (2011).

[12] F.Aharonian, *et al.* (LHAASO Coll.), Phys.Rev. **D 104**, 062007 (2021).

[13] Bai,L., Liu,H., on behalf of LHAASO Coll., ICRC2022, DOI: 10.22323/1.395.0246 (2022)

[14] Zhang,H., He, H., Feng, C., Phys.Rev. **D 106**, 123028 (2022).

[15] J. Matthews, Astroparticle Physics 22, 387 (2005).

[16] T. Gaisser, T. Stanev, S. Tilav, Frontiers of Physics **8**, 748 (2013).

[17] Liping Wang *et al.*, submitted to Phys.Rev. **D**, (2023).

[18] Gao,W., *et al.*, on behalf of LHAASO Coll., ICRC2022, DOI: 10.22323/1.395.0351 (2022).

https://doi.org/10.1051/epjconf/202328302002

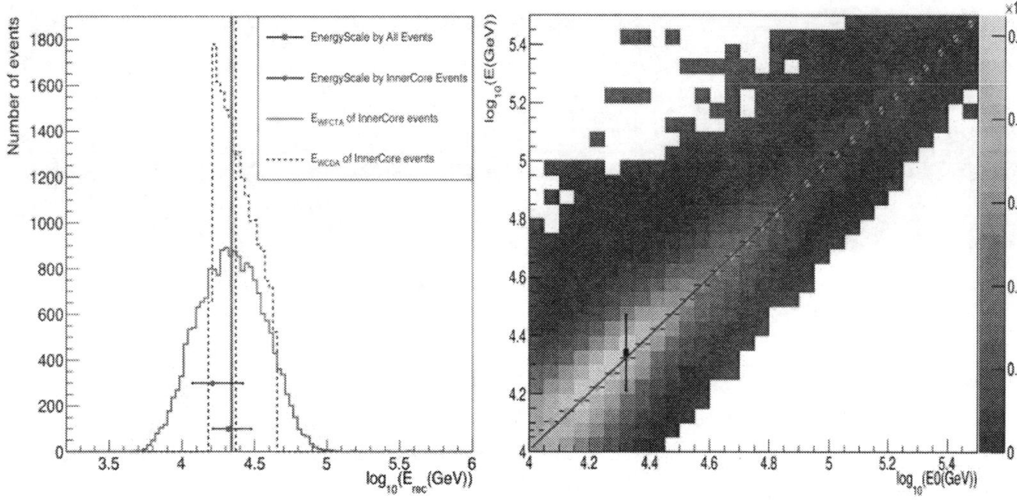


**Figure 1.** The energy scale measurement by WCDA (dotted histogram and the red dot with the horizontal error bar of 30% in the left panel) and propagation to WFCTA (solid histogram in the left panel). The dotted and solid vertical lines indicate the median energies[12]. The energy response function of the WFCTA is shown in the right panel. The black dot is the calibrated energy of 21.0 TeV with the 30% uncertainty shown as the error bar.

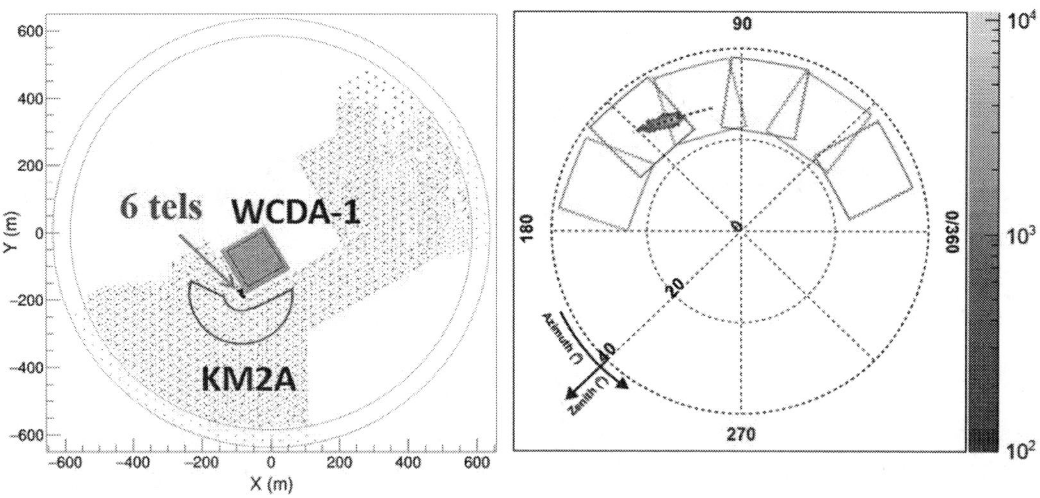


**Figure 2.** The configuration of LHAASO hybrid operation for cosmic ray observation before April 2021. The layout of all detectors is shown in the leftt panel. The FoV of the 6 WFCTs is shown in the right panel in the polar-coordinate system where the radius indicates the zenith angle and the polar angle in counterclockwise indicates the azimuth angle. Each square indicates the corresponding FoV of the telescope. There are overlaps between the adjacent FoV's. An image of a shower is also shown in the panel.

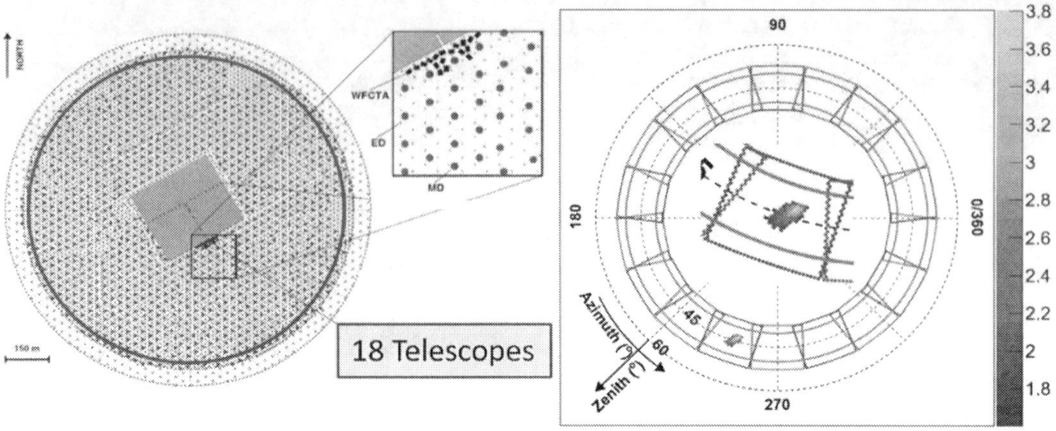


**Figure 3.** The configuration of LHAASO hybrid operation for cosmic ray observation after July 2021. The layout of all detectors is shown in the left panel. The FoV of the 18 WFCTs is shown in the right panel in the same coordinate system as in Figure 2. It is clearly seen that telescopes form a boundary-less FoV in the azimuthal direction between zenith angles 37° and 53°. Two zenith angle contour lines in pink indicate the event selection criteria images having heads between them, are well measured showers. The FoV of the telescope that has a shower image recorded is enlarged and put in the middle of the panel as an inset.

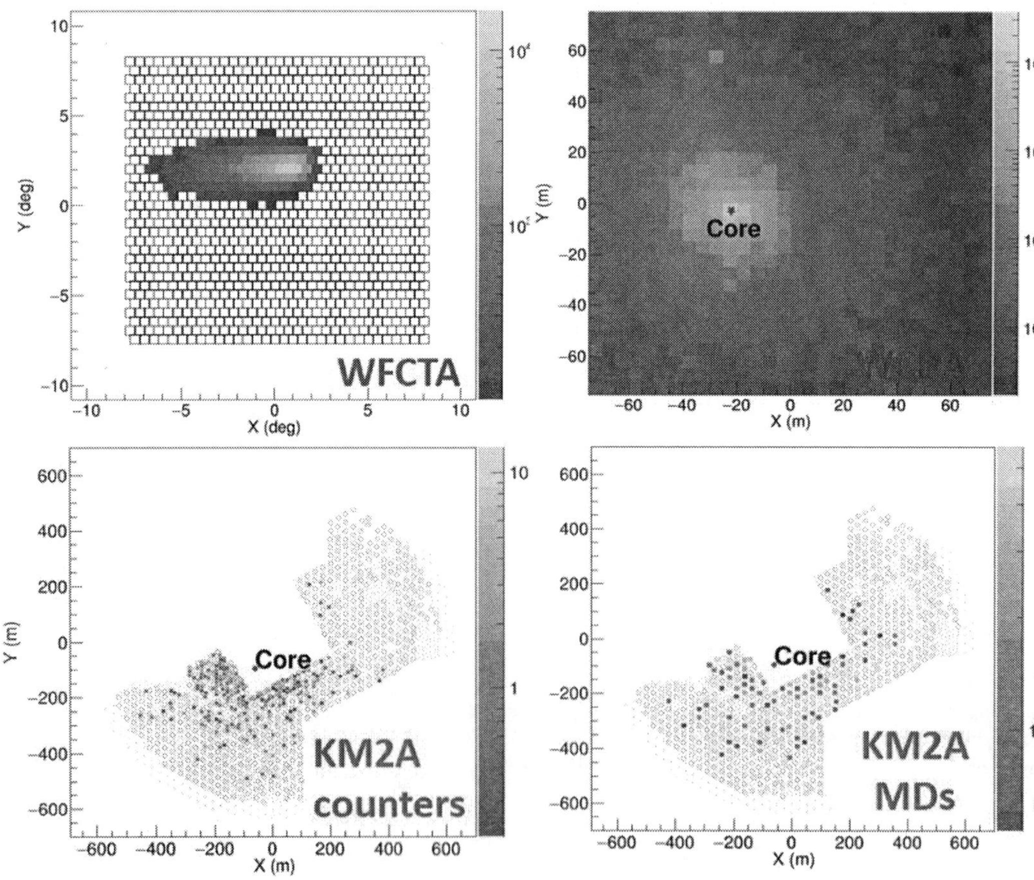


**Figure 4.** A typical hybrid event recorded in Phase-1.

 https://doi.org/10.1051/epjconf/202328302002

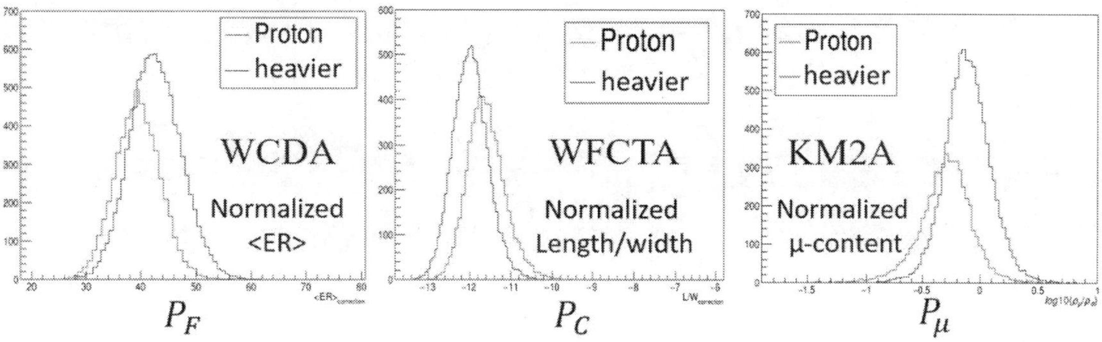


**Figure 5.** Primary species sensitive shower parameters measured by WCDA ($P_F$), WFCTA ($P_C$) and KM2A ($P_\mu$), respectively. Distributions of those parameters of showers induced by protons are in red. The same distributions of showers induced by Helium and heavier nuclei are in blue, according to the composition model by Gaisser *et al.*[16].

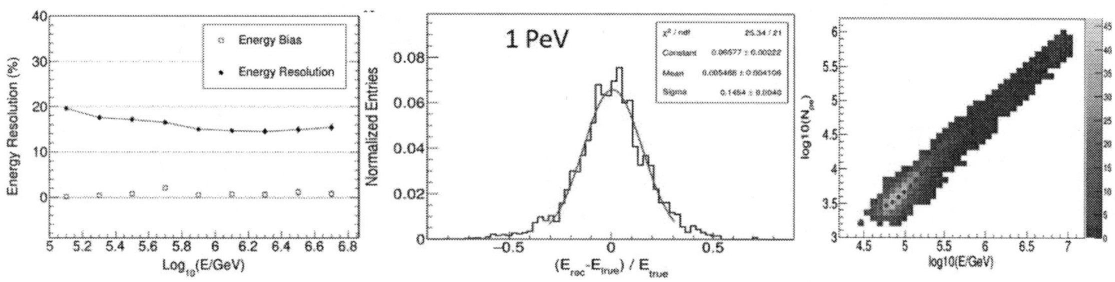


**Figure 6.** Energy reconstruction for pure proton samples. In the left panel, the shower energy resolution and systematic shift are shown over the energy from 100 Te V to 5 PeV. In the middle panel, the energy resolution function, i.e. the distribution of relative deviation between the reconstructed energy $E_{rec}$ and input energy $E_{true}$, at 1 PeV. In the right panel, the energy response function for showers at impact parameters from 120 to 130 m from the telescope. The energy estimator is the total charge measured in the shower image in unit of number photo-electrons ($N_{pe}$).

 https://doi.org/10.1051/epjconf/202328302002

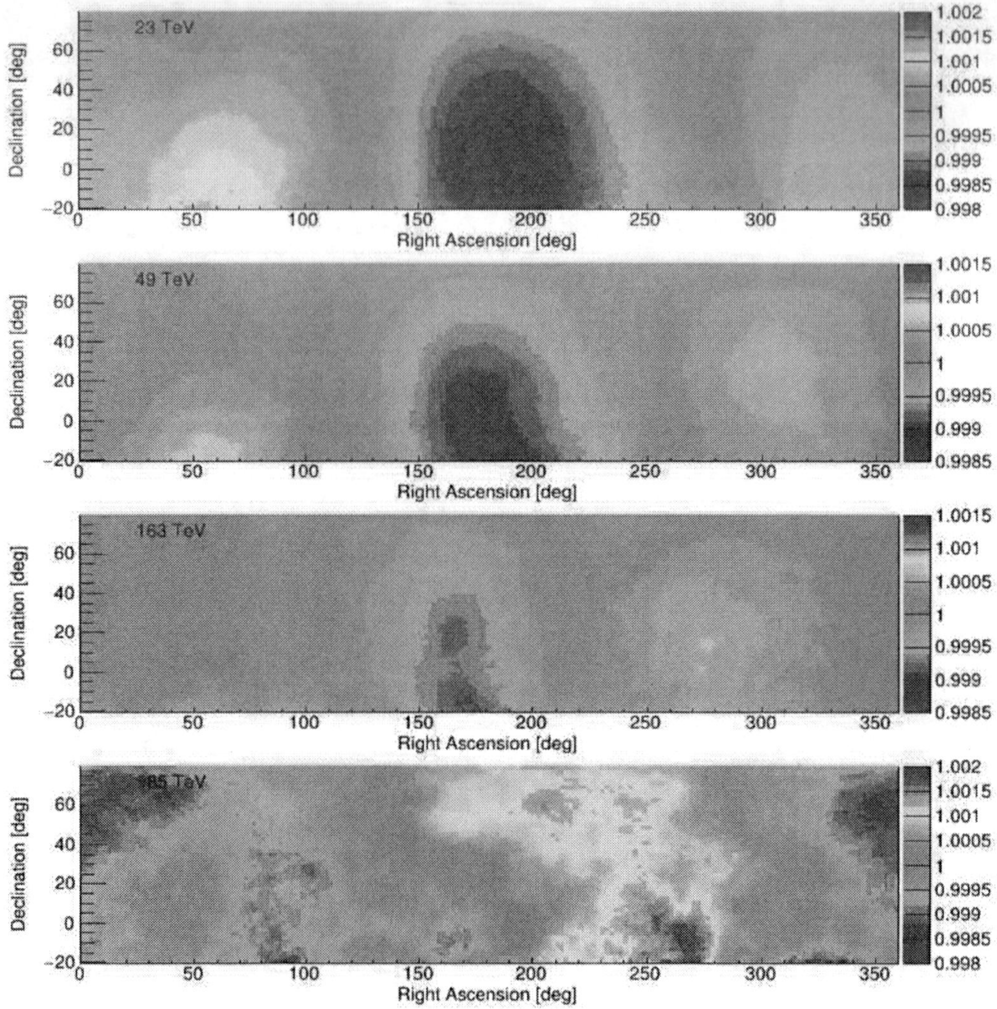


**Figure 7.** Anisotropy of flux of CRs at energies from 23 TeV to 1 PeV measured by KM2A[18].

# The energy spectrum of ultra-high energy cosmic rays measured at the Pierre Auger Observatory and the Telescope Array

*D. R.* Bergman[1], *O.* Deligny[2], *F.* Fenu[3], *T.* Fujii[4], *K.* Fujita[5], *J. H.* Kim[1], *I.* Lhenry-Yvon[2], *I.* Mariş[6], *Q.* Luce[7], *M.* Roth[3], *F.* Salamida[8,9], *Y.* Tsunesada[4,10], and *V.* Verzi[11,*] for the Pierre Auger[12,**] and the Telescope Array[13,***] Collaborations

[1]University of Utah, High Energy Astrophysics Institute, Salt Lake City, Utah, USA
[2]CNRS/IN2P3, IJCLab, Université Paris-Saclay, Orsay, France
[3]Karlsruhe Institute of Technology (KIT), Institute for Astroparticle Physics, Karlsruhe, Germany
[4]Graduate School of Science, Osaka Metropolitan University, Osaka, Osaka, Japan
[5]Institute for Cosmic Ray Research, The University of Tokyo, Kashiwa, Chiba, Japan
[6]Université Libre de Bruxelles (ULB), Brussels, Belgium
[7]Karlsruhe Institute of Technology (KIT), Institute for Experimental Particle Physics, Karlsruhe, Germany
[8]Universitá dell'Aquila, Dipartimento di Scienze Fisiche e Chimiche, L'Aquila, Italy
[9]INFN Laboratori Nazionali del Gran Sasso, Assergi (L'Aquila), Italy
[10]Nambu Yoichiro Institute of Theoretical and Experimental Physics, Osaka Metropolitan University, Osaka, Osaka, Japan
[11]INFN, Sezione di Roma "Tor Vergata", Roma, Italy
[12]Observatorio Pierre Auger, Av. San Martín Norte 304, 5613 Malargüe, Argentina
[13]Telescope Array Project, 201 James Fletcher Bldg, 115 S. 1400 East, Salt Lake City, UT 84112-0830, USA

**Abstract.** In the study of cosmic rays, the measurement of the energy spectrum of the primaries is one of the main issues and provides fundamental information on the most energetic phenomena in the Universe. At ultra-high energies, beyond $10^{18}$ eV, the cosmic rays are studied by the two largest observatories built so far, the Pierre Auger Observatory and the Telescope Array. Both observatories are based on a hybrid design and reported a measurement of the energy spectrum using the high duty cycle of the surface detector and the calorimetric estimation of the energy scale provided by the fluorescence detector.

The differences among the reported spectra are scrutinized by a working group made by members of the Pierre Auger and Telescope Array Collaborations. The two measurements have been found well in agreement below $10^{19}$ eV while, at higher energies, they show an energy-dependent difference that is beyond the systematic uncertainties associated to the energy scale.

In this contribution we review the status and perspectives of the working group activities including new studies aiming at addressing the impact on the flux measurement at the highest energies of potential biases in the estimation of the shower size.

## 1 Introduction

Ultra-high-energy cosmic rays (UHECRs) are atomic nuclei arriving from outer space with energies beyond $10^{18}$ eV, the highest-energy particles observed in nature. Their energy spectrum, the differential flux of particles, is the basic experimental information because its absolute scale and its shape provide information on the acceleration mechanisms, the spatial distribution of the sources, and propagation of cosmic rays in the inter-galactic space. UHECRs arrive at the Earth very rarely, with a flux that decreases with energy, reaching less than one event per square kilometer per century at $10^{20}$ eV, and therefore their observation require huge detection area and long observation times.

The Pierre Auger Observatory (Auger) and the Telescope Array (TA) are the two largest observatories of UHECRs built so far. They have been in operation fo more than a decade in the Southern and Northern hemispheres, respectively. Auger [1] is located near the small town of Malargüe in the province of Mendoza (Argentina) at a latitude of about 35.2° S. It has a surface detector (SD) of 1600 water Cherenkov detectors placed on a triangular grid with 1500 m spacing that extends over 3000 km$^2$. TA [2, 3] is located near Delta, Millard County, Utah, U.S. at the latitude 39.3° N. Its SD has 507 scintillation counters on a square grid with 1.2 km spacing covering an area of 700 km$^2$. Both observatories follow a so-called hybrid approach, because the SD measurements are combined with the ones performed by a fluorescence detector (FD). With the FD it is possible to reconstruct the longitudinal profile of the shower and to obtain a calorimetric estimation of the shower energy. In this way, it is possible to measure the energy spectrum with the high exposure of

*e-mail: valerio.verzi@roma2.infn.it
**e-mail: spokespersons@auger.org, full author list available at https://www.auger.org/archive/authors_2022_10.html
***Full author list: http://www.telescopearray.org/research/collaborators

https://doi.org/10.1051/epjconf/202328302003

the SD and with an energy scale that, being set by the FD, it is largely independent of air shower simulations and of assumptions on hadronic interaction models.

The differences among the spectra measured at the two observatories are scrutinized by a joint working group that was formed in 2012 and that has reported its studies in the UHECR and ICRC conference series [4, 5]. In this contribution, we briefly report the results of the activities of the working group revising the details of the Auger and TA data analysis and focusing on the discrepancy between the spectra at the highest energies. In particular, we discuss new studies of potential biases in the estimation of the shower size measured with the SD that can't be corrected by the calibration performed with the FD energies.

## 2 The experimental methods

Despite the TA and Auger observatories having both adopted the hybrid detection technique, there are several differences in the analysis methods to determine the energy scale and to estimate the energy spectrum.

In both observatories, the signals detected by the SD stations are fitted with a lateral density function. The fit provides the position of the shower core and the lateral density function evaluated at a some distance from the core is used to get an energy estimator that is calibrated against the FD energies. This distance is 1000 m for Auger and 800 m for TA and the resulting energy estimators are called $S(1000)$ and $S(800)$, respectively. The calibration procedures against the FD energies developed by the two Collaborations are rather different as detailed below.

The Auger analysis is performed selecting showers with zenith angles $\theta < 60°$ and with energies greater than $2.5 \times 10^{18}$ eV and these conditions ensure that the trigger efficiency is 100%. In this way $S(1000)$ can be corrected for attenuation effects using the empirical procedure of the so-called Constant Intensity Cut (CIC) method [6]. The resulting energy estimator, $S_{38}$, is the zenith-angle independent energy estimator and can be thought of as the signal, $S(1000)$, that the shower would have produced at a zenith angle of 38°. Then, $S_{38}$ is calibrated against the FD energies using a power-law relationship $E_{\rm FD} = A\, S_{38}^{B}$ where the two parameters $A$ and $B$ are fitted to the data ($B \approx 1.03$). The Auger analysis is described in detail in [7].

The full trigger efficiency for the TA SD is attained at $10^{18.8}$ eV for showers with zenith angles $\theta < 55°$. Therefore, in order to estimate the shower energy below this energy threshold, the CIC method can't be used and one has to calculate the attenuation effects using Monte-Carlo (MC) simulations. For a given value of $S(800)$ and $\theta$, a MC lookup table is used to estimate the shower energy ($E_{\rm TBL}$). The TA MC simulations use the CORSIKA software package and the showers are generated according the QGSJetII-03 hadronic interaction model assuming proton primaries. Then, $E_{\rm TBL}$ is calibrated against the FD energies ($E_{\rm FD}$) using a linear relationship among $E_{\rm TBL}$ and $E_{\rm FD}$. The calibration fit provides a normalization factor of the MC energies equal to 1/1.27. For further details on TA energy determination see [8].

A detailed analysis of the systematic uncertainties in the energy scale of TA is reported in [10] and for Auger in [11]. The results are summarized in table 1. The total uncertainty for TA and Auger is 21% and 14%, respectively, and they are almost energy independent. The main difference in the uncertainties is related to the fluorescence yield. Auger uses the recent and precise measurements of the Airfly experiment [12, 13] while TA uses the measurement of the absolute yield made by Kakimoto et al. in 1996 [14], and the wavelength spectrum measured more recently by FLASH [15]. Another important difference is related to the atmosphere, and in particular to the determination of the aerosols that in Auger are measured every hour of data taking while, in TA, the same average aerosol profile is used to reconstruct all FD events. It is worth nothing that, besides the contribution to the uncertainty being not so large, the invisible energy also plays an important role in the determination of the energy scale. The invisible energy is the non-calorimetric part of the FD energy estimation being associated to the high-energy muons and neutrinos that do not deposit their energies in the atmosphere. In TA it is estimated using MC simulations of proton showers and it amounts to about 7% of the total shower energy, while the Auger estimation is significantly larger, at the level of 14%, as it incorporates the muon number excess measured with the water Cherenkov detectors [16].

**Table 1.** Systematic uncertainties in the energy scale for TA [10] and Auger [11]. The mild energy dependence of the uncertainties is reported for Auger.

| Systematic uncertainties in the energy scale | | |
|---|---|---|
| | TA | Auger |
| Fluorescence Yield | 11% | 3.6% |
| Atmosphere | 11% | 3.4%–6.2% |
| FD Calibration | 10% | 9.9% |
| FD Reconstruction | 9% | 6.5%–5.6% |
| Invisible Energy | 5% | 3%–1.5% |
| Other contributions | | 5% |
| Total | 21% | 14% |

The other important ingredient for the determination of the energy spectrum is the exposure. The Auger analysis is performed in an energy range in which the SD is fully efficient, and therefore the calculation of the exposure reduces to a geometrical calculation plus the knowledge of the live-time of the array. In TA, since the analysis is extended at energies where the detector is not fully efficient, the exposure related to the geometry and live-time of the array has to be folded with the trigger and reconstruction efficiency that are estimated using MC simulations. Another ingredient is the energy resolution that must be known in order to account for the distortion of the spectrum shape given by the migration of events between neighboring energy bins. Resolution corrections are in general not so large and are precisely estimated.

For this contribution, we use the Auger data set presented in [7], whose exposure is 60 400 km$^2$ sr yr, and the TA data set presented in [9], whose exposure ranges from

**Table 2.** Parameters relevant for the measurements of the energy spectrum at the Auger Observatory [7] and the Telescope Array [9].

| | Telescope Array | Auger |
|---|---|---|
| data period | 11/05/2008 – 11/05/2019 | 01/01/2004 – 31/08/2018 |
| energy threshold | $10^{18.2}$ eV | $10^{18.4}$ eV |
| zenith angle | $\theta < 45°$ | $\theta < 60°$ |
| declination band | $-6° < \delta < 90°$ | $-90° < \delta < 24.8°$ |
| full trigger efficiency | $> 10^{18.8}$ eV | $> 10^{18.4}$ eV |
| exposure above full trigger efficency | $\approx$ 10 000 km$^2$ sr yr | 60 400 km$^2$ sr yr |
| number of events $E > 10^{19}$ eV ($10^{20}$ eV) | 3292 (13) | 16 737 (15) |
| SD energy resolution ($10^{19}$ eV – $10^{20}$ eV ) | 21% – 15% | 11% – 8% |
| FD energy resolution | 19% | 7.4% |
| uncertainty in the energy scale | 21% | 14% |

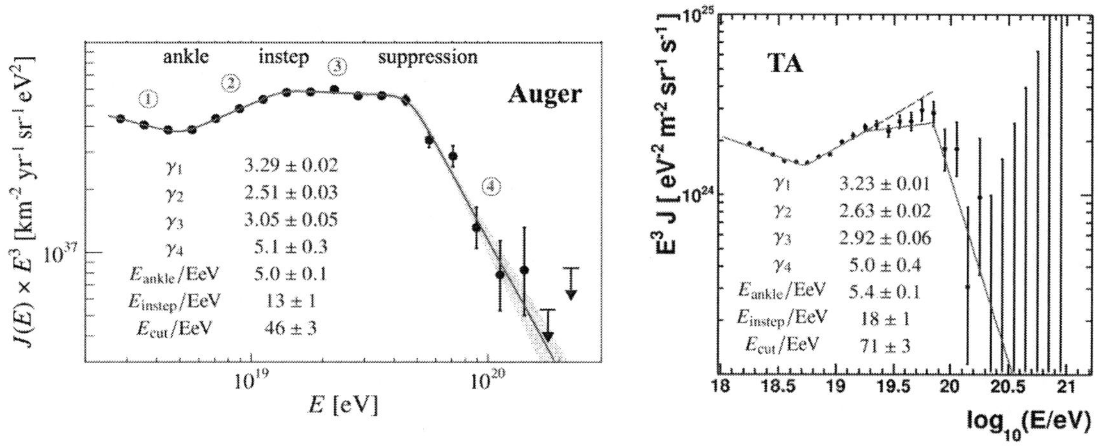


**Figure 1.** Energy spectrum and spectral features measured by Auger [7] and TA [9]. Only statistical uncertainties are shown. For a rough estimation of the systematic uncertainties in the energy that defines the transition points one can consider the total uncertainty in the energy scale reported in table 1 with the remark that it would be fully correlated for the different inflection points.

770 km$^2$ sr yr at $10^{18.2}$ eV up to $\approx$ 10 000 km$^2$ sr yr above $10^{18.8}$ eV. Other information and relevant parameters for the determination of the energy spectrum at the two observatories are reported in table 2.

## 3 Comparison of the spectra measured at the two observatories

The measurements of the energy spectrum performed by Auger [7] and TA [9] are shown in figure 1. They are presented multiplying the flux by the third power of the energy (corresponding to the central value of the energy bin $\Delta \log_{10} E = 0.1$) in order to better visualize their shape. The spectrum is characterized by both Collaborations with four broken power laws, i.e. a sequence of four power laws with fast transitions of the spectral index in three inflection points. The latter are identified as the *ankle*, the *suppression* at highest energies and the *instep*, the new steepening recently reported just above $10^{19}$ eV. The spectral features are in remarkable agreement with some tension that emerges at the highest energies. It is worth noting that the flux parameterization is the simplest possible model that describes well the data. A more realistic evolution of the spectral index with energy is likely more complex than simple broken power laws, as it depends by many factors, such as the production rate in the sources, the evolution of mass composition with energy and propagation effects [17]. Further statistics is needed to address more precisely the shape of the energy spectrum [7].

The two spectra are compared in the top-left panel of figure 2. The TA data points are systematically higher than the Auger ones with a discrepancy that is larger at the highest energies. Such discrepancy can be interpreted as a difference in the energy scale of the two observatories. This is because the uncertainty in the energy scale gives the dominant contribution to the uncertainty in the spectrum normalization. As a first approximation one has $\Delta J/J \sim (\gamma - 1)\, \Delta E/E$ where $\gamma$ is the spectral index. With the typical uncertainties affecting the energy scale, it is likely to observe differences in the measured spectra that can be as large as several tens of percentage. Other contri-

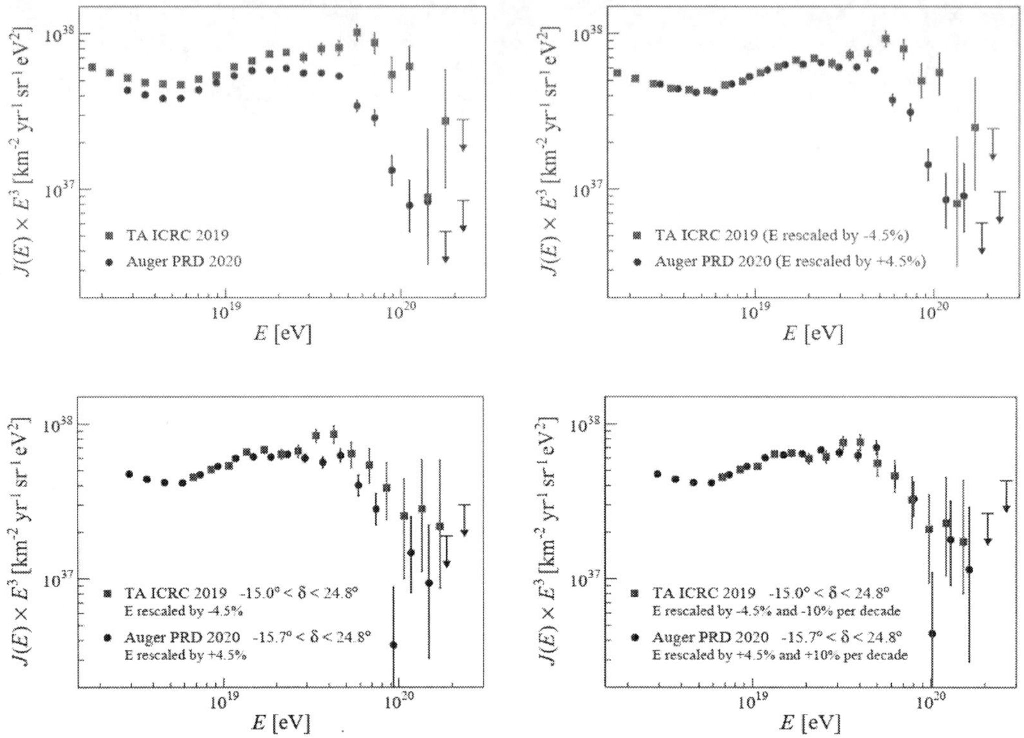


**Figure 2.** Energy spectrum measured by Auger [7] and TA [9] in the full declination band (upper-left), in the full band after a rescaling of the energy by an overall 9% factor (upper-right), in the common band with the same energy rescaling factor (bottom-left) and in the common band once the energy scale is furtherly rescaled in an energy dependent way by 20%/decade defined as $\Delta E/E = \left[\pm 4.5 \pm 10\log_{10}(E/10^{19}\text{ eV})\right]\%$.

butions to the uncertainty in the flux are in general much smaller (see e.g. [7]).

In the top-right panel of figure 2 we show how the spectra can be put in agreement up to about $10^{19}$ eV introducing an overall energy independent energy shift of 9% (−4.5% for TA and +4.5% for Auger). The discrepancy persisting at the highest energies can be recovered only introducing an energy dependent energy shift. However, interpreting the spectral difference only as an energy scale issue may not be fully correct. This because the two detectors observe different parts of the celestial sphere and at the highest energies, where the deflection of the CRs in the galactic and inter-galactic magnetic fields are expected to be not so large, the energy spectrum may depend on the sources that fall within the field of view of the detector.

A more correct comparison of the two spectra at the highest energies is done by limiting the measurements to the declination band accessible by both observatories. This is shown in the bottom-left panel of figure 2. Here, the TA spectrum is calculated using the selection criteria adopted for the analyses aiming to search the anisotropy signals in the CRs arrival directions. This allows us to extend the zenith angle range from $\theta < 45°$ to $\theta < 55°$ and therefore to decrease the minimum detectable declination from −6° to −15°, allowing to have a larger overlap with the Auger field of view. For this selection criteria the analysis is limited at the energies larger than $10^{18.8}$ eV. In the figure we show how the discrepancy at the highest energies persists even in the common band: from the agreement attained at $\approx 10^{19}$ eV thanks to the energy rescaling factor determined in the full band around the ankle, the difference between then flux becomes up to about 70% at $10^{20}$ eV. As shown in the bottom-right panel of figure 2, such energy dependent difference can be explained introducing a further energy dependent energy shift that amounts to 20% per decade (−10%/decade for TA and +10%/decade for Auger). We have verified that the same results are obtained when the analysis accounts for the different shapes of the directional exposure of the two observatories [18].

The statistical significance of the energy-dependent energy shift has been evaluated using the following method: thousands of independent spectra with the TA statistics in the common band have been generated simulating events according to the functional shape that describes the Auger spectrum (both shape and normalization) and taking into account resolution effects. For each generated spectrum we estimate the energy dependent energy shift to reach full agreement with Auger obtaining a distribution centered in 0 and with an RMS of 6%. The distribution allows us to reject the scenario in which the 20%/decade energy shift is due to a statistical fluctuation with a significance of about $3\sigma$.

## 4 Toward understanding the differences between the energy scales of the two observatories

In this section we discuss the energy shifts determined in the previous section in the light of the systematic uncertainties in the energy scales. The discussion is done separately for the constant energy shift derived below $10^{19}$ eV and for the additional energy dependent energy shift at higher energies because, as we will see, they need an analysis of different types of systematic uncertainties.

### 4.1 Understanding the 9% energy shift below $10^{19}$ eV

The overall energy shift of 9% that brings the Auger and TA spectra in agreement in the energy region below $10^{19}$ eV is of course fully consistent with the uncertainties in the energy scales. It is worth noting that an even better agreement can be achieved if the fluorescence events collected at the two observatories would be reconstructed using the same model of the fluorescence yield and invisible energy (see [4] and references therein). Introducing the Airfly absolute yield in the TA reconstruction would lower the energy by 20%. Such energy shift is reduced to −14% if also all the other Airfly parameters describing the wavelength spectrum and quenching effects are introduced in the analysis. On the other hand, the invisible energy correction of Auger would increase the TA energies by 7%. Therefore, the combined effect of using the same model of fluorescence yield and invisible energy would lower the 9% energy shift to a value well below 5%. This value is surprisingly low when compared with the uncertainty in the energy scales obtained subtracting (in quadrature) the contributions from the fluorescence yield and invisible energy (from table 1 one can estimate 13% for Auger and 17% for TA). The good agreement between the energy scales suggests that the systematic uncertainties in the fluorescence events reconstruction, and in particular the ones on the absolute calibration of the telescopes, are well under control. On this respect, a remarkable test has been done by TA using a linear accelerator installed in front of one of the fluorescence telescopes [19]. The facility allows to mimic a cosmic ray air shower and it provides an effective test of the combined effect of the fluorescence yield and of the absolute calibration of the telescopes. The test performed by TA has a precision of 7.9% and has shown that data are fully consistent with the FD absolute calibration when the Airfly model is used in the analysis [20].

### 4.2 Understanding the additional 20%/decade energy shift determined above $10^{19}$ eV

The understanding of the additional 20%/decade energy shift needed to bring in agreement the spectra above $10^{19}$ eV is much more complicated than the one at lower energies, both due to the lack of statistics especially for hybrid events and because the two Collaborations perform the energy calibration with methods that are substantially different and therefore affected by different kinds of systematics.

As seen in Sect. 2, the Auger Collaboration calibrates the SD signal using a power law and therefore the systematics in the FD energies play a crucial role even at the higher energies. The analysis of the hybrid events benefits from enough statistics and good reconstruction performances. As seen in figure 3, Auger has almost 600 hybrid events above $10^{19}$ eV and the distribution of the ratio of SD to FD energies is rather narrow, with an RMS of 12% that is the result of the combined effect of the SD and FD energy resolutions (see table 2). These features guarantee that the SD energies are well aligned to the FD ones up to the highest energies: the statistical uncertainty in the calibration curve is 0.3% at $10^{19}$ eV to reach its maximum of about 1% at $10^{20}$ eV [7], therefore introducing only a small uncertainty that can't explain the difference

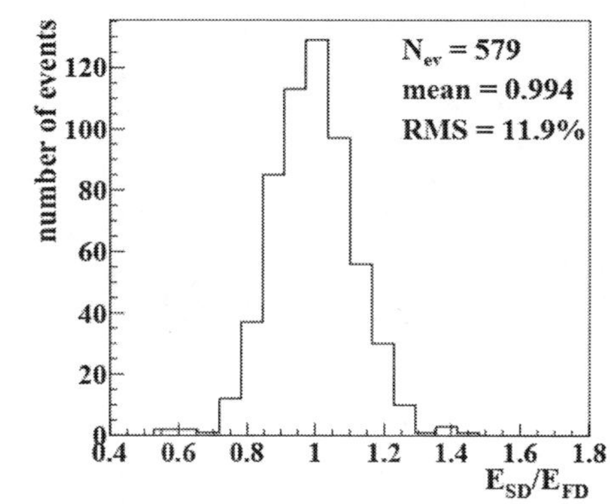


**Figure 3.** Ratio of the SD to FD energies for the Auger hybrid events above $10^{19}$ eV [7].

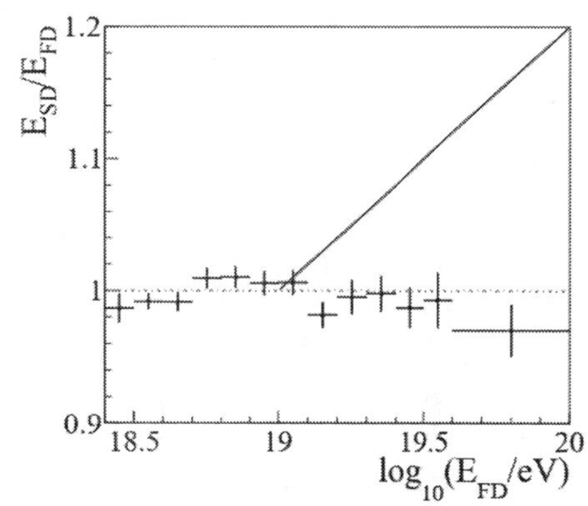


**Figure 4.** Average ratio of the SD to FD energies in bins of energies for the Auger hybrid events [7]. The red line shows the 20%/decade energy shift above $10^{19}$ eV needed to explain the difference between the Auger and TA spectra.

between the TA and Auger spectra. This is also shown in figure 3, where the 20%/decade energy shift (red line) is compared with the average ratio of the SD to FD energies calculated in bins of energy. It is then clear that for Auger it is important to address precisely the energy dependence of the systematics in the FD energies. As shown in table 1, the systematics depend only little on energy and can't explain the 20%/decade energy shift. Even in the case of the aerosols under an extreme scenario of their underestimation, the energy dependence of the bias would be rather small, below 3%/decade [21].

As seen in Sect. 2, TA calibrates the SD energy estimator through an overall normalization factor ($E_{\rm FD} = E_{\rm TBL}/1.27$). The study of possible energy-dependent reconstruction biases estimated comparing the SD and FD energies as a function of energy leads to a ($-1\% \pm 9\%$)/decade energy shift [21], and it is on some extent limited by the low TA hybrid statistics at higher energies. This result can't be significantly affected by the energy dependence of the FD systematics because, like in Auger, it is also small in TA [21]. Therefore, it is natural to investigate possible energy biases arising from the reconstruction of the SD energy estimator that could emerge at the higher energies as they can't be corrected by the TA calibration procedure (because it provides one factor that is the same at all energies). It is worth noting that such kinds of bias could also be relevant for Auger even though the calibration fit provides also the slope ($B$) of the power law ($E_{\rm FD} = A\, S_{38}^{B}$).

The central point of the SD reconstruction is the choice of the lateral density function (LDF) and of the distance from the shower core at which the energy estimator is evaluated. Auger and TA use an LDF whose shape is fixed to a predetermined average parameterization. This because, given the large spacing of the arrays, the LDF is sampled in only few points and its shape can't be determined on an event-by-event basis (the LDF depends on how the shower has developed in the atmosphere). This causes large uncertainties in the normalization of the LDF (proxy of the shower energy) and leads to define the energy estimator using the LDF evaluated far away from the core were fluctuations of the signals are quite small (see [22] and references therein). It turns out that the fluctuations are minimal at a so-called "optimal distance" ($r_{\rm opt}$) whose value primarily depends on the array geometry, with little dependence on the energy or zenith angle of the shower or choice of the LDF [22].

The LDF used by Auger is a modified Nishimura–Kamata–Greisen function (see [23] and references therein). Its shape is parametrized empirically from data as a function of $S(1000)$ and of the zenith angle analyzing those events where the LDF can be sampled with a large enough number of stations to provide a sufficiently high lever-arm. The energy estimator is evaluated at $r_{\rm opt}$ = 1000 m [22]. In figure 5 we present a recent cross-check of the LDF fluctuations, evaluated repeating several times the LDF fit of Auger simulated events where the slope parameters are changed in each fit according to their uncertainty. As we can see, the fluctuations are minimal at 1000 m, confirming the early results reported in [22]. The fluctuations are a few percent at $r_{\rm opt}$ and increase to beyond 10% at 500 m from $r_{\rm opt}$. Evaluating the energy estimator minimizing the uncertainties related to the LDF is important to ensure that the energy estimator is a good proxy of the shower energy. The risk in workink with larger uncertainties is that we may have an energy estimator that, once it is corrected by attenuation effects, can't be properly calibrated using a power law (something difficult to detect with the limited hybrid statistics), therefore with the risk of introducing energy biases to which the energy spectrum is very sensitive. The Auger collaboration has estimated that the uncertainties related to the $S(1000)$ reconstruction affect the measured spectrum by less than 3% [7].

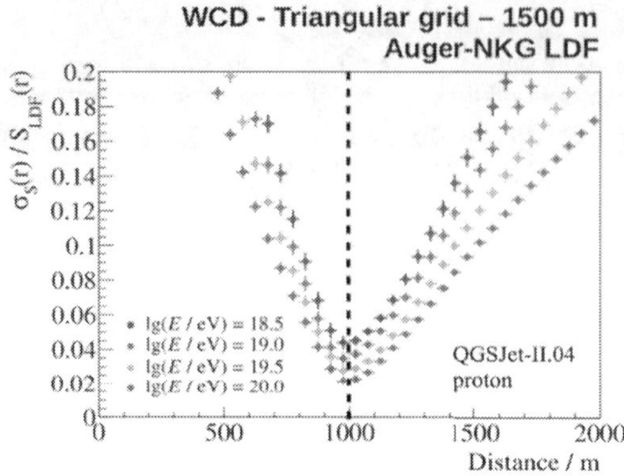


**Figure 5.** Fluctuations of the signal estimated with the Auger LDF arising from the uncertainty in the slope parameters.

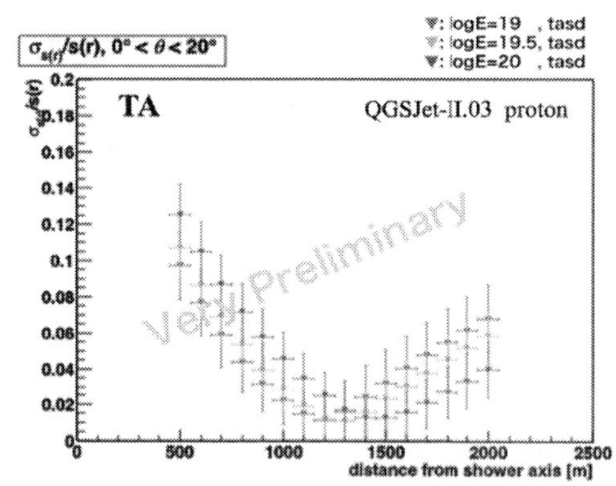


**Figure 6.** Fluctuations of the signal estimated with the TA LDF arising from the uncertainty in the slope parameters.

The TA LDF is the same as the one that was used by the AGASA experiment (see [8] and references therein),[1] with the slope parameter that depends only on the zenith angle. The energy estimator is obtained taking the signal at 800 m from the core because this distance minimizes the difference between the LDF of proton and iron showers. The fluctuations of the LDF for the TA events calculated changing the slope by its uncertainty are shown in

[1] Like for TA, the AGASA detectors were scintillators.

figure 6, and one can see that 800 m does not coincide with the "optimal distance" of 1300 m at which the fluctuations are minimal. The difference between the fluctuations at 1300 m and 800 m is not so large (from 2% to ≈ 6%) and therefore one can expect that the two energy estimators ($S(800)$ and $S(1300)$) provide similar performances.

In order to address this point, the TA collaboration has developed an analysis similar to the one performed by Auger. Selecting the events above $10^{19}$ eV where the array is fully efficient, $S(800)$ and $S(1300)$ are corrected for the attenuation effects using the CIC method (figure 7 shows the attenuation curves of $S(1300)$). The resulting zenith angle independent energy estimators are calibrated against the FD energies using a power law relationship through a $\chi^2$ minimization (see figure 8, where $S_{35}$ is the shower size at the zenith angle of 35°). The two energies derived from $S(800)$ and $S(1300)$ agree very well, at the 4% level as shown in figure 9, demonstrating that there is not a significant worsening of the performances of the energy reconstruction when the shower size is not estimated at $r_{\rm opt}$.

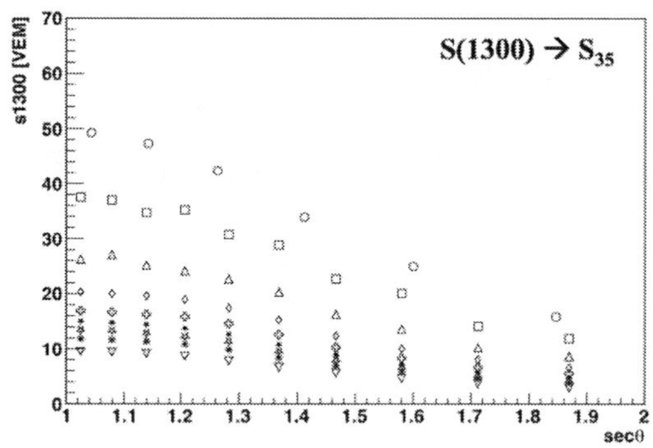


**Figure 7.** Attenuation curves of $S(1300)$ for different intensities.

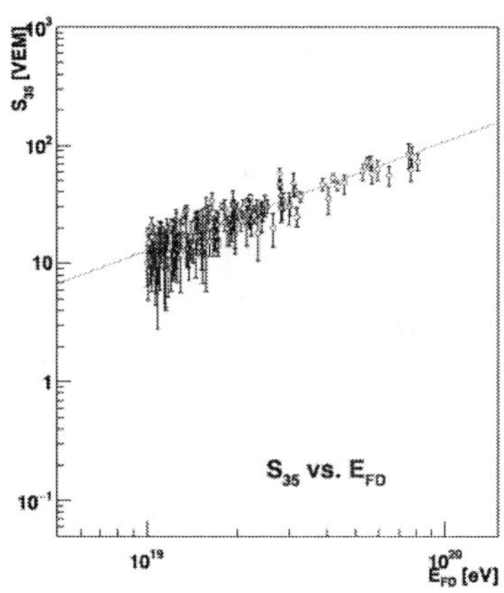


**Figure 8.** Energy calibration of $S_{35}$ derived from $S(1300)$.

It has been checked that the two energy estimations are in fair agreement with the standard TA energy calculation ($E_{\rm standard}$) that is used to measure the energy spectrum. The distribution of $\ln(E_{\rm standard}/E(S_{800}))$ has a mean value of −4.2% and an RMS of 9.4%, while for $\ln(E_{\rm standard}/E(S_{1300}))$ the mean is −3.9% and the RMS is 8.2%. It is interesting to note that $E_{\rm standard}$ is obtained from $S_{800}$ but the RMS is smaller when compared with the energy estimation performed at $r_{\rm opt}$ ($E(S_{1300})$). This demonstrates some effectiveness of the MC lookup table in providing good performances in the energy reconstruction. However, no firm conclusions useful to explain the difference between the TA and Auger spectra can be derived from the comparison of $E_{\rm standard}$ with the CIC based energy estimations, as the performances of their energy calibration are limited by the small hybrid statistics above $10^{19}$ eV and resolution effects (see figure 10).

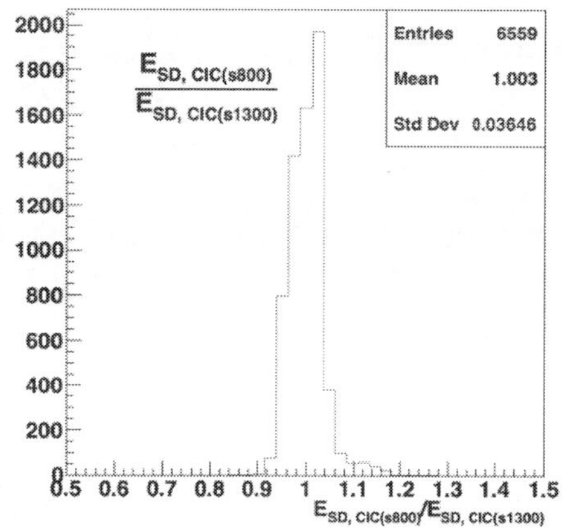


**Figure 9.** Comparison among the energies reconstructed starting from $S(800)$ and $S(1300)$.

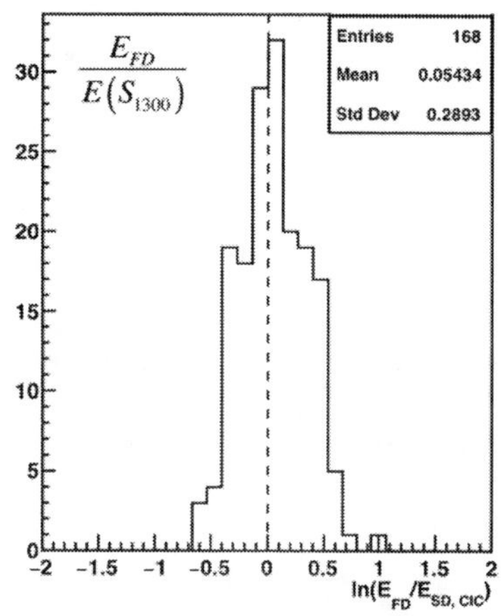


**Figure 10.** Ratio of the SD to FD energies for the TA hybrid events above $10^{19}$ eV. The SD energy is estimated from $S(1300)$.

We conclude this section showing in figure 11 another measurement of the energy spectrum obtained using the Auger SD events inclined at large zenith angles ($\theta > 60°$) [24]. These events require a completely different and more complex reconstruction technique [25] with respect to the one used for the vertical events with $\theta < 60°$ [23]. In fact, at large zenith angles, the signals

https://doi.org/10.1051/epjconf/202328302003

are dominated by muons that make long paths in the atmosphere and the geomagnetic field, separating laterally positively and negatively charged particles, destroys the circular symmetry of the LDF making impossible to define an energy estimator like $S(1000)$. The energy estimator is a normalization factor of simulated muon density maps that are parametrized as function of the zenith and azimut angles of the shower. In this way the attenuation effects are estimated using MC simulations, contrary to what is done for $S(1000)$ where a data-driven approach (CIC) is used. The normalization factor is then fitted to the data and, like for $S_{38}$, it is calibrated against the FD energies using a power law relationship. As can be seen in figure 11, the Auger spectra obtained with inclined and vertical events agree rather well and this suggests that the systematic uncertainties related to the shower size reconstruction are well under control.

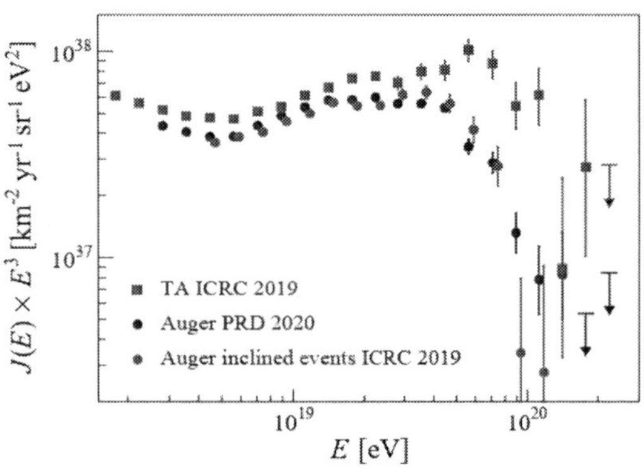


**Figure 11.** Auger energy spectrum obtained from the events inclined at large zenith angles ($\theta > 60°$) [24] compared with the measurements discussed in this proceeding.

## 5 Summary and future perspectives

We have reviewed and compared the results of the energy spectra measured at the Pierre Auger Observatory and at the Telescope Array. It has been shown that there is a good agreement among the spectral features, including the *instep* one that has been discovered only recently, modulo a difference in the energy scales that can be inferred from the different normalization of the fluxes.

The energy offset needed to explain the differences in the flux up to $10^{19}$ eV is 9.5% and it is fully consistent with the uncertainties in the energy scales. Moreover, the offset can be significantly reduced to a value below 5% if the two experiments would use the same model for the fluorescence yield and invisible energy. The fluorescence yield and the invisible energy can be considered as a sort of external parameters of the reconstruction of FD events and the choice of the particular model by the two Collaborations is sometimes made for historical reasons. The convergence toward the use of the same models in both experiments would have obvious benefits for both the Collaborations and the community.

The understanding of the offset between the spectra at energies above $10^{19}$ eV is much more complicated. In addition to the 9.5% energy shift, it is necessary to introduce an energy dependent energy shift of 20%/decade. The analysis is performed in the declination band visible by both experiments ensuring that the observed differences are caused by instrumental effects and are not of astrophysical origin. We have excluded that the energy dependent energy shift is due to a statistical fluctuation with a significance of $3\sigma$. An even higher significance has been obtained with the larger data set used by the joint WG on the study of the arrival directions of CRs [26].

We have presented an in-depth discussions of the systematic uncertainties without finding anything useful to explain the 20%/decade energy shift. The matter is very complicated because the energy dependence of the systematic uncertainties can be related to subtle details of the event reconstruction and of the calibration procedure, details that are on some extent difficult to understand and to quantify.

We have shown that in the case of Auger it is very important to understand the energy dependence of the FD systematics. The calibration fit is performed using a power law relationship and the analysis benefits of a large enough hybrid statistics and the good energy resolution with the result that the SD energies are well aligned to the FD ones. The energy dependence of the FD uncertainties are in general quite small and can't explain the 20%/decade energy shift.

We have seen that the TA SD energy estimator is the shower energy obtained from a MC lookup table and for its calibration it is enough to consider only an energy-independent rescaling factor. By construction, the procedure can't ensure a perfect alignment of the SD energies to the FD ones over the full energy range and therefore the energy dependence of the FD systematics are less important in comparison to the Auger analysis. We have verified that potential non linearity effects in the SD energy estimator arising from the fact that the shower size is not estimated at the distance from the core that minimizes the uncertainties in the LDF are quite small. The test has been performed using the data-driven approach of Auger (CIC and energy calibration with a power law) and the resulting energies have been found in fairly agreement with the ones obtained using the MC lookup table. However, no firm conclusions useful to understand the 20%/decade energy shift can be drawn as the performances of the energy calibration fit are on some extent limited by the small TA hybrid statistics and resolution effects.

Finally, for the first time we have discussed another measurement of the spectrum performed by Auger with the events inclined at large zenith angles. The measurement agree quite well with the one performed using the vertical events (zenith angles < 60°) used for the comparison with the TA spectrum. The reconstruction of vertical and inclined events are quite different and have a different sensitivity to several physical and instrumental effects (e.g. different sensitivity of the energy estimator to the primary mass and no saturated stations in inclined events). Therefore, the good consistency between the spectra provides an indication that the systematics related to the SD reconstruction are well under control.

The joint WG is strongly engaged into continue the activities to understand the differences between the spectra. This work is of primary importance because the combination of Auger and TA data allows us to study UHECRs with full sky coverage, and a proper combination of the data requires the cross-calibration of the energy scales and therefore the understanding of their systematic uncertainties.

An interesting perspective to continue and improve the joint activity is to make the comparison in the common band including the Auger spectrum obtained with the inclined events. This will allow the common band to be extended up to the declination of 44.8° (from the current 24.8°) and therefore to increase significantly both the Auger and TA statistics. The statistics can be also increased adding more years of data taking. Concerning the energy scale it is important that the Collaborations will refine their study of the energy dependence of the systematic uncertainties associated to both the FD ans SD reconstruction. Another interesting perspective is to refine the techniques used for the energy calibration fit in order to address more precisely the consistency between SD and FD energies and to facilitate the comparison of the energy scales. On a longer time scale, a significant improvement of the joint analyses will be possible thanks to the current upgrade of the observatories, with the scintillators of AugerPrime [27] that are installed on the top of the water Cherenkov detectors, and with the larger exposure of TA×4 [28].

## References

[1] Pierre Auger Collaboration, Nucl. Instrum. Meth. A **798**, 172 (2015).

[2] Telescope Array Collaboration, Nucl. Instrum. Meth. A **689**, 187 (2013).

[3] Telescope Array Collaboration, Nucl. Instrum. Meth. A **676**, 54 (2012).

[4] V. Verzi, D. Ivanov and Y. Tsunesada, Prog. Theor. Exp. Phys. **2017**, 12A103 (2017).

[5] Y. Tsunesada (Pierre Auger and Telescope Array Collaborations), Proc. Sci. **ICRC2021**, 337 (2021).

[6] J. Hersil, I. Escobar, D. Scott, G. Clark, and S. Olbert, Phys. Rev. Lett. **6**, 22 (1961).

[7] Pierre Auger Collaboration, Phys. Rev. D **102**, 062005 (2020).

[8] Telescope Array Collaboration, Astrophys. J. **768**, L1 (2013).

[9] D. Ivanov (Telescope Array Collaboration), Proc. Sci. **ICRC2019**, 298 (2019).

[10] Y. Tsunesada et al., in Proc. 32nd ICRC 2011, Beijing, China, 1227.

[11] V. Verzi (Pierre Auger Collaboration), in Proc. 33rd ICRC 2013, Rio de Janeiro, Brazil, arXiv:1307.5059.

[12] AIRFLY Collaboration, Astropart. Phys. **42**, 90 (2013).

[13] AIRFLY Collaboration, Astropart. Phys. **28**, 41 (2007).

[14] F. Kakimoto *et al.*, Nucl. Instrum. Meth. A **372**, 527 (1996).

[15] R. Abbasi *et al.*, Astropart. Phys. **29**, 77 (2008).

[16] Pierre Auger Collaboration, Phys. Rev. D **100**, 082003 (2019).

[17] Pierre Auger Collaboration, Phys. Rev. D **125**, 121106 (2020).

[18] T. AbuZayyad et al. (Pierre Auger and Telescope Array Collaborations), JPS Conf. Proc. **19**, 011003 (2018).

[19] T. Shibata et al., Nucl. Instrum. Methods A **597**, 61 (2008).

[20] B. Shin *et al.* (Telescope Array Collaboration), Proc. Sci. **ICRC2015**, 640 (2015).

[21] T. AbuZayyad et al. (Pierre Auger and Telescope Array Collaborations), EPJ Web of Conferences **210**, 01002 (2019).

[22] D. Newton, J. Knapp and A.A. Watson, Astropart. Phys. **26**, 414 (2007).

[23] Pierre Auger Collaboration, JINST **15**, P10021 (2020).

[24] V. Verzi (Pierre Auger Collaboration), Proc. Sci. **ICRC2019**, 450 (2019).

[25] Pierre Auger Collaboration, JCAP **08**, 019 (2014).

[26] F. Urban et al. (Pierre Auger and Telescope Array Collaborations), these proceedings.

[27] C. Bérat (Pierre Auger Collaboration), these proceedings.

[28] E. Kido (Telescope Array Collaboration), these proceedings.

# The energy spectrum of cosmic rays above 6 PeV as measured at the Pierre Auger Observatory

*Quentin* Luce[1], on behalf of the Pierre Auger Collaboration[2,*]

[1]Institute for Experimental Particle Physics, Karlsruhe Institute of Technology (KIT), Karlsruhe, Germany
[2]Observatorio Pierre Auger, Av. San Martín Norte 304, 5613 Malargüe, Argentina
Full author list: https://www.auger.org/archive/authors_2022_10.html

**Abstract.** Since 2004, the Pierre Auger Collaboration has measured, with unprecedented precision, one of the most important features of ultra-high-energy cosmic rays, the energy spectrum. Located in the Southern hemisphere, the Observatory comprises an array of 1660 water-Cherenkov detectors covering 3000 km² overlooked by 27 fluorescence telescopes. Five sets of measurements have been used to reconstruct the spectrum from 6 PeV to beyond 100 EeV. The highest-energy events, recorded by surface detectors on a 1500 m triangular grid, are reconstructed differently depending on their inclination (vertical events below 60° and inclined ones above 60°). A cross-check of these measurements is made using 'hybrid events', in which there are simultaneous detections at both the fluorescence detectors and at least one surface detector. Events with energies below 3 EeV have been studied using a nested array with a spacing of 750 m and using the Cherenkov light recorded with three high-elevation telescopes. In this contribution, updated methods for the reconstruction of all events are discussed, together with their associated uncertainties. A combination of all data sets is reported to construct a spectrum above 6 PeV to the highest energies. A detailed discussion of the spectral features will be presented.

## 1 Introduction

More than a century ago, cosmic rays have been observed for the first time and decades ago, the first cosmic ray with an energy larger than 10 EeV has been detected. However, at the highest energies ever observed, the origin of these extraordinary particles is still a source of debate and unknowns among the scientific community. One of the keys to unlock the path to the origin of ultra-high-energy cosmic rays (UHECRs) is to reconstruct the energy spectrum. Thus, experiments have been built or are still operating to measure as precisely as possible the energy spectrum of cosmic rays. When it comes to the highest energies ever observed on Earth, a precise measurement of the spectrum has been reported by the Pierre Auger Observatory [1].

The Pierre Auger Observatory is, at the time when this proceeding is written, the largest ground-based observatory ever built dedicated to the indirect detection of cosmic rays through their interaction in the atmosphere of the Earth and the creation of Extensive Air-Showers (EAS). It is located in the Southern hemisphere, in the province of Mendoza, Argentina, at an altitude around 1400 m above sea level. It consists in two main parts: a Surface Detector (SD) with 1600 water-Cherenkov detector stations, deployed on a triangular grid with a 1500 m spacing, covering an area larger than 3000 $\mathrm{km}^2$. The SD is overlooked by 24 telescopes located at the edges of the SD, thus forming the Fluorescence Detector (FD) and giving the main characteristic of the Pierre Auger Observatory: its hybrid design. Both detectors, SD and FD, have been designed to work independently from each other, thus offering the possibility to simultaneously observe the lateral and longitudinal profiles of a shower initiated by a cosmic ray. While FD is working only during moonless and clear nights, reducing its duty cycle ~15%, SD is operating 24 hours per day in almost all weather conditions. Thanks to this, the Pierre Auger Observatory is able to collect a large statistic of events with the SD, for which the calibration in energy is ensured by a use of a subset of events that can be reconstructed by both detectors at the same time, as it will be described later.

Initially designed to observe UHECRs with an energy above 3 EeV to get a precise measurements of the ankle at ~5 EeV and the flux at the highest energies with ≳50 EeV, the Pierre Auger Observatory experienced several upgrades. In particular, an extension towards lower energies with the aim to observe the region of the second knee at ~0.1 EeV. For the SD, this extension consists in building inside the existing array, a denser region with 61 water-Cherenkov detectors with a spacing of 750 m and covering an area of 24 $\mathrm{km}^2$. This infilled array is then overlooked by 3 High-Elevation Auger Telescopes (HEAT), observing the higher altitudes and detecting events dominated by the Cherenkov emission. Several detection and reconstruction techniques have thus been used to reconstruct the energy spectrum of cosmic rays from 6 PeV to the highest energies. Alongside with the data sets used, these techniques will be described below.

*e-mail: spokespersons@auger.org

## 2 Five data sets from two types of detectors

To reconstruct the energy spectrum of cosmic rays starting at 6 PeV, several instruments and methods have been applied. Thus, not less than five different data sets are used by the Pierre Auger Collaboration: two data sets for the FD, the *hybrid* and *Cherenkov*, and three data sets for the SD, *SD1500-vertical* for events with a zenith angle $\theta$ below 60°, *SD1500-inclined* for events with $60° < \theta < 80°$ and observed using the stations distant from each other by 1500 m, and *SD750* for events observed with the denser array. In the following, main characteristics of the data sets and their reliability will be presented.

### 2.1 Fluorescence Detector

The fluorescence detector of the Pierre Auger Observatory is designed to observe the emission of the fluorescence in the atmosphere due to the passage of the particles, giving access to the longitudinal profile of the shower. Four sites with six telescopes each are thus placed at the four corners of the SD and are observing the sky at elevations from 0° to 30°. At one of these sites, three other telescopes are constituting the extension of the observations at lower energies, HEAT and are overlooking the infilled region of the SD. They are observing the sky at an higher elevation, from 30° to 60°, detecting showers developing at higher altitudes. From these two sets of telescopes, observation of events dominated by fluorescence light or Cherenkov emission lead to the construction of two data sets called *hybrid* and *Cherenkov*, respectively.

*Hybrid data set*

As its name is indicating, the first data set constructed from the observations with the FD is using the hybrid design of the observatory. In an event dominated by the fluorescence light, the shower is seen from the side and a line of triggered pixels is observed on the camera of at least one telescope. The details about the telescopes and its triggering system are presented in Ref. [2]. The known pointing direction in the sky of the triggered pixels is used to determine the shower-detector plane containing the axis of the shower and the location of the considered telescope. Then the timing information of each pixel is used to fit the inclination of the shower axis inside the shower-detector plane. At the energies considered by the hybrid data set, the observed FD events is combined with the detection of the shower by at least one station of the SD on the ground. The timing information and position of the station are then used to improve the constraints on the reconstructed geometry of the shower (location of the core on ground, zenith, and azimuth angles) and thus greatly improving the accuracy of the geometrical reconstruction. With the geometry determined, the next step consists in the reconstruction of the longitudinal profile of the shower to estimate the energy of the cosmic rays. This step requires to take into account the detector efficiency, the fluorescence and Cherenkov yields, and attenuation and scattering in the atmosphere of the emitted photons. The last point is possible

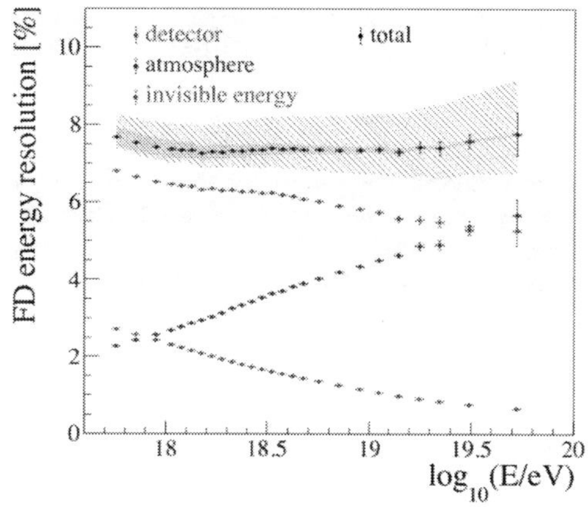


**Figure 1.** Energy resolution of the Fluorescence Detector for the hybrid data set. Contributions from the detector effects (detection and reconstruction), the atmosphere, and the correction for invisible energy are reported. Details are presented in Ref. [5].

thanks to the extensive monitoring of the atmosphere for each night of the measurements. Once this profile is constructed from the data of the telescopes, it is fitted with a Gaisser-Hillas function, which integral gives an estimate of the calorimetric energy deposited in the atmosphere. However, for the total deposited energy, we need to correct the energy (~15% of the total at $10^{18}$ eV and decreasing to ~12% at $10^{20}$ eV) for the contribution carried away by muons and neutrinos. The details about the correction of the invisible energy are presented in Ref. [3] but it is worth mentioning that this correction is based on a data-driven analysis.

Selection criteria are applied to guarantee the quality of the spectrum resulting from the use of this data set. Thus only events with $\theta < 60°$, with good reconstruction of the geometry and the longitudinal profiles are selected. The depth of the maximum development of the shower also has to be observed in the field of view of the detector. Cuts on the atmospheric conditions for which the information on the aerosol content is known, on the energy resolution, etc., complete the selection process [4]. Moreover, given the trigger probability of a single station of the SD and the systematic uncertainty on the exposure, the measurements of the energy spectrum is performed for energies above $10^{18}$ eV. For the reconstruction of the spectrum from the hybrid data, 14675 events are selected in the period from 01/2007 to 12/2017.

The resolution of the FD energy for these hybrid events is driven by three contributions: the detector (calibration of the telescopes and reconstruction of the profile), the knowledge of the atmosphere at the time of the observation (molecules and aerosols concentration), and the invisible energy. The results of the study performed in Ref. [5] are reported in Fig. 1. The total resolution of the FD energy is stable with the energy of the cosmic rays at a level of around 7 to 8%.

 https://doi.org/10.1051/epjconf/202328302004

*Cherenkov data set*

In contrast to the previous data set, in which the light collected by the telescopes is dominated by the fluorescence emission, the second data set constructed from the FD measurements considers events in which the Cherenkov emission is dominating and pointing directly towards the telescopes. While this allows for an extension of the observations to lower energies, it comes at the cost of having no counterparts from the SD because of the low distance between the FD detector and the location of the core of the shower. Thus the previous hybrid reconstruction procedure cannot be applied anymore for the data recorded at these energies at the HEAT telescopes. The reconstruction of the geometry of the shower is then performed using the Profile-Constrained Geometry Fit (PCGF). All the geometries compatible with the timing information of each triggered pixel are tested and for each one of them the longitudinal profile is reconstructed assuming a Gaisser-Hillas function. The geometry that gives the best expectation of the profile of energy deposit in the atmosphere is then selected.

To decrease the energy threshold of events recorded by the FD and reconstructed with a PCGF method, two data sets have to be distinguished. The first one consists of events for which the full trigger procedure of the telescopes is satisfied. However, this trigger procedure has been designed to reduce the number of events dominated by Cherenkov light [6]. To monitor the evolution of the HEAT telescopes, the Pierre Auger Observatory is randomly storing 10% of events that do not pass the last step of the trigger chain. These events, called minimum-bias events, lead to increase of the statistics of events at the lowest energies considered and thus decrease the energy threshold to $10^{15.8}$ eV and a selection of 123159 events in the period from 06/2012 to 12/2017. As for the hybrid data sets, a selection is applied to the events to minimize the introduction of biases. Cuts on the quality of the reconstruction using the PCGF method, requiring that the maximum of the shower is in the field of view of the telescopes, requiring that the minimum fraction of the Cherenkov light is above 50% etc., are applied. This criteria enable us to reduce the energy bias below 5% and ensure a resolution evolving from 12% at $10^{15.8}$ eV to 6% at $10^{18}$ eV.

## 2.2 Surface Detector

The 1660 detectors of the surface array, whether they are deployed on the triangular grid with a spacing of 1500 m or 750 m, are water-Cherenkov detectors sensitive to both the muonic and electromagnetic components of the showers. With the increase of the inclination of the shower, the electromagnetic component gets more and more absorbed in the atmosphere and asymmetry effects due to the geomagnetic field have to be taken into account to perform an unbiased reconstruction of the characteristics of the cosmic ray initiating the shower. To reconstruct the events observed using such detectors, two techniques have been developed, thus splitting the events in two categories: vertical for which $\theta < 60°$ and inclined for $60° < \theta < 80°$.

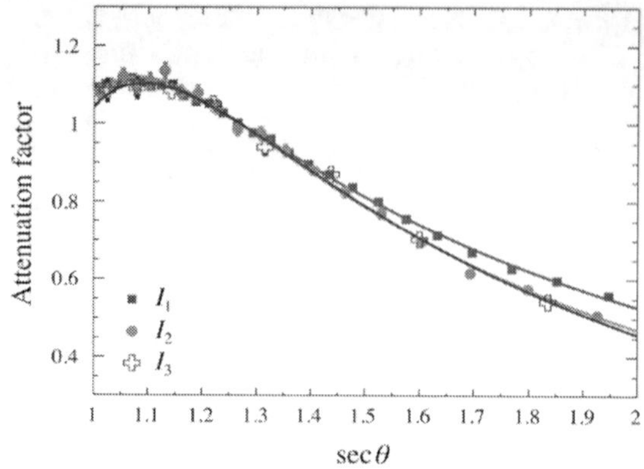


**Figure 2.** Attenuation curves of $S(1000)$ as a function of sec $\theta$ for three different intensities corresponding to the following energies: $3\times10^{18}$ eV (red), $8\times10^{18}$ eV (green) and $2\times10^{19}$ eV (blue). Details are presented in Ref. [11].

However this distinction is relevant only for the 1500 m array. The size of the infilled region, 24 km$^2$, and the spacing between the detectors, allows a fully triggered detection of EAS only with zenith angles up to 40°.

*Vertical data sets*

Covering large arrays at small costs implies a sparse deployment of the detectors, thus the lateral profile of the shower observed on the ground is sampled only by a finite number of detectors for which the time and amplitude of the signal they observe is extracted. While for the reconstruction of the geometry of the shower, i.e. the location of the core on the ground and the arrival direction (zenith and azimuth angles), we exploit the timing at which the stations are triggered by the passage of the particles, the first step of the reconstruction of the energy of the vertical data of the SD1500 and SD750 data corresponds to the reconstruction of the Lateral Distribution Function (LDF) [7, 8]. Thus, using the amplitude of the signal of each station participating in an event, the signal at a particular distance from the shower axis is extracted. In the case of the SD1500, this optimal distance is 1000 m and it has been shown in Ref. [9] that the fluctuations of the signals at this distance are minimal. This signal, $S(1000)$, is then the more accurate estimator of the shower size and by extension, of the energy of the primary. Recent studies of the impact of this procedure, especially the choice of one optimal distance, are reported in Ref. [10]. For the SD750, the same procedure is applied, extracting $S(450)$, the signal at 450 m. However, these estimators of the shower size are dependent on the inclination of the shower since with an increase of the inclination, the electromagnetic component is more and more absorbed. To overcome this, the Constant Intensity Cut (CIC) method is applied, assuming that the flux of cosmic rays is isotropic above a defined threshold in energy. Based on that assumption, for which the validity has been scrutinized in Ref. [11], attenuation curves are established at different intensities (~energies) of cosmic rays as shown in Fig. 2 for SD1500. The estimator of the shower size is reduced by a factor 2 when comparing $S(1000)$ at 0° and 60°. According to the model parametrised from these curves, $S(1000)$ and $S(450)$ are corrected in such a way that the energy estimators are defined as $S_{38}$ and $S_{35}$, i.e. the signals that would be observed if inclination of the shower was 38° and 35° for SD1500 and SD750, respectively. Finally, it is important to mention that these energy estimators are then corrected to take into account effects from the atmospheric conditions at the time of the observations [12] and effects from the geomagnetic field [13]. Thanks to these procedures and the detection methods deployed in the field of the observatory, the SD array is fully efficient for vertical events for energies above $10^{18.4}$ eV with a selection of 215030 events in time period from 01/2004 to 08/2018. For the SD750 array, the extension to the lower energies is first assured by the smaller distance between two adjacent detectors but, on top of that, the use of triggers dedicated to the detection of low-energy events ensure a full efficiency above $10^{17}$ eV [14]. In the period from 01/2014 to 08/2018, a total of 545090 events were selected to reconstruct the spectrum of cosmic rays from the SD750.

*Inclined data set*

For the more inclined showers, $\theta > 60°$, the signals observed on the ground are dominated by the muonic component. Moreover, at such inclinations, the circular symmetry around the shower axis is broken by attenuation and geometrical effects and by the influence of the geomagnetic field on the propagation of muons. While the geometry can be reconstructed using the timing of the station, the procedure described in the previous section cannot be applied anymore. For such events, instead of a LDF maps of the density of muons on the ground are built from Monte-Carlo simulations of EAS [15]. Under the assumption that the normalisation of this distribution of muons depends only on the energy of the shower, the scaling parameter $N_{19}$ is used as an estimator of the energy of the shower. It corresponds to the scaling factor needed to reconstruct the map of the density of muons for a simulated shower initiated by a cosmic ray with an energy of $10^{19}$ eV to the data. The dependency on the zenith and azimuth angles of the showers is thus incorporated in the density map, leaving the estimator of the energy $N_{19}$ free of such dependencies. These methods of detection and reconstruction ensure a full efficiency of SD1500 for the inclined events above $10^{18.6}$ eV with the selection of 24209 events over a period of acquisition from 01/2004 to 08/2018.

*Calibration in energy*

To convert the estimators of the energy, $S_{38}$, $S_{35}$, and $N_{19}$ (for the SD1500-vertical, SD750, and SD1500-inclined, respectively), the hybrid design of the Pierre Auger Observatory is playing a central role, making it possible to perform a calibration in a data-driven way, free of any assumptions about the mass composition of the cosmic rays or about the models of hadronic interactions. For each of the three SD data sets, a subset of events has also been observed and reconstructed by the FD. Applying a set of selection criteria on the quality of the reconstruction for both the SD and FD, on the knowledge of the atmospheric conditions at the time of the observation, etc., the energy calibration of the SD estimators is performed comparing the SD estimators with the calorimetric energy obtained from the FD reconstruction. The calibration of the SD1500 data sets (vertical and inclined) is ensured by a comparison with the standard FD telescopes while the one of the SD750 is also counting on the use of the HEAT data. This procedure is shown in Fig. 3. For each data set, the correlation with the FD energy, is fitted using

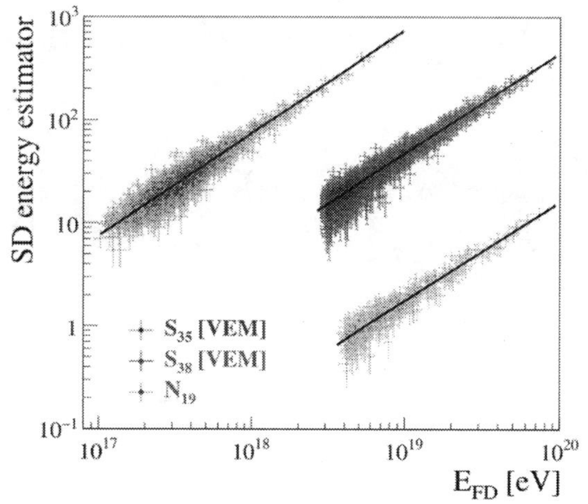


**Figure 3.** Calibration functions: correlation between the FD energy and the SD energy estimator for the three SD data sets. The black lines represent the best fit for each data set. Figure taken from Ref. [6].

$$E_{FD} = A\,\hat{S}^{\,B}, \tag{1}$$

where $\hat{S}$ is the considered estimator of the energy and the parameters $A$ and $B$ are left as free parameters. The results of the fit is reported in Fig. 3 by the black lines.

The evaluation of the energy resolution of the SD data sets is performed using the same subsets of events seen by both detectors and thus building the distributions of ratio $E_{SD}/E_{FD}$ as a function of energy. This ratio is shown in Fig. 4 for the SD1500-vertical data set at different energies. Knowing the FD energy resolution [5], ~7 to 8%, the SD energy resolution is estimated decreasing from ~20% to less than 10% in the energy range of each data set. The bias is inferred using the same ratio for the SD1500 vertical and inclined data sets, while for the SD750 data set a use of MC simulations complete the study at the lowest energies considered [8, 11, 16].

The systematic uncertainties in the energy scale of the Pierre Auger Observatory are derived the same as for the energy resolution, using hybrid events [5]. The main contributions to the systematic uncertainties are coming from the uncertain knowledge of the atmosphere (evolving with

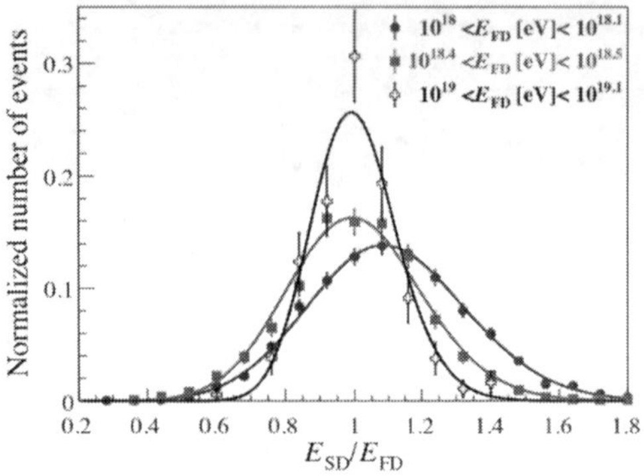


**Figure 4.** Ratio distribution of the Surface Detector and Fluorescence Detector energies for three different energy ranges. Details can be found in Ref. [11].

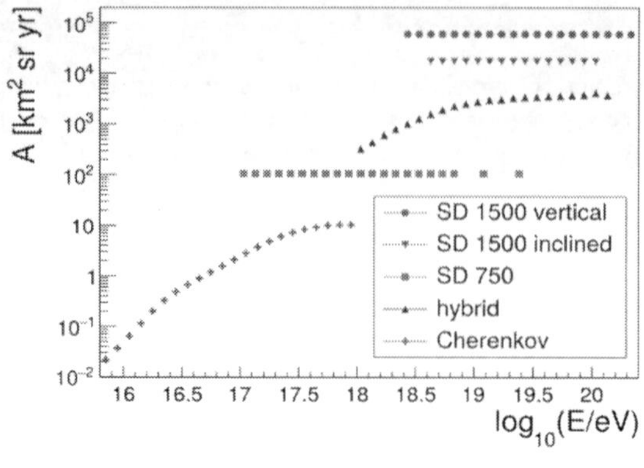


**Figure 5.** Exposure of the surface and fluorescence detectors.

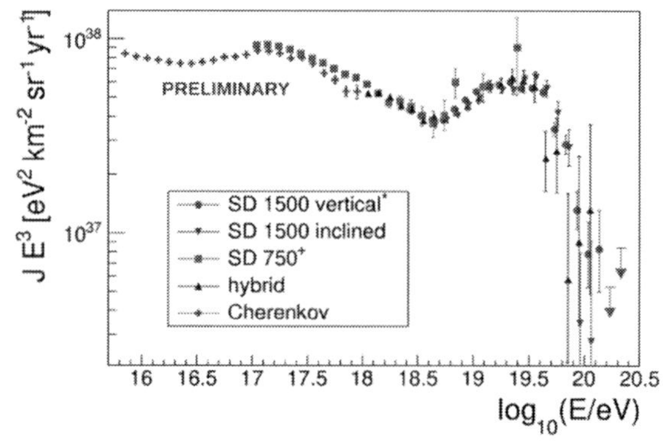


**Figure 6.** Five spectra of the Pierre Auger Observatory. Taken from Ref. [6].

energy from 3.4% to 6.2%), the calibration of the FD telescopes (9.9%), and the method to reconstruct the longitudinal profile (decreasing with energy from 6.5% to 5.6%). Adding the uncertainties arising from the invisible energy, fluorescence yield, and the total uncertainties in the energy scale results in ~14%.

## 3 The combined spectrum and a new feature

For each of the five data sets previously described, the raw spectrum of the cosmic rays is obtained by dividing the rate of events in a particular bin of energy by the width of this bin and its corresponding exposure.

*Exposure*

In the case of the three data sets constructed from the SD, the computation of the exposure is reduced to a geometrical problem. Indeed, at energies above the full-efficiency threshold, the total exposure is calculated by integrating in time the aperture of all working cells of the surface array, i.e. hexagonal cells in the case of the triangular grid. This calculation is possible using the monitoring database of the SD, and because for each event in the SD1500-vertical, -inclined, and SD750 data set, it is required that the six stations around the station recording the highest amplitude of the signal are active at the time of the detection. The exposure, computed in such a way, is then independent of any MC assumptions and of the energy. Its uncertainties are kept below 3%. The results are presented in Fig. 5. Thanks to its large area of detection and its more than 15 years of data taking, the Pierre Auger Observatory has reached exposures of 60400 $km^2$ sr yr, 17500 $km^2$ sr yr, and 105 $km^2$ sr yr for the SD1500-vertical, -inclined, and SD750 data sets, respectively.

The computation of the exposure for the FD data sets is less straightforward and has to be performed simulating a large number of MC events that are taking into account the status of the atmosphere and of the telescopes [17]. Due to the constraints inherent in the use of MC simulations, the mass composition chosen is following the one derived from the Global Spline Fit model [18] and simulations assuming pure composition are then used to derive the associated uncertainties related to the lack of knowledge of the true composition reaching Earth at the highest energies. The resulting exposure is thus dependent on the energy of the cosmic rays and is ~2200 $km^2$ sr yr at $10^{19}$ eV and ~10 $km^2$ sr yr at $10^{17.5}$ eV for the hybrid and Cherenkov data sets, respectively. The uncertainties arising from an exposure computed using MC simulations are described in Ref. [6].

*Five spectra*

The raw spectra are suffering from the distortions induced by the finite energy resolution and the shape of the spectrum, causing a bin-to-bin migration of events. Taking into account the detector effects and assuming a spectrum following a multiply-broken power law, a correction factor is established as a function of the energy and applied to the raw spectra. The resulting spectra are shown in Fig. 6 after normalizing (independently of the energy) the spectra by <1%, +5%, −2%, <1%, and +7% for the SD1500-

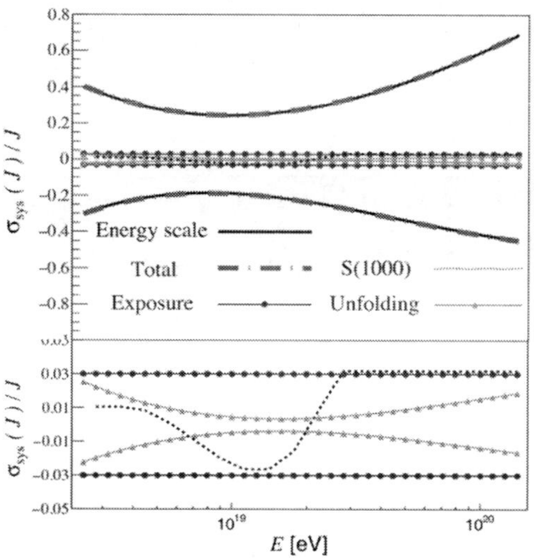


**Figure 7.** Systematic uncertainties in the energy spectrum as a function of the energy. Contributions from the energy scale, exposure, $S(1000)$ reconstruction, and unfolding procedure are presented. More details can be found in Ref. [11]

vertical, -inclined, SD750, hybrid, and Cherenkov spectra, respectively. These normalization values are obtained applying the combination method that will be described later in this section. The discrepancies between all the different reconstructed spectra is well encompassed by the uncertainties in the energy scale (~15% for the Cherenkov data set, ~14% for the other data sets). Moreover, it is interesting to note that the uncertainty in the energy scale is the main source of uncertainties in the reconstruction of the spectra observed by the Pierre Auger Observatory. In Fig. 7, the analysis of all possible sources of uncertainties in the spectrum is reported for the SD1500-vertical data set [11]. Systematic uncertainties coming from the reconstruction of the energy estimator, computation of the exposure, or forward-folding procedure are below 3% and thus well below the uncertainties in the energy scale. The domination of the uncertainties in the energy scale in the total uncertainty of the spectrum is found also for all the other data sets used [4, 6, 8, 16]. It is worth noting that at the lowest energies of the spectrum reconstructed from the Cherenkov data set, the uncertainty are dominated by the uncertainties in the exposure due to the assumptions made with the use of simulations.

The combination of all the five spectra is performed using a forward-folding approach, beginning with the evaluation of the uncorrelated uncertainties between the different data sets, a shift in exposure and in the energy calibration is introduced to modify a given spectrum relative to the others. For each spectrum, a bin-to-bin migration of events is taken into account considering the reconstruction biases and resolutions of each data set. The shape of the expected spectrum is assumed as a sequence of six power-laws written as

$$J(E) = J_0 \left(\frac{E}{10^{16}\,\mathrm{eV}}\right)^{-\gamma_0} \prod_{i=0}^{4} \left[1 + \left(\frac{E}{E_{ij}}\right)^{\frac{1}{\omega_{ij}}}\right]^{(\gamma_i-\gamma_j)\omega_{ij}}, \quad (2)$$

where $J_0$ is the normalization parameter, $\gamma_i$ are the spectral indices, and $E_{ij}$ the energies at which the transitions occur. The width of the transition between two power-law regions is set by the parameters $\omega_{ij}$ fixed as $\omega_{01} = \omega_{12} = 0.25$ and $\omega_{23} = \omega_{34} = 0.05$ (half of the width of one logarithmic bin in energy). Thus a combined likelihood, written as the product of Poissonian terms, is applied to determine for each spectrum the best shift (constrained by a Gaussian penalization included in the likelihood) in exposure and in energy and the best set of parameters is selected to get the combined spectrum shown in Fig. 8.

The combined spectrum is fitted using Eq. (2) and the best-fit parameters are reported in Tab. 1 with the transition width fixed as mentioned previously. This spectrum is now covering five orders of magnitude in energy, from $6\times10^{15}$ eV to the highest energies ever observed. Positions of the low-energy ankle at $2.8\times10^{16}$ eV, firstly reported in Ref. [19], of the second knee at $1.58\times10^{17}$ eV, of the ankle at $5\times10^{18}$ eV and of the suppression at $4.7\times10^{19}$ eV are thus reported. With the increase of the number of events collected by the Pierre Auger Observatory, a new feature has recently been reported at an energy of $1.4\times10^{19}$ eV and is named *in-step* [11]. The interpretation of all these features is now under scrutiny to understand the mechanisms behind the acceleration of the cosmic rays. Such interpretation is presented in Ref. [20] using the data about the mass composition of the cosmic rays. For example, in the scenario explored, the in-step feature appears as a transition between helium-like nuclei and CNO-like nuclei.

## 4 Conclusion: a glimpse into the future

The continuous observation of UHECR by the Pierre Auger Observatory in the last 18 years allows the reconstruction of the energy spectrum with an unprecedented precision. However, the story of the energy spectrum is not planned to end soon. Associated to its longevity, the Observatory is constantly improved. To reinforce the measurements at the lowest energies, an array of 19 water-Cherenkov detectors has been deployed in the region of the SD750 with a spacing between two adjacent detectors reduced to 433 m [21]. It is then expected, from its measurements, to build a sixth energy spectrum of cosmic rays starting at $6\times10^{16}$ eV for vertical events ($\theta < 45°$).

Moreover, on top of each water-Cherenkov detector, a scintillator is added to improve the sensitivity of the SD to the mass composition of UHECRs [22]. Radio antennas, muon counters etc. complete the upgrade of the observatory. The knowledge of the mass of the cosmic rays on an event-by-event basis, would constitute a remarkable asset to build spectra for different mass primaries and thus, to help constraining various scenarios of acceleration and propagation of cosmic rays.

**Table 1.** Best-fit parameters of the combined energy spectrum. Results are shown with their statistical and systematical uncertainties.

| | | |
|---|---|---|
| | | $\gamma_0 = 3.09 \pm 0.01 \pm 0.10$ |
| low-energy ankle | $E_{01} = (2.8 \pm 0.3 \pm 0.4)\times10^{16}$ eV | $\gamma_1 = 2.85 \pm 0.01 \pm 0.05$ |
| second knee | $E_{12} = (1.58 \pm 0.05 \pm 0.2)\times10^{17}$ eV | $\gamma_2 = 3.283 \pm 0.002 \pm 0.10$ |
| ankle | $E_{23} = (5.0 \pm 0.1 \pm 0.8)\times10^{18}$ eV | $\gamma_3 = 2.54 \pm 0.03 \pm 0.05$ |
| instep | $E_{34} = (1.4 \pm 0.1 \pm 0.2)\times10^{19}$ eV | $\gamma_4 = 3.03 \pm 0.05 \pm 0.10$ |
| suppression | $E_{45} = (4.7 \pm 0.3 \pm 0.6)\times10^{19}$ eV | $\gamma_5 = 5.3 \pm 0.3 \pm 0.1$ |
| | $J_0 = (8.34 \pm 0.04 \pm 3.40)\times10^{-11}\mathrm{km^{-2}sr^{-1}yr^{-1}eV^{-1}}$ | |

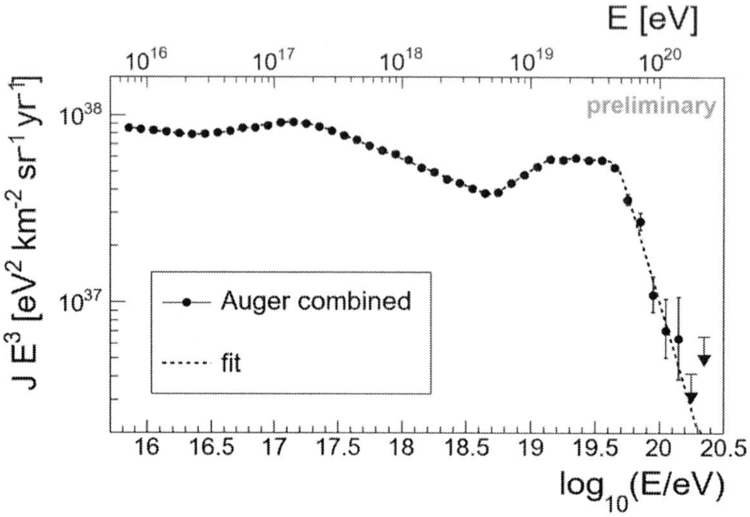


**Figure 8.** Combined energy spectrum, multiplied by $E^3$ and best fit function, presented in [6].

Such scenarios are already tested combining the energy spectrum as presented in this proceeding with the mass composition extracted from the measurement of the shower maximum. A recent update of this analysis performed by the Pierre Auger Collaboration has been reported in Ref. [23]. The transition between Galactic and extra-galactic cosmic rays is thus scrutinized, as well as the end of the spectrum interpreted as a combination of a rigidity-dependent acceleration and propagation effects.

Finally it is not possible to track back the origins of the UHECRs without analyzing the arrival directions of the cosmic rays. Anisotropy studies at different scales have been conducted and led to the discovery of a dipole above 8 EeV and an indication of the Starburst galaxies as a possible sources of UHECRs. The combination of all the characteristics of the cosmic rays: energy, mass, and arrival direction is constituting the next step in solving the question of the origins of such particles and is already under way [24].

## References

[1] A. Aab et al. [Pierre Auger Coll.], *Nucl. Instrum. Meth. A* **798** (2015) 172-213, [1502.01323].

[2] J. Abraham et al. [Pierre Auger Coll.], *Nucl. Instrum. Meth. A* **620** (2010) 227-251, [0907.4282].

[3] A. Aab et al. [Pierre Auger Coll.], *Phys. Rev. D* **100** (2019) 082003, [1901.08040].

[4] M.A. Settimo [Pierre Auger Coll.], *Eur. Phys. J. Plus* **127** (2012) 87, [1208.6574].

[5] B. Dawson [Pierre Auger Coll.], *PoS* (ICRC2019) 231.

[6] V. Novotný [Pierre Auger Coll.], *PoS* (ICRC2021) 324.

[7] A. Aab et al. [Pierre Auger Coll.], *JINST* **15** P10021 (2020), [2007.09035].

[8] A. Aab et al. [Pierre Auger Coll.], *Eur. Phys. J.* C **81** 966 (2021), [2109.13400].

[9] D. Newton, J. Knapp and A.A. Watson, *Astropart. Phys.* **26** 414 (2007), [astro-ph/0608118].

[10] V. Verzi [Pierre Auger Coll., Telescope Array Coll.], these proceedings.

[11] A. Aab et al. [Pierre Auger Coll.], *Phys. Rev. D* **125** 121106 (2020), [2008.06486].

[12] A. Aab et al. [Pierre Auger Coll.], *JINST* **12** (2017) P02006, [1702.02835].

[13] P. Abreu et al. [Pierre Auger Coll.], *JCAP* **11** (2011) 022, [1111.7122].

[14] A. Coleman [Pierre Auger Coll.], *PoS* (ICRC2019) 225.

[15] A. Aab et al. [Pierre Auger Coll.], *JCAP* 08 (2014) 019, [1407.3214].

[16] A. Aab et al. [Pierre Auger Coll.], *JCAP* 08 (2015) 049, [1503.07786].

[17] P. Abreu et al. [Pierre Auger Coll.], *Astropart. Phys.* **34** (2011) 368, [1010.6162].

[18] H.P. Dembinski et al., *PoS* (ICRC2017) 533.

[19] W.D. Apel et al., *Astropart. Phys.* **36** (2012) 183[1206.3834]

[20] A. Aab et al. [Pierre Auger Coll.], *Phys. Rev. L* **102** 062005 (2020), [2008.06488].

[21] G. Silli [Pierre Auger Coll.], *PoS* (ICRC2021) 224.

[22] A. Aab et al. [Pierre Auger Coll.], [1604.03637].

[23] E. Guido [Pierre Auger Coll.], *PoS* (ICRC2021) 311.

[24] T. Bister [Pierre Auger Coll.], *PoS* (ICRC2021) 368.

# Energy Spectrum Measured by the Telescope Array Surface Detectors

*Jihyun* Kim[1*], *Dmitri* Ivanov[1], *Charles* Jui[1], and *Gordon* Thomson[1] for the Telescope Array Collaboration

[1]High Energy Astrophysics Institute and Department of Physics and Astronomy, University of Utah, Salt Lake City, Utah 84112-0830, USA

**Abstract.** Located in the west desert of Utah, USA, the Telescope Array experiment is the largest ultra-high energy cosmic ray observatory in the northern hemisphere. It consists of two types of detectors: scintillator surface detectors (SDs) and air fluorescence detectors (FDs). A total of 507 SDs consisting of two-layer plastic scintillation counters is deployed with 1.2 km spacing, making measurements over an area of approximately 700 km$^2$. There are 3 FD stations, having 38 fluorescence telescopes viewing 3°–31°in elevation, overlooking the SD array. In this work, we update the Telescope Array energy spectrum as measured by the SD array for 14 years. We verify the linearity in TA SD energy reconstruction using three different energy comparisons: Monte Carlo comparison, FD/SD comparison, and Constant Intensity Cut method. The spectral features of ankle, softening, and GZK cutoff are seen in 14 years of TA SD data. Also, we consistently observe that the cutoff energy depends on declination.

## 1 Introduction

Ultra-high energy cosmic rays (UHECRs), extremely energetic ($E > 10^{18}$ eV) charged particles flying into Earth's atmosphere from outer space, have been investigated as a window of exploring the Universe. The traditional main observables of UHECRs are mass composition, arrival direction, and energy spectrum. Comprehensive understandings of those observables all together are necessary to elucidate the nature and origin of UHECRs. It is essential to investigate the energy spectrum of UHECRs since the spectral features of cosmic rays contain information about the sources and about cosmic ray propagation. The high energy cutoff at $\sim 10^{19.8}$ eV, the energy predicted by Greisen, Zatsepin, and Kuzmin (GZK) [1, 2], was first observed by the High Resolution Fly's Eye experiment (HiRes) [3] and later confirmed by Telescope Array (TA) experiment [4]. In the southern hemisphere, the Pierre Auger Observatory (Auger) [5] observed the cutoff at a slightly lower energy. Although the GZK cutoff mechanism is pion photoproduction due to cosmic microwave background photons striking cosmic ray protons, the high energy cutoff might be caused by the maximum energy of sources. The ankle, a hardening of the spectrum at $\sim 10^{18.7}$ eV, can be interpreted as due to $e^+e^-$ pair production [6] if cosmic rays are protons. In the southern hemisphere, the Auger collaboration has a different interpretation [7]. In this work, we investigate energy spectrum features using the most up-to-date data measured by the TA surface detector array.

*e-mail: jihyun@cosmic.utah.edu

## 2 Telescope Array Experiment

The Telescope Array (TA) experiment [8, 9] is the largest UHECR observatory in the northern hemisphere, which is designed to observe the extensive air shower induced by a primary particle hitting Earth's atmosphere. It is located near the west desert of Utah, USA (39.3° N, 112.9° W, 1400 m above sea level). We deployed a total of 507 surface detectors (SDs) consisting of two-layer plastic scintillation counters with 1.2 km spacing, making measurements over an area of approximately 700 km$^2$. This SD array is overlooked by three fluorescence detector (FD) stations having a total of 38 fluorescence telescopes viewing 3°–31°in elevation above the SD array. The deployment layout is depicted in Figure 1.

## 3 Reconstruction of Events Observed by Surface Detectors

When a primary particle of UHECRs collides a nucleus of Earth's atmosphere, a cascade of millions of subatomic particles, an extensive air shower (EAS), is initiated and develops along the track of the primary particle. The profile of the EAS is reconstructed as follows. First, we locate the core position and determine the arrival direction of the primary particle using SD counter position and timing by using the modified Linsley shower-shape function fit [10]. Next, we fit counter signal size to find its lateral distribution [11, 12]. The signal size at 800 m, S800, is the density of shower particles at a lateral distance of 800 m from the shower axis is used as an energy estimator. Using S800 and zenith angle, the energy of the primary particle is determined from the *look-up table*, $E_{\mathrm{TBL}}$, which is produced from the SD Monte Carlo simulation using CORSIKA

https://doi.org/10.1051/epjconf/202328302005

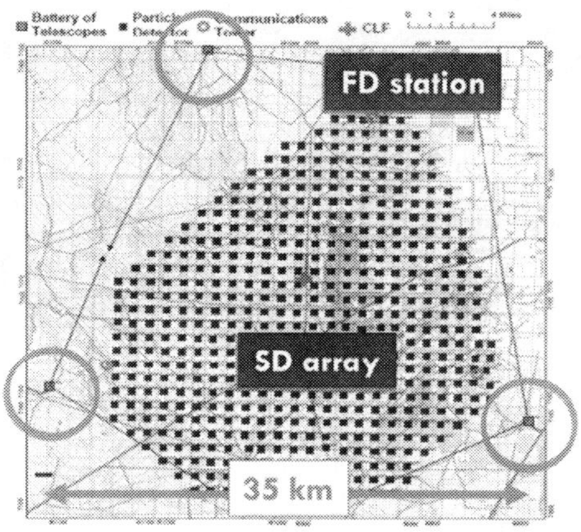


**Figure 1. Layout of the Telescope Array (TA) experiment.** The black squares are the TA surface detector (SD) counters, and the gray squares are the TA fluorescence detector (FD) stations. The arrows from the FD stations represent the each FD's viewing ranges. The distance scale of two southern FD stations is approximately 35 km.

software package [13] with the QGSJetII-03 hadronic interaction model assuming proton primaries [14]. Lastly, we compare these Monte Carlo determined energies with FD's calorimetric energies using hybrid events. The scale factor is found to be 1.27, and the final energy is obtained by $E_{\mathrm{Final}} = E_{\mathrm{TBL}}/1.27$.

## 4 Descriptions on Dataset

In this work, we used a dataset measured by the TA SD array from May 11, 2008 up to May 10, 2022; i.e. 14 years of SD data. The selection criteria employed in this analysis are as follows: 1. Each event must include at least five SD counters.
2. The reconstructed primary zenith angle must be less than 45°.
3. The reconstructed event core must be more than 1200 m from edge of the array.
4. Both the geometry and lateral distribution fits must have $\chi^2$/degree of freedom value less than 4.
5. The angular uncertainty estimated by the geometry fit must be less than 5°.
6. The fractional uncertainty in S800 estimated by the lateral distribution fit must be less than 25%.

Monte Carlo based on CORSIKA software package is used for resolution and exposure calculations. The energy and angular resolutions of 14 years of TA SD data are summarized below.

- $E \geq 10^{19.0}$ eV: 20% in energy and 1.4° in angular,
- $10^{18.5}$ eV $\leq E < 10^{19.0}$ eV: 29% in energy and 2.1° in angular,
- $10^{18.0}$ eV $\leq E < 10^{18.5}$ eV: 34% in energy and 2.4° in angular.

## 5 Results

### 5.1 Linearity in Energy Reconstruction

Figure 2 shows the linearity in energy reconstruction with three different approaches. The left panel shows the ratio between reconstructed energy and thrown energy in Monte Carlo simulation. The same reconstruction program for data analysis is used for this analysis. The linearity is observed in this comparison. The middle panel shows the ratio between reconstructed energy measured by SD and FD as a function of energy. For higher energy bins the number of events is not sufficient but linearity is observed. The right panel shows the comparison of energies determined by the TA's standard SD reconstruction method and the Constant Intensity Cut method [15]. The result agrees within ~3% level at the TA's highest energy [16]. The results from these three different approaches are consistent with each other showing the linearity in the energy reconstruction above $10^{19}$ eV.

### 5.2 Energy Spectrum and Features

Figure 3 shows the energy spectrum derived from the 14 years of TA SD data with the selection criteria described in Section 4. The data is represented by black points with error bars and a fit to the broken power law is shown with the black solid line. Here, we fit to the broken power law having two breakpoints. The first break point, the ankle feature, is found at $E_{\mathrm{ankle}} = 10^{18.68\pm0.01}$ eV with spectral indices of $p1 = -3.27 \pm 0.02$ before the ankle and $p2 = -2.69 \pm 0.02$ after the ankle. The second break point, the GZK cutoff feature, is obtained at $E_{\mathrm{GZK}} = 10^{19.78\pm0.04}$ eV with the spectral index of $p3 = -4.47 \pm 0.41$ after the GZK cutoff.

It is known that there are differences in the cutoff energies depending on declination in TA's filed of view from the previous energy spectrum analysis results [17]. Using the 14-year TA SD dataset, we confirm that the cutoff energy depends on declination. In the lower declination band, between -15.7° and 24.8° (the band seen by both TA and Auger), the cutoff energy is fit to be $10^{19.65\pm0.002}$ eV. In the upper declination band, $\delta \geq 24.8°$, the cutoff energy is fit to be $10^{19.84\pm0.02}$ eV. This energy difference has a local significance of about $5\sigma$. The declination dependence feature is shown in the right panel of Figure 3.

Recently, Auger reported evidence for a new spectral feature, a softening of the energy spectrum at $10^{19.11\pm0.03}$ eV [18]. Following that, we searched for the new feature in the northern hemisphere data, including HiRes, TA FD, and TA SD. The same softening feature in the northern hemisphere has been observed but at the energy of $10^{19.25\pm0.03}$ eV [19]. Here, we investigate this feature using the 14 years of SD data. The softening feature is consistently observed, and the fit to the broken power law shows the softening feature appears at energy of $10^{19.22\pm0.08}$ eV. The probability to observe such a feature in the spectrum by chance is excluded at $4.0\sigma$ level. The detailed results of spectral features using three break points are shown with the black fitting line in Figure 4.

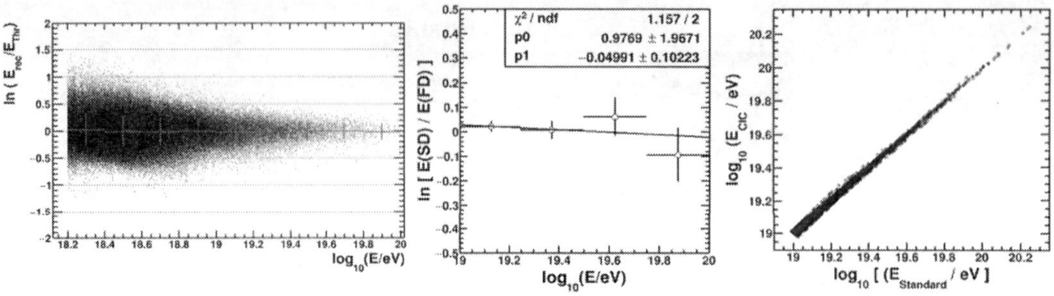


**Figure 2. Linearity in energy reconstruction.** The ratio between reconstructed energy and thrown energy in Monte Carlo simulation (left), the ratio between reconstructed energy measured by SD and FD (middle), and the comparison of energies determined by the TA's standard SD reconstruction method and the Constant Intensity Cut method (right) as a function of energy show consistent results of the linearity in the energy reconstruction.

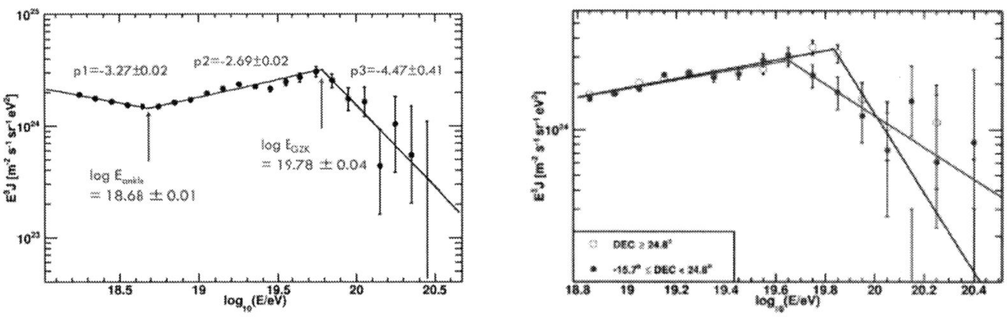


**Figure 3. Energy spectrum using 14-year SD data and declination dependence. (Left)** The data is represented with the black points with error bars, and a fit to the broken power law is shown with the black solid line. Here, we fit to the broken power law having two breakpoints, where p1, p2 are the spectral indices before and after the ankle, respectively, $E_{\text{ankle}}$ is the energy of the ankle, $E_{\text{GZK}}$ is the energy of the GZK cutoff, and p3 is the spectral index after the GZK cutoff. **(Right)** The data in the lower declination ($-15.7° \leq \delta < 24.8°$) are marked with the red dots overlaid with the red fitting line. The data in the higher declination ($\delta \geq 24.8°$) and its fitting result are shown with the open black circles superimposed and the black line, respectively.

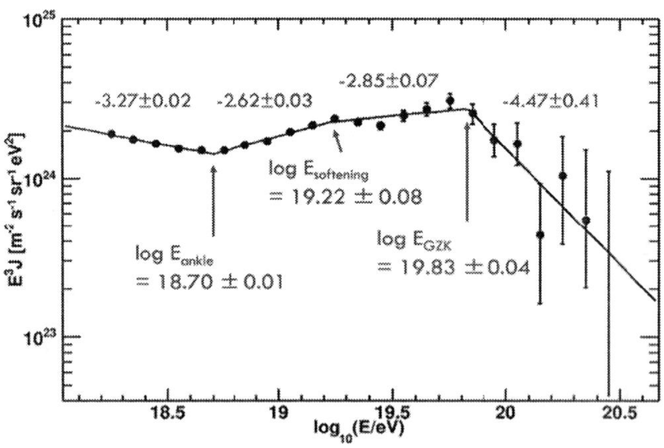


**Figure 4. Softening spectral feature in 14-year SD data.** The data is represented with the black points with error bars, and a fit to the broken power law having three break points is shown with the black solid line. The detailed results are written on the energy spectrum, including spectral indices and break energies.

 https://doi.org/10.1051/epjconf/202328302005

## 6 Summary

The TA SD array has been steadily collecting data since May 11, 2008. In this work, we used 14 years of TA SD data and derived energy spectrum. First, we validated Monte Carlo simulation carefully by comparing it with the distribution of the data. We concluded that TA SD energy reconstruction is robust from three different energy comparisons: FD/SD comparison, Monte Carlo, and Constant Intensity Cut method. The TA SD energy spectrum shows three spectral features—ankle, softening (*instep/shoulder*), and GZK cutoff—with 14 years of data. In addition, the declination dependence of the cutoff energy was consistently seen in the up-to-date dataset.

## References

[1] Greisen, K., Physical Review Letters, **16**, 183 (1966).

[2] Zatsepin, G. T. and Kuz'min V. A., Soviet physics. JETP letters, **4**, 114 (1966).

[3] Abbasi, R. U., *et al.*, Physical Review Letters, **100**, 101101 (2008).

[4] Abu-Zayyad, T., *et al.*, Astrophysical Journal Letters, **768**, L1 (2013).

[5] Abraham, J., *et al.*, Physical Review Letters, **101**, 061101 (2008).

[6] Berezinsky, V., *et al.*, Physical Review D, **74**, 043005 (2006).

[7] Aab, A., *et al.*, Physics Letters B, **762**, 288 (2016).

[8] Abu-Zayyad, T., *et al.*, Nuclear Instruments Methods in Physics Research Section a-Accelerators Spectrometers Detectors and Associated Equipment, **689**, 87 (2012).

[9] Tokuno, H., *et al.*, Nuclear Instruments Methods in Physics Research Section a-Accelerators Spectrometers Detectors and Associated Equipment, **676**, 54 (2012).

[10] Teshima, M., *et al.*, Journal of Physics G Nuclear Physics, **12**, 1097 (1986).

[11] Takeda, M., *et al.*, Physical Review Letters, **81**, 1163 (1998).

[12] Takeda, M., *et al.*, Astroparticle Physics, **19**, 447 (2003).

[13] Heck, D., Schatz, G., Thouw, T., Knapp, J., and Capdevielle, J. N., CORSIKA: A Monte Carlo Code to Simulate Extensive Air Showers, Tech. Rep. 6019, FZKA.

[14] Ostapchenko, S., Nuclear Physics B Proceedings Supplements, **151**, 143 (2006).

[15] Hersil, J., Escobar, I., Scott, D., Clark, G., and Olbert, S., Physical Review Letters, **6**, 22 (1961).

[16] Ivanov, D., for the Telescope Array Collaboration, in Proc. UHECR 2018, Paris, France (2018).

[17] Ivanov, D., for the Telescope Array Collaboration, PoS(ICRC2019)298 (2019).

[18] A. Aab *et al.*, Physical Review Letters, **125**, No.12, 121106 (2020).

[19] Ivanov, D., *et al.*, for the Telescope Array Collaboration, PoS(ICRC2021)341 (2021).

# Measurement of Cosmic-ray Energy Spectrum with the TALE Detector in Hybrid Mode

*Hitoshi* Oshima[1,*], *Keitaro* Fujita[1], *Shoichi* Ogio[1], *Takashi* Sako[1], and *Telescope Array Collaboration*

[1]Institute for Cosmic Ray Research, the University of Tokyo

**Abstract.** The TA Low-energy Extension (TALE) experiment extends the energy range observed by the TA experiment to below $10^{16}$ eV. We aim to study the transition from galactic to extragalactic cosmic rays. The TALE detector is a hybrid apparatus composed of fluorescence telescopes and surface detectors. The surface detectors are arranged to be suitable for hybrid energy spectrum measurements in the low-energy region. We measured the energy spectrum with the 392 hours observation data of the TALE hybrid detector. This energy spectrum measurement will play an important role in understanding the transition from cosmic rays of galactic origin to those of extragalactic origin.

## 1 Introduction

More than 100 years have passed since the discovery of cosmic rays, but the origin of ultra-high energy cosmic rays (UHECRs) have not been revealed yet. Revealing the UHECR origins is one of the most important topics of cosmic-ray physics. In the energy region below $10^{18.7}$ eV, galactic cosmic rays, which originate within the Milky Way Galaxy and extragalactic cosmic rays, which originate outside the galaxy are in competition with each other. The energy spectrum shows a fine structure, especially in the region of galactic cosmic rays, which reflects various physics such as the characteristics and acceleration limits of galactic cosmic-ray sources and confinement by the galactic magnetic field. It is essential to observe the cosmic-ray energy spectrum and measure the composition over a wide energy range. We will cover the wide energy range of five orders of magnitude from $10^{15}$ eV to $10^{20}$ eV by both the Telescope Array (TA) [1, 2] and the TA Low-energy Extension (TALE) [3, 4] experiments. We aim to measure the acceleration limit of galactic cosmic rays through the measurements of composition and energy spectrum, and transition from galactic to extragalactic cosmic rays using the TALE detector.

## 2 TALE detector and data samples

TA is the largest cosmic-ray detector in the Northern hemisphere, located in Millard County, Utah, USA. The TALE experiment extends the energy range observed by the TA experiment to lower energies. The TALE detector is suitable to study the energy spectrum structure and the associated change in the composition of cosmic rays below $10^{18.6}$ eV. The TALE detector is a hybrid apparatus composed of scintillation surface detectors (SDs) covering an area of approximately 20 $km^2$ and fluorescence telescopes (FDs) overlooking the SDs, as shown in Fig. 1. The TALE SDs are placed at 400 m and 600 m spacing for low-energy thresholds. In this study, we analyzed data sample between November 2018 and February 2020 corresponding to 392 hours of operation. These data are collected using both the FDs and the SDs.

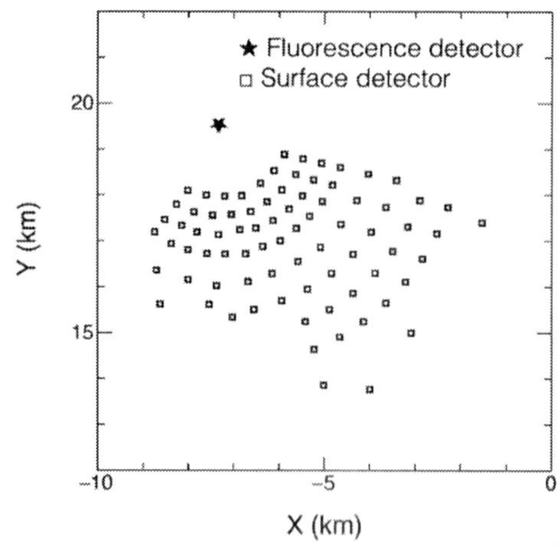


**Figure 1.** Physical location of TALE detector. The TALE detector is a hybrid apparatus composed of surface detectors and fluorescence detectos.

## 3 Monte Carlo simulation

Detector aperture is evaluated using a Monte Carlo (MC) simulation. The MC simulations used in this work comprises of air showers and includes the detector response.

*e-mail: oshima@icrr.u-tokyo.ac.jp

 https://doi.org/10.1051/epjconf/202328302006

Air-shower simulation is performed with CORSIKA [5], using EPOS LHC [6] as the hadronic interaction model. Iron-induced air showers are generated with primary energies ranging from $10^{16.3}$ eV to $10^{18.5}$ eV. These showers were thrown with an isotropic zenith angle distribution between 0° and 70° and a uniform azimuthal distribution between 0° and 360°. Particle information such as position and momentum at the detector level are used to get the energy deposit in each SD with GEANT4 [7–9]. The response of the SD electronics is also taken into account [10]. The FD simulation includes fluorescence and Cherenkov photon generations, telescope optics [11], detector calibration [12], and the response of the electronics [13, 14].

## 4 Data analysis

The analysis is based on hybrid data collected by the TALE FDs and SDs. We used events which are triggered by the FDs and have at least one SD with a signal above three minimum ionizing particles.

Image of the air shower observed by FDs is referred to as the event track. The following conditions are required to select event tracks that have high quality reconstruction: 1) there is no saturated PMT in the track; 2) the depth of maximum, $X_{max}$, of the air shower is within the field of view of the FDs; and 3) the $\chi^2$ in the fitting of longitudinal development of the air shower is less than 100.

As the air shower develops, both the atmospheric fluorescence and the Cherenkov light are emitted. Cherenkov light is the dominant component below $10^{17.0}$ eV, while the fluorescence component increases with increasing energy, becoming the main component above $10^{17.5}$ eV. Events with the fraction of fluorescence light above 75% are defined as fluorescence events, while the others are defined as Cherenkov events. Several filtering criteria are then applied to select shower-induced track-like events. Regarding the Cherenkov events, the following conditions are required: 1) event duration is above 0.1 $\mu$s; 2) the number of PMTs constituting the track is at least 10; 3) the average number of photo-electrons/PMT is grater than 50; and 4) the track length is grater than 6.5°. Concerning the fluorescence events, the number of photo-electrons is required at least 2000 per event. In addition to these filtering conditions, the zenith angle acceptance was set to $\theta < 60°$, while the energy range was set between $10^{16.6}$ eV and $10^{18.5}$ eV.

The measured energy spectrum is not an exact representation of the true energy spectrum due to the detector efficiency and resolution. Therefore, an unfolding is performed to correct the smearing between the true spectrum and the measured spectrum. In this analysis, we used the Iterative unfolding method proposed by D'Agostini [15, 16]. This unfolding method is based on Bayes' theorem and the unsmearing can be written as

$$C'_i = \sum_{j=1}^{N_m} U_{ij} E_j^{data}, \quad (1)$$

where $C'_i$ is a number of true events in bin $i$, $N_m$ is a number of bins of measured spectrum, $U_{ij}$ is an unsmearing matrix, and $E_j^{data}$ is a number of events in measured bin $j$. Figure 2 shows a resolution map between true and measured energies. The resolution map is evaluated by the MC simulation. Unsmearing matrix can be calculated by the resolution map. Figure 3 shows true, measured, and unfolded spectra in the MC simulation. The unfolded spectrum is obtained through the unsmearing process of the measured spectrum based on the unsmearing matrix. The unsmearing process is successfully performed and the unfolded spectrum is consistent with the true spectrum.

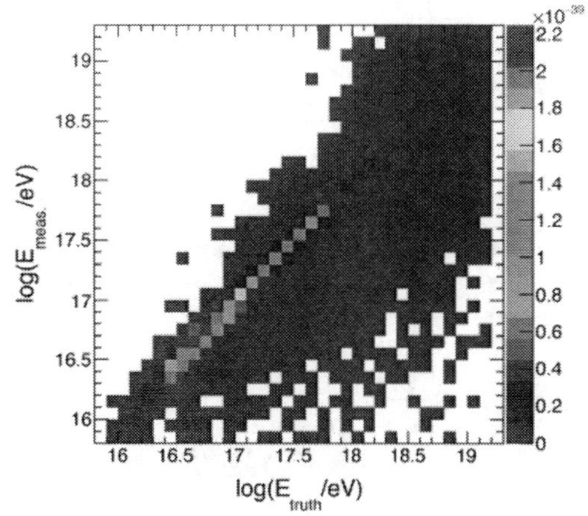


**Figure 2.** Resolution map between true and measured energies. The resolution map is evaluated by the MC simulation.

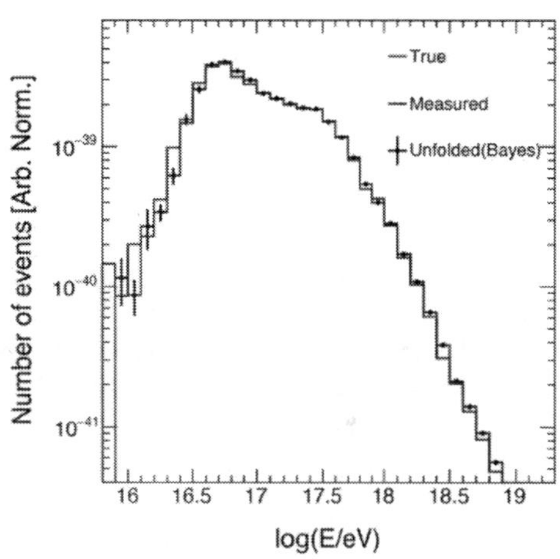


**Figure 3.** True, measured, and unfolded energy spectra. The unfolded spectrum is obtained through the unsmearing process of the measured spectrum based on the unsmearing matrix. The unsmearing is successfully performed and the unfolded spectrum is consistent with the true spectrum.

The product of the effective aperture and the observation time is called the effective exposure. Figure 4 shows the effective exposure used in this analysis. The effective exposure is evaluated by the MC simulation. Energy spec-

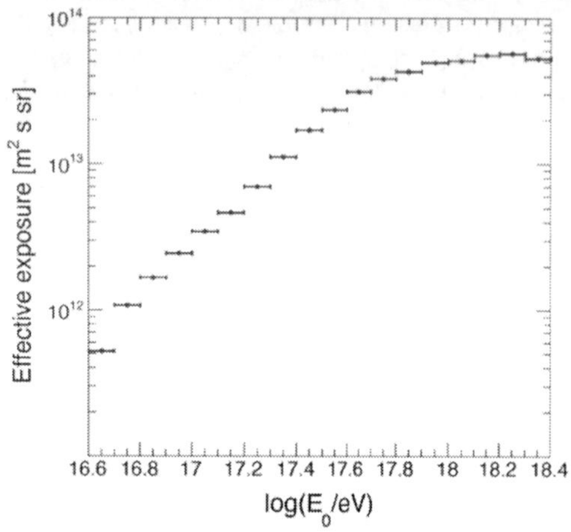


**Figure 4.** Effective exposure in this analysis. The effective exposure is expressed as a function of the true energy. This exposure is evaluated by the MC simulation.

trum is expressed as

$$J(E_i) = \frac{\sum_{j=1}^{N_m} U_{ij} N_j^{sel}}{A\Omega(E_i) \cdot T \cdot \Delta E_i}, \quad (2)$$

where $J(E_i)$ is the energy spectrum, $U_{ij}$ is the unsmearing matrix, $N_{sel}$ is the number of selected events, $A\Omega(E_i)$ is an effective aperture, $T$ is an observation time, and $\Delta E_i$ is a width of the $i$-th bin.

## 5 Results

Figure 5 shows the distributions of the reconstructed shower geometries which are core positions $X$ and $Y$, zenith angle, azimuth angle, $\Psi$ angle, and $R_p$. The $\Psi$ angle is the angle of the shower axis with respect to the direction of the center of the shower track and $R_p$ is the shower impact parameter to the detector. The measured distributions are consistent with the MC predictions.

Figure 6 shows the energy distribution of the events. The measured energy is consistent with the MC prediction. The energy resolution is 8.7% which is evaluated by the MC simulation and it is consistent with previous studies [17].

Figure 7 shows the result of energy spectrum obtained in this study, as well as that reported by TALE FD mono observation [3]. In this analysis, the TALE hybrid observation data was used. The difference of the TALE hybrid and the TALE FD mono observation results in the energy range of $10^{17.0}$ eV to $10^{17.5}$ eV can be explained by the fluctuations due to the small statistics in this analysis. The analysis with more statistics is on-going and the expected statistics will increase around four times. Note that the MC studies with other hadronic interaction model and compositions are also work in progress.

## 6 Summary

It is important to measure the acceleration limit of galactic cosmic rays through the measurements of composition and energy spectrum, and transition from galactic to extragalactic cosmic rays. To study those physical topics, it is essential to observe the cosmic-ray energy spectrum and measure the composition over a wide energy range. Combining the TA and TAx4 as well as TALE observation data, the energy spectrum will be measured over five orders of magnitude from $10^{15}$ eV to $10^{20}$ eV. Therefore, we observed and analyzed the cosmic-ray events using the TALE detector. We checked the fundamental information, such as core positions, and measure the energy spectrum with the TALE hybrid mode. In this work, we presented the preliminary result with a part of the TALE hybrid observation corresponding to the 392 hours observation data. However, the analysis with more statistics is on-going and the expected statistics will increase around four times. This energy spectrum measurement will play an important role in understanding the transition from cosmic rays of galactic origin to those of extragalactic origin.

## References

[1] H. Kawai *et al.* (TA Collaboration), J. Phys. Soc. Jpn. Suppl. A **78**, 108 (2009).

[2] T. Abu-Zayyad *et al.* (TA Collaboration), Astrophys. J. Lett. **768**, L1 (2013).

[3] R. U. Abbasi *et al.* (TA Collaboration), Astrophys. J. **865**, 74 (2018).

[4] R. U. Abbasi *et al.* (TA Collaboration), Astrophys. J. **909**, 178 (2021).

[5] D. Heck, G. Schatz, T. Thouw, J. Knapp, and J.N. Capdevielle, Tech. Rep. FZKA **6019** (1998).

[6] T. Pierog, Iu. Karpenko, J. M. Katzy, E. Yatsenko, and K. Werner, Phys. Rev. C **92**, 034906 (2015).

[7] S. Agostinelli *et al.*, Nucl. Instrum. Methods Phys. Res., Sect. A **506**, 250 (2003).

[8] J. Allison *et al.*, IEEE Trans. Nucl. Sci. **53**, 270 (2006).

[9] J. Allison *et al.*, Nucl. Instrum. Methods Phys. Res., Sect. A **835**, 186 (2016).

[10] T. Abu-Zayyad *et al.* (TA Collaboration), Nucl. Instrum. Methods Phys. Res. A **689**, 87 (2012).

[11] H. Tokuno *et al.* (TA Collaboration), Nucl. Instrum. Methods Phys. Res. A **676**, 54 (2012).

[12] H. Tokuno *et al.* (TA Collaboration), Nucl. Instrum. Methods Phys. Res. A **601**, 364 (2009).

[13] A. Taketa *et al.* (TA Collaboration), Proceedings of the 29th International Cosmic Ray Conference, Pune, India, vol. **8**, 209 (2005).

[14] Y. Tameda *et al.* (TA Collaboration), Nucl. Instrum. Methods Phys. Res. A **609**, 227 (2009).

[15] G. D'Agostini, Nucl. Instrum. Methods Phys. Res., Sect A **362**, 487 (1995).

[16] G. D'Agostini, arXiv:1010.0632

[17] K. Fujita, Ph.D. thesis, Osaka City University (2022).

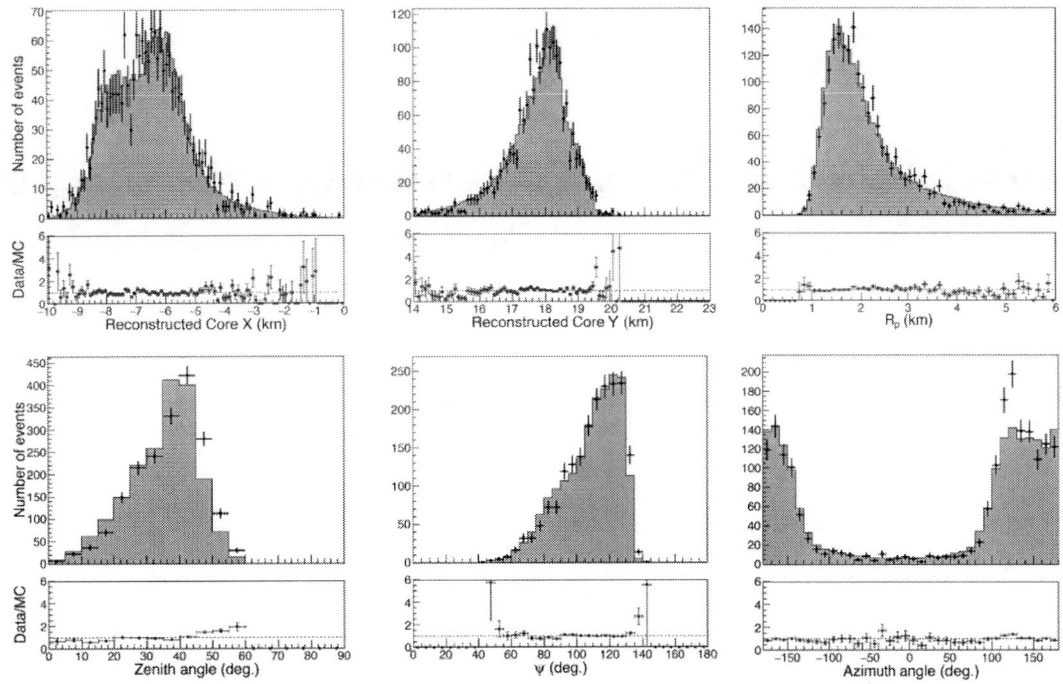


**Figure 5.** Distributions of reconstructed shower geometries. The data are shown by marker points and the MC prediction are shown by histogram. The MC predictions are normalized with the number of events in the data.

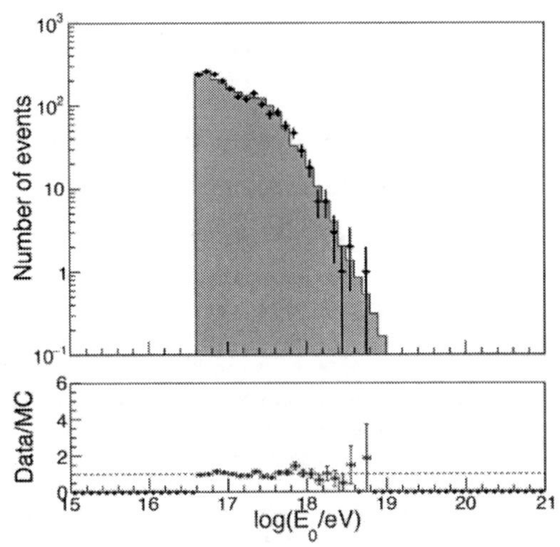


**Figure 6.** Energy distribution. The data are shown by marker points and the MC prediction are shown by histogram. The MC predictions are normalized with the number of events in the data.

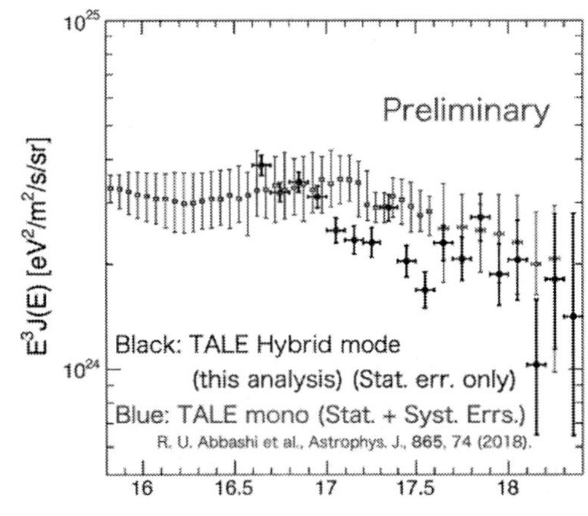


**Figure 7.** The result of energy spectrum. The spectrum obtained in this analysis is shown by black marker points, while that reported by TALE FD mono observation is shown by blue ones.

# Measurements of Cosmic Ray Mass Composition with the IceCube Neutrino Observatory

*Matthias* Plum[1,*] for the IceCube Collaboration

[1]South Dakota School of Mines and Technology, Rapid City, SD 57701 USA

**Abstract.** The IceCube Neutrino Observatory is a multi-component detector at the South Pole. Besides studying high-energy neutrinos, it is capable of measuring high-energy cosmic rays from PeV to EeV. This energy region is thought to cover the transition from galactic to extragalactic sources of cosmic rays. The observatory consists of the deep in-ice IceCube array, which measures the high-energy (≥ 500 GeV) muonic component, and the IceTop surface array, which is sensitive to the electromagnetic and low-energy muonic part of an air shower. The primary energy and the mass composition can be measured simultaneously by applying statistical methods including modern machine-learning techniques to reconstruct cosmic ray air showers. In this contribution, we will discuss recent improvements to the reconstruction techniques, the mass composition sensitivity, and an outlook on future improved measurements with the full surface scintillator/radio array to mitigate snow accumulation and measure the air shower maximum $X_{max}$ using imaging air-Cherenkov telescopes IceAct.

## 1 Introduction

High-energy cosmic rays are observed indirectly through the extensive air showers they initiate in the Earth's atmosphere. By measuring the electromagnetic and muonic shower components simultaneously, the mass and energy of the primary cosmic ray can be studied. However, the interpretation of such measurements in terms of the properties of the primary nucleus relies on detailed simulations of the air shower development.

## 2 IceCube and IceTop

IceCube [1] is a cubic-kilometer astroparticle detector installed in the ice at the geographic South Pole between depths of 1450 m and 2450 m, completed in 2010. The detector setup is shown in Figure 1 together with a schematic of the air shower detection capabilities of the surface and the in-ice component. It consists of 86 detector strings instrumented with 60 digital optical modules (DOMs). The deep in-ice detector IceCube, designed for high-energy neutrino astronomy, is complemented by the square-kilometer surface array IceTop [3], designed to measure the secondary particles from cosmic ray air showers. IceTop consists of 81 surface stations composed of two tanks separated by 10 m. Each tank is an ice-Cherenkov detector with two DOMs, one operating at low gain and the other at high gain, to maximize the dynamic range of the detector. The IceTop and the IceCube arrays can study astroparticle events independently or in coincidence. The IceTop tank signal is dominated by the electromagnetic component of incoming cosmic ray air showers, which allows a reconstruction of the primary energy and the geometry of the air showers. The in-ice array, on the other hand, can only measure the high-energy muon component of the air showers, which allows for the direct study of TeV-muon bundles. Together both detector arrays provide a unique environment for the study of the properties of primary cosmic rays over a wide energy range from PeV to EeV.

*e-mail: matthias.plum@icecube.wisc.edu

## 3 IceTop-IceCube Cosmic Ray Mass Composition

The combination of IceTop and IceCube events provides a good means to measure the primary mass composition. Using the tank signals, the core position and direction of the shower axis are reconstructed by fitting the signal times. A primary energy proxy is obtained by fitting the following function to the signal charges as a function of radial distance $R$ to the shower axis:

$$S(R) = S_{125} \cdot \left(\frac{R}{125\,\mathrm{m}}\right)^{-\beta-\kappa\cdot\log_{10}\left(\frac{R}{125\,\mathrm{m}}\right)} \tag{1}$$

where $\kappa = 0.303$ as obtained from simulations (Sibyll2.1 [4]), and the free fit parameters $\beta$ and $S_{125}$ represent the slope of the function and the signal strength at a reference distance of 125 m respectively. An example of the air shower reconstruction in the spatial and time domain is shown in Figure 2. As snow accumulation above the tanks causes a fraction of the shower particles to be absorbed, the reconstruction modifies the expectations from Eq. 1 to account for the additional snow attenuation. $S_{125}$ is highly correlated to the primary particle's energy as

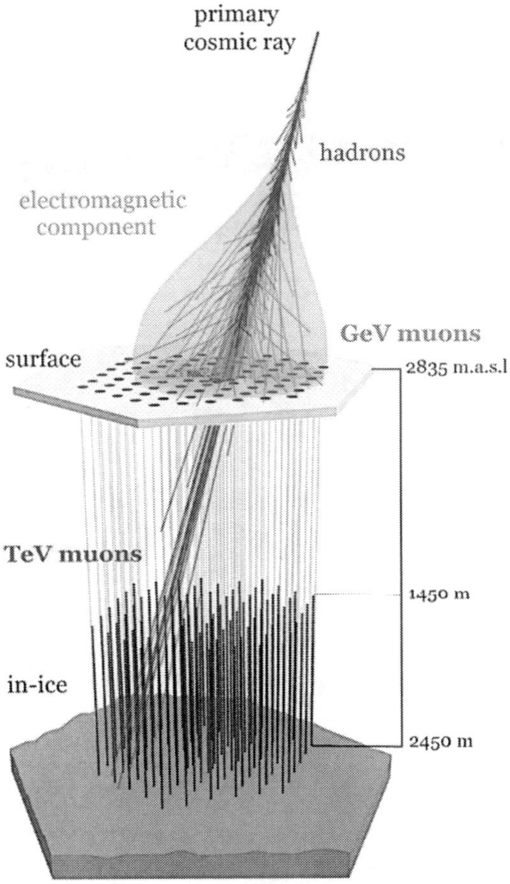


Figure 1: Air shower measurements of the IceCube Neutrino Observatory with the in-ice IceCube and surface IceTop array [2].

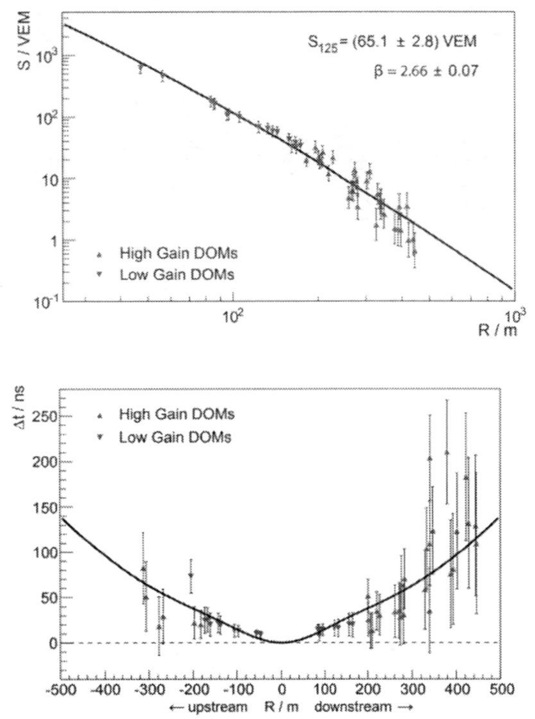


Figure 2: Example of the IceTop air shower reconstruction [3]. (Top) Lateral fit to signals from 25 triggered stations with a reconstructed shower size $S_{125} = (65.1 \pm 2.8)$ VEM. (Bottom) Time residuals with respect to a plane perpendicular to the shower direction. "Upstream" and "downstream" refer to tanks being hit before and after the shower core reaches the ground.

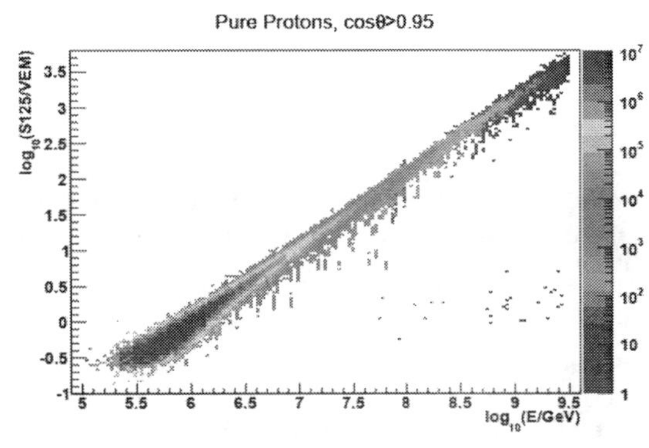


Figure 3: Correlation between $S_{125}$ and primary energy in IceTop for proton simulations [5] using Sibyll2.1 [4].

shown in Figure 3. The shower axis from IceTop reconstruction is used to determine the shower axis from the high-energy muon bundle in the in-ice IceCube detector. With this axis, the energy loss of the muon bundle can be measured as a function of slant depth. An example of the reconstructed energy deposit profile is shown in Figure 4. This energy deposit profile is evaluated at a reference slant depth of 1500 m. As shown in Figure 5, the energy loss at this depth is highly mass composition sensitive. As high-energy muons have a higher probability of losing larger energy clumps in stochastic processes, which produce spikes in the energy deposit profile. The number of those stochastic losses is counted as they increase the composition sensitivity of the analysis. Combining reconstructed detector observables from IceTop ($S_{125}$, zenith) and IceCube ($dE/dX_{1500\,m}$, number of high energy stochastic losses) using a dedicated Neural Network (NN) reconstruction, the energy and the mass composition of the primary cosmic ray are studied. The NN is trained to deliver an estimate for each event's energy and mass. While the

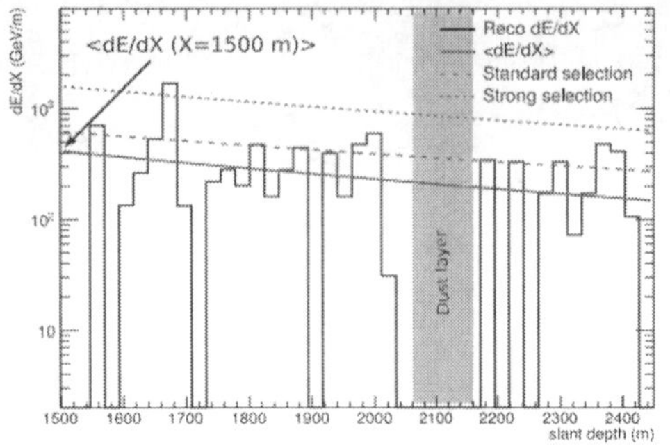


Figure 4: Energy loss reconstruction of an example air shower [5] event, where the solid red line demonstrates the average energy loss fit, the dashed red and dotted red lines indicate the standard and strong stochastics selection, respectively. The gray band is the approximate location of the dust layer for the slant depth of this particular event.

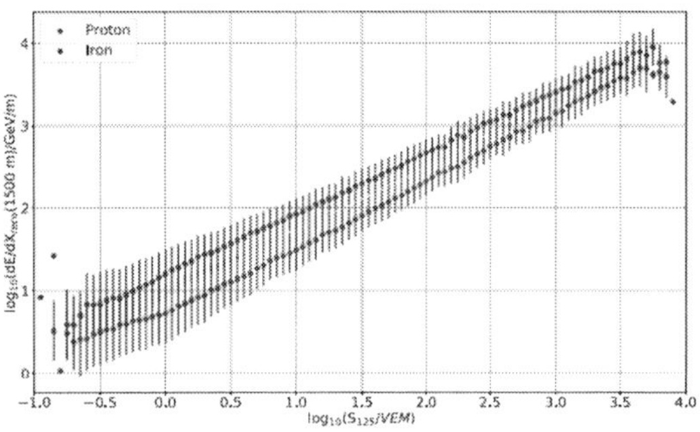


Figure 5: The composition sensitivity of the reconstructed energy loss parameter (dE/dX) at 1500 m slant depth for a given $S_{125}$ for H and Fe simulation [5] using Sibyll2.1 [4]. The error bars indicate the 1-sigma width of the (dE/dX) distribution for a given $S_{125}$.

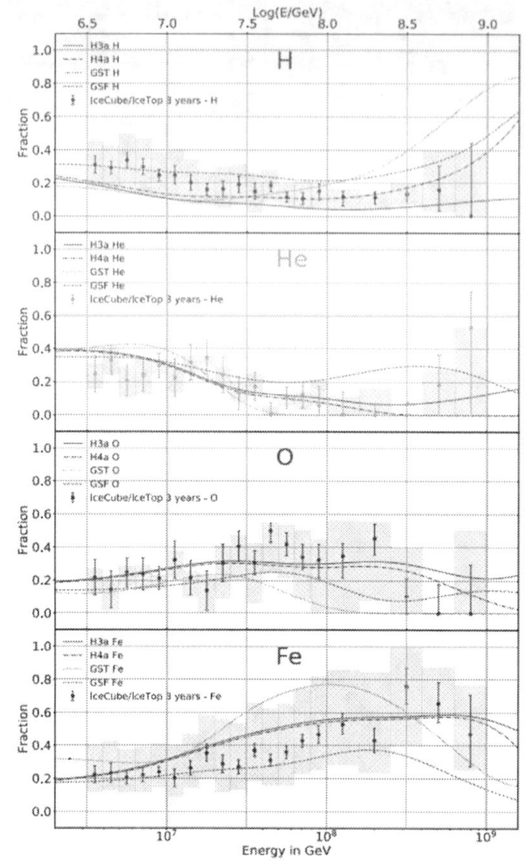


Figure 6: Fractions for the four elementary mass groups (H, He, O and Fe) [5] including the total detector systematic compared with various cosmic ray models (H3a and H4a [6]) and phenomenological experimental fits (GST [7] and GSF [8]). As a hadronic interaction model, Sibyll2.1 [4] is used in the simulated dataset.

energy output is a good estimator of the true energy, the mass output yields a broad distribution for a given element. Therefore, the composition can only be obtained statistically for the whole event sample by fitting distribution templates of four elementary mass groups (H, He, O and Fe) to the NN mass output[5]. In Figure 6, the mass composition fractions for these elementary groups as a function of the reconstructed energy are shown in comparison to CR propagation models. The elemental fraction results agree well with the prediction of the phenomenological models of the H3a and H4a [6] in this energy range. There are some clear differences in the shape of the fractions, especially the Fe fraction, in comparison to the prediction by the phenomenological data fits, Gaisser-Stanev-Tilav (GST) [7] fit and Global Spline Fit (GSF) [8]. However, inside the statistical and systematical uncertainties, none of these models can be ruled out. The calculated mean logarithmic mass $\langle \ln A \rangle$ also shows the predicted behavior of an increasingly heavier mass composition with increasing energies up to about 100–200 PeV, as shown in Figure 7. Given the statistical and systematic uncertainties, the reconstructed data beyond this point are consistent with either constant or decreasing mass composition. As shown in Figure 8, the absolute scale of the mass composition is dependent on the choice of the hadronic interaction model used for the simulations. The hadronic interaction model Epos-LHC [9] with a higher muon multiplicity leads to heavier mass composition measurements.

## 4 High-Energy Muon Multiplicity

Similar to the analysis in section 3, nearly vertical coincident events can be used to directly measure the TeV-muon multiplicity. This analysis [12] is based on a Neural Network (NN) reconstruction to simultaneously reconstruct

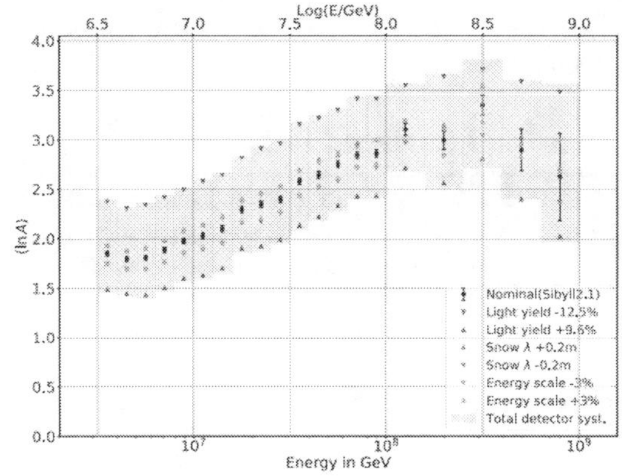


Figure 7: Calculated mean logarithmic mass $\langle \ln A \rangle$ with the total detector uncertainty(grey) from the combination of the light yield of the in-ice detector, snow correction and energy scale uncertainty [5].

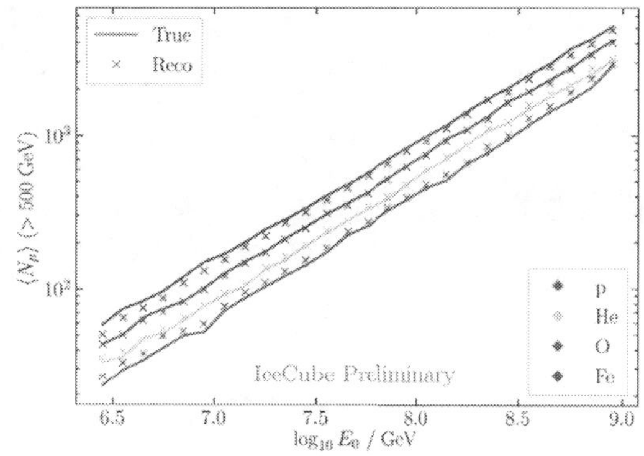


Figure 9: Average muon multiplicity $\langle N_\mu \rangle$ (> 500 GeV) as a function of primary energy in simulation obtained using the neural network determination of $N_\mu$ and $E_0$ compared to the true values for several elementary groups [12].

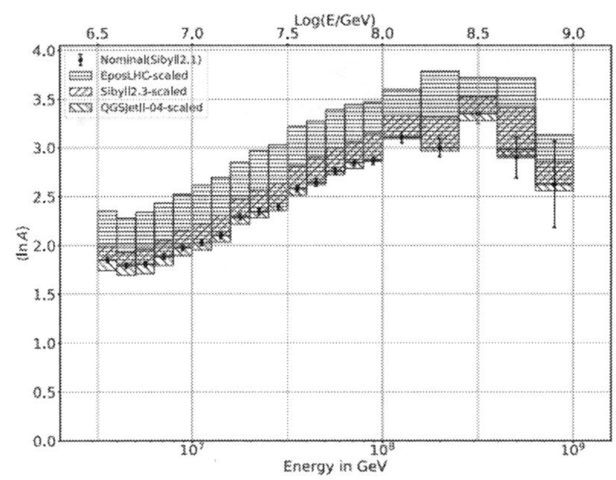


Figure 8: Hadronic interaction model uncertainty range on $\langle \ln A \rangle$ relative to the base model for Epos-LHC [9] (blue), Sibyll2.3 [10] (red) and QGSJetII-04 [11] (green) [5].

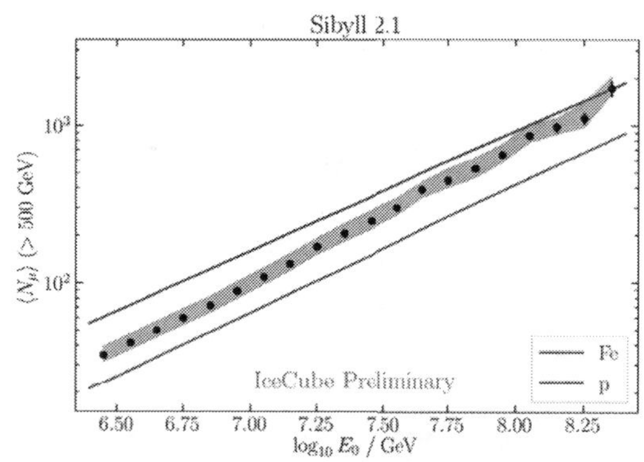


Figure 10: $\langle N_\mu \rangle$ with energy larger than 500 GeV as a function of primary energy [12], obtained from experimental data with Sibyll2.1 [4] simulation. Error bars indicate the statistical uncertainty and the band indicates the total systematic uncertainty.

the primary energy and the number of TeV-muons in the bundle. This NN uses inputs from IceTop, the shower size $S_{125}$ (Eq. 1), the zenith angle, and from IceCube, the reconstructed energy loss profile. The full energy loss profile is transformed into a vector of 20 m segments for a recurrent neural network (RNN) layer. The combination of the RNN layer with the input variables from IceTop and several dense layers returns reconstructed values for $E_0$ and $N_\mu$. For training, testing and validation, air shower simulations from 4 elementary groups (H, He. O and Fe) are used. The event-by-event neural network predictions are used to calculate the average multiplicity $\langle N_\mu \rangle$ per energy bin. As this reconstruction method shows a mass-dependent bias, dedicated simulation studies are used to derive correction factors, which are applied to compensate for neural network biases. In Figure 9, the reconstructed $\langle N_\mu \rangle$ is shown as a function of the primary energy $E_0$ for 4 elementary groups. This Figure demonstrates the small composition-depended biases of this method. A further simulation study is used to derive a correction for these biases [12]. The corrected experimental data are shown in Figure 10 in comparison to Sibyll2.1 simulations. For a better qualitative comparison, the measurements are re-scaled to the so-called "z-scale" [13]. In the "z-scale", the reconstructed muon number is normalized to the expected difference between the iron and proton simulation of a given hadronic interaction model. The re-scaling is defined as

$$z = \frac{\ln\langle N_\mu \rangle - \ln\langle N_\mu \rangle_p}{\ln\langle N_\mu \rangle_{Fe} - \ln\langle N_\mu \rangle_p}. \quad (2)$$

The re-scaling results for different hadronic interaction models are shown in Figure 11 in comparison to cosmic

https://doi.org/10.1051/epjconf/202328302007

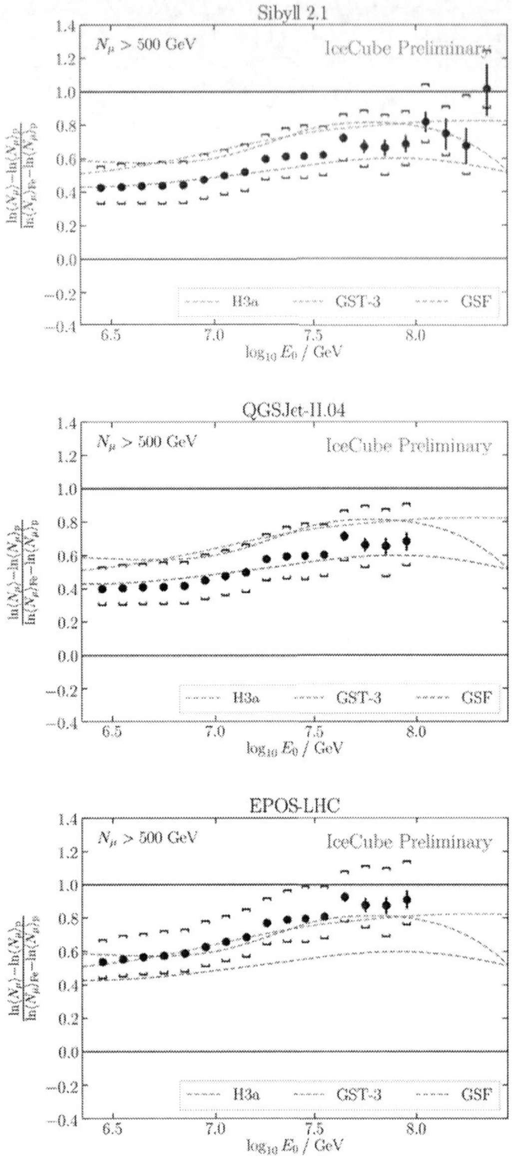


Figure 11: $z$-value results (Eq. 2) [12] with different hadronic interaction model assumptions, compared to expectations from mass composition models. Error bars indicate the statistical uncertainty and the brackets indicate the total systematic uncertainty.

ray models. Depending on the chosen hadronic interaction model some of the cosmic ray models describe the evolution of the data as a function of the reconstructed energy better than the other models. However, inside the statistical and systematical uncertainties, the results cannot rule out any of the expectations from the different mass composition models when assuming Sibyll 2.1 [4], QGSJet-II.04 [11] and EPOS-LHC [9] hadronic interaction models. Overall, these results show over the whole energy range the expected number of muons and do not see a muon deficit associated with other cosmic ray experiments[14].

## 5 IceTop - Surface Enhancement Detectors

Upgrades of the IceCube Neutrino Observatory are going to improve the mass composition sensitivity and reduce systematic uncertainties. The full surface array footprint of IceTop is currently been enhanced by additional surface stations, each consisting of 8 scintillator panels and 3 radio antennas [15]. A prototype of this new surface station is installed and taking successful data [16–18]. Additionally, 2 working prototypes of imaging air-Cherenkov telescopes IceAct [19–21] are installed at the South Pole.

### 5.1 IceTop Surface Enhancement

As the IceTop tanks get covered with additional snow every year, which is over time increasing the energy threshold, the surface scintillator panels are designed to mitigate the effect of snow attenuation as they are installed and kept above the snow level. A picture of a prototype scintillator panel at the South Pole is shown in Figure 12 (Top). These panels will also increase the sensitivity to study the surface muon density [22] of the air showers and add more observables for the study of the primary mass composition. The surface radio antennas are triggered by the scintillator stations. A picture of the radio antennas is shown in Figure 12 (Bottom left). These antennas will allow studying the air shower development of high-energy events inside the atmosphere, directly measuring the air shower maximum $X_{max}$ for inclined air showers at the higher end of the observatories energy range. The preliminary results of a Monte Carlo study to estimate the $X_{max}$ sensitivity are shown in Figure 13.

### 5.2 IceAct Telescopes

The IceAct telescopes are designed to withstand the environment of the South Pole, with a wide field of view of 12° to measure air showers inside the atmosphere at the lower end of the energy range from TeV to PeV. A picture of one of the IceAct prototypes is shown in Figure 12 (Bottom right). The measurement of air showers inside the atmosphere provides a complementary measurement to the surface array and in-ice response. This will allow independent reconstruction as well as improvement of the coincident analysis by providing additional detector observables. By using machine learning reconstruction, the air shower geometry, primary energy, and $X_{max}$ can be measured simultaneously by using a single telescope. Improvements to this reconstruction are achieved by stereo observations by several telescopes or by hybrid measurement together with the other detector components. A preliminary graph neural network reconstruction using a single IceAct telescope [24] from the image for $X_{max}$ is shown in Figure 14.

 https://doi.org/10.1051/epjconf/202328302007

Figure 12: Surface enhancement prototype detectors - (Top) Surface scintillator panel [16–18], (Bottom left) Surface Radio antenna [17, 18] (Bottom right) Imaging air-Cherenkov telescope IceAct [19–21]. Pictures by Yuya Makino, IceCube/NSF

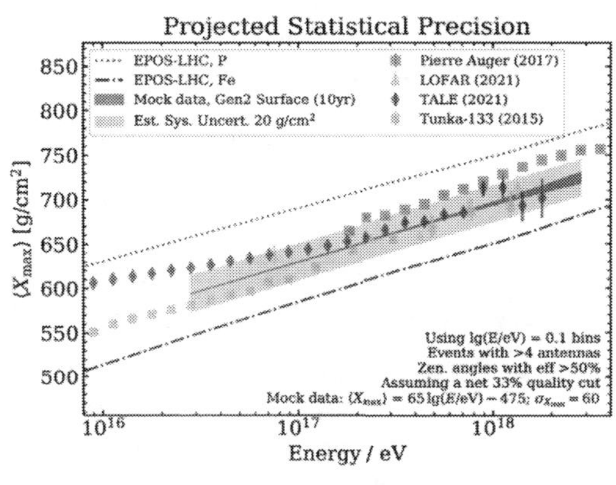


Figure 13: Projected precision from surface radio antennas for a measurement of $X_{max}$ using 10 years of mock data and an IceCube-Gen2 detector [23].

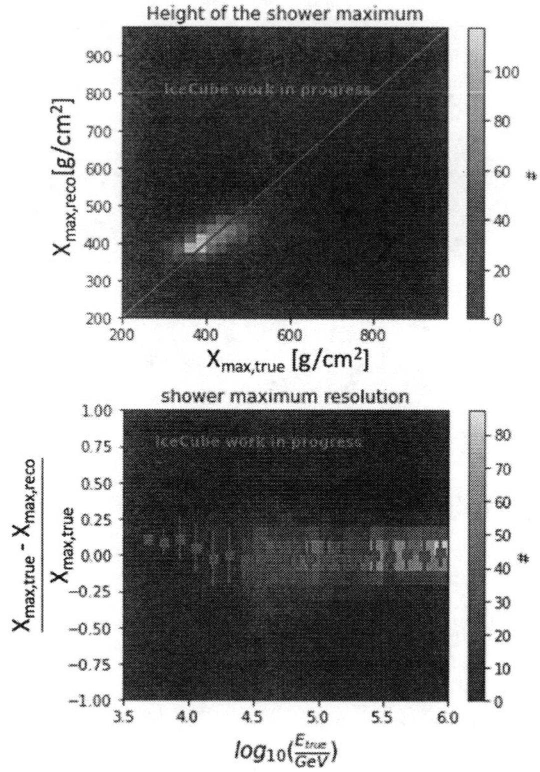


Figure 14: Preliminary IceAct graph neural network reconstruction result on the air shower maximum, trained with different elementary groups (H, He, N, Al and Fe) [24] using Sibyll2.1 [4] simulations. (Top) $X_{max,true}$ vs. $X_{max,reco}$. (Bottom) Relative $X_{max}$ resolution as a function of energy.

### 5.3 Future Outlook

In the near future, the IceCube Neutrino Observatory plans to upgrade the detector to instrumentalize an 8 times bigger volume with optical sensors and install a new shallow in-ice radio array to detect ultra-high energy (>100 PeV) neutrinos, called IceCube-Gen2 [25]. This new detector will also include an expanded surface array utilizing the already established new detector components, a scintillator panel array with radio antennas and a small IceAct telescope array. The schematic for this new detector layout and the IceTop surface enhancement is shown in Figure 15. With this larger detector, the upper energy bound for all analyses will be increased allowing cosmic ray studies to a few EeV with high enough statistics. With the increased number of surface/in-ice coincidences over a large energy range further tests of the hadronic interaction models can be performed to study the muon multiplicity. An improved in-ice detector component, a better in-situ calibration by new calibration devices, and new reconstruction techniques, like [26], will allow also provide a better mass

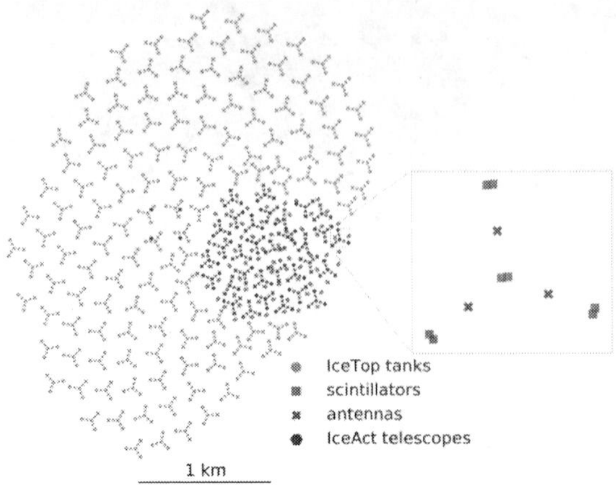


Figure 15: Layout of the currently under construction IceTop surface enhancement array (darker colors) and the proposed surface component of the future IceCube-Gen2. Modified from [23].

composition resolution and smaller systematic uncertainties.

## 6 Summary

The analyses presented here show the scientific capabilities of the IceCube Neutrino Observatory as a unique cosmic ray detector for mass composition in the energy range from PeV to EeV. The evolution of the mass composition fractions as a function of energy from the elementary groups as well as the measurements of TeV-muon multiplicities show reasonable agreement between the data from IceCube and IceTop and the prediction from cosmic ray models in this energy range. However, the measurements cannot discriminate any models yet. In the near future, these results will be improved by the addition of new detector components. The planned IceCube-Gen2 will further enhance the measurements of the primary cosmic ray mass composition by the increased detector size and improved new detector components.

## References

[1] M.G. Aartsen et al. (IceCube), JINST **12**, P03012 (2017)

[2] S. Verpoest, D. Soldin, S. De Ridder (IceCube), PoS **ICRC2021**, 357 (2021)

[3] R. Abbasi et al. (IceCube), Nucl. Instr. and Meth. A **700**, 188 (2013)

[4] E. Ahn, R. Engel, T. Gaisser, P. Lipari, T. Stanev, Physical Review D **80**, 94003 (2009)

[5] M.G. Aartsen et al. (IceCube), Phys. Rev. D **100**, 082002 (2019)

[6] T.K. Gaisser, Astroparticle Physics **35**, 801 (2012)

[7] T.K. Gaisser, T. Stanev, S. Tilav, Front. Phys. (Beijing) **8**, 748 (2013)

[8] H.P. Dembinski, R. Engel, A. Fedynitch, T. Gaisser, F. Riehn, T. Stanev, PoS **ICRC2017**, 533 (2018)

[9] T. Pierog, I. Karpenko, J.M. Katzy, E. Yatsenko, K. Werner, Phys. Rev. C **92**, 034906 (2015)

[10] F. Riehn, R. Engel, A. Fedynitch, T.K. Gaisser, T. Stanev, PoS **ICRC2015**, 558 (2016)

[11] N.N. Kalmykov, S.S. Ostapchenko, A. Pavlov, Nucl. Phys. B Proc. Suppl. **52** (1997)

[12] S. Verpoest (IceCube), PoS **27th European Cosmic Ray Symposium**, 074 (2022)

[13] H.P. Dembinski et al., EPJ Web Conf. **210**, 02004 (2019)

[14] J. Albrecht et al., Astrophys. Space Sci. **367**, 27 (2022)

[15] A. Haungs (IceCube), EPJ Web Conf. **210**, 06009 (2019)

[16] M. Kauer, T. Huber, D. Tosi, C. Wendt (IceCube), PoS **ICRC2019**, 309 (2019)

[17] M. Oehler, R. Turcotte (IceCube), PoS **ICRC2021**, 225 (2021)

[18] A. Coleman, H. Dujmovic, M. Oehler (IceCube), PoS **ICRC2021**, 314 (2021)

[19] M. Aartsen, et al. (IceCube), JINST **15**, T02002–T02002 (2020)

[20] K. Andeen, M. Schaufel, J. Auffenberg (IceCube), PoS **ICRC2019**, 179 (2019)

[21] L. Paul, M. Schaufel, M. Plum (IceCube), PoS **ICRC2021**, 276 (2021)

[22] R. Abbasi et al. (IceCube), Phys. Rev. D **106**, 032010 (2022)

[23] A. Coleman et al., Astropart. Phys. **147**, 102794 (2023)

[24] L. Paul, T. Bretz, J. Hewitt, A. Zink (IceCube) (2022), `https://doi.org/10.5281/zenodo.6354743`

[25] M.G. Aartsen et al. (IceCube-Gen2), J. Phys. G **48**, 060501 (2021)

[26] P. Koundal (IceCube), PoS **27th European Cosmic Ray Symposium**, 085 (2022)

# Testing the Compatibility of the Depth of the Shower Maximum Measurements performed at Telescope Array and the Pierre Auger Observatory

## Auger-TA Mass Composition Working Group Report

*Douglas* Bergman[1,*], *Jose* Bellido[2], *Vitor* de Souza[3], *Ralph* Engel[4], *Zane* Gerber[1], *JiHyun* Kim[1], *Eric* Mayotte[5], *Olena* Tkachenko[4], *Michael* Unger[4], and *Alexey* Yushkov[6] for the Telescope Array[7] and Pierre Auger[8,**] Collaborations

[1]University of Utah, High Energy Astrophysics Institute, Salt Lake City, UT, USA
[2]University of Adelaide, Adelaide, S.A., Australia
[3]Universidade de São Paulo, Instituto de Física de São Carlos, São Carlos, SP, Brazil
[4]Karlsruhe Institute of Technology, Institute for Astroparticle Physics, Karlsruhe, Germany
[5]Colorado School of Mines, Golden, CO, USA
[6]Institute of Physics of the Czech Academy of Sciences, Prague, Czech Republic
[7]Telescope Array Project, 201 James Fletcher Bldg, 115 S 1400 East, Salt Lake City, UT 84112-0830, USA. Full author list: http://www.telescopearray.org/research/collaborators
[8]Observatorio Pierre Auger, Av. San Martín Norte 304, 5613, Malargüe, Argentina. Full author list: http://www.auger.org/archive/authors_2022_10.html

**Abstract.** The Telescope Array and the Pierre Auger Observatory estimate the composition of ultra-high-energy cosmic rays by observing the distribution of depths of air-shower maxima, $X_{\text{max}}$. Both experiments directly observe the longitudinal development of air showers using fluorescence telescopes with surface particle detectors used in conjunction to provide precision in determining air-shower geometry. The two experiments differ in the details of the analysis of events, so a direct comparison of $X_{\text{max}}$ distributions is not possible. The Auger – Telescope Array Composition Working Group presents their results from a technique to compare $X_{\text{max}}$ measurements from Auger with those of Telescope Array. In particular, the compatibility of the first two moments of the $X_{\text{max}}$ distributions of Auger with the data from the Black Rock Mesa and Long Ridge detectors of the Telescope Array is tested for energies above $10^{18.2}$ eV. Quantitative comparisons are obtained using air-shower simulations of four representative species made using the Sibyll 2.3d high-energy interaction model. These are weighted to fit the fractional composition seen in Auger data and reconstructed using the Telescope Array detector response and analysis methods.

## 1 Context and History

The Telescope Array (TA) observatory [1] and the Pierre Auger Observatory [2] (Auger) both measure the nuclear composition of ultra-high-energy cosmic rays (UHECRs) by observing the distribution of the depths of shower maxima ($X_{\text{max}}$) of extensive air showers reconstructed in hybrid mode, combining surface detector (SD) and fluorescence detector (FD) measurements. The conclusions and interpretations made from these measurements have differed between the two observatories, which has lead to confusion amongst outside observers. One source of this confusion may be that TA and Auger employ different strategies in selecting the data sets used in the measurements. Auger selects events with analysis cuts designed to minimize bias in $X_{\text{max}}$ acceptance and reconstruction, and then corrects the resulting $X_{\text{max}}$ moments for remaining biases. TA on the other hand selects all well-reconstructed events and then models the biases using Monte Carlo (MC) simulations. This difference in event selection strategy makes direct comparison of $X_{\text{max}}$ distributions problematic.

Beginning with UHECR2012 [3], the Auger-TA Mass Composition Working Group has tried to assess the degree of agreement between Auger and TA $X_{\text{max}}$ measurements. The procedure for comparison involves making a representative breakdown of the Auger $X_{\text{max}}$ distributions into several single species fractions using a given high-energy interaction model, then using the representative fractions and high-energy interaction model in the TA MC simulation. The result is interpreted as the $X_{\text{max}}$ distributions TA *would have measured* given the $X_{\text{max}}$ distributions actually measured at Auger. These distributions, specifically their first and second moments, are compared to the actual TA measurements to assess the degree of agreement. In practice, the four representative nuclear species are taken to be hydrogen (H), helium (He), nitrogen (N), and iron (Fe). The relative fractions of these species in energy bins found to represent the Auger data is called the AugerMix.

*e-mail: bergman@physics.utah.edu
**e-mail: auger_spokespersons@fnal.gov

 https://doi.org/10.1051/epjconf/202328302008

At UHECR2014 [4], a comparison of the AugerMix to TA Middle Drum hybrid data was made, just comparing $\langle X_{\max}\rangle$. This showed agreement within systematic uncertainties. At UHECR2016 [5] and UHECR2018 [6], the AugerMixes using both QGSJetII-04 [7] and EPOS-LHC [8] high-energy interaction models were compared to TA Black Rock and Long Ridge hybrid data using both $\langle X_{\max}\rangle$ and $\sigma(X_{\max})$. The comparison showed agreement within systematic uncertainties for $\langle X_{\max}\rangle$ but tension at energies above $10^{18.8}$ eV in $\sigma(X_{\max})$. Due to problems in the implementation of EPOS-LHC within CORSIKA [9] at that time, the EPOS-LHC result was actually performed by re-weighting QGSJetII-04 distributions.

While QGSJetII-04 works well in simulating the response of the TA SD (and FD) to air showers, it cannot describe Auger measurements because it predicts broader distributions than observed. It is thus useful to use another high-energy interaction model. We show here the result of using the Sibyll 2.3d [10] high-energy interaction model in both making the AugerMix and in simulating the result of that mix in the TA detector.

## 2 The AugerMix with Sibyll 2.3d

Auger hybrid $X_{\max}$ data [11] were binned by energy into $X_{\max}$ histograms. The energy bins were chosen to match those used by TA, which are constrained by the TA statistics. The bin edges in $\log_{10}(E/\mathrm{eV})$ are (18.2, 18.3, 18.4, 18.5, 18.6, 18.7, 18.8, 18.9, 19.0, 19.2, 19.4, 19.9). In each of the energy bins, $X_{\max}$ distributions from Conex [12] simulations of single species (H, He, N, Fe) using Sibyll 2.3d were created. These single-species distributions are then modified by the $X_{\max}$ acceptance and resolution following [13] and used as templates in fitting the observed Auger distribution into a set of fractions. While there are correlations in the fit fractions, only the best-fit fractions are used in this analysis.

The fit AugerMix fractions using Sibyll 2.3d as a function of energy are shown in Figure 1. These fractions describe the Auger measured data as the weighted sum of the templates can reproduce both $\langle X_{\max}\rangle$ and $\sigma(X_{\max})$ as shown in Figure 2.

## 3 Generating the TA distributions

The AugerMix fractions were used to generate $X_{\max}$ distributions as they would have been seen in the TA detector. Single-species MC sets for each of H, He, N, and Fe were generated using CORSIKA v7.7402 with Sibyll 2.3d as the high-energy interaction model. In total, 250 events for each species in each tenth-of-a-decade bin from $10^{18}$ eV to $10^{20.6}$ eV were generated using CORSIKA/Sibyll with optimal thinning [14]. The events were dethinned using the Stokes method [15] and the response of TA SDs placed every 6 m in a grid around the shower core was calculated using GEANT [16] to create a "tile file" for each simulated event. These events were then sampled at random locations and with random azimuthal orientations about the TA detector area with sufficient sampling to produce

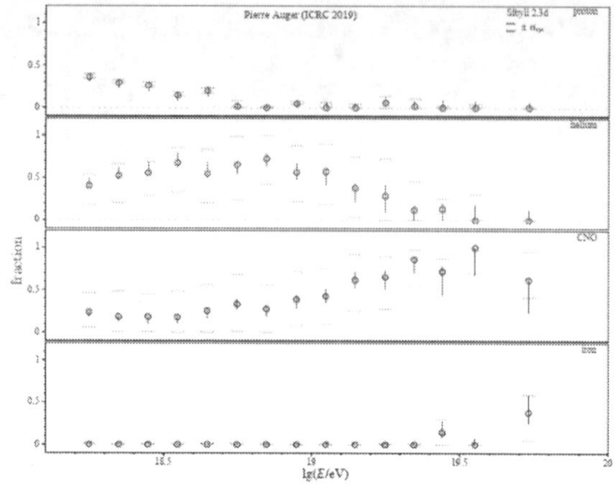


**Figure 1.** The fractions of H, He, N, and Fe that best fit the Auger $X_{\max}$ distributions, using templates produced by the Sibyll 2.3d high-energy interaction model.

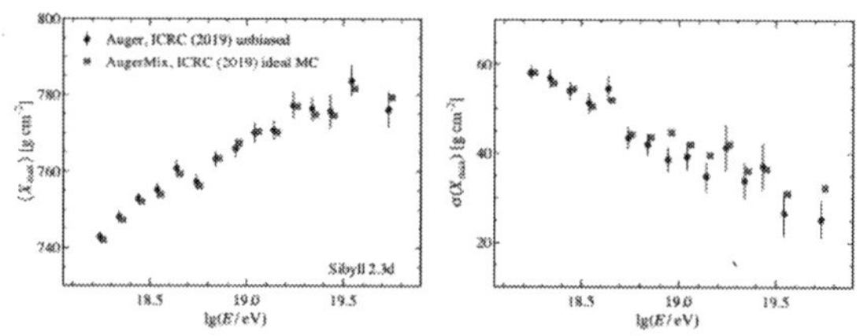


**Figure 2.** Comparison of $\langle X_{\max}\rangle$ (left) and $\sigma(X_{\max})$ (right) between the sum of AugerMix weighted templates from Sibyll 2.3d and the measured Auger $X_{\max}$ distributions.

the measured flux. The response of the TA FD was also simulated at this time from the stored longitudinal shower information from CORSIKA. This produces a spectral hybrid composition data set for each individual species. This data is in the same format as the collected TA hybrid data, and the same program of analysis cuts is applied to the simulated data that is used with the measured data.

The $X_{\max}$ distributions resulting from this analysis are combined using the AugerMix fractions as weights. As each single-species set is a spectral set using the same flux, this is the same as selecting events according to the AugerMix at the "thrown" level of MC.

## 4 Comparison of the AugerMix to TA Data

The results of the comparison between TA and the Auger⊗TA results using the Sibyll 2.3d AugerMix are shown in Figure 3. The $\langle X_{\max}\rangle$ comparison shows agreements within systematic uncertainties throughout the compared energy range, with some slight tension at the lowest energies. However, $\sigma(X_{\max})$ shows some disagreement between TA and Auger⊗TA in the upper half of the $10^{18}$ eV decade; the TA results show a roughly constant width throughout the decade, while the Auger⊗TA results show a continual narrowing of the distribution.

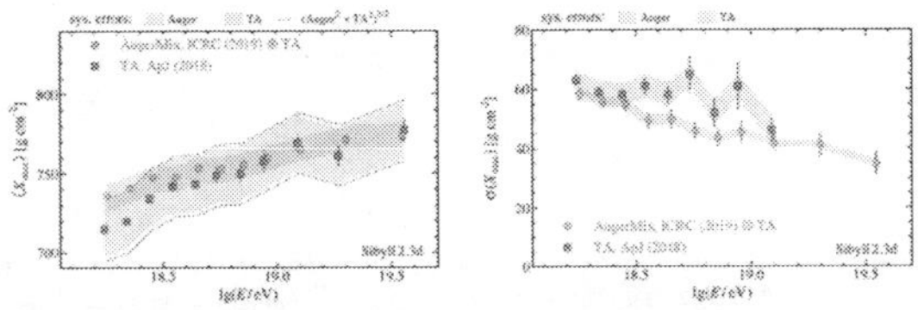


**Figure 3.** Comparison of $\langle X_{\max}\rangle$ (left) and $\sigma(X_{\max})$ (right) between measured TA data distributions (blue squares) and the Auger⊗TA results (red circles) using Sibyll 2.3d.

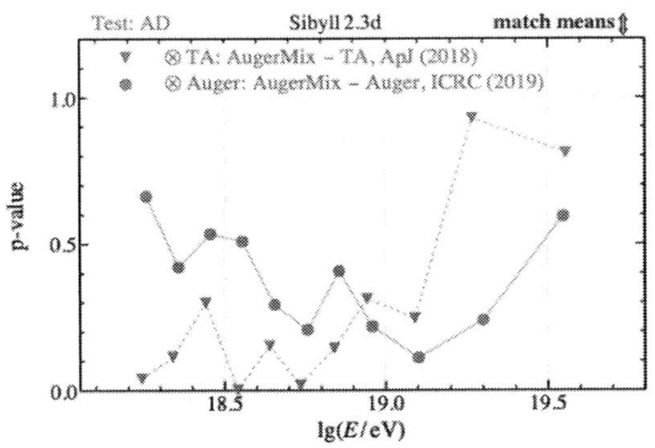


**Figure 4.** $p$-values from Anderson-Darling comparisons after removing differences in mean between TA and Auger⊗TA $X_{\max}$ distributions and between Auger and the AugerMix distributions. $p$-values should be roughly uniformly distributed between 0 and 1 for distributions drawn from the same underlying distribution. The very low $p$-values in the TA-Auger⊗TA comparison at 18.55 and 18.75 are indicative of a poor match.

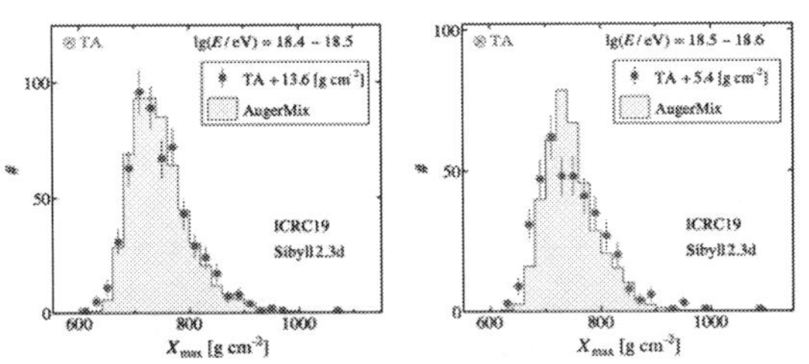


**Figure 5.** Example of a comparison between TA and Auger⊗TA with a good AD $p$-values (left) with $18.4 < \log_{10}(E/\mathrm{eV}) < 18.5$ and one with a bad $p$-value (right) with $18.5 < \log_{10}(E/\mathrm{eV}) < 18.6$.

To compare the shape of the distributions directly, we perform an Anderson-Darling test [17] on the two distributions in each energy bin. The test is made after removing the difference in means between the two distributions. In Figure 4 we plot the $p$-values of the test as a function of the energy for both the Auger⊗TA comparison to TA, and for the AugerMix comparison to the Auger data. Examples of distributions that match and that do not match are shown in Figure 5

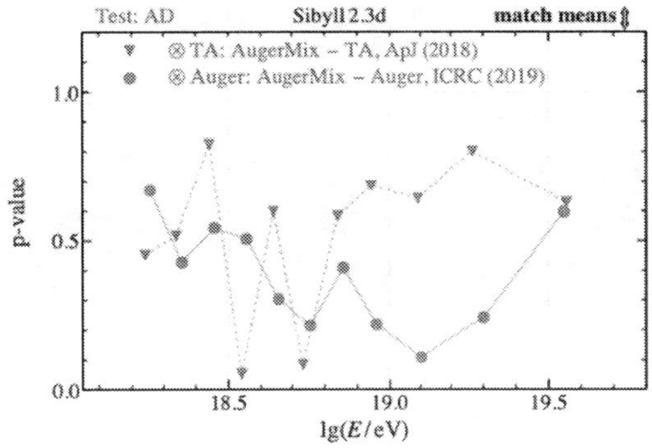


**Figure 6.** $p$-values from Anderson-Darling comparisons after removing differences in mean between TA and Auger⊗TA $X_{\max}$ distributions and between Auger and the AugerMix distributions, and smearing the Auger⊗TA $X_{\max}$ distributions by an additional 18.9 g/cm$^2$.

One systematic effect that was not considered in these comparisons, was the variation in atmospheric aerosol levels on a night-by-night basis. The MC simulations use a constant vertical aerosol optical depth (VAOD) value of 0.04, whereas the data was collected with varying VAODs values, which have a mean of 0.04. The systematic effect of this variation may add up to 18.9 g/cm$^2$ to the measured width of the $X_{\max}$ distributions. If we smear the Auger⊗TA distributions by and additional 18.9 g/cm$^2$ and redo the Anderson-Darling comparisons as before, we get a set of $p$-values consistent with both distributions coming from the same parent distribution. The values from this comparison are shown in Figure 6. It should be noted that this additional smearing of the $X_{\max}$ distributions does not significantly ameliorate the differences in widths seen in Figure 3. The additional 18.9 g/cm$^2$ will be added in quadrature to the significantly larger intrinsic widths of the Auger⊗TA data.

## 5 Discussion

We have made a comparison of the cosmic ray composition measured by Auger as it would have been seen at TA. This was performed using a set of representative fractions of single species $X_{\max}$ distributions generated using Sibyll 2.3d to simulate the response of the TA detector, also using Sibyll 2.3d, and compare the resulting distribution to the actual data distributions measured in TA. The result shows that the Auger⊗TA $\langle X_{\max}\rangle$ measurements agree within systematic uncertainties. However, there is disagreement in the width $\sigma(X_{\max})$ at energies between $10^{18.5}$ and $10^{19}$ eV. This difference in widths is similar to that seen previously in Auger⊗TA comparisons preformed using both QGSJetII-04 and EPOS-LHC models. The difference in widths can be slightly ameliorated, but not removed, by smearing the Auger⊗TA distributions by an additional factor of 18.9 g/cm$^2$ to account for variations in atmospheric conditions which effect TA data but are not present in these simulations.

We do not yet claim, however, to have discovered a robust difference in the $X_{\max}$ distributions between TA and Auger.

## References

[1] H. Kawai et al. (Telescope Array), Nucl. Phys. B Proc. Suppl. **175-176**, 221 (2008)

[2] A. Aab et al. (Pierre Auger), Nucl. Instrum. Meth. A **798**, 172 (2015), 1502.01323

[3] E. Barcikowski, J. Bellido, J. Belz, Y. Egorov, S. Knurenko, V. de Souza, Y. Tameda, Y. Tsunesada, M. Unger (Pierre Auger, Telescope Array, Yakutsk), EPJ Web Conf. **53**, 01006 (2013), 1306.4430

[4] R. Abbasi et al. (Pierre Auger, Telescope Array), JPS Conf. Proc. **9**, 010016 (2016), 1503.07540

[5] W. Hanlon, J. Bellido, J. Belz, S. Blaess, V. de Souza, D. Ikeda, P. Sokolsky, Y. Tsunesada, M. Unger, A. Yushkov (Pierre Auger, Telescope Array), JPS Conf. Proc. **19**, 011013 (2018)

[6] A. Yushkov, J. Bellido, J. Belz, V. de Souza, W. Hanlon, D. Ikeda, P. Sokolsky, Y. Tsunesada, M. Unger (Pierre Auger, Telescope Array), EPJ Web Conf. **210**, 01009 (2019), 1905.06245

[7] S. Ostapchenko, Phys. Rev. D **83**, 014018 (2011), 1010.1869

[8] T. Pierog, I. Karpenko, J.M. Katzy, E. Yatsenko, K. Werner, Phys. Rev. C **92**, 034906 (2015), 1306.0121

[9] D. Heck, J. Knapp, J.N. Capdevielle, G. Schatz, T. Thouw (1998)

[10] F. Riehn, R. Engel, A. Fedynitch, T.K. Gaisser, T. Stanev, Phys. Rev. D **102**, 063002 (2020), 1912.03300

[11] A. Yushkov (Pierre Auger), PoS **ICRC2019**, 482 (2020)

[12] T. Bergmann, R. Engel, D. Heck, N.N. Kalmykov, S. Ostapchenko, T. Pierog, T. Thouw, K. Werner, Astropart. Phys. **26**, 420 (2007), astro-ph/0606564

[13] A. Aab et al. (Pierre Auger), Phys. Rev. D **90**, 122006 (2014), 1409.5083

[14] M. Kobal (Pierre Auger), Astropart. Phys. **15**, 259 (2001)

[15] B.T. Stokes, R. Cady, D. Ivanov, J.N. Matthews, G.B. Thomson, Astropart. Phys. **35**, 759 (2012), 1104.3182

[16] S. Agostinelli et al. (GEANT4), Nucl. Instrum. Meth. A **506**, 250 (2003)

[17] T.W. Anderson, D.A. Darling, J. Am. Stat. Assoc. **49**, 765 (1954)

# Cosmic ray mass composition measurement with the TALE hybrid detector

*Keitaro* Fujita[1,*] and for the Telescope Array collaboration

[1]Institute for Cosmic Ray Research, the University of Tokyo

**Abstract.** The Telescope Array (TA) located in the State of Utah in the US is the largest ultra-high energy cosmic rays observatory in the northern hemisphere. The Telescope Array Low-energy Extension (TALE) detector was constructed to study the transition of cosmic rays from Galactic to extra-galactic origin. The TALE detector consists of a Fluorescence Detector (FD) station with 10 high elevation telescopes located at the TA Middle Drum FD Station (itself made up of 14 FD telescopes), and a Surface Detector (SD) array made up of 80 scintillation counters, including 40 with 400 m spacing and 40 with 600 m spacing. We have continued stable observation with hybrid mode since 2017. In this contribution, we present the latest result of the cosmic ray mass composition measurement using almost 4 years of TALE hybrid data.

## 1 Introduction

The Telescope Array (TA) cosmic ray observatory is the largest hybrid cosmic ray detector in the northern hemisphere in Millard Country, Utah St. The main part of the experiment consists of a surface detector (SD) array that is overlooked by 3 fluorescence detector (FD) stations. The TA SD consists of 507 scintillation counters with 1.2 km spacing and covering a total of $\sim$ 700 km$^2$ area on the ground. The three TA FD stations are located at Black Rock Mesa (BRM), Long Ridge (LR), and Middle Drum (MD), and these are overlooking the area of SD array. The Telescope Array Low-energy Extension(TALE), located at the north part of the TA experiment site, is aimed at measuring the energy spectrum and the mass composition of very high energy cosmic rays above $10^{16}$ eV. The TALE detector consists of one FD station with ten fluorescence telescopes and an array of 80 scintillation surface detectors, which were deployed to cover a total area of approximately 20 km$^2$. All 10 telescopes were refurbished from components previously used by HiRes [1], and view 31° to 59° in elevation, directly above the field of view of the MD telescopes. The TALE FD began operation in 2013 at the MD station. On the other hand, the TALE SD consists of 40 scintillation counters with 400 m and 40 counters with 600 m spacing, and started observation from 2017. In addition, an external trigger from the TALE FD to the TALE SD to detect low energy cosmic rays, so-called hybrid trigger system, was installed in 2018. The full details of the TALE detectors are found in [2] [3]. In this contribution, we report on the performance of the TALE hybrid detector and present preliminary results of mass composition measurements using 4 years of the TALE hybrid data.

*e-mail: kfujita@icrr.u-tokyo.ac.jp

## 2 Event Reconstruction

The reconstruction process for the hybrid events is composed of the following 3 steps: the PMT selection for FD side, the determination of shower detector plane (SDP), which is the plane including the shower axis and the FD location, and the profile constrained geometry fit (PCGF) [4]. At first PMTs to be used in the reconstruction are selected by rejecting those that are spatially and temporally isolated from the shower image. After the good PMTs are selected, the SDP is determined from the pointing direction vectors of the selected PMTs. Once the SDP is determined, we perform the PCGF reconstruction that simultaneously reconstruct the shower geometry and the shower profile. So far the PCGF reconstruction is applied to the FD monocular data, and we perform it in hybrid data. Technically, for each given $\psi$ angle, which is the shower inclination angle in the SDP as shown in Fig. 1, the shower geometry is calculated by a time vs angle fit that uses the pointing directions and timings of the PMTs and the one SD. This SD information provides us with more accurate shower geometry than monocular mode. Then the shower profile is fitted in given shower geometry using the Gaisser-Hillas parameterization formula [5]

$$N(x) = N_{\rm max}\left(\frac{x - X_0}{X_{\rm max} - X_0}\right)^{\frac{X_{\rm max}-X_0}{\lambda}} \exp\left(\frac{X_{\rm max} - x}{\lambda}\right), \quad (1)$$

where $N(x)$ is the number of charged particles at a given slant depth, $x$, $X_{\rm max}$ is the depth of shower maximum, $N_{\rm max}$ is the maximum number of particles at $X_{\rm max}$, $X_0$ is the depth of the first interaction, and $\lambda$ is the interaction length of shower particles. For each trial, a combined $\chi^2_{\rm com} = \chi^2_{\rm geo} + \chi^2_{\rm pfl}$ is evaluated, and the best expectation of the shower geometry and energy deposit profile from a cosmic ray shower is chosen. All reconstructed events are processed to apply the quality cuts summarized in Table. 1. The obtained shower parameter resolutions energies above

$10^{16.5}$ eV are $\sim 3$ % in $R_p$, $\sim 1°$ in $\psi$ angle, $\sim 30$ g/cm$^2$ in $X_{\mathrm{max}}$ and $\sim 10$ % in energy (Fig. 2, 3).

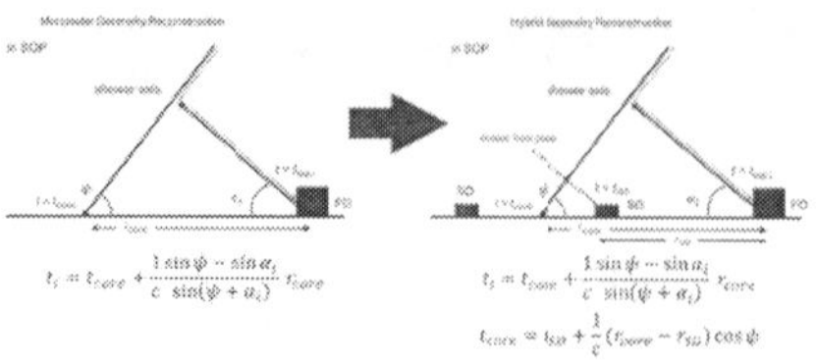

Figure 1: The schematics of the monocular and the hybrid shower geometry reconstruction. The relations between the measured values, which are $t_{\mathrm{exp,i}}$, $\alpha_i$ and the fitting parameters, which are $t_{\mathrm{core}}$, $r_{\mathrm{core}}$ and $\psi$. In the hybrid analysis, the two observable, $t_{\mathrm{SD}}$ and $r_{\mathrm{SD}}$, are added to the relation of the monocular analysis, and as a result the number of the fitting parameter is reduced to two and the geometry determination accuracy is improved.

| Variable | CL | FL |
|---|---|---|
| No saturated PMTs in FD | applied | |
| $X_{\mathrm{max}}$ bracketing cut | applied | |
| Angular track-length [deg] | track > 6.5° | - |
| Event duration [ns] | > 100 ns | - |
| # of PMTs | > 10 | - |
| # of Photo-electrons / # of PMTs | > 50 | - |
| # of Photo-electrons | - | > 2000 |

Table 1: Quality cuts applied in this work. Here we define fluorescence events as fractional contribution to the total signal of Fluorescence Light (FL) > 0.75, and Cherenkov Light (CL) events as fractional contribution to the total signal of FL $\leq$ 0.75.

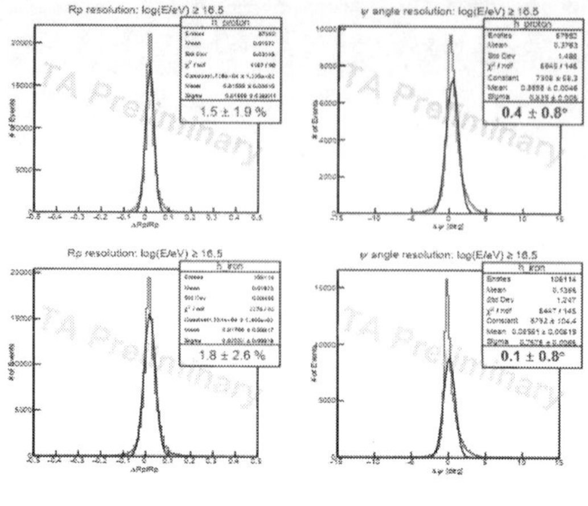


Figure 2: Reconstruction resolutions of the impact parameter $R_p$, the shower inclination angle in the SDP $\psi$, respectively. Top panels are shown for proton MC case and bottom panels are shown for iron MC case. The black curve is a Gaussian fit to the distribution of the uncertainty.

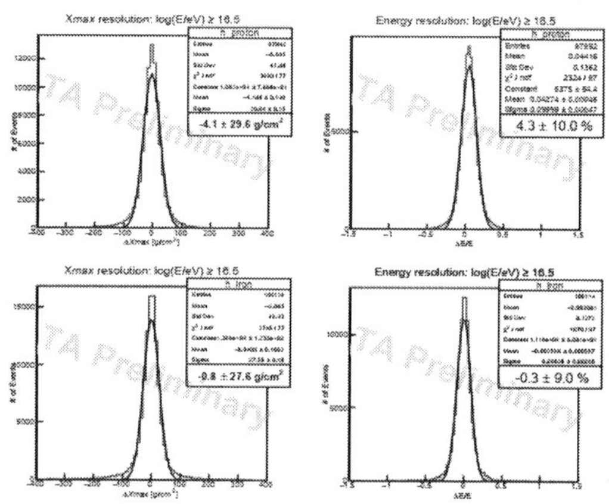


Figure 3: Reconstruction resolutions of the shower maximum $X_{\mathrm{max}}$, and the shower energy $E$, respectively. Top panels are shown for proton MC case and bottom panels are shown for iron MC case. The black curve is a Gaussian fit to the distribution of the uncertainty.

## 3 Composition Analysis

We presented the preliminary results of a measurement of the cosmic rays mass composition in the energy range of $10^{16.5}$ - $10^{18.4}$ eV. The result of the mean of the shower maximum, $\langle X_{\mathrm{max}} \rangle$, and the width of the observed $X_{\mathrm{max}}$ distributions, $\sigma(X_{\mathrm{max}})$ as a function of the logarithmic shower energy are presented in Fig. 4. For the comparison, the pure proton and pure iron predictions calculated by our Monte-Carlo simulation are also shown beside the observed ones. In the left panel of Fig. 4, the observed elongation rate shows clearly a break, where the energy is just above $10^{17}$ eV. A linear fit has been rejected with a $p-$value of $2.5 \times 10^{-7}$. The elongation rate before the break energy is $16 \pm 5$ g/cm$^2$/decade and after the break energy is $97 \pm 4$ g/cm$^2$/decade, while the pure proton and iron ones are $68 \pm 2$ g/cm$^2$/decade and $62 \pm 2$ g/cm$^2$/decade, respectively. On the other hand, the $\sigma(X_{\mathrm{max}})$ is compatible or wider than pure proton composition in whole energies. These results indicate that the mass composition at around $10^{17}$ eV, i.e. around the well known 2nd knee in the cosmic ray spectrum, is consistent with mixed composition, and the average mass of cosmic ray increases up to the break energy, then changing to lighter composition with increasing energies.

The left panel of Fig. 5 is a comparison of $\langle X_{\mathrm{max}} \rangle$ with other fluorescence measurements, with systematic error denoted by gray band. The contributions of the various source on the $X_{\mathrm{max}}$ measurement are listed in Table. 2. For

https://doi.org/10.1051/epjconf/202328302009

further comparison with particle detection based experiments, we display a mean logarithmic mass plot. From the observed $X_{\max}$, the mean $\ln A$ can be calculated by:

$$\langle \ln A \rangle = \frac{X_{\max}^{\text{data}} - X_{\max}^{\text{proton}}}{X_{\max}^{\text{iron}} - X_{\max}^{\text{proton}}} \cdot \ln A_{\text{iron}}, \tag{2}$$

where $X_{\max}^{\text{data}}$ is the mean $X_{\max}$ observed by experiments, $X_{\max}^{\text{proton/iron}}$ are the mean $X_{\max}$ for the proton and the iron primaries obtained by MC simulation, and $\ln A_{\text{iron}}$ is the natural logarithm of the iron atomic mass. The right panel of Fig. 5 shows the $\langle \ln A \rangle$ as a function of energy. The $\langle \ln A \rangle$ values measured by the TALE hybrid detector are shown as black dots with systematic error denoted by gray band.

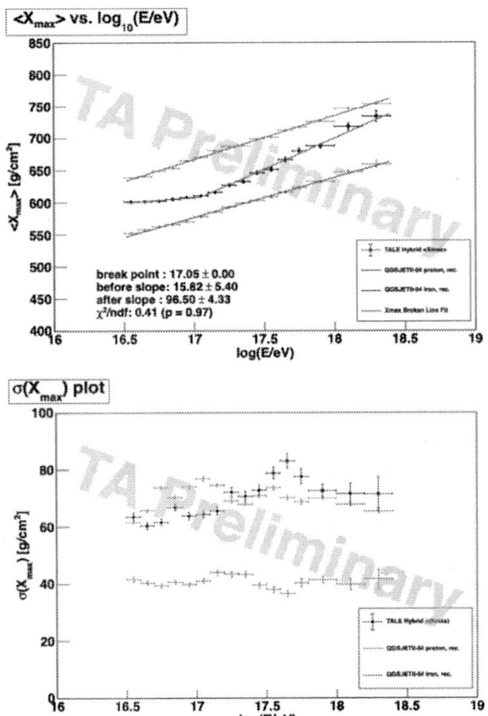


Figure 4: Top: $\langle X_{\max} \rangle$ as a function of shower energy, measured by using 4 years of the TALE hybrid data. Bottom: $\sigma(X_{\max})$ as a function of shower energy. For both panel, the proton and iron MC rails are also shown.

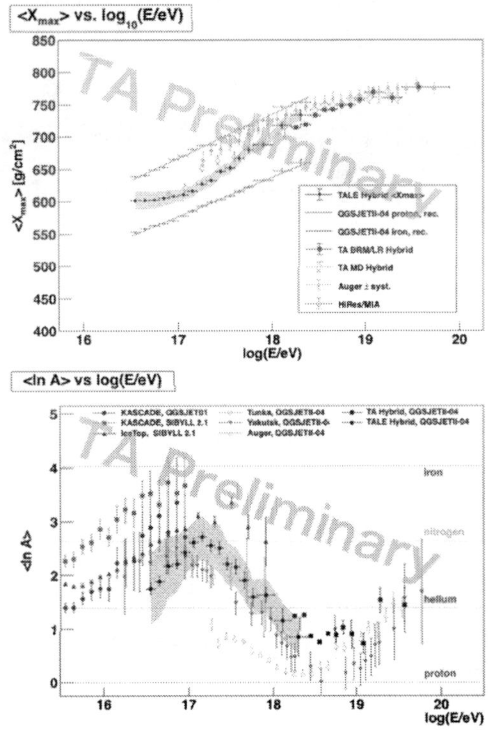


Figure 5: Top: Comparison of $\langle X_{\max} \rangle$ as a function of shower energy with Auger [6], TA [7, 8], HiRes/MIA [9] measurements. Bottom: Comparison of $\langle \ln A \rangle$ as a function of energy with various measurements. For comparison, two interpretations by KASCADE [10], IceTop [11], Tunka [12], Yakutsk [13], Auger [6], and 8.5 yrs TA BRM/LR hybrid [7] results are shown. The gray band in both figures represents the systematic uncertainty on $X_{\max}$ measurement.

| Sources | $X_{\max}$ |
|---|---|
| Photonic Scale | 5 g/cm$^2$ |
| Relative Time of FD and SD | 3.5 g/cm$^2$ |
| Fluorescence yield | 5 to 1 g/cm$^2$ |
| Cherenkov model | 11 to 3 g/cm$^2$ |
| Atmosphere | 1.4 g/cm$^2$ |
| Missing energy | 3 g/cm$^2$ |
| Total | 13.8 to 7.1 g/cm$^2$ |

Table 2: Summary of systematic uncertainties on the $X_{\max}$ measurements. Lines with multiple entries represent the values at the low and high end of the considered energy range ($\simeq 10^{16.5}$ eV and $\simeq 10^{18.5}$ eV, respectively).

## 4 Conclusion

We report on the preliminary result of the mass composition measurement obtained by using 4 years of TALE hybrid data. In this contribution, we present the measured $\langle X_{\max}\rangle$ and $\sigma(X_{\max})$ as a function of primary energy. The $\langle X_{\max}\rangle$ elongation rate shows a change in the slope at the energy just above $\sim 10^{17}$ eV. This break in the elongation rate is likely correlated with the observed break in the cosmic ray energy spectrum by the TALE FD monocular measurement [3].

## Acknowledgement

The Telescope Array experiment is supported by the Japan Society for the Promotion of Science(JSPS) through Grants-in-Aid for Priority Area 431, for Specially Promoted Research JP21000002, for Scientific Research (S) JP19104006, for Specially Promoted Research JP15H05693, for Scientific Research (S) JP19H05607, for Scientific Research (S) JP15H05741, for Science Research (A) JP18H03705, for Young Scientists (A) JPH26707011, and for Fostering Joint International Research (B) JP19KK0074, by the joint research program of the Institute for Cosmic Ray Research (ICRR), The University of Tokyo; by the Pioneering Program of RIKEN for the Evolution of Matter in the Universe (r-EMU); by the U.S. National Science Foundation awards PHY-1607727, PHY-1712517, PHY-1806797, PHY-2012934, and PHY-2112904; by the National Research Foundation of Korea (2017K1A4A3015188, 2020R1A2C1008230, & 2020R1A2C2102800) ; by the Ministry of Science and Higher Education of the Russian Federation under the contract 075-15-2020-778, IISN project No. 4.4501.18, and Belgian Science Policy under IUAP VII/37 (ULB). This work was partially supported by the grants ofThe joint research program of the Institute for Space-Earth Environmental Research, Nagoya University and Inter-University Research Program of the Institute for Cosmic Ray Research of University of Tokyo. The foundations of Dr. Ezekiel R. and Edna Wattis Dumke, Willard L. Eccles, and George S. and Dolores Doré Eccles all helped with generous donations. The State of Utah supported the project through its Economic Development Board, and the University of Utah through the Office of the Vice President for Research. The experimental site became available through the cooperation of the Utah School and Institutional Trust Lands Administration (SITLA), U.S. Bureau of Land Management (BLM), and the U.S. Air Force. We appreciate the assistance of the State of Utah and Fillmore offices of the BLM in crafting the Plan of Development for the site. Patrick A. Shea assisted the collaboration with valuable advice and supported the collaboration's efforts. The people and the officials of Millard County, Utah have been a source of steadfast and warm support for our work which we greatly appreciate. We are indebted to the Millard County Road Department for their efforts to maintain and clear the roads which get us to our sites. We gratefully acknowledge the contribution from the technical staffs of our home institutions. An allocation of computer time from the Center for High Performance Computing at the University of Utah is gratefully acknowledged.

## References

[1] J.H. Boyer, B.C. Knapp, E.J. Mannel, M. Seman, Nucl. Instrum. Meth. A **482**, 457 (2002)
[2] S. Ogio (Telescope Array), PoS **ICRC2019**, 375 (2019)
[3] R.U. Abbasi et al. (Telescope Array), Astrophys. J. **865**, 74 (2018), 1803.01288
[4] R.U. Abbasi et al. (HiRes), Phys. Rev. Lett. **100**, 101101 (2008), astro-ph/0703099
[5] T.K.Gaisser, A.M.Hillas, *Reliability of the Method of Constant Intensity Cuts for Reconstructing the Average Development of Vertical Showers* (1977), Vol. Proceedings of 15th International Cosmic Ray Conference (Plovdiv, Bulgaria) 8
[6] A. Yushkov, PoS **ICRC2019**, 482 (2019)
[7] R.U. Abbasi et al. (Telescope Array), Astrophys. J. **858**, 76 (2018), 1801.09784
[8] R.U. Abbasi et al. (Telescope Array), Astropart. Phys. **64**, 49 (2015), 1408.1726
[9] T. Abu-Zayyad, K. Belov, D.J. Bird, J. Boyer, Z. Cao, M. Catanese, G.F. Chen, R.W. Clay, C.E. Covault, H.Y. Dai et al., The Astrophysical Journal **557**, 686 (2001)
[10] T. Antoni et al. (KASCADE), Astropart. Phys. **24**, 1 (2005), astro-ph/0505413
[11] M.G. Aartsen et al. (IceCube), Phys. Rev. D **100**, 082002 (2019), 1906.04317
[12] V.V. Prosin et al., EPJ Web Conf. **121**, 03004 (2016)
[13] S. Knurenko, I. Petrov, Adv. Space Res. **64**, 2570 (2019), 1908.01508

# The depth of the shower maximum of air showers measured with AERA

*Bjarni* Pont[1] for the Pierre Auger Collaboration[2,*,**]

[1]Department of Astrophysics/IMAPP, Radboud University, P.O. Box 9010, NL-6500 GL Nijmegen, The Netherlands
[2]Observatorio Pierre Auger, Av. San Martín Norte 304, 5613 Malargüe, Argentina

**Abstract.** The Auger Engineering Radio Array (AERA), as part of the Pierre Auger Observatory, is an array of radio antennas probing the nature of ultra-high energy cosmic rays at energies around the transition from Galactic to extra-galactic origin. It measures the MHz radio emission of extensive air showers produced by cosmic rays hitting our atmosphere. The elemental composition of cosmic rays is a crucial piece of information in determining what the sources of cosmic rays are and how cosmic rays are accelerated. This composition can be obtained from the mass-sensitive parameter $X_{max}$, the depth of the shower maximum. We reconstruct $X_{max}$ with a likelihood analysis, by comparing the measured radio footprint on the ground to an ensemble of footprints from Monte-Carlo CORSIKA/CoREAS air shower simulations. We compare our $X_{max}$ reconstruction with fluorescence $X_{max}$ measurements on a per-event basis, a setup unique to the Pierre Auger Observatory, and show the methods to be fully compatible. Furthermore, we extensively validate our reconstruction by identifying and correcting for systematic uncertainties. We determine the resolution of our method as a function of energy and reach a precision better than 15 g cm$^{-2}$ at the highest energies. With a bias-free set of around 600 showers, we find a light to light-mixed composition at energies between $10^{17.5}$ to $10^{18.8}$ eV, also in agreement with the Auger fluorescence measurements.

## 1 Introduction

The detection of ultra-high energy cosmic rays (UHECRs) is only feasible through indirect detection of extensive air showers. The footprints of the showers, covering extended areas, allow for detection with an array of detectors. Using the captured information the properties of air showers be reconstructed, which can tell us about the cosmic rays and their sources. One key property to differentiate between source models is the particle type, or mass, of the primary cosmic ray. The mass itself however is not a direct observable, but this information can, for example, be extracted from the depth of the shower maximum ($X_{max}$). In general, the development of an air shower occurs higher in the atmosphere for a heavy particle than for a light particle, because the heavy particle can be considered a superposition of multiple light particles, each with a fraction of the total energy. These lower-energy particles finish their cascades earlier and hence $X_{max}$ will be shallower for heavier cosmic rays. In this work, we will present results on extracting $X_{max}$ using the radio emission of air showers, compare it with simulations fluorescence measurements, and interpret the results in the context of the mass composition of cosmic rays.

## 2 AERA and the Pierre Auger Observatory

The Pierre Auger Observatory [1], located near the town of Malargüe in Argentina, is aimed at detecting UHECRs up to the highest energies. It covers an area of 3000 km$^2$, making it the largest of its kind in the world. The main components are an array of 1660 water-Cherenkov detectors (WCDs) making up the surface detector (SD) and 27 fluorescence detectors (FD) surrounding and looking over the SD. Located in the west of the surface array is also the Auger Engineering Radio Array (AERA) [2], an array of radio antennas covering roughly 17 km$^2$. AERA measures the radio emission produced in air showers at frequencies of 30 to 80 MHz and is sensitive to air showers with energies above roughly $10^{17}$ eV (limited by the radio noise floor) and up to roughly $10^{19}$ eV (exposure-limited by its surface area and current operation time). The combination of radio and various other detection techniques makes the Pierre Auger Observatory an ideal place to study both the properties of cosmic rays and the systematics of their detection techniques.

## 3 Reconstruction and Resolution of $X_{max}$

The reconstruction of $X_{max}$ with AERA [3, 4] makes use of simulation-fitting, based on [5], where the radio signals from a measured air shower are compared to the signals from an ensemble of simulated air showers in order to derive $X_{max}$. After quality cuts, a set of 2153 air showers with a high-quality reconstruction of shower core position, arrival direction, and energy are selected. For each of these showers, these properties are given as input to the air-shower simulation code in order to create representative simulations. In these simulations $X_{max}$ will vary, both

*e-mail: spokespersons@auger.org
**Full author list at https://www.auger.org/archive/authors_2022_10.html

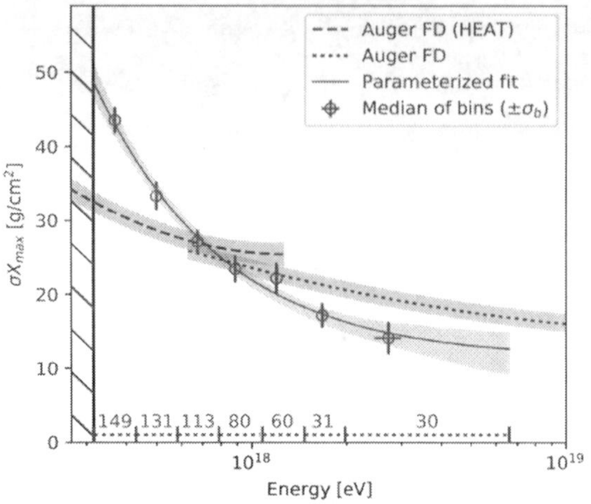


**Figure 1.** Median resolution of the AERA $X_{\max}$ reconstruction (green points) and fitted function (green line, Eq. 1). Inset at the bottom are the sizes of the energy bins and the number of events per bin. For comparison also the $X_{\max}$ resolution for the Auger FD is shown (black lines).

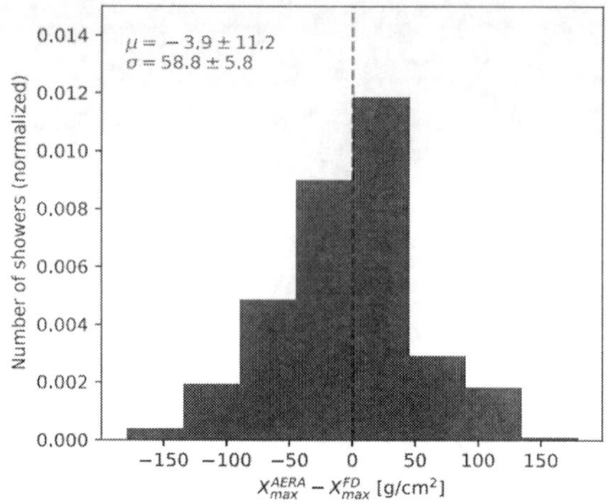


**Figure 2.** Distribution of differences between AERA and FD reconstructed $X_{\max}$ values for the set of hybrid showers. Inset at the top are the mean and width of the differences as determined with bootstrap resampling.

because of the primary particle that initiated the shower, and because of the stochastic nature of shower development. Hence, we simulate 12 showers induced by iron nuclei and 15 induced by protons to get full coverage of the values of $X_{\max}$ that the shower could have realistically had (according to the models). Comparison of the simulations to the measured shower then allows us to extract the best-matching $X_{\max}$ value. For the simulations, we use the air-shower simulation code Corsika (v7.7100) [6] where the radio signals are calculated with CoREAS [7]. The high-energy interactions are modelled with QGSJetII-04 [8] and the low-energy ones with UrQMD 1.3cr [9]. Furthermore, for each of the measured air showers that simulations are made for we provide an atmospheric model of the date and time of the event using the Global Data Assimilation System (GDAS) atmospheric model [10].

We then select only the showers with energies above $10^{17.5}$ eV where the trigger of the SD, which we use for the triggering of the radio stations, becomes fully efficient [11]. Crucially, we select from this the showers that are free from selection bias, caused by the radio detection and reconstruction, by evaluating if each of the showers could have been reconstructed had they arrived with any other $X_{\max}$ value [3, 4]. Showers that do not satisfy this criterion are rejected. After all quality and selection cuts a sample of 594 shower with reconstructed $X_{\max}$ is obtained.

The $X_{\max}$ reconstruction method also provides an estimation of the uncertainty of the reconstructed $X_{\max}$ values, obtained by evaluating the $X_{\max}$ reconstruction accuracy of simulated air showers with all noise and detector effects included. Figure 1 shows the median resolution of $X_{\max}$ as a function of shower energy (green points), for the sample of 594 showers with energies above $10^{17.5}$ eV. To these points, we fit a function (green line), inspired by the resolution in the energy of electromagnetic calorimeters [12]

$$\sigma X_{\max} = a \cdot \sqrt{\frac{10^{18}\text{eV}}{E}} \oplus b \cdot \frac{10^{18}\text{eV}}{E} \oplus c, \quad (1)$$

where $a = 13.8 \pm 6.5\,\text{g cm}^{-2}$, $b = 12.8 \pm 2.1\,\text{g cm}^{-2}$, and $c = 11.3 \pm 4.5\,\text{g cm}^{-2}$ are the fitted parameters and $\oplus$ indicates the quadratic sum. We achieve a resolution of $X_{\max}$ of better than $15\,\text{g cm}^{-2}$ at the highest energies. At lower energies the resolution decreases due to a decreasing signal-to-noise ratio (leading to relatively lower quality signals) and decreasing shower footprint size (leading to relatively less stations per event). For comparison, the Auger FD resolution of $X_{\max}$ (grey lines) [13] is plotted, showing the radio technique to be very competitive.

## 4 Comparison to Hybrid Fluorescence Data

For 53 of the air showers in our data set $X_{\max}$ has also been measured simultaneously with the FD [14]. This number is limited by the requirement of the fluorescence technique to have dark and clear nights and the relatively large distance between AERA and the FD combined with FD field-of-view cuts. These 53 showers, however, are extremely valuable. They allow for event-to-event comparison of $X_{\max}$ , which gives a handle on understanding systematic uncertainties between the two methods independent of selection effects (unlike comparisons of $X_{\max}$ distributions). Fig. 2 shows the differences in $X_{\max}$ as measured by FD and AERA. Inset at the top of the figure are the mean and width of the differences. For this calculation, each event is weighted by its combined uncertainty on $X_{\max}$: $w = (\sigma^2_{X^{\text{AERA}}_{\max}} + \sigma^2_{X^{\text{FD}}_{\max}})^{-0.5}$. The width of the distribution, $58.8 \pm 5.8\,\text{g cm}^{-2}$, matches within uncertainties with the combined resolution of AERA and the FD (53.3 ±

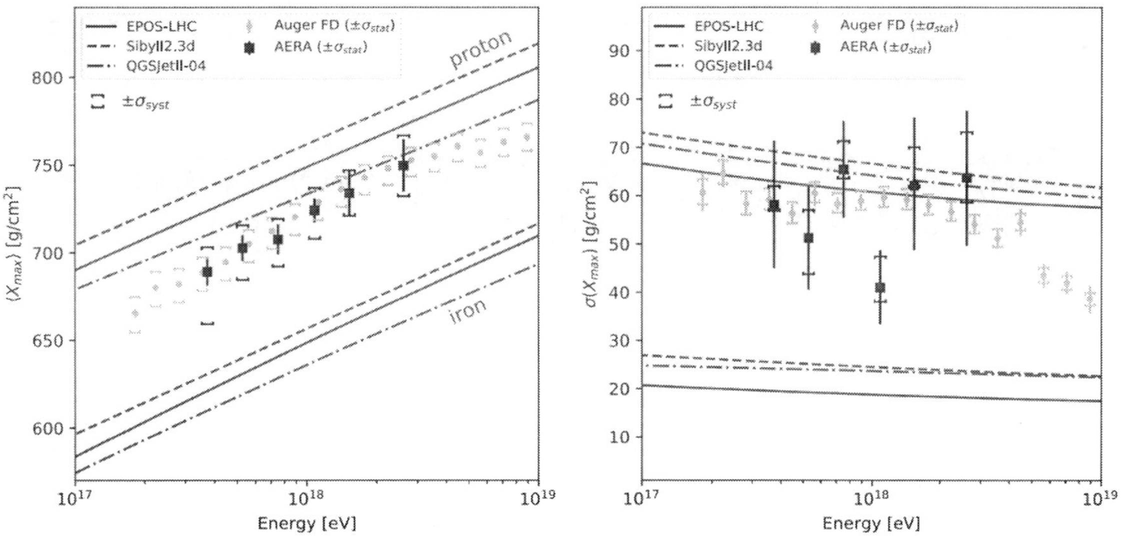


**Figure 3.** Mean (left) and width (right) of $X_{\rm max}$ for our measurements (black markers), Auger FD measurements [13] (gray markers), and pure proton and iron nuclei composition from air shower simulations using three different hadronic interaction models (red and blue lines) [13]. Statistical uncertainties are shown with error bars and systematic uncertainties (see also Fig. 4) with caps.

$5.7\,\mathrm{g\,cm^{-2}}$). The mean difference of $-3.9 \pm 11.2\,\mathrm{g\,cm^{-2}}$ shows there is no significant bias between the two methods and sets a limit on systematic uncertainties between the reconstructions in this energy range (predominantly at the lower energy range of AERA, between $10^{17.5}$ and $10^{18.0}$ eV). Furthermore, the compatibility provides an independent check on the fluorescence technique and gives a new handle on investigating shower physics. This new constraint on the $X_{\rm max}$ scale might prove useful in the context of the muon deficit in air shower simulation codes with respect to measurements [15].

## 5 Moments of the $X_{\rm max}$ Distribution

We show the two central moments of the $X_{\rm max}$ distribution in Fig. 3 for the set of 594 showers with high-quality $X_{\rm max}$ (black markers). For the interpretation of these results, we show the predictions from air shower simulations with three different hadronic interaction models for a composition of only protons or only iron nuclei (red and blue lines, respectively). AERA finds a relatively light composition in the energy range of $10^{17.5}$ to $10^{18.8}$ eV, in agreement with the Auger FD measurements [13].

## 6 Systematic Uncertainties on the Moments of $X_{\rm max}$

For the calculation of the two central moments of the $X_{\rm max}$ distribution, we account for various sources of systematic uncertainty. The contributions are shown in Fig. 4 as a function of energy. There are several small contributions from the models that we use to simulate our showers; the choice of hadronic interaction model has been shown to vary $\langle X_{\rm max}\rangle$ by $5\,\mathrm{g\,cm^{-2}}$ [16] and the GDAS atmospheric model has a systematic uncertainty of $2.2\,\mathrm{g\,cm^{-2}}$ [16, 17]. Furthermore, our SD energy measurements have a systematic uncertainty, which we propagate to $\langle X_{\rm max}\rangle$, resulting in $2.9\,\mathrm{g\,cm^{-2}}$ (and $0.3\,\mathrm{g\,cm^{-2}}$ for $\sigma(X_{\rm max})$ ).

Next, we account for the effects of our $X_{\rm max}$ reconstruction method (red bars in Fig. 4). The reconstruction bias of $X_{\rm max}$ for our method depends on $X_{\rm max}$ itself, hence we quantify, by using the reconstruction of $X_{\rm max}$ with simulated showers, how $\langle X_{\rm max}\rangle$ and $\sigma(X_{\rm max})$ vary for compositions ranging from pure proton to pure iron nuclei. The resulting range in $X_{\rm max}$ we quote as, admittedly conservative, systematic uncertainty. We note that when comparing our $X_{\rm max}$ results to a particular composition measurement (e.g., from a different detector) the AERA systematic uncertainties would be much reduced. For example, for the composition as measured by the Auger FD the total AERA systematic uncertainties would reduce to about $\pm 10\,\mathrm{g\,cm^{-2}}$ at all energies.

For the acceptance cuts in Sec. 3 we determined whether AERA would have been able to detect and reconstruct our measured showers if they would have had a different $X_{\rm max}$. To be more precise, we required that at least 90% of a Gumbel $X_{\rm max}$ distribution [18, 19] for both proton and iron nuclei showers would have been reconstructed. This calculation was performed using the simulation set we have for each measurement [3]. A requirement of 100% would unnecessarily cut many events because they might not have been detected if they had a very unlikely $X_{\rm max}$ value. To still account for this effect we calculate the effect of the non-unity acceptance on the two central moments. This results in less than $4\,\mathrm{g\,cm^{-2}}$ on $\langle X_{\rm max}\rangle$ and $5\,\mathrm{g\,cm^{-2}}$ on $\sigma(X_{\rm max})$ (blue bars in Fig. 4).

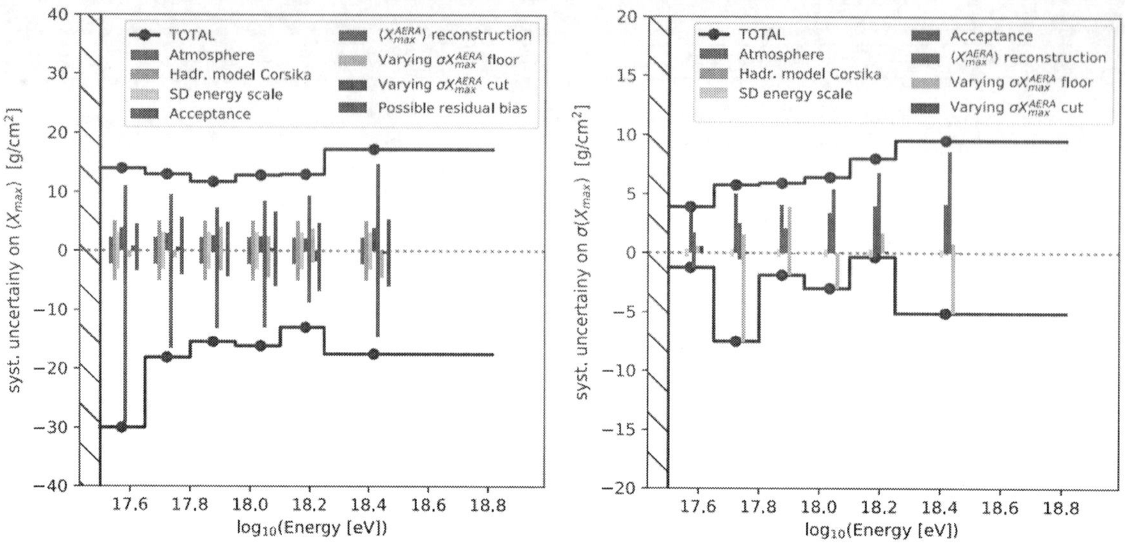

**Figure 4.** Contributions (bars) to the total systematic uncertainty (black lines) on $\langle X_{max}\rangle$ (left) and $\sigma(X_{max})$ (right) for six energy bins centred on the mean energy of each bin (black circles).

We also account for systematic uncertainties from the calculation of the central moments. Because of the relatively low number of events per energy bin these calculations can be sensitive to single outliers. We quantify this with two contributions: the first accounts for events with very small uncertainties (i.e., events where the uncertainty on the $X_{max}$ uncertainty estimation is an under-fluctuation). This can have an especially significant effect on the second moment of $X_{max}$ where one has to remove the detector resolution from the width of the $X_{max}$ distribution. Hence, we perform the calculations of the moments with lower limits for the $X_{max}$ uncertainty, ranging between 5 and 15 g cm$^{-2}$, and quote the range of the moments as a systematic uncertainty (see orange bars in Fig. 4). The second effect is for the few events with very large uncertainty (larger than 80 g cm$^{-2}$). Here we repeat the calculations with an event cut on the $X_{max}$ uncertainty at 80 g cm$^{-2}$ and quote the difference as systematic uncertainty (cyan bars). This latter effect is minor since there are only a few events with higher uncertainties.

Finally, to check for any residual biases that have not been accounted for by the previous contributions we check for any dependencies in $\langle X_{max}\rangle$ as a function of geometry parameters such as arrival direction (zenith and azimuth angle), shower core position, and angle to the geomagnetic field. These represent the effects on acceptance from, respectively, the asymmetric directional sensitivity of the radio antennas, the irregular spacing of radio antennas, and the dependence of the radio emission strength on the angle between the magnetic field and the shower axis. It is conceivable that these effects could lead to part of the showers never being seen at all (even though we checked with the acceptance that each of our showers that we do see would have been seen at any $X_{max}$). We have checked the trends of $\langle X_{max}\rangle$ as a function of these geometry parameters and find a maximal residual bias of 3 to 10 g cm$^{-2}$ (depending on shower energy) is allowed within statistical uncertainties. To be conservative we include this as a systematic uncertainty (see green bars in Fig. 4).

## 7 Conclusions

We have presented the results of the measurement of $X_{max}$ with AERA at the Pierre Auger Observatory. We observe a rather light composition at energies between $10^{17.5}$ and $10^{18.8}$ eV, the energy range where the origin of cosmic rays is expected to transition from Galactic to extragalactic. With our results, we also show compatibility with Auger fluorescence measurements of the central moments of $X_{max}$ with a set of 594 air showers and show the compatibility of $X_{max}$ measurements of individual showers with a set of 53 events measured simultaneously with both detectors. This is strong independent support for the validity of both techniques. Furthermore, we have shown our method is able to reach a precision of $X_{max}$ reconstruction better than 15 g cm$^{-2}$ at high energies, competitive with other techniques.

## References

[1] A. Aab *et al.* (Pierre Auger Collaboration), *Nucl. Instrum. Meth. A* **798** (2015) 172–213
[2] P. Abreu *et al.* (Pierre Auger Collaboration), *J. Instrum.* **7** (2012) P10011
[3] B. Pont, Ph.D. thesis, Radboud University, 2021.
[4] B. Pont [for the Pierre Auger Collaboration], PoS(ARENA2022)044 (in preparation)
[5] S. Buitink *et al.*, *Nature* **531** (2016) 70
[6] D. Heck *et al.*, *FZKA Tech. Umw. Wis. B* **6019** (1998)

[7] T. Huege *et al.*, *AIP Conf. Proc.* **1535** (2013) 128-132
[8] S. Ostapchenko, *Nucl. Phys. B-Proc. Sup.* **151** (2006) 143-146
[9] M. Bleicher, *J. Phys. G* **25** (1999) 1859–1896
[10] P. Mitra *et al.*, *Astropart. Phys.* **123** (2020) 102470
[11] P. Abreu *et al.* (Pierre Auger Collaboration), *Eur. Phys. J. C* **81** (2021) 966
[12] C. W. Fabjan, F. Gianotti, *Rev. Mod. Phys* **75** (2003) 1243
[13] A. Yushkov [for the Pierre Auger Collaboration], PoS(ICRC2019)482
[14] J. Bellido [for the Pierre Auger Collaboration], PoS(ICRC2017)506
[15] A. Aab *et al.* (Pierre Auger Collaboration), *Phys. Rev. Lett.* **126** (2021) 152002
[16] S. Buitink *et al.*, *Nature* **531** (2016) 70
[17] P. Mitra *et al.*, *Astropart. Phys.* **123** (2020) 102470
[18] S. Petrera, personal communication (October 30, 2020).
[19] M. de Domenico *et al.*, *J. Cosmol. Astropart. P.* **07** (2013) 050

# A machine learning approach for mass composition analysis with TALE-SD data

*Ryuhei* Arimura[1,*] on behalf of the Telescope Array Collaboration
(a complete list of authors can be found at the end of the proceedings)

[1] *Graduate School of Science, Osaka City University, Osaka, Japan*

**Abstract.** The TALE experiment is a Telescope Array Low-energy Extension constructed to observe cosmic rays with energies down to $10^{16.5}$ to clarify the origin of the second knee and the energy of the galactic to extragalactic CRs transition. TALE consists of 10 high-elevation fluorescence detectors and 80 scintillation counters in an area of 21 km$^2$. The key of data interpretation is the mass composition of cosmic rays, and we report on a machine learning approach of mass composition analysis that utilizes waveform data of TALE scintillation counters.

## 1 Introduction

### 1.1 Telescope Array experiment

The Telescope Array (TA) experiment is the largest cosmic ray observatory in the northern hemisphere designed to detect ultra high energy cosmic rays is deployed in Millard County, Utah, USA. It mainly observes ultra high energy (above ~ $10^{18}$ eV) cosmic rays, using both Fluorescence Detectors (FD) and scintillator Surface Detectors (SD). There are 507 SDs with about 700 km$^2$ effective detection area and the 3 FD stations, which are called Black Rock Mesa (BRM), Long Ridge (LR) and Middle Drum (MD). The FD telescopes cover the sky above the SD array from 3° to 31° in elevation. The general map of the TA detectors is shown in Fig. 1a.

### 1.2 Telescope Array Low-energy Extension (TALE)

The Telescope Array Low-energy Extension (TALE) detectors are designed for the energy threshold of the experiment to be well below $10^{16.5}$ eV. We call the bend in the energy spectrum around $10^{17}$ eV as the "second knee" and consider that the feature suggests a galatic-to-extragalactic transition of cosmic ray origin. The motivation of the TALE experiment is to clarify the origin of the feature measuring the energy spectrum and mass composition of cosmic rays around $10^{17}$ eV. To observe lower energy showers, TALE utilizes 10 FDs with elevation angles higher than that of TA-FD, from 31° up to 59°. In addition 80 SDs with denser grids are also installed near the MD site. The full details of the detectors may be found in [1]. For this analysis, we report on a machine learning approach of mass composition analysis that utilizes waveform data of TALE-SD array. The layout of TALE detectors is shown in Fig. 1b.

*e-mail: m21sa003@st.osaka-cu.ac.jp

## 2 TALE-SD mass composition analysis

### 2.1 Monte Carlo (MC) simulation and Event selection criteria

We generate cosmic ray air showers using CORSIKA-based MC simulation code developed for TA [2] in this analysis. QGSJETII-04 [3] and Geant 4 are used for air shower simulation and detector response simulation, respectively. The MC input parameters are given in Table 1. The simulated air shower is reconstructed using TALE-SD software [4]. Only events that pass the event selection criteria shown in Table 2a below are used in the analysis. The number of Monte-Carlo events after the selection are 17,121 events (proton) and 17,262 (iron), respectively.

**Table 1:** MC simulation dataset

| primary | proton | iron |
|---|---|---|
| MC simulation | CORSIKA | |
| interaction model | QGSJETII-04 | |
| primary energy | $E = 10^{18.0}$ eV (fixed) | |
| zenith angle | $\theta = 30.0°$ (fixed) | |
| azimuthal angle | $0° \leqq \varphi \leqq 360°$ | |
| TALE-SDs | 78SDs, 2019/10/01-2020/01/29 | |
| core position | refer to Fig. 2 | |
| Number of generated event | 100,000 | 100,000 |
| Number of selected event | 17,121 | 17,262 |

### 2.2 Composition-sensitive parameters

The advantages of the TALE-SD array are its high statistics and high uniform sensitivity compared to FD. If we analyze mass composition analysis using TALE-SD data, we can apply the result to cosmic ray energy spectrum and

https://doi.org/10.1051/epjconf/202328302011

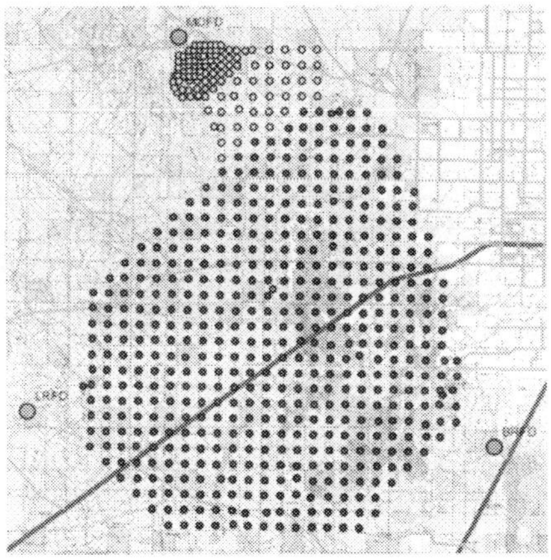


(a)

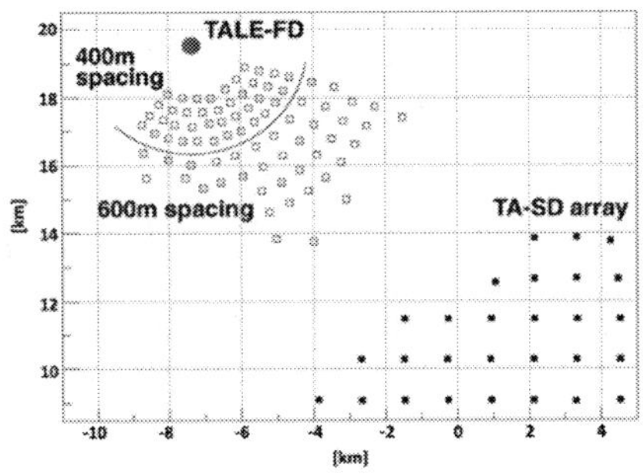


(b)

**Figure 1:** (a)General map of the TA experiment site. The locations of the TA-SDs are shown as red circles and the locations of the three TA-FD stations are indicated by light blue circles. Yellow circles show the location of the TALE-SDs. The TALE-FD station is located at the MD site. (b)The layout of the TALE detectors. Open square boxes (□) represent the locations of the TALE-SDs and blue circle (•) correspond to the MD / TALE-FD station. Black boxes (■) repesent the locations of the TA-SDs.

anisotropy analysis.
A schematic view of an air shower arriving at the surface is shown in Fig. 3. As shown in this figure, air showers arrive at the surface with 3 characteristics: lateral distribution, curvature, and thickness. Using TALE-SD software and simulated proton and iron air showers, we search and extract 22 parameters [5, 6] that exhibit 3 characteristics. The parameters' histograms are shown in Fig. 4.

**Table 2:** (a)Event selection criteria (b)$N_{\text{thickness}}$ selection (SD selection)

(a)

| |
|---|
| $N_{\text{SD}} \geqq 5,\ N_{\text{thickness}} \geqq 1$ |
| $\chi^2_{\text{geometry}}/\text{d.o.f.} \leqq 4,\ \chi^2_{\text{ldf}}/\text{d.o.f.} \leqq 2$ |
| $\left(\sigma_\theta^2 + \sin^2\theta\,\sigma_\phi^2\right)^{0.5} \leqq 2.5$ deg. |
| $\sigma_{S_{600}}/S_{600} \leqq 0.25$ |
| $0° \leqq \theta_{\text{rec.}} \leqq 45°$ |

(b)

| |
|---|
| $t_{\text{recorded waveform}} \leqq 2.56\,\mu\text{s}$ (128 bin) |
| $N_{\text{bin}}(\geqq 15$ FADC count) is more than 2. |
| $N_{\text{bin}}(\geqq 45$ FADC count) is more than 1. |
| No saturation in recorded signal (upper/lower) within the range of 400 m $\leqq r \leqq$ 700 m. |

https://doi.org/10.1051/epjconf/202328302011

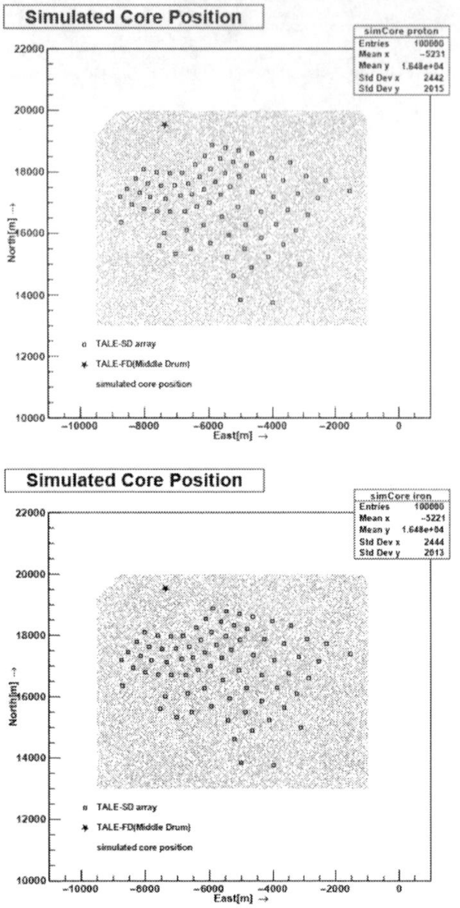


**Figure 2:** Simulated core position distribution for TALE-SD array. □, ★ and •/• indicates TALE-SD array, TALE-FD and simulated core positions, respectively. red one is (a)proton and blue one is (b)iron. "Entries" indicates the number of generated events.

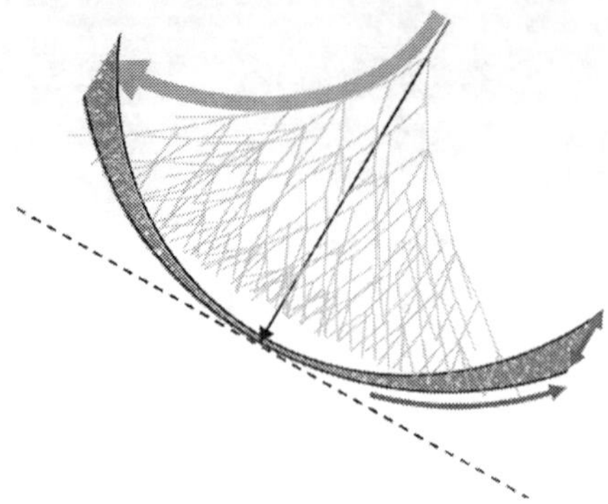

**Figure 3:** A schematic view of the development of an air shower around the surface. Dashed line and black arrow are shower plane and shower axis, respectively. Purple, red and blue arrow indicates "lateral distribution", "thickness" and "curvature" of air showers, respectively.

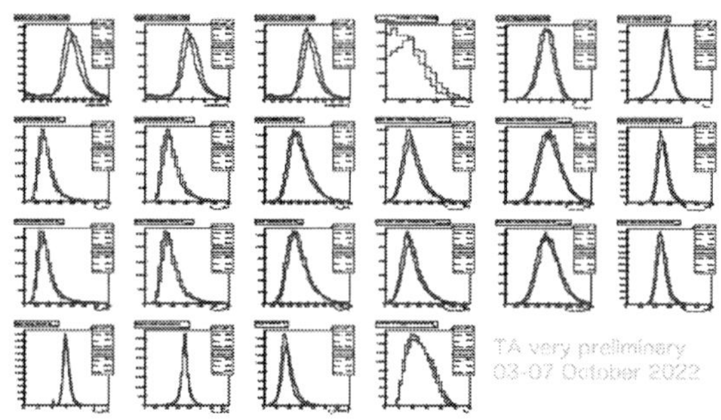


**Figure 4:** 22 parameters by the TALE-SD array histograms. Proton MC is shown with red lines and iron MC is shown with blue them. "Entries" indicates the number of events.

## 3 Machine learning approach

### 3.1 Neural network details

We have developed a machine learning model to take advantage of the high-statistics TALE-SD data for mass composition analysis. The machine learning model that we use is a binary classification to discriminate between proton and iron cosmic ray events by MC simulation as shown the Fig. 5.

The input vector $\alpha$ consists of 22 normalized parameters. Output value $\beta$ is a numerical values from 0 to 1, with 0 being labeled as proton and 1 as iron. The output value $\beta$ is determined to be proton if it is less than 0.5, and iron if it is greater than 0.5.

The machine learning model details are shown in Table 6.

**Figure 5:** Machine learning model outline

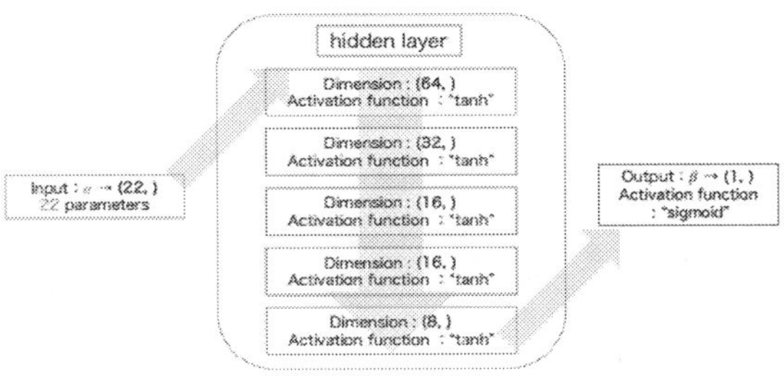


We use two activation functions (a hyperbolic tangent and a sigmoid). To eliminate learning bias, the same number of events are prepared for proton and iron air showers. All data are split into training, validation, and test data at a ratio of 8:1:1.

**Figure 6:** Machine learning model details

| primary | proton | iron |
|---|---|---|
| All data | 17120 | 17120 |
| Training data | 13696 | 13696 |
| Validation data | 1712 | 1712 |
| Test data | 1712 | 1712 |
| Optimizer | Adam | |
| Loss function | Cross-Entropy Error (CEE) | |

### 3.2 The prediction of the machine learning model

The performance of the trained machine learning model was evaluated using all test data. The result is shown in Fig. 7. The red line represents the model's prediction on test data where the answer is proton ($\beta = 0$), and the blue line represents the model's prediction on test data where the answer is iron ($\beta = 1$). Classification matrix for all test data are shown in Table 3. The percentage of those that predicted that the answer is proton is 68.9%, and the percentage of those that predicted that the answer is iron is 65.4%. The overall accuracy is determined by the average of these values, which is 67.1%.

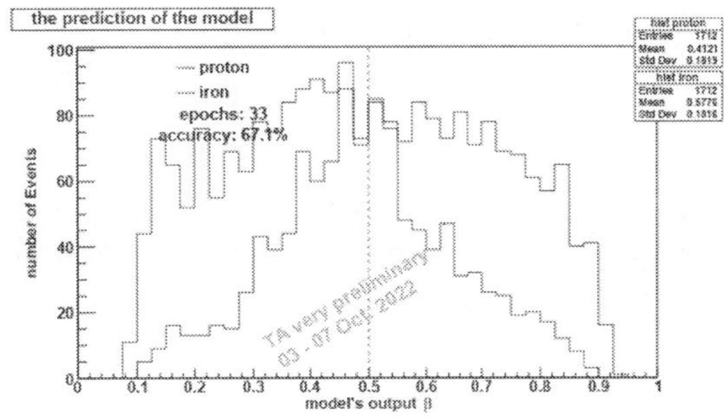


**Figure 7:** The response distribution of the test data to the machine learning model. The purple dashed line is drawn at $\beta = 0.5$.

## 4 Conclusion

We search and extract 22 parameters that depend on primary cosmic ray for cosmic ray mass composition analysis using TALE-SD array. Using these parameters, we have developed a binary classification machine learning model to discriminate proton and iron cosmic rays by MC simulation. The classification accuracy of the machine learning model is currently 67.1%. To improve the accuracy, it is planned to add information for each detector, search for new parameters, and use a Graph Neural Network.

## Acknowledgements

The Telescope Array experiment is supported by the Japan Society for the Promotion of Science(JSPS) through Grants-in-Aid for Priority Area 431, for Specially Promoted Research JP21000002, for Scientific Research (S) JP19104006, for Specially Promoted Research JP15H05693, for Scientific Research (S) JP19H05607, for Scientific Research (S) JP15H05741, for Science Research (A) JP18H03705, for Young Scientists (A) JPH26707011, and for Fostering Joint International Research (B) JP19KK0074, by the joint research program of the Institute for Cosmic Ray Research (ICRR), The University of Tokyo; by the Pioneering Program of RIKEN for the Evolution of Matter in the Universe (r-EMU); by the U.S. National Science Foundation awards PHY-1607727, PHY-1712517, PHY-1806797, PHY-2012934, and PHY-2112904; by the National Research Foundation of Korea (2017K1A4A3015188, 2020R1A2C1008230, & 2020R1A2C2102800) ; by the Ministry of Science and Higher Education of the Russian Federation under the contract 075-15-2020-778, IISN project No. 4.4501.18, and Belgian Science Policy under IUAP VII/37 (ULB). This work was partially supported by the grants ofThe joint research program of the Institute for Space-Earth Environmental Research, Nagoya University and Inter-University Research Program of the Institute for Cosmic Ray Research of University of Tokyo. The foundations of Dr. Ezekiel R. and Edna Wattis Dumke, Willard L. Eccles, and George S. and Dolores Doré Eccles all helped with generous donations. The State of Utah supported the project

**Table 3:** Classification matrix for all test data in Fig. 7. The percentage of cosmic-ray events considered to be of proton origin if the output value $\beta$ is less than 0.5, and of iron origin if the output value is greater than 0.5, is shown in brackets.

| answer \ prediction | proton($0 \leqq \beta < 0.5$) | iron($0.5 \leqq \beta \leqq 1$) | total |
|---|---|---|---|
| proton($0 \leqq \beta < 0.5$) | 1180 (68.9%) | 532 (31.1%) | 1712 |
| iron ($0.5 \leqq \beta \leqq 1$) | 593 (34.6%) | 1119 (65.4%) | 1712 |
| total | 1773 | 1651 | 3424 |

through its Economic Development Board, and the University of Utah through the Office of the Vice President for Research. The experimental site became available through the cooperation of the Utah School and Institutional Trust Lands Administration (SITLA), U.S. Bureau of Land Management (BLM), and the U.S. Air Force. We appreciate the assistance of the State of Utah and Fillmore offices of the BLM in crafting the Plan of Development for the site. Patrick A. Shea assisted the collaboration with valuable advice and supported the collaborations efforts. The people and the officials of Millard County, Utah have been a source of steadfast and warm support for our work which we greatly appreciate. We are indebted to the Millard County Road Department for their efforts to maintain and clear the roads which get us to our sites. We gratefully acknowledge the contribution from the technical staffs of our home institutions. An allocation of computer time from the Center for High Performance Computing at the University of Utah is gratefully acknowledged.

## References

[1] S. Ogio, for the Telescope Array collaboration, "Telescope Array Low-energy Extension (TALE) hybrid", *PoS* **ICRC2019** (2019) 375.

[2] B.T. Stokes, R. Cady, D. Ivanov, J.N. Matthews and G.B. Thomson, "Dethinning extensive air shower simulations" *Astroparticle Physics* **35** (2012) 759-766.

[3] S. Ostapchenko, "Monte Carlo treatment of hadronic interactions in enhanced Pomeron scheme: QGSJET-II model", *Phys. Rev. D* **83** (2011) 014018.

[4] K. Sato, for the Telescope Array collaboration, "Cosmic ray energy spectrum in the 2nd knee region measured by the TALE-SD array", *PoS* **ICRC2021** (2021) 395.

[5] P. Abreu et al., "A Search for Photons with Energies Above $2 \times 10^{17}$ eV Using Hybrid Data from the Low-Energy Extensions of the Pierre Auger Observatory", *Astrophys. J.* **424** (1994) 491.

[6] R. U. Abbasi et al., "Mass composition of ultrahigh-energy cosmic rays with the Telescope Array Surface Detector data", *Phys. Rev. D* **99** (2019) 022002.

## Full Authors List: Telescope Array Collaboration

R.U. Abbasi[1], T. Abu-Zayyad[1,2], M. Allen[2], Y. Arai[3], R. Arimura[3], E. Barcikowski[2], J.W. Belz[2], D.R. Bergman[2], S.A. Blake[2], I. Buckland[2], B.G. Cheon[4], M. Chikawa[5], T. Fujii[6], K. Fujisue[5], K. Fujita[3], R. Fujiwara[3], M. Fukushima[5], G. Furlich[2], N. Globus[7*], R. Gonzalez[2], W. Hanlon[2], N. Hayashida[8], H. He[7], K. Hibino[8], R. Higuchi[5], K. Honda[9], D. Ikeda[8], N. Inoue[10], T. Ishii[9], H. Ito[7], D. Ivanov[2], H. Iwakura[11], A. Iwasaki[3], H.M. Jeong[12], S. Jeong[12], C.C.H. Jui[2], K. Kadota[13], F. Kakimoto[8], O. Kalashev[14], K. Kasahara[15], S. Kasami[16], S. Kawakami[3], K. Kawata[5], I. Kharuk[14], E. Kido[7], H.B. Kim[4], J.H. Kim[2**], J.H. Kim[2], S.W. Kim[12], Y. Kimura[3], I. Komae[3], Y. Kubota[11], V. Kuzmin[14***], M. Kuznetsov[14,17], Y.J. Kwon[18], K.H. Lee[12], B. Lubsandorzhiev[14], J.P. Lundquist[2,19], H. Matsumiya[3], T. Matsuyama[3], J.N. Matthews[2], R. Mayta[3], I. Myers[2], S. Nagataki[7], K. Nakai[3], R. Nakamura[11], T. Nakamura[20], T. Nakamura[11], Y. Nakamura[11], A. Nakazawa[11], E. Nishio[16], T. Nonaka[5], H. Oda[3], S. Ogio[5], M. Ohnishi[5], H. Ohoka[5], Y. Oku[16], T. Okuda[21], Y. Omura[3], M. Ono[7], A. Oshima[22], S. Ozawa[23], I.H. Park[12], M. Potts[2****], M.S. Pshirkov[14,24], J. Remington[2], D.C. Rodriguez[2], C. Rott[2,12], G.I. Rubtsov[14], D. Ryu[25], H. Sagawa[5], N. Sakaki[5], T. Sako[5], N. Sakurai[3], K. Sato[3], T. Seki[11], K. Sekino[5], P.D. Shah[2], Y. Shibasaki[11], N. Shibata[16], T. Shibata[5], J. Shikita[3], H. Shimodaira[5], B.K. Shin[25], H.S. Shin[5], D. Shinto[16], J.D. Smith[2], P. Sokolsky[2], B.T. Stokes[2], T.A. Stroman[2], K. Takahashi[5], M. Takamura[26], M. Takeda[5], R. Takeishi[5], A. Taketa[27], M. Takita[5], Y. Tameda[16], K. Tanaka[28], M. Tanaka[29], Y. Tanoue[3], S.B. Thomas[2], G.B. Thomson[2], P. Tinyakov[14,17], I. Tkachev[14], H. Tokuno[30], T. Tomida[11], S. Troitsky[14], R. Tsuda[3], Y. Tsunesada[3,31], S. Udo[8], T. Uehama[11], F. Urban[32], D. Warren[7], T. Wong[2], M. Yamamoto[11], K. Yamazaki[22], K. Yashiro[26], F. Yoshida[16], Y. Zhezher[5,14], and Z. Zundel[2]

[1] *Department of Physics, Loyola University Chicago, Chicago, Illinois 60660, USA*
[2] *High Energy Astrophysics Institute and Department of Physics and Astronomy, University of Utah, Salt Lake City, Utah 84112-0830, USA*
[3] *Graduate School of Science, Osaka Metropolitan University, Sugimoto, Sumiyoshi, Osaka 558-8585, Japan*
[4] *Department of Physics and The Research Institute of Natural Science, Hanyang University, Seongdong-gu, Seoul 426-791, Korea*
[5] *Institute for Cosmic Ray Research, University of Tokyo, Kashiwa, Chiba 277-8582, Japan*
[6] *The Hakubi Center for Advanced Research and Graduate School of Science, Kyoto University, Kitashirakawa-Oiwakecho, Sakyo-ku, Kyoto 606-8501, Japan*
[7] *Astrophysical Big Bang Laboratory, RIKEN, Wako, Saitama 351-0198, Japan*
[8] *Faculty of Engineering, Kanagawa University, Yokohama, Kanagawa 221-8686, Japan*
[9] *Interdisciplinary Graduate School of Medicine and Engineering, University of Yamanashi, Kofu, Yamanashi 400-8511, Japan*
[10] *The Graduate School of Science and Engineering, Saitama University, Saitama, Saitama 338-8570, Japan*
[11] *Academic Assembly School of Science and Technology Institute of Engineering, Shinshu University, Nagano, Nagano 380-8554, Japan*
[12] *Department of Physics, SungKyunKwan University, Jang-an-gu, Suwon 16419, Korea*
[13] *Department of Physics, Tokyo City University, Setagaya-ku, Tokyo 158-8557, Japan*
[14] *Institute for Nuclear Research of the Russian Academy of Sciences, Moscow 117312, Russia*
[15] *Faculty of Systems Engineering and Science, Shibaura Institute of Technology, Minato-ku, Tokyo 337-8570, Japan*
[16] *Department of Engineering Science, Faculty of Engineering, Osaka Electro-Communication University, Neyagawa-shi, Osaka 572-8530, Japan*
[17] *Service de Physique Thorique, University Libre de Bruxelles, Brussels 1050, Belgium*
[18] *Department of Physics, Yonsei University, Seodaemun-gu, Seoul 120-749, Korea*
[19] *Center for Astrophysics and Cosmology, University of Nova Gorica, Nova Gorica 5297, Slovenia*
[20] *Faculty of Science, Kochi University, Kochi, Kochi 780-8520, Japan*
[21] *Department of Physical Sciences, Ritsumeikan University, Kusatsu, Shiga 525-8577, Japan*
[22] *College of Engineering, Chubu University, Kasugai, Aichi 487-8501, Japan*
[23] *Quantum ICT Advanced Development Center, National Institute for Information and Communications Technology, Koganei, Tokyo 184-8795, Japan*
[24] *Sternberg Astronomical Institute, Moscow M.V. Lomonosov State University,*

*Moscow 119991, Russia*
[25] *Department of Physics, School of Natural Sciences, Ulsan National Institute of Science and Technology, UNIST-gil, Ulsan 689-798, Korea*
[26] *Department of Physics, Tokyo University of Science, Noda, Chiba 162-8601, Japan*
[27] *Earthquake Research Institute, University of Tokyo, Bunkyo-ku, Tokyo 277-8582, Japan*
[28] *Graduate School of Information Sciences, Hiroshima City University, Hiroshima, Hiroshima 731-3194, Japan*
[29] *Institute of Particle and Nuclear Studies, KEK, Tsukuba, Ibaraki 305-0801, Japan*
[30] *Graduate School of Science and Engineering, Tokyo Institute of Technology, Meguro, Tokyo 152-8550, Japan*
[31] *Nambu Yoichiro Institute of Theoretical and Experimental Physics, Osaka Metropolitan University, Sugimoto, Sumiyoshi, Osaka 558-8585, Japan*
[32] *CEICO, Institute of Physics, Czech Academy of Sciences, Prague 182 21, Czech Republic*

* Presently at: University of Californa - Santa Cruz and Flatiron Institute, Simons Foundation

** Presently at: Argonne National Laboratory, Physics Division, Lemont, Illinois 60439,USA

*** Deceased

**** Presently at: Georgia Institute of Technology, Physics Department, Atlanta, Geogia 30332,USA

# Testing Model Predictions of Depth of Air-Shower Maximum and Signals in Surface Detectors using Hybrid Data of the Pierre Auger Observatory

*Jakub* Vícha[1,*] for Pierre Auger Collaboration[2,**]

[1]Institute of Physics of the Czech Academy of Sciences
[2]Observatorio Pierre Auger, Av. San Martín Norte 304, 5613 Malargüe, Argentina
for full-author list see https://www.auger.org/archive/authors_2022_10.html

**Abstract.** We present a method for testing the predictions of hadronic interaction models and improving their consistency with observed two-dimensional distributions of the depth of shower maximum, $X_{\rm max}$, and signal at the ground level as a function of zenith angle. The method relies on the assumption that the mass composition is the same at all zenith angles, while the atmospheric shower development and attenuation depend on composition in a correlated way. In the present work, for each of the three leading LHC-tuned hadronic interaction models, we allow a global shift $\Delta X_{\rm max}$ of the predicted shower maximum, which is the same for every mass and energy, and a rescaling $R_{\rm Had}$ of the hadronic component at the ground level which is constant with the zenith angle. We apply the analysis to 2297 events reconstructed with both the fluorescence and surface detectors of the Pierre Auger Observatory with energies $10^{18.5-19.0}$ eV and zenith angles below 60°. Given the modeling assumptions made in this analysis, the best fit reaches its optimum value when shifting the $X_{\rm max}$ predictions of hadronic interaction models to deeper values and increasing the hadronic signal. This change in the predicted $X_{\rm max}$ scale alleviates the previously identified model deficit in the hadronic signal (commonly called the muon puzzle) but does not fully remove it. Because of the size of the adjustments $\Delta X_{\rm max}$ and $R_{\rm Had}$ and the large number of events in the sample, the statistical significance of need for these adjustments is large, greater than $5\sigma_{\rm stat}$, even for the combination of the systematic experimental shifts within $1\sigma_{\rm sys}$ that are the most favorable for the models.

## 1 Introduction

The cosmic rays of ultra-high energy (above $10^{18}$ eV) are still of unknown origin and uncertain mass composition despite the huge effort of the community for more than 50 years since their discovery. The main complications are their very low flux and uncertainties coming from the extrapolation of hadronic interactions that take place in the development of air-showers initiated by the primary cosmic rays. The Pierre Auger Observatory [1] provides an unprecedented large amount of high-quality air showers reconstructed using two detection techniques - Surface Detector (SD) and Fluorescence Detector (FD), providing an excellent opportunity to test predictions of hadronic interaction models.

The SD measurement of the ground signal estimates the shower size using the signal expected at 1000 m from the shower core, $S(1000)$ [2]. In this work, the signal $S(1000)$ is assumed to be composed of the hadronic $S_{\rm Had}$ and electromagnetic $S_{\rm em}$ components. The signal $S_{\rm Had}$ is induced by muons, and electromagnetic (EM) particles from muon decays and low-energy neutral pions according to the 4-component shower universality description [3],[4]. The signal $S_{\rm em}$ is induced by EM particles originating from high-energy neutral pions. The muon component attenuates less with the zenith angle ($\theta$) than the EM component and its size is a measure of the primary

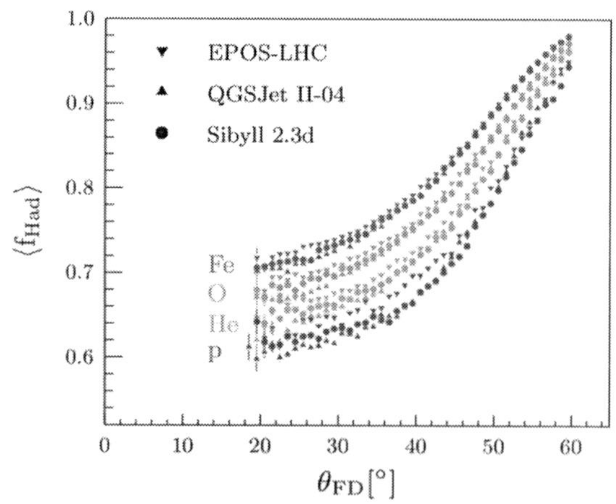


Figure 1: Average fraction of the hadronic signal at 1000 m, $f_{\rm Had} = S_{\rm Had}/S(1000)$, as a function of the zenith angle reconstructed by the FD. Signals from four primaries (colors) of energies $10^{18.5} - 10^{19.0}$ eV and three models of hadronic interactions (markers) are shown.

*e-mail: vicha@fzu.cz
**e-mail: spokespersons@auger.org

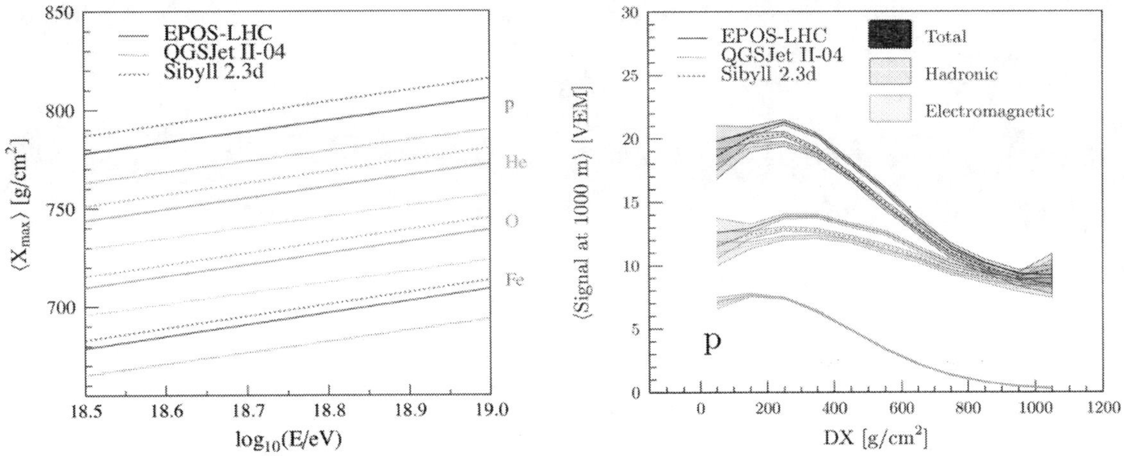


Figure 2: Left: Energy evolution of the mean of $X_{max}$ distribution predicted for three hadronic interaction models and four primary species. The model Sibyll 2.3d predicts on average by ≈7 g/cm² and ≈22 g/cm² deeper $\langle X_{max} \rangle$ than the models Epos-lhc and Qgsjet II-04, respectively. Right: Decomposition of the total ground signal (black) into the hadronic (braun) and EM (turquoise) parts as predicted to depend on the distance of $X_{max}$ to the ground ($DX$). The ground signal at 1000 m from the shower core is simulated for responses at the Pierre Auger Observatory to showers generated using three hadronic interaction models in case of protons of energies $10^{18.5} - 10^{19.0}$ eV.

mass. As a consequence, the fraction of the hadronic signal to $S(1000)$ is increasing with $\theta$ and mass of the primary (protons p, and He, O, Fe nuclei), see Fig. 1.

The FD measurement of the deposited energy on the shower depth provides a precise estimation of the shower energy ($E_{FD}$) and depth of the shower maximum ($X_{max}$) [5]. The latter quantity is another measure of the mass of primary particle initiating the shower.

In this work, we use the showers reconstructed in both SD and FD to analyse simultaneously the mass composition of primary particles and deficiencies of Monte Carlo (MC) predictions on $S(1000)$ and $X_{max}$. We test in this way predictions of three models of hadronic interactions Epos-lhc [6], Qgsjet II-04 [7], Sibyll 2.3d [8] in the energy range around the ankle region [2].

### 1.1 Data Selection and Observables

We select 2297 events detected with FD energies $10^{18.5}$ – $10^{19.0}$ eV and $\theta$ within 60° during the period 1. 1. 2004 - 31. 12. 2018. These events survive high-quality selection criteria used to derive the SD energy spectrum [2] and FD analysis of $X_{max}$ [5, 9]. These events are divided into 5 zenith-angle ranges containing nearly the same and sufficiently large number of events.

The observables $X_{max}^{Ref}$ and $S^{Ref}(1000)$ contain corrections for the energy evolution of $X_{max}$ and $S(1000)$ using the FD energy as

$$S^{Ref}(1000) = S(1000) \cdot \left(\frac{E^{Ref}}{E_{FD}}\right)^{1/B}, \qquad (1)$$

and

$$X_{max}^{Ref} = X_{max} + D \cdot \log_{10}\left(\frac{E^{Ref}}{E_{FD}}\right), \qquad (2)$$

where $B = 1.031$ is the SD energy calibration parameter [2] and the elongation rate of a single primary $D = 58$ g/cm² is taken as the average value over the four primary particles and the three hadronic interaction models. The reference energy is set to $10^{18.7}$ eV.

## 2 Adjustments to Monte Carlo Predictions

There are persistent inconsistencies in the mass interpretation of $X_{max}$ measurements and ground-signal measurements [10], [11]. All these interpretations are based on the assumption of $X_{max}$ scale, $\langle X_{max} \rangle$ at given energy, predicted by a given model of hadronic interactions and driving this way an assumption on the mass composition. Another assumption in these analyses is a tension between the data and simulations coming only from the lack of hadronic (muon) signal in the MC predictions.

Recently, a novel method considering insufficiencies not only in the predictions of ground signal, but also in the predicted $X_{max}$ scale was applied to the data of the Pierre Auger Observatory [12]. The motivation for such a generalized approach is indicated on the left panel of Fig. 2 where the differences in model predictions of $\langle X_{max} \rangle$ are approximately energy and primary-mass independent. On the right panel of Fig. 2, the difference in the ground signal predicted by different models of hadronic interactions is stemming from the different scales of hadronic signal. The EM part of signal is very universal within the applied shower-universality approach to the ground signal.

### 2.1 Fitting Method in Nutshell

In this work, we show the application of this novel method to the data of the Pierre Auger Observatory considering

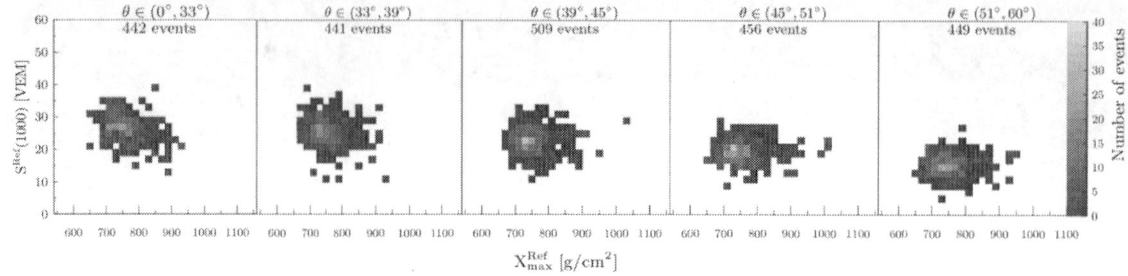


Figure 3: The two-dimensional distributions of $S(1000)$ and $X_{max}$ for data measured by the Pierre Auger Observatory in the energy range $10^{18.5} - 10^{19.0}$ eV and divided into five zenith-angle ($\theta$) bins.

Table 1: Minimal values of the Likelihood-ratio expression from Eq.(3) for the best fits of three hadronic interaction models allowing different adjustments to the MC templates.

| $\mathscr{L}_{min}$ | EPOS-LHC | QGSJET II-04 | SIBYLL 2.3d |
|---|---|---|---|
| none | 2097.410 | 4650.070 | 2581.530 |
| $\Delta X_{max}$ | 764.090 | 1746.945 | 1060.795 |
| $R_{Had}$ | 518.146 | 720.086 | 550.954 |
| $R_{Had}$ & $\Delta X_{max}$ | 479.456 | 515.055 | 480.611 |

simultaneously freedom in the mass composition and the predicted hadronic and $X_{max}$ scale. The freedom in the predicted hadronic signal is assumed to be a constant function of the zenith angle on the contrary to the more general assumptions made in [12], where more details can be found.

In short, we perform a composition fit simultaneously to five two-dimensional (2D) distributions of $X_{max}^{Ref}$ and $S^{Ref}(1000)$ corresponding to five zenith-angle ranges between 0° and 60°, see Fig. 3, with a combination of MC templates for four primary species (p, He, O, Fe). A constant freedom in the predicted $X_{max}$ scale ($X_{max} = X_{max} + \Delta X_{max}$) and hadronic scale ($S_{Had} = S_{Had} \cdot R_{Had}$) in these templates is considered. The effect of the change of $X_{max}$ on the ground signal is incorporated through the separate effects on the hadronic and EM signals estimated from the parameterized evolution of the mean ground signal parts with the distance of $X_{max}$ to the ground ($DX = 880\ \mathrm{g/cm^2}/\cos(\theta) - X_{max}$). In this way, the total ground signal in a given $\theta$-range is estimated to be modified by about 7% at most in case of a change of $X_{max}$ by 50 g/cm$^2$.

The likelihood-ratio expression that is minimized in the method for a given model of hadronic interactions is of the form

$$\forall n_{jz} > 0: \quad \mathscr{L} = \sum_z \sum_j \left( C_{jz} - n_{jz} + n_{jz} \cdot \ln \frac{n_{jz}}{C_{jz}} \right),$$
$$\forall n_{jz} = 0: \quad \mathscr{L} = \sum_z \sum_j C_{jz}, \qquad (3)$$

with the sums running over the 2D-bins $j$ and five $\theta$-bins $z$. The number of showers measured in a 2D-bin $j$ and a $\theta$-bin $z$ is denoted by $n_{jz}$ and the predicted number of MC showers for the same bin by $C_{jz}$. These MC predictions at the reconstruction level are obtained from parameterization of MC templates for each of the three models of hadronic interactions.

## 3 Results

### 3.1 Data Description

The resulting minimal likelihood-ratio value for different adjustments on the MC predictions is listed in Tab. 1 for all three hadronic interaction models. This value is a measure how well the observed data are described by the adjusted MC templates. A large improvement in the data description is obtained considering freedom in the predicted hadronic scale and further significant improvement is achieved combining with the freedom in the predicted $X_{max}$ scale as well. This improvement of data description is further demonstrated in Figs. 4, 5 and 6 for no adjustments to MC predictions, adjusted hadronic scale and adjusted hadronic and $X_{max}$ scales, respectively. These figures include projected distributions of the ground signal and $X_{max}$, together with their zenith-dependent correlation $r_G$ [13]. The improvement, especially in case of comparison Figs. 5 and 6, is also quantified by p-values of a $\chi^2$ test. For all three models, the overall p-value of 2D data description using freedom in both hadronic and $X_{max}$ scales is at a level of several percent based on MC-MC tests.

### 3.2 Fitted Parameters

The most likely values of adjustments to the MC templates that describe the Pierre Auger Observatory data are shown

 https://doi.org/10.1051/epjconf/202328302012

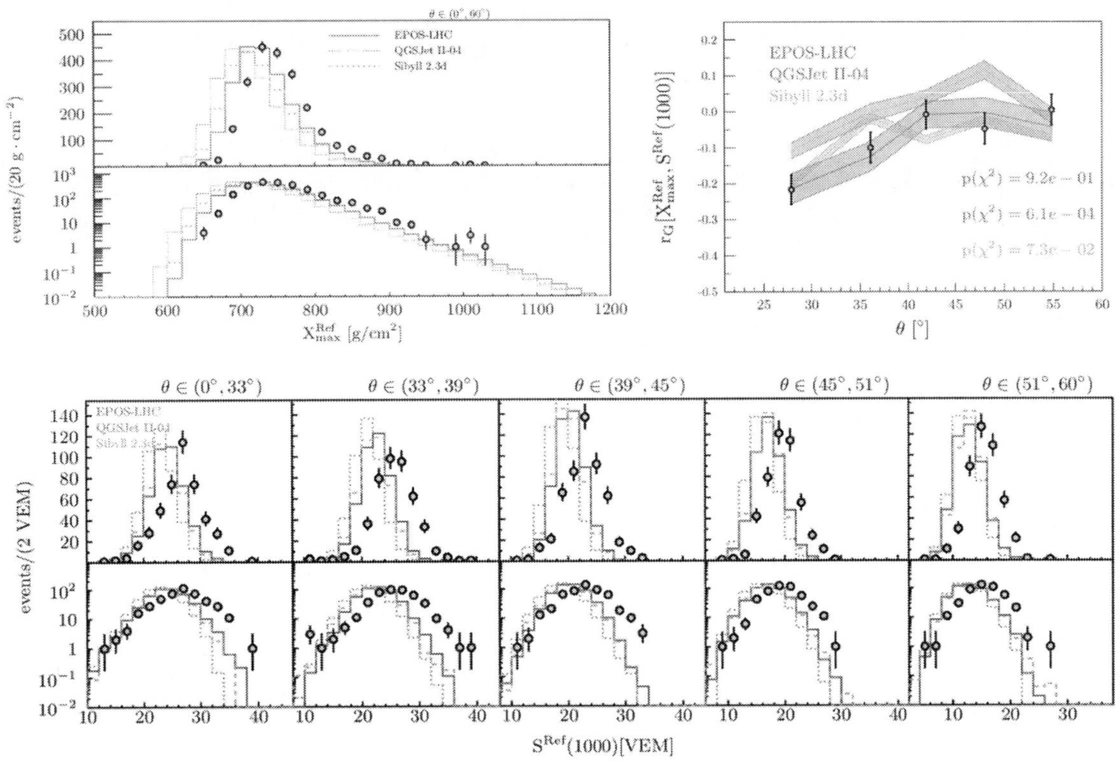


Figure 4: Description of data (black) by the composition fits without any modification of the MC templates. The projected $X_{max}$ distribution (top left), projected $S(1000)$ distributions for all five zenith-angle ($\theta$) ranges and the zenith-angle dependence of the correlation between $X_{max}$ and $S(1000)$ are shown for the three models of hadronic interactions (lines).

on the left panel of Fig. 7. The measured data are best described when the MC predictions on $X_{max}$ are on average deeper by ≈20, 50 and 30 g/cm$^2$ for Epos-lhc, QgsJet II-04 and Sibyll 2.3d, respectively. In case of the predicted hadronic signal, an increase by 15-25% is simultaneously needed for all three hadronic interaction models. The adjusted $X_{max}$ scales have a consequence of heavier mass compositions preferred by the fits to describe the measured data, see the right panel of Fig. 7, than in case of the mass-composition fits to the measured $X_{max}$ distributions with the unmodified predictions on $X_{max}$ by the hadronic interaction models as in [14, 15] in the corresponding energy range. As our method removes the main differences in the predictions of models of hadronic interactions by introducing freedom in hadronic and $X_{max}$ scales, the fitted primary fractions show smaller model dependence of the primary fractions than in case of the $X_{max}$ distribution fits with the unmodified predictions.

### 3.3 Systematic Uncertainties

There are 4 dominant sources of systematic uncertainties: energy scale (±14%), $X_{max}$ measurement ($^{+8}_{-9}$ g/cm$^2$), $S(1000)$ measurement (±5%) and biases of the method. The individual contributions of these systematic effects to the MC adjustments are shown in Fig. 8 together with their quadratic sums.

The systematic effects on the energy correction parameters ($B$, $D$, $\beta$) and the long-term effects have negligible contribution compared to the above-mentioned systematic uncertainties as well as the effect of idealization of Auger detectors in the simulations.

### 3.4 Tests of Models

To quantify the significance of the assumed adjustments to the model predictions, we scan all possible linear combinations of the three experimental systematic uncertainties (energy scale, $X_{max}$ and $S(1000)$ measurements). The change of attenuation of total ground signal is correlated with the systematic change of energy scale due to different energy dependence of EM ($S_{EM} \propto E_{FD}$) and hadronic ($S_{Had} \propto E_{FD}^{\beta}$) parts of the signal. The parameter $\beta = 0.92$, expressing the growth of hadronic component with energy, is chosen in accordance with [16]. Therefore there is a degeneration between the systematic change of energy scale and change of attenuation of the hadronic signal (EM signal is assumed universal in this method, see the right panel of Fig. 2). The correct way to estimate the significance of a need for the assumed MC adjustments is through the more

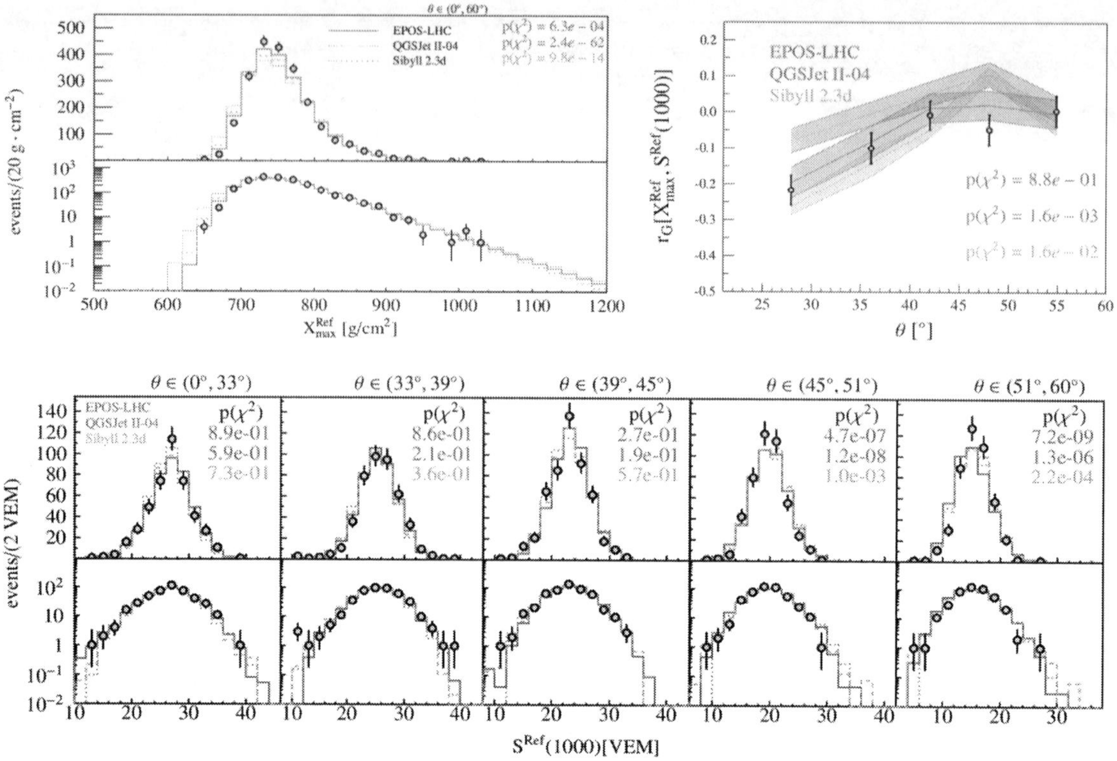


Figure 5: Composition fits with a freedom in $R_{\rm Had}$, see caption of Fig. 4.

general approach as in [12] considering the zenith dependence of $R_{\rm Had}$. Even for the most favorable combination of experimental systematic uncertainties for each model of hadronic interaction, the statistical significance of a need for these MC adjustments is higher than $5\sigma$.

## 4 Conclusions

The combined measurements of cosmic rays with energies between $10^{18.5}$ eV and $10^{19.0}$ eV using the Surface Detector and the Fluorescence Detector of the Pierre Auger Observatory provide statistics large enough to apply a novel method allowing a complex testing of the air-shower observables predicted by the hadronic interaction models and simultaneously fit the fractions of primary particles. The results of the method are in tension with the predictions (i.e. $X_{\rm max}$ and $S_{\rm Had}$ scales) of hadronic interaction models Epos-lhc, Sibyll 2.3d and Qgsjet II-04 by more than $5\sigma$ with much higher significance in case of the Qgsjet II-04. The best description of the data is obtained when the $X_{\rm max}$ scale of the MC predictions is deeper by about 20 g/cm$^2$, 30 g/cm$^2$ and 50 g/cm$^2$ for Epos-lhc, Sibyll 2.3d and Qgsjet II-04, respectively. The adjusted $X_{\rm max}$ scales are shown on the left panel of Fig. 9 together with measured data. At the same time, the differences between the hadronic interaction models in the fitted mass composition fractions are decreased and the mass composition is found heavier than using the unmodified predictions of the three hadronic interaction models. For such heavier mass compositions, the "muon problem" of the hadronic interaction models is alleviated with respect to the previous studies, observing ≈15-25% deficit of the hadronic component of simulated ground signal for all three studied models.

## Acknowledgements

This work was supported by the Ministry of Education, Youth and Sports of the Czech Republic – Grant No. LTT18004, LM2018102 and CZ.02.1.01/0.0/0.0/16_013/0001402.

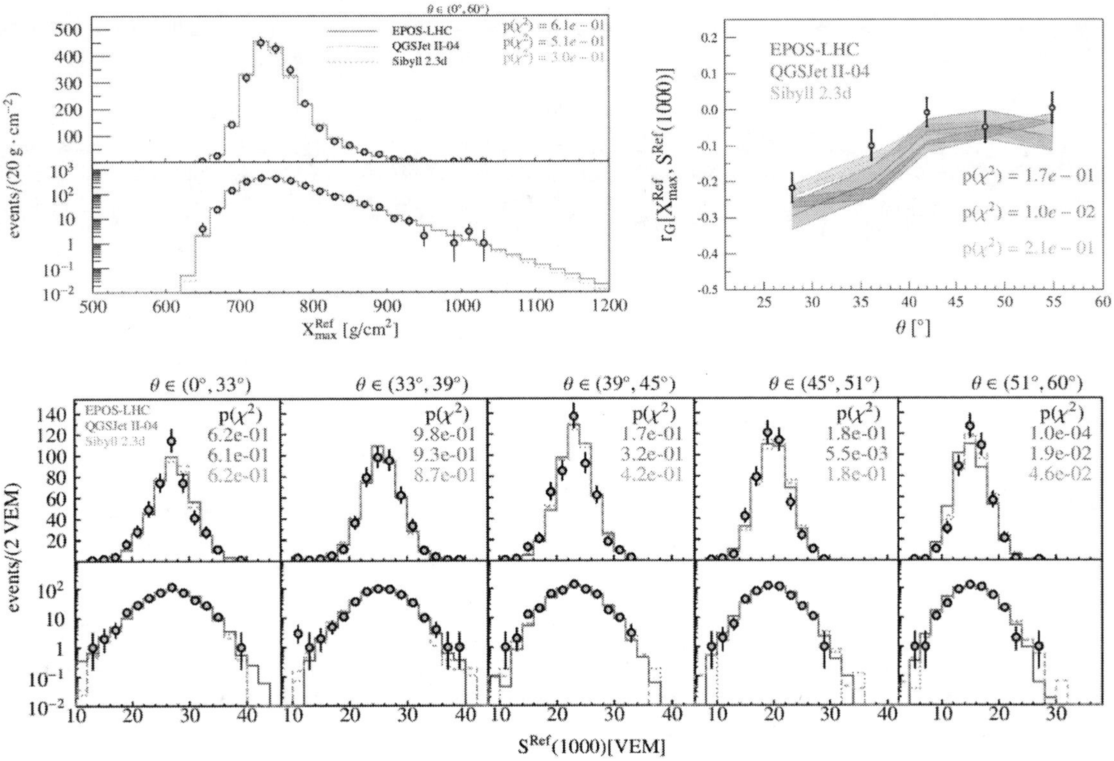


Figure 6: Composition fits with a freedom in $R_{\mathrm{Had}}$ and $\Delta X_{\mathrm{max}}$, see caption of Fig. 4.

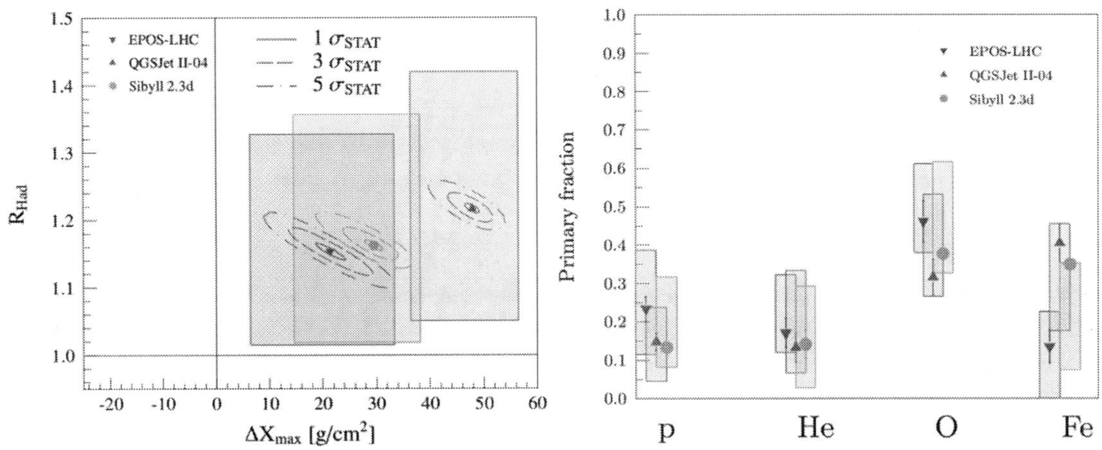


Figure 7: Best-fit values of the fitted parameters $R_{\mathrm{Had}}$ and $\Delta X_{\mathrm{max}}$ (left) and primary fractions (right) for all three models of hadronic interactions. The systematic uncertainties are depicted by gray bands.

https://doi.org/10.1051/epjconf/202328302012

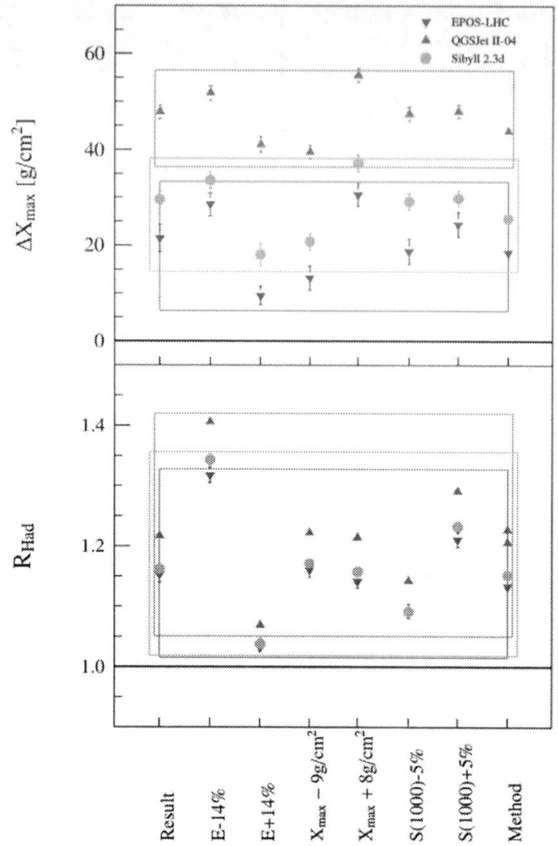


Figure 8: Individual systematic contributions (points) to the total systematic uncertainty (band) of $R_{\text{Had}}$ and $\Delta X_{\text{max}}$.

## References

[1] A. Aab et al. (Pierre Auger Collaboration), Nucl. Instrum. Methods Phys. Res. A **798**, 172 (2015)

[2] A. Aab et al. (Pierre Auger Collaboration), Phys. Rev. D **102**, 062005 (2020)

[3] M. Ave, R. Engel, M. Roth, A. Schulz, Astropart. Phys. **87**, 23 (2017)

[4] M. Ave, M. Roth, A. Schulz, Astropart. Phys. **88**, 46 (2017)

[5] A. Aab et al. (Pierre Auger Collaboration), Phys. Rev. D **90**, 122005 (2014)

[6] T. Pierog, I. Karpenko, J.M. Katzy, E. Yatsenko, K. Werner, Phys. Rev. C **92**, 034906 (2015)

[7] S. Ostapchenko, Phys. Rev. D **83**, 014018 (2011)

[8] F. Riehn, R. Engel, A. Fedynitch, T.K. Gaisser, T. Stanev, Phys. Rev. D **102**, 063002 (2020)

[9] A. Yushkov (Pierre Auger Collaboration) (2019), PoS(ICRC19)482

[10] A. Aab et al. (Pierre Auger Collaboration), Phys. Rev. Lett. **117**, 192001 (2016)

[11] A. Aab et al. (Pierre Auger Collaboration), Eur. Phys. J. C **210**, 751 (2020)

[12] J. Vícha (Pierre Auger Collaboration) (2021), PoS(ICRC2021)310

[13] R.A. Gideon, R.A. Hollister, Journal of the American Statistical Association **82**, 656 (1987)

[14] A. Aab et al. (Pierre Auger Collaboration), Phys. Rev. D **90**, 122006 (2014)

[15] J. Bellido (Pierre Auger Collaboration) (2018), PoS(ICRC19)482

[16] K.H. Kampert, M. Unger, Astropart. Phys. **35**, 660 (2012)

[17] C.J.T. Peixoto (Pierre Auger Collaboration) (2019), PoS(ICRC19)440

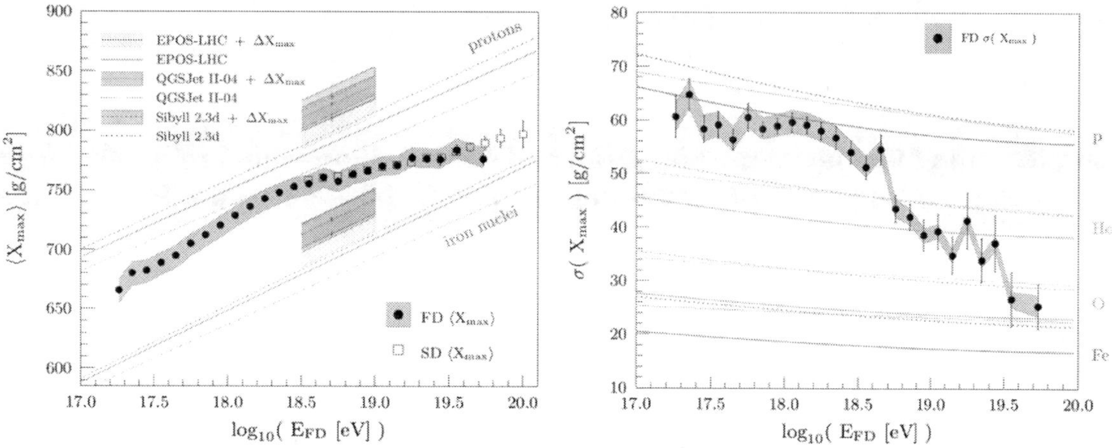


Figure 9: The energy evolution of the mean $X_{\max}$ (left) and the standard deviation of $X_{\max}$ (right) measured by the Pierre Auger Observatory (in black) using FD [9] and SD [17]. The adjusted $X_{\max}$ scales of MC predictions (left panel) obtained using the fit results in the energy range $10^{18.5}-10^{19.0}$ eV are shown in blue and red for protons and iron nuclei, respectively, with the bands corresponding to the systematic uncertainties. The original predictions by hadronic interaction models are depicted by lines without bands.

# Interpreting the cosmic ray spectrum and composition measurements across the ankle and up to the highest energies with the data of the Pierre Auger Observatory

*Eleonora* Guido[1,*] on behalf of the Pierre Auger Collaboration[2,**]

[1]Department für Physik, Universität Siegen, Walter-Flex-Str. 3, 57068 Siegen, Germany
[2]Observatorio Pierre Auger, Av. San Martín Norte 304, 5613 Malargüe, Argentina
Full author list: https://www.auger.org/archive/authors_2022_10.html

**Abstract.** In this work we present a combined fit of the energy spectrum and mass composition data above $6{\times}10^{17}$ eV as measured by the Pierre Auger Collaboration and we show that our data can be reproduced by the superposition of two (or more) components, either Galactic or extragalactic. We use a simple astrophysical model with two extragalactic components, one dominant at high energy and the other at low energy, whose superposition generates the so-called "ankle" feature. We find that the component dominating the low (high) energy region is required to be emitted from the sources with a very soft (hard) spectrum. In our best fit scenario the sources contributing above the ankle emit a mixed and increasingly heavier mass composition, whereas a mix of protons and intermediate-mass nuclei is observed below the ankle, which reach the Earth almost unaffected by the propagation in the intergalactic space. The possible presence of a Galactic population providing the intermediate-mass contribution at low energy is also explored.
In order to evaluate our capability to constrain astrophysical models, we finally discuss the impact on the fit results of the main experimental systematic uncertainties and of the assumptions on the quantities affecting propagation, both in the atmosphere and in the intergalactic medium.

## 1 Introduction

The existence of ultra-high-energy cosmic rays (UHECRs), the ones reaching Earth with energies above $\sim 10^{18}$ eV, was proven in the early 1960s, but their origin and the acceleration mechanism governing them are still unsettled. The large ground-based experiments built in the past few decades, like the Pierre Auger Observatory, are providing very high data quality and setting the basis for shedding light on such open questions. The energy spectrum of cosmic rays exhibits several features in the highest-energy region: a sharp hardening, known as the ankle, is observed at $\sim 10^{18.7}$ eV; a new feature, dubbed the instep, at $\sim 10^{19.1}$ eV, could be attributed to the interplay of light-to-intermediate nuclei [1]; finally, the flux suppression above $\sim 10^{19.7}$ eV may be due to energy losses during the propagation of UHECRs [2, 3], to the limited maximum energy the sources acceleration can reach, or possibly to a combination of the two effects. The mass composition [4, 5] is estimated by the distributions of depth of maximum development of the showers $X_{\max}$: below the ankle it is given by a mix of protons and medium-mass nuclei, whereas above the ankle it gets increasingly heavier and less mixed, suggesting a superposition of alternating groups of elements with a steep rigidity-dependent cutoff and a progressively heavier mass (a similar sequence, known as Peters cycle [6], is associated to the knee of the cosmic-ray spectrum around $10^{15.3}$ eV).

The quest for the origin of UHECRs is strongly related to the investigation of the energy region between the second knee and the ankle, where the transition from Galactic to extragalactic cosmic rays occurs. In the region around the ankle, the dominance of a light-to-intermediate Galactic contribution can be excluded by the low level of anisotropy in the distribution of arrival directions [7]. Consequently, the protons at and below the ankle must have an extragalactic origin and may originate either from the interactions of heavier nuclei in the source environment [8, 9] or from another population of sources; as concerns the medium-mass nuclei observed around the ankle, they should be provided by an additional component, which can be either Galactic or extragalactic. Above 8 EeV, the observation of a dipole pointing away from the Galactic centre suggest an extragalactic origin for UHECRs [10–12].

In this analysis [13] we aim to constrain the physical parameters related to the energy spectrum and the mass composition of particles escaping the environments of extragalactic sources (after acceleration and in-source propagation). We simultaneously fit a simple astrophysical model to both the energy spectrum and the mass composition data measured at the Pierre Auger Observatory, considering energies from $10^{17.8}$ eV. In a previous publication [14], a model consisting of one single population of

*guido@hep.physik.uni-siegen.de
**spokespersons@auger.org

extragalactic sources was fitted to the data above the ankle ($E > 10^{18.7}$ eV). Here, with the aim of interpreting the ankle feature as the superposition of different components, we assume the presence of one (or more) additional contribution(s) at low energies, either Galactic or extragalactic. Further details can be found in Ref. [13, 15].

## 2 The astrophysical model

In our model each extragalactic component originates from a population of identical sources uniformly distributed in the comoving volume, except for a local overdensity for distances smaller than $\sim 30$ Mpc, with a minimum source distance of 1 Mpc, which takes into account that the Milky Way belongs to a group of galaxies embedded on the Local Sheet. For each extragalactic population of sources the spectrum of particles escaping from the source environment is given by the superposition of the contributions of $n \leq 5$ representative nuclear species $A$, chosen among $^{1}$H, $^{4}$He, $^{14}$N, $^{28}$Si, $^{56}$Fe, ejected according to a power-law spectrum with a rigidity-dependent broken exponential cutoff. The generation rate $\tilde{Q}_A(E)$, defined as the number of nuclei with mass $A$ ejected per unit of energy, volume and time, is then given by:

$$\tilde{Q}_A(E) = \tilde{Q}_{0A} \left(\frac{E}{E_0}\right)^{-\gamma} \begin{cases} 1, & E \leq Z_A R_{\rm cut}; \\ \exp\left(1 - \frac{E}{Z_A R_{\rm cut}}\right), & E > Z_A R_{\rm cut}, \end{cases} \tag{1}$$

where $Z_A$ is the atomic number of each species $A$, and $\tilde{Q}_{0A}$ is the generation rate at a reference energy $E_0$, set to an arbitrary value lower than the energy cutoff of protons.

The free parameters for each component are the spectral index $\gamma$, the rigidity cutoff $R_{\rm cut}$, and $n$ partial normalisations $\tilde{Q}_{0A}$. To immediately get a more physically meaningful information about the mass composition at the sources, in our fit results the mass fractions are expressed in terms of fractions $I_A$ of the total source emissivity $\mathcal{L}_0$ of each population at redshift $z = 0$, defined as $\mathcal{L}_0 = \sum_A \int_{E_{\rm min}}^{\infty} E\, \tilde{Q}_A(E)\, {\rm d}E$:

$$I_A = \frac{\int_{E_{\rm min}}^{\infty} E\, \tilde{Q}_A(E)\, {\rm d}E}{\mathcal{L}_0} \tag{2}$$

starting from the fit energy threshold $E_{\rm min} = 10^{17.8}$ eV.

We also consider the possible presence of a Galactic component at Earth, which is modelled as a power law with $\gamma_{\rm Gal} \sim 3$ modified by a simple exponential cutoff. [1] We performed the fit by making different assumptions on its mass composition to find the one which best describes our data. The normalisation $J_0^{\rm Gal}$ at $E_0^{\rm Gal} = 10^{16.85}$ eV and the rigidity cutoff $R_{\rm cut}^{\rm Gal}$ are free parameters of the fit.

The energy spectrum and mass composition of the particles escaping from the environment of extragalactic sources are modified during their propagation in the intergalactic medium by the adiabatic energy losses and the interactions with background photons, from both the cosmic microwave background (CMB) and the extragalactic background light (EBL). We take into account these effects by using SimProp [16] simulations, where the uncertain quantities are treated with phenomenological models: we model photodisintegrations via the cross sections computed by Talys [17–19] with the settings described in Ref. [20], or the ones from the Puget, Stecker and Bredekamp (PSB) [21, 22] model, and the EBL with the Gilmore [23] or Domínguez [24] model. The observed energy spectrum $J_{\rm obs}(E')$ is thus obtained by integrating the contributions of all the sources weighted by the redshift and modified by the effects of interactions with radiation photons.

Since a direct measurement of the mass composition is not possible on an event-by-event basis, we use the distribution of $X_{\rm max}$ as an estimator of the mass distribution in each energy bin. Such a conversion depends on the choice of hadronic interaction model (HIM), which is thus another source of uncertainty. In this work, we use the HIMs Epos-LHC [25], QGSJet II-04 [26] and Sibyll 2.3d [27].

## 3 Fit procedure and reference results

### 3.1 The fit procedure

We use a measurement of the energy spectrum in $\log_{10}(E/{\rm eV})$ bins of 0.1 width from $10^{17.8}$ eV to $10^{20.2}$ eV, obtained with the data collected over 15 years with the Surface Detector Array of the Observatory [28]. As for the $X_{\rm max}$ distributions, measured by the fluorescence telescopes, we consider $\log_{10}(E/{\rm eV})$ bins of 0.1 from $10^{17.8}$ eV to $10^{19.6}$ eV and one additional larger bin containing all events with greater energies; each $X_{\rm max}$ distribution is binned in intervals of 20 g cm$^{-2}$ [29]. In the fit we minimise the deviance $D = -\ln(L/L_{\rm sat})$, a generalised $\chi^2$, where $L$ is the likelihood of our model and $L_{\rm sat}$ that of a model which perfectly describes the data. It consists of two terms, $D_J$ and $D_{X_{\rm max}}$. The first is for the energy spectrum and is a product of Gaussian distributions. The latter is a product of multinomial distributions used for the fit of the $X_{\rm max}$ distributions; they are modelled as Gumbel distribution functions [30], whose parameters depend on the hadronic interaction model.

The best-fit parameter values for each scenario are those with which the total deviance $D = D_J + D_{X_{\rm max}}$ attains its minimum value $D_{\rm min}$, which we locate using the Minuit package [31]; the statistical uncertainties on the spectral parameters correspond to the half extent of the 1D profile in the parameter space where $D \leq D_{\rm min} + 1$, as computed using the `MINOS` routine of Minuit; the uncertainties on the emissivity and on the mass fractions are computed with Monte Carlo simulations.

All the reference results are obtained by using Talys for the photodisintegration cross sections, the Gilmore model for the EBL spectrum and evolution, and Epos-LHC as the HIM. Other combinations of models will be discussed in Section 4.2. In order to focus on the simplest case, in this section we assume a flat cosmological evolution for the

[1] Since the spectral index and the cutoff energy are nearly degenerate with each other and this component is non-negligible only in a rather narrow energy range, the exact value for the slope does not affect the results.

| | Scenario 1 | | Scenario 2 | |
|---|---|---|---|---|
| Galactic contribution (at Earth) | pure N | | — | |
| $J_0^{\mathrm{Gal}}/(\mathrm{eV}^{-1}\,\mathrm{km}^{-2}\,\mathrm{sr}^{-1}\,\mathrm{yr}^{-1})$ * | $(1.06 \pm 0.04)\times10^{-13}$ | | — | |
| $\log_{10}(R_{\mathrm{cut}}^{\mathrm{Gal}}/\mathrm{V})$ | $17.48 \pm 0.02$ | | — | |
| EG components (at the escape) | LE | HE | LE | HE |
| $\mathcal{L}_0/(10^{44}\,\mathrm{erg\,Mpc}^{-3}\,\mathrm{yr}^{-1})$ ** | $6.54 \pm 0.36$ | $5.00 \pm 0.35$ | $11.35 \pm 0.15$ | $5.07 \pm 0.06$ |
| $\gamma$ | $3.34 \pm 0.07$ | $-1.47 \pm 0.13$ | $3.52 \pm 0.03$ | $-1.99 \pm 0.11$ |
| $\log_{10}(R_{\mathrm{cut}}/\mathrm{V})$ | $>19.3$ | $18.19 \pm 0.02$ | $>19.4$ | $18.15 \pm 0.01$ |
| $I_{\mathrm{H}}$ (%) | 100 (fixed) | $0.0 \pm 0.0$ | $48.7 \pm 0.3$ | $0.0 \pm 0.0$ |
| $I_{\mathrm{He}}$ (%) | — | $24.5 \pm 3.0$ | $7.3 \pm 0.4$ | $23.6 \pm 1.6$ |
| $I_{\mathrm{N}}$ (%) | — | $68.1 \pm 5.0$ | $44.0 \pm 0.4$ | $72.1 \pm 3.3$ |
| $I_{\mathrm{Si}}$ (%) | — | $4.9 \pm 3.9$ | $0.0 \pm 0.0$ | $1.3 \pm 1.3$ |
| $I_{\mathrm{Fe}}$ (%) | — | $2.5 \pm 0.2$ | $0.0 \pm 0.0$ | $3.1 \pm 1.3$ |
| $D_J$ ($N_J$) | 48.6 (24) | | 56.6 (24) | |
| $D_{X_{\max}}$ ($N_{X_{\max}}$) | 537.4 (329) | | 516.5 (329) | |
| $D$ ($N$) | 586.0 (353) | | 573.1 (353) | |

* defined at $E_0^{\mathrm{Gal}} = 10^{16.85}$ eV.
** from $E_{\min} = 10^{17.8}$ eV.

**Table 1.** Best-fit parameters obtained in the two reference scenarios. Scenario 1: a Galactic contribution of pure nitrogen, a low-energy extragalactic component of pure protons (LE), and a high-energy extragalactic component with a mixed mass composition (HE). Scenario 2: two mixed extragalactic components (LE and HE) overlapping in the ankle energy region.

extragalactic sources, whereas the effect of other choices of source evolution are investigated in Section 5.

### 3.2 The reference results

We consider two possible scenarios in the reference configuration. The best-fit parameter values thus obtained are listed in Table 1.

In Scenario 1 we assume an extragalactic population with a mixed mass composition dominating at high energies ("HE"), plus an additional extragalactic population of pure protons dominating at low energies ("LE"). The LE extragalactic component could e.g. arise from the photo-disintegration of HE cosmic rays in the environment of their sources [9]. The heavier nuclei below the ankle are instead assumed to originate from a Galactic population. We found that a Galactic component at Earth of pure nitrogen, extending up to a relatively high energy $Z\,R_{\mathrm{cut}}^{\mathrm{Gal}} \approx 2\times10^{18}$ eV, provides the best fit to the data, whereas heavier compositions with no nitrogen result in deviances $D \gtrsim 1000$. Hence, here we only show the results obtained in the case of a composition of pure nitrogen.

Alternatively, in Scenario 2 we describe our data by assuming that the ankle feature is due to the superposition of two extragalactic components, one dominating at LE and the other at HE. We assume that the two components are both ejected according to energy spectra described by Eq. (1) but with different parameter values, since they are reasonably associated to two different populations of sources. We are here implicitly assuming that a possible Galactic contribution is subdominant in the considered energy range.

The generation rate at the sources estimated in Scenario 1 and in Scenario 2 is shown for each component and each ejected nuclear species in the left panel of Fig. 1 and Fig. 2, respectively. In the right panel of the same figures we report also the corresponding contributions at the top of the atmosphere grouped according to mass number. For Scenario 1 the first two moments of the $X_{\max}$ distributions along with their best-fit function are shown in Fig. 3; very similar results are obtained in Scenario 2.

A common finding between the two proposed scenarios is that the HE component requires very hard spectra with low rigidity cutoffs and intermediate mass compositions, while the LE component requires much steeper spectra.

In order to reproduce the very pronounced spectral features of the measured energy spectrum and the rather narrow $X_{\max}$ distributions, a negative spectral index is estimated for the HE component. However, it is important to stress that the UHECR spectrum escaping the source environment may differ from the accelerated one due to energy-dependent effects of interactions and diffusion in the source environments. Besides, also the effect of intergalactic magnetic field, here neglected, may cause a suppression of the flux at low energies [32], making the observed spectrum harder than the actual one escaping from the sources.

As concerns the LE component, if for example the assumption of identical sources is relaxed and different maximal energies are taken into account, the effective en-

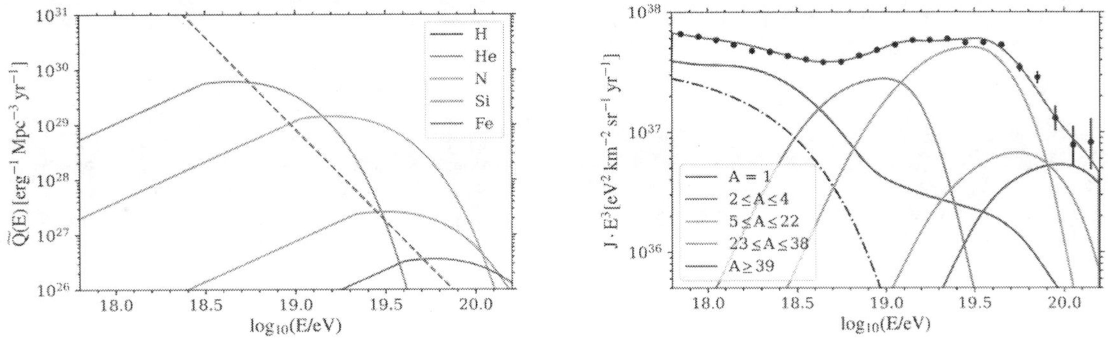


**Figure 1.** Scenario 1. *Left:* The generation rate at the extragalactic sources for each representative mass; the LE and HE contributions are shown as dashed and solid lines, respectively. *Right:* The corresponding Galactic (dot-dashed line) and extragalactic (grouped according to mass number) contributions to the energy spectrum at the top of atmosphere.

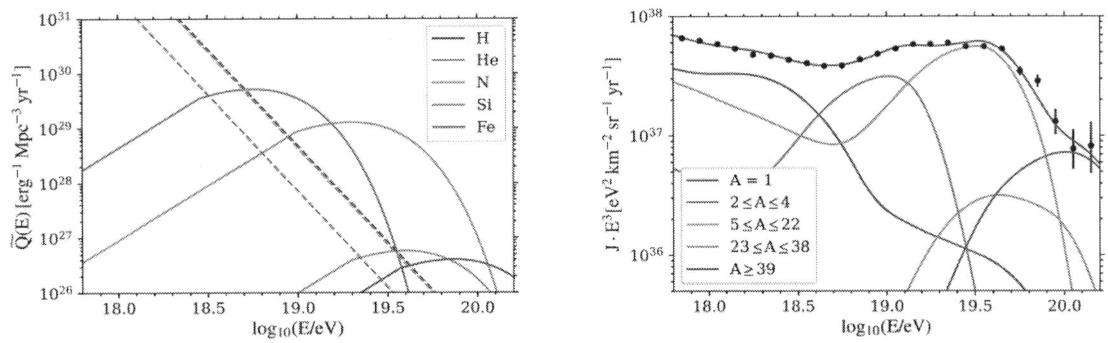


**Figure 2.** Scenario 2. *Left:* The generation rate at the extragalactic sources for each representative mass; the LE and HE contributions are shown as dashed and solid lines, respectively. *Right:* The corresponding extragalactic (grouped according to mass number) contributions to the energy spectrum at the top of atmosphere.

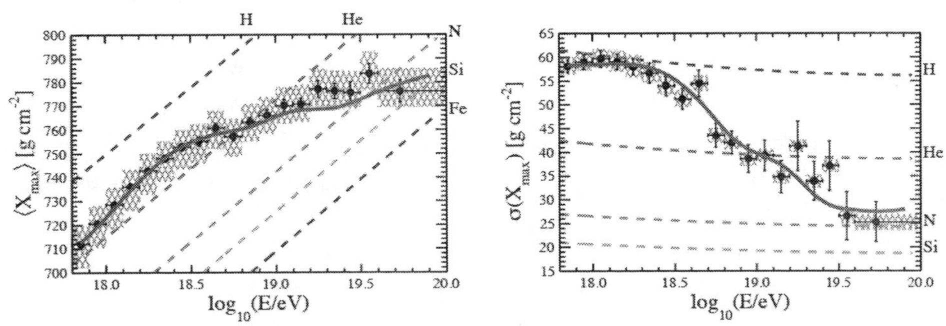


**Figure 3.** First two moments of the $X_{max}$ distributions as predicted by the best-fit results in Scenario 1, along with the measured values and the predictions for pure compositions of various nuclear species according to Epos-LHC (dashed lines).

ergy spectrum obtained by integrating over them would be steeper than the one of each individual source, as demonstrated in Ref. [33], and this could be a possible explanation for the very soft estimated energy spectrum.

In terms of mass composition at LE, we find that the data can be described by a mix of nitrogen and hydrogen in both scenarios, their relative contributions respectively decreasing and increasing with energy, which confirms what found in Ref. [34] about the needed mixture at the ankle. In the case of Scenario 1, the dominance of Galactic iron in the region below the ankle is disfavoured. Hence, models proposing a medium-mass contribution e.g. from explosions in the winds of Wolf-Rayet-like stars [35] would better describe our data than models foreseeing a dominance of Galactic iron in the region below the ankle, like the one originally proposed by Hillas [36].

The HE rigidity cutoff found as a result of the fit suggests that the maximum energy emitted at the sources is not high enough to entirely attribute the suppression at the highest energies to propagation effects. As concerns the LE component, the fit is degenerate with respect to $R_{\rm cut}^{\rm LE}$ for values $\gg 10^{19.5}$ V, thus fixing this parameter to any arbitrarily higher value provides the same best-fit results. Such a degeneracy can be explained by the fact that the estimated energy spectrum of this component is very steep, and hence it is rapidly suppressed even in the absence of an exponential cutoff, making the energy range where this component is the dominant one rather narrow and the fit insensitive to the details of its shape.

The two scenarios provide very similar results in terms of deviance values. The scenario with two extragalactic mixed components provides a better fit of the $X_{\rm max}$ measurements but a worse description of the very pronounced features in the energy spectrum. One way in which a Galactic and an extragalactic below-ankle medium-mass composition would differ is in their distribution of arrival directions, which are not considered in this work, hence firm conclusions about a favoured scenario could be drawn only after a further investigation of the Galactic-to-extragalactic transition region.

## 4 Effect of the systematic uncertainties

The effect of the systematic uncertainties on the fit results is here shown only for Scenario 2, since the two scenarios are nearly equivalent and the effects of different assumptions about the distribution and evolution of extragalactic sources and the propagation in intergalactic space are expected to be less noticeable in Scenario 1.

### 4.1 Experimental uncertainties

The energy scale and the $X_{\rm max}$ scale are the most important sources of experimental systematic uncertainties. For the energy scale, an energy independent uncertainty $\Delta E/E = 14\%$ is adopted in the whole considered energy region [28]. As concerns the systematic uncertainties on the measured $X_{\rm max}$ values, they are asymmetric and slightly energy-dependent, ranging from 6 to 9 g/cm$^2$ [37].

Regarding the energy scale uncertainty, we follow the same approach used in Ref. [14] and we shift all the measured energies by one systematic standard deviation in each direction. As concerns the $X_{\rm max}$ scale uncertainty, since the correlations between the lowest and the highest energy bins can go down to $\sim 0.6$, we add two nuisance parameters, $a$ and $b$, to the fit to allow different shifts in $X_{\rm max}$ at different energies. The first two eigenvectors of the covariance matrix define two functions of energy, $v_1(E_i)$ and $v_2(E_i)$; all the $X_{\rm max}$ distributions are shifted by a quantity $a\,v_1(E) + b\,v_2(E)$, and an additional term given by $a^2 + b^2$ is added to the deviance. The parameter $a$ shifts all the $X_{\rm max}$ distributions in the same direction by an energy-dependent amount, whereas $b$ has an opposite effect on the high-energy and the low-energy distributions.

The variations on the predicted fluxes at Earth are shown in the left panel of Fig. 4. The rather large uncertainty on the predicted total fluxes (brown band) is mainly due to the $\pm 14\%$ shifts in the energy scale, which significantly affects only the estimated source emissivities, whereas the $X_{\rm max}$ uncertainties have a stroger impact on the predictions at Earth and on the deviance value .

### 4.2 Uncertainties from models

We explored effect of the propagation models on the fit results by repeating the fit considering different combinations of them from those used in the reference configuration. As regards the photodisintegration, we tested the PSB model, that neglects photodisintegration channels in which alpha particles rather than single nucleons are ejected. Besides, as concerns the EBL spectrum and evolution, we tested also the Domínguez model, which has a higher spectral energy density in the far infrared than the Gilmore one. Regarding the HIM, we verified that QGSJet II-04 cannot properly describe our data ($D \gtrsim 1000$ in all cases), and is thus excluded from this analysis. Instead of fixing a single HIM, we allow for the possibility to describe our data with an intermediate model between Epos-LHC and Sibyll 2.3d by introducing an additional nuisance parameter $\delta_{\rm HIM}$, limited between 0 and 1. In this way each HIM-dependent Gumbel parameter is interpolated as alpha as $\alpha_{\rm HIM} = \delta_{\rm HIM}\,\alpha_{\rm Epos\text{-}LHC} + (1 - \delta_{\rm HIM})\,\alpha_{\rm Sibyll\,2.3d}$, so that $\delta_{\rm HIM} = 0$ corresponds to "pure" Sibyll 2.3d and $\delta_{\rm HIM} = 1$ to "pure" Epos-LHC. The effect of such variations on the predicted fluxes at Earth is shown in the right panel of Fig. 4.

Regardless of the propagation models configuration, our data appear to be better described by pure Epos-LHC or by intermediate models much closer to Epos-LHC than to Sibyll 2.3d: the HIM choice is then the dominant uncertainty among the ones from models in terms of predictions at Earth.

As concerns the propagation models effects, even if the impact on the deviance and on the predicted fluxes at the Earth is smaller, some changes in the best fit parameters at the sources are observed, which are in agreement with what is expected to compensate the differences in the propagation to produce similar fluxes at the Earth. Further details and the parameters values can be found in Ref. [13].

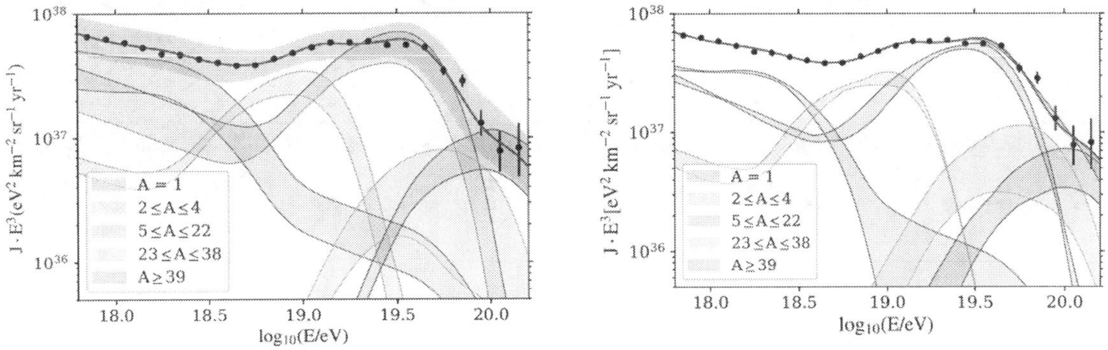


**Figure 4.** The combined effect of the experimental uncertainties (*left*) and of the uncertainties from models (*right*) on the all-particle energy spectrum and on the partial contributions of the different mass groups at Earth.

However, it is worth stressing that the impact of changing the propagation models on the deviance and on the predicted fluxes at Earth is encompassed by the effect of the experimental systematic uncertainties.

## 5 Effect of the cosmological evolution of sources

We repeated the fit considering, for each population of sources, three different models for the cosmological evolution of the source emissivity, parameterised as $\propto (1+z)^m$, namely $m = +5, +3$ and $-3$, in addition to the no-evolution ($m = 0$) case considered so far. Each one of these choices can be associated to different classes of sources [13].

In this analysis we consider an evolution $(1+z)^m$ with constant $m$ in the entire considered redshift range (up to $z = 10$), but we have verified that other possibilities for the behaviour of the evolution at $z > 1$ have only a small impact on the LE component, not affecting our main conclusions, and a completely negligible effect on the HE component. This can be explained by the fact that practically all nuclei reaching us with energies $\geq 10^{17.8}$ eV originate from $z \lesssim 3$, and in particular those with $E \gtrsim 10^{18.4}$ eV from $z \lesssim 1$.

Since the LE and HE populations might be accelerated in different classes of sources, they could have different source evolutions. Hence, we consider all sixteen possible pairs of evolutions among $m \in \{-3, 0, +3, +5\}$. Our results in terms of total deviance are summarised in Fig. 5.

A positive (negative) evolution means that particles were on average accelerated longer ago (more recently) than in the no-evolution scenario, and hence had the time to undergo more (fewer) interactions in intergalactic space. The effects are more noticeable for the HE population, as interactions are more frequent at high energies. In the case of a strong positive ($m = 5$) evolution of the HE component, corresponding to the last column in Fig. 5, its secondary flux at ankle energies exceeds the observed all-particle spectrum, so that no good fit of the data is possible ($D \sim 1000$), regardless of the source evolution of the LE

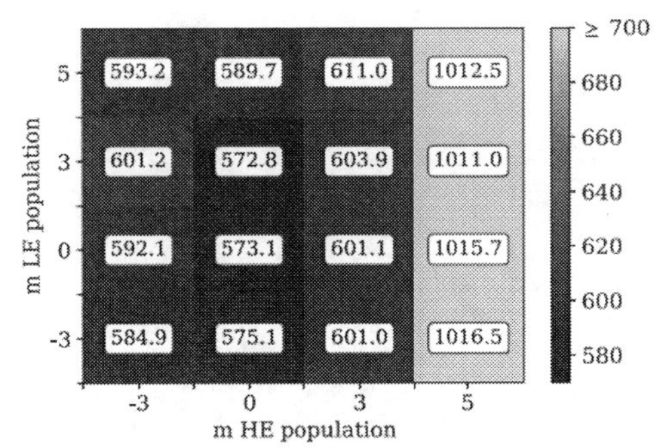


**Figure 5.** The total deviance is colour-coded as in the right bar and shown for all the possible combinations of source evolution of the two populations. The values of $m$ are shown on the axes.

component. On the other hand, the scenarios with no evolution for the HE population appear to be favoured overall, though acceptable fits can also be found with a weak evolution ($m = \pm 3$).

Additional constraints on the cosmological evolution of sources may be obtained by considering the flux of the expected cosmogenic neutrinos associated to the best-fit results of each chosen source evolution scenario. In fact, cosmogenic neutrinos do not undergo any interactions during their propagation, except for adiabatic energy losses due to the expansion of the Universe and flavour oscillations, so they can reach us even from very high redshifts, providing complementary information to that available from UHECR measurements.

In general, the contribution of the HE population to the flux of expected neutrinos is negligible, regardless of its cosmological evolution: due to its rather low rigidity cutoff, the pion photoproduction interactions cannot occur on CMB photons, but only on the EBL ones, which contributes to the neutrino flux to a lesser extent because of the much greater interaction length. As a consequence,

 https://doi.org/10.1051/epjconf/202328302013

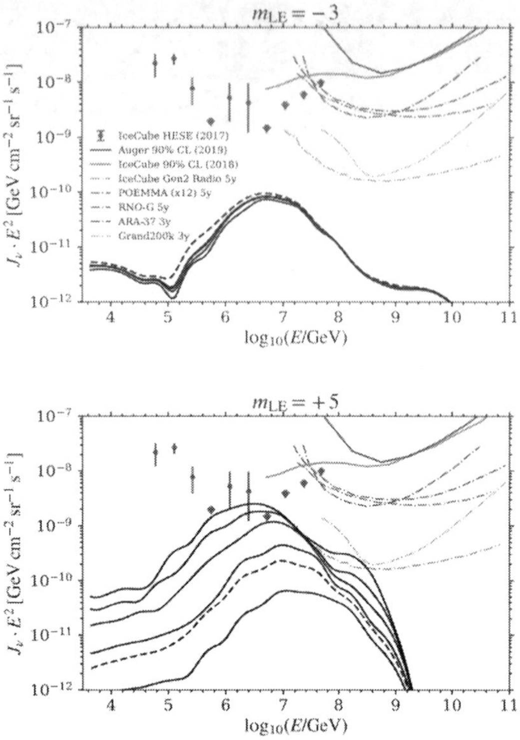


**Figure 6.** The predicted fluxes of neutrinos (single flavour) corresponding to the best fit results obtained by assuming a source evolution with $m = -3$ (top) and $m = 5$ (bottom) for the LE component; the HE component has no source evolution ($m = 0$). The observed IceCube HESE flux, the current upper limits from IceCube and Auger (solid lines), and the predicted sensitivities of future detectors (dot-dashed lines) are also shown for reference.

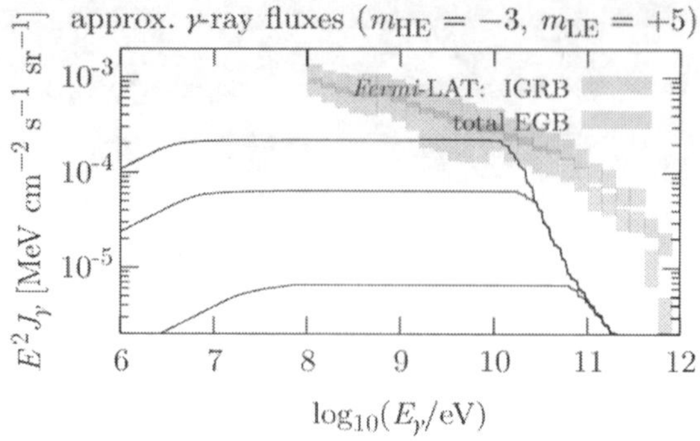


**Figure 7.** Fluxes at Earth of gamma-ray cascades produced in UHECR propagation for $m_{HE} = -3$ and $m_{LE} = 5$ when assuming $z_{max} = 1$, 3 and 5, compared to *Fermi*-LAT measurements from Ref. [40].

the expected neutrino fluxes are entirely provided by the LE population of sources. In the case of a flat or negative source evolution for the LE component, the expected neutrino fluxes are well below the current observations and the future detectors sensitivity; the case with $m = -3$ is shown in the top panel of Fig. 6. On the other hand, the largest fluxes of neutrinos are predicted when assuming a strong positive source evolution for the LE component, as shown in the bottom panel of Fig. 6. In the plots of Fig. 6 the black solid curves represent the fluxes corresponding (from the bottom to the top) to $z_{max} = 1, 2, 3, 4$ and 5, assuming a constant $m$ value in the whole redshift range; the dashed black curve shows the expected fluxes corresponding to $z_{max} = 3$ with a flat source evolution above $z = 1$. The maximum rigidity of the LE component and its uncertainties have also a strong impact on the neutrino flux and in particular on the peak at around $10^9$ GeV, corresponding to the UHECR interactions with CMB photons.

It is worth noting that future neutrino detectors will provide an improved sensitivity to cosmogenic neutrinos at energies above $10^8$ GeV. As shown in Fig. 6, our predictions in the cases of positive source evolutions would be constrained by the most stringent future limits, provided by the next-generation detector upgrade of IceCube [38] and by planned detectors. We can conclude that in the near future some additional constraints on the source properties may be set, for example excluding some source evolutions for the LE component and/or limiting the possible values of its rigidity cutoff.

Finally, another messenger of potential interest is the flux of gamma-ray cascades originated from photo-hadronic interactions of UHECRs in intergalactic space, which develop until all the secondaries have $E \lesssim 100\,\mathrm{GeV}$. A rough estimate of the resulting gamma-ray fluxes at $E \lesssim 100\,\mathrm{GeV}$ can be obtained by applying the analytical approximation from Ref. [39] to electrons and photons produced in SimProp simulations. In this case, both the LE and the HE component have a non-negligible contribution, as the Lorentz factor threshold for electron–positron pair production by UHECRs on CMB photons is two orders of magnitude lower than for pion production. The results are compared to *Fermi*-LAT measurements [40] and the only scenarios in tension with the data would be the ones where the HE component has a strong positive evolution and $z_{max} \gtrsim 2$, or the LE component has a strong positive evolution and $z_{max} \gtrsim 4$, as shown for example in Fig. 7. However, as shown in Fig. 5 the former are already excluded by the deviance of our combined fit, and as shown in Fig. 6 the latter may also result in amounts of cosmogenic neutrinos within the reach of future planned detectors.

## 6 Conclusions

In this analysis we have performed a fit of the energy spectrum and mass composition data across the ankle as measured at the Pierre Auger Observatory. We considered the hypothesis of two extragalactic components, in presence or not of a secondary Galactic contribution, and we showed that they reasonably succeed to reproduce the ankle feature, whose sharpness would be hardly described by other scenarios. As concerns the newly observed feature at $\sim 10^{19}$ eV (the 'instep'), it is here given by the interplay of light-to-intermediate nuclei.

Despite the fact that a definite conclusion on the presence of a subdominant Galactic flux cannot be reached, our results show that its end is compatible with the data only if it is composed by medium-mass nuclei. An extension of the combined fit to even lower energies will be more effective to investigate the transition from Galactic to extragalactic cosmic rays, increasing the lever-arm with the use of composition data from HEAT (High Elevation Auger Telescopes). Composition results below $10^{17.8}$ eV have been already reported in a preliminary analysis [4]. An update of the $X_{\max}$ analysis in the whole energy range is currently in progress and its results are expected to push remarkably the sensitivity of the combined fit studies in the transition region.

A central part of this analysis is the detailed treatment of the systematic uncertainties and the investigation of their effect on the fit results. The possible systematic uncertainties from both experimental and model sources do not spoil our main conclusions, even though they have a non-negligible impact on the deviance value and they affect the fit parameters.

Thanks to the inclusion in our fit of data from the below-ankle region, some constraints on the cosmological evolution of the two populations can be set: very strong evolutions would cause a flux of secondary particles at the ankle exceeding the observed spectrum, even in presence of a negligible contribution from the LE component in that region, and are thus excluded by our fit. Besides, for some of the considered scenarios the predictions of cosmogenic neutrino fluxes might reach the sensitivity range of the next-generation detectors and could be thus excluded in the near future.

Further insight on the possible sources of UHECRs can be gained by extending the combined fit to include the arrival directions information to the spectrum and composition data. The results of a preliminary analysis were shown in Ref. [41].

In this analysis, the mass composition data do not extend to energies where the suppression occurs, because of the limited duty cycle of the FD. The interpretation of the suppression in the flux as due to propagation effects or to the maximum energy reached in the sources can provide some constraints on the sources of UHECRs and their properties. In the near future, mass composition estimates will be obtained through $X_{\max}$ and the muon content of showers by using machine learning techniques on SD data [42, 43]. Furthermore, the Pierre Auger Observatory is currently undergoing an upgrade, AugerPrime [44, 45], that includes the deployment of scintillators on top of the SD stations to better distinguish the muonic and electromagnetic content of the showers. This will improve the measurement of the mass composition at the highest energies, allowing us to investigate the possible presence of a sub-dominant light contribution and extend up to the suppression region to perform an analysis similar to the one presented here with much larger statistics.

## References

[1] Pierre Auger Collaboration, *Phys. Rev. Lett.* **125** (2020) 121106.
[2] K. Greisen, *Phys. Rev. Lett.* **16** (1966) 748.
[3] G. T. Zatsepin, V. A. Kuzmin, *JETP Lett.* **4** (1966) 78.
[4] J. Bellido for the Pierre Auger Collaboration, *PoS (ICRC2017)*, **301** (2017) 506.
[5] Pierre Auger Collaboration, *Phys. Rev.* **D 90** (2014) 122006.
[6] B. Peters, *Il Nuovo Cimento* **22** (1961) 800.
[7] Pierre Auger Collaboration, *Astrophys. J.* **762** (2013) L13.
[8] N. Globus, D. Allard and E. Parizot, *Phys. Rev.* **D 92** (2015) 021302.
[9] M. Unger, G . R. Farrar and L. A. Anchordoqui, *Phys. Rev.* **D 92** (2015) 123001.
[10] Pierre Auger Collaboration, *Science* **357** (2017) 1266.
[11] Pierre Auger Collaboration, *Astrophys. J.* **891** (2020) 142.
[12] R. de Almeida for the Pierre Auger Collaboration, *PoS(ICRC2021)* **395** (2021) 335.
[13] Pierre Auger Collaboration, submitted to *JCAP* (2022), [arXiv:2211.02857v1].
[14] Pierre Auger Collaboration, *JCAP* **04** (2017) 038.
[15] E. Guido for the Pierre Auger Collaboration, *PoS(ICRC2021)* **395** (2021) 311.
[16] R. Aloisio et al., *JCAP* **11** (2017) 009.
[17] A. J. Koning, S. Hilaire and M. C. Duijvestijn, *American Institute of Physics Conference Series*, (2005) 1154.
[18] A. J. Koning and D. Rochman, *Nuclear Data Sheets*, **13** (2012) 2841.
[19] A. J. Koning, S. Hilaire and S. Goriely, *TALYS User Manual*.
[20] R. Alves Batista et al., *JCAP* **10** (2015) 063.
[21] J.L. Puget, F.W. Stecker and J.H. Bredekamp, *Astrophys. J.* **205** (1976) 638.
[22] F.W. Stecker and M.H. Salamon, *Astrophys. J.* **512** (1999) 521.
[23] R. Gilmore, R. Somerville, J. Primack and A. Domínguez, *Mon. Not. Roy. Astron. Soc.* **422** (2012) 3189.
[24] A. Domínguez et al., *Mon. Not. Roy. Astron. Soc.* **410** (2011) 2556.
[25] T. Pierog et al., *Phys. Rev.* **C 92** (2015) 034906.
[26] S. Ostapchenko, *Phys. Rev.* **D 83** (2011) 014018.
[27] F. Riehn et al., *Phys. Rev.* **D 102** (2020) 063002.
[28] Pierre Auger Collaboration, *Eur. Phys. J. C* **81** (2021) 966.
[29] A. Yushkov for the Pierre Auger Collaboration, *PoS(ICRC2019)*, **358** (2019) 482.
[30] M. De Domenico et al., *JCAP* **07** (2013) 050.
[31] F. James and M. Roos, *Comput. Phys. Commun.* **10** (1975) 343.
[32] S. Mollerach and E. Roulet, *Phys. Rev.* **D 101** (2020) 103024.
[33] M. Kachelrieß and D.V. Semikoz, *Phys. Lett.* **B 634** (2006) 143.
[34] Pierre Auger Collaboration, *Phys. Lett.* **B 762** (2016) 288.
[35] S. Thoudam et al., *A&A* **595** (2016) A33.
[36] M. Hillas, *J. Phys. G: Nucl. Part. Phys.* **31** (2005) R95.
[37] Pierre Auger Collaboration, *Phys. Rev.* **D 90** (2014) 122005.
[38] IceCube Collaboration, *Astro2020 decadal survey* (2019) [arXiv:1911.02561 [astro-ph.HE]].
[39] V. Berezinsky and O. Kalashev, *Phys. Rev. D* **94** (2016) 023007.
[40] *Fermi*-LAT Collaboration, M. Ackermann et al., *Astrophys. J.* **799** (2015) 86.
[41] T. Bister for the Pierre Auger Collaboration, *PoS(ICRC2021)*, **395** (2021) 368.
[42] Pierre Auger Collaboration, *J. Instrum.* **16** (2021) P07019.
[43] Pierre Auger Collaboration, *J. Instrum.* **16** (2021) P07016.
[44] Pierre Auger Collaboration, [arXiv:1604.03637].
[45] A. Castellina for the Pierre Auger Collaboration, *EPJ Web Conf.* **210** (2019) 06002.

 https://doi.org/10.1051/epjconf/202328302014

# Model-independent Evidence for an Increase in the Mean Mass of Cosmic Rays above 3 EeV

A A Watson[1]

[1] School of Physics and Astronomy, University of Leeds, Leeds, UK

**Abstract.** Measurements of the Elongation Rate of the depth of shower maximum above 1 EeV are reviewed. There is evidence, from four independent estimates of this rate, made in the two hemispheres using three different techniques, for a decrease in the Elongation Rate above ~3 EeV, as first discovered by the Pierre Auger Collaboration over 15 year ago. Unless there is a dramatic change in the hadronic physics above this energy, the mean mass of the primary cosmic rays must increase as a function of energy, well into the decade beyond 10 EeV. To estimate the mass, the use of hadronic models is required, the accuracy of which remains uncertain. However, the possibility of a dramatic change in the hadronic physics appears unlikely, and would be inconsistent with data from the Auger Collaboration on the mass composition in the range 3 to 10 EeV, and on the anisotropy of arrival directions above 8 EeV. Both of these conclusions are insensitive to uncertainties in the shower models. Some remarks are made about the belief, still firmly held by some, that the highest-energy cosmic rays are dominantly protons.

## 1 Introduction

The talk, on which this paper is based, was prepared in the hope that it would re-activate discussion of the topic of the Elongation Rate at high energies within the Auger/Telescope Array (TA) Working Group on Mass Composition. In fact, during UHECR2022, Bergman [1] presented evidence showing that measurements of the depth of shower maximum by the TA group are now considered as being consistent with the mixed mass observed by the Auger Collaboration. Thus, significant progress has been made, and, although the conclusions are restricted to energies below ~10 EeV, and apply only for some hadronic models, a major step forward appears to have been made: the joint publication promised by the Working Group is therefore eagerly awaited.

Some of what follows is a replay of arguments set out in [2] in which the claim made by Sokolsky and D'Avignon [3] that the Elongation Rates measured in the Northern and Southern Hemispheres are different was refuted. Additionally, the likelihood of a change in the hadronic physics, and the evidence against it are briefly discussed, and the long-standing belief of some that protons dominate at the highest energies is examined.

The TA Collaboration has not yet made comparisons of their $X_{max}$ results with model predictions beyond 10 EeV as they hold the view that ~300 events above 10 EeV are too few to make reliable deductions [4]. It was argued in [2] that *only* TA data from the Black Rock and Long Ridge detectors should be used in such studies, and this is what has now been adopted in [1, 4]. In this TA data set, there are 3330 events above 1 EeV: the measurements [4] are shown in figure 1, alongside those from a recent report from the Auger Collaboration [5].

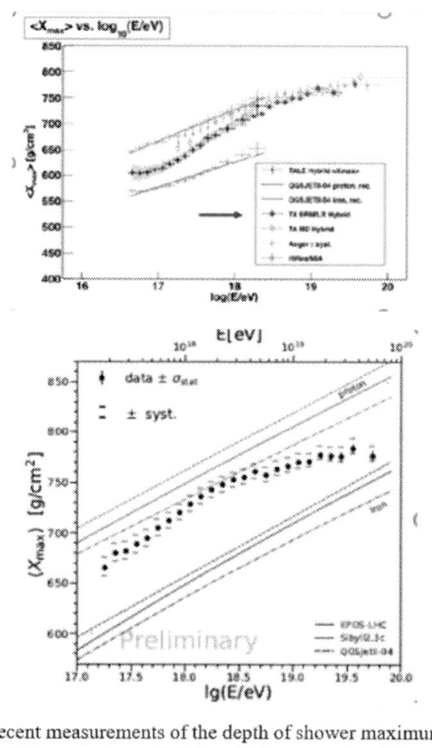


**Fig.1**. Recent measurements of the depth of shower maximum from the Telescope Array Collaboration (upper plot) [4], and from the Auger Collaboration (lower plot figure) [5]. Note that different models have been used by the Collaborations for the comparisons of data with models. The QGSJet II-04 model gives a poor description of some features of data from the Auger Observatory. For the Auger data, the Elongation Rate above 3 EeV is $26 \pm 2.5$ g cm$^{-2}$/decade.

Conclusions about primary mass drawn from plots such as those of figure 1 are dependent on the hadronic model assumed. This is a long-standing, and continuing, problem, as illustrated in figure 2 where data from the Fly's Eye detector are compared with the predictions from different models. The upper-plot is what was shown in [6], nearly 30 years ago, when the QCD Pomeron model was 'state-of-the-art'. This model was used to draw the conclusion that a light component, usually assumed to be protons, was becoming dominant at the highest energies.

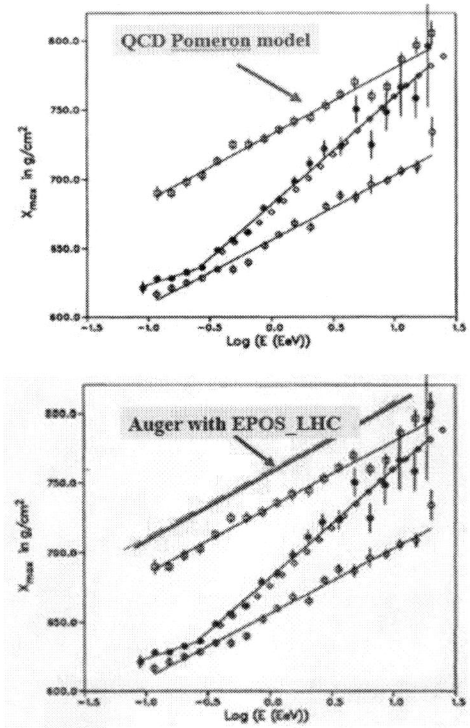


**Fig.2.** Data from [6] are shown in both plots. The different conclusions that can be drawn according to which model is adopted are evident (see discussion in text).

This plot has, I believe, had profound influence on the thinking of many people, particularly phenomenologists. In the lower plot, the same data are compared with the predictions of a current hadronic model (EPOS-LHC) from which larger values $X_{max}$ are found. These predictions shown were made with the selection procedures used to derive values of $X_{max}$ from the data of the Auger Observatory, and would differ by a few g cm$^{-2}$ from what is shown were the Fly's Eye selection procedures adopted instead. However, the important point is that the conclusion would be the same: the mass derived is strongly dependent on the hadronic model selected, and, above ~0.1 EeV, deductions from LHC data require extrapolation.

In [2], it was pointed out that the break in the Elongation Rate deduced from the *ensemble* of data amassed from projects carried out in Utah over the years [3], had been discussed previously [7]. The break is also seen by the Yakutsk group, using measurements of the lateral distribution air-Cherenkov radiation, and in measurements using the risetimes of the signals in the water-Cherenkov tanks of the Auger Observatory. These three results are summarised in figure 3. A detailed discussion of these plots can be found in [2].

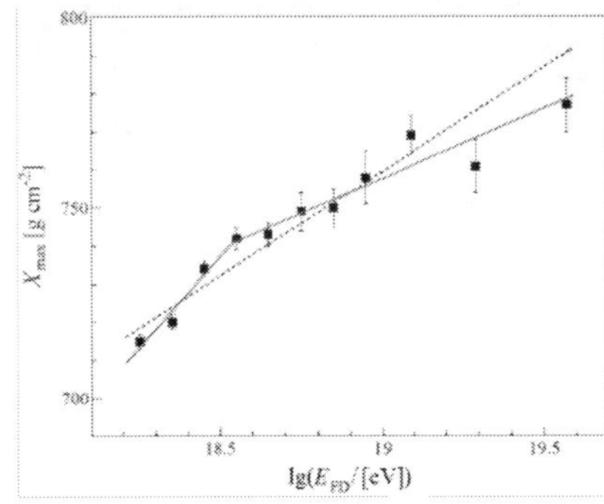


**Fig.3a.** $X_{max}$ vs Energy from the Black Rock/Long Ridge Observations by the Telescope Array Collaboration. The dashed line is a fit to the points with a single straight line (Elongation Rate = 54 ± 3 g cm$^{-2}$/decade). The full line is a fit with a break and the best-fit straight lines found simultaneously. The Elongation Rate above the break at 3 EeV is 37 ± 5 g cm$^{-2}$/decade. See [2] for further discussion.

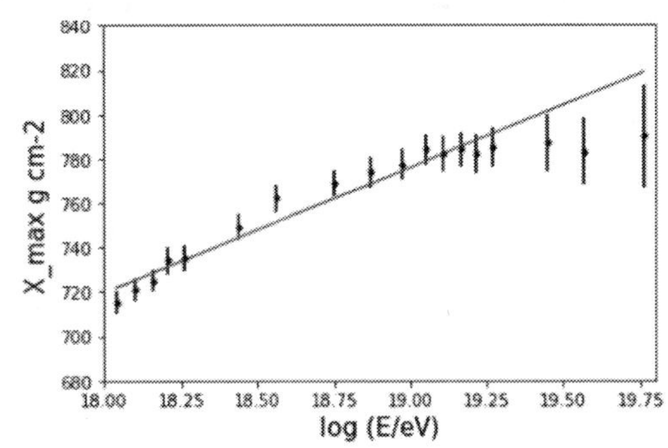


**Fig.3b.** Data of $X_{max}$ vs log E/eV from Yakutsk. The Elongation Rate above 3 EeV is 28 ± 7 g cm$^{-2}$/decade. See [2] for details.

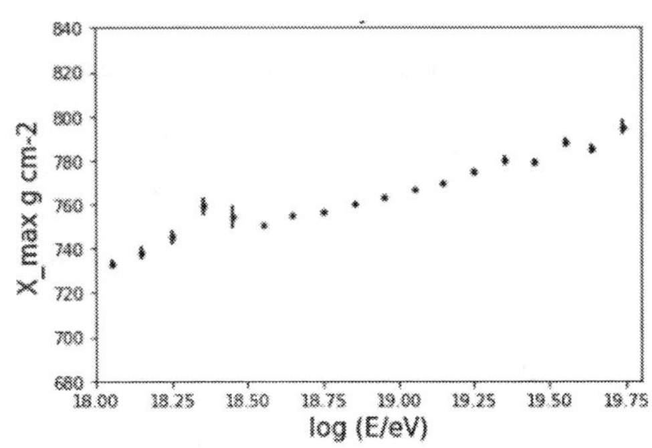


**Fig.3c.** Estimates of $X_{max}$ as a function of energy made at the Auger Observatory using the risetime technique. The Elongation Rate above 3 EeV is 33.2 ± 0.7 g cm$^{-2}$/decade. Further details are given in [2].

 https://doi.org/10.1051/epjconf/202328302014

The weighted averages of the Elongation Rates (ER2) above 3 EeV for the two hemispheres are:

Northern hemisphere:

TA and Yakutsk: ER2 = 34 ± 4 g cm$^{-2}$/decade

Southern hemisphere:

Two independent measurements from the Auger Observatory: ER2 = 32.7 ± 0.7 g cm$^{-2}$/decade.

It is evident that the claim advanced in [3] that the Elongation Rates are different in the two hemispheres is not supported.

It also follows that the mean mass of the primary particles increases as the cosmic-ray energy rises, unless there is a dramatic change in features of the hadronic interactions above the centre-of-mass energies reached at the LHC (equivalent to ~0.1 EeV for p-p collisions in cosmic rays). Exactly what mean mass is deduced depends strongly on the choice of models (see figures 1 and 2). While the conclusion drawn here is certainly not original (see for example [8]), it is worth restating, as it does not depend on assumptions about hadronic physics. Also, there remain many who do not accept this conclusion.

The situation with regard to the best model to adopt remains fluid. For example, at this meeting, we learned from Pierog [9] that it may be necessary to include the production of quark-gluon plasmas in the calculations. Of course, a dramatic rise in the p-p cross-section and multiplicity of pions resulting from collisions can also be invoked to enable a proton to mimic a heavy nucleus.

## 2 The possibility of changes in the hadronic physics

The idea that features observed in cosmic-ray data have their origins in changes in hadronic physics, rather than providing evidence of an astrophysical feature, has a long history. In 1958, Kulikov and Khristiansen [10] reported a break in the number spectrum of extensive air showers, now known as the knee. They proposed that this was evidence for an astrophysical feature, specifically a demonstration in favour of an extragalactic origin of cosmic rays above $10^{16}$ eV. However, interpretations of this measurement, and of later confirmatory ones, in terms of a change in hadronic physics persisted for over 40 years (e.g. [11]). It was not until 2002, that the KASCADE group ruled out this possibility, *without resorting to models,* by demonstrating that the number spectrum of muons steepened around the knee, but only in showers that were rich in electrons (i.e. shower produced by light nuclei) [12]. A similar approach was later used by the KASCADE-Grande Collaboration to show that there is a second knee at around 0.1 EeV, associated with the loss of heavy nuclei from the primary beam [13]. I find such *model-independent conclusions* very compelling: trends are demonstrated that can be interpreted rather easily with an elementary understanding of the principles of shower development.

Also, one should be alert to the over-confidence of theorists. Up until 1973, major authorities, including Heisenberg, asserted that the proton-proton cross-section would remain constant with energy beyond a few GeV. This was a strongly held view, based on Regge Pole theory, that the p-p cross-section had reached a finite asymptotic value. Of course, a study at the CERN ISR proved this theoretical prediction to have been incorrect (see figure 4, taken from [14]).

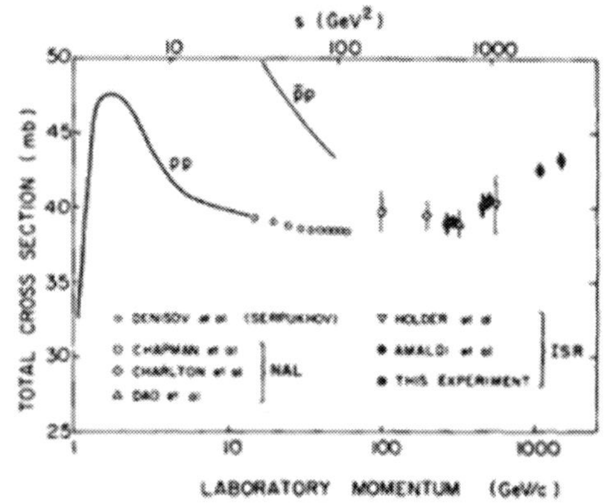


**Fig.4**. Rise in the total p-p cross-section as reported in [14].

The unexpected nature of this result has been discussed by Amaldi [15], while in [16] Matthiae, who was heavily involved in developing the 'Roman pots' used for these measurements, describes the rise in the cross-section as *'a startling discovery'*. There had been earlier claims by Grigorov et al. [17] and by Yodh et al. [18] of a rise in the nucleon-air cross-section deduced by comparing the rate of unaccompanied nucleons at high altitude with the flux of protons at the top of the atmosphere. However, it was difficult to assess the systematic uncertainties in these pioneering cosmic-ray experiments, particularly because of limited knowledge of the primary spectrum of protons. Indications of a rise in the p-air cross-section were certainly present, with the results in reasonable accord with the accelerator results.

Whether such surprises await us above 1 EeV is, of course, unknowable. At a recent conference [19], Unger reported on work carried out by Engel, Ulrich and himself, in which it was argued that the change in the p-air cross-section by the factor of two required to explain the change in the Elongation Rate, if protons persist, is very extreme. The minimum change required for the pp cross-section to obtain such a large value for the p-air cross-section with a Glauber model of nucleon-nucleus scattering is about a factor of three (at a centre-of-mass energy within about an order of magnitude of that reached at the LHC). Romanopoulos, Pavliou and Tomaras [20] have recently discussed the same problem.

There is also experimental evidence against such a change. Firstly, in the region of the ankle, 3 to 10 EeV, the Auger Collaboration [21] has presented an analysis showing that the correlation between the depth of shower maximum and the signal in the water-Cherenkov detectors supports a mixed composition with the atomic mass greater than 4, thus disfavouring claims of a pure composition in this energy range. This conclusion does not depend on different models of the hadronic physics.

Secondly, it may well be that evidence from anisotropy studies, along with better a understanding of the magnetic fields lying between the sources and the earth, will make it possible to rule out such a possibility in a rather direct manner. I see this line of attack as being as fruitful as that used in [12], which ruled out explanations of the knee in terms of changes in hadronic physics. At present, features such as the dipole measured in the distribution of arrival directions by the Auger Collaboration above 8 EeV, with very high significance [22], are found to incompatible with a pure proton composition [23]. These experimental results [21 and 22] should be taken as a clear refutation of a pure proton composition.

## 3 The Belief in Protons at the Highest Energies

The idea that the highest-energy cosmic rays are dominantly protons has enormously influenced thinking about the origin of these particles for over 60 years. As long ago as 1955, the distinguished cosmic ray and particle physicist, G Cocconi wrote [24]:

*"We remain with the dilemma: protons versus heavy nuclei. A clear cut decision cannot be reached yet. I believe that up to the highest energies the protons are the most abundant in the primary cosmic rays. …………………… However, I must confess that a leak proof test of the protonic nature of the primaries at the highest energies does not exist. This is a very important problem. Experimentally it is quite a difficult problem."*

The final sentence has proved to be a prophetic understatement. Also, it is worth keeping in mind that, in 1955, the detection of particles with energies above 0.1 EeV had barely been achieved.

At that time, estimates of the primary energy were made using a calorimetric approach, assessing the energies carried by the electrons and muons separately, and by including ionisation losses, which dominate, in both the atmosphere and below ground level. Estimates based on models of particle interactions were just being developed. Particles above 0.1 EeV were soon to be observed, principally through the efforts of the MIT group [25, 26]. The view of Linsley on the primary mass, based on observations with the small muon detector included in 8 km$^2$ array at Volcano Ranch, was that the muon distribution above ~0.1 EeV implied that either protons or nuclei of the Fe group dominate, though he also had arguments that appeared to favour protons [26]. The Volcano Ranch results were for the range 0.1 EeV to 1 EeV.

That the proton idea became so embedded in the thinking of the cosmic ray community is rather alarming. There has been an almost Orwellian group-think, possibly largely stimulated by the results from the Fly's Eye experiment [6] shown in figure 2. That result, dependent as it was on assumptions on the choice of hadronic model, appears to have carried more weight than other indications from astrophysical ideas about acceleration.

Cocconi's view was based on a number of observations and the theories that were available at a time when there was even deep suspicion, from many low-energy cosmic-ray physicists, as to the reality of the shower phenomenon. The idea that a shower might be induced by several lower energy photons arriving simultaneously was seriously considered. Cocconi [24] used the lack of lateral spread of the nucleonic component in showers as an argument against heavy primaries. Also, Greisen's view [27], that if Fermi's theory for an acceleration mechanism was correct, then lighter nuclei would be favoured, as their cross-section for collision was smaller, would almost surely have been well-known to Cocconi who was working along side Greisen at Cornell.

The idea that the higher energy particles might be of extragalactic origin, first discussed in [10], led Gerasimova and Rozenthal [28] to point out that nuclei with energies of $10^{16}$ eV/nucleon would be subject to photodisintegration in intergalactic space, and, with the photon fields assumed by them, to the claim that most heavy nuclei would disappear. However, Ginzburg and Syrovatskii argued that the photon density adopted was a least an order of magnitude lower than assumed in [28] so that heavy nuclei would survive [29]. After the theory of diffusive shock acceleration became widely accepted, Hillas [30], in the famous 'Hillas plot', pointed out that the highest energy attained was proportional to Z, the atomic number of the nucleus of interest. Rigidity is a key parameter.

Evidence for the feature in the elongation rate discussed in section 1 began to emerge in data obtained by the Auger Collaboration even before construction of the instrument was completed. In 2007, on the basis of ~1000 events with E > 2 EeV, it was pointed that there was evidence for the mean mass becoming heavier as the energy grew [31]. This indication was reinforced in 2010 when results from approximately twice as many events were reported [32].

That these results were in contradiction with conclusions drawn from the successive projects in Utah (Fly's Eye, HiRes - and later Telescope Array) has clearly been part of the problem, and certainly key people in the field, such as V S Berezinsky and E Waxman, were extraordinarily reluctant to consider other than pure protons at the highest energies. Around the time of the publication of [8], in which results based on over 5000 events above 2 EeV are presented, I was invited to a meeting of acceleration specialists at Dunsink, Ireland. I had been charged to set out the experimental situation and I described in critical detail the evidence available. I recall being told by M Lemoine that he preferred to try to find ways of accelerating protons to 100 EeV rather than iron nuclei as it was a more challenging problem. True of course, but not a way to help our understanding, particularly as theorists generally have much more influence than do the gatherers of the data.

Of course, the belief in a dominance of protons at the highest energies also leads to more encouraging fluexes for those planning the detection of UHE neutrinos.

## 4 Take Home Messages

1. There is no evidence for any difference in the Elongation Rate as observed in the two hemispheres.

2. The break in the Elongation Rate above 3 EeV povides *model-free* evidence that the mean mass of the primary particles becomes heavier as the energy increases.

3. Claims that protons are in fact dominant at high energies because of the increase in the p-air cross-section and the multiplicity of pions from such collisions are in disagreement with experimental evidence on the arrival directions of cosmic rays above 8 EeV and on the model-independent evidence for A>4 in the ankle region.

4. The belief that protons dominate at the highest energies should be abandoned.

I am grateful to Michael Unger for clarifying information about the rise in the p-p cross-section at the highest energies.

## References

I have used PICRC for ICRC Proceedings, giving the place and the date. Proceedings for most of the biennial International Cosmic Ray Conferences can now be readily accessed at https://ads.harvard.edu/pubs/icrc_proceedings.html.

1. D. Bergman, This Conference: 'Testing the Compatibility of the Depth of Shower Maximum Measurements performed at the Telescope Array and the Pierre Auger Observatory' (2022)

2. A. A. Watson, Journal of High Energy Astrophysics **33** 13 (2022) (arXiv 2112.06525)

3. P. Sokolsky and R. D'Avingon, JETP **134** 459 (2022) (arXiv 2110.09588)

4. H. Sagawa, Proceedings of 21$^{st}$ ISVHECRI Symposium, Ooty (arXiv 2209.0359) (2022)

5. A. Yushkov for Pierre Auger Collaboration, PICRC Madison 2019 POS(ICRC2019) 482

6. D. J. Bird et al., Phys. Rev. Letters **71** 3401 1993

7. A. A. Watson, Proceedings of UHECR2018, Paris: https://indico.in2p3.fr/event/17063/(arXiv: 1906.06676)

8. A. Aab et al., Pierre Auger Collaboration, Phys. Rev. D **90** 122006 (2014)

9. T. Pierog, This Conference: 'Hadronic Interaction Studies at the LHC' (2022)

10. G. V. Kulikov and G. B. Khristiansen, J. Exp. Theo. Phys. **35** 635 (1958)

11. S. I. Nikolsky, PICRC Hamburg 2001 p 1389

12. T. Antoni et al., Astroparticle Physics **16** 273 (2002)

13. KASCADE Grande Collaboration, Phys. Rev. Letters **107** 171104 (2011)

14. U. Amaldi et al., Physics Letters **44B** 112 (1973)

15. U. Amaldi, in '60 Years of CERN Experiments and Discoveries', Editors H Schopper and L Di Lella, World Scientific (2015)

16. G. Matthiae, CERN publication ISSN 1793 – 1339 **section 4.7**, p 112 (2017)

17. N. L. Grigorov et al. PICRC London 1965 **Vol 2** 860

18. G. B. Yodh, Y. Pal and J. S. Trefil, Phys. Rev. Letters **28** 1005 (1972)

19. M. Unger, R. Engel and R. Ulrich, Cosmic Rays and Neutrinos in the Multi-messenger Era: Paris December 2020 https://indico.in2p3.fr/event/20789/contributions/83757/.

20. S. Romanopoulos, V. Pavliou and T. Tomaras, arXiv 2206.14847

21. A. Aab et al. (Pierre Auger Collaboration), Physics Letters **762B** 288 (2016)

22. A. Aab et al. (Pierre Auger Collaboration), Science **357** 1266 (2017)

23. C. Ding, N. Globus and G. R. Farrar, ApJ Letters **913** L13 (2021)

24. G. Cocconi, PICRC Guanajuato 1955 p 168

25. J. Linsley, L. Scarsi and B. Rossi, Phys. Rev. Letters **6** 485 (1961)

26. J. Linsley, PICRC Jaipur 1963 **Vol 4** 77. In answer to a question by Colgate (p 86), Linsley pushes protons rather less strongly than is argued in J Linsley and L Scarsi, Phys. Rev. Letters **9** 123 (1962). In that work, a discrimination between protons and iron was attempted based on a measurement of shower development at large angles. Protons appeared to be favoured in this energy range. While I do not recall Linsley ever withdrawing the 1962 claim, he was a man who was particularly careful in his use of words, and the 1963 remarks may well have represented his more mature view.

27. K. Greisen, In 'Progress in Cosmic Ray Physics', North Holland Vol III p 3, Editor J G Wilson (1956)

28. N. M. Gerasimova and I. L. Rozenthal, JETP **14** 350 (1962)

29. V. L. Ginzburg and S I Syrovatskii, 'The Origin of Cosmic Rays', Pergamon Press p 128 (1964)

30. A. M. Hillas, Ann. Rev. Astr. and Astrophys **22** 425 (1984)

31. M. Unger for the Pierre Auger Collaboration, PICRC Merida 2007 **Vol 4** 373

32. J. Abraham et al. Pierre Auger Collaboration, Phys. Rev. Letters **104** 091101 2010.

# A detailed presentation of the highest-energy cosmic rays recorded at the Pierre Auger Observatory

*Mario* Buscemi[1,2,*] for the Pierre Auger Collaboration[3**]

[1]Università di Catania, Dipartimento di Fisica e Astronomia "Ettore Majorana", Catania, Italy
[2]INFN, Sezione di Catania, Catania, Italy
[3]Observatorio Pierre Auger, Av. San Martín Norte 304, 5613 Malargüe, Argentina
Full author list: https://www.auger.org/archive/authors_2022_10.html

**Abstract.** A catalog that contains details of the highest-energy cosmic rays, recorded by the Pierre Auger Collaboration between 1 January 2004 and 31 December 2020, is presented. Data from 100 air showers, generated by particles having energies in the range 78 EeV to 166 EeV, are described, together with nine other very energetic events used in the energy calibration. The catalog has been created to demonstrate the quality of the data that underlie measurements reported by the Collaboration, and to make the details of these events available for scrutiny. After a brief description of the techniques used for data acquisition and reconstruction, the contents of the catalog will be described. Some events within the catalog will be discussed in detail.

## 1 Introduction

The Pierre Auger Observatory, located in Malargüe, in the Province of Mendoza (Argentina), is largest cosmic ray observatory in the world and is used to study the extensive air-showers produced by cosmic rays above $\sim 10^{17}$ eV [1]. The Observatory features an array of 1600 water-Cherenkov particle detector (the Surface Detector, SD) spread over 3000 km$^2$ on a 1500 m triangular grid, overlooked by 24 air-fluorescence telescopes (the Fluorescence Detector, FD). The Observatory is at a mean altitude of about 1400 m, corresponding to an atmospheric overburden of about 875 g/cm$^2$. Data-taking started on 1 January 2004 with 154 water-Cherenkov detectors and one fluorescence detector in operation. Installation was completed in June 2008 and running has been on-going since that date. At the beginning of operation, in 2005, the Auger Collaboration presented a list of the 10 most energetic events detected until then, with the aim to illustrate the quality of the information that they contained [2]. After the conclusion of Phase I of operation, occurred on 31 December 2020, the Pierre Auger Collaboration released a catalog to demonstrate the quality of the data that lie behind measurements of the energy spectrum, the mass composition and the distribution of arrival directions of the highest-energy cosmic rays, and to make the details of these events available for scrutiny [3].

## 2 The Catalog

The catalog contains the 100 most energetic events recorded using the Surface Detectors of the Pierre Auger Observatory and included in the data set used in the analysis of the arrival directions of events above 32 EeV [4]. In addition nine hybrid events used in the energy-calibration procedure are presented. The energy of the particles described in the catalog is in the range 78 to 166 EeV. The catalog is described in a paper accecpeted for publication by the Astrophysical Journal Supplement Series, and in available online event browser containing details of all 109 events [5]. For each shower, the catalog contains the footprint at ground, the footprint in a plane perpendicular to the arrival direction, the lateral distribution function (LDF), and the time delays in each triggered WCD with respect to a plane shower front. Moreover the calibrated waveforms from Flash Analog-to-Digital converters (FADCs) associated with each of the three photomultipliers in the water-Cherenkov detectors (WCDs) are provided. Some peculiar FADC traces are described in detail. The events are identified with a catalog number (#*N*), and by an identifier, PAOddmmyy, that indicates the day, month and year of detection. A typical event of the catalog is shown in Figure 1.

### 2.1 Vertical events

The reconstruction of the arrival direction and of the size of an extensive air-shower is obtained from the magnitudes and temporal distributions of signals detected in each SD station. Different methods are used to reconstruct events with a zenith angle $\theta < 60°$, vertical events, and with $\theta > 60°$, inclined events. As the zenith angle increases, the shower loses its circular symmetry, due to the effect of the geomagnetic field, and to geometrical and attenuation effects acting as the particles cross the atmosphere. The methods used to reconstruct the inclined events are

*e-mail: mario.buscemi@ct.infn.it
**e-mail: spokespersons@auger.org

https://doi.org/10.1051/epjconf/202328302015

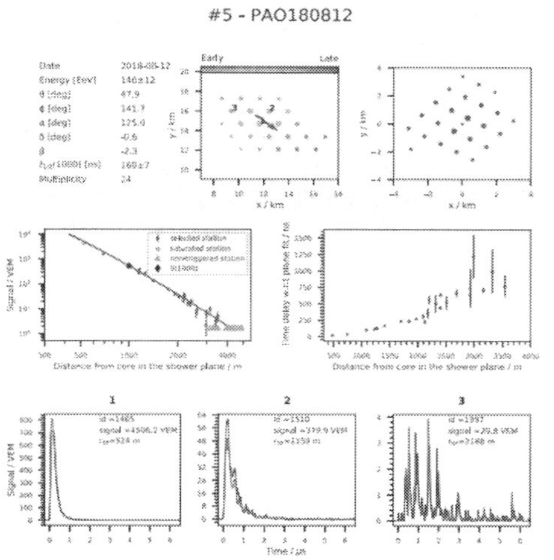


**Figure 1.** An event of the catalog, PAO180812 (#5). Top panel: the reconstructed features of the shower (left); the footprint at ground (middle) and in a plane perpendicular to the shower arrival direction (right). Middle panel: the lateral distribution function (left) and the time delays with respect to a plane shower front (right). Bottom panel: FADC traces of three water-Cherenkov detectors (different lines correspond to the three calibrated photomultipliers).

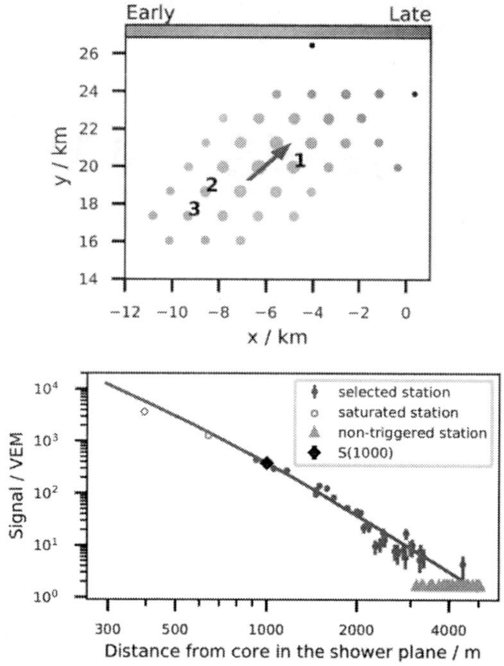


**Figure 2.** *Top*: Footprint on the ground of the most energetic shower detected at the Pierre Auger Observatory, PAO191110 (#1), with 34 WCDs triggered. Black dots correspond to stations that were triggered randomly. *Bottom*: The lateral distribution function of the same event. Open circles indicate saturated stations.

described in detail in [6]. The estimator of the shower size is found by fitting the observed signals to a lateral distribution function. A modified NKG function [7], [8], [9] is adopted for the procedure:

$$S_{\rm LDF}(r) = S(1000)\left[\left(\frac{r}{r_{\rm opt}}\right)\left(\frac{r+r_{\rm s}}{r_{\rm opt}+r_{\rm s}}\right)\right]^{\beta}$$

with $r_{\rm s}$ fixed at 700 m. The slope factor, $\beta$, changes from about $-2.6$ at $\theta = 0°$ to about $-1.9$ at 60°. The value $r_{\rm opt}$ depends on the spacing of the array, and is the distance at which the uncertainty in the signal size due to shower-to-shower fluctuations is minimized. For a 1500 m spaced array, as the Pierre Auger Observatory, $r_{\rm opt}$ is close to 1000 m. The size of the shower at 1 km from the core, $S(1000)$, is the energy estimator.

The LDF of the most energetic event ever detected at the Pierre Auger Observatory is shown in Figure 2 together with the shower footprint at ground. This shower, PAO191110 (#1), triggered 34 stations over an area of $\sim(13\times 6)\,\mathrm{km}^2$. The energy of the shower is 166 EeV, the zenith angle is 58.6° . This event has two saturated stations. These events are extremely rare, only three events in the full data sample have two detectors that are saturated simultaneously.

### 2.2 Inclined Events

Events with zenith angles > 60° are important to enhance the exposure of the Observatory by 30%, and to extend sky coverage to regions that would otherwise be inaccessible. The procedures developed to analyze these events differ from those used to reconstruct vertical showers, and are discussed in detail in [10]. Above 70° most of the particles at detector level are energetic muons accompanied by an electromagnetic component which makes up 25% of the signal beyond ~1 km from the core and around 30% within 1 km. The number of triggered stations above 60° increases with $\sec\theta$. For a shower with energy 30 EeV the average number is ~25 at 60°, while at 80° it is ~45. The method used for reconstruction is based on fitting the signal pattern recorded to what is predicted from modeling the shower development. The muon density scales with energy as $\rho_\mu(r) \propto E^\alpha$ with $\alpha$ in the range 0.90 to 0.95. The expected density of muons at the ground is given by

$$\rho_\mu(r) = N_{19}\,\rho_{\mu,19}(r,\theta,\phi),$$

where $N_{19}$ is chosen by convention as a measure of the shower size using a reference shower model and comparing the signals to those expected from simulated showers of 10 EeV with the same arrival direction. Simulations have shown that $\rho_{\mu,19}(r,\theta,\phi)$, at fixed zenith and azimuth angle, varies by only about 5% for changes in the energy (between $10^{18}$ and $10^{20}$ eV) and mass of the primary particle [11].

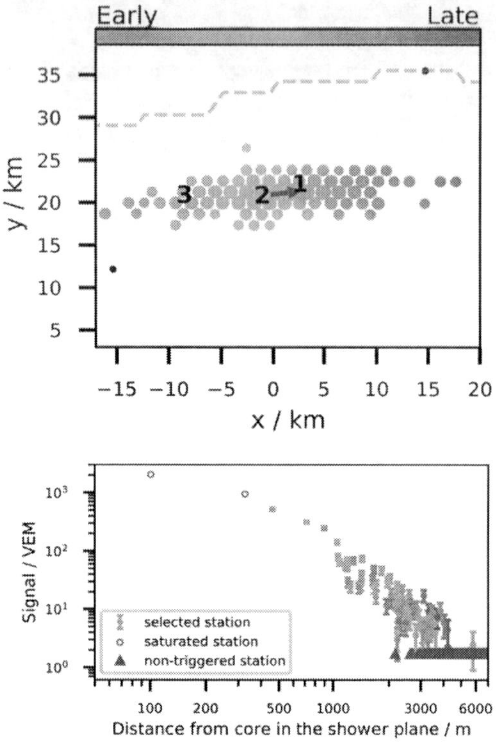


**Figure 3.** *Top*: Footprint on the ground of the most-energetic inclined shower detected at the Pierre Auger Observatory, PAO150926 (#17). *Bottom*: The distribution of the recorded signals as a function of the distance of the same event.

Some features of the most-energetic inclined event, PAO150926 (#17), are shown in Figure 3. The reconstructed energy of the shower is (113 ± 14) EeV and the zenith angle is $\theta = 77.2°$. The shower triggered 75 WCDs in an elongated pattern on the ground, over an area close to $(35 \times 6)\,\mathrm{km}^2$. Since the distribution of the integrated signal on the ground loses the near-rotational symmetry characteristic of vertical events, it cannot be described by a single rotationally-symmetric function.

## 2.3 Hybrid Events

An event that is acquired simultaneosly by an FD telescope and at least by one SD station is called hybrid. Measurements of the fluorescence and Cherenkov emission, through the FD, give details of the longitudinal development of air showers in the atmosphere, with the determination of the depth at which the deposition of energy is greatest, the shower maximum, $X_{\mathrm{max}}$. This is a key measurement for mass estimation (see [12] and [13]). A subsample of hybrid events can be fully reconstructed by both the FD hybrid reconstruction and, independently, by the SD reconstruction. These events are used for the calibration of the SD energy estimator with the calorimetric measurement performed by the FD. The catalog contains the ten most energetic events of the calibration subsample.

The camera of each telescope contains 440 photomultipliers (pixels). The energy deposited in the slant-depth is estimated from the light flux collected by each pixel in 100 ns interval. The resulting profile is fitted using the universal shower profile function described in [14],

$$f(X) = \left(\frac{\mathrm{d}E}{\mathrm{d}X}\right)_{\mathrm{max}} \left(1 + \frac{R}{L}(X - X_{\mathrm{max}})\right)^{1/R^2} \exp\left(-\frac{X - X_{\mathrm{max}}}{RL}\right),$$

where $f(X)$ is the energy deposit at the slant-depth $X$ and $(\mathrm{d}E/\mathrm{d}X)_{\mathrm{max}}$ is the energy deposit at shower maximum. $R$ and $L$ are shape parameters loosely constrained in the fit to the average of measured values [15]. The universal shower profile function is a modified version of the Gaisser-Hillas functional form [16].
The energy of each event is determined by integrating the area defined by the longitudinal profile, $f(X)$, with the addition of the energy that is carried into the ground largely by muons and neutrinos and that is not deposited in the atmosphere [17]. This invisible energy amounts to 20% of the at 0.1 EeV and 12% at 100 EeV. Above 10 EeV, the energy is determined with a systematic uncertainty of ~14% [15]. The methods by which data from the surface detectors are calibrated to obtain the energies of the primaries are described in [18]. Hybrid events are used for the calibration of both vertical showers and inclined events.
The energy resolution, at the highest energies, is ~8% for vertical events [19] and 12% for inclined ones.

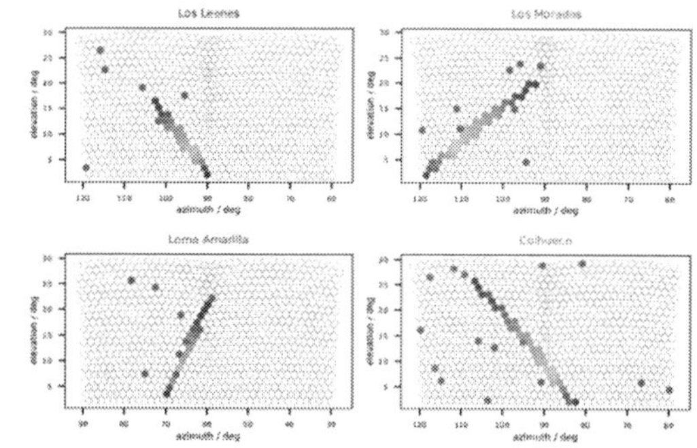


**Figure 4.** The camera views at the four sites where the event PAO100815, #84, was detected. The colors (violet to red) represent the time (early to late) at which the light reaches each pixel. Dark grey pixels indicate pixels that were triggered randomly.

In the most energetic hybrid event, PAO100815 (#84), the fluorescence light was detected at all four FD stations. The camera views of all telescopes at the four sites where the event was detected are shown in Figure 4. The reconstructed profiles of the energy deposition in the atmosphere are shown in Figure 5-Bottom. The lateral distribu-

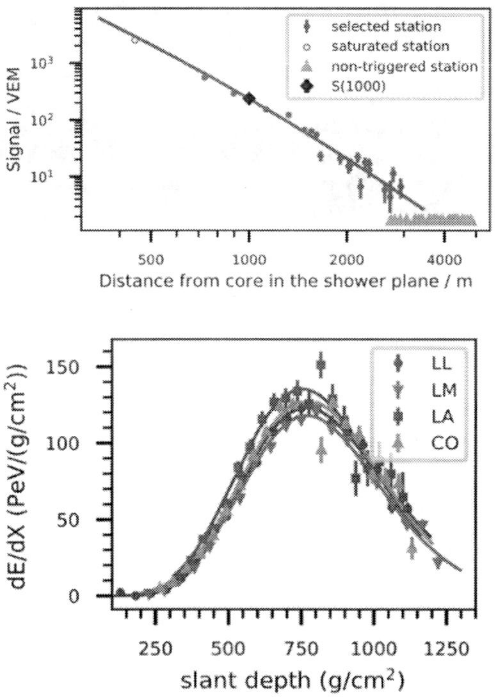


**Figure 5.** *Top*: The lateral distribution function of the most energetic hybrid event detected at the Pierre Auger Observatory, PAO100815, #84. *Bottom*: Profiles of the energy deposits reconstructed in each FD site. The solid lines correspond to fits to the profiles of the energy deposition using the universal shower profile function.

tion of signals is shown in Figure 5-Top. The energy estimate from the determination of $S(1000)$ is $(82 \pm 7)$ EeV. It is consistent with that from the fluorescence measurements of $(85 \pm 4)$ EeV.

### 2.4 Typical FADC traces

For each event, the catalog includes the FADC traces of the three photomultipliers of all triggered stations. The FADCs have 10-bit dynamic range and 40 MHz sampling rate. From the FADC information, the amplitude and time structure of each signal are obtained. Some typical signals are shown in Figure 6.

A typical signal produced a 1 km from the core in a vertical event is shown in Figure 6-a. The early parts of the signals are produced mainly by muons, while the tails are populated with broader signals due to the electromagnetic component of the shower. In an inclined event, electromagnetic particles are mostly absorbed in the atmosphere and the signals at the ground are produced almost entirely by muons, that reach a station in about 200 ns producing narrow FACD traces (Figure 6-b).
When the core of a shower falls close to a detector, within about 500 m, the large amount of particles generates a signal that can saturate the dynamic range of the electronics. An example is shown in Figure 6-c. A peculiar signal, with a spike at about 6 µs, is shown in Figure 6-d. Such signals are due to a contribution from direct light reaching one photomultiplier and are likely caused by the passage of a particle close to the location of the photomultiplier.

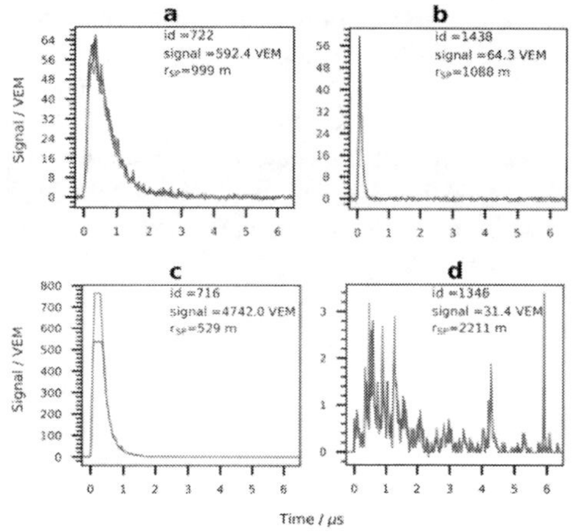


**Figure 6.** Some typical calibrated FADC traces of the WCDs. The different colors indicate the signals from the photomultipliers in each detector. In detectors #716 and #1346 two of the three photomultipliers were operational.

## 3 Conclusions

After the conclusion of Phase I of operation, the Pierre Auger Collaboration released a catalog containing the most energetic events recorded using the detectors of the Observatory. The catalog is described in a paper, and an event browser containing details of all events is available online.

## References

[1] A. Aab et al. (Pierre Auger Coll.), Nucl. Instrum. Meth. A **798**, 172 (2015)
[2] M. Ave Pernas (Pierre Auger Coll.), Proc. of the 29$^{th}$ International Cosmic Ray Conference p. 283 (2005)
[3] A. Abdul Halim et al. (Pierre Auger Coll.) (2022), `arXiv:2211.16020`
[4] P. Abreu et al. (Pierre Auger Coll.), Astrophys. J. **935**, 170 (2022)
[5] `https://opendata.auger.org/catalog/`
[6] A. Aab et al. (Pierre Auger Coll.), JINST **15**, P10021 (2020)
[7] K. Greisen, *Progress in Cosmic Ray Physics*, Vol. 3 (North-Holland Publishing, 1956)
[8] K. Greisen, Annual Review of Nuclear Science **10**, 63 (1960)
[9] K. Kamata, J. Nishimura, Progress of Theoretical Physics Supplement **6**, 93 (1958)

[10] A. Aab et al. (Pierre Auger Coll.), JCAP **08**, 019 (2014)

[11] H. Dembinski, P. Billoir, O. Deligny, T. Hebbeker, Astroparticle Physics **34**, 128 (2010)

[12] J. Abraham et al. (Pierre Auger Coll.), Nucl. Instrum. Meth. A **620**, 227 (2010)

[13] A. Aab et al. (Pierre Auger Coll.), Phys. Rev. D **90**, 122005 (2014)

[14] S. Andringa, R. Conceicao, M. Pimenta, Astropart. Phys. **34**, 360 (2011)

[15] B. Dawson (Pierre Auger Coll.), PoS **ICRC2019**, 231 (2020)

[16] T.K. Gaisser, A.M. Hillas, *Reliability of the Method of Constant Intensity Cuts for Reconstructing the Average Development of Vertical Showers*, in *International Cosmic Ray Conference* (1977), Vol. 8, p. 353

[17] A. Aab et al. (Pierre Auger Coll.), Phys. Rev. D **100**, 082003 (2019)

[18] H.P. Dembinski, B. Kégl, I.C. Mariş, M. Roth, D. Veberič, Astroparticle Physics **73**, 44 (2016)

[19] A. Aab et al. (Pierre Auger Coll.), Phys. Rev. D **102**, 062005 (2020)

 https://doi.org/10.1051/epjconf/202328302016

# Single source scenario describing the very end of the cosmic ray energy spectrum

*Alena* Bakalová[1,*], *Jakub* Vícha[1], and *Petr* Trávníček[1]

[1]FZU - Institute of Physics of the Czech academy of sciences, Na Slovance 1999/2, Prague, Czech Republic

**Abstract.** The energy spectrum of cosmic rays is steeply falling with a suppression of the flux at the highest energies caused by energy losses during propagation or by reaching the maximum energy of cosmic accelerators. The energy spectrum at the highest energies is currently measured with high precision by two experiments, Telescope Array in the Northern hemisphere and the Pierre Auger Observatory in the Southern hemisphere. In this work, we study the possibility of explaining the shape of the energy spectrum above $10^{19.5}$ eV measured by the Pierre Auger Observatory with a single source. Using CRPropa 3 numerical simulations of cosmic ray propagation in the Universe we show possible features of such a single source that would be able to create a shape of the energy spectrum compatible with the measurement including limitations on the source distance, spectral index and mass composition.

## 1 Introduction

Ultra high energy cosmic rays (UHECRs) are currently measured by two major experiments; Telescope Array in the Northern hemisphere and the Pierre Auger Observatory in the Southern hemisphere. At the highest energies, the two experiments both measure a strong suppression of the flux (above $\approx 5 \cdot 10^{19}$ eV), nevertheless, the measured spectral index differs between these two observatories [1]. The steep suppression of the cosmic ray flux might be explained by either the GZK cutoff caused by particle interaction with mainly cosmic microwave background (CMB) or by reaching the maximum acceleration capabilities of astrophysical sources in our UHECR horizon, or a combination of both. However, a disagreement between the two observatories might be caused by the ability of the two experiments to observe different sources on the sky in the Northern and Southern hemispheres that are able to accelerate cosmic rays up to the highest energies.

In this work we investigate the possibility of explaining the very end of the energy spectrum of cosmic rays (above $10^{19.5}$ eV) as measured by the Pierre Auger Observatory by a single source scenario. We use Monte Carlo simulations of cosmic-ray propagation and study different features of the potential single source including its distance, spectral index or mass composition. We are looking for such features of the source that would create an energy spectrum compatible with the measurements. We apply additional requirements on the mass composition on Earth based on measurements by the Pierre Auger Observatory. Older results of this research were previously introduced in [2]. In this proceeding we report new results obtained with updated data of the energy spectrum by the Pierre Auger Observatory and new simulations.

*e-mail: bakalova@fzu.cz

## 2 Simulated data

### 2.1 Simulation of cosmic ray propagation

Simulations of cosmic ray propagation are performed using CRPropa 3 [3]. One-dimensional mode of propagation is used, where only particle interactions are taken into account and no deflections in magnetic fields are applied. Four types of particles are simulated; proton, helium, nitrogen and iron nuclei each with statistics of 10,000 particles. Particles undergo interactions with cosmic microwave background (CMB) and extragalactic background light (EBL). The Gilmore12 model of EBL [4] is used in our simulations. The interactions of cosmic rays with CMB and EBL include photo-pion production, electron pair production and photodisintegration in case of cosmic rays heavier than proton. Other processes included in the simulations are nuclear decay of unstable nuclei and cosmological redshift accounting for the expansion of the Universe. The mass composition on the source is investigated in discrete steps of 10% in the relative fraction of individual particle species.

A single source is considered at distances from 3 Mpc up to 100 Mpc with a step of 1 Mpc up to 20 Mpc and a step of 10 Mpc further out. The energy spectrum on the source is simulated in the energy range from $10^{19.5}$ eV up to $10^{20.5}$ eV as a power law with rigidity dependent exponential cutoff as

$$\frac{\mathrm{d}N}{\mathrm{d}E} = E^{-\gamma} \cdot f_{\mathrm{cut}}\,, \tag{1}$$

where $\gamma$ is the spectral index and $f_{\mathrm{cut}}$ is the rigidity exponential cutoff function defined for a given value of rigidity cutoff $R_{\mathrm{cut}}$ as

$$f_{\rm cut} = \begin{cases} 1 & (E < ZeR_{\rm cut}) \\ \exp\left(1 - \frac{E}{ZeR_{\rm cut}}\right) & (E > ZeR_{\rm cut}) \end{cases}, \quad (2)$$

where $Z$ is the proton number of the particle species, see e.g. [5]. The spectral indices of the energy spectrum are considered in the range $\gamma \in (1, 3)$ with a step of 0.5. We investigate the rigidity cutoff from $\log_{10}(R_{\rm cut}/{\rm V}) = 20.5$ with discrete steps of 0.1 down to $\log_{10}(R_{\rm cut}/{\rm V}) = 18.5$. For lower $R_{\rm cut}$ values the energy spectrum above $10^{19.5}$ eV is dominated by the exponential behavior of the generated energy spectrum for all particle species considered.

### 2.2 Comparison of MC with data

We compare the energy spectra obtained from our simulations with the energy spectrum of cosmic rays above $10^{19.5}$ eV measured by the Pierre Auger Observatory [6]. We take into account the statistical errors of the measured spectrum and statistical errors obtained from the simulations. We compare the two energy spectra based purely on the shape of the measured spectrum. For that purpose, we rescale the measured values of the energy spectrum in such a way that the value in the lowest energy bin is equal to 1 and the simulated energy spectrum is rescaled in the same manner. Therefore, the values are equal in the first energy bin and we calculate a reduced $\chi^2$ for the rest of the energy bins. This is done separately for each spectrum from simulations obtained from a source with given distance, spectral index, mass composition and rigidity cutoff. The agreement between data and simulations is based on two conditions. The first condition is given by $\chi^2/ndf \le 2$ and the second condition is that the simulated energy spectrum must contain particles in all investigated energy bins.

The measured energy spectrum is not the only information we have about UHECRs, therefore, we can further restrict the single source features. Additional condition is put on the mass composition of the cosmic rays arriving on the Earth based on measurements by the Pierre Auger Observatory [7, 8]. We take into account interpretation of the data measured by the Pierre Auger Observatory using EPOS-LHC and Sibyll 2.3c models of hadronic interactions. QGSJet II-04 model is not taken into account as it suggests nonphysical values of the variance of mean $X_{\rm max}$. For that reason, we restrict ourselves to cases with the mean mass composition on the Earth above $10^{19.5}$ eV heavier than nitrogen, e.i. $\langle \ln A \rangle > \ln(14)$ and the variance of the mean $\ln A$ to be smaller than 0.5. Further support for heavy composition at the highest energies was also shown in [9].

## 3 Possible properties of a single source at the highest energies

When comparing the measured energy spectrum with the simulated energy spectra produced by a single source with different parameters of spectral index, mass composition, rigidity cutoff and source distance without the condition on heavy mass composition above $10^{19.5}$ eV a large set of possible features of a single source is obtained. Figure 1 shows the minimal value of $\chi^2/ndf$ for different combinations of rigidity cutoff and spectral index of the source. For spectral index $\gamma = 3$ the energy spectrum can be described well in the whole range of investigated $R_{\rm cut}$ values. In case of low spectral indices ($\gamma = 1.0$ and $\gamma = 1.5$), the energy spectrum above $10^{19.5}$ eV can be described well only for a narrow set of rigidity cutoff values in the range $R_{\rm cut} \approx (10^{19.0} - 10^{19.6})$ V.

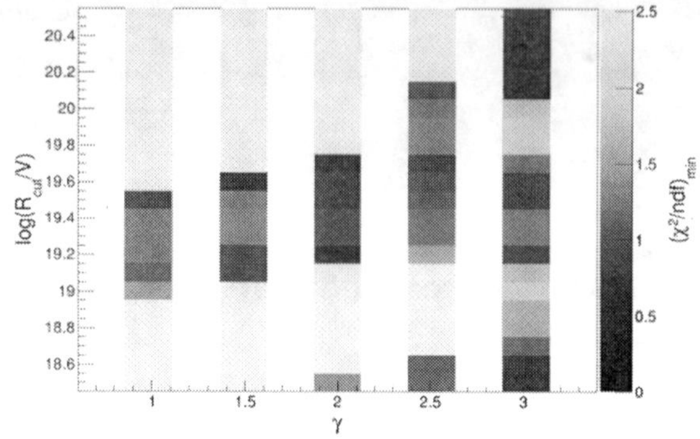


**Figure 1.** Minimal value of $\chi^2/ndf$ between the simulated and measured energy spectra for different combinations of spectral indices and rigidity cutoff values.

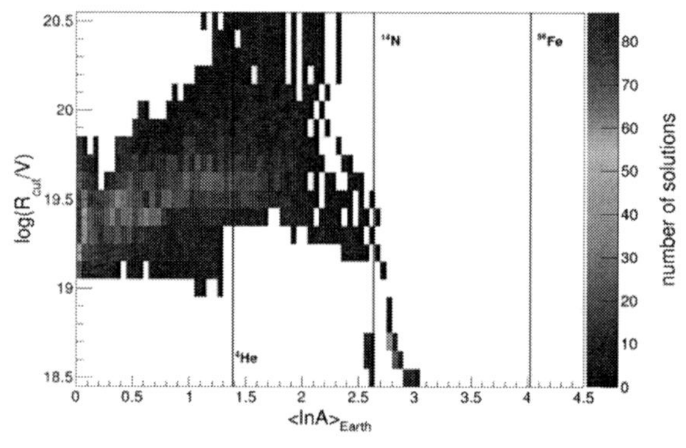


**Figure 2.** Number of solutions for different values of $R_{\rm cut}$ with respect to mean mass composition on the Earth.

The number of solutions, satisfying condition on the compatibility of the energy spectrum with measurement, for different $R_{cut}$ values with respect to the mean mass composition on the Earth are depicted in Figure 2. For the whole investigated phase space of parameters no possible solutions of the source features are found to obtain a very heavy composition (close to iron) on the Earth. A large number of solutions can be found in the region of mass composition close to proton for rigidity cutoff values between $R_{\rm cut} \sim 10^{19.1}$ V and $R_{\rm cut} \sim 10^{19.8}$ V. With increasing mass composition on the Earth, rigidity cutoff values of the source are allowed up to the maximum investigated value $R_{\rm cut} = 10^{20.5}$ V. However, the solutions of the heav-

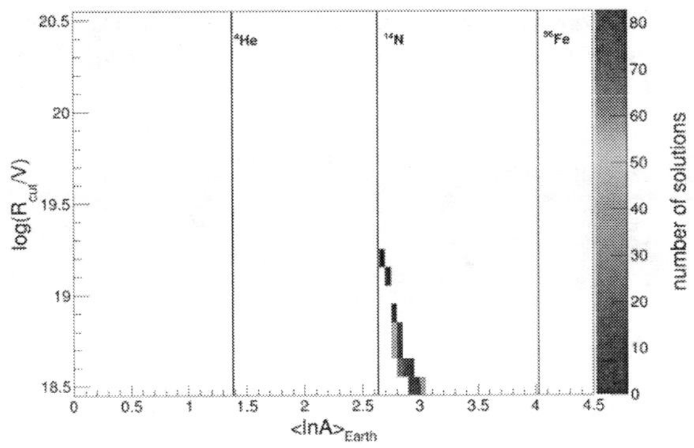


**Figure 3.** Number of solutions for different values of $R_{cut}$ with respect to mean mass composition on the Earth after applying additional mass composition cuts.

iest mass compositions found are only for rather smaller values of $R_{cut}$.

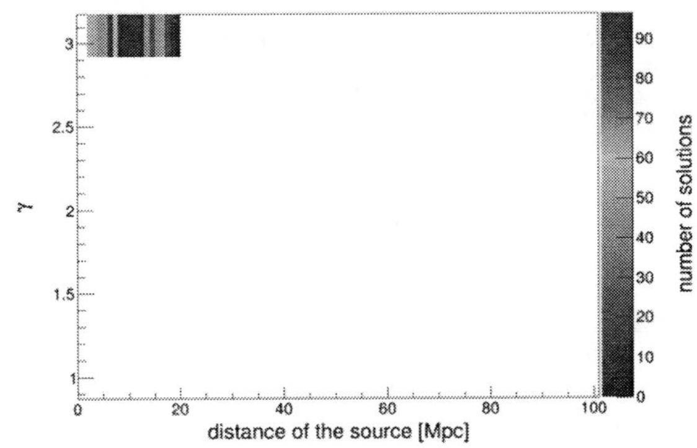


**Figure 4.** Number of solutions for different values of spectral indices $\gamma$ with respect to distance of the source after applying additional mass composition cuts.

Without the additional conditions on mass composition a large set of combinations of single source features is allowed in order to explain the shape of the measured energy spectrum. However, restricting ourselves to the cases of $\langle \ln A\rangle > \ln(14)$ and $\sigma(\ln A) < 0.5$ significantly restricts the possible single source features. The number of solutions for different values of $R_{cut}$ with respect to mean mass composition on the Earth after applying additional mass composition cuts are shown in Figure 3. Only sources with $R_{cut} \leq 10^{19.3}$ V are able to describe both the energy spectrum and mass composition well. Such low values of $R_{cut}$ are needed in order to sufficiently suppress the presence of light elements in the energy spectrum above $10^{19.5}$ eV. The number of remaining solutions are plotted for different spectral indices with respect to distance of the source in Figure 4. This exhibits that a single source describing well both the energy spectrum and mass composition of cosmic rays above $10^{19.5}$ eV measured by the Pierre Auger Observatory should have a spectral index $\gamma \approx 3$ and its distance should be within 20 Mpc from the Earth. Sources at distances farther are problematic as the longer trajectory causes disintegration of the heavy elements thus pushing the $\langle \ln A\rangle >$ lower than required in this work.

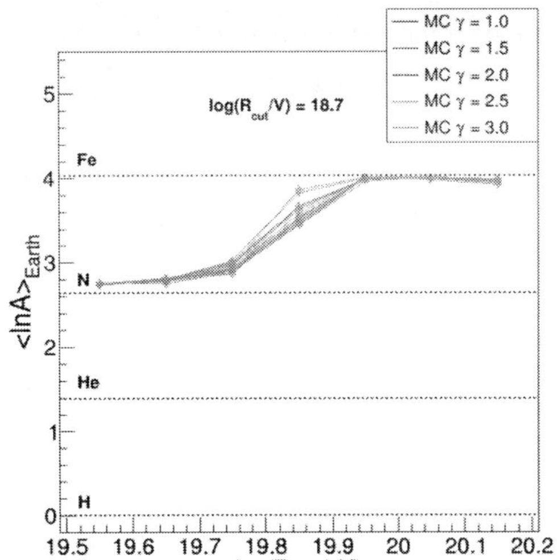


**Figure 5.** Evolution of the mean mass with energy for solutions found for $R_{cut} = 10^{18.7}$ V satisfying restrictions put on energy spectrum and mass composition from the measurements.

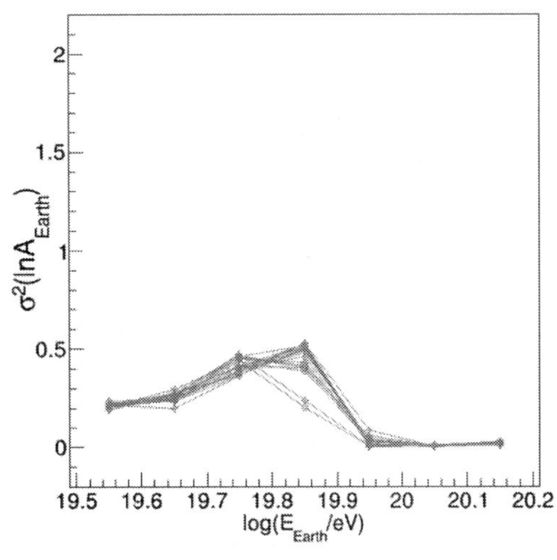


**Figure 6.** Evolution of the variance of mass with energy for solutions found with $R_{cut} = 10^{18.7}$ V satisfying restrictions put on energy spectrum and mass composition from the measurements.

An example of the evolution of the mean mass with the energy and the variance of the mass with energy for the solutions found for $R_{cut} = 10^{18.7}$ V are demonstrated in Figure 5 and Figure6, respectively. The mean mass composition grows with energy from a composition little heavier than nitrogen at $E = 10^{19.5}$ eV up to a composition close to pure iron at the highest energies. The small mixing of the elements can be seen in Figure 6 showing a small variance of the mean mass.

## 4 Conclusions

We investigated the possibility of explaining the energy spectrum measured by the Pierre Auger Observatory above $10^{19.5}$ eV by a single source. Numerical simulations of the energy spectrum created by a single source of various properties including distance, spectral index, mass composition and rigidity cutoff are compared with the measured energy spectrum above $10^{19.5}$ eV by the Pierre Auger Observatory. Additional condition is put on the mean mass composition $\langle \ln A \rangle$ of the cosmic-ray mix and on the variance of the mean mass $\sigma(\ln A)$ based on current measurements by the Pierre Auger Observatory. We show that a single source scenario is able to create an energy spectrum and mass composition compatible with measurements. In order to reproduce well both the energy spectrum and mass composition, such a source should be within 20 Mpc from the Earth. Note that this is an "effective" distance of the source as deflections of cosmic rays in magnetic fields are not taken into account in the simulations. Moreover, such a source should have a spectral index $\gamma \approx 3$ and rigidity cutoff of the source should be $R_{\rm cut} \leq 10^{19.3}$ V.

## Acknowledgement

This work was supported by: Czech Republic – Grant No. MEYS LTT18004, LM2018102, CZ.02.1.01/0.0/0.0/16_013/0001402.

## References

[1] T. AbuZayyad, O. Deligny, D. Ikeda, D. Ivanov, I. Lhenry-Yvon, I. Mariş, D. Mockler, T. Nonaka, M. Roth, F. Salamida et al., European Physical Journal Web of Conferences **210**, 01002 (2019)

[2] A. Bakalova, J. Vicha, P. Travnicek, 37th International Cosmic Ray Conference **ICRC2021**, 363 (2022)

[3] R. Alves Batista, A. Dundovic, M. Erdmann et al., Journal of Cosmology and Astroparticle Physics **2016**, 038 (2016)

[4] R.C. Gilmore, R.S. Somerville, J.R. Primack, A. Domínguez, Monthly Notices of the Royal Astronomical Society **422**, 3189 (2012), 1104.0671

[5] A. Aab et al. (Pierre Auger Collaboration), Journal of Cosmology and Astroparticle Physics **2017**, 038 (2017), 1612.07155

[6] A. Aab et al. (Pierre Auger Collaboration), Physical Review D **102**, 062005 (2020), 2008.06486

[7] A. Yushkov, 36th International Cosmic Ray Conference **ICRC2019**, 482 (2019)

[8] C.J. Todero Peixoto, 36th International Cosmic Ray Conference **ICRC2019**, 440 (2019)

[9] J. Vícha, arXiv e-prints arXiv:2209.00744 (2022), 2209.00744

# Magnetic fields and UHECR propagation

*Katia* Ferrière[1,*]

[1]IRAP, Université de Toulouse, CNRS, 9 avenue du Colonel Roche, BP 44346, 31028 Toulouse Cedex 4, France

**Abstract.** I review the main observational tools used to detect and measure magnetic fields in the interstellar medium of our Galaxy (considering both the Galactic disk and the Galactic halo), in the interstellar medium of external galaxies, and in the extragalactic medium. In each case, I present a summary of the most important results obtained with the different tools, trying to lay the emphasis on the magnetic field properties that are relevant to the propagation of ultra-high–energy cosmic rays (UHECRs).

## 1 Introduction

The existence of interstellar magnetic fields was first predicted as a means to confine and accelerate Galactic cosmic rays. Back in 1937, Hannes Alfvén[1] correctly pointed out that the confinement of cosmic rays in our Galaxy implied "the existence of a magnetic field in interstellar space". Twelve years later, Enrico Fermi [1] went one step further and argued that "cosmic rays are originated and accelerated primarily in the interstellar space of the Galaxy by collisions against moving magnetic fields." He also added, with surprising accuracy for the time, that "the magnetic field in the dilute matter is of the order of magnitude of $5 \times 10^{-6}$ gauss, while its intensity is probably greater in the heavier clouds." Thus, it appears that the history of interstellar magnetic fields is intimately linked to the history of cosmic rays.

Since cosmic rays are electrically charged particles, their trajectories are deflected by the interstellar magnetic field. This deflection is measured by the Larmor radius,

$$r_g = \frac{E}{q\,B}\,\sin\alpha\,, \qquad (1)$$

where $E$ is the energy of the particle, $q$ its electric charge, $B$ the magnetic field strength, and $\alpha$ the particle pitch angle (angle between the particle velocity and the magnetic field).

For a 1 GeV ($10^9$ eV) cosmic ray,

$$\begin{aligned} r_g &\simeq (0.2\,\mathrm{AU}) \left(\frac{B}{1\,\mu\mathrm{G}}\right)^{-1} \sin\alpha \\ &\simeq (10^{-6}\,\mathrm{pc}) \left(\frac{B}{1\,\mu\mathrm{G}}\right)^{-1} \sin\alpha\,, \end{aligned} \qquad (2)$$

which is considerably smaller than both the dominant scale of the interstellar magnetic field, $L_B$, and the scale height of the Galactic disk, $H_G$. As a result, the cosmic ray has a mostly helical motion along magnetic field lines, with negligible drift, though with possible diffusion due to small-scale magnetic fields and collisions.

In contrast, for a 1 EeV ($10^{18}$ eV) cosmic ray,

$$r_g \simeq (1\,\mathrm{kpc}) \left(\frac{B}{1\,\mu\mathrm{G}}\right)^{-1} \sin\alpha\,, \qquad (3)$$

which is roughly comparable to both $L_B$ and $H_G$. As a result, the motion of the cosmic ray is moderately deflected by the interstellar magnetic field, such that its trajectory is neither straight nor helical.

Throughout this paper, the magnetic field vector is denoted by $\vec{B}$, its component in the plane of the sky (PoS) by $\vec{B}_\perp$, and its component along the line of sight (LoS) by $B_\parallel$. Furthermore, Galactic longitude and latitude are denoted by $l$ and $b$, respectively.

In Sect. 2, I review the main observational methods employed to detect and measure interstellar magnetic fields in our own Galaxy, and I summarize what each of these methods has taught us about their typical strength, their direction/orientation, their spatial distribution, and their overall topology. In brief, the linear polarization of starlight and of dust thermal emission yields the orientation of $\vec{B}_\perp$ in the general, dusty interstellar medium (ISM) (Sect. 2.1). Synchrotron emission, in combination with data or assumptions on the relativistic-electron density, is a tracer of $\vec{B}_\perp$ in the general, cosmic-ray filled ISM (Sect. 2.2). Faraday rotation measures, in conjunction with the observed or modeled distribution of thermal electrons, lead to $B_\parallel$ in ionized regions (Sect. 2.3). Exploiting the Faraday rotation of the Galactic diffuse synchrotron emission makes it possible to perform Faraday tomography (Sect. 2.4). Zeeman splitting [e.g., 2] and the Goldreich-Kylafis effect [3, 4] also offer important diagnostic tools of interstellar magnetic fields, but since they are not directly relevant to cosmic-ray propagation, I do not discuss them in this paper.

*e-mail: katia.ferriere@irap.opm.eu

[1]"Versuch zu einer Theorie über die Entstehung der kosmischen Strahlung". II. Zeitschrift FüR Physik, 107, 579-588. DOI: 10.1007/BF01339294

In Sect. 3, I present the main observational results obtained for interstellar magnetic fields in external galaxies (Sect. 3.1) and for extragalactic magnetic fields (Sect. 3.2).

## 2 Interstellar magnetic fields in our Galaxy

### 2.1 Dust polarization

Interstellar dust grains generally have irregular, elongated shapes, which, in a directional stellar radiation field, cause them to feel radiative torques. These torques have two important effects [see, e.g., 5–7]. First, they spin up the grains to suprathermal rotation about their short axes. Second, they gradually bring the grain spin axes into alignment with the local interstellar magnetic field. As a result, dust grains tend to line up with their long axes perpendicular to the local magnetic field.

The magnetically aligned dust grains collectively act as a polarizing filter for starlight. Since they preferentially scatter and absorb the starlight component that is polarized parallel to their long axes (as seen in the PoS), the starlight that passes through is linearly polarized parallel to $\vec{B}_\perp$. In consequence, the measured polarization orientation of starlight directly gives the orientation of $\vec{B}_\perp$.

In addition to polarizing starlight, the magnetically aligned dust grains also emit infrared/submm thermal radiation in a dichroic manner. Since the emissivity is strongest for the radiation component that is polarized parallel to the grains' long axes (as seen in the PoS) [8], dust thermal emission is linearly polarized perpendicular to $\vec{B}_\perp$. Thus, the measured polarization orientation of dust thermal emission, rotated by 90°, corresponds to the orientation of $\vec{B}_\perp$.

If $\vec{B}$ varies along the LoS and/or across the observed sky area, polarization data, from either starlight or dust thermal emission, smoothed in the PoS, provide the orientation of the PoS component of the ordered magnetic field, $\vec{B}_{\rm ord}$. For future reference, we note that the *ordered* magnetic field, $\vec{B}_{\rm ord}$, discussed in Sects. 2.1 and 2.2, differs from the *regular* magnetic field, $\vec{B}_{\rm reg}$, discussed in Sect. 2.3, in the sense that $\vec{B}_{\rm ord}$ has a given orientation, but not necessarily a given direction, whereas $\vec{B}_{\rm reg}$ has both a given orientation and a given direction[2].

Polarization data not only give the orientation of $\vec{B}_\perp$, but they also make it possible to estimate, under certain conditions, the inclination of $\vec{B}$ to the PoS [see, e.g., 9, and references therein]. This is because the polarization fraction, i.e., the ratio of polarized intensity to total intensity, varies as $\cos^2 \gamma_B$, where $\gamma_B$ is the inclination angle of $\vec{B}$ to the PoS. In particular, the signal has maximum polarization when $\vec{B}$ lies in the PoS ($\gamma_B = 0$), while the signal is completely unpolarized when $\vec{B}$ lies along the LoS ($\gamma_B = 90°$). The difficulty here is that the polarization fraction depends not only on $\gamma_B$, but also on the optical properties of dust grains and on their alignment efficiency, and it is generally not straightforward to isolate the contribution from $\gamma_B$.

Starlight polarimetry of a large number of stars within a few kpc of the Sun shows that $\vec{B}_{\rm ord}$ in the nearby Galactic disk is horizontal, i.e., parallel to the Galactic plane, and nearly azimuthal [10]. In the Galactic vicinity of the Sun, $\vec{B}_{\rm ord}$ is found to be oriented along the axis $(l, b) \simeq (83°, 0°)$[3] [11]. The horizontal orientation of $\vec{B}_{\rm ord}$ in the nearby Galactic disk appears clearly in the all-sky map of starlight polarization from the compilation by [12] (see Figure 1). It also emerges from this map that $\vec{B}$ does not remain horizontal at high Galactic latitudes. Most strikingly, $\vec{B}_\perp$ appears to follow approximately the shape of the North Polar Spur (NPS), which extends from the Galactic plane at $l \approx +20°$ nearly all the way up to the north Galactic pole. More detailed polarization maps of the north and south Galactic polar regions reveal a significant asymmetry between both hemispheres, probably due to the Local Bubble and other local features [13].

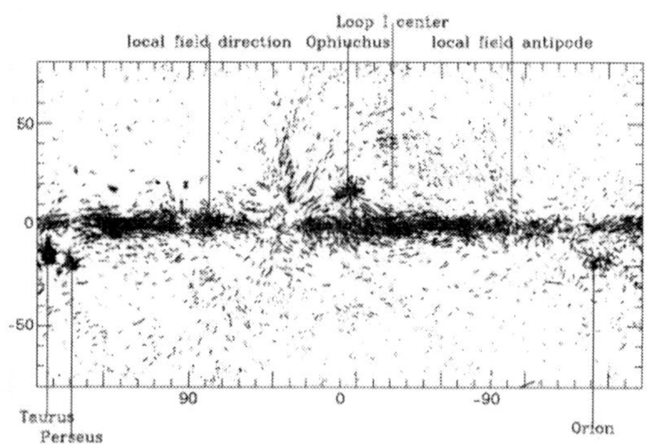


**Figure 1.** All-sky map in Galactic coordinates (with the Galactic center in the middle, longitude $l$ increasing to the left, and latitude $b$ increasing upward) of the polarization half-vectors of 8 662 stars from the compilation by [12]. The polarization half-vectors directly trace the orientation of $\vec{B}_\perp$.

Starlight polarization measurements, combined with stellar distances (e.g., from *Gaia*), open the way to probe the orientation of $\vec{B}_\perp$ in 3D. The optical polarization survey PASIPHAE is expected to soon offer a large enough data set to perform such a tomographic analysis in a meaningful way [14].

Polarization measurements of the dust thermal emission provide a more complete and more detailed view of the orientation of $\vec{B}_\perp$ across the sky, though without any information on its LoS variations. The all-sky 353 GHz polarization map obtained with the *Planck* satellite [15] (see Figure 2) is generally consistent with the starlight polarization map displayed in Figure 1. In the Galactic disk, $\vec{B}_{\rm ord}$ is again found to be horizontal, and this appears to

[2] The total magnetic field, $\vec{B}$, is composed of a regular field, $\vec{B}_{\rm reg}$, and a fluctuating field, $\vec{B}_{\rm fluc}$: $\vec{B} = \vec{B}_{\rm reg} + \vec{B}_{\rm fluc}$. The fluctuating field, in turn, has an isotropic component, $\vec{B}_{\rm fluc,iso}$, and an anisotropic (or directional) component, $\vec{B}_{\rm fluc,dir}$: $\vec{B}_{\rm fluc} = \vec{B}_{\rm fluc,iso} + \vec{B}_{\rm fluc,dir}$. The ordered field, $\vec{B}_{\rm ord}$, is composed of the regular field and the anisotropic fluctuating field: $\vec{B}_{\rm ord} = \vec{B}_{\rm reg} + \vec{B}_{\rm fluc,dir}$. Thus, the total magnetic field can also be regarded as the superposition of the ordered field and the isotropic fluctuating field: $\vec{B} = \vec{B}_{\rm ord} + \vec{B}_{\rm fluc,iso}$.

[3] Faraday rotation studies, which give the sign of $B_\parallel$, indicate that $\vec{B}_{\rm ord}$ actually points toward $(l, b) \simeq (83°, 0°)$, not in the opposite direction (see Sect. 2.3).

remain true out to large distances from the Sun. At high Galactic latitudes, the distorsion of $\vec{B}_\perp$ toward the NPS is similar to that observed in the optical.

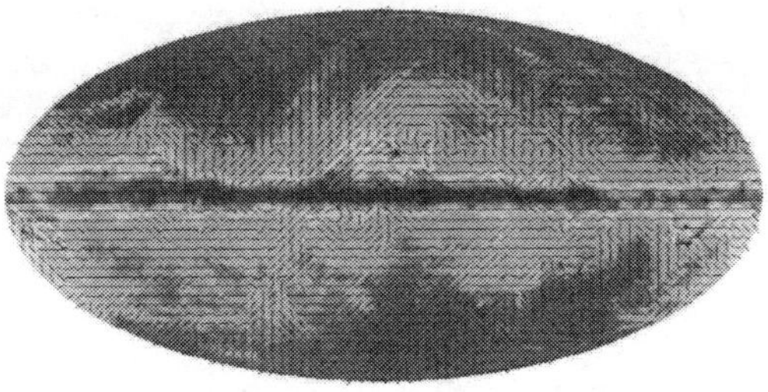

**Figure 2.** All-sky map in Galactic coordinates (with the Galactic center in the middle) of the total intensity of the 353 GHz dust thermal emission measured by *Planck* [15]. The small bars show the polarization half-vectors rotated by 90° to trace the orientation of $\vec{B}_\perp$.

## 2.2 Synchrotron emission

Synchrotron emission is produced by relativistic electrons gyrating about magnetic field lines. The synchrotron emissivity at frequency $\nu$ due to a power-law energy spectrum of relativistic electrons, $f(E) = K_e\,E^{-\gamma}$, is given by

$$\mathcal{E}_\nu = \mathrm{fc}(\gamma)\,K_e\,B_\perp^{\frac{\gamma+1}{2}}\,\nu^{-\frac{\gamma-1}{2}}\,, \tag{4}$$

where fc($\gamma$) is a known function of $\gamma$ and $B_\perp$ is the strength of $\vec{B}_\perp$ [16]. The synchrotron total intensity is obtained by integrating the synchrotron emissivity along the LoS:

$$I_\nu = \int \mathcal{E}_\nu\,ds\,\cdot \tag{5}$$

Low-frequency radio maps of the Galactic synchrotron emission can be used to model the spatial distribution of the interstellar magnetic field, from the large Galactic scales down to the smallest scales resolved by radio telescopes. This modeling requires knowing the relativistic-electron spectrum, which can be derived either from cosmic-ray propagation models or from gamma-ray observations. Alternatively, one sometimes resorts to the double assumption that (1) relativistic electrons represent a fixed fraction of the cosmic-ray population and (2) cosmic rays and magnetic fields are in (energy or pressure) equipartition. While the second assumption can find some rough justification at large scales,[4] there is no guarantee that it holds at small scales. Nevertheless, direct measurements of the interstellar magnetic field [18] and of the Galactic cosmic-ray ion and electron spectra [19] by the *Voyager* spacecraft indicate that, in the very local ISM, magnetic fields and cosmic rays are indeed close to (pressure) equipartition, with a total magnetic field strength $B \approx 5\,\mu$G.

[4]The argument usually put forward to justify the equipartition assumption is that energy equipartition corresponds almost exactly to the condition that the total (cosmic-ray + magnetic) energy is minimum [17].

An important property of synchrotron emission is that it is linearly polarized perpendicular to $\vec{B}_\perp$, so that information can also be gained on the orientation of $\vec{B}_\perp$. If the observing frequency happens to be low enough to be affected by Faraday rotation (see Sect. 2.3), the received polarized signal must somehow be "de-rotated" in order to retrieve the true orientation of $\vec{B}_\perp$. In addition, if $\vec{B}$ has a fluctuating component, the contributions from isotropic magnetic fluctuations to the polarized intensity cancel out, leaving only the contribution from the ordered magnetic field, $\vec{B}_{\rm ord} = \vec{B}_{\rm reg} + \vec{B}_{\rm fluc,dir}$ (see footnote 2). Thus, while the synchrotron *total* intensity yields the strength of the *total* magnetic field, the synchrotron *polarized* intensity yields the strength and the orientation of the *ordered* magnetic field (both in the PoS). In the Galactic vicinity of the Sun, the ratio of ordered to total magnetic field strength turns out to be $\frac{B_{\rm ord}}{B} \approx 0.6$ [20]. Together with $B \approx 5\,\mu$G, this ratio implies $B_{\rm ord} \approx 3\,\mu$G.

The reference all-sky map of the Galactic synchrotron emission remains the all-sky 408 MHz radio continuum emission map of [21] (displayed in Figure 3). Based on this map, [22] constructed an axisymmetric model of the spatial distribution of the Galactic synchrotron emissivity. When combined with the equipartition assumption, this model implies that the total magnetic field strength has a radial scale length $\simeq$ 12 kpc and a vertical scale height $\simeq$ 4.5 kpc near the horizontal position of the Sun [23].

**Figure 3.** All-sky map in Galactic coordinates (with the Galactic center in the middle) of the total intensity of the 408 MHz radio continuum emission (presumably pure synchrotron emission) from the Bonn 408 MHz All-Sky Survey, using data obtained with the Jodrell-Bank 76 m, Effelsberg 100 m and Parkes 64 m radio telescopes. Figure credit: MPIfR/Patricia Reich.

Synchrotron emission at 408 MHz suffers strong depolarization by differential Faraday rotation along the LoS. As a result, the polarized intensity of the 408 MHz emission is very weak and only representative of the nearby ISM. To detect synchrotron emission in polarization from deeper in the Galaxy, one must turn to higher frequencies. Figure 4 shows an all-sky map of the polarized intensity measured by WMAP at 23 GHz – a frequency little affected by Faraday rotation. The two features that clearly stand out are the strongly polarized NPS in the middle and Fan region on the left. The orientation of $\vec{B}_\perp$ inferred from synchrotron polarization is in broad agreement with the orientation inferred from dust polarization (see Figures 1 and 2 in Sect. 2.1).

 https://doi.org/10.1051/epjconf/202328303001

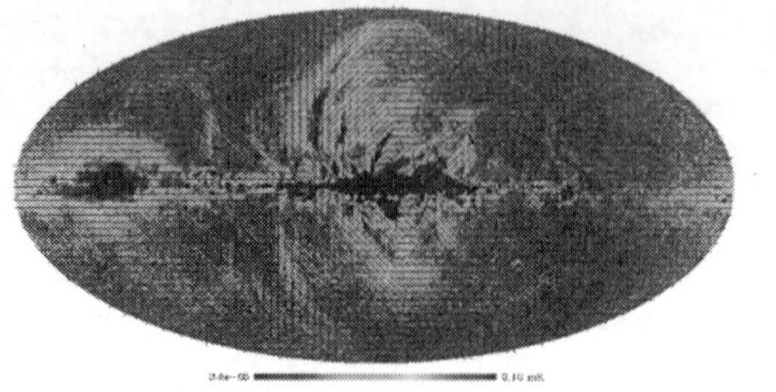

**Figure 4.** All-sky map in Galactic coordinates (with the Galactic center in the middle) of the polarized intensity of the 23 GHz continuum emission measured by WMAP, with the polarization half-vectors rotated by 90° to trace the orientation of $\vec{B}_\perp$. Figure credit: Tess Jaffe.

### 2.3 Faraday rotation

In ionized regions of the ISM, the interstellar magnetic field can be probed with Faraday rotation measures of Galactic pulsars and extragalactic sources of linearly polarized radio waves [24].

Faraday rotation of a linearly polarized radio wave occurs when the wave propagates through a magneto-ionized medium. This rotation results from the interaction of the wave with the thermal electrons, which systematically gyrate in a right-handed sense about magnetic field lines. The angle by which the orientation of polarization rotates is given by

$$\Delta\psi = \mathrm{RM}\,\lambda^2\,, \tag{6}$$

where $\lambda$ is the observing wavelength and

$$\mathrm{RM} = C \int_0^L n_e\, B_\parallel\, ds \tag{7}$$

is the so-called rotation measure, with $C$ a numerical constant, $n_e$ the thermal-electron density, $B_\parallel$ the LoS component of the magnetic field, and $L$ the path length from the source to the observer. In practice, the rotation measure of a given radio source can be determined by measuring the polarization orientation of the incoming radiation at at least two different wavelengths.

Rotation measures have now been obtained for about 1 200 Galactic pulsars and 45 000 extragalactic radio point sources.

The big advantage of Galactic pulsars is that one can measure, in addition to their rotation measures, their dispersion measures,

$$\mathrm{DM} = \int_0^L n_e\, ds\,, \tag{8}$$

and their distances, $L$ (with some uncertainty). The ratio of RM to DM directly yields the $n_e$-weighted average value of $B_\parallel$ between the considered pulsar and the observer. By combining the average values of $B_\parallel$ toward pulsars with their measured distances, one can in principle map out the large-scale 3D distribution of $B_\parallel$.

The big advantage of extragalactic radio point sources is that they are much more numerous than pulsars, such that they give access to finer details in the magnetic field distribution. The fact that they lie beyond the Galaxy is both an asset and a drawback, in the sense that their rotation measures also probe the outskirts of the Galaxy, beyond the pulsar region, but they are contaminated by contributions from the host galaxy and from the intervening intergalactic medium. Another limitation is that extragalactic sources have no dispersion measures, so their rotation measures must be complemented with a model for the spatial distribution of thermal electrons.

Rotation-measure studies have provided invaluable information on the strength, direction, and spatial configuration of the interstellar magnetic field in ionized regions. Low-latitude radio sources (both Galactic pulsars and extragalactic point sources) have been used to probe the Galactic disk (see Figure 5), while radio sources across the entire sky have been used to probe the Galactic halo (see Figure 6 for extragalactic point sources and Figure 7 for Galactic pulsars).

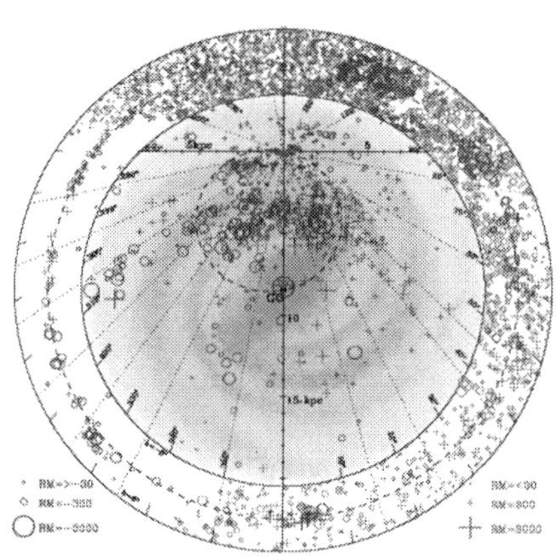

**Figure 5.** Top view of the Galactic disk showing the RMs of low-latitude ($|b| < 8°$) radio sources: Galactic pulsars observed with the Parkes 64 m and Green-Bank 100 m radio telescopes (in the inner disk) and extragalactic point sources from the literature (in the outer ring) [25]. Positive (negative) RMs, wich correspond to a magnetic field pointing on average toward (away from) the observer, are plotted in red (blue). The symbol size is proportional to $|\mathrm{RM}|^{0.5}$. The background yellow pattern is an artist's impression of the Galactic structure.

The regular magnetic field, $\vec{B}_{\rm reg}$, has a strength $B_{\rm reg} \simeq 1.5\,\mu$G in the Galactic vicinity of the Sun, compared to $B_{\rm fluct} \sim 5\,\mu$G for the fluctuating magnetic field [29]. $B_{\rm reg}$ increases toward the Galactic center, to $\gtrsim 3\,\mu$G at Galactocentric radius $r = 3$ kpc [30], i.e., with an exponential scale length $\lesssim 7.2$ kpc. Moreover, $B_{\rm reg}$ decreases away from the Galactic midplane, with an exponential scale height estimated at $\approx 2$ kpc, on average over Galactic quadrants Q1 and Q2, above and below the midplane [27].

In the Galactic disk, $\vec{B}_{\rm reg}$ is nearly horizontal and generally dominated by its azimuthal component. In the Galactic vicinity of the Sun, $\vec{B}_{\rm reg}$ points toward $(l, b) \simeq (82°, 0°)$ [31], in very good agreement with the orientation of $\vec{B}_{\rm ord}$ inferred from starlight polarimetry (see Sect. 2.1).

https://doi.org/10.1051/epjconf/202328303001

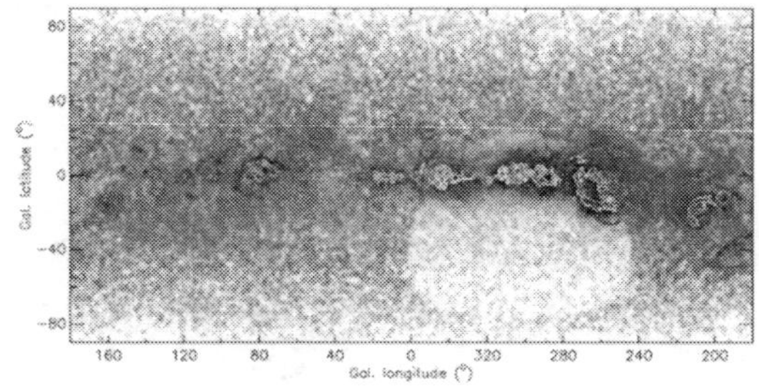


**Figure 6.** All-sky map in Galactic coordinates (with the Galactic center in the middle) of the RMs of ≃ 45 000 extragalactic radio point sources from the NVSS ($\delta > -40°$) and S-PASS ($\delta < 0°$) surveys [26]. Positive (negative) RMs are plotted in red (blue).

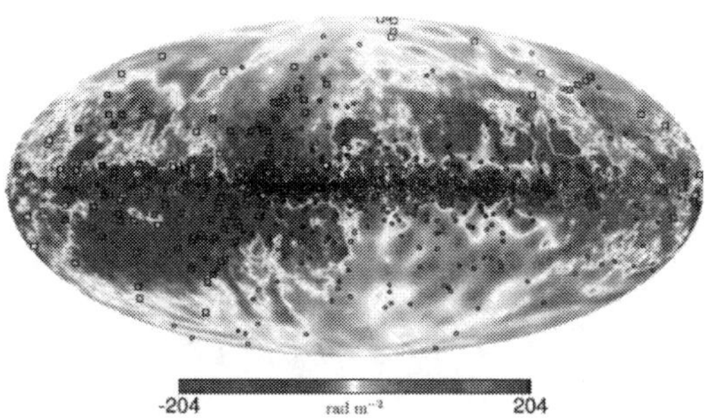


**Figure 7.** All-sky map in Galactic coordinates (with the Galactic center in the middle) of the RMs of 1 133 Galactic pulsars from the current pulsar catalogue (1.59) [27]. The squares correspond to the 137 new RMs obtained with the LOFAR High-Band Antenna (HBA) at ≃ (110 − 190) MHz. The background represents the all-sky Galactic Faraday depth (defined by Eq. (9)) reconstructed from the RMs of extragalactic sources [28]. Positive (negative) RMs are plotted in red (blue).

$\vec{B}_{\rm reg}$ reverses direction at least a couple of times with decreasing Galactocentric radius, but the exact number and radial locations of the field reversals are still controversial. These field reversals have often been interpreted as evidence that $\vec{B}_{\rm reg}$ is bisymmetric (azimuthal wavenumber $m = 1$), while an axisymmetric ($m = 0$) field would be expected from dynamo theory. In reality, [32] showed that neither the axisymmetric nor the bisymmetric picture is consistent with the existing pulsar rotation measures, and they concluded that $\vec{B}_{\rm reg}$ must have a more complex pattern. In general, $\vec{B}_{\rm reg}$ is described as a spiral magnetic field with a small pitch angle. A possible model of $\vec{B}_{\rm reg}$ in the Galactic disk, derived by [33], is sketched in Figure 8.

In the Galactic halo, $\vec{B}_{\rm reg}$ probably has a significant vertical component, $(B_{\rm reg})_z$. Near the horizontal position of the Sun, [34] obtained $(B_{\rm reg})_z \simeq -0.14\,\mu$G above the Galactic midplane ($z > 0$) and $(B_{\rm reg})_z \simeq +0.30\,\mu$G below the midplane ($z < 0$), whereas [35] obtained $(B_{\rm reg})_z \simeq 0.00\,\mu$G toward the north Galactic pole and $(B_{\rm reg})_z \simeq +0.31\,\mu$G toward the south Galactic pole. In contrast to the situation prevailing in the Galactic disk, $\vec{B}_{\rm reg}$ shows no sign of reversal with decreasing Galactocentric radius.

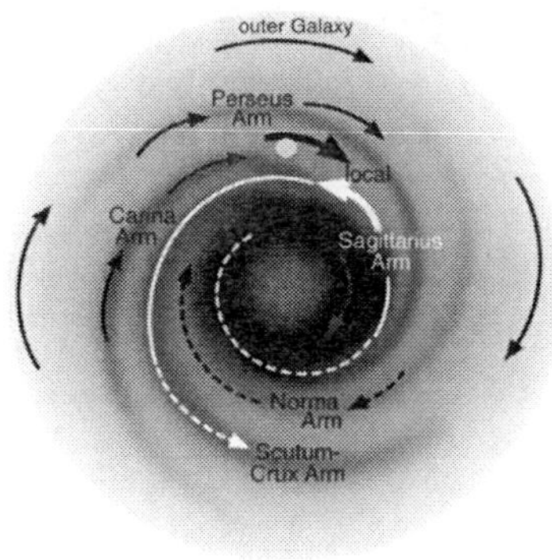


**Figure 8.** Sketch of the large-scale configuration of $\vec{B}_{\rm reg}$ in the Galactic disk from the model of [33]. $\vec{B}_{\rm reg}$ runs clockwise (black arrows) in most of the disk and counterclockwise (white arrows) in a restricted region of reversed field. The dashed arrows are more uncertain than the solid arrows. The yellow dot indicates the position of the Sun.

However, the horizontal component of $\vec{B}_{\rm reg}$ clearly reverses direction between the northern and southern Galactic hemispheres, which suggests that $\vec{B}_{\rm reg}$ is antisymmetric (odd symmetry) with respect to the midplane. [36] found that $\vec{B}_{\rm reg}$ is slightly more likely to be bisymmetric than axisymmetric. They also showed that $\vec{B}_{\rm reg}$ can be described as an upward/downward spiraling magnetic field (in the northern/southern hemisphere), formed by an azimuthal field plus an X-shape poloidal field (as observed in the halos of external spiral galaxies; see Sect. 3.1). A field line from a possible model of $\vec{B}_{\rm reg}$ in the Galactic halo, derived by [36], is plotted in Figure 9.

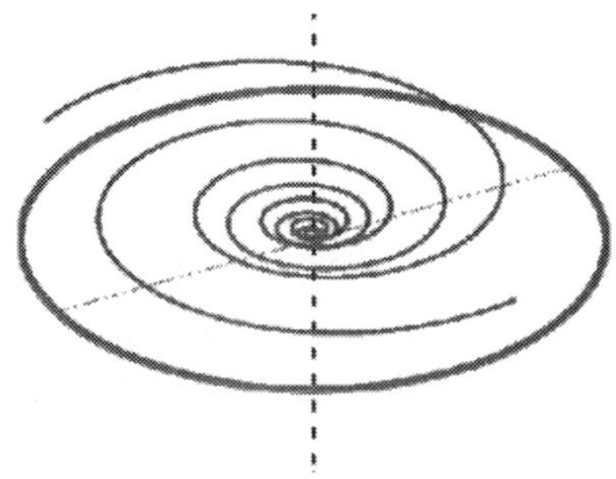

**Figure 9.** One field line of $\vec{B}_{\rm reg}$ in the Galactic halo from a model of upward spiraling magnetic field in the Galactic halo [36]. Figure credit: Philippe Terral.

### 2.4 Faraday tomography

An important limitation of the observational methods described in Sects. 2.1–2.3 is that they provide only LoS-integrated quantities, with no details on how the integrants vary along the LoS. For instance, the synchrotron intensity (Eq. 5) measured in a given direction tells us only about the total amount of synchrotron emission produced along the entire LoS through the Galaxy, with no information on the local value of $\mathcal{E}_\nu$ (Eq. 4). Similarly, the rotation measure of a given radio source (Eq. 7) tells us only about the total amount of Faraday rotation incurred along the LoS between the source and the observer, with no information on the local value of $n_e\, B_\parallel$.

A powerful and promising approach to probe the 3D structure of the interstellar magnetic field is now being increasingly utilized. This approach is also based on Faraday rotation, but instead of considering the Faraday rotation of the linearly polarized radiation from a background radio source (as explained in Sect. 2.3), the idea is to exploit the Faraday rotation of the synchrotron radiation from the Galaxy itself (discussed in Sect. 2.2).

In the case of a background radio source, i.e., when the regions of synchrotron emission and Faraday rotation are spatially separated, the Faraday rotation angle, $\Delta\psi$, varies linearly with $\lambda^2$, and the rotation measure, RM, is defined as the slope of the linear relation between $\Delta\psi$ and $\lambda^2$ (see Eq. 6). Hence, RM is a purely observational quantity, which can be meaningfully measured only for a background radio source and which can then be related to the physical properties ($n_e$ and $B_\parallel$) of the foreground Faraday-rotating medium through Eq. (7).

In contrast, when the radio source is the Galaxy itself, the regions of synchrotron emission and Faraday rotation are spatially mixed. In that case, $\Delta\psi$ no longer varies linearly with $\lambda^2$ and the very concept of rotation measure becomes meaningless. However, one may resort to the more general concept of Faraday depth, defined as

$$\Phi(z) \equiv C \int_0^z n_e\, B_\parallel\, ds\,, \tag{9}$$

where all symbols have the same meaning as in Eq. (7) and $z$ is the LoS distance from the observer[5] [37, 38]. $\Phi(z)$ has basically the same formal expression as RM (Eq. 7), but it differs from RM in the sense that it is a truly physical quantity, which can be defined at any point of the ISM, independent of any background radio source. $\Phi(z)$ simply corresponds to the LoS depth, $z$, measured in terms of Faraday rotation – in much the same way as optical depth corresponds to LoS depth measured in terms of opacity.

When synchrotron emission and Faraday rotation are mixed along the LoS, the polarized intensity measured at a given wavelength $\lambda$, is the superposition of the polarized emission produced at every LoS distance $z$, i.e., at every Faraday depth $\Phi$, and Faraday-rotated by an angle $\Delta\psi = \Phi\, \lambda^2$ (from Eq. (6) with RM replaced by $\Phi$). Since the Faraday rotation angle, $\Delta\psi$, varies with wavelength, the polarized intensities measured at different wavelengths correspond to different combinations of all the LoS contributions and, therefore, provide different pieces of information. Thus, the idea is to measure the polarized intensity at a large number of different wavelengths and to convert its variation with $\lambda^2$ into a variation with $\Phi$. Mathematically, this can be done through a Fourier transform.

[5]Here, the symbol $z$ refers to the distance from the observer, whereas in Sect. 2.3, $z$ refers to the vertical component in Galactic coordinates, i.e., the distance from the Galactic midplane.

By applying this rotation measure synthesis to all the pixels of a given area in the sky, one obtains a so-called Faraday cube, i.e., a 3D map of the synchrotron polarized emission in Faraday space, $(\alpha, \delta, \Phi)$ or $(l, b, \Phi)$. An example is shown in Figure 10.

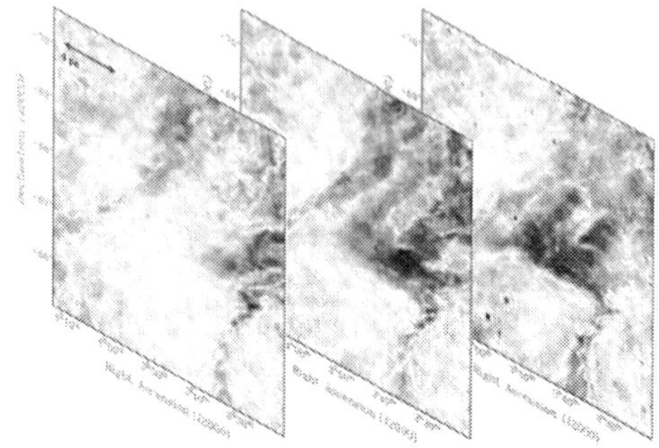

**Figure 10.** Illustration of Faraday tomography. Three slices, at Faraday depths $\Phi_1 = -2.0$ rad m$^{-2}$, $\Phi_2 = -1.5$ rad m$^{-2}$, and $\Phi_3 = -1.0$ rad m$^{-2}$, through the Faraday cube obtained by [39] toward the Fan region, based on data from the LOFAR HBA at (115 – 178) MHz. Figure credit: Marta Alves.

In practice, Faraday tomography can be used to separate synchrotron-emitting regions located at different Faraday depths along the LoS and to estimate their respective polarized synchrotron intensities, which in turn can lead to the strength and the orientation of their $\vec{B}_\perp$ (see Sect. 2.2). Faraday tomography can also be used to uncover intervening Faraday screens and to estimate their Faraday thicknesses, which in turn can lead to their $B_\parallel$. The method is particularly interesting when the uncovered Faraday screens can be identified with known gaseous structures, because it then offers a new way of probing their magnetic fields.

# 3 Magnetic fields outside our Galaxy

## 3.1 Interstellar magnetic fields in external galaxies

Interstellar magnetic fields in external galaxies are detected and measured with the same tools as those employed for our own Galaxy, namely, synchrotron emission, Faraday rotation, and, more recently, dust polarized emission. Observations of external galaxies are complementary to observations of our Galaxy, because they directly provide a bird's eye view of their large-scale magnetic fields, whereas our position within our Galaxy makes it difficult to disentagle the contributions from the large-scale Galactic magnetic field and from small-scale nearby magnetic fluctuations.

Radio observations of external galaxies have systematically detected synchrotron emission from spiral, barred, flocculent (with disks, but no spiral arms), and irregular galaxies [see, e.g., 40, for a detailed review].

Measurements of the synchrotron total intensity, interpreted with the equipartition assumption (see Sect. 2.2), indicate that these galaxies host interstellar magnetic fields with total field strengths $B \sim$ a few $\mu$G. The total fields tend to be strongest in spiral arms and in bars, with $B \sim (20 - 30)\,\mu$G, and in central starburst regions, with $B \sim (50 - 100)\,\mu$G.

Measurements of the synchrotron polarized intensity reveal the existence of ordered magnetic fields in spiral, barred, and flocculent galaxies. The ordered fields are generally strongest in interarm regions, with $B_{\rm ord} \sim (10 - 15)\,\mu$G. In face-on galaxies, $\vec{B}_{\rm ord}$ describes a spiral pattern, with $\vec{B}_{\rm ord}$ approximately parallel to the gaseous spiral arms (when present). The prototype example, displayed in Figure 11, is the whirlpool galaxy, M 51. In edge-on galaxies, $\vec{B}_{\rm ord}$ appears to be horizontal in the disk and X-shaped in the halo. This is illustrated in Figure 12, with the galaxy NGC 891.

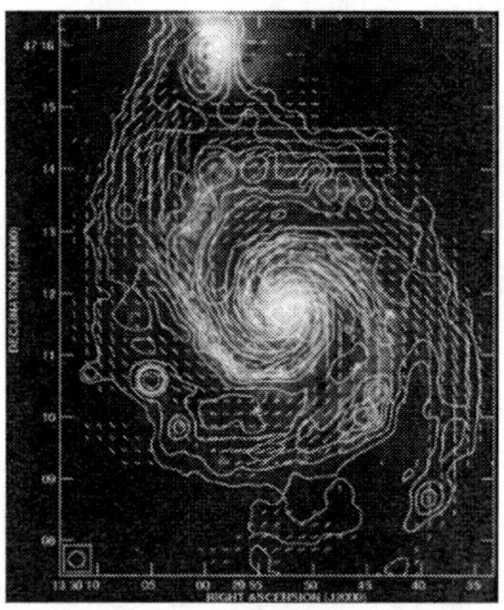

**Figure 11.** Total intensity contours of the $\lambda\,6$ cm radio continuum emission from the face-on spiral galaxy M 51 observed with the VLA and the Effelsberg 100 m radio telescope. The background is an optical image from the HST. The small bars show the polarization half-vectors rotated by 90° to trace the orientation of $\vec{B}_\perp$, with no correction for Faraday rotation [41].

It is very likely that elliptical galaxies also possess $\mu$G interstellar magnetic fields. However, these fields are generally much more difficult (or even impossible) to detect through synchrotron emission, because of a lack of relativistic electrons in the absence of active star formation [43]. Nevertheless, one can hold the following theoretical reasoning: since elliptical galaxies have strong interstellar turbulence, but a very low rotation rate, they must be subject to a fluctuating dynamo – rather than the mean-field $\alpha\omega$ dynamo that operates in spiral galaxies – and this dynamo must produce purely fluctuating magnetic fields, i.e., magnetic fields without an ordered component [44, 45].

As explained in Sect. 2.2, synchrotron emission is sensitive to the orientation, but not the direction, of the magnetic field. This is why synchrotron observations carry information on the ordered magnetic field, $\vec{B}_{\rm ord}$, but not on the regular magnetic field, $\vec{B}_{\rm reg}$. The only way to separate out the contribution from $\vec{B}_{\rm reg}$ is to resort to Faraday rotation measures. In the case of external galaxies, the rotation measures of interest are those of background radio sources or those of the galaxies themselves.

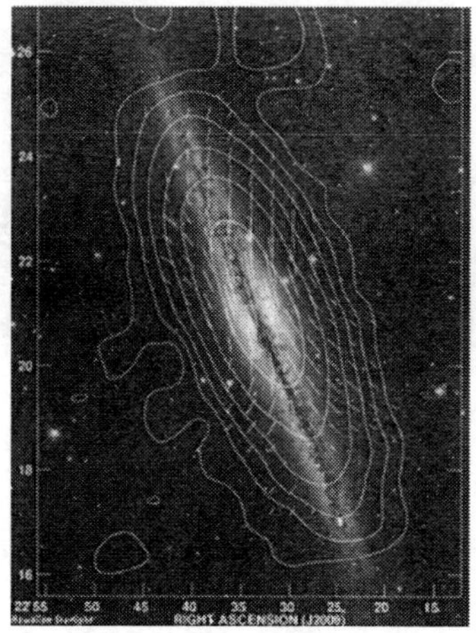

**Figure 12.** Total intensity contours of the $\lambda\,3.6$ cm radio continuum emission from the edge-on spiral galaxy NGC 891 observed with the Effelsberg 100 m radio telescope. The background is an optical image from the CFHT. The small bars show the polarization half-vectors rotated by 90° to trace the orientation of $\vec{B}_\perp$, with no correction for Faraday rotation [42]. Copyright: MPIfR Bonn.

Large-scale patterns in the RM maps of external galaxies reveal the presence of regular magnetic fields. The azimuthal structure of $\vec{B}_{\rm reg}$ in galactic disks, inferred from the large-scale azimuthal variation of RM, displays a variety of behaviors: in some cases, $\vec{B}_{\rm reg}$ is dominated by one azimuthal (generally axisymmetric) mode; in other cases, a superposition of 2 or 3 azimuthal modes are needed to describe $\vec{B}_{\rm reg}$ [see, e.g., Table 2 in 40]. The vertical structure of $\vec{B}_{\rm reg}$ and its potential (even or odd) symmetry with respect to the midplane are more difficult to establish. Indications of even symmetry have been found for a handful of nearby galaxies [40].

The CHANG-ES (Continuum HAlos in Nearby Galaxies – an EVLA Survey) project to observe 35 nearby edge-on galaxies at 6 GHz and 1.6 GHz [46] has provided the most comprehensive data set to date on magnetic fields in galactic halos. The polarization maps of the observed galaxies confirm and generalize the earlier finding that $\vec{B}_{\rm ord}$ has an X-shape pattern in the halos of spiral galaxies [47]. In addition, the sensitivity and broad bandwidth of the observations have made it possible to perform Faraday tomography. The resulting RM maps unambiguously show that the ordered magnetic field, $\vec{B}_{\rm ord}$, is partly (perhaps mostly) a regular field, $\vec{B}_{\rm reg}$ (see footnote 2), which extends over several kpc and has a typical scale $\gtrsim 1$ kpc [47].

Interstellar magnetic fields in external galaxies are now also being observed through the polarization of the far-infrared (FIR) dust thermal emission. For instance, the magnetic fields of the face-on galaxy M 51

and the edge-on galaxy NGC 891 were observed at 154 $\mu$m with the High-resolution Airborne Wideband Camera Plus (HAWC+) on the Stratospheric Observatory for Infrared Astronomy (SOFIA) [48].
For M 51, the FIR polarization map (Figure 13) is in very good agreement with its synchrotron counterpart (Figure 11): here, too, the inferred $\vec{B}_{\rm ord}$ follows approximately the material spiral arms. The FIR polarization fraction is much lower in the FIR-bright central region, possibly as a result of turbulent depolarization (across the beam and along the LoS).
For NGC 891, the FIR polarization map (Figure 14) is also consistent with its synchrotron counterpart (Figure 12), especially in the disk where the inferred $\vec{B}_{\rm ord}$ is again nearly horizontal. There is tentative evidence for a vertical $\vec{B}_{\rm ord}$ off the disk, but the FIR polarized emission from the halo is too weak to reveal a possible X-shape $\vec{B}_{\rm ord}$. The FIR polarization fraction is very low in a few places, probably where one is looking down spiral arms and viewing the magnetic field almost head-on.

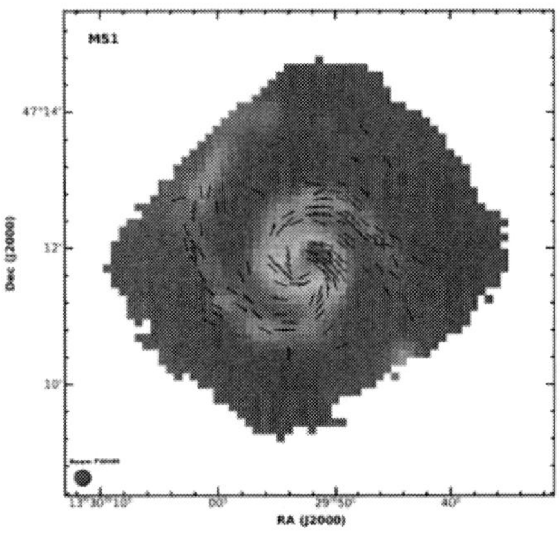


**Figure 13.** Total intensity map of the $\lambda$ 154 $\mu$m FIR continuum emission from the face-on spiral galaxy M 51 observed with SOFIA/HAWC+. The small bars show the polarization half-vectors rotated by 90° to trace the orientation of $\vec{B}_\perp$ [48].

New FIR polarimetric mapping observations of nearby galaxies, including M 51 [49], were recently carried out as part of the SALSA (Survey on extragALactic magnetiSm with SOFIA) Legacy Program. The first data release, for a set of 14 nearby (< 20 Mpc) galaxies, observed with SOFIA/HAWC+ from 53 $\mu$m to 214 $\mu$m, was presented by [50]. The results of SALSA are complementary to those of radio polarimetric observations. Both sets of results will eventually be combined to produce a comprehensive empirical picture of the kpc-scale structure and strength of magnetic fields in external galaxies.

### 3.2 Extragalactic magnetic fields

Magnetic fields have also been detected outside galaxies – certainly in the intergalactic medium of clusters of galaxies, probably in filaments of the cosmic web, and tentatively in cosmic voids.

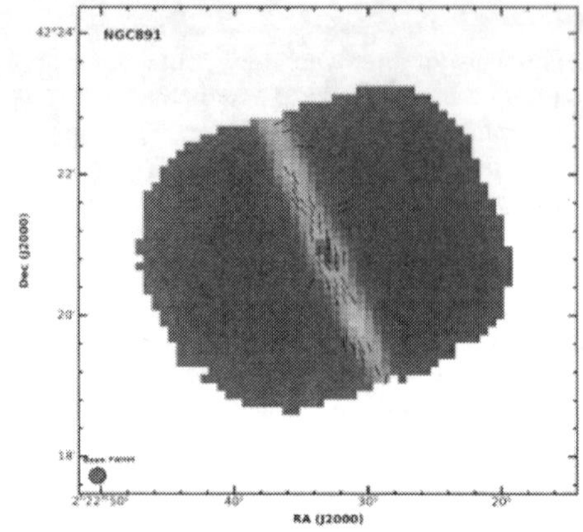


**Figure 14.** Total intensity map of the $\lambda$ 154 $\mu$m FIR continuum emission from the edge-on spiral galaxy NGC 891 observed with SOFIA/HAWC+. The small bars show the polarization half-vectors rotated by 90° to trace the orientation of $\vec{B}_\perp$ [48].

Intergalactic magnetic fields in clusters of galaxies are mostly detected through diffuse radio synchrotron emission from the intergalactic medium (IGM) and through Faraday rotation of background or embedded radio galaxies [see, e.g., [51] for a comprehensive review of early results and [52] for a review of more recent results]. A small fraction of the clusters (about 50 clusters according to [52]) display extended diffuse synchrotron emission, which shows no correlation with the individual galaxies, but appears to be associated with the IGM. It is likely that the other clusters also emit synchrotron radiation, but their surface brightnesses are probably too low to enable detection with present-day radio telescopes.

Clusters observed in radio synchrotron emission have led to a classification into radio halos, radio relics, and mini-halos according to the morphology and location of their radio emission.
Radio halos are roundish structures extending over $\gtrsim$ 1 Mpc in the centers of certain clusters undergoing violent merger events. Their measured synchrotron total intensities, together with the assumption that the intergalactic magnetic field is homogeneous and in equipartition with cosmic rays (see Sect. 2.2), lead to magnetic field strengths $\sim (0.1 - 1)\,\mu$G. The non-detection of polarization in most cases indicates either beam depolarization in the absence of an ordered magnetic field or Faraday depolarization due to mixing of thermal and relativistic electrons (see Sect. 2.4).
Radio relics are elongated structures with sizes $\gtrsim$ 1 Mpc near the outskirts of certain clusters, where they are probably produced by shock compression during merger events. Their equipartition magnetic field strengths are typically $\sim (0.5 - 2)\,\mu$G. In contrast to radio halos, they are strongly and coherently polarized, and the inferred magnetic field orientations are generally along their long axes.
Mini-halos extend over $\lesssim$ 0.5 Mpc around dominant radio galaxies in the centers of relaxed, cooling-core clus-

ters. Their equipartition magnetic field strengths can reach $\sim 10\,\mu$G.
To illustrate, Figure 15 shows an image of the radio halo in the galaxy cluster A 665, while Figure 16 shows an image of the pair of radio relics in the merging galaxy cluster ZwCl 2341.1+0000.

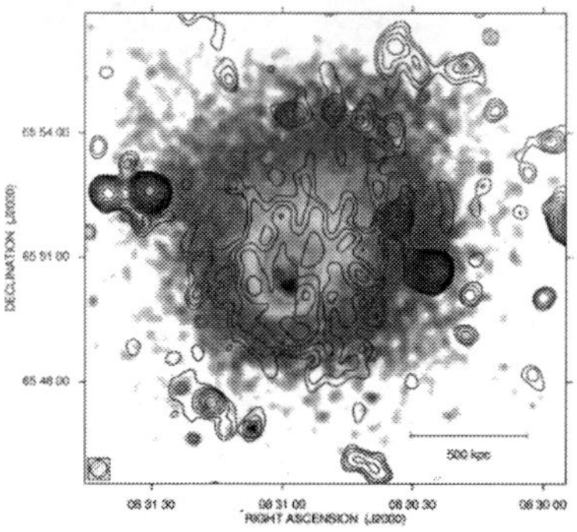


**Figure 15.** Radio halo in the central region of the galaxy cluster A 665. The contours show the total intensity of the 1.4 GHz radio continuum emission measured with the VLA. The background is an X-ray image from *Chandra* in the (0.8 – 4) keV band [53].

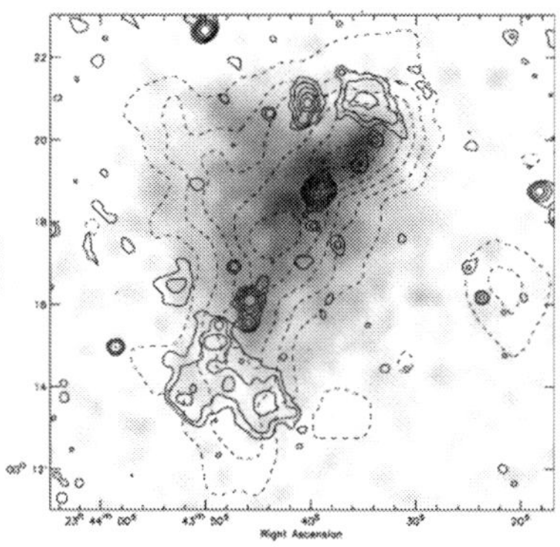

**Figure 16.** Pair of radio relics at the outskirts of the merging galaxy cluster ZwCl 2341.1+0000. The contours show the total intensity of the 610 MHz radio continuum emission measured with the GMRT. The background is an X-ray image from *Chandra* in the (0.5 – 3) keV band [54].

Rotation measures of background or embedded radio galaxies, combined with estimates of both the thermal-electron density in the IGM and the dominant scales of magnetic fluctuations, provide useful complementary information on intergalactic magnetic fields. The thermal-electron density is generally inferred from the measured X-ray surface brightnesses of clusters or taken from a generic model. The dominant magnetic fluctuation scales are estimated from the patchiness of the observed RM distributions. The resulting magnetic field strengths are $\sim (1 - 10)\,\mu$G, with values as high as $\sim 30\,\mu$G in cooling cores.

At scales even larger than galaxy clusters, there have been detections of extended diffuse synchrotron emission from filaments of the cosmic web [55, 56], indicating the likely existence of extracluster magnetic fields. [55] compared their observations to MHD simulations and found that their detected radio sources were similar to the brightest patches of diffuse synchrotron emission, which correspond to the densest and hottest portions of cosmic filaments, in the simulations. From this, they concluded that the underlying magnetic fields must be on average $\sim (20 - 50)$ nG strong. For illustration, an image of the radio synchrotron ridge connecting the merging galaxy clusters A 399 and A 401 within a cosmic filament [56] is displayed in Figure 17.

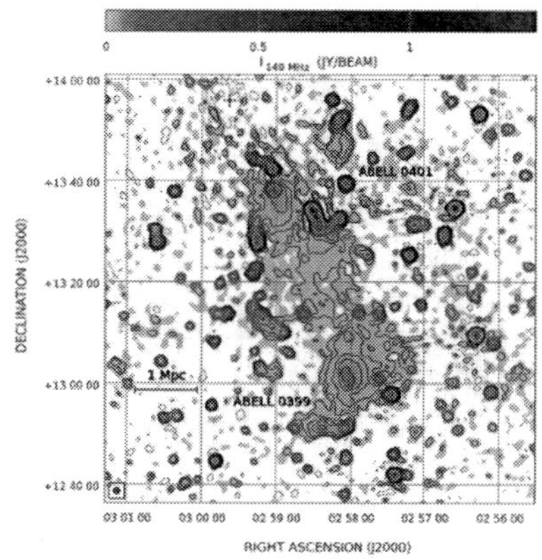


**Figure 17.** Ridge of radio synchrotron emission in a filament of the cosmic web, between the merging galaxy clusters A 399 and A 401. Color and contours show the total intensity of the 140 MHz radio continuum emission measured with LOFAR [56].

Finally, voids of the cosmic web have been suggested to host extremely weak magnetic fields. The evidence is indirect and based on an interpretation of the measured $\gamma$-ray emission from blazars. From the non-detection of GeV echos in the long-term GeV-TeV light curves of the MRK 421 blazar, [57] derived a lower limit to the magnetic field strength in cosmic voids of $B \gtrsim 10^{-20.5}$ G. A more constraining estimate was later derived by [58], who relied on the sizes of the GeV halos around stacked low-redshift blazars to find $B \sim (10^{-17} - 10^{-15})$ G.

## References

[1] E. Fermi, Physical Review **75**, 1169 (1949)
[2] R.M. Crutcher, B. Wandelt, C. Heiles, E. Falgarone, T.H. Troland, ApJ **725**, 466 (2010)
[3] P. Goldreich, N.D. Kylafis, ApJL **243**, L75 (1981)
[4] P. Goldreich, N.D. Kylafis, ApJ **253**, 606 (1982)
[5] B.T. Draine, J.C. Weingartner, ApJ **470**, 551 (1996), `arXiv:astro-ph/9605046`
[6] A. Lazarian, T. Hoang, MNRAS **378**, 910 (2007), `0707.0886`

[7] T. Hoang, A. Lazarian, MNRAS **388**, 117 (2008), 0707.3645
[8] R.H. Hildebrand, QJRAS **29**, 327 (1988)
[9] Planck Collaboration, P.A.R. Ade, N. Aghanim, D. Alina, M.I.R. Alves, G. Aniano, C. Armitage-Caplan, M. Arnaud, D. Arzoumanian, M. Ashdown et al., A&A **576**, A105 (2015), 1405.0872
[10] D.S. Mathewson, V.L. Ford, MmRAS **74**, 139 (1970)
[11] C. Heiles, ApJ **462**, 316 (1996)
[12] C. Heiles, AJ **119**, 923 (2000), arXiv:astro-ph/9910303
[13] A. Berdyugin, V. Piirola, P. Teerikorpi, A&A **561**, A24 (2014)
[14] V. Pelgrims, G.V. Panopoulou, K. Tassis, V. Pavlidou, A. Basyrov, D. Blinov, E. Gjerløw, S. Kiehlmann, N. Mandarakas, A. Papadaki et al., arXiv e-prints arXiv:2208.02278 (2022), 2208.02278
[15] Planck Collaboration, P.A.R. Ade, N. Aghanim, D. Alina, M.I.R. Alves, C. Armitage-Caplan, M. Arnaud, D. Arzoumanian, M. Ashdown, F. Atrio-Barandela et al., A&A **576**, A104 (2015), 1405.0871
[16] V.L. Ginzburg, S.I. Syrovatskii, ARAA **3**, 297 (1965)
[17] G. Miley, ARAA **18**, 165 (1980)
[18] L.F. Burlaga, N.F. Ness, ApJ **784**, 146 (2014)
[19] A.C. Cummings, E.C. Stone, B.C. Heikkila, N. Lal, W.R. Webber, G. Jóhannesson, I.V. Moskalenko, E. Orlando, T.A. Porter, ApJ **831**, 18 (2016)
[20] R. Beck, Space Science Reviews **99**, 243 (2001), arXiv:astro-ph/0012402
[21] C.G.T. Haslam, C.J. Salter, H. Stoffel, W.E. Wilson, A&AS **47**, 1 (1982)
[22] K. Beuermann, G. Kanbach, E.M. Berkhuijsen, A&A **153**, 17 (1985)
[23] K. Ferrière, ApJ **497**, 759 (1998)
[24] L. Spitzer, *Physical processes in the interstellar medium* (Wiley, Somerset, 1978)
[25] J.L. Han, R.N. Manchester, W. van Straten, P. Demorest, ApJS **234**, 11 (2018), 1712.01997
[26] D.H.F.M. Schnitzeler, E. Carretti, M.H. Wieringa, B.M. Gaensler, M. Haverkorn, S. Poppi, MNRAS **485**, 1293 (2019), 1902.09556
[27] C. Sobey, A.V. Bilous, J.M. Grießmeier, J.W.T. Hessels, A. Karastergiou, E.F. Keane, V.I. Kondratiev, M. Kramer, D. Michilli, A. Noutsos et al., MNRAS **484**, 3646 (2019), 1901.07738
[28] N. Oppermann, H. Junklewitz, M. Greiner, T.A. Enßlin, T. Akahori, E. Carretti, B.M. Gaensler, A. Goobar, L. Harvey-Smith, M. Johnston-Hollitt et al., A&A **575**, A118 (2015), 1404.3701
[29] R.J. Rand, S.R. Kulkarni, ApJ **343**, 760 (1989)
[30] J.L. Han, R.N. Manchester, A.G. Lyne, G.J. Qiao, W. van Straten, ApJ **642**, 868 (2006), arXiv:astro-ph/0601357
[31] J.L. Han, R.N. Manchester, G.J. Qiao, MNRAS **306**, 371 (1999), arXiv:astro-ph/9903101
[32] H. Men, K. Ferrière, J.L. Han, A&A **486**, 819 (2008), 0805.3454
[33] C.L. Van Eck, J.C. Brown, J.M. Stil, K. Rae, S.A. Mao, B.M. Gaensler, A. Shukurov, A.R. Taylor, M. Haverkorn, P.P. Kronberg et al., ApJ **728**, 97 (2011), 1012.2938
[34] A.R. Taylor, J.M. Stil, C. Sunstrum, ApJ **702**, 1230 (2009)
[35] S.A. Mao, B.M. Gaensler, M. Haverkorn, E.G. Zweibel, G.J. Madsen, N.M. McClure-Griffiths, A. Shukurov, P.P. Kronberg, ApJ **714**, 1170 (2010), 1003.4519
[36] P. Terral, K. Ferrière, A&A **600**, A29 (2017), 1611.10222
[37] B.J. Burn, MNRAS **133**, 67 (1966)
[38] M.A. Brentjens, A.G. de Bruyn, A&A **441**, 1217 (2005), astro-ph/0507349
[39] C.L. Van Eck, M. Haverkorn, M.I.R. Alves, R. Beck, A.G. de Bruyn, T. Enßlin, J.S. Farnes, K. Ferrière, G. Heald, C. Horellou et al., A&A **597**, A98 (2017), 1612.00710
[40] R. Beck, AApR **24**, 4 (2015), 1509.04522
[41] A. Fletcher, R. Beck, A. Shukurov, E.M. Berkhuijsen, C. Horellou, MNRAS **412**, 2396 (2011), 1001.5230
[42] M. Krause, *Magnetic Fields and Star Formation in Spiral Galaxies*, in *Revista Mexicana de Astronomia y Astrofisica Conference Series* (2009), Vol. 36 of *Revista Mexicana de Astronomia y Astrofisica Conference Series*, pp. 25–29, 0806.2060
[43] R. Beck, R. Wielebinski, in *Planets, Stars and Stellar Systems. Volume 5: Galactic Structure and Stellar Populations*, edited by T.D. Oswalt, G. Gilmore (2013), Vol. 5, p. 641
[44] D. Moss, A. Shukurov, MNRAS **279**, 229 (1996)
[45] A. Seta, L.F.S. Rodrigues, C. Federrath, C.A. Hales, ApJ **907**, 2 (2021), 2012.02329
[46] J. Irwin, R. Beck, R.A. Benjamin, R.J. Dettmar, J. English, G. Heald, R.N. Henriksen, M. Johnson, M. Krause, J.T. Li et al., AJ **144**, 43 (2012), 1205.5694
[47] M. Krause, J. Irwin, P. Schmidt, Y. Stein, A. Miskolczi, S. Carolina Mora-Partiarroyo, T. Wiegert, R. Beck, J.M. Stil, G. Heald et al., A&A **639**, A112 (2020), 2004.14383
[48] T.J. Jones, J.A. Kim, C.D. Dowell, M.R. Morris, J.L. Pineda, D.J. Benford, M. Berthoud, D.T. Chuss, D.A. Dale, L.M. Fissel et al., AJ **160**, 167 (2020), 2008.07897
[49] A.S. Borlaff, E. Lopez-Rodriguez, R. Beck, R. Stepanov, E. Ntormousi, A. Hughes, K. Tassis, P.M. Marcum, L. Grosset, J.E. Beckman et al., ApJ **921**, 128 (2021), 2105.09315
[50] E. Lopez-Rodriguez, S.A. Mao, R. Beck, A.S. Borlaff, E. Ntormousi, K. Tassis, D.A. Dale, J. Roman-Duval, K. Subramanian, S. Martin-Alvarez et al., ApJ **936**, 92 (2022), 2205.01105

[51] F. Govoni, L. Feretti, International Journal of Modern Physics D **13**, 1549 (2004), astro-ph/0410182
[52] V. Vacca, M. Murgia, F. Govoni, T. Enßlin, N. Oppermann, L. Feretti, G. Giovannini, F. Loi, Galaxies **6**, 142 (2018)
[53] V. Vacca, M. Murgia, F. Govoni, L. Feretti, G. Giovannini, E. Orrù, A. Bonafede, A&A **514**, A71 (2010)
[54] R.J. van Weeren, H.J.A. Röttgering, J. Bagchi, S. Raychaudhury, H.T. Intema, F. Miniati, T.A. Enßlin, M. Markevitch, T. Erben, A&A **506**, 1083 (2009), 0908.0728
[55] V. Vacca, M. Murgia, F. Govoni, F. Loi, F. Vazza, A. Finoguenov, E. Carretti, L. Feretti, G. Giovannini, R. Concu et al., MNRAS **479**, 776 (2018), 1804.09199
[56] F. Govoni, E. Orrù, A. Bonafede, M. Iacobelli, R. Paladino, F. Vazza, M. Murgia, V. Vacca, G. Giovannini, L. Feretti et al., Science **364**, 981 (2019), 1906.07584
[57] K. Takahashi, M. Mori, K. Ichiki, S. Inoue, H. Takami, ApJL **771**, L42 (2013), 1303.3069
[58] W. Chen, J.H. Buckley, F. Ferrer, PRL **115**, 211103 (2015), 1410.7717

# 2022 report from the Auger-TA working group on UHECR arrival directions

*A.* di Matteo, *L.* Anchordoqui, *T.* Bister, *R.* de Almeida, *O.* Deligny, *L.* Deval, *G.* Farrar, *U.* Giaccari, *G.* Golup, *R.* Higuchi, *J.* Kim, *M.* Kuznetsov, *I.* Mariş, *G.* Rubtsov, *P.* Tinyakov, and *F.* Urban for the Pierre Auger [1,*] and Telescope Array collaborations [2]

[1]Observatorio Pierre Auger, Av. San Martín Norte 304, 5613 Malargüe, Argentina
[2]Telescope Array Project, 201 James Fletcher Bldg, 115 S. 1400 East, Salt Lake City, UT 84112-0830, USA

**Abstract.** After over 60 years, the powerful engines that accelerate ultra-high-energy cosmic rays (UHECRs) to the formidable energies at which we observe them from Earth remain mysterious. Assuming standard physics, we expect UHECR sources to lie within the local Universe (up to a few hundred Mpc). The distribution of matter in the local Universe is anisotropic, and we expect this anisotropy to be imprinted on the distribution of UHECR arrival directions. Even though intervening intergalactic and Galactic magnetic fields deflect charged UHECRs and can distort these anisotropies, some amount of information on the distribution of the sources is preserved. In this proceedings contribution, we present the results of the joint Pierre Auger Observatory and Telescope Array searches for (a) the largest-scale anisotropies (the harmonic dipole and quadrupole) and (b) correlations with a sample of nearby starburst galaxies and the 2MRS catalogue tracing stellar mass within 250 Mpc. This analysis updates our previous results with the most recent available data, notably with the addition of 3 years of new Telescope Array data. The main finding is a correlation between the arrival directions of $12.1\%^{+4.5\%}_{-3.1\%}$ of UHECRs detected with $E \geq 38$ EeV by Auger or with $E \gtrsim 49$ EeV by TA and the positions of nearby starburst galaxies on a $15.1^{\circ}{}^{+4.6^{\circ}}_{-3.0^{\circ}}$ angular scale, with a $4.7\sigma$ post-trial significance, up from $4.2\sigma$ obtained in our previous study.

## 1 Introduction

Ultra-high-energy cosmic rays (UHECRs) are particles which impact the Earth's atmosphere with energies $E \geq 1\,\text{EeV} = 10^{18}\,\text{eV} \approx 0.16\,\text{J}$. Over 60 years after their discovery, we still do not know what are the sources of UHECRs and what is the mechanism that accelerates them to such formidable energies [1, 2]. To a smaller extent, we also still do not know what UHECRs are, namely what is their mass composition. What we do know is that the highest energy rays are most likely extra-Galactic. This can be inferred from the fact that their arrival direction distribution does not correlate with the matter density in the Milky Way [3–5].

At the highest energies, UHECRs lose energy or disintegrate because of their interactions with cosmological background photons. This significantly reduces their propagation length, which is limited to a few hundred Mpc at a few tens of EeV and rapidly decreases further at higher energies. At such distances the local Universe is highly anisotropic, the more so the closer we look. Therefore, we expect that the distribution of UHECR arrival directions is also anisotropic. However, charged UHECRs on their way to the Earth encounter intervening extragalactic and Galactic magnetic fields which deflect their trajectories, thereby obfuscating the true position of their sources, blurring and distorting their original anisotropy. The strength of magnetic deflections depends on $ZB/E$, where $Z$ is the UHECR electric charge, $B$ the strength of the magnetic field, and $E$ is the UHECR energy. Therefore, at higher energies we expect the anisotropic distribution of sources to be better preserved than at low energies because UHECRs travel more rectilinearly. Moreover, at higher energies the UHECR propagation horizon is smaller and the original anisotropy is more pronounced. However, at higher energies the number of UHECRs is also much smaller, thereby increasing the shot-noise contribution for anisotropy searches. Large-scale anisotropies, for example the harmonic dipole or quadrupole, corresponding to anisotropies on angular scales of 180° and 90°, respectively, are expected to be much less affected by strong magnetic deflections of even some tens of degrees. On the other hand, associating events with specific sources is very hard unless the magnetic deflections, particularly the stochastic ones, are small enough. It is therefore fruitful to try to detect anisotropies at different energy scales.

The two largest cosmic-ray detector arrays in the world are the Pierre Auger Observatory (Auger) in Argentina and the Telescope Array (TA) in Utah, USA. Neither observatory can see the full sky: TA observes the full northern celestial hemisphere plus declinations $\delta > -15.7°$ below the equator; Auger observes the southern hemisphere plus northern declinations up to $\delta < +44.8°$. Without full-sky coverage, it is not possible to measure the dipole and quadrupole unless assumptions about higher multipoles

*e-mail: spokespersons@auger.org; full author list: https://auger.org/archive/authors_2022_10.html

are made. Moreover, there are comparably bright starburst galaxies (SBGs) in both hemispheres, and with partial-sky coverage we would miss some of them. It is therefore crucial to combine the data from both observatories in order to maximize their data yield. This is the motivation behind the establishment of the TA–Auger working group on UHECR arrival directions.

The first, and thus far only, anisotropy detected with more than $5\sigma$ significance was a 6.5% dipole at $E \geq 8$ EeV reported by Auger [4] with approximately 32 000 events, which has since grown in amplitude to 7.3% and in significance to $6.6\sigma$ [5] (with about 44 000 events). Hints of a medium-scale anisotropy have been reported as a correlation between the arrival directions of UHECRs and the positions of nearby SBGs on a $15.5^{\circ}{}^{+5.3^{\circ}}_{-3.2^{\circ}}$ angular scale, disfavouring isotropy at $4.2\sigma$ post-trial significance, using a joint Auger-TA dataset [6] (for the Auger-only analysis see [7, 8]).

## 2 The data sets

The Pierre Auger Observatory [9] is a hybrid detector of UHECRs located in the Southern hemisphere in Argentina at a latitude of −35.2°. It consists of a surface array of 1660 water-Cherenkov detectors covering an area of approximately 3000 km$^2$, overlooked by the fluorescence detector composed of 27 fluorescence telescopes. The detector has been taking data since January 2004. In this work, we use the dataset described in Ref. [10], consisting of events detected by the surface detector array from 2004 Jan 01 to 2020 Dec 31. The geometrical exposure is 95 700 km$^2$ yr sr for Auger vertical events (zenith angles $\theta < 60°$) and 26 300 km$^2$ yr sr for Auger inclined events ($60° \leq \theta < 80°$). This corresponds to 39 691 events with $E_{\text{Auger}} \geq 8.53$ EeV and 2635 events with $E_{\text{Auger}} \geq 32$ EeV.

The Telescope Array [11] is a hybrid detector of UHECRs located in the Northern hemisphere in Utah, USA at a latitude of 39.3°. It is taking data since May 2008. The surface detector of TA consists of 507 plastic scintillator detectors covering an area of about 700 km$^2$. The fluorescence detector of TA is composed of 38 fluorescence telescopes arranged in 3 towers overlooking the surface detector area. In this work, we use the events with the zenith angles $\theta < 55°$ detected by the TA surface detectors array from 2008 May 11 to 2022 May 10, namely three years more compared to the previous analyses [6, 12]. The effective exposure is 18 000 km$^2$ yr sr for TA events: this corresponds to 6014 events with $E_{\text{TA}} \geq 10$ EeV and 395 events with $E_{\text{TA}} \geq 40.5$ EeV.

## 3 The cross-calibration

UHECR energy measurements are affected by sizeable systematic uncertainties (±14% for Auger, ±21% for TA). If not corrected, a mismatch between energy scales can yield a spurious dipole. For instance, imagine that events with $E_{\text{true}} = 10$ EeV are reconstructed as $E_{\text{rec}} = 9$ EeV by Auger and as $E_{\text{rec}} = 11$ EeV by TA. If we collect all events with $E_{\text{rec}} \geq 10$ EeV, then events with $E_{\text{true}} = 10$ EeV are included in the TA data set but not in the Auger one; therefore, the flux would appear to be larger in the North compared to the South. It is therefore necessary to cross-calibrate the two data sets before searching for anisotropies. This can be done from the data themselves without any extra assumptions on the nature of the energy determination systematics by comparing the data of the two observatories in the equatorial band where their exposures overlap [13].

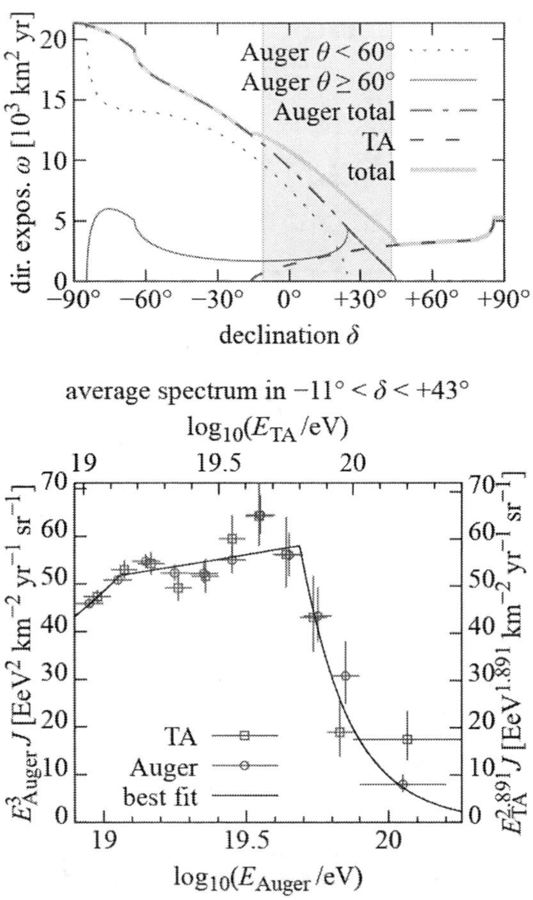


**Figure 1.** Top: the directional exposures of the datasets, with the declination band used for the energy cross-calibration highlighted. Bottom: the energy cross-calibration fit ($\chi^2/n_{\text{dof}} = 20.7/14$; $p = 0.11$).

In this contribution we follow the recipe outlined in [12] (which itself follows the earlier works [13–15]) and update it using the latest available data sets. We simultaneously fit a twice-broken power-law model for the energy spectrum and a power-law model for the $E_{\text{Auger}} \leftrightarrow E_{\text{TA}}$ conversion to the data in a common declination band as shown in figure 1:

$$E_{\text{Auger}} = \hat{E}e^{\alpha}(E_{\text{TA}}/\hat{E})^{\beta} \text{ , and } E_{\text{TA}} = \hat{E}e^{-\alpha/\beta}(E_{\text{Auger}}/\hat{E})^{1/\beta},$$

where $\hat{E} = 10\,\text{EeV}$, $\alpha = -0.159 \pm 0.012$ and $\beta = 0.945 \pm 0.016$. The fit returns $\chi^2/n = 20.7/14$ with $n = 14$ degrees of freedom, corresponding to a $p$-value of 0.11. Note that in this fit we used data sets optimized for anisotropy studies at $E_{\text{TA}} \geq 10\,\text{EeV}$: the result must not be extrapolated to lower energies or used outside of the scope of this analysis.

## 4 Latest results

### 4.1 Large-scale anisotropies: dipoles and quadrupoles

The flux $\Phi(\hat{\mathbf{n}})$ of UHECRs as a function of the arrival direction $\hat{\mathbf{n}}$ can be expanded into spherical harmonics $Y_{\ell m}(\hat{\mathbf{n}})$ as

$$\begin{aligned}\Phi(\hat{\mathbf{n}}) &= \sum_{\ell=0}^{+\infty}\sum_{m=-\ell}^{+\ell} a_{\ell m} Y_{\ell m}(\hat{\mathbf{n}}) \\ &= \Phi_{\text{avg}} \times \left(1 + \mathbf{d}\cdot\hat{\mathbf{n}} + \frac{1}{2}\hat{\mathbf{n}}\cdot\mathsf{Q}\hat{\mathbf{n}} + \cdots\right),\end{aligned}$$

where the coefficients $a_{\ell m}$ represent anisotropies on scales $O(180°/\ell)$. The contribution of the two lowest non-trivial harmonics $\ell = 1, 2$, the dipole and quadrupole, can be rewritten in terms of a dipole vector $\mathbf{d}$ and the symmetric traceless quadrupole tensor $Q_{ij}$ as $\mathbf{d} = \sqrt{3}(a_{11}\hat{\mathbf{x}} + a_{1-1}\hat{\mathbf{y}} + a_{10}\hat{\mathbf{z}})/a_{00}$, and $Q_{xx} - Q_{yy} = 2\sqrt{15}a_{22}/a_{00}$, $Q_{xz} = \sqrt{15}a_{21}/a_{00}$, $Q_{yz} = \sqrt{15}a_{2-1}/a_{00}$, $Q_{zz} = 2\sqrt{5}a_{20}/a_{00}$, $Q_{xy} = \sqrt{15}a_{2-2}/a_{00}$ (the remaining components of $Q_{ij}$ are obtained from its symmetry and zero trace conditions).

The dipole amplitude $|\mathbf{d}|$ and the quadrupole amplitude $|\mathsf{Q}|$ are relatively insensitive to magnetic fields, providing some information about sources: coherent deflections mainly affect the directions of $\mathbf{d}, \mathsf{Q}$ rather than their amplitudes, and small-scale random deflections only attenuate the $2^\ell$-pole by a factor $O\left(\exp(-\ell^2\theta_{\text{turb}}^2/2)\right)$, where $\theta_{\text{turb}}$ is the average deflection accumulated from the scattering in turbulent magnetic fields, so that most of the $|\mathbf{d}|$ and a sizeable fraction of the $|\mathsf{Q}|$ should survive (see also ref. [16]).

Using a full-sky data set wherein the exposure $\omega(\hat{\mathbf{n}})$ is non-zero everywhere, each $a_{\ell m}$ can be independently estimated in a way that is unbiased regardless of any assumption made about higher multipoles [12], as $\hat{a}_{\ell m} = \sum_k Y_{\ell m}(\hat{\mathbf{n}}_k)/\omega(\hat{\mathbf{n}}_k)$ where the sum runs over all $k$ UHECR events.

Our results are listed in table 1: we do not find any deviations from isotropy other than a weakly energy-dependent dipole towards a direction far away from the Galactic Center and a hint of a quadrupole at the highest energies roughly along the Supergalactic Plane. Compared to the previous analysis [12] the dipole component $d_y = 5.0\% \pm 1.1\% \pm 0.0\%$ has become more prominent, viz. $d_y = 4.8\% \pm 1.1\% \pm 0.0\%$, while most uncertainties have slightly decreased. The reconstructed strength of large-scale anisotropies remains at the low edge of a range of model expectations such as those of refs. [17, 18], suggesting a medium to heavy mass composition.

### 4.2 Medium-scale anisotropies: correlations with nearby galaxies

In order to search for medium-scale anisotropies we need to focus on the highest energies, where the magnetic deflections are expected to be smaller and the UHECR propagation horizon shrinks (thus, fewer candidate sources remain, making the anisotropy from the distribution of the sources more pronounced). However, the amount of statistics available at those energies is dramatically lower, making "blind" searches very unlikely to succeed. We therefore perform targeted searches based on two different catalogues, following our previous study [6]. The first is a list of 44 113 galaxies of all types at distances $1\text{ Mpc} \leq D < 250\text{ Mpc}$, based on the 2MASS catalogue with distances from HyperLEDA and weights assumed proportional to the near-infrared flux in the $K$-band (2.2 µm). The second is a list of 44 starburst galaxies (SBGs) at distances $1\text{ Mpc} \leq D < 130\text{ Mpc}$, taken from Ref. [19] except that we removed the SMC and LMC (these are dwarf irregular galaxies, not SGBs, as inferred from the fact that their infrared-to-radio flux ratio is much lower than all other objects of the list), and we added the Circinus galaxy with data from the Parkes telescope ($\alpha = 213.29°$, $\delta = -65.34°$, $D = 4.21$, $S_{1.4\text{ GHz}} = 1.50$ Jy); these galaxies were assigned weights proportional to their radio flux at 1.4 GHz, see also Ref. [8].

We define the test statistic

$$\begin{aligned}\text{TS}(\psi, f, E_{\min}) &= 2\ln\frac{L(\psi, f, E_{\min})}{L(\psi, 0, E_{\min})}, \\ L(\psi, f, E_{\min}) &= \prod_{E_i \geq E_{\min}} \frac{\Phi(\hat{\mathbf{n}}_i; \psi, f)\omega(\hat{\mathbf{n}}_i)}{\int_{4\pi} d\Omega\, \Phi(\hat{\mathbf{n}}; \psi, f)\omega(\hat{\mathbf{n}})},\end{aligned}$$

where $\omega(\hat{\mathbf{n}})$ is the combined directional exposure of the two data sets, and the flux model is

$$\Phi(\hat{\mathbf{n}}; \psi, f) = f\Phi_{\text{signal}}(\hat{\mathbf{n}}; \psi) + (1 - f)\Phi_{\text{background}},$$

where the contribution of each source is a von Mises–Fisher distribution and the background is isotropic (normalized to 1 over the whole sphere):

$$\begin{aligned}\Phi_{\text{signal}}(\hat{\mathbf{n}}; \psi) &= \frac{1}{\sum_j w_s}\sum_j w_s \frac{\psi^{-2}}{4\pi\sinh\psi^{-2}}\exp\left(\psi^{-2}\hat{\mathbf{n}}_s\cdot\hat{\mathbf{n}}\right); \\ \Phi_{\text{background}} &= \frac{1}{4\pi}.\end{aligned}$$

Here $E_i$ and $\hat{\mathbf{n}}_i$ are the energy and arrival direction of the $i$-th event; $w_s$ and $\hat{\mathbf{n}}_s$ are the weight and position of the $s$-th source candidate, and $\psi$ is the root-mean-square deflection per transverse dimension (i.e. the total r.m.s. deflection is $\sqrt{2} \times \psi$). The analysis is repeated using energy thresholds of $32\text{ EeV}, 33\text{ EeV}, \ldots, 80\text{ EeV}$ on the Auger scale, corresponding to $40.5\text{ EeV}, 41.9\text{ EeV}, \ldots, 106.8\text{ EeV}$ on the TA scale.

In this work, we neglect the energy losses undergone by cosmic rays, because, given the distance distributions of the objects we are considering, the effect of energy losses on the results can be presumed to be relatively small (in the case of all galaxies) or negligible (in the case of

**Table 1.** Our measurements of the dipole and quadrupole moments. The first uncertainty is statistical, the second is due to the uncertainty in the cross-calibration of energy scales. Values in *italics* **(bold)** are locally significant at $\geq 2\sigma$ ($\geq 4\sigma$).

| $E_{\text{Auger}}$ [EeV]<br>$E_{\text{TA}}$ [EeV] | [8.53, 16)<br>[10, 19.49) | [16, 32)<br>[19.49, 40.5) | [32, +∞)<br>[40.5, +∞) |
|---|---|---|---|
| $d_x$ [%] | $-0.2 \pm 1.1 \pm 0.0$ | $+0.9 \pm 1.9 \pm 0.0$ | $-4.4 \pm 3.7 \pm 0.1$ |
| $d_y$ [%] | $\boldsymbol{+5.0 \pm 1.1 \pm 0.0}$ | $\mathit{+4.4 \pm 1.9 \pm 0.0}$ | $\mathit{+10.0 \pm 3.5 \pm 0.0}$ |
| $d_z$ [%] | $-3.0 \pm 1.3 \pm 1.2$ | $-8.4 \pm 2.2 \pm 1.3$ | $+3.3 \pm 4.4 \pm 3.5$ |
| $Q_{xx} - Q_{yy}$ [%] | $-4.3 \pm 4.6 \pm 0.0$ | $+12.9 \pm 8.1 \pm 0.0$ | $\mathit{+39.7 \pm 15.0 \pm 0.0}$ |
| $Q_{xz}$ [%] | $-2.7 \pm 2.7 \pm 0.0$ | $+4.1 \pm 4.7 \pm 0.0$ | $+4.9 \pm 9.7 \pm 0.1$ |
| $Q_{yz}$ [%] | $-4.3 \pm 2.7 \pm 0.0$ | $-8.3 \pm 4.6 \pm 0.1$ | $+12.8 \pm 9.1 \pm 0.3$ |
| $Q_{zz}$ [%] | $+0.5 \pm 3.1 \pm 1.5$ | $+4.5 \pm 5.4 \pm 1.5$ | $+22.0 \pm 10.3 \pm 4.1$ |
| $Q_{xy}$ [%] | $+1.3 \pm 2.3 \pm 0.0$ | $-0.6 \pm 4.0 \pm 0.1$ | $+4.0 \pm 7.8 \pm 0.1$ |

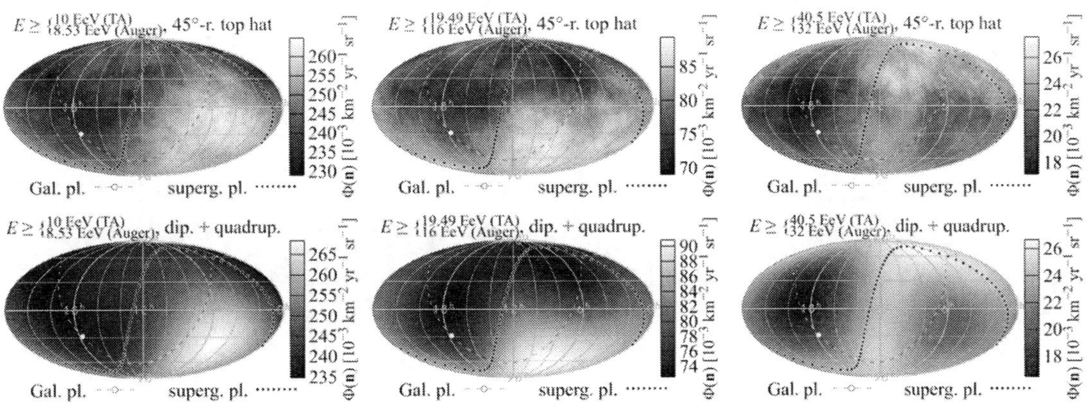


**Figure 2.** The observed flux in the three energy bins used for the large-scale anisotropies (top panels), and the reconstructed directional flux truncated to $\ell \leq 2$ (bottom panels).

SBGs). This enables us to be agnostic about the UHECR injection properties, thus reducing the number of model parameters for this search.

We present the result for the TS in figure 3. The most significant correlation we found gives TS = 31.1 with the SGB catalogue, using $E_{\min} = 38\,\text{EeV}$ on the Auger scale (49 EeV on the TA scale), $\psi = 15.1°^{+4.6°}_{-3.0°}$, and $f = 12.1\%^{+4.5\%}_{-3.1\%}$; the observed flux and the model prediction are compared in figure 4. The post-trial significance of this is $4.7\sigma$.[1] This is a significant increase compared to the previous analysis [6], wherein the TS = 27.2 corresponded to a post-trial significance of $4.2\sigma$; the Auger-only analysis presented in [10] obtained $TS = 25.0$ and a post-trial significance of $4.0\sigma$. The correlation with the catalogue of all galaxies is much weaker at TS = 14.3 using $E_{\min} = 40\,\text{EeV}$ on the Auger scale (51 EeV on the TA scale), $\psi = 28°^{+11°}_{-12°}$, and $f = 41\%^{+29\%}_{-18\%}$, with a post-trial global significance of $2.7\sigma$—this has decreased compared to our previous study in which the TS = 16.2 returned a global significance of $2.9\sigma$.

[1]The post-trial significance is found accounting for the scans over all three parameters: $\psi$, $f$ and $E_{\min}$.

## 5 Summary and Outlook

In this work we have updated the full-sky searches for large-scale and medium-scale anisotropies in the distribution of UHECR arrival directions, making use of the latest data sets available, namely 17 years of data taken with the Pierre Auger Observatory (the same used in our previous analyses) and 14 years of data taken with the Telescope Array (three years more compared to our previous study). Using the same methods outlined in [6, 12] the most important findings are: (1) a mild increase in the significance of the $d_y$ dipole component which grew from $d_y = 4.8\% \pm 1.1\% \pm 0.0\%$ to $d_y = 5.0\% \pm 1.1\% \pm 0.0\%$; (2) a marked increase in the post-trial statistical significance of the correlation with SBGs, which in this study is found to peak at $4.7\sigma$ (cf. $4.2\sigma$ in the previous analysis) using $E_{\min} = 38\,\text{EeV}$ on the Auger scale (49 EeV on the TA scale), with r.m.s. deflection parameter $\psi = 15.1°^{+4.6°}_{-3.0°}$, and correlated flux fraction $f = 12.1\%^{+4.5\%}_{-3.1\%}$.

Besides the obvious increase in statistics, our results complement single-hemisphere analyses in two qualitative ways. In the case of large-scale anisotropies, having full-sky coverage means that we do not need to assume anything about the UHECR flux, and the first moments of the harmonic flux decomposition can be recovered without bias. In the case of medium-scale anisotropies, we are able

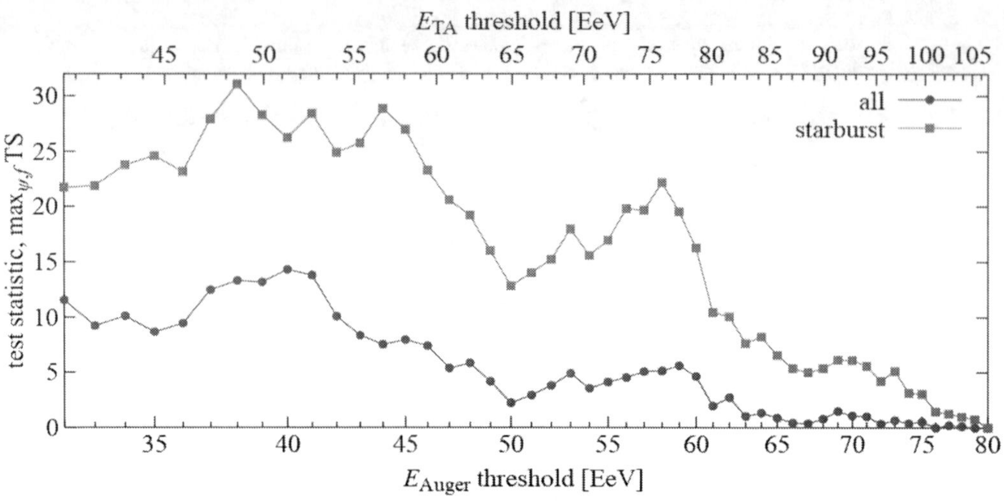


**Figure 3.** The TS as a function of the energy scale for the SBG catalogue (green) and the all-galaxies catalogue (violet).

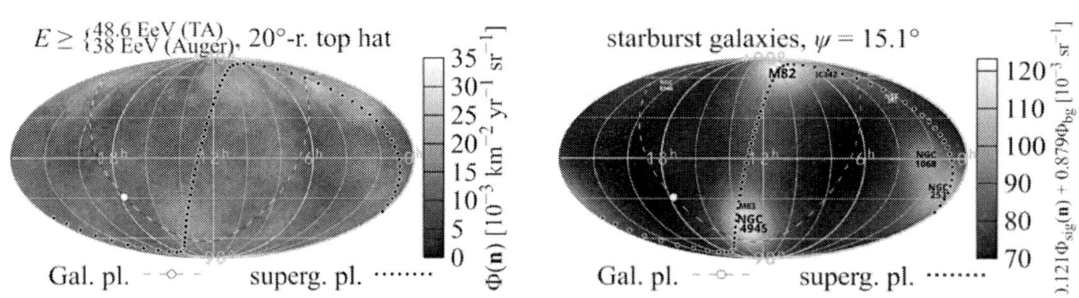


**Figure 4.** The observed UHECR flux above 38 EeV on the Auger scale (49 EeV on the TA scale) (left), compared to the starburst galaxy model predictions (right).

to observe all comparable sources; for example in the case of SBGs, the two most UHECR-bright sources, NGC4945 and M82, are in opposite hemispheres.

The astrophysical interpretation of the UHECR–SBG association is complicated by our incomplete knowledge of intergalactic and Galactic magnetic fields as well as the UHECR mass composition. Indeed, in our analysis, in order to reduce statistical penalties, the TS was based on a simple model that does not take into account the energy losses of UHECRs (which depend on their mass composition), coherent magnetic deflections, and the possibility of several anisotropic classes of sources at once. In order to estimate their effects, and therefore to better interpret our results, we are generating simulated sets of data based on a variety of scenarios, which we will subject to the same analyses as the observational data to test which simulations return similar values for $\psi, f,$ TS as the data.

TA is undergoing an upgrade (TA×4) which will increase its area by a factor of 4, and which will rapidly reduce statistical uncertainties in the northern hemisphere—currently a bottleneck for our combined analyses. New scintillation and radio detectors are being added to the existing water-Cherenkov and fluorescence detectors (the AugerPrime upgrade), which will reduce statistical and systematic uncertainties on UHECR masses. Moreover, improved mass estimation from new analysis techniques (e.g., machine learning) are being developed [20, 21]. The combination of these will allow us to study mass-dependent anisotropies, potentially allowing us to disentangle the effects of magnetic deflections from the distribution of UHECR sources, for instance by selecting high-rigidity samples with smaller magnetic deflections. To conclude, given the statistical significance of the results presented here, in particular the 4.7$\sigma$ correlation of UHECRs with the SBG catalogue, it is imperative that the two observatories continue running through this decade in order to decidedly confirm this correlation.

## References

[1] R. Alves Batista et al. (2021), 2110.10074
[2] L.A. Anchordoqui, Phys. Rept. **801**, 1 (2019), 1807.09645

 https://doi.org/10.1051/epjconf/202328303002

[3] R.U. Abbasi et al., Astropart. Phys. **86**, 21 (2017), 1608.06306

[4] A. Aab et al. (Pierre Auger), Science **357**, 1266 (2017), 1709.07321

[5] R. de Almeida et al. (Pierre Auger), PoS **ICRC2021**, 335 (2021)

[6] A. di Matteo et al. (Telescope Array, Pierre Auger), PoS **ICRC2021**, 308 (2021)

[7] A. Aab et al. (Pierre Auger), Astrophys. J. Lett. **853**, L29 (2018), 1801.06160

[8] P. Abreu et al. (Pierre Auger), PoS **ICRC2021**, 307 (2021)

[9] A. Aab et al. (Pierre Auger), Nucl. Instrum. Meth. A **798**, 172 (2015), 1502.01323

[10] P. Abreu et al. (Pierre Auger), Astrophys. J. **935**, 170 (2022), 2206.13492

[11] T. Abu-Zayyad et al. (Telescope Array), Nucl. Instrum. Meth. A **689**, 87 (2013), 1201.4964

[12] P. Tinyakov et al. (Telescope Array, Pierre Auger), PoS **ICRC2021**, 375 (2021)

[13] A. Aab et al. (Telescope Array, Pierre Auger), Astrophys. J. **794**, 172 (2014), 1409.3128

[14] A. di Matteo, O. Deligny, K. Kawata, R.M. de Almeida, M. Mostafá, E. Moura Santos, H. Sagawa, P. Tinyakov, I. Tkachev, N. Toshiyuki, JPS Conf. Proc. **19**, 011020 (2018)

[15] J. Biteau et al. (Pierre Auger, Telescope Array), EPJ Web Conf. **210**, 01005 (2019), 1905.04188

[16] B. Eichmann, T. Winchen, JCAP **04**, 047 (2020), 2001.01530

[17] A. di Matteo, P. Tinyakov, Mon. Not. Roy. Astron. Soc. **476**, 715 (2018), 1706.02534

[18] C. Ding, N. Globus, G.R. Farrar, Astrophys. J. Lett. **913**, L13 (2021), 2101.04564

[19] C. Lunardini, G.S. Vance, K.L. Emig, R.A. Windhorst, JCAP **10**, 073 (2019), 1902.09663

[20] R.U. Abbasi et al. (Telescope Array), Phys. Rev. D **99**, 022002 (2019), 1808.03680

[21] A. Aab et al. (Pierre Auger), JINST **16**, P07019 (2021), 2101.02946

# Update on the indication of a mass-dependent anisotropy above $10^{18.7}$ eV in the hybrid data of the Pierre Auger Observatory

*Eric* Mayotte[1,2,*] and *Thomas* Fitoussi[3] for the Pierre Auger Collaboration[4,†]

[1]Colorado School of Mines, Department of Physics, Golden, CO, USA
[2]Bergische Universität Wuppertal, Department of Physics, Wuppertal, Germany
[3]Karlsruhe Institute of Technology (KIT), Institute for Astroparticle Physics, Karlsruhe, Germany
[4]Observatorio Pierre Auger, Av. San Martín Norte 304, 5613, Malargüe, Argentina.
Full author list: https://www.auger.org/archive/authors_2022_10.html

**Abstract.** We test for an anisotropy in the mass of arriving cosmic-ray primaries associated with the galactic plane. The sensitivity to primary mass is obtained through the depth of shower maximum, $X_{\max}$, extracted from hybrid events measured over a 14-year period at the Pierre Auger Observatory. The sky is split into distinct on- and off-plane regions using the galactic latitude of each arriving cosmic ray to form two distributions of $X_{\max}$, which are compared using an Anderson-Darling 2-samples test. A scan over roughly half of the data is used to select a lower threshold energy of $10^{18.7}$ eV and a galactic latitude splitting at $|b| = 30°$, which are set as a prescription for the remaining data. With these thresholds, the distribution of $X_{\max}$ from the on-plane region is found to have a $9.1 \pm 1.6^{+2.1}_{-2.2}$ g/cm$^2$ shallower mean and a $5.9 \pm 2.1^{+3.5}_{-2.5}$ g/cm$^2$ narrower width than that of the off-plane region and is observed in all telescope sites independently. These differences indicate that the mean mass of primary particles arriving from the on-plane region is greater than that of those from the off-plane region. Monte Carlo studies yield a $5.9\times10^{-6}$ random chance probability for the result in the independent data, lowering to a $6.0 \times 10^{-7}$ post-penalization random chance probability when the scanned data is included. Accounting for systematic uncertainties leads to an indication for anisotropy in mass composition above $10^{18.7}$ eV with a $3.3\,\sigma$ significance. Furthermore, the result has been newly tested using additional FD data recovered from the selection process. This test independently disfavors the on- and off-plane regions being uniform in composition at the $2.2\,\sigma$ level, which is in good agreement with the expected sensitivity of the dataset used for this test.

## 1 Introduction

The Pierre Auger Observatory is currently the largest observatory dedicated to studying cosmic rays with energies in the EeV range, so-called ultra-high-energy cosmic rays, *UHECR* [1]. To do this, it uses both an array of particle detectors on the Earth's surface, the Surface Detector, *SD* [2], and an array of fluorescence telescopes monitoring the atmosphere above the SD, the Fluorescence Detector, *FD* [3]. The highest quality data set of the Observatory is made up of UHECR events which have simultaneously been measured by both the FD and SD, so-called *hybrid events*. In a hybrid event reconstruction, the geometry of the shower axis is highly constrained by combining the triggered pixel geometry/timing from the FDs and the high confidence core location/timing provided by the SDs. This results in an angular resolution for the pointing direction of the shower axis of better than 0.5° [4], and a resolution on the location of the shower core of 50 m [5].

The low uncertainty geometric reconstruction provided by the hybrid method allows the evolution of the intensity of UV fluorescence light measured by the FD to be inverted to model the *shower profile*, which is the number of charged particles in the air shower as a function of the amount of matter it has traversed, the slant depth, $X$. From the shower profile, the slant depth at which the maximum development of the shower occurs, $X_{\max}$, can be extracted. $X_{\max}$ is closely related to the mass of the primary cosmic ray which induced the air-shower, but is subject to large fluctuations meaning that it can not be used on a shower-by-shower basis to determine primary mass. However, if collected with sufficient statistics, the first and second moments of distributions of $X_{\max}$, $\langle X_{\max}\rangle$ and $\sigma(X_{\max})$ respectively, can be used to make high certainty estimations of the mean mass of the UHECR events used to form that particular $X_{\max}$ distribution [6].

Up until recently [7], $X_{\max}$ derived from hybrid measurements has been used to study the average composition of the cosmic-ray sky as whole, rather than being used to compare the mean compositions of different subsets of the sky. This choice was likely driven by the relatively sparse statistics available in hybrid studies due to the upper-limit of exposure available to them being set by the relatively low 14 % up-time of FDs. The possibility of splitting the data set into subsamples was therefore limited by the need to maintain sufficient statistics to say something useful about primary composition. However, the Pierre Auger

*e-mail: emayotte@mines.edu
†e-mail: spokespersons@auger.org

Observatory has now collected more than 14-years of FD data, and tens of thousands of high-quality hybrid measurements. With this quantity of data, the sky can be split into different regions and the mean mass of UHECRs arriving from them can be studied. This new reality then prompts two questions:

1. Are the systematic uncertainties which trend with event arrival direction for hybrid reconstruction low enough to allow the mean mass arriving from different regions of the sky to be meaningfully compared?

2. Is there a reason to expect that different regions of the sky may display differing compositions due to astrophysical causes?

Question 1) will be explored in section 4. For question 2) it is clear that the opportunity exists. The flux above the ankle at ~5 EeV [8] is mixed in composition and has long been thought (now confirmed) to be extragalactic in origin [9]. Furthermore, it definitively displays anisotropy above 8 EeV [10]. Additionally, as was nicely put by Alan Watson in 1990:

> "*... the Larmor radius of a proton of* $10^{18}$ *eV in a 3 $\mu$G field is about 400 pc, comparable to the thickness of the galactic disk. It follows, therefore, that, if the bulk of cosmic rays are protons, anisotropies associated with the magnetic field structure of the galactic disk might appear as the energy increases.*" - [11]

Indeed, there were hints of such a spectral feature starting somewhere around $10^{18.5}$ eV for directions within 30° of the galactic plane [11, 12]. Unfortunately, such an excess so far does not appear to be significant in the data of current experiments and therefore has not been given much attention since those initial publications.

Now it is known that the flux above 1 EeV is best described as an evolving mix of light-, intermediate-, and high-mass primaries [13, 14]. Due to the galactic magnetic field, *GMF*, the different mass components present at any given energy will be deflected to different degrees as they travel from their extragalactic sources to Earth. This mass dependent deflection suggests that an anisotropy associated with the structure of the GMF would kick in for increasingly heavier components as energies climb. It is therefore distinctly possible that an anisotropy associated with the galactic plane could arise in the higher mass components of the flux at some energy in the EeV range.

What follows below are specific tests for such a mass-dependent anisotropy associated with the galactic plane using hybrid data of the Pierre Auger Observatory collected between 2004 and the end of 2018. To avoid repeating the contents of the ICRC 2021 proceedings on this result [7] in its entirety, the contents of this proceeding will aim to include components of the analysis that could not be fit in the eight pages allotted in that publication. Therefore, while this proceeding will cover the details of the analysis, increased space will be given to the cross checks and studies of the systematics uncertainties of the analysis. Additionally, a new test on an independent hybrid data set, recovered from the quality cuts, will also be discussed.

## 2 Reconstruction and selection

The reconstruction methods, selection cuts, and core analysis of $X_{max}$ distributions used here are largely the same as those described in [15]. A rigorous description of these methods can be found in [6]. Other than the fiducial field-of-view cut, *FidFoV*, which is treated below, a detailed description of the methods will not be provided here.

The important differences from [15] are that the minimum energy for inclusion in the data set has been raised to $E > 10^{18.4}$ eV, and that, as described below in subsection 2.1, the event acceptance, reconstruction bias, and $X_{max}$ resolution are now treated based on the arrival direction of each event. The lower limit of $10^{18.4}$ eV has been chosen as above this energy, the composition is well mixed and expected to be primarily of extragalactic origin [16]. The period over which data has been collected has also been slightly expanded to span events observed between 2004-12-01 and 2018-12-31, yielding 7572 high-quality events. A further subdivision of the data is necessary to test for the hypothesized anisotropy. Following the results of the scan shown in Figure 5, the data is split into the events with a galactic latitude, $|b| \leq 30°$, the *on-plane* sample, and, $|b| > 30°$, the *off-plane* sample.

#### *Fiducial field-of-view selection*

To ensure a good reconstruction of $X_{max}$ with the FD, $X_{max}$ itself should be directly observed. The telescopes of the FD have a field-of-view, *FoV*, which is vertically constrained. Additionally, some showers in more vertical geometries will not reach $X_{max}$ before impacting the ground. These factors together lead to a geometric and $X_{max}$ dependence for what events end up in the analyzed data set. If unaccounted for, this *$X_{max}$ acceptance*, will inevitably bias a composition study based on FD $X_{max}$ data. At the Observatory, this $X_{max}$ acceptance is addressed primarily through mitigation of the effect with the fiducial field-of-view, *FidFoV*, cut. The FidFoV cut constrains the FD detector volume to only event geometries where the expected range of $X_{max}$ values would be visible in the FD FoV. As can be seen in Figure 1, this changes the natural FD $X_{max}$ acceptance (gray) to one which is unbiased from ~600 to ~900 g/cm$^2$, which spans the typical range of observed event $X_{max}$ values. Remaining effects on rare events outside this range are corrected using parameterizations as a function of $X_{max}$ and primary energy.

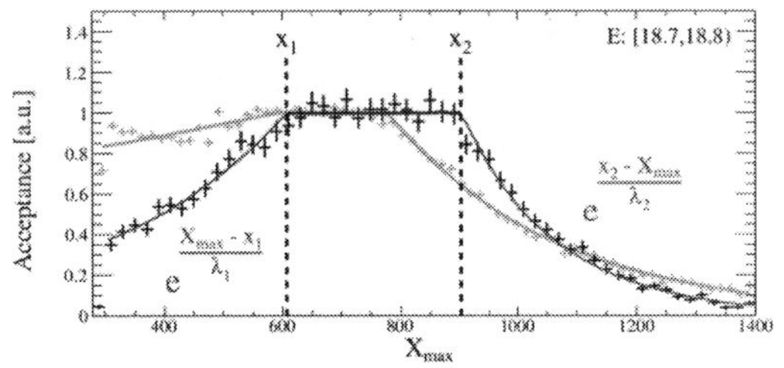


Figure 1: FD $X_{max}$ acceptance before (gray) and after (black) FidFoV cuts. A 4-variable parameterization of the post-FidFoV acceptance is shown in red. The range of $X_{max}$ values with unbiased sampling is shown in blue.

### 2.1 Distributions of $X_{max}$ and arrival direction

After measurement, reconstruction, and selection, the observed $X_{max}$ distribution does not quite represent the true $X_{max}$ distribution of all cosmic rays landing within the Observatory. This is due to energy dependent biases on the reconstruction of $X_{max}$ ($B$), the resolution on $X_{max}$ of the hybrid reconstruction method ($R$), and the residual effects of the $X_{max}$ acceptance ($A$). Since the location of an event and its inclination with respect to the observing fluorescence telescope plays a role in the magnitude of $A$, $R$, and $B$, they have an inherent geometric dependence. If unaccounted for, this could bias this study. Distributions of the key geometric relationships between events and the FD can be seen in Figure 2 for on- and off-plane regions. It is clear that there is little difference between the two regions, so $A$, $R$, and $B$ are expected to also be similar.

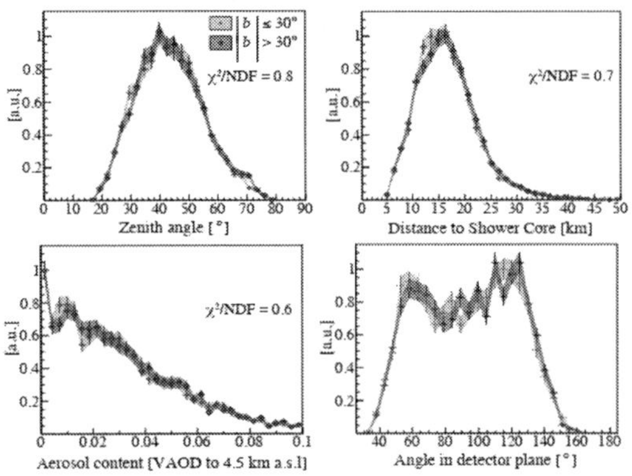

Figure 2: Geometries of the on- and off-plane samples.

To explicitly verify $A$, $R$, and $B$ similarity on and off the galactic plane, CONEX [17] is used to generate showers with Sibyll-2.3c [18]. The showers are then isotropically thrown into detector simulations which include the time-dependent state of the FD and SD from 2004 through 2018. These simulations, therefore, mimic the measurement conditions, trigger efficiency, and up-time of the real data, accurately modelling the exposure and geometries of events arriving from all parts of the sky [19]. Two sets of these simulations are produced, one formed from showers generated with a flat sampling of $X_{max}$ between 300 and 1500 g/cm$^2$, the *flat-MC*, and one formed from an equal number of proton, helium, nitrogen, and iron simulations which are then weighted to their abundances observed in data as reported in [20], the *mixed-MC*. These simulated event sets are then subjected to the same reconstruction and selection techniques used on the real data, so that $A$, $R$, and $B$ are accurately included in them.

The flat-MC is split into on- and off-plane subsamples which are then used to extract the functional form of $A$ in $0.1 \log_{10}(E/\mathrm{eV})$ energy bins using the method illustrated in Figure 1. The form of $A$ is extracted by leveraging the flatly sampled $X_{max}$ generation of the Monte Carlo, as, once the plateau of the distribution is normalized to one, the height of each bin represents the acceptance of events with $X_{max}$ values in that range. The acceptance is then fit

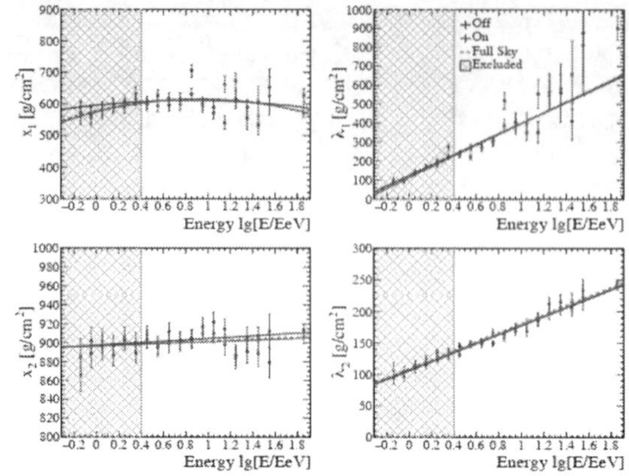

Figure 3: The $X_{max}$ acceptance parameterizations for the on- and off-plane sky regions from Monte Carlo.

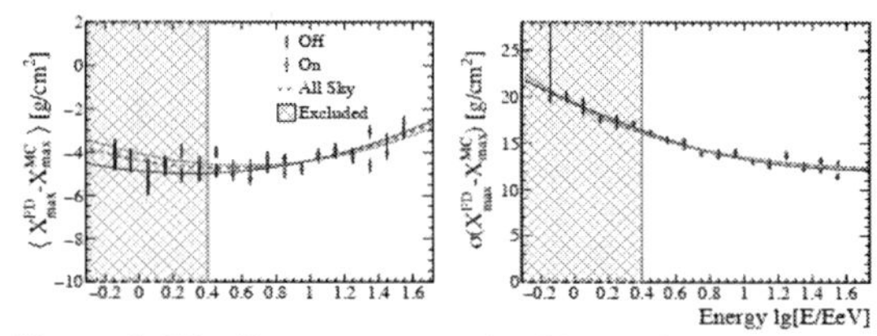

Figure 4: The $X_{max}$ reconstruction bias and resolution parameterizations for the on- and off-plane sky regions from Monte Carlo. Note: only the detectors and reconstruction $X_{max}$ resolution is shown. Other effects lowering the resolution are included as specified in [6].

with the 4-component parameterization illustrated in Figure 1. The energy evolution of $x_1, \lambda_1, x_2$, and $\lambda_2$ is then parameterized separately with 2D polynomials for the on- and off-plane subsamples, resulting in Figure 3. From Figure 3, it is clear that above $10^{18.4}$ eV there is no statistically significant difference in $X_{max}$ acceptance between the on- and off-plane regions. Even so, the region-specific parameterizations of $A$ are used to correct the 1.4 % of events with partial $X_{max}$ acceptance using the up-weighting technique outlined in [6]. Uncertainties in these parameterizations result in the systematic uncertainties on the first and second moments specified in Table 1.

The mixed-MC is likewise split into on- and off-plane subsamples which are then used to extract the energy evolution of $B$ and $R$ for each region. $B$ and $R$ are extracted by forming a distribution of the difference between the FD reconstructed value of $X_{max}$ and the Monte Carlo truth value in $0.1 \log_{10}(E/\mathrm{eV})$ energy bins. From these, the mean reconstruction bias, $B = \langle X_{max}^{FD} - X_{max}^{MC} \rangle$, and the $X_{max}$ resolution, $R = \sigma\left(X_{max}^{FD} - X_{max}^{MC}\right)$, are extracted in each energy bin. Again, the evolution of each is parameterized with a 2D polynomial. Figure 4 shows that $B$ and $R$ for the two regions are found to agree within errors. These however are also corrected for separately. Fit uncertainties again result in the systematic uncertainties outlined in Table 1. At this point, the $X_{max}$ distributions and moments from the on- and off-plane regions in each energy bin can be compared without bias from selection and reconstruction.

## 3 Testing for anisotropy

The specific hypothesis to be tested is whether, above some energy threshold, $E_{\mathrm{th}}$, the mean composition of UHECRs coming from directions near to the galactic plane is significantly higher in mass than those arriving further from it. This is to be tested using $X_{\max}$ as a mass sensitive parameter. Typically, $X_{\max}$ based composition analyses leverage the first two moments of $X_{\max}$ distributions binned in energy, to comment on primary mass. This approach, however, does not lend itself well to quantifying the significance of a result testing the above statement. Instead, a test statistic, $TS$, which quantifies the degree of dissimilarity between the $X_{\max}$ distributions in the two regions in a single value is preferred. For this, the returned value from the Anderson-Darling two-sample homogeneity test [21], *AD-test*, has been selected as it scales with the dissimilarity of the tested distributions. The AD-test has good sensitivity to the full width of a distribution [22], and has more power than the Kolmorogov-Smirnov test while remaining robust against false positives [23].

To use the AD-test and $X_{\max}$ for this purpose, two modifications are required. First, a single $TS$ comparing all events in each region above $E_{\mathrm{th}}$ is desired. So, all events with $E \geq E_{\mathrm{th}}$ in the on- and off-plane samples separately need to be collected into a common on-plane distribution and a common off-plane distribution. To do this, the natural evolution of $X_{\max}$ with energy needs to be removed so that spectral features in the flux do not influence the result. Therefore, we define an energy-normalized $X_{\max}$ value

$$X'_{\max} = X_{\max} - \underbrace{\left(649 + 63.1\,Z_{18} + 1.97\,Z_{18}^2\right)}_{\text{EPOS-LHC elongation rate for iron}}, \qquad (1)$$

where $Z_{18} = \log_{10}(E_{\mathrm{rec}}/\,\mathrm{EeV})$. The last term in Equation 1 is the natural energy evolution of mean $X_{\max}$ for iron primaries as predicted by EPOS-LHC [24]*. Second, the $X'_{\max}$ distribution of an on-plane sample populated with primaries which are on average heavier than those in the off-plane sample will display a lower mean and a narrower width than that of the off-plane $X'_{\max}$ distribution. Since the null hypothesis is that there is either no composition difference or a heavier off-plane sample, a $TS$ sensitive to the ordering of the $X'_{\max}$ distributions is required†. The AD-test is insensitive to ordering, so it is modified to

$$TS = \begin{cases} AD : \langle X'_{\max}\rangle^{\mathrm{on}} < \langle X'_{\max}\rangle^{\mathrm{off}} \\ -3 \ : \text{else} \end{cases}, \qquad (2)$$

where $AD$ is the result of the AD-test comparing the on- and off-plane distributions, and $-3$ is selected as it is well below the minimum of the AD-test.

*Scan for energy and galactic latitude thresholds*

A scan has been used to select the optimal on/off splitting latitude, $b_{\mathrm{split}}$, and minimum energy, $E_{\mathrm{th}}$, as uncertainties in GMF models and source distributions make other approaches impractical. In this scan, each trial $[E_{\mathrm{th}}, b_{\mathrm{split}}]$ pair is used to form on- and off-plane subsets and the $TS$ is extracted. To preserve the statistical strength of the sparse FD data set, a coarse scan of 5° steps in $|b|$ from 20° to 35° and $0.1\log_{10}(E/\mathrm{eV})$ steps in energy from 18.4 to $19.4\log_{10}(E/\mathrm{eV})$ is used. The scan is performed on the data set from [6], which includes events through Dec 31st 2012. At the time of writing, this *scan data set* represents 54 % of the analyzed events. The remaining 46 % of events, the *post-scan data set*, is reserved as blind.

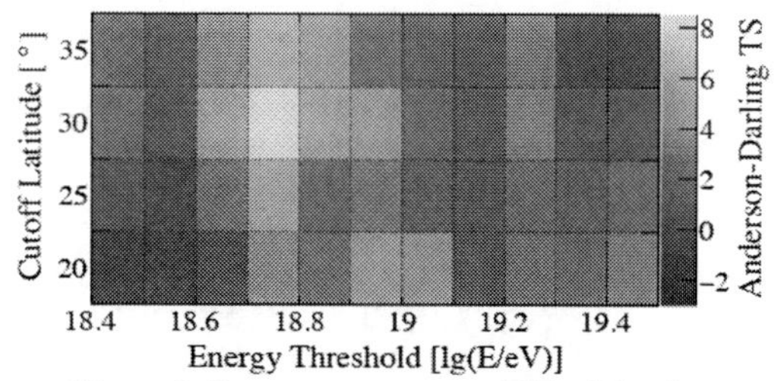


Figure 5: Parameter scan over 54% of the data.

Interestingly, as shown in Figure 5, all tested pairs result in $\langle X'_{\max}\rangle^{\mathrm{on}} < \langle X'_{\max}\rangle^{\mathrm{off}}$. An optimal $[E_{\mathrm{th}}, b_{\mathrm{split}}]$ of $[10^{18.7}\,\mathrm{eV}, 30°]$ was found with a $TS = 8.4$. The selected $[E_{\mathrm{th}}, b_{\mathrm{split}}]$ is applied as a prescription to the post-scan data set, which independently confirms the result with a $TS = 12.6$, for a total $TS = 21.0$ for the full data set.

*Statistical significance*

The chance probability of the observed TS occurring with in an isotropic sky is tested using Monte Carlo methods on randomized skies derived from the real data. To form each randomized sky, the arrival direction is first decoupled from the energy and $X'_{\max}$ values of each event. These are then randomly re-paired to create a new sky which maintains the real $X_{\max}$, energy, and sky exposure distributions, but has a scrambled arrival direction/composition pairing. The above analysis is then used to extract a $TS$ from each sky which is compared to the result in data. Skies which display more extreme on-/off-plane differences than those observed in data are tallied and used to calculate the probability of an isotropic sky generating the observed $TS$. The results of this procedure are shown in Figure 6.

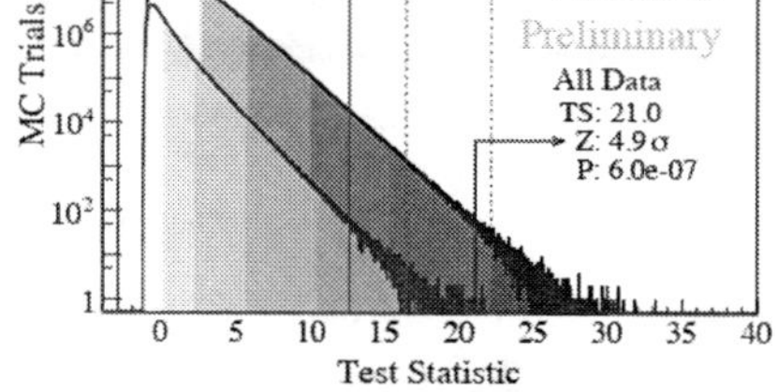


Figure 6: The Monte Carlo determination of the post-scan (red) and all-data (blue) significance with 1 and 10 billion randomized skies, respectively.

*Choice of hadronic interaction model varies result by $\sim 0.02\,\mathrm{g/cm^2}$.

†Modifying the test to also require $\sigma(X'_{\max})^{\mathrm{on}} < \sigma(X'_{\max})^{\mathrm{off}}$ would be more restrictive, but conservatively has not been applied.

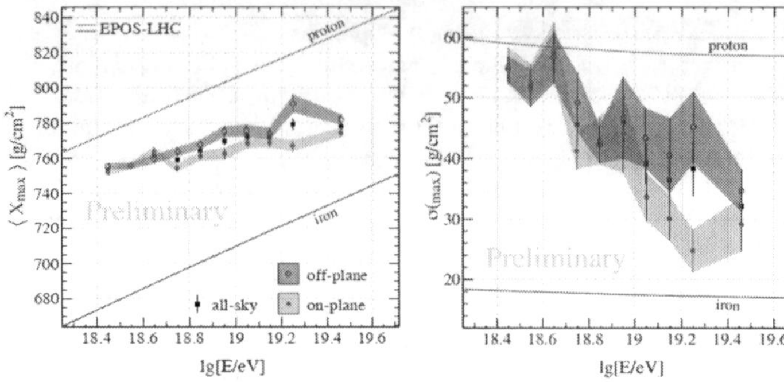


Figure 7: The first (left) and second (right) moments of the $X_{max}$ distributions from on- and off-plane regions.

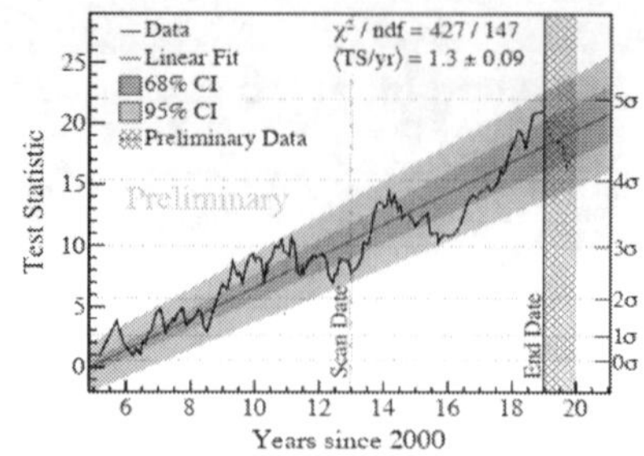


Figure 8: The time evolution of the TS with significance indicated on the right. The shaded region is preliminary data.

For the blind, post-scan data set, the prescribed $[E_{\rm th}, b_{\rm split}]$ pair is used to split each randomized sky into on- and off-plane samples and a $TS$ is extracted. In one billion random skies, only 5865 resulted in a more extreme $TS$ than the 12.6 observed in data. This indicates a chance probability of $5.87 \times 10^{-6}$ which corresponds to $4.4\,\sigma$.

To calculate the significance of the result when the scan- and post-scan data sets are combined, the entire analysis chain, including the scan, is duplicated. In each random sky, 54 % of the data is used to scan for the $[E_{\rm th}, b_{\rm split}]$ pair which results in the most extreme result, fully penalizing for the scan. These values are then used to split all data in the random sky into on- and off-plane subsamples and the $TS$ for the sky is extracted. From 10 billion random skies, only 5964 resulted in a more extreme $TS$ than the 21.0 observed in data. This indicates to a chance probability of $5.96 \times 10^{-7}$ which corresponds to $4.9\,\sigma$. The strong penalization of the scanned data is evident as the additional 54 % of the data (with $\Delta\langle X'_{\rm max}\rangle = 8.5$ g/cm$^2$) only resulted in an 11 % increase of the significance of the observation.

*$X_{max}$ moments and trends*

To illustrate the difference in composition on and off the plane, the first two moments of the $X_{\rm max}$ distribution in each 0.1 $\log_{10}(E/{\rm eV})$ energy bin has been plotted in Figure 7 for both regions. Above $10^{18.7}$ eV there is a clear separation in $\langle X_{\rm max}\rangle$ for all energy bins. Most energy bins also display a separation in $\sigma(X_{\rm max})$. Heavier primaries are expected to, on average, have a shallower $X_{\rm max}$ and lower shower-to-shower fluctuations. Therefore the correlated difference seen here indicates that, for this data sample, primaries from the on-plane region have a higher mean mass than that of the off-plane region above $10^{18.7}$ eV.

To evaluate the degree to which fluctuation plays a role in the observed result, the growth of the $TS$ over time has been plotted in Figure 8. The time evolution of the signal is consistent with linear growth at a rate of 1.3 $TS$ yr$^{-1}$. This behavior is in line with expectations for a real difference in mean mass between the subsamples. The shaded region of Figure 8 shows preliminary data from 2019. These reconstructions were not subject to a validated reconstruction chain and may change. Still, when added, a $3.7/4.4\,\sigma$ (post-scan/all data) statistical significance is expected. The best fit rate of growth of 1.3 $TS$ yr$^{-1}$ remains unchanged.

## 4 Systematic uncertainties

Figure 2 shows that there is very little difference between the on- and off-plane regions in the local reference frame of the detector. The data sets from the two regions also consist of events measured with the same instrumentation in the same location, and have been reconstructed with the same methods. Because systematic uncertainties are derived from measurement effects in the local frame, these similarities between the on- and off-plane samples mean that the majority of uncertainty sources outlined in [6] will apply equally to both regions and thereby cancel in a comparison. Furthermore, from the acceptance, resolution, and bias studies in subsection 2.1, the two regions are also free from selection and reconstruction biases.

Table 1: Systematic uncertainties on the difference in the means, $\Delta\langle X_{\rm max}\rangle$, and widths, $\Delta\,\sigma(X_{\rm max})$, of the on- and off-plane $X_{\rm max}$ distributions.

| Source | Uncertainty [g/cm$^2$] of | |
|---|---|---|
| | $\Delta\langle X_{\rm max}\rangle$ | $\Delta\,\sigma(X_{\rm max})$ |
| $A$ correction | $^{+1.14}_{-0.71}$ | $^{+2.37}_{-1.61}$ |
| $B$ correction | ±0.36 | ±0.01 |
| $R$ correction | 0 | $^{+1.78}_{-0.24}$ |
| Seasonal | $^{+1.00}_{-1.53}$ | $^{+1.19}_{-1.23}$ |
| Instrumentation | ±1.41 | ±1.41 |
| Sum in Quadrature | $^{+2.10}_{-2.23}$ | $^{+3.49}_{-2.48}$ |

To test for potential systematic effects derived from uncertainties in the $A$, $B$, and $R$ corrections, possible seasonal effects, and small differences between the instrumentation at different fluorescence telescope sites, *FD-sites*, several studies were performed. All permutations of the uncertainties in the $A$, $B$, and $R$ corrections were evaluated and the maximum changes in $\Delta\langle X'_{\rm max}\rangle$ and $\Delta\sigma(X'_{\rm max})$ were recorded. To look for on-/off-plane biases associated with instrumentation differences, events seen by two or more FD-sites were used to compare the $X_{\rm max}$ reconstructions of each site. No significant biases were found, and the instrumentation-derived systematic uncertainties from [6] were adopted. To check for systematics derived from a combination of the seasonal dependence of the FD exposure and normal yearly variation of the atmospheric quality, $\langle X'_{\rm max}\rangle$ and $\sigma(X'_{\rm max})$ for the on- and off-plane regions

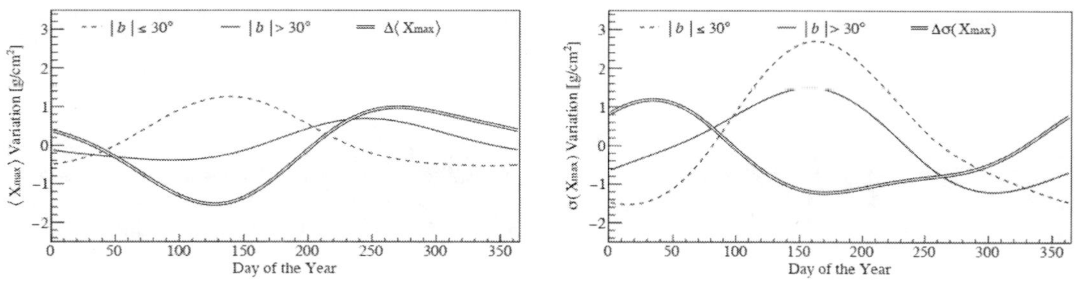


Figure 9: Seasonal fluctuation of the first (left) and second (right) $X_{max}$ moments for the on-plane (blue) and off-plane (red) samples. The double black line shows the difference between these curves. The maximum and minimum values of these differences are taken as the systematic uncertainty on $\Delta\langle X'_{max}\rangle$ and $\Delta\sigma(X'_{max})$.

were tracked over the course of the year, and is shown in Figure 9. The maximum exposure-weighted difference between the two regions at any point of the year has been adopted as the systematic uncertainty due to seasonal effects. These uncertainties and the total systematic uncertainty on $\Delta\langle X'_{max}\rangle$ and $\Delta\sigma(X'_{max})$ are listed in Table 1.

*Confidence level considering systematic uncertainties*

The observed $\Delta\langle X'_{max}\rangle$ of $9.1 \pm 1.6\,g/cm^2$ is 4.1 times larger than the $2.2\,g/cm^2$ systematic uncertainty listed in Table 1. The observed $\Delta\sigma(X'_{max})$ of $5.9 \pm 2.9\,g/cm^2$ is 2.4 times larger than its $2.5\,g/cm^2$ systematic uncertainty. This means it is unlikely that the result could be entirely due to systematic effects. However, the systematic uncertainties in $\Delta\langle X'_{max}\rangle$ and $\Delta\sigma(X'_{max})$ may increase the likelihood of an extreme result occurring in data. To quantify the result significance taking possible systematic effects into account, a two step approach is taken. First, the on-/off-plane difference is reduced by adding a value sampled from a Gaussian distribution with $\mu = 2.2\,g/cm^2$ and $\sigma = 2.5\,g/cm^2$ to the on-plane sample. Then, the AD-test is applied to the resulting on- and off-plane distributions. Repeating this process 1 million times results in a mean TS of $11.3 \pm 0.5$ ‡. If these values are converted to significances using the data from Figure 6, this corresponds to at least $3.3\,\sigma$. Conservatively, to include systematic effects, this lower bound of $3.3\,\sigma$ is adopted as the confidence level of the result.

*Cross-check: Results by zenith angle and FD-site*

If astrophysical in nature, the difference in composition of UHECR arriving from the on- and off-plane regions should be independently observed by each FD-site and in all zenith angle ($\theta$) ranges. Figure 10 shows that the difference in $X_{max}$ on and off the plane is indeed present in all zenith angles bins and is also observed by all FD-sites independently. Even more stringently, when the response of each FD-site is split in $\cos^2\theta$ bins, it appears in 22 out of 28 bins. This independent observation at all sites and zeniths is a strong confirmation of the stereo study described above, showing that detector systematics can not play a large role in the result. Furthermore, because the FD-sites have FoVs differing by 90° on average, each sees the galactic plane at a different local geometry and time during the year, making it unlikely that some unidentified detector, reconstruction, or atmospheric effect is causing the observed anisotropy.

‡Treating the other side of the systematic errors in the same way results in a $TS$ of $31.8 \pm 1.1$ ($6.3\,\sigma$).

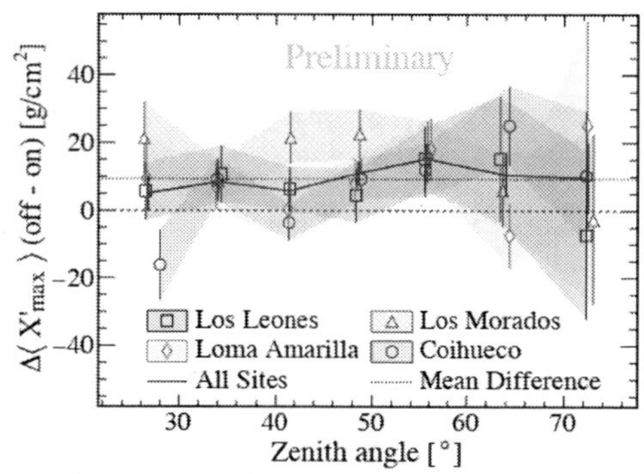


Figure 10: $\Delta\langle X'_{max}\rangle$ by FD-site and zenith bin.

## 5 Independent Test of the Result

To cross-check the validity of the previous results, an independent data set is needed. The selection presented in section 2 aims to select only high-quality events. In particular, the *FidFoV* cut is needed to ensure that the distribution of selected events covers the expected range of $X_{max}$ values with an unbiased acceptance. The events rejected by the FidFoV cut are still high-quality and perfectly well reconstructed. In this section, a second data set is built from the events removed in this FidFoV cut to form an *out-FidFoV* data set which is 82.9 % as large as the data set already used in this analysis, the *in-FidFoV* data set.

Figure 11 represents the $X_{max}$ acceptance of this out-FidFoV data set in different energy ranges. For all energies, the efficiency maximized for low $X_{max}$ values and decreases as $X_{max}$ increases. Indeed, without the FidFoV selection, deep showers tend to be under-represented [6]. Thus, the new data set is made out of shallower $X_{max}$ events. As consequence, the $\langle X_{max}\rangle$ distribution of the new data set is on average 10 $g/cm^2$ shallower. Despite this bias, if the difference between on- and off-plane is astrophysical, it should also appear in the out-FidFoV data set.

To test this hypothesis, this new data set is divided into on- and off-plane samples using the $E_{th} = 10^{18.7}$ eV,

 https://doi.org/10.1051/epjconf/202328303003

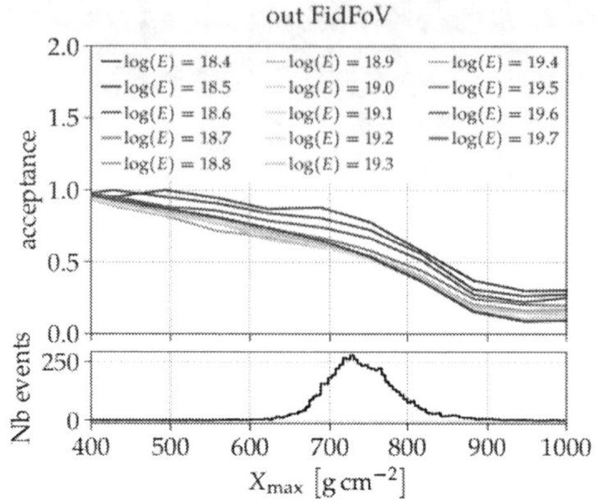


Figure 11: Top panel: acceptance for out-*FidFoV* data set split in $0.1 \log_{10}(E/\text{eV})$ energy bins. The energy in the legend is the lower energy edge of the bin, excepting 19.7 which corresponds to $\log(E/\text{eV}) \geq 19.7$. Lower panel: the $X_{max}$ distribution of events for the full energy range.

$b_{split} = 30°$ splitting determined by the scan. The resulting on/off distributions show a somewhat smaller $\Delta\langle X_{max}\rangle$ of ~5g/cm$^2$, which is only 55 % of what was obtained with the in-FidFoV data set. The AD-test returns a $TS = 1.8$. The post-trial significance of this test is re-evaluated with the method in section 3, using two million randomized MC trials generated from the out-FidFoV data set. The corresponding $TS$ distribution is represented by the orange histogram in Figure 12. The red dashed line depicts the AD-test obtained for the data set. The corresponding significance is $\sim 2.2\sigma$, which gives a probability of 0.03 that this would result from an isotropic sky. The significance seen in this sample is, therefore, much lower than with the in-FidFoV data set.

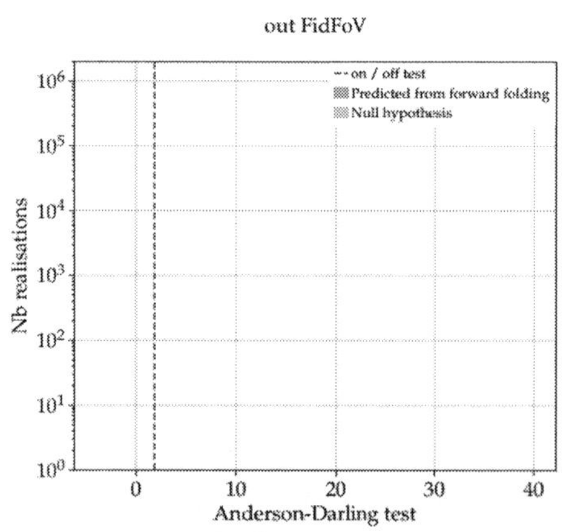


Figure 12: Anderson-Darling test from randomized skies (orange) and from forward folding (blue) for the out-*FidFoV* data set. The red dashed line indicates the value obtained for data in the on/off test.

The lower significance of the out-FidFoV sample begs the question, why is there such a large difference between the in-FidFoV and out-FidFoV data sets? The bottom panel of Figure 11 shows the distribution of $X_{max}$ values from events. Clearly the bulk of the distribution lies in a region where the acceptance is decreasing quickly. It is possible that this could reduce the sensitivity to a difference in composition. To test this hypothesis, two million mock data sets have been generated from the on- and off-plane $X_{max}$ distributions of the in-FidFoV data set. The non-flat acceptance and lower $X_{max}$ resolution of the out-FidFoV data set has then been forward folded onto these samples to create out-FidFoV mock data sets which assume the difference in-FidFoV is real. The AD-test is then computed for each mock data set. The blue histogram represents the corresponding $TS$ distribution. It peaks exactly at the value observed with the out-FidFoV data set, showing that the out-FidFoV is indeed less sensitive to the tested anisotropy. Overall, the independent test with the out-FidFoV shows that the on/off separation is present in both data sets, and that the difference seen in the out-FidFoV sample is consistent with the magnitude of the difference seen in the in-FidFoV sample.

## 6 Mapping the UHECR sky in composition

To aid interpretation of the latitude-dependent difference in composition, a test statistic quantifying the relative difference in $X_{max}$ between different parts of the sky is mapped in Figure 13, for UHECR primaries with $E \geq 10^{18.7}$ eV. To produce this map from the in-FidFoV data set, first the requirement $E \geq 10^{18.7}$ eV is imposed. Then, because small portions of the sky are analysed, in contrast to the on/off study, each event has its $B$, $R$, and $A$ corrected for based on its arriving declination instead of its arriving galactic latitude. This is because local geometry has a time-independent relationship with arrival declination. This means $B$, $R$, and $A$ can be corrected equally well for each direction in the sky*.

At this point, a top-hat sampling is used to collect all events with arrival directions within 30° of a point $(\ell, b)$ into an *in-hat* sample. All other events are placed in an *out-hat* sample. The distributions of $X'_{max}$ for the in-hat and out-hat samples are then compared using Welch's t-test [25]:

$$TS = \frac{\langle X'^{\,in}_{max}\rangle - \langle X'^{\,out}_{max}\rangle}{\sqrt{\left(\sigma\left(X'^{\,in}_{max}\right)/\sqrt{N^{in}}\right)^2 + \left(\sigma\left(X'^{\,out}_{max}\right)/\sqrt{N^{out}}\right)^2}}, \quad (3)$$

where $N^{in}$ and $N^{out}$ are the event counts for the in- and out-hat samples respectively§. This procedure is repeated for top-hats centered on each point in a 5° by 5° galactic latitude and longitude grid. The result is shown in Figure 13, which illustrates the relative composition of UHECRs with $E \geq 10^{18.7}$ eV arriving from each point in the sky.

In Figure 13, positive $TS$ values (red) indicate that events within 30° of that point have a lighter mean mass than the rest of the sky. Negative values (blue) indicate that events within 30° of that point have a heavier mean mass than the rest of the sky. An excess of heavy particles

*Using declination-dependent corrections changes the on/off comparison only by +0.1 g/cm$^2$ and increases systematic uncertainties.

§Because Welch's t-test considers event statistics, the FD arrival direction-dependent exposure is naturally treated through its use.

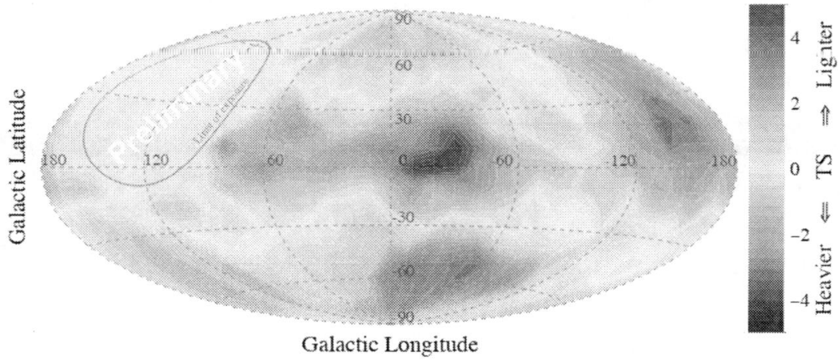


Figure 13: Sky map of comic ray composition for $E \geq 10^{18.7}$ eV

within 30° of the galactic plane is visible. This can not be due to detector systematics as they would be declination dependent and appear as radial patterns centered on $\ell = -57°, b = -27°$ due to the geographic location of the Observatory.

## 7 Conclusions and Outlook

There is an apparent difference in the mean mass of primaries with energies greater than $10^{18.7}$ eV that arrive from within 30° of the galactic plane. This has been observed at least at the $3.3\,\sigma$ level in the standard hybrid data set used for $X_{\rm max}$-based composition analyses. It has now been independently confirmed with an additional $2.1\,\sigma$ significance in a second hybrid data set formed from high quality events cut by a selection aimed at reducing the bias caused by the $X_{\rm max}$-dependent event acceptance. The combined significance of these two results has not yet been evaluated. Further tests of the on-/off-plane difference are being planned using analyses of data from the SD and will be reported elsewhere.

Currently, this result should be considered to primarily provide a new verification of a mixed composition above the ankle as it is clear no such difference could be observed in a flux with a single mass component. Though the analysis provides a possible indication that the galactic magnetic field may have an observable impact on mass-dependent anisotropies, the result found in this analysis does not necessarily support a causal relationship with galactic structures. The differing horizons of different nuclear species at a given energy could also result in composition-dependent anisotropic patterns [26]. It is important, however, to note there is significant tension with models [27]. Alternative scenarios are being explored along these lines of thought.

## References

[1] A. Aab et al. (Pierre Auger), Nucl. Instrum. Meth. A **798**, 172 (2015), 1502.01323
[2] I. Allekotte et al. (Pierre Auger), Nucl. Instrum. Meth. A **586**, 409 (2008), 0712.2832
[3] J. Abraham et al. (Pierre Auger), Nucl. Instrum. Meth. A **620**, 227 (2010), 0907.4282
[4] C. Bonifazi (Pierre Auger), Nucl. Phys. B Proc. Suppl. **190**, 20 (2009), 0901.3138
[5] M. Mostafa, Nucl. Phys. B Proc. Suppl. **165**, 50 (2007), astro-ph/0608670
[6] A. Aab et al. (Pierre Auger), PRD **90**, 122005 (2014)
[7] E. Mayotte et al. (Pierre Auger), PoS **ICRC2021**, 321 (2021)
[8] A. Aab et al. (Pierre Auger), Phys. Rev. Lett. **125**, 121106 (2020), 2008.06488
[9] J. Linsley (1963), ICRC 8 77
[10] A. Aab et al. (Pierre Auger), Science **357**, 1266 (2017)
[11] A.A. Watson, Nucl. Phys. B Proc. Suppl. **22**, 116 (1991)
[12] J. Szabelski, J. Wdowczyk, A.W. Wolfendale, J. Phys. G **12**, 1433 (1986)
[13] Pierre Auger Collaboration, PRD **90**, 122006 (2014)
[14] P. Abreu et al. (Pierre Auger), PoS **ICRC2021**, 311 (2021)
[15] A. Yushkov (Pierre Auger), PoS **ICRC2019**, 482 (2020)
[16] A. Aab et al. (Pierre Auger), JCAP **04**, 038 (2017)
[17] T. Bergmann et al., Astropart. Phys. **26**, 420 (2007)
[18] F. Riehn et al., PoS **ICRC2015**, 558 (2016)
[19] P. Abreu et al. (Pierre Auger), Astropart. Phys. **34**, 368 (2011)
[20] J. Bellido (Pierre Auger), PoS **ICRC2017**, 506 (2018)
[21] T.W. Anderson, D.A. Darling, Ann. Math. Statist. **23**, 193 (1952)
[22] F.W. Scholz, M.A. Stephens, JASA **82**, 918 (1987)
[23] S. Engmann, D. Cousineau, JAQM **6** (2011)
[24] T. Pierog et al., PRC **92**, 034906 (2015)
[25] B.L. Welch, Biometrika **29**, 350 (1938)
[26] C. Ding, N. Globus, G.R. Farrar, Astrophys. J. Lett. **913**, L13 (2021), 2101.04564
[27] D. Allard, J. Aublin, B. Baret, E. Parizot, Astron. Astrophys. **664**, A120 (2022), 2110.10761

https://doi.org/10.1051/epjconf/202328303004

# UHECR anisotropy and extragalactic magnetic fields with the Telescope Array

*Mikhail* Kuznetsov[1,2,*] and *Peter* Tinyakov[2,**] for the Telescope Array collaboration

[1]Institute for Nuclear Research, Moscow 117312, Russia
[2]Université Libre de Bruxelles, CP225, bvd. du Triomphe, 1050 Bruxelles, Belgium

**Abstract.** We study the energy-dependent distribution of ultra-high energy cosmic ray arrival directions with respect to luminous matter in the local Universe. We use a specially designed test statistic (TS) that is robust to uncertainties of the galactic magnetic field. We generate realistic mock UHECR sets assuming various injected compositions, and different strengths of the extragalactic magnetic field (EGMF). Applying the TS to both mock sets and the Telescope Array Surface Detector data we constrain, for a given EGMF strength, the UHECR injected mass composition at energies above 10 EeV. At highest energies, higher than 100 EeV, only heavy composition — iron or at least silicon — is compatible with the data, irrespective of the EGMF strength. We then compare the obtained results with the direct Telescope Array fluorescence measurements of the UHECR mass composition. Requiring that the TA composition measurements are compatible with the arrival direction distribution allows us to constrain the parameters of the EGMF. It appears that light composition, measured by TA at around 10 EeV, is compatible with the arrival directions if the EGMF has strength of order 1 nG.

## 1 Introduction

Ultra-high energy cosmic rays (UHECR) are particles of cosmic origin with extreme energies in excess of $10^{18}$ eV that bombard the Earth atmosphere producing extensive air showers observed either by large arrays of particle detectors on the ground, or by the telescopes that detect the fluorescence light emitted by the shower, or by both, the latter being referred to as the hybrid mode. Despite several decades of observations, the nature and origin of these particles still remains unknown, undoubtedly due to the indirect character of the existing detection techniques.

The key questions that remain unsolved are the mass composition of the incident cosmic rays, and their sources. Both observational techniques — the ground array and the fluorescent telescopes — allow for a direct reconstruction of the arrival direction and, to a lesser extent, of energy of the incident particle. However, in both cases the mass of the incident particle has to be inferred from indirect observables through complicated modeling of the air shower development. This only works so far with large uncertainties, and on a statistical rather than event-by-event basis.

The arrival directions of UHECR are measured with good accuracy and are basically free from systematic errors. However, they do not necessarily point back to the UHECR sources because of the deflections in the Galactic and extragalactic magnetic fields. These magnetic fields are not known well enough to reconstruct true directions to the sources. The unknown charges of particles make the problem even more complicated, so the two problems are entangled.

Although the UHECR accelerators are not yet identified, their global distribution must follow the large-scale structure (LSS) of the ordinary matter. The latter is inhomogeneous at scales $\lesssim$ 100 Mpc which is a typical size of the UHECR collection region at the highest energies. Depending on the magnitude of deflections, these inhomogeneities may have an imprint on the distribution of UHECR arrival directions over the sky, which may be possible to identify in the data. This may give information on typical deflections, from which both the cosmic ray composition and the magnetic fields may potentially be constrained.

A way to quantify these arguments has been recently proposed in Ref. [1] which is based on measuring a typical UHECR deflection angle with respect to the LSS and comparing it to the predictions of various composition models. In this contribution we apply this method to the most recent TA data. This allows us to estimate or to constrain from below, depending on the energy, the typical UHECR deflections. Using these results, even after accounting for possible uncertainties, some light composition models can be ruled out. We will also see that, if the direct TA measurements of the composition are included, the conclusions about the strength of the extragalactic magnetic field can be made.

## 2 Experiment and data

Telescope Array [2, 3] is a hybrid detector of UHECR located at 39.3° N, 112.9° W in Utah, USA. Its surface detector (SD) consists of an array of 507 scintillator stations placed in a square grid with the 1.2 km spacing covering the area of $\sim$ 700 km$^2$. The atmosphere above the surface

*e-mail: mkuzn@inr.ac.ru
**e-mail: petr.tiniakov@ulb.be

array is overlooked by 38 fluorescence telescopes arranged in three stations. The detector is taking data in a hybrid mode since May 2008. The duty cycle of the SD is about 95% [4].

In this analysis we use the TA SD data collected during 14 years of operation from 2008-05-11 to 2022-05-10. We use the standard TA data set with the reconstruction optimized for anisotropy analysis; the corresponding quality cuts are described in Refs. [4, 5]. It contains events with energy $E \geq 10$ EeV and zenith angle $z \leq 55°$. The angular resolution of these events is estimated as 1.5° [6]. The energy resolution is 18% [4, 5], while the systematic uncertainty of the energy scale is estimated as 21% [7].

In addition to standard cuts, we also eliminate the events that may have been induced by lightnings [8, 9]. The latter are taken from the Vaisala lightning database compiled by the U.S. National Lightning Detection Network [10]. We correlated the list of the lightning events detected within 15 miles from the Central Laser Facility of the TA during the full time of TA operation with the list of TA events and removed all the TA events that occur within 10 minutes before or after the lightnings. This cut was shown to reduce the total exposure by less than 1% [11]. The total number of events in the resulting set is 5966; of these 19 events have energies exceeding 100 EeV.

## 3 Analysis

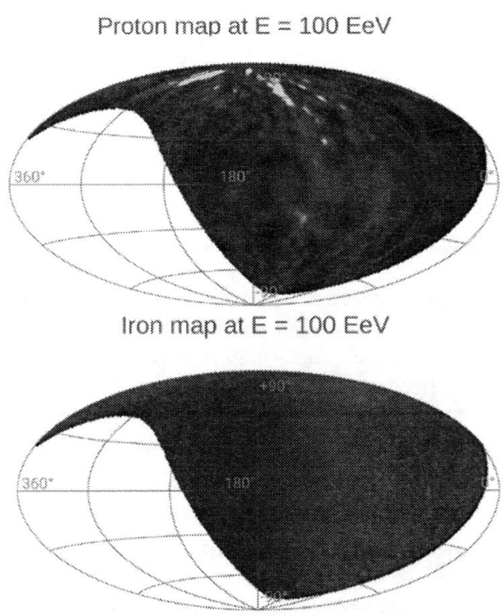


**Figure 1.** Examples of flux maps used for mock UHECR sets generation.

We implement the analysis method in three steps. First, for each injected mass composition model considered we generate a large mock set of realistic UHECR events, represented by their directions and energies. Then, we quantify the mean deflection of a given event set with respect to the Large Scale Structure of the Universe (LSS) using the specially designed test-statistics (TS) which is robust to the uncertainties of the magnetic fields. Finally, we compute the TS for each mock event set and for the experimental data set and quantify the compatibility of each mass composition model with the data.

Let us describe the analysis steps in more detail. To generate realistic mock UHECR event sets, we first build a state-of-the-art model of the UHECR flux. Basically, we set all the model parameters but nuclei injected fractions to some reasonable values. The effect of the uncertainties in these parameters can be estimated in the very end of the analysis, by varying them for typical composition models. We left this issue for further studies.

For the flux model we assume the UHECR sources to trace the luminous matter distribution in the local Universe. To achieve this we attribute an equal intrinsic UHECR flux for each galaxy in a complete volume-limited sample. Technically, we use a highly-complete flux-limited galaxy sample. We derive it from the 2MRS galaxy catalog [12] in which we are cutting out galaxies with mag > 12.5 and those located closer than 5 Mpc and farther than 250 Mpc from the Earth. To compensate the observational selection effects of the flux-limited sample we assign a larger flux to more distant galaxies. We assume that the sources beyond 250 Mpc are distributed uniformly and have same mean density as the sources within this distance. The procedure of the catalog construction is described in Ref. [13]. This source model successfully represent all the scenarios with sufficiently numerous sources (source number density $n \gg 10^{-5}$ Mpc$^{-3}$). The sensitivity of our method for the models with source densities of order $10^{-5}$ Mpc$^{-3}$, that are not yet ruled out [14], will be given in further studies.

We specify the flux model further by building a separate flux map for each injected primary. We use 5 benchmark pure injected compositions: protons (p), helium (He), oxygen (O), silicon (Si) and iron (Fe). We set the injection spectrum for each primary by making a separate fit to the TA and Auger observed spectra [15]. Namely, we set power law spectra with indices −2.55, −2.20, −2.10 and without cut-off for p, He and O, respectively; power law with index −1.50 and with sharp cut-off at 280 EeV for Si; power law with index −1.95 and with sharp cut-off at 560 EeV for Fe. For each primary we take into account the effect of absorption in interstellar medium. The flux of secondary particles produced during propagation of injected primary through the intergalactic medium is taken into account for He and O but neglected for Si and Fe, see Ref. [15] details.

We also simulate the deflections in magnetic fields with the account of particle charge $Z$ and its energy $E$. We take into account the particles deflection in regular [16] and random [17] galactic magnetic field (GMF) components. The deflections in the extra-galactic magnetic field (EGMF) are neglected for the basic flux model, this corresponds to magnetic field strength $B_{\rm EGMF} \ll 1$ nG for the correlation length $\lambda \sim 1$ Mpc. We also build a separate flux model with non-negligible deflections in EGMF to test its impact on results.

In order to make mock event sets similar to data sets we also take into account the experimental effects: the energies of the events are generated according to the observed TA spectrum [18] and we add effects of TA experiment energy resolution and exposure to the simulations. The examples of two flux maps used for mock event set simulations are shown in Fig. 1.

The test-statistics we use is based on the UHECR flux maps similar to those used for the mock event sets generation, but with smaller number of free parameters. Namely we use the same source catalog and the flux is attenuated as protons with $\sim E^{-2.5}$ injection spectrum without cut-off. The only free parameter of the TS is the magnitude of the uniform smearing of sources defined at the energy 100 EeV: $\theta_{100}$.

For each given value of $\theta_{100}$ we build a set of maps $\Phi_k(\theta_{100}, \mathbf{n})$ where $\mathbf{n}$ is the direction in the sky and $k$ is the number of energy bin. The smearing of each map scales as $100\ \mathrm{EeV}/E_k$. The test statistics $\mathrm{TS}(\theta_{100})$ for a given event set with directions $\mathbf{n}_i$ is defined in a following way:

$$\mathrm{TS}(\theta_{100}) = -2 \sum_k \left( \sum_i \ln \frac{\Phi_k(\theta_{100}, \mathbf{n}_i)}{\Phi_{\mathrm{iso}}(\mathbf{n}_i)} \right), \qquad (1)$$

where the sum run over the events in the set $i$ and energy bins $k$. For the particular analysis we use logarithmic energy bins of 0.05 decade width. The TS is normalized with a factor $\Phi_{\mathrm{iso}}(\mathbf{n}_i) = \Phi(\infty, \mathbf{n}_i)$ that corresponds to an isotropic distribution of events. This TS was first introduced in our study [1], where it was shown that for a large number of events it is distributed around its minimum according to $\chi^2$-distribution with one d.o.f.

The position of the TS minimum $\theta_{100}^{\mathrm{min}}$ for each event set has a meaning of the energy-rescaled average event deflection with respect to the sources in LSS. Therefore, for a large enough mock event set of a given composition the TS should have a deep and narrow minimum and the respective value $\theta_{100}^{\mathrm{min}}$ characterizes this composition model. These values could then be compared with the $\mathrm{TS}(\theta_{100})$ distribution for the data sets.

## 4 Results

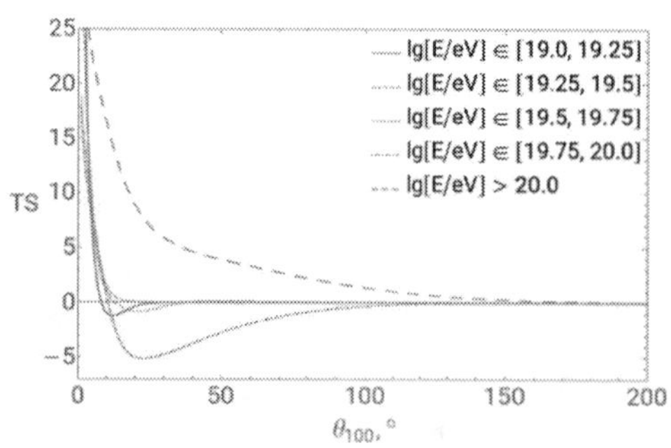


**Figure 2.** The test statistics $TS(\theta_{100})$ calculated the data in five energy bins as indicated on the plot.

The test statistics Eq. 1 calculated for the data in five energy bins is shown in Fig. 2 as a function of $\theta_{100}$. Note, that the plotted bins are logarithmic and have 0.25 decade width, i.e. each of them contain 5 technical energy bins used for computation of Eq. 1. We limit the range of $\theta_{100}$ by 200° as set by the size of the TA field of view; beyond this value we consider configurations to be essentially isotropic. It can be seen that in all energy bins the small values of $\theta_{100}$ are excluded, the strongest exclusion arising from the highest energy bin $E > 10^{20}$ eV. The low energy bins show indications of minima at $10 - 30°$ which are not significant (recall that 1-$\sigma$ interval corresponds to the change of the TS by 1). The minimum is more pronounced in the energy bin $10^{19.75} - 10^{20}$ eV where the TA data exhibit the hot spot of the size $\sim 25°$ matching the position of the minimum in this bin. Surprisingly, the highest-energy bin $E > 10^{20}$ eV shows no minimum and prefers instead a nearly isotropic distribution with deflections $\theta_{100} \gtrsim 50°$ at $2\sigma$ CL. The positions of the minima of the TS (where they exist) and the 1- and 2-$\sigma$ confidence intervals are shown in Fig. 3.

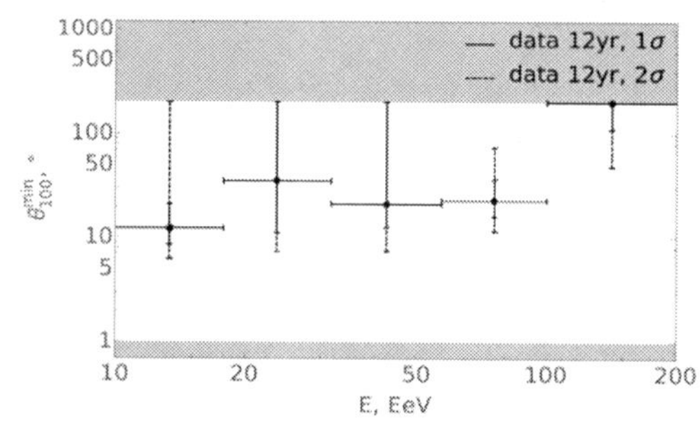


**Figure 3.** The positions of the TS minima, as well as the corresponding 1- and 2-$\sigma$ intervals, for the actual TA data in 5 energy bins.

In order to interpret this result in terms of composition models we generate, for each given composition, a large mock set of events as we describe in previous Section. For each of these sets we calculate the TS in exactly the same way as for the data. Since the mock sets are very large, the statistical uncertainties in the positions of the TS minima are negligible. The comparison of the model predictions with the data is shown in Fig. 4. The cases where the minimum for the model does not exist (like Si and Fe in low energy bins) are shown by upward arrows.

It is tempting to conclude from Fig. 4 that predominantly light compositions are excluded, but this is not so straightforward without examining mixed compositions in various combinations, which we do not attempt in this contribution. Let us instead focus on the highest-energy bin $E > 10^{20}$ eV where conclusions can be drawn from already existing results. As can be seen from Fig. 4, in this energy bin no pure composition fits the data, and even the heaviest — pure iron — is disfavored by the data at about $2\sigma$ level. The same is clearly true for an arbitrary mixture. We conclude therefore that the isotropy of the data at $E > 10^{20}$ eV indicates a very heavy composition at these

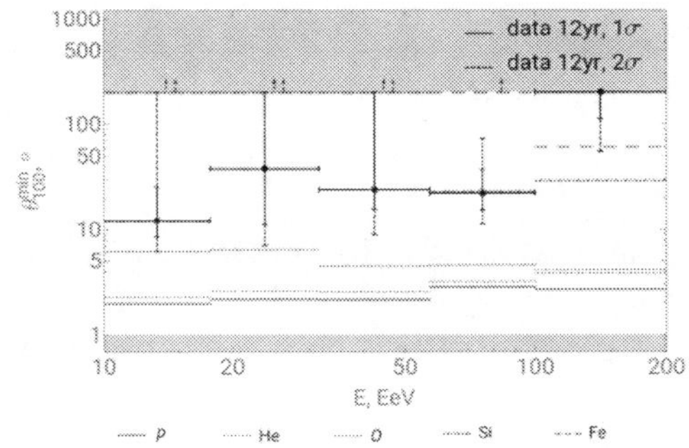


**Figure 4.** Comparison of the pure composition models (color lines as indicated on the plot) with the TA data (black crosses). Upward arrows indicate model predictions indistinguishable from isotropy.

energies. It should be stressed that this result is obtained *with the account of the deflections in the GMF*. The latter is not well known; it can be checked however that the conclusion is robust: changing the GMF model to the one of Ref. [19] affects the model predictions only slightly, and thus the conclusion remains. We will see shortly that assuming a non-negligible extragalactic MF also does not change the picture.

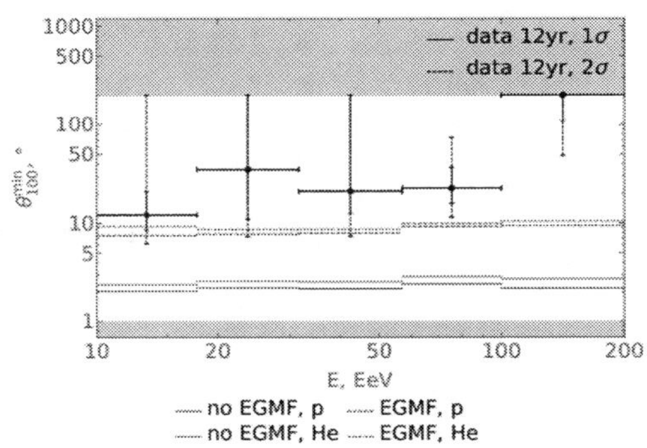


**Figure 5.** Comparison of the model predictions in the case of pure light composition (p, He) with the data, with zero and with maximum EGMF.

Let us now focus on the low energy bins. Here it is not possible to conclude that the composition should be heavy without the detailed study of the mixed composition models. It may appear that the data still exclude the light p-He mix. This conclusion, however, is not robust with respect to possible effects of the EGMF. While the EGMF is not known, there exist experimental upper bounds at the level of $10^{-9}$ G assuming maximum correlation length of 1 Mpc [20]. If the EGMF is pushed to this extreme value, the model predictions change. This change is shown in Fig. 5 where the predictions for pure proton and helium compositions are compared to the data in two cases: zero EGMF, and maximum EGMF satisfying the current limits. one can see that inclusions of deflections in the EGMF can make the light composition models compatible with the TA data.

This argument can be inverted if extra information concerning the composition is included. In the energy bin $10^{19}-10^{19.25}$ eV there exist a TA measurement of the composition from the $X_{\max}$ distribution of the fluorescence detector (FD) data [21]. This measurement indicates a light composition consisting of the proton-helium mix. If this measurement is combined with the above measurement of event deflections, together they imply that the EGMF should be large, at its current experimental upper-limit.

It remains to be checked whether the large EGMF, as large as indicated by the combined TA SD and FD data, changes the conclusion about a heavy composition at the highest energies. From Fig. 5 one can see that in the highest energy bin the light elements are still not deflected enough to explain the isotropy of the events. Fig. 6 shows how the predictions for the pure heavy composition (Si, Fe) change upon the inclusion of the maximum EGMF. In the highest energy bin, the predicted deflection for iron become compatible with the data, while those for silicone become marginally compatible. The conclusion about heavy composition at the highest energies remains, therefore, unchanged.

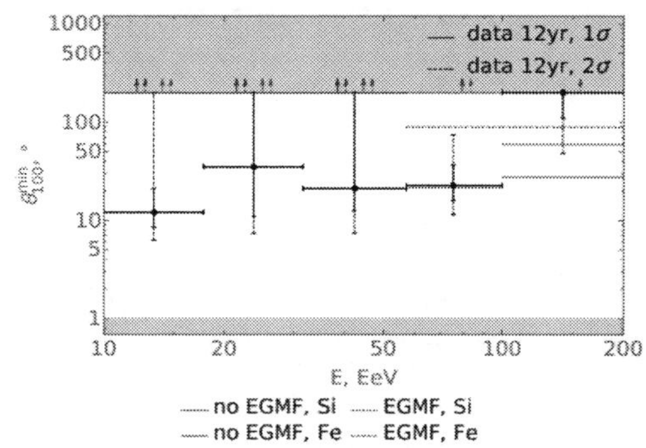


**Figure 6.** Comparison of the model predictions in the case of pure heavy composition (Si, Fe) with the data, with zero and with maximum EGMF.

## 5 Conclusions

To summarize, we have introduced a quantitative measure of the overall cosmic ray deflections in a given data set with respect to the large-scale structure on the Universe, $\theta_{100}^{\min}$ defined in terms of Eq. 1. We then calculated this quantity for the most recent TA SD data set in 5 energy bins at $E > 10^{19}$ eV. The result is shown in Fig. 3. This quantity can be compared to the predictions of different composition models. Despite the uncertainties inherent in these models, two conclusions can be robustly deduced:

- In the highest energy bin $E > 10^{20}$ eV only the heaviest composition — iron if no EGMF is assumed, or possibly silicone in the case of a maximum EGMF — gives deflections large enough to be compatible with the measured value.

- In the lowest energy bin $10^{19} - 10^{19.25}$ eV, the independently measured (from the TA FD data) light composition is only compatible with the observed large deflections if a maximum allowed EGMF is assumed.

We see therefore that the observed high isotropy of the UHECR and, consequently, their large deflections, indicate the heavy composition of UHECR at highest energies, and large, at the level of existing experimental limit, extragalactic magnetic fields.

## Acknowledgements

The Telescope Array experiment is supported by the Japan Society for the Promotion of Science(JSPS) through Grants-in-Aid for Priority Area 431, for Specially Promoted Research JP21000002, for Scientific Research (S) JP19104006, for Specially Promoted Research JP15H05693, for Scientific Research (S) JP19H05607, for Scientific Research (S) JP15H05741, for Science Research (A) JP18H03705, for Young Scientists (A) JPH26707011, and for Fostering Joint International Research (B) JP19KK0074, by the joint research program of the Institute for Cosmic Ray Research (ICRR), The University of Tokyo; by the Pioneering Program of RIKEN for the Evolution of Matter in the Universe (r-EMU); by the U.S. National Science Foundation awards PHY-1607727, PHY-1712517, PHY-1806797, PHY-2012934, and PHY-2112904; by the National Research Foundation of Korea (2017K1A4A3015188, 2020R1A2C1008230, & 2020R1A2C2102800) ; by IISN project No. 4.4501.18, and Belgian Science Policy under IUAP VII/37 (ULB). This work was partially supported by the grants of The joint research program of the Institute for Space-Earth Environmental Research, Nagoya University and Inter-University Research Program of the Institute for Cosmic Ray Research of University of Tokyo. The foundations of Dr. Ezekiel R. and Edna Wattis Dumke, Willard L. Eccles, and George S. and Dolores Doré Eccles all helped with generous donations. The State of Utah supported the project through its Economic Development Board, and the University of Utah through the Office of the Vice President for Research. The experimental site became available through the cooperation of the Utah School and Institutional Trust Lands Administration (SITLA), U.S. Bureau of Land Management (BLM), and the U.S. Air Force. We appreciate the assistance of the State of Utah and Fillmore offices of the BLM in crafting the Plan of Development for the site. Patrick A. Shea assisted the collaboration with valuable advice and supported the collaboration's efforts. The people and the officials of Millard County, Utah have been a source of steadfast and warm support for our work which we greatly appreciate. We are indebted to the Millard County Road Department for their efforts to maintain and clear the roads which get us to our sites. We gratefully acknowledge the contribution from the technical staffs of our home institutions. An allocation of computer time from the Center for High Performance Computing at the University of Utah is gratefully acknowledged. The work on the EGMF analysis was supported by the Russian Science Foundation, grant No. 22-12-00253

## References

[1] M.Y. Kuznetsov, P.G. Tinyakov, JCAP **04**, 065 (2021), 2011.11590

[2] T. Abu-Zayyad et al. (Telescope Array), Nucl. Instrum. Meth. A **689**, 87 (2013), 1201.4964

[3] H. Tokuno et al., Nucl. Instrum. Meth. **A676**, 54 (2012), 1201.0002

[4] T. Abu-Zayyad et al. (Telescope Array), Astrophys. J. **768**, L1 (2013), 1205.5067

[5] Telescope Array Collaboration, arXiv e-prints (2014), 1403.0644

[6] T. Abu-Zayyad et al. (Telescope Array), Astrophys. J. **757**, 26 (2012), 1205.5984

[7] R.U. Abbasi et al. (Telescope Array), Astropart. Phys. **80**, 131 (2016), 1511.07510

[8] R. Abbasi et al., J. Geophys. Res. Atmos. **123**, 6864 (2018), 1705.06258

[9] R. Abbasi et al. (Telescope Array), Physics Letters A **381**, 2565 (2017)

[10] K. Cummins, M.J. Murphy, IEEE Trans. **51**, 499 (2009)

[11] R.U. Abbasi et al. (Telescope Array), Astropart. Phys. **110**, 8 (2019), 1811.03920

[12] J.P. Huchra et al., Astrophys. J. Suppl. **199**, 26 (2012), 1108.0669

[13] H.B. Koers, P. Tinyakov, Mon. Not. Roy. Astron. Soc. **399**, 1005 (2009), 0907.0121

[14] P. Abreu et al. (Pierre Auger), JCAP **05**, 009 (2013), 1305.1576

[15] A. di Matteo, P. Tinyakov, Mon. Not. Roy. Astron. Soc. **476**, 715 (2018), 1706.02534

[16] M.S. Pshirkov, P.G. Tinyakov, P.P. Kronberg, K.J. Newton-McGee, Astrophys. J. **738**, 192 (2011), 1103.0814

[17] M.S. Pshirkov, P.G. Tinyakov, F.R. Urban, Mon. Not. Roy. Astron. Soc. **436**, 2326 (2013), 1304.3217

[18] D. Ivanov, PoS **ICRC2015**, 349 (2016)

[19] R. Jansson, G.R. Farrar, Astrophys. J. **757**, 14 (2012), 1204.3662

[20] M. Pshirkov, P. Tinyakov, F. Urban, Phys. Rev. Lett. **116**, 191302 (2016), 1504.06546

[21] R. Abbasi et al. (Telescope Array), Astrophys. J. **858**, 76 (2018), 1801.09784

# Updates on the Hotspot and the Perseus-Pisces supercluster Excess Observed by the Telescope Array Experiment

*Jihyun* Kim[1*], *Dmitri* Ivanov[1], *Kazumasa* Kawata[2], *Hiroyuki* Sagawa[2], and *Gordon* Thomson[1] for the Telescope Array Collaboration

[1]High Energy Astrophysics Institute and Department of Physics and Astronomy, University of Utah, Salt Lake City, Utah 84112-0830, USA
[2]Institute for Cosmic Ray Research, University of Tokyo, Kashiwa, Chiba 277-8582, Japan

**Abstract.** The Telescope Array (TA) experiment, the largest observatory studying ultra-high energy cosmic rays in the northern hemisphere, has reported evidence for two medium-scale anisotropies. The first, known as the TA hotspot, is an excess in the arrival direction distribution for events with energies greater than $5.7 \times 10^{19}$ eV. More recently, an additional excess of events with energies greater than $10^{19.4}$ eV appearing in the direction of the Perseus-Pisces supercluster has been named the PPSC excess. In this presentation, we will update the status of the TA hotspot and the PPSC excess results using the most recent data measured by the TA surface detector array.

## 1 Introduction

Ultra-high-energy cosmic rays (UHECRs) are energetic charged particles originating from outer space that impinge on Earth's atmosphere with energies above $10^{18}$ eV. There have been many efforts to elucidate their origin; however, it is still challenging since cosmic magnetic fields are ubiquitous in the universe and bend UHECR trajectories from their sources. However, for higher energy cosmic rays having $E > 10^{19.5}$ eV, they have still reasonable deflections to search for their sources by studying their arrival direction distribution.

Anisotropy studies reveal patterns in the arrival direction distribution that may be indicative of UHECR's source locations. In 2014, the Telescope Array (TA) experiment reported anisotropy in the arrival direction distribution of UHECR events with energy greater than 57 EeV (1 EeV = $10^{18}$ eV), called the TA hotspot [1]. By using 5 years of data collected by the TA surface detector array, we found an excess of events in the Ursa Major constellation, using an oversampling analysis with an intermediate angular scale of 20°. Using the Li-Ma analysis method, the maximum significance of excess compared to the isotropic background was calculated to be $5.1\sigma$ at the position of (146.7°, 43.2°) in equatorial coordinates. The global significance of this excess was estimated to be $3.4\sigma$ by Monte Carlo simulations. A number of sources have been suggested, including M82, Mrk 180 [2], Mrk 421 [3], and filaments of galaxies connected to the Virgo cluster [4]. However, it is necessary to get higher statistics to discern the nature of the TA hotspot.

Recently, TA observed an additional excess of events with energy greater than $10^{19.4}$ eV. While investigating the energy spectrum mismatch between TA and Pierre Auger Observatory, we found excesses in the energy spectrum in energy bins of $10^{19.4}$ eV, $10^{19.5}$ eV, and $10^{19.6}$ eV. We examined their arrival direction distributions to understand what causes these excesses in the energy spectrum and found the new excess of events appearing in the direction of the Perseus-Pisces supercluster (PPSC), which has been named the PPSC excess [5, 6].

Here, we update the TA hotspot and the PPSC excess using the most up-to-date data measured by the TA surface detector array.

## 2 Telescope Array Experiment

The TA experiment is the largest UHECR observatory in the northern hemisphere. It is located at 39.3° N 112.9° W in the west desert of Utah, USA. Its elevation is 1400 m above sea level, which is perfectly suitable for observing the maximum development of air showers. TA is a hybrid experiment, using a combination of ground array and air-fluorescence techniques. A total of 507 surface detectors (SDs) are deployed with 1.2 km spacing over an area of approximately 700 km$^2$. Each SD counter consists of two-layers of plastic scintillator to sample the footprint of the air shower when it reaches Earth's surface [7]. Three fluorescence detector (FD) stations overlook the SD array using 38 fluorescence telescopes viewing 3°–31° in elevation. The FDs measure the ultra-violet light generated as the shower passes through the gas of the atmosphere [8].

## 3 Description of the Dataset

In this analysis, we use data measured by the TA SD array that has the highest statistical power. The data acquisition

*e-mail: jihyun@cosmic.utah.edu

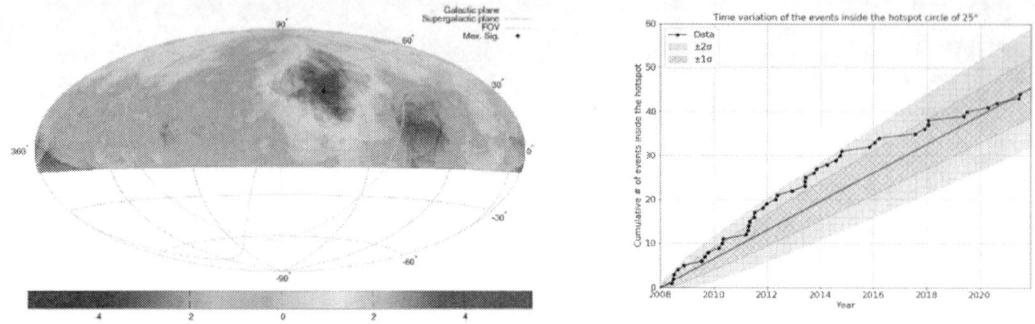


**Figure 1. The sky map of the TA hotspot and the increase rate of events inside the hotspot circle. (Left)** It is shown that the Li-Ma significance using a 25°-circle angular window in equatorial coordinates for 14 years of SD events with energies greater than 57 EeV. The black diamond indicates the maximum Li-Ma significance position at (144.0°, 40.5°). The color code indicates an excess (red) and a deficit (blue) of events compared to isotropy. **(Right)** The black dots represent the cumulative number of events inside the hotspot circle of 25° from the most significant excess to the isotropy at (144.0°, 40.5°). The blue solid line indicates the estimated event rate inside the hotspot. The light blue and gray bands show $\pm1\sigma$ and $\pm2\sigma$ deviations from the linear increase rate.

period is between May 11, 2008, and May 10, 2022. The selection criteria employed in this analysis are as follows:
1. Each event must include at least five SD counters.
2. The reconstructed primary zenith angle must be less than 55°.
3. Both the geometry and lateral distribution fits must have $\chi^2$/degree of freedom value less than 4.
4. The angular uncertainty estimated by the geometry fit must be less than 5°.
5. The fractional uncertainty in S800 estimated by the lateral distribution fit must be less than 25%.
6. The largest signal counter surrounded by four working counters, there must be one working counter to the left, right, down, up on the grid of the largest signal counter, but they do not have to be immediate neighbors of the largest signal counter.

The reconstructed SD energy is normalized to the FD calorimetric energy using hybrid events [9]. For the hotspot analysis, we adopt the same energy threshold and selection criteria that were used in the original hotspot paper [1]. The energy threshold is 57 EeV and the selection criteria are looser than those described above to include more events at higher energies while keeping reasonable energy and angular resolutions. For the PPSC excess analysis, the energy threshold is set to $10^{19.4}$ eV, where an excess in the energy spectrum was observed. The total number of events used for the hotspot analysis is 205 and the energy and angular resolutions of the events are 10% and 1.0°, respectively. On the other hand, the total number of events used for the PPSC excess analysis is 1060, and energy and angular resolutions of the events are 20% and 1.5°, respectively [10].

## 4 Methods and Results

We conduct oversampling searches to investigate intermediate scale anisotropies in the data sets. Oversampling calculations are done at grid points in equatorial coordinates, having 0.1° steps in right ascension (0°–360°) and declination (−10°–90°). Specifically, at each grid point, we sum over the number of events in a given angular scale circle for the data, defined as $N_{\rm on}$. ($N_{\rm off} = N_{\rm tot} - N_{\rm on}$) Then, we generate $10^5$ events assuming an isotropic flux taking into account the geometrical exposure $g(\theta) = \sin\theta\cos\theta$ as a function of zenith angle ($\theta$) because the detection efficiency for this energy range is ~100% regardless of zenith angle $\theta$. Using these isotropic background events, we sum over the number of events in the given angular scale circle in the same manner we did for the data, which defines $\alpha = N_{\rm iso,on}/N_{\rm iso,off}$ for exposure ratio of *on* to *off*. Then, by comparing the oversampling of observed data to the isotropic background events, the statistical significance of the excess of events at each grid point is calculated by the following equation [11]:

$$S_{\rm LM} = \sqrt{2}\left[N_{\rm on}\ln\left(\frac{(1+\alpha)N_{\rm on}}{\alpha(N_{\rm on}+N_{\rm off})}\right) + N_{\rm off}\ln\left(\frac{(1+\alpha)N_{\rm off}}{N_{\rm on}+N_{\rm off}}\right)\right]^{1/2}. \quad (1)$$

### 4.1 TA Hotspot

In 14 years of TA SD data, we take 205 events with energies greater than 57 EeV and then examine their arrival direction distribution using the Li-Ma method with 25° oversampling angular size, which gave the maximum excess of events in the previous analysis using 12 years of data [5]. The most significant excess to the isotropy was found at (144.0°, 40.5°) in equatorial coordinates. Inside the 25°-circle from the position, 44 out of 205 events are detected, whereas only 16.9 events are expected from the isotropic distribution. The Li-Ma significance is calculated to be 5.1$\sigma$. The left panel of Figure 1 shows the sky map with the Li-Ma significance indicating excess (red) and deficit (blue) of events compared to the isotropic sky. The black diamond represents the maximum Li-Ma significance position. We estimate how often such excess could appear

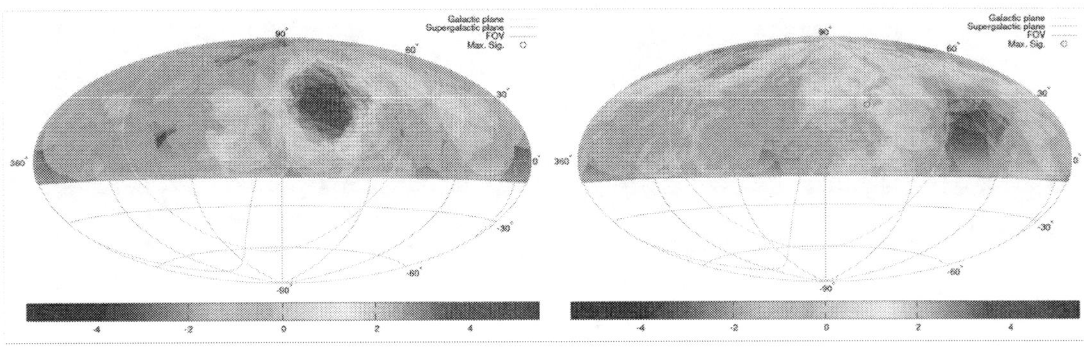

**Figure 2. The sky map of the TA hotspot for independent data sets.** It is shown that the Li-Ma significance using a 25°-circle angular window in equatorial coordinates for first 5-year events and the last 9-year events with energies greater than 57 EeV. The open circles indicate the maximum Li-Ma significance position at (144.0°, 40.5°) obtained from the 14-year data. **(Left)** For the first 5-year events, which is the period the TA hotspot was observed for the first time, the local maximum Li-Ma significance is calculated to be $5.0\sigma$. **(Right)** For the last 9-year events, the local maximum Li-Ma significance is calculated to be $2.5\sigma$, which still indicates an excess of events in the hotspot direction.

anywhere in TA's field of view by Monte Carlo simulation. The chance probability is estimated to be $7.4 \times 10^{-4}$, corresponding to $\sim 3.2\sigma$.

In addition, we examine the time variation of the hotspot by counting the cumulative number of events inside the hotspot as a function of time. The result is shown on the right panel of Figure 1. Inside the 25°-circle from the maximum excess position of (144.0°, 40.5°), the cumulative number of observed events is shown with $\pm 1\sigma$ and $\pm 2\sigma$ bands deviations from a linear increase rate. We conclude that the increase rate of the events inside the hotspot circle is consistent with a constant within $2\sigma$ fluctuation.

We also investigate the excess of events by dividing the whole data set into two periods for independent analysis: the first 5 years and the last 9 years. We split data into the first 5 years and the remaining because the first 5-year data was the one we found the hotspot in 2014 [1]. Figure 2 shows the sky maps of independent analysis. In the first 5-year data, 22 out of 72 events are observed inside the 25°-circle from the maximum excess position of (144.0°, 40.5°), taken from the 14-year result, whereas 5.2 events are expected from the isotropy. The local maximum Li-Ma significance is calculated to be $5.0\sigma$. In the last 9-year data, 22 out of 133 events are detected inside the oversampling circle while 11.6 events are expected. The local maximum Li-Ma significance is calculated to be $2.5\sigma$, which still indicates an excess of events in the hotspot direction.

### 4.2 Perseus-Pisces Supercluster Excess

Here, we revisit the PPSC excess using 14 years of TA SD data. By adopting the same methodology used for the original hotspot study [1], we conduct the oversampling analysis with 20°-radius angular windows. The data selection criteria are described in Section 3, which are tighter than those for the hotspot analysis since we have more events at lower energies and need to maintain good energy and angular resolution.

In 14 years of TA SD data, we have 1060, 685, and 413 events with energy greater than $10^{19.4}$ eV, $10^{19.5}$ eV, and $10^{19.6}$ eV, respectively. The oversampling analysis results are summarized as follows: the Li-Ma significances are calculated to be $3.8\sigma$ at (17.4°, 36.0°), $3.8\sigma$ at (19.0°, 35.1°), and $3.5\sigma$ at (19.7°, 34.6°) for $E \geq 10^{19.4}$ eV, $E \geq 10^{19.5}$ eV, and $E \geq 10^{19.6}$ eV, respectively. The sky maps of the PPSC excess for each energy threshold are shown in Figure 3. We observe consistent excesses of events in the direction of the PPSC.

We examine whether another excess of events is observed close to other nearby supercluster locations. We choose all the similar major structures to the PPSC in TA's field of view within 150 Mpc, which is the GZK horizon assuming proton primaries: Virgo cluster (17 Mpc), PPSC (70 Mpc), Coma supercluster (90 Mpc), Leo supercluster (135 Mpc), and Hercules supercluster (135 Mpc). Figure 3 (d) shows their center locations overlaid with the excess map. The data do not show an excess at any of the locations of other nearby major structures other than the PPSC. None of them have Li-Ma significances greater than $1\sigma$. The PPSC is a unique and significant structure in TA's field of view because it is the closest supercluster other than the Local supercluster where we reside and its location is next to the Local Void [12, 13], where the magnetic field strength is presumed to be weaker than other structures in the cosmic web.

It is suggestive that the excess in the data falls in the direction of the PPSC. We estimate how often this happens by chance. First, we generate many Monte Carlo event sets, each containing the same number of events as the data. We throw Monte Carlo events isotropically according to the acceptance of the TA SD. We conduct the same oversampling analysis for each Monte Carlo event set using the Li-Ma method. After that, we count as successes the number of sets where the point of maximum

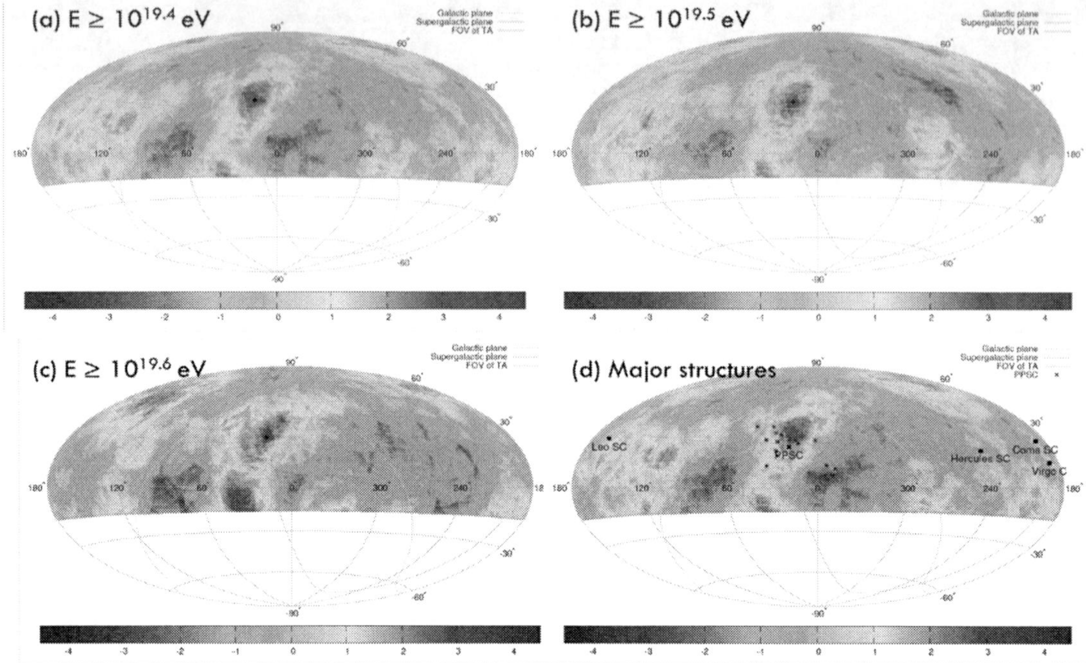


**Figure 3. The sky maps of the PPSC excess using Hammer projection.** It is shown that the Li-Ma significance using a 20°-circle angular window in the equatorial coordinates for different energy thresholds: (a) $E \geq 10^{19.4}$ eV, (b) $E \geq 10^{19.5}$ eV, and (c) $E \geq 10^{19.6}$ eV. In addition, (d) shows the nearby major large-scale structures with the Li-Ma significance map with $E \geq 10^{19.4}$ eV. The color code indicates an excess (red) and a deficit (blue) of events compared to isotropy at each grid point.

Li-Ma significance is at least as significant as in the data, and also occurs at least as close to the PPSC as in the data: ($S_{\rm mc} \geq S_{\rm obs}$) and ($\theta_{\rm mc} \leq \theta_{\rm obs}$). Finally, we calculate the chance probability of having equal or higher excess as close to the PPSC as in the data. We obtain for $3.1\sigma$, $3.2\sigma$, and $3.0\sigma$ for $E \geq 10^{19.4}$ eV, $E \geq 10^{19.5}$ eV, and $E \geq 10^{19.6}$ eV, respectively. We also perform the same calculations for the chance probability of having equal or higher excess as close to any of the nearby major structures we discussed above as PPSC. The results are $2.5\sigma$, $2.6\sigma$, and $2.4\sigma$ for $E \geq 10^{19.4}$ eV, $E \geq 10^{19.5}$ eV, and $E \geq 10^{19.6}$ eV, respectively. We conclude that the results indicate that a cosmic ray source may exist in the direction of the PPSC.

## 5 Summary

In this work, intermediate scale anisotropy studies were conducted using 14 years of TA SD data. We have persistent evidence for the TA hotspot at the highest energies, $E > 57$ eV. The global significance of such excess appearing by chance anywhere in TA's field of view is estimated to be $3.2\sigma$. The new excess in slightly lower energy events having $E \geq 10^{19.4}$ eV in the direction of the Perseus-Pisces supercluster has been identified. The chance probability of having an excess as close to the Perseus-Pisces supercluster as the data is estimated to be $3.2\sigma$.

## References

[1] Abbasi, R.U., *et al.*, Astrophysical Journal Letters, **790**, L21 (2014).

[2] He, H.-N., *et al.*, Physical Review D, **93**, 043011 (2016).

[3] Fang, K., Fujii, T., Linden, T., and Olinto, A.V., Astrophysical Journal **794**, 126 (2014).

[4] Kim, J., Ryu, D., Kang, H., Kim, S., and Rey, S.-C., Science Advances **5** eaau8227, (2019).

[5] Kim, J., Ivanov, D, Kawata, K., Sagawa, H., and Thomson, G., PoS(ICRC2021)328 (2021)

[6] Abbasi, R.U., *et al.*, https://arxiv.org/abs/2110.14827 (2021)

[7] Abu-Zayyad, T., *et al.*, Nuclear Instruments Methods in Physics Research Section a-Accelerators Spectrometers Detectors and Associated Equipment, **689**, 87 (2012).

[8] Tokuno, H., *et al.*, Nuclear Instruments Methods in Physics Research Section a-Accelerators Spectrometers Detectors and Associated Equipment, **676** 54 (2012).

[9] Abu-Zayyad, T., *et al.*, Astrophysical Journal Letters, **768**, L1 (2013).

[10] Abbasi, R.U., *et al.*, Astrophysical Journal, **862**, 91 (2018).

[11] Li, T.-P. and Y.-Q. Ma, Astrophysical Journal Letters, **272**, 317 (1983).
[12] Courtois, H.M., et al., Astronomical Journal, **146**, 69 (2013).
[13] Tully, R.B., *et al.*, Astrophysical Journal, **880**, 24 (2019).

# Flux predictions in the transition region incorporating the effects from propagation of cosmic rays in the Galactic magnetic field

*Alex* Kääpä[1,★], *Karl-Heinz* Kampert[2], and *Julia* Becker Tjus[1]

[1] *Ruhr-Universität Bochum, 44780 Bochum, Germany*
[2] *Bergische Universität Wuppertal, Gaußstraße 20, 42119 Wuppertal, Germany*

**Abstract.** Galactic cosmic rays (GCRs) and (anisotropically injected) extragalactic cosmic rays (EGCRs) are both affected by the Galactic magentic field (GMF) on their voyage to Earth at energies pertaining to the transition from GCRs to EGCRs, such that their flux, composition and arrival directions are modified. GCRs increasingly leak from the Galaxy with rising energy, leading to a flux suppression. The flux modification imposed on EGCRs is more complex, but may exhibit (subtle) spectral breaks depending on the direction and nature of the injected anisotropy.
Using a full Monte Carlo approach with CRPropa and making realistic and minimal assumptions about the injected GCR and EGCR fluxes, we derive a prediction of the total all-particle flux in the transition region. We find that it cannot account for the flux measured by various cosmic ray experiments in this energy range. This calls for the need of an additional component to the flux in the transition region.

## 1 Introduction

The energy at which the dominant contribution to the cosmic ray (CR) flux transitions from Galactic to extragalactic cosmic rays (GCRs and EGCRs, respectively) remains an open question to this day. Numerous models have been shown to account for certain or most features of CR observables, such as the flux, composition and anisotropy (e.g. [1–3]). Two features of the energy spectrum, the *knee*[1] [4] and the *ankle*[2] [5], are of particular importance, as they set the upper and lower bound of the energy range within which this transition is believed to occur for virtually all models[3] ([1–3] mentioned above). This energy range will henceforth be referred to as the *transition region* (or *shin* when referring to the energy spectrum [4]). The rather wide range of proposed origins of these features, however, is exemplary of the substantial uncertainties that still govern our understanding of the processes included in the creation & acceleration, propagation (including deflections in magnetic fields and interactions), and even the detection & identification of CRs.

Major progress has been made on all fronts, however, constraining models, and even strongly disfavouring certain, once popular, ones [1]. On top of that, the development of improved numerical methods and increased computing power have lead to unequalled precision in the description of various processes the CR undergoes from source to observer. One aspect that is particularly relevant for the transition region is the propagation in the Galactic magnetic field (GMF).

While the GMF itself is still poorly constrained, recent models have, for the first time, been fit to observational data, that additionally employ a more complex topology and structure [6]. Especially noteworthy is the separation of the GMF into several components, which are each relevant for different propagation regimes. They include a coherent field which is responsible for large-scale deflections in the ballistic regime, as well as a random component that leads to turbulent motion in the diffusive regime. Propagation studies with such GMF models are particularly useful in the energy range pertaining to transition region, as CR propagation changes from diffusive to ballistic within this range.

This work is based on results from simulations studying the propagation of GCRs and EGCRs in the GMF [7, 8]. Therein, significant features were found for CR observables, such as the flux, composition and anisotropy, that could be traced back to the change in propagation regimes from diffusive to ballistic propagation. The objective here is to make a flux prediction based on these features that can be compared to data (i.e. the energy spectrum). The outline is as follows: In section 2, we will outline the propagation effects and resulting spectral features for both GCRs and EGCRs. In section 3, further considerations and assumptions that are needed to make a flux prediction are discussed. In section 4, the resulting total (i.e. involving both GCR and EGCR) all-particle (i.e. consisting of a mixed composition CR nuclei) energy spectrum is pre-

★e-mail: $\mu$ alex.kaeaepae@rub.de

[1] It describes a steepening of the spectrum at around $4 \cdot 10^{15}$ eV.

[2] It refers to a spectral hardening at roughly $3 \cdot 10^{18}$ eV.

[3] Dipole anisotropy data, more precisely the direction to which the phase points depending on CR energy, provides the strongest support for this notion.

sented and discussed. Finally, conclusions are drawn and an outlook is given in section 5.

## 2 Propagation of GCRs and EGCRs in the GMF in the transition region

The objective of this work is to derive a realistic prediction for the *total*[4] all-particle[5] CR flux in the transition region, by extrapolating two individual contributions, one tuned to the flux up to the knee (representing GCRs), the other to the flux beyond the ankle (representing EGCRs). The defining feature of this prediction is its incorporation of propagation effects that the GMF imposes on GCRs and well as EGCRs, based on [7, 8]. With this prediction, we wish to make inferences about the prospects of describing the shin without invoking any additional contributions. For the latter objective, an outline of said studies is in order. On top of giving the reader an overview of the observed features in CR observables, particularly the flux, their framing in the context of the change in propagation regimes is crucial for drawing valid conclusions from the comparison of our prediction within the CR spectrum.

### 2.1 Summary of earlier simulation studies

Our propagation studies were performed with the established and commonly used Monte-Carlo-based software framework for CR propagation *CRPropa3* [9]. We performed forward-tracking studies, where we simulated in the order of $10^9$ particles with rigidities ranging from $10^{16}$–$10^{20}\,V$, by tracking them individually through the GMF, modeled by the JF12 field[6] [10]. For all simulations, particles are injected with a flat injection spectrum[7] $R^{-1}$. The range of injection positions (and directions) that define the *source* varies depending on the intended class of CRs (i.e. of Galactic or extragalactic origin). The sources of GCRs all lie in the Galactic plane (GP)[8]. The distances of EGCR sources to the Galaxy exceed Galactic distances by a factor of 100 or more, meaning that the bulk of the voyage of EGCRs to Earth takes place outside our region of interest. On the one hand, this means that we can neglect this part of their propagation in the context of our studies, such that their points of injection are located at the edge of the volume that defines the Galaxy[9]. On the other hand, such a projection of the source to the Galactic shell requires an adaptation of the injection direction range, depending on the desired properties of the sources. One goal of our studies was to be able to easily rescale our results across as wide a range of parameters as possible. As a result, the distribution of sources was to mimic isotropy as closely as possible. This entails a continuous and uniform distribution of sources, as well as an isotropic injection direction distribution[10]. Particles are tracked until they satisfy certain break conditions (i.e. diverging propagation times, or leaving the propagation volume), or if they enter or leave the boundary of some specific predetermined region that defines the *observer*. Depending on the nature of our analysis, two classes of observers are defined. In the context of illustrating the propagation effects arising from the change in propagation regimes, the GP[11], which plays an instrumental role in giving rise to said effects, constitutes the observer. For determining the effects this has on CR observables, the observer represents Earth[12].

The theoretical starting point of our studies is the fact that the transition region also marks a shift in propagation regimes from diffusive to ballistic for CRs propagating within the GMF. This shift can best be illustrated using the particle's *gyro radius* $r_g = \frac{R}{B}$. $R$ refers to the CR's *rigidity*, and is defined as the ratio between its total energy $E$ and electric charge $q(= Ze$, $Z$ constituting its atomic number)[13]. $B$ denotes the strength of (the perpendicular component of) the magnetic field. The typical scale of $B$ for the GMF (particularly in the GP region) is $\approx 3\,\mu$G. Translating the energy range of the transition region into its corresponding rigidity range, the lower bound is set by the heaviest CR nucleus, iron, at knee energies, whose gyro radius is $r_g \lesssim 1$ pc. The size of the Galaxy is in the order tens of kpc. Hence, the particle can undergo a very large number of gyrations, and in the presence of strong turbulent fields (which the GMF has), the motion is chaotic. The upper end of the rigidity range is set by the proton energy beyond the ankle, yielding $r_g \sim 10$ kpc, i.e. of similar scale as the size of the Galaxy. Propagation is no longer turbulent in this case, as the particle escapes the volume before it can undergo a large number of gyrations. All of these facts are illustrated in Fig. 1, depicting the gyro radius $r_g$ as a function of rigidity $R$ for various magnetic field strengths $B$, along with relevant length scales (horizontal lines).

[4] In this work, this term always refers to the combination of both GCR and EGCR contributions.

[5] We assume a mixed composition of CRs, i.e. contributions from several different CR species (i.e. nuclei), across the entire measured energy range. Nuclear fractions vary with energy owed to both their differing charges (that affect their rigidities and, hence, nature of propagation) and masses (that affect they interaction cross sections). All contributions cumulatively give rise to the total flux.

[6] The JF12 field is to-date the default GMF model, owing its prominence to the fact that it results from a fit to actual data of relevant observables. In addition, it is made up of both a regular (or ordered), as well as two random components (striated and turbulent), making it particularly well-suited for our purposes.

[7] This choice is primarily owed to the fact that arising features in CR observables are best visually discernible this way.

[8] The GP is parameterised by a disk centered at Galactic latitude $b = 0$ with a radius of 20 kpc

[9] Following the default volume within which the JF12 field is defined, this constitutes a 20 kpc shell centered around the Galactic centre.

[10] For EGCRs, with the Galactic edge being a shell, the injection directions follow a Lambertian distribution to account for the aforementioned projection of the sources to the shell.

[11] The GP as observer is to be modified w.r.t to its parametrisation as a source. On the one hand, it has a non-zero width (100 pc) to account for the condition that the observer be a volume. On the other hand, in the study of EGCRs, its radius is reduced to 19 kpc to eliminate an artefact that occurs due to the discontinuous cut-off of the GMF to zero at the edge of the Galaxy (which is most pronounced at the GP).

[12] Earth is parametrised via a shell centered at Earth's position in the Galaxy. Several shells of varying radius, ranging from 5 pc to 1 kpc, are included in each simulation, to be sensitive to artefacts arising from their non-zero size.

[13] Technically, this definition, $R = \frac{E}{q}$, is only valid in the ultra-relativistic limit, where $\gamma = \frac{E}{mc^2} \gg 1$, and the particle's total energy is essentially equal to its kinetic energy $T$.

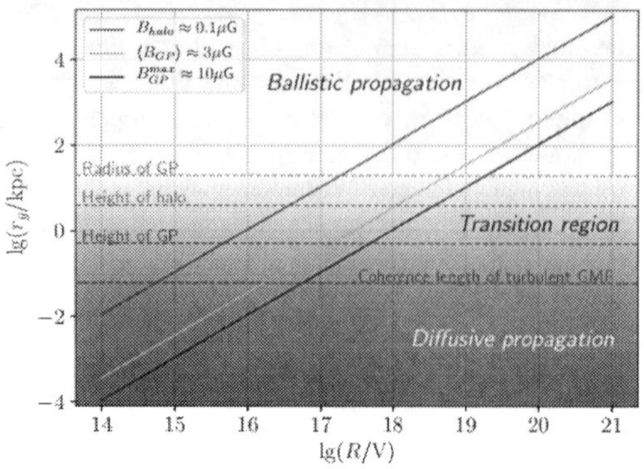


Figure 1: Gyro radius $r_g$ as a function of rigidity $R$ for various magnetic field strengths $B$ (indicated in legend), along with relevant length scales (horizontal lines).

The dependence of $r_g$ on the rigidity $R$, rather than just $E$, exemplifies one way in which the nature of propagation varies for mono-energetic CRs across nuclear species. It also shows that inferences on the deflections in a magnetic field of one CR species are transferable to any other species, by scaling the energy via the ratio of their charges to match the respective rigidity. In fact, we made use of this fact in our studies and performed our simulations to the propagation of protons[14].

The shift from highly turbulent/chaotic motion (diffusive propagation) to largely ordered deflections of a few degrees (ballistic propagation) leads to large-scale propagation effects concerning the evolution the spatial CR distribution. A decisive role here is played by the GP, as it encompasses the region with the strongest fields, while at the same time being confined to a narrow volume of the Galaxy, particularly in north/south direction ($\lesssim$ 1 kpc height versus $\approx$ 20 kpc radius [10]). GCRs that originate therein, will experience *confinement* within the GP at lowest rigidities with propagation being diffusive. EGCRs approaching the GP from outside the Galaxy will observe a large positive gradient in the field strength, and consequently experience strongly increasing deflections; once they start pointing away from the GP, they observe the same gradient in reverse, and their deflection quickly weakens. Effectively, they are reflected, or *shielded*, from the GP. If they do manage to penetrate into the GP, they are trapped therein, i.e. experiencing similar confinement as GCRs do. With other words, EGCRs experience two *counteracting* effects.

The aforementioned effects are qualitatively illustrated in Fig. 2, which show the evolution of the number of GCRs (left, a) and EGCRs (right, b) residing in the GP for various rigidity bins, which are colour-coded. Confinement is discernible via the "depletion times" (the intersection with the x-axis where all CRs have escaped), and shielding via the initial (EG)CR count (the intersection of the lines with the y-axis). Both effects decrease as rigidity increases.

The resulting effect on the CR flux depends on their origin. A decreasing confinement as CR rigidity increases, i.e. a quicker escape from the GP, results in a decreasing probability of reaching a designated region of the GP, including Earth. Hence, the flux of GCRs, which only experience confinement due to the GMF, decreases with rigidity. This can be seen in their rigidity spectrum depicted in Fig. 3. The spectrum flattens at highest rigidities, indicating that propagation becomes ballistic and the probability of reaching Earth no longer depends on rigidity, and is determined by the geometric fractions of the observer w.r.t the source. For EGCRs, the same considerations apply for the EGCRs confined in the GP. Since shielding leads to a lower probability of reaching Earth, the combination of both effects determines the resulting flux. In fact, their contributions cancel exactly, meaning that the flux is conserved across the entire rigidity range. This result is a consequence of Liouville's theorem [11, 12], which our studies effectively serve to confirm.

While the flux of isotropically injected EGCRs is unaffected by the GMF, modifications to the spectrum may occur in the case of anisotropic injection. Since the EGCR flux is not necessarily isotropic upon arrival at the Galaxy, our propagation studies included the injection of anisotropic EGCR fluxes. Two scenarios were explored, one representing the largest-scale, the other the smallest-scale anisotropy: a dipole and a point-source. With the help of the *Galactic lensing* scheme [13, 14] and using custom lenses[15], several iterations of these scenarios were studied, each varying the phase (of the dipole) or direction (of the point source), respectively. For both scenarios, the injected EGCR flux is modified by the GMF. In the case of an injected dipole, this takes the form of a subtle change in the spectral slope. For point sources, the flux modification is significantly more complex, ranging from changes in the slope, similar to the dipole scenario, to the occurrence of spectral breaks, depending on the source direction. An exemplary plot of such a break is given in Fig. 4, showing the modified rigidity spectrum for an injection from the direction of Centaurus A. To understand this complexity, consider that modifications to the shape of an injected anisotropic EGCR flux come about due to rigidity-dependent changes in the magnetic transparency of the GMF. The transparency of the GMF regarding some injection direction, which we henceforth call "directional magnetic transparency", is determined by the trajectory that the CR undergoes. Due to the turbulent fields, small deviations from that trajectory quickly diverge, such that the CR does not hit Earth in the first place, making the GMF opaque concerning said injection direction. That the directional transparency exhibits such strong rigidity-

[14] This is the most straightforward choice of nuclear species, as the determination of the CR rigidity from its energy is the simplest with $Z = 1$ (i.e. rigidity and energy are identical, save the differing dimension between the two quantities of $\frac{[R]}{[E]} = 1/e$)

[15] The custom lenses were generated from the inverted injection and arrival directions of $10^8$ anti-protons that are injected isotropically from Earth and backtracked to the edge of the Galactic shell. Time-reversibility of trajectories guarantees that these directions represent the arrival and injection directions of isotropically injected protons into the Galaxy.

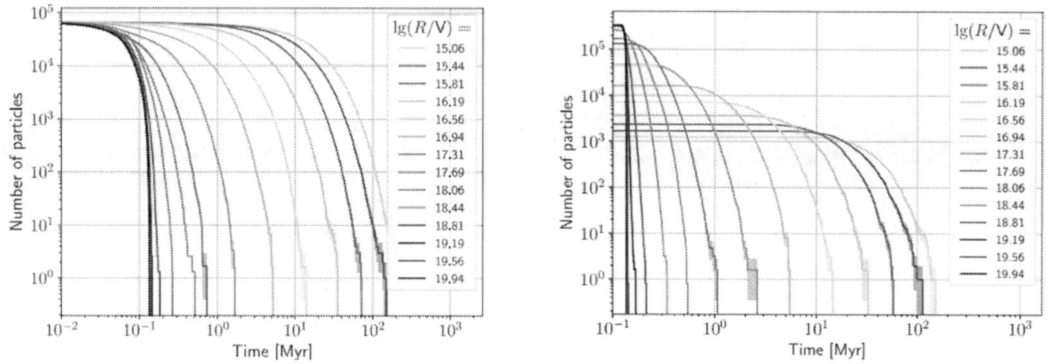


(a) GCRs

(b) EGCRs

Figure 2: Evolution of the CR count in the GP for several rigidity bins indicated in legend (from [8]).

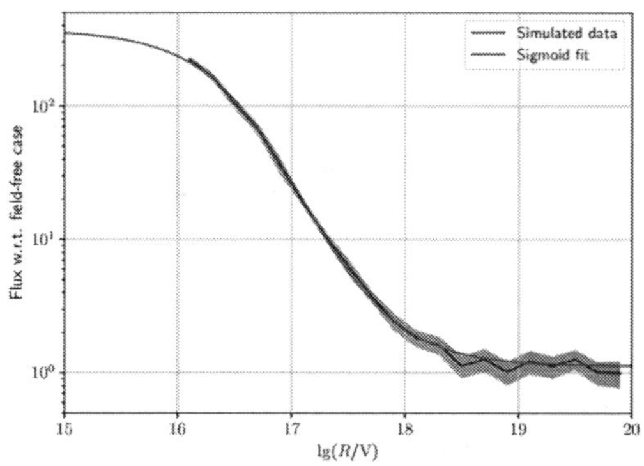


Figure 3: Modification of the rigidity spectrum of GCRs imposed by the GMF.

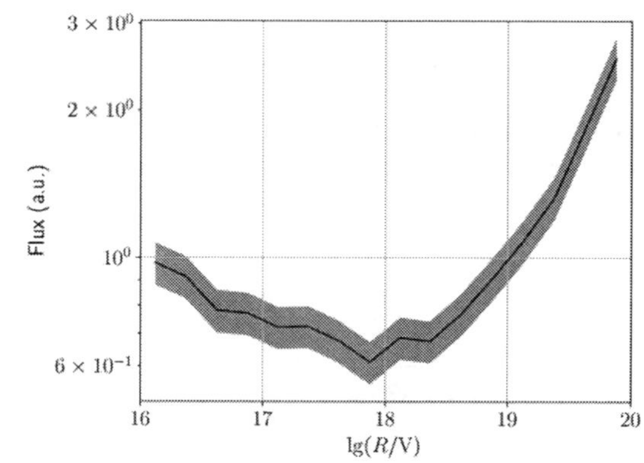


Figure 4: Modification of the rigidity spectrum of EGCRs originating from Centaurus A imposed by the GMF (from [8]).

dependence is reflective of the complex structure of the GMF.

## 2.2 Adaptation towards a concrete description of propagation effects

The key goal of this work is to ascertain in what way the aformentioned propagation effects affect the CR flux in the transition region. To this end, the source distribution for both GCRs and EGCRs must be specified, such that we obtain a concrete description of the flux modifications that the GMF imposes on both classes of CRs.

For GCRs, this implies a re-weighting of the uniform source distribution used in our studies to a more realistic distribution. Since supernova remnants (SNRs) are considered to be the prime source candidate of GCRs, performing our re-weighting according to their (galactocentric) distribution is a natural choice. Following [15], the re-weighting function becomes:

$$w(r) = \frac{1}{25.0544}\left(\frac{r}{8.5}\right)^{1.09} \cdot \exp\left(-3.87 \cdot \frac{r-8.5}{8,5}\right). \quad (1)$$

The distribution of EGCR sources follows the positions of the ten brightest galaxies of the "starburst catalogue", a catalogue of galaxies the Pierre Auger Collaboration compiled for analysis of possible source candidates of ultra-high energy CRs (or UHECRs) [16]. We distribute our sources according to the positions of the 10 brightest galaxies from said catalogue, with their relative contributions matching the relative brightness assigned in the analysis. On top of that, we added a distance- and rigidity-dependent smearing $\Delta \propto \frac{d}{R}$ to the flux to account for the diffusive propagation in the intergalactic magnetic fields. A skymap of the injected flux for a fixed (arbitrary) rigidity in given in Fig. 5.

https://doi.org/10.1051/epjconf/202328303006

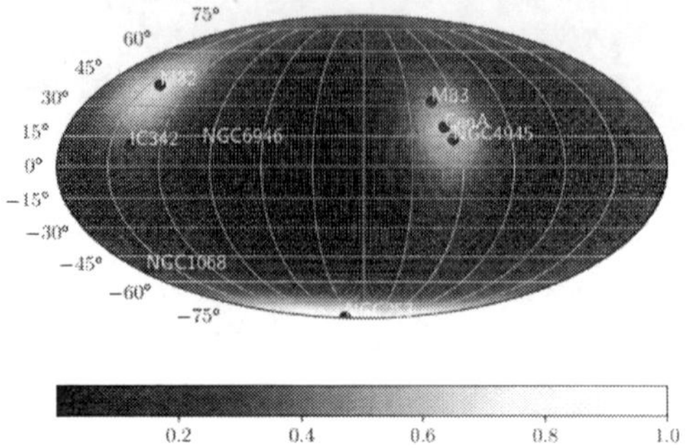


Figure 5: Skymap of the injected EGCR flux following the distribution of the 10 brightest galaxies from the starburst catalogue, with their relative contributions matched to their given brightness, as well as rigidity- and distance-dependent smearing.

The resulting rigidity spectra, indicating the concreting flux modifications imposed by the GMF, are shown in Fig. 6 for both GCRs (left, a), and EGCRs (right, b). Besides larger fluctuations, due to larger contributions of further-away sources, the radially dependent SNR distribution hardly affects the spectral shape of GCR flux. The EGCR spectrum takes on a rather complex structure. At lowest and highest rigidities, the spectrum is flat, i.e. it has the same spectral index as the injected spectrum, but undergoes both a hardening, occurring at roughly $10^{18}$ V, followed by a steepening at about $10^{18.5}$ V. Assuming the accuracy of the choice of distribution of EGCR sources and of the JF12 field to model the GMF, this indicates that the ankle may partly arise from propagation effects. While being an intriguing result, the determination of both the EGCR sources and the GMF is subject to large uncertainties, so this assumption is rather spurious. Nevertheless, we have obtained a quantitative description of the rigidity-dependent modifications that the GMF introduces to the flux of GCRs and EGCRs originating from a fixed, but at the same time realistic and generic, distribution of sources. These shall be captured in the quantity $m_i(R)$, where $i$ = GCR, EGCR. This lays the foundation for obtaining the final flux prediction.

## 3 Towards a flux prediction

The propagation studies outlined in section 2 show that the fluxes of GCRs and (anisotropically injected) EGCRs are modified by the GMF, the origin of which can be traced to a shift in propagation regimes. By specifying the source distributions, we have arrived at a full quantitative prediction of the modifications that the GMF imposes on the flux of GCRs and EGCRs as a function rigidity.

For a complete flux prediction, further parameters that impact the CR flux have to be considered. The ones we included all specify source properties: an intrinsic rigidity-dependent cut-off in the source output (e.g. via a maximum acceleration rigidity of the sources) which we denote by $c_i(R)$; the fraction of each injected CR species, signified by $f_i^j$; the injected spectral slope of each nucleus $\gamma_i^j$. As for the flux modification $m_i$, $i$ = GCR, EGCR, and $j$ denotes the individual nuclei contributing to the flux (specified below). These quantities are tuned in order to reproduce the spectral data outside of the transition region, with the GCR flux to match the energy range around the knee, and the EGCR flux to match the energy range beyond the ankle. The two individual energy spectra $J_i(E)$ are then combined to obtain the total, combined all-particle energy spectrum $J(E)$. This spectrum extrapolated to the transition region is our flux prediction that we set out for, and that can be compared with spectral data.

The motivation to introduce a rigidity-dependent cut-off $c_i(R)$ to the spectrum is similar for both classes of CRs, namely to account for a specific spectral feature. A large consensus exists that GCR flux is tightly intertwined with the knee of the spectrum, though its origin is still disputed. One popular description of the knee is that it marks the maximum rigidity that Galactic sources, e.g. the shock front of (SNRs) can accelerate CRs to. Since the propagation-related flux suppression of GCRs occurs for rigidities exceeding the energy of the knee for protons, including an additional cut-off of this kind is necessary to be able to describe the knee feature in the first place. A cut-off to the EGCR spectrum is invoked to recover the high-energy cut-off in the energy spectrum at highest energies. Its origin is also not settled, though evidence that the composition is becoming heavier at highest energies points to a underlying rigidity-dependent process, like a maximum acceleration rigidity of cosmic accelerators [17]. Since the knee exhibits a gradual change of the spectral index, as opposed to a hard break for the high-energy cut-off, we introduced a smooth cut-off ($c_{\rm GCR}(R) = \exp(1 - R/R_{\rm cut}^{\rm GCR})$ to the GCR spectrum, and a broken cut-off ($c_{\rm EGCR}(R) = \exp(R/R_{\rm cut}^{\rm EGCR})$ for $R > R_{\rm cut}^{\rm EGCR}$, $c_{\rm EGCR}(R) = 1$ otherwise) to the EGCR spectrum.

While the position of the break can be determined via the aforementioned cut-offs, the shape that the spectrum takes immediately following said break is primarily determined by the injected composition[16]. Thus, the spectral trend directly following the knee and the high-energy cut-off, respectively, serves to determine the injected compositions of the GCR and EGCR sources. We modeled the composition via a four-component approach where the flux is made up of the nuclei Hydrogen ($j = 1, Z = 1, A = 1$), Helium ($j = 2, Z = 2, A = 4$), Oxygen ($j = 3, Z = 8, A = 16$) and Iron ($j = 4, Z = 26, A = 56$), each representing individual mass groups (protons, light, intermediate, heavy) [18]. This categorisation has the benefit of keeping the composition simple, yet approximately capturing the main features CR transport that affect the composition[17]. For both GCRs and EGCRs, all mass groups

[16] Since the cut-off is rigidity-dependent, its onset is shifted further to the right in energy, the heavier the injected composition is.

[17] The most noteworthy of these is that interactions and deflections are sensitive to different properties of the nucleus and hence modify the flux

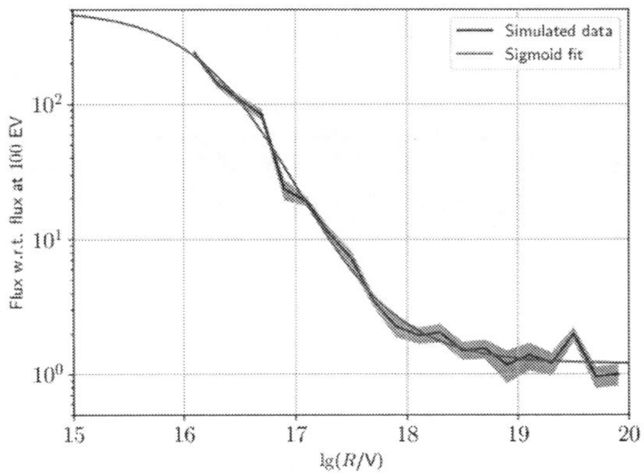


(a) GCRs

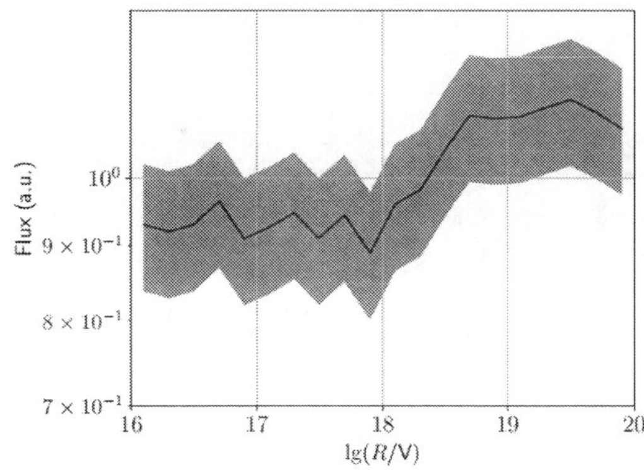


(b) EGCRs

Figure 6: Rigidity spectra indicating the flux modifications imposed by the GMF.

are injected with the same spectral index and are set equal to the slope of the measured spectrum prior to the knee, i.e. $\gamma \equiv \gamma_i^j = -2.7\ \forall i, j$ [4]. The parameters we alter to reproduce the aforementioned regions of the spectrum are the fractions of each mass group, by which their spectra are scaled.

All in all, the total, combined energy spectrum that we wish to match, and later compare, with spectral data has the following functional form:

$$J(E) = \sum_i J_i(E) = \sum_i \sum_j J_{i,j}(E), \tag{2}$$

$$J_{i,j}(E) = J_0 \cdot f_j^i\, E^\gamma \cdot m_i\left(\frac{E}{Z_j}\right) \cdot c_i\left(E, R_{\text{cut}}^i \cdot Z_j\right). \tag{3}$$

$J_0$ is a normalisation factor, and $f_j^i, R_{\text{cut}}^i$ are free variables, which we tune to match the spectrum up to and around the knee for $J_{\text{GCR}}$ and the region beyond the ankle for $J_{\text{EGCR}}$.

## 4 Results and discussion

The individual energy spectra $J_{\text{GCR}}$ and $J_{\text{EGCR}}$ are shown in Fig. 7. The spectra of the individual components $J_{i,j}$ are also depicted. The rather smooth change in slope around the knee is partly reproduced via roughly equal fractions of light, intermediate and heavy nuclei, which cumulatively roughly match the proton fraction. Matching the shape of the spectrum above the ankle and particularly the slope of the high-energy cut-off requires a dominant light component, with a small but significant proton fraction. Intermediate to heavy mass groups influence the spectral shape only at the highest energies of the cut-off, where the flux is the lowest, and the spectral data carries the highest degree of uncertainty. Hence, their contribution is negligible when describing the CR flux in the transition region. Such fractions are reasonably in line with other composition studies, given the qualitative nature of their determination.

The combined spectrum, overlain on the measured energy spectrum, is depicted in Fig. 8. As already suggested in the individual spectral plots, the fluxes of both GCRs and EGCRs, do not follow the trend of the measured flux into the transition region. Each individual spectrum undershoots the measured spectrum from their respective side, such that their combined contribution is consistently too small. The GCR flux drops off too rapidly, whereas the break in the EGCR spectrum is not sufficiently pronounced. The difference between the measured and predicted flux, the "missing" flux, is indicated in black. This suggests the necessity of some additional component to account for this difference, a notion that has already been discussed in other contexts [1, 19].

When considering the robustness of this finding, prospects of improving our prediction, particularly $m_i(R)$, are worth discussing. The steep suppression of the GCR flux appears to be a necessary consequence of the propagation effects in the GMF outlined in section 2. Their gradual escape from the GP and the resulting suppression of their flux is inexorably linked to the rigidity range in question and the strength of the (turbulent) field through which they propagate. The propagation effects involving EGCRs are subject to more uncertainty. Since the flux that is injected into the Galaxy strongly affects the nature of the flux modification after propagation in the GMF, the parameters determining said flux are worth further scrutinising. They

of a mixed composition of CRs differently. While deflections depend on rigidity and scale with the charge number Z, interactions depend on the energy per nucleon, thereby scaling with the mass number A. Since $A \approx Z/2$ for most nuclei except Hydrogen, interaction-related modifications to the spectrum that occur at the same energy as some rigidity-dependent modification for protons will be shifted to the right in energy by a factor of two compared to the latter. In the standard log-log-representation of the spectrum, this shift is most striking for Helium, where we have a fourfold increase of A compared to protons, whereas a two-fold increase is less pronounced for the intermediate to heavy mass range whose shift in energy with Z is already sizeable.

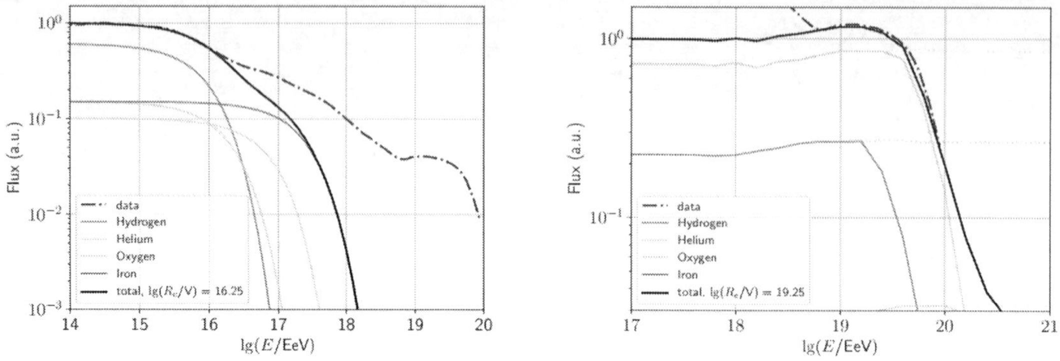


(a) GCRs (b) EGCRs

Figure 7: Predicted all-particle energy spectra for each individual class of CRs.

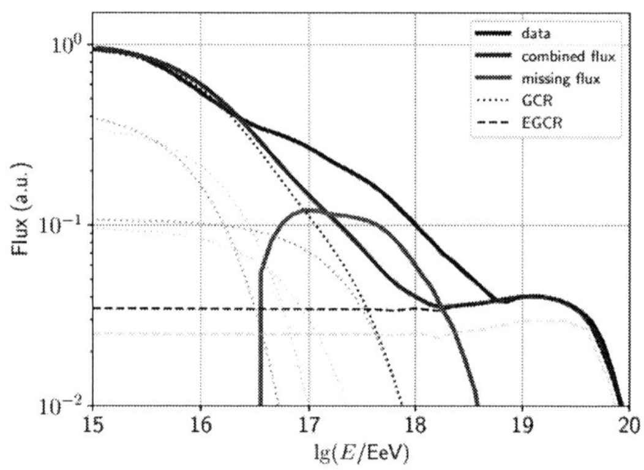


Figure 8: Total predicted all-particle energy spectrum.

include the source distribution, and the smearing of the fluxes originating from each. Concerning the sensitivity of $m_i(R)$ to the latter, we varied the smearing to twice and half the value chosen for this prediction, and found that this has no large effect on the overall trend of the resulting flux modification, and mainly affects statistics[18]. Besides the nature of the incoming flux, the turbulent component of the GMF plays an instrumental role in the modification of the EGCR flux. It does so by governing the amount of smearing of the incoming EGCR flux and, effectively, the Galactic transparency. Both the angular scale of regions of higher or lower transparency, as well as their amplitudes, are affected.

Since the flux modification of both GCRs and EGCRs depends chiefly on the turbulent field strength, exploring the effect of higher or lower turbulence levels naturally presents itself. Such a discussion is further motivated due to the general uncertainty of our current GMF. With regard to the JF12 field specifically, there is, in fact, a long-standing suspicion that the strength of its turbulent component is overestimated. Other publications have found hints of this, pertaining to the deflection of EGCRs [20], average grammage of GCRs [3], or different treatments of relevant observables [21, 22]. Further, two observations from our studies support this notion. For one, the Galactic residence times of GCRs (in our case estimated where the count has decreased by one half) is slightly high at lowest rigidities with $t_{\rm res} \approx 100\,\mathrm{Myr}$ [8, 23]. The residence time scales with the degree of confinement in the GP which, in turn, are determined by the strength of the turbulent component. The second observation is found in our study of anisotropically injected EGCRs. They also contained dipole anisotropy studies, where the dipole amplitude was determined for several rigidity bins. At highest rigidities, where a significant dipole has been measured by the Pierre Auger Observatory [24, 25], even the amplitude of the most significant dipole among the point sources can barely account for the experimental measured value of 8 %. This tension only seems more pertinent in light of the injected fluxes in our studies being *purely* anisotropic (i.e. no isotropic fraction). The observed amplitudes, therefore, are better understood as upper limits and are, thus, expected to be even smaller for non-zero isotropic fractions to the injected EGCR flux. This tension can be remedied, if the random field strength is in fact overestimated, since this would imply that the smearing of directions by the GMF that lowers the dipole amplitude is too strong.

Smaller turbulence levels weaken the confinement of GCRs and decrease the smearing of anisotropically injected EGCRs. The flux suppression of GCRs would commence at lower rigidities, potentially eliminating the need for a rigidity-dependent cut-off from a phenomenological point of view. The suppression of the GCR flux is thence less steep than our prediction suggests, and part of the "missing" flux we find may be accounted for by it. Anisotropies in the EGCR flux are better preserved,

[18] Additional considerations such as these to appear in a publication that is still in preparation at the time of writing these proceedings.

which may also increase the angular resolution and amplitude of the directional transparency of the GMF. While the sensitivity to the injected anisotropy would likely increase, spectral breaks arising in the EGCR flux may become more pronounced, improving the description of the transition region also at highest energies.

Whether a reduction of the turbulent field strength would eliminate the need of an additional component altogether is not clear. For this to be the case, however, the rigidity-dependent escape of GCRs would have to be sufficiently slow to adequately flatten the resulting flux suppression, and the Galactic transparency of EGCRs would have to exhibit a rigidity-dependence (from the injection directions of interest), such that the modified EGCR flux possesses a spectral break at roughly the same position as does our prediction, while, at the same time, being sufficiently steep left of the break. Further, Galactic sources would be required to accelerate CRs to energies comfortably exceeding the knee, such that the lack of a cut-off is justified. In this context, we also wish to refer to [3], where a weaker turbulence level of the JF12 was analysed and found to fully describe the transition region.

For completeness, we briefly outline the effects of larger turbulence levels. On top of being poorly motivated, this would result in the GCR flux overshooting around the knee due to their increased confinement in the GP, which would have to be compensated by a smaller cut-off rigidity $R_{\mathrm{cut}}$. While this would loosen the requirements on the energy output of Galactic sources, the suppression of the Galactic flux beyond the knee would only become steeper. Anisotropic EGCR fluxes would be smeared out more strongly and flux modifications would likely become even less pronounced.

## 5 Conclusion

In this work, we detail the procedure for obtaining a flux prediction of GCRs and EGCRs based on propagation studies of both classes of CRs in the GMF [7, 8]. Said prediction is constructed with the objective of relying on a minimal set of additional assumptions pertaining to the distribution and properties of the respective sources. To this end, the GCR sources are distributed according to the galactocentric distribution of SNRs [15], the EGCRs sources follow the distribution of ten brightest galaxies in the "Starburst catalogue" from [16]. For both classes of CRs, a four-component composition is injected from the sources with equal spectral indices for all mass groups that set to the value of the measured energy spectrum below the knee. In addition, a rigidity-dependent cutoff is introduced for both classes with differing shapes in order to match the smoothness of the breaks in the measured spectrum that the are aimed to produce (knee, and high-energy cutoff, respectively). The cut-off rigidity $R^{i}_{\mathrm{cut}}, i =$ GCR, EGCR, and the nuclear fractions $f^{i}_{j}$, $j =$ H, He, O, Fe, are free parameters that are varied to match the position and subsequent spectral shape following the aforementioned breaks.

We find that the obtained flux underpredicts the measured flux across the entire transition region, suggesting the need for an additional component to fill the unaccounted for flux. We argue that this discrepancy between prediction and data follows rather strictly from the propagation effects themselves. Motivated by the presumed overestimation of the strength of the turbulent component of the JF12, the effects of lower turbulence levels are briefly explored. We conclude that this could alleviate some of the aforementioned tension that we find. The prospect of this eliminating the need to invoke additional components entirely is addressed by alluding to the rather strict set of requirements this places on the nature of the propagation effects in the GMF (and on the energy output of Galactic sources).

As a final point, it is worth noting that our prediction makes an important contribution, despite being based on a presumably systematically biased GMF model (i.e. one overestimating the strength of the random component of the GMF). This holds true because the JF12 model constitutes a GMF model that results from fits to observational data. Modifying the random component ad hoc, as was done in [3], has the drawback of not fulfilling said criterion anymore. Given the ongoing work on creating updated and improved GMF models (see e.g. [22]), an outlook on further research presents around this results itself, namely re-investigating the propagation we identified using on an alternative GMF model with an improved determination of the random field component, ideally one that is specifically based on the JF12 model.

## Acknowledgements

The simulations were carried out on the Pleiades cluster at the University of Wuppertal, which was supported by the Deutsche Forschungsgemeinschaft (DFG). Further financial support by the German Ministry of Research and Education (Verbundforschung Astroteilchenphysik) is gratefully acknowledged.

## References

[1] S. Thoudam, J.P. Rachen, A. van Vliet, A. Achterberg, S. Buitink, H. Falcke, J.R. Hörandel, Astron. Astrophys. **595**, A33 (2016), `1605.03111`

[2] N. Globus, D. Allard, E. Parizot, Phys. Rev. D **92**, 021302 (2015), `1505.01377`

[3] G. Giacinti, M. Kachelrieß, D.V. Semikoz, Phys. Rev. D **91**, 083009 (2015), `1502.01608`

[4] J. Becker Tjus, L. Merten, Phys. Rept. **872**, 1 (2020), `2002.00964`

[5] A. Aab et al. (Pierre Auger), Phys. Rev. D **102**, 062005 (2020), `2008.06486`

[6] T.R. Jaffe, Galaxies **7**, 52 (2019), `1904.12689`

[7] A. Kääpä, K.H. Kampert, E. Mayotte, PoS **ICRC2021**, 004 (2022)

[8] A. Kääpä, *Propagation in the Galactic magnetic field – Effects on the spectrum, composition and*

*anisotropy of Galactic and extragalactic cosmic rays* (University of Wuppertal, 2022)

[9] R. Alves Batista, A. Dundovic, M. Erdmann, K.H. Kampert, D. Kuempel, G. Müller, G. Sigl, A. van Vliet, D. Walz, T. Winchen, JCAP **05**, 038 (2016), 1603.07142

[10] R. Jansson, G.R. Farrar, ApJ **757**, 14 (2012), 1204.3662

[11] J. Liouville, Journal de Mathématiques Pures et Appliquées pp. 342–349 (1838)

[12] T.K. Gaisser, R. Engel, E. Resconi, *Cosmic Rays and Particle Physics: 2nd Edition* (Cambridge University Press, 2016), ISBN 978-0-521-01646-9

[13] H.P. Bretz, M. Erdmann, P. Schiffer, D. Walz, T. Winchen, Astropart. Phys. **54**, 110 (2014), 1302.3761

[14] M. Erdmann, G. Müller, M. Urban, M. Wirtz, Astropart. Phys. **85**, 54 (2016), 1607.01645

[15] D.A. Green, Mon. Not. Roy. Astron. Soc. **454**, 1517 (2015), 1508.02931

[16] A. Aab et al. (Pierre Auger), Astrophys. J. Lett. **853**, L29 (2018), 1801.06160

[17] A. Aab et al. (Pierre Auger), Phys. Rev. Lett. **125**, 121106 (2020), 2008.06488

[18] A. Aab et al. (Pierre Auger), JCAP **04**, 038 (2017), [Erratum: JCAP 03, E02 (2018)], 1612.07155

[19] L. Merten, C. Bustard, E.G. Zweibel, J. Becker Tjus, Astrophys. J. **859**, 63 (2018), 1803.08376

[20] G.R. Farrar, M.S. Sutherland, JCAP **05**, 004 (2019), 1711.02730

[21] R. Adam et al. (Planck), Astron. Astrophys. **596**, A103 (2016), 1601.00546

[22] M. Unger, G. Farrar, EPJ Web Conf. **210**, 04005 (2019), 1901.04720

[23] N.E. Yanasak, M.E. Wiedenbeck, R.A. Mewaldt, A.J. Davis, A.C. Cummings, J.S. George, R.A. Leske, E.C. Stone, E.R. Christian, T.T. von Rosenvinge et al., ApJ **563**, 768 (2001)

[24] A. Aab et al. (Pierre Auger), Science **357**, 1266 (2017), 1709.07321

[25] A. Aab et al. (Pierre Auger), Astrophys. J. **891**, 142 (2020), 2002.06172

EPJ Web of Conferences **283**, 03007 (2023)
*UHECR 2022*
https://doi.org/10.1051/epjconf/202328303007

# Search for the large-scale cosmic-ray anisotropies using the TA and TALE surface detector arrays

*Yusuke* Kimura[1], *Toshihiro* Fujii[1,2,*], *Shoichi* Ogio[3], and *Yoshiki* Tsunesada[1,2] for the Telescope Array Collaboration

[1]Graduate School of Science, Osaka Metropolitan University, Sumiyoshi, Osaka, Japan
[2]Nambu Yoichiro Institute of Theoretical and Experimental Physics, Osaka Metropolitan University, Sumiyoshi, Osaka, Japan
[3]Institute for Cosmic Ray Research, University of Tokyo, Kashiwa, Chiba, Japan

**Abstract.** Origins of ultra-high-energy cosmic rays (UHECRs) are still largely unknown. In particular, the "2nd knee" around $10^{17}$ eV could be related to a transition of origins from the galactic to extra-galactic sources. In this scenario, one would expect the large-scale anisotropies of UHECRs to change over the 2nd knee region. We search for large-scale anisotropies using data taken from surface detector (SD) arrays of the Telescope Array experiment (TA), the largest cosmic-ray detector in the northern hemisphere, and from its low energy extension, the Telescope Array Low-energy Extension (TALE). In this contribution, we report on preliminary results in the search for the large-scale cosmic-ray anisotropies with 11-years TA SD and two-years TALE SD data.

## 1 Introduction

Deciphering origins of ultra-high-energy cosmic rays (UHECRs) is one of the most important question in a field of astroparticle physics. Since the cosmic rays are charged particles, they are deflected by the galactic and extragalactic magnetic fields. With small deflections of UHECRs by galactic and intergalactic magnetic fields, $\sim 5°Z(E/50\,\mathrm{EeV})^{-1}$ where $Z$ is the charge of nuclei [1], a small-intermediate scale anisotropy is predicted at the highest energies. The deflections by magnetic fields are enlarged at lower energies below 10 EeV, resulting in a large-scale anisotropy, which can be approximated as a dipole [2–4].

In 2017 the Pierre Auger Collaboration reported the observation of a significant large-scale anisotropy in the arrival directions of cosmic rays above 8 EeV, indicating a significant dipole structure of 4.7% amplitude in a projection of the right ascension with a $5.2\sigma$ significance [5]. An enhancement of the dipole amplitude above 4 EeV and results down to 0.03 EeV are also reported [6, 7]. These results are agreement with an extragalactic origin of UHECRs. The Telescope Array Collaboration reported a consistent result with both the Auger reported dipole and isotropic distribution due to the current statistics [8].

In this proceedings we search for an energy dependence of a dipole structure at lower energy below 8 EeV using the Telescope Array experiment, and from its low energy extension.

*e-mail: toshi@omu.ac.jp

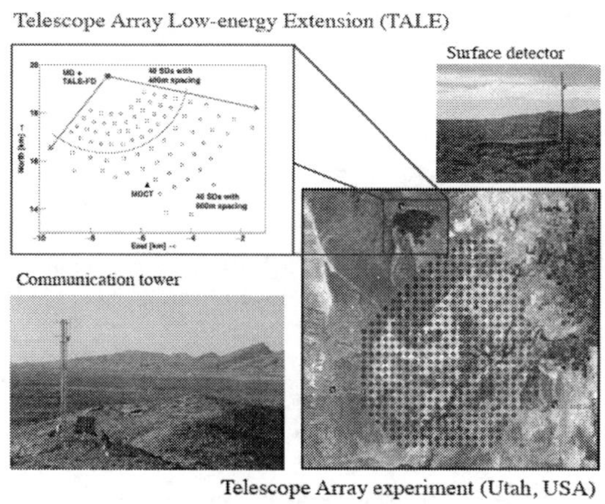


**Figure 1.** The detector configuration of TA SD and TALE SD. The square grid is used in the TA SD deployment, while the hybrid detection efficiency is maximized for the arrangement of TALE SD. The picture of communication tower and individual surface detector are displayed.

## 2 Telescope Array experiment (TA) and Telescope Array Low-energy Extension (TALE)

The Telescope Array experiment (TA) is the largest cosmic-ray detector in the Northern hemisphere, located near the city of Delta, Utah, USA (39.30° North and 112.91° West at ~1400 m above sea level [9]. The surface detector array (SD) consists of 507 plastic scintillators of 3 m$^2$ area deployed in a square grid with a 1.2 km spac-

https://doi.org/10.1051/epjconf/202328303007

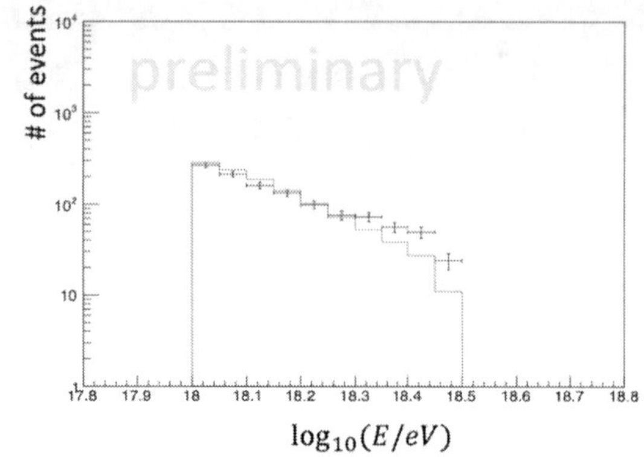


**Figure 2.** Energy distribution of TALE SD during two-years data (black plots) and the MC prediction using proton of QGSJetII-04 model under the assumption of isotropic distribution (red histogram)

ing. The total effective area is approximately 700 km$^2$. Additional surface detectors, designed to provide a four-fold increase in effective area, named TAx4, are now being deployed and partly in operation [10]. Total 80 additional SDs with 600 m and 400 m spacings are installed to be sensitive to lower energies below 3 EeV than TA SD, named Telescope Array Low-energy Extension (TALE). The total effective area of TALE is 21 km$^2$. Figure 1 shows the detector configuration of TA SD and TALE SD. The TA SD and TALE SD are overlooked by fluorescence detectors, which are used for determination of the calorimetric energy of UHECR from energy deposited in the atmosphere during its development [11]. The obtained energy is calibrated to the calorimetric energy measured by the fluorescence detectors by a factor of 1/1.27 for TA SD [12] and of 1/1.08 for TALE SD.

## 3 Large-scale anisotropy search with TALE around 2 EeV

Since the TALE SD data taking was started from 2019, two years data from October 2nd, 2019 to September 28th 2021 are used for the large-scale anisotropy search around 2 EeV. The obtained result is compared to time dependent detailed detector simulation under assumptions of the isotropic distribution and primary proton of QGSJetII-04, which is the post-LHC interaction model. We found no significant non-uniformities in the sidereal time and anti-sidereal time distributions.

Figure 2 shows the reconstructed energy distribution of TALE SD, compared to the Monte Carlo (MC) expectation with the normalization to the total number of the observed events with the TALE SD. The total 1122 events are recorded in the energy range from 1 EeV to 3 EeV with TALE SD data during two-years data. Figure 3 shows residual intensities as a function of the right ascension with a fitted curve by a formula of $r_\alpha \cos(\alpha + \phi_\alpha)$ where $\alpha$ is the right ascension, $r_\alpha$ and $\phi_\alpha$ are amplitude and

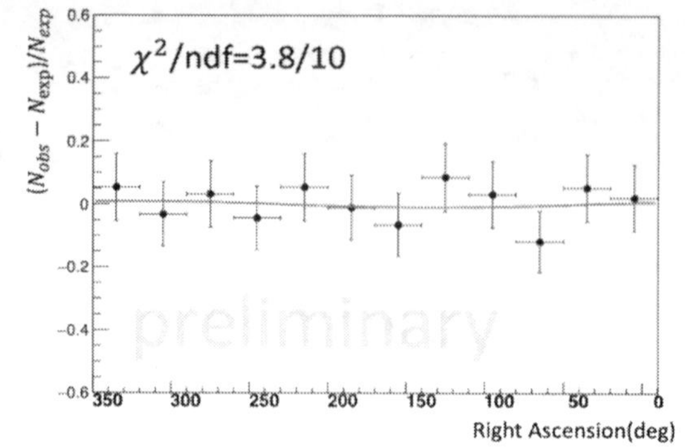


**Figure 3.** Residual intensities vs right ascension from two-years of TALE SD data (black points). The red curve shows a fitted result by $r_\alpha \cos(\alpha + \phi_\alpha)$

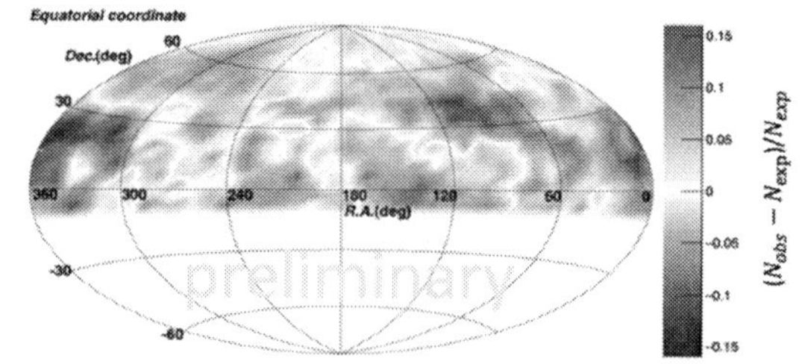


**Figure 4.** Skymap of residual intensities with TALE SD in equatorial coordinates using a top-hat function of 45° over-samplings

phase of the dipole structure, respectively. The fitted amplitude is 0.009±0.040 with a phase of 327°±267°. The obtained amplitude is consistent with the isotropic distribution. Also we found no significant non-uniformities in the sidereal time and anti-sidereal time distributions.

Figure 4 shows residual intensities measured with the TALE data in equatorial coordinates using a top-hat funtion of 45° over-samplings. There is no significant dipole structure seen under the limited statistics.

## 4 Large-scale anisotropy search with TA around 6 EeV

As the TA SD data taking is continued over 11 years, the 11-years data from May 2008 to May 2019 are analyzed for the large-scale anisotropy around 6 EeV. There were 8810 events surviving in a range from 4 EeV to 8 EeV as shown in Figure 5, compared to the time-dependent detailed MC simulation using primary proton of QGSJetII-03 under the assumption of isotropic distribution. We confirmed no significant non-uniformities in the sidereal time and anti-sidereal time distributions.

Figure 6 shows residual intensities of 11-years TA SD data below 8 EeV in equatorial coordinates with a fitted curve of the dipole component. The fitted amplitude is 0.044±0.016 with a phase of 225°±20°. The $\chi^2$/ndf

 https://doi.org/10.1051/epjconf/202328303007

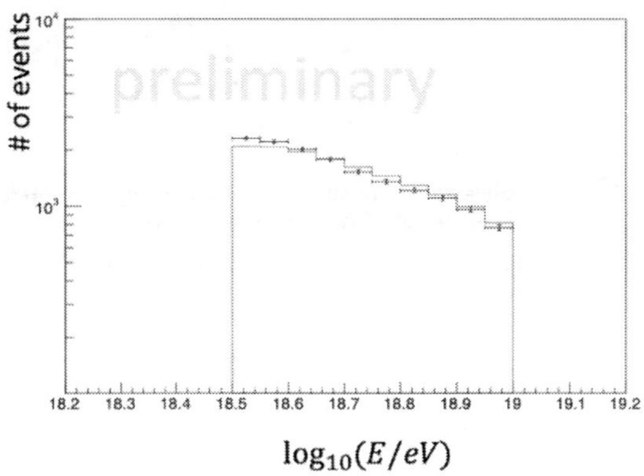


**Figure 5.** Energy distribution measured by the TA SD during 11 years (black points), compared to the time dependent detailed MC simulations (red histogram)

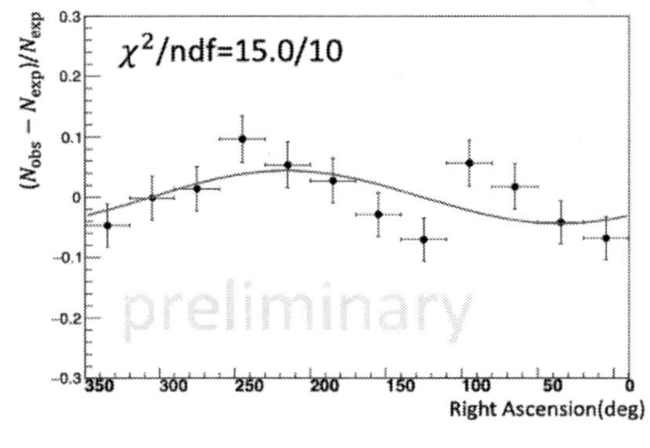


**Figure 6.** Residual intensities vs right ascension analyzed from 11-years of TA SD data of residual intensities. The red curve shows a fitted result by $r_a \cos(\alpha + \phi_a)$

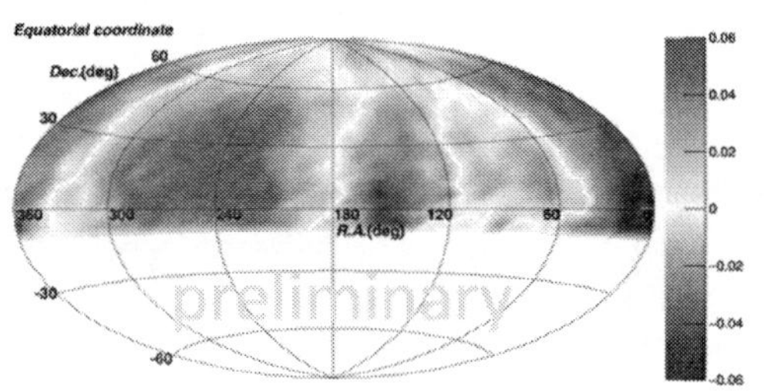


**Figure 7.** Preliminary residual intensities of 11-years TA SD data below 8 EeV in equatorial coordinates

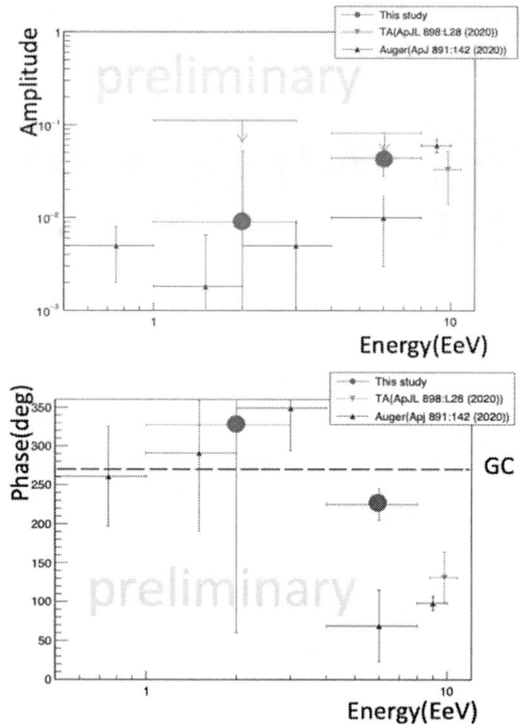


**Figure 8.** The measured amplitude with their upper limits of 99% confidence level and phase using both TALE SD and TA SD. The obtained results are compared with the result reported from Pierre Auger Collaboration [6, 7]

against the isotropic distribution is 21.5/12. Figure 7 shows a skymap of the residual intensities in equatorial coordinates. Although a small enhancement toward the galactic center of ~266° is seen in Figure 6 and Figure 7, the measured amplitude is not significant due to the current statistics.

Because of no significant results reported from the dipole search using both TALE SD and TA SD, the upper limits of 99% confidence level are evaluated. The measured amplitudes with their upper limits and phases are indicated in Figure 8, compared to the already reported result for the Pierre Auger Collaboration [6, 7]. The upper limits of TALE and TA are consistent with the Auger result. Further statistic of UHECR is required for the detailed check with these results.

## 5 Summary

The large-scale anisotropies below 8 EeV were searched using data-sets of TALE SD during two years data and TA SD during 11 years. Since there are no significant results seen in the current limited statistics, the upper bounds on the amplitude of dipole structure are derived from this results of TALE SD and TA SD. Further data-taking of these

SD arrays are essential to understand the origin and nature of UHECRs.

## Acknowledgements

The Telescope Array experiment is supported by the Japan Society for the Promotion of Science(JSPS) through Grants-in-Aid for Priority Area 431, for Specially Promoted Research JP21000002, for Scientific Research (S) JP19104006, for Specially Promoted Research JP15H05693, for Scientific Research (S) JP19H05607, for Scientific Research (S) JP15H05741, for Science Research (A) JP18H03705, for Young Scientists (A) JPH26707011, and for Fostering Joint International Research (B) JP19KK0074, by the joint research program of the Institute for Cosmic Ray Research (ICRR), The University of Tokyo; by the Pioneering Program of RIKEN for the Evolution of Matter in the Universe (r-EMU); by the U.S. National Science Foundation awards PHY-1607727, PHY-1712517, PHY-1806797, PHY-2012934, and PHY-2112904; by the National Research Foundation of Korea (2017K1A4A3015188, 2020R1A2C1008230, & 2020R1A2C2102800) ; by the Ministry of Science and Higher Education of the Russian Federation under the contract 075-15-2020-778, IISN project No. 4.4501.18, and Belgian Science Policy under IUAP VII/37 (ULB). This work was partially supported by the grants ofThe joint research program of the Institute for Space-Earth Environmental Research, Nagoya University and Inter-University Research Program of the Institute for Cosmic Ray Research of University of Tokyo. The foundations of Dr. Ezekiel R. and Edna Wattis Dumke, Willard L. Eccles, and George S. and Dolores Doré Eccles all helped with generous donations. The State of Utah supported the project through its Economic Development Board, and the University of Utah through the Office of the Vice President for Research. The experimental site became available through the cooperation of the Utah School and Institutional Trust Lands Administration (SITLA), U.S. Bureau of Land Management (BLM), and the U.S. Air Force. We appreciate the assistance of the State of Utah and Fillmore offices of the BLM in crafting the Plan of Development for the site. Patrick A. Shea assisted the collaboration with valuable advice and supported the collaboration's efforts. The people and the officials of Millard County, Utah have been a source of steadfast and warm support for our work which we greatly appreciate. We are indebted to the Millard County Road Department for their efforts to maintain and clear the roads which get us to our sites. We gratefully acknowledge the contribution from the technical staffs of our home institutions. An allocation of computer time from the Center for High Performance Computing at the University of Utah is gratefully acknowledged.

## References

[1] J. Bray, A. Scaife, Astrophys. J. **861**, 3 (2018), 1805.07995

[2] D. Harari, S. Mollerach, E. Roulet, Phys. Rev. D **92**, 063014 (2015), 1507.06585

[3] A. di Matteo, P. Tinyakov, Mon. Not. Roy. Astron. Soc. **476**, 715 (2018), 1706.02534

[4] B. Eichmann, T. Winchen, JCAP **04**, 047 (2020), 2001.01530

[5] A. Aab et al. (Pierre Auger), Science **357**, 1266 (2017), 1709.07321

[6] A. Aab et al. (Pierre Auger), Astrophys. J. **868**, 4 (2018), 1808.03579

[7] A. Aab et al. (Pierre Auger), Astrophys. J. **891**, 142 (2020), 2002.06172

[8] R.U. Abbasi et al. (Telescope Array), Astrophys. J. Lett. **898**, L28 (2020), 2007.00023

[9] T. Abu-Zayyad et al. (Telescope Array), Nucl. Instrum. Meth. **A689**, 87 (2013), 1201.4964

[10] E. Kido et al. (Telsecope Array), EPJ Web Conf. **210**, 06001 (2019)

[11] H. Tokuno et al., Nucl. Instrum. Meth. **A676**, 54 (2012), 1201.0002

[12] T. Abu-Zayyad et al. (Telescope Array), Astrophys. J. **768**, L1 (2013), 1205.5067

# Sensitivity of the combined fit of energy spectrum, shower depth distributions, and arrival directions at the Pierre Auger Observatory

*Teresa* Bister[1,2*] for the Pierre Auger Collaboration[3**]

[1]Institute for Mathematics, Astrophysics and Particle Physics (IMAPP), Radboud Universiteit, Nijmegen, Netherlands
[2]Physics Institute IIIA, RWTH Aachen University, Otto-Blumenthal-Str., 52074 Aachen, Germany
[3]Observatorio Pierre Auger, Av. San Martín Norte 304, 5613 Malargüe, Argentina

**Abstract.** The Pierre Auger Observatory measures several characteristics of ultra-high-energy cosmic rays (UHECRs), most importantly the energy spectrum, the distribution of maximum shower depths and the arrival directions. We use all three observables for a combined fit, in which parameters of possible UHECR sources can be constrained. The astrophysical model used in the fit consists of homogeneously distributed background sources as well as an adaptable contribution from a nearby source population. For this, the catalogs of starburst galaxies and active galactic nuclei are used which show an indication for a correlation with the UHECR arrival directions. The signal fraction, as well as the size of a rigidity-dependent magnetic field blurring, are part of the fit parameters, along with the parameters describing the source emission. In this work, we present an astrophysical simulation containing an energy-dependent contribution from starburst galaxies, which simultaneously describes the energy spectrum, shower depth distributions, and arrival directions measured by the Pierre Auger Observatory. On this simulation, the discrimination power of the method regarding the differentiation of source catalogs is demonstrated, and the expected statistical significance of the result is investigated.

## 1 Introduction

The astrophysical model established in this work consists of homogeneous background sources, in analogy to [1], and an additional part coming from catalogs of candidate sources, specifically starburst galaxies (SBGs) and jetted active galactic nuclei ($\gamma$-AGNs). These catalogs have been used in previous works by the Pierre Auger Collaboration [2–4], where a correlation between the arrival directions of ultra-high-energy cosmic rays (UHECRs) and the catalog sources was investigated. The significance of the correlation is currently $4.0\sigma$ for the SBGs and $3.1\sigma$ for the $\gamma$-AGNs [4].

Contrary to [2–4], in this work we will not only use the arrival directions, but also the energy spectrum and the shower depth distributions as observables. Additionally, the whole energy dependency of the observables will be modeled and compared to the data in each energy bin. Together, this not only allows us to reconstruct parameters of the source emission by including the full propagation in the model, but it also increases the sensitivity of the correlation analysis. The expected sensitivity will be investigated in this work using data-similar simulations.

## 2 Astrophysical model and combined fit

The analysis and the astrophysical model have been described before in [5], so here we will only present a short overview. In a future publication, all details regarding the calculation of the model will be discussed in detail. The source injection for both homogeneous background sources and catalog sources is described by a Peters cycle as in [1]:

$$J_{\rm inj}(E_{\rm inj}, A_{\rm inj}) = J_0 \cdot a_A \cdot \Big(\frac{E_{\rm inj}}{10^{18}\,{\rm eV}}\Big)^{-\gamma} f_{\rm cut}\Big(\frac{E_{\rm inj}}{Z_A R_{\rm cut}}\Big) \quad (1)$$

with a broken exponential cut-off function $f_{\rm cut}$. Here, we use 5 representative elements (H, He, Ni, Si, Fe) with fractions $a_A$. The integral fractions $I_A$ (defined as emissivity fractions above a common rigidity threshold) which can be calculated from $a_A$, the spectral index $\gamma$ and the rigidity cutoff $R_{\rm cut}$ are free model parameters.

To describe the propagation we use a propagation database calculated with CRPropa3 [6]. For the background sources, the calculation of the arriving species at Earth is similar as in [1]. The source evolution is described as $\propto (1+z)^m$ with the redshift $z$, where $m = 3.4$ for the SBGs and $m = 5.0$ for the $\gamma$-AGNs [7]. For the candidate sources, the model takes into account the effect of the source distances, the flux weights (both taken from the catalog), and the direction (regarding the observatory's exposure into that direction).

The contributions of the background and the catalog sources are weighted with another free model parameter, the signal fraction $f_0$. Its value is defined at the energy bin $(10^{19.5} - 10^{19.6})$ eV. The energy dependency of the catalog contribution, however, is produced by propagation effects. This will in general lead to an increase of the catalog con-

*e-mail: teresa.bister@ru.nl
**e-mail: spokespersons@auger.org;
full author list: https://auger.org/archive/authors_2022_10.html

https://doi.org/10.1051/epjconf/202328303008

tribution with the energy as the catalog sources are closer to Earth than the background.

The modeling of the observables on Earth is described for the energy spectrum and shower depth distributions in [1]. For the arrival directions, we consider a rigidity-dependent blurring of the form $\delta = \frac{\delta_0}{R/10\ \mathrm{EV}}$ with the free model parameter $\delta_0$. Then, the arrival directions distribution $S$ expected from the catalog is assembled by adding Fisher distributions $F$ for each element $A_{\mathrm{det}}$ from each source $C$ in each energy bin $E_{\mathrm{det}}$, and weighting them with the corresponding contribution $p$ calculated from the propagation:

$$S(E^e_{\mathrm{det}}, \mathrm{pix}^p) = \sum_{k,s} \left( p_{\mathrm{sig}}(E^e_{\mathrm{det}}, A^k_{\mathrm{det}}, C^s) \cdot F(\mathrm{pix}^p | R^{e,k}_{\mathrm{det}}, C^s) \cdot \mathcal{E}(\mathrm{pix}^p) \right. \quad (2)$$

Here, $\mathcal{E}(\mathrm{pix}^p)$ represents the observatory exposure in each pixel $\mathrm{pix}^p$. The total probability density function (pdf) is then calculated by weighting the catalog part with the background $B = \mathcal{E}(\mathrm{pix}^p)$ using the signal fraction. An example of arrival direction pdfs in two energy bins is shown in Fig. 1. Here, it is visible how the blurring decreases while the catalog contribution rises with the energy. Additionally, the effect of propagation is visible as the flux contribution of further away sources e.g. on the Galactic north decreases with the energy.

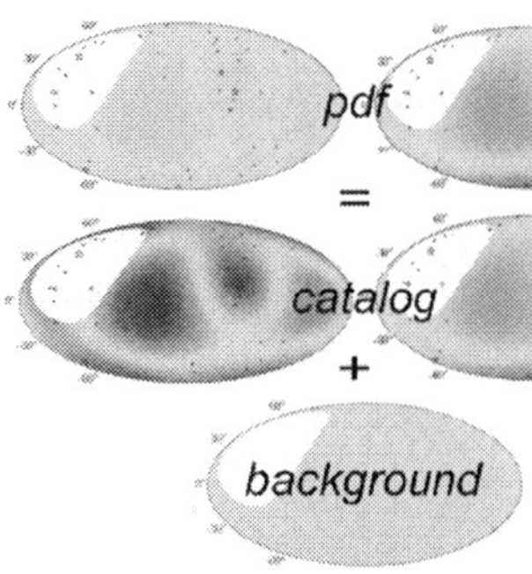


**Figure 1.** Setup of the total arrival directions pdf (*upper*) in Galactic coordinates for the astrophysical model for a representative small energy bin (*left*) and one at larger energy (*right*). The pdf is assembled from a part $S$ coming from the source catalog (*middle*) and a part $B$ from the homogeneous background (*lower*). The weighting of the two parts is determined by propagation effects and an adaptable signal fraction $f_0$ (see text).

We use a Markov-Chain-Monte-Carlo sampler to reconstruct the posterior distributions of the model parameters, and a maximum-likelihood optimizer to find the best-fit. The log-likelihood function $\log \mathcal{L}_{\mathrm{tot}}$ is a sum of the log-likelihoods for the energy spectrum, the shower depth distributions (both as in [1]), and the arrival directions. The likelihood of the arrival directions is calculated by comparing the measured (or simulated) arrival directions $v^{e,p}$ in each energy bin $E^e_{\mathrm{det}}$ with the modeled pdf:

$$\mathcal{L}_{\mathrm{AD}} = \prod_e \prod_p \mathrm{pdf}^{e,p}(v^{e,p}) \quad (3)$$

## 3 Benchmark simulation

To test the sensitivity of the fit method, we apply it to data-like benchmark simulations with SBGs as sources. For that, we select model parameters in a way that the simulated observables resemble the measurements by the Pierre Auger Observatory as well as possible. Specifically, we adapt the signal fraction and blurring until we find a simulation that can reproduce the test statistic $\mathrm{TS}_{\mathrm{AD-only}}$ (where AD-only is always used to denote the analysis from [2–4] based solely on the arrival directions) observed on the measured data. This is depicted in Fig. 2. It is visible how the simulations (200 representations) resemble the test statistic on the measured data for both SBG and $\gamma$-AGN hypothesis. The maximum test statistic, obtained for the best energy threshold, is stated in the figure. Even though in this simulation SBGs are set as the simulated truth, the test statistic of the $\gamma$-AGN model is still sizeably large. This is because some of the strongest sources in both catalogs are located in similar directions. It has to be noted that the statistical fluctuations between the simulations are extremely large, which comes mostly from fluctuations of the individual UHECR arrival directions.

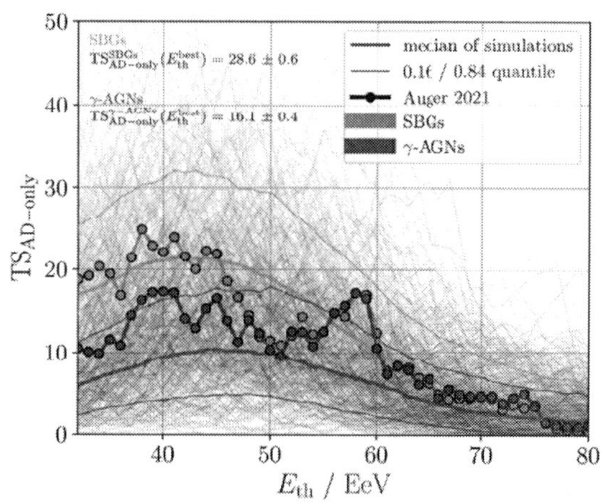


**Figure 2.** Test statistic of the arrival direction stacking analysis following [2–4] applied on simulations containing SBGs as sources, as a function of the energy threshold. The thin lines depict individual simulations, the thick line the median. The data from [3] is indicated with round markers. The maximum test statistic (mean and uncertainty on the mean over 200 representations) is given in the upper left. It is larger for the SBG model (= simulated truth, blue) than for the $\gamma$-AGN model ($\neq$ simulated truth, purple).

The simulated true parameters which have been inserted in order to reproduce the measured observables on Earth are given in tab. 1. Thus, with a signal fraction of

| $\gamma$ | $\log_{10}(R_{\mathrm{cut}}/\mathrm{V})$ | $I_{\mathrm{H}}$ | $I_{\mathrm{He}}$ | $I_{\mathrm{N}}$ | $I_{\mathrm{Si}}$ | $I_{\mathrm{Fe}}$ | $f_0$ | $\delta_0/°$ |
|---|---|---|---|---|---|---|---|---|
| −3.22 | 18.09 | 0% | 15% | 77% | 5.4% | 2.6% | 15% | 20° |

**Table 1.** Values of the model parameters used as simulated truth

around 15% at $10^{19.5}$ eV coming from the SBG catalog and a magnetic field blurring of 20° for a 10 EV particle, the simulations resemble not only the measured energy spectrum and shower depth distributions measured by the Pierre Auger Observatory, but can also in median reproduce the energy dependency of the arrival direction test statistic from [3] for both SBG and $\gamma$-AGN catalogs.

## 4 Fit result

As a first test of the fit method we apply it to one specific representation of the benchmark simulations. When the SBG model (=simulated truth) is used, the simulated true parameters from tab. 1 can be reconstructed very well, and useful uncertainties are provided by the posterior distributions (not depicted). The fitted energy spectrum and shower depth distributions are shown in Fig. 3. Here, it is visible how the simulated observables (markers) are reproduced well by the model (lines). The contribution by the SBG catalog is depicted as a dashed line. It is visible how the catalog mostly contributes at higher energies.

When the $\gamma$-AGN model ($\neq$ simulated truth) is applied to the benchmark simulation, the observables cannot be described as well. The best-fit signal fraction is very close to zero, indicating that the $\gamma$-AGN catalog does not contribute to the simulated data.

The quality of the fit can be compared via the test statistic, which in our case is defined as 2 times the negative total likelihood ratio between the model and a reference model with the same background source evolution, but signal fraction $f_0 = 0$. The values are given in tab. 2. The fit can clearly identify the SBG catalog as the simulated truth by a much larger total test statistic. Here, the arrival directions are the most important part for the source differentiation. This is because the energy dependency of the arrival directions of the SBG catalog cannot be described with the $\gamma$-AGN catalog, leading to the very small value of $\mathrm{TS}_{\mathrm{ADs}}^{\gamma\text{-AGN}} = 2.6$.

| | SBG model | | $\gamma$-AGN model | |
|---|---|---|---|---|
| | *example* | *mean* | *example* | *mean* |
| $\mathrm{TS}_{\mathrm{tot}}$ | 35.6 | $35.3 \pm 0.8$ | 0.4 | $1.9 \pm 0.2$ |
| $\mathrm{TS}_E$ | 9.8 | $4.3 \pm 0.4$ | 2.8 | $-0.2 \pm 0.1$ |
| $\mathrm{TS}_{X_{\mathrm{max}}}$ | −0.6 | $-0.91 \pm 0.03$ | 0.6 | $-0.13 \pm 0.02$ |
| $\mathrm{TS}_{\mathrm{ADs}}$ | 26.4 | $31.8 \pm 0.8$ | 2.6 | $2.3 \pm 0.2$ |

**Table 2.** Test statistic for the example simulation depicted in Fig. 3, and mean and uncertainty on the mean of the test statistic over the 200 simulations shown in Fig. 2 and Fig. 4 (same values as stated in the figures).

## 5 Sensitivity estimation

To estimate the sensitivity more reliably, we apply the fit method to all 200 representations of the benchmark simulation depicted in Fig. 2. Then, we calculate the mean and uncertainty on the mean[1] of the test statistic, which is also

[1] Note that even though the uncertainties on the mean are reasonably small, the standard deviation is very large for both $\mathrm{TS}_{\mathrm{AD-only}}$ and $\mathrm{TS}_{\mathrm{tot}}^{\mathrm{SBG}}$ as is visible in Fig. 2 and Fig. 4.

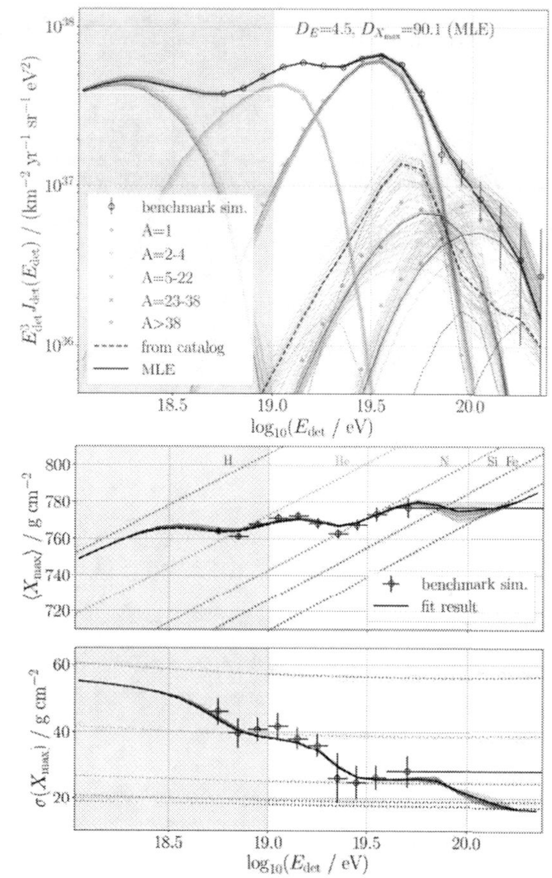


**Figure 3.** Energy spectrum (*upper*) and first two moments of the shower depth distributions (*lower*). Markers with Poisson errors represent the benchmark simulation containing SBGs as sources, thick lines the best-fit model, and thin lines indicate the uncertainties via several draws from the posterior distributions. The dashed line corresponds to the fitted contribution of the SBG catalog to the total flux. The deviance $D$ is stated in the top figure.

provided in tab. 2. Here, it is again visible how the arrival directions have the largest discrimination power. When comparing the mean arrival directions test statistic of the combined fit to the one obtained with the stacking analysis using only energy-independent arrival directions [2–4] (values given in Fig. 1), it is clearly visible how the $\gamma$-AGN model ($\neq$ simulated truth) can be discarded with a much larger certainty using the combined fit:

$$\mathrm{TS}_{\mathrm{AD-only}}^{\gamma\text{-AGN}} = 16.1 \pm 0.4 \gg \mathrm{TS}_{\mathrm{AD}}^{\gamma\text{-AGN}} = 2.3 \pm 0.2 \quad (4)$$

Also, the test statistic of the simulated true SBG model has increased even more:

$$\mathrm{TS}_{\mathrm{AD-only}}^{\mathrm{SBG}} = 28.6 \pm 0.6 < \mathrm{TS}_{\mathrm{AD}}^{\mathrm{SBG}} = 31.8 \pm 0.8 \quad (5)$$

The energy dependency of the total test statistic is shown in Fig. 4. Here, one can see how the most important contribution to the test statistic comes from the energy bins

 https://doi.org/10.1051/epjconf/202328303008

around $10^{19.7}$ eV. This is due to the interplay of the rising catalog contribution with the energy, and the shrinking statistics at the highest energies.

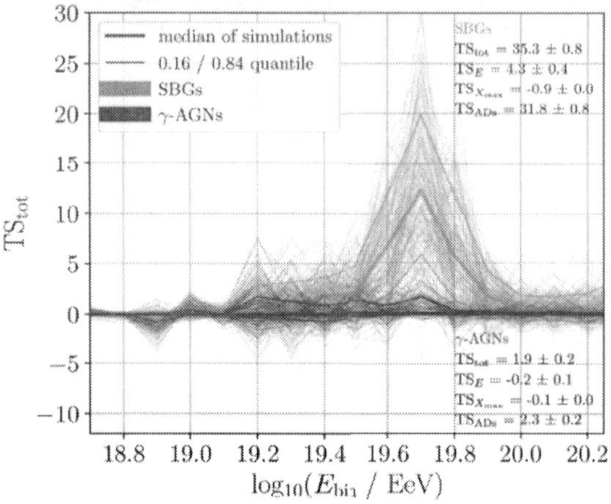


**Figure 4.** Total test statistic of the combined fit as a function of the energy bin. Note that differently to Fig. 2 the sum over all energy bins gives the total test statistic, which is stated on the right. The total test statistic of the SBG model (= simulated truth, blue) is much larger than that of the $\gamma$-AGN model ($\neq$ simulated truth, purple).

The energy spectrum plays a minor role in the model differentiation, but still has a positive impact in the identification of the correct catalog. The shower depth distributions on the other hand contribute almost nothing to the sensitivity. This is because of the very small statistic at the highest energies due to the measurement with fluorescence telescopes (which is considered in the simulated data). This limits the differentiation power of the shower depth observable in the energy range which is most important as this is where the catalog sources contribute most.

As a final test we also looked into the best-fit values for the signal fraction and magnetic field blurring to compare them for the two catalogs. This is depicted in Fig. 5. It is visible how for the simulated true SBG model, both parameters are reconstructed reliably. When the $\gamma$-AGN model is applied, however, the estimated signal fraction $f_0$ is always close to zero as expected. For a signal fraction of zero, the blurring parameter $\delta_0$ is not reconstructable.

## 6 Conclusion

In this contribution we have demonstrated the sensitivity of the combined fit of energy spectrum, shower depth distributions, and arrival directions. We have presented benchmark simulations that resemble the energy spectrum and shower depth distributions measured by the Pierre Auger Collaboration. Additionally, we showed that simulations with a signal fraction of around 15% at $10^{19.5}$ eV from the SBG catalog and a blurring of 20° for a particle with a rigidity of 10 EV can reproduce the energy dependency of the measured arrival directions.

On these simulations, the differentiation power regarding the source catalog is largely enhanced compared to previous analyses using only the energy-independent arrival directions. Also, no scan of the best energy threshold is necessary (which would decrease the significance), as the whole energy dependency of the observables is captured by the astrophysical model. Additionally, a reliable reconstruction of the parameters of the source emission, as well as the signal fraction and magnetic field blurring, is possible with the combined fit.

A publication which applies the method to the measured data of the Pierre Auger Observatory is in preparation.

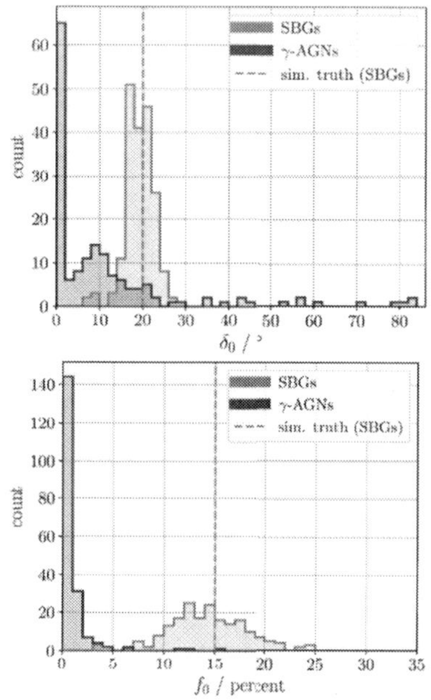


**Figure 5.** Histogram of the fitted magnetic field blurring $\delta_0$ (*upper*) and signal fraction $f_0$ (*lower*) for both the SBG model (blue) and the $\gamma$-AGN model (purple) applied on the simulations containing SBGs as sources. The simulated truth is indicated by the blue dashed line.

## References

[1] A. Aab et al. (The Pierre Auger Collaboration), JCAP **04**, 038 (2017)

[2] A. Aab et al. (The Pierre Auger Collaboration), ApJL **853**, L29 (2018)

[3] J. Biteau for the Pierre Auger Collaboration, PoS **ICRC2021**, 307 (2021)

[4] P. Abreu et al. (The Pierre Auger Collaboration) ApJ **935** 170 (2022)

[5] T. Bister for the Pierre Auger Collaboration, PoS **ICRC2021**, 368 (2021)

[6] R. Alves Batista et al., JCAP **1605** 038 (2016)

[7] R. Alves Batista, R. M. de Almeida, B. Lago, K. Kotera, JCAP **01** 002 (2019)

# A Bayesian source association analysis of UHECRs: Impact of the Galactic magnetic field and composition

*Keito* Watanabe[1,2,*], *Anatoli* Fedynitch[3,4,**], *Francesca* Capel[1,***], and *Hiroyuki* Sagawa[4,****]

[1]Max Planck Institute for Physics, Föhringer Ring 6, Munich, 80805, Germany
[2]I. Physikalisches Insitut, Universität zu Köln, Zülpicher Str. 77, 50937 Köln, Germany
[3]Institute of Physics, Academia Sinica, Taipei City, 11529, Taiwan
[4]Institute for Cosmic Ray Research, the University of Tokyo, 5-1-5 Kashiwa-no-ha, Kashiwa, Chiba 277-8582, Japan

**Abstract.** We present a statistical analysis of the association between UHECRs and proposed astrophysical sources. Our approach is based on the Bayesian hierarchical framework presented in Capel & Mortlock 2019, but with notable extensions. Using CRPropa3, we now include the lensing effect of the Galactic magnetic field and explore the impact of heavier compositions. This analysis directly connects to the physics of UHECR propagation so that each detected event is allowed different possible deflections and energy-loss horizons based on its measured energy, arrival direction and the corresponding uncertainties. In this way, we can easily interpret the connection to proposed sources in a physical way. We verify our approach using simulated data and then present our results with views of the Northern and Southern skies thanks to publicly available data from both the Telescope Array experiment and Pierre Auger Observatory.

## 1 Introduction

The quest to determine the origins of ultra-high-energy cosmic rays (UHECRs) is an ongoing problem, due to the complex physical processes involved in their injection, acceleration, and propagation, as well as the technical challenges in their detection. In particular, point-source analyses cannot accurately determine the sources of UHECRs due to their rigidity-dependent deflections from the extra-Galactic and Galactic magnetic fields (EGMF and GMF, respectively). While several attempts have been made to associate sources with UHECRs through their detected positions, no consensus for exact sources have yet been established (see recent reviews, e.g. Anchordoqui 2018 [1] and Batista et al. 2019 [2]).

Due to the large number of parameters and uncertainties involved, it is challenging to construct a comprehensive statistical model from UHECR observations. Nevertheless, this can be approached within the framework of Bayesian hierarchical modeling. Earlier work has made progress in this direction, with Watson [3], Soiaporn [4], and Kanin & Mortlock [5] developing Bayesian models that focused on using the UHECR arrival directions. Capel & Mortlock 2019 (CM19 hereafter) [6] extended these models by presenting a joint model including both UHECR energies and arrival directions, yielding a source association fraction of up to $\sim 0.2$ with a set of starburst galaxies proposed from Ref. [7] (SBG hereafter) and the public UHECR events from the Pierre Auger Observatory (PAO) [8] as reported in Ref. [9] (PAO14 hereafter).

In this paper, we report on the progress in quantifying the impact of the GMF and UHECR composition on source-UHECR associations by further developing the joint model presented in CM19. In Section 2 and 3, we describe the physical and statistical model used to include GMF and composition effects and to perform the Bayesian source association analysis. In Section 4, we apply our model to the PAO14 data and to the public data from the Telescope Array experiment (TA) [10] reported in Ref. [11] (TA15 hereafter) with the SBG catalogue as above. We show the individual source-UHECR association probabilities and also the posterior distribution of the source association fractions to describe the impact of the GMF and composition. In Section 5, we discuss our obtained results. Finally, in Section 6, we summarize on the current status of our work and outline plans for the future. The package used in our work is available online.[1]

## 2 The Physical Model

The model is based on the physical model presented in CM19.

### 2.1 Source

We assume a population of steady-state sources with equal particle luminosity $Q$ such that the UHECR flux $F_k = Q/4\pi D_k^2$ only depends on the source distance $D_k$ for each

*e-mail: watanabe@ph1.uni-koeln.de
**e-mail: anatoli@gate.sinica.edu.tw
***e-mail: capel@mpp.mpg.de
****e-mail: hsagawa@icrr.u-tokyo.ac.jp

[1]https://github.com/cescalara/fancy

source $k$. We assume that the sources produce UHECRs of the same composition with a power law spectrum, $dN/dE \propto E^{-\alpha}$, where $\alpha$ is the spectral index. We also include an isotropic background flux $F_0$ for UHECRs that originate from sources not present in the considered catalogue.

### 2.2 Energy Losses

As an intermediate stage in our approach, we describe the energy losses that the UHECRs undergo through their propagation using the loss length. Under the continuous energy loss approximation, we evaluate the arrival energies of the UHECRs by the following first order differential equation:

$$\frac{\mathrm{d}E}{\mathrm{d}z} = -\frac{E}{L_{\mathrm{loss}}(E,z)} \tag{1}$$

where $z$ is the redshift of the UHECRs and the loss length $L_{\mathrm{loss}}(E,z)$ is described by losses due to photopion production ($\beta_\pi$), electron pair production ($\beta_{e^\pm}$), adiabatic expansion of the universe ($\beta_{\mathrm{adi}}$) and photodisintegration ($\beta_\gamma$) processes:

$$L^{-1}_{\mathrm{loss}}(E,z) = \frac{\mathrm{d}t}{\mathrm{d}z}(\beta_\pi(E,z) + \beta_{e^\pm}(E,z) + \beta_{\mathrm{adi}}(E,z) + \beta_\gamma(E,z)). \tag{2}$$

Note that, unlike CM19, which assumes a pure proton UHECR composition, we consider photodisintegration processes as they are relevant for UHECR compositions with Z > 1. A preliminary implementation of this is performed using the loss length as given in the propagation modules provided by CRPropa3 [12]. It is assumed that the energy loss rate is described by the average number of nucleons lost for all possible disintegration channels.

### 2.3 EGMF Deflections

The magnetic deflections from the EGMF experienced by a UHECR with energy $E$, charge $Z$, travelling at a distance $D$ from Earth are described by an analytical approach, with a root-mean-square (RMS) angular scale as such:

$$\bar{\theta} \approx 2.3^\circ Z \left(\frac{E}{50\,\mathrm{EeV}}\right)^{1} \left(\frac{\bar{B}}{1\,\mathrm{nG}}\right) \left(\frac{D}{10\,\mathrm{Mpc}}\right)^{1/2} \left(\frac{l_c}{1\,\mathrm{Mpc}}\right)^{1/2} \tag{3}$$

where $\bar{B}$ is the RMS EGMF strength and $l_c$ is the field coherence length [4].

### 2.4 Detection

As in CM19, we model the detected energies $\hat{E}$ from a Gaussian distribution, given the reported reconstruction uncertainty $f_E$ and "true" arrival energies $E$. Similarly, we model the detected arrival directions $\hat{\omega}$ using a von-Mises-Fisher (vMF) distribution [13] as in CM19:

$$p(\hat{\omega} \mid \omega, \kappa_d) = \frac{\kappa_d}{4\pi \sinh \kappa_d} \exp(\kappa_d \hat{\omega} \cdot \omega) \tag{4}$$

where $\omega$ is reported "true" arrival direction and $\kappa_d$ is the concentration parameter, inversely related to the angular reconstruction uncertainty $\sigma_\omega$. The exposure of PAO and TA is calculated using the formalism provided in CM19.

### 2.5 GMF Deflections

Unlike CM19, which describes both EGMF and GMF deflections using Eq. 3, we use a backtracking approach to describe the GMF deflections. We backtrack each UHECR with a detected arrival direction $\hat{\omega}$ (sampled by Eq. 4) using CRPropa3, assuming a GMF model postulated by Jansson & Farrar [14] (JF12 hereafter). Fig. 1 shows the skymap of the deflected UHECRs from the TA15 dataset [11] using this approach, assuming a composition of protons.

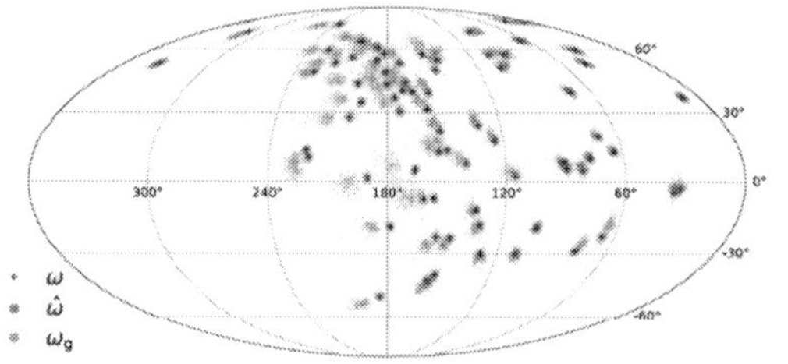


**Figure 1.** Skymap of deflected UHECRs ($\omega_g$, red) using our approach, assuming all UHECRs are protons. The true arrival directions ($\omega$, black) provided from the TA15 dataset are sampled via Eq. 4 ($\hat{\omega}$, blue) and then deflected under the GMF model from JF12. The skymap is in galactic coordinates with the Galactic anti-center at the origin.

From the arrival direction at the galactic boundary $\omega_g$, we can extend the vMF distribution in Eq. 4 by replacing $\kappa_d$ with a modified parameter $\kappa_{\mathrm{GMF}}$, which now includes the effects of the GMF. The concentration parameter $\kappa_{\mathrm{GMF}}$ is determined by numerically solving

$$\int_\Omega \mathrm{d}\omega\, p(\omega_g \mid \hat{\omega}, \kappa_{\mathrm{GMF}}) = \frac{1 - e^{-\kappa_{\mathrm{GMF}}(1-\hat{\omega}\cdot\omega_g)}}{1 - e^{-2\kappa_{\mathrm{GMF}}}} = \mathrm{P} \tag{5}$$

where $\Omega$ is the cone of solid angle subtended by the contour of the region containing P = 0.683, analogous to the "$1\sigma$" region of a Gaussian distribution [4]. The new concentration parameter $\kappa_{\mathrm{GMF}}$ is then considered in the likelihood function for the arrival directions, which takes into account the GMF deflections by modifying the width of the distribution (see Sec. 3).

## 3 The Statistical Model

We use a Bayesian hierarchical model with Stan, a state-of-the-art platform for statistical modeling and computation [15]. The parameters are fitted using a Markov Chain Monte Carlo sampling method with the Hamiltonian Monte Carlo algorithm. Through this approach, we can encompass more propagation physics into our model and construct a "bubble" of possible origins for each UHECR based on its energy and arrival direction.

### 3.1 Likelihood Function

In our analysis, we follow the likelihood function as presented in CM19. However, to take the GMF deflections into account, we replace the concentration parameter $\kappa_d$ by $\kappa_{\mathrm{GMF}}$, which now includes the effects of the GMF deflections. This is reflected in the probability distribution that describes the probability of observing a UHECR with arrival direction $p(\hat{\omega} \mid \bar{\omega}_k, \bar{\theta})$ (Eq. (A5) in CM19), which is subsequently contained in the likelihood function. In doing so, we assume that the detection uncertainties are smaller than the impact GMF deflections, which holds for the cases considered here.

### 3.2 Priors

To describe a more accurate representation of the EGMF strength, we consider a narrower prior for $\bar{B}$ compared to CM19. In our analysis, we consider a normal prior with $\mu_{\bar{B}} = 1$ nG with a spread of $\sigma_{\bar{B}} = 3$ nG, motivated by observational constraints on the EGMF strength, e.g. Refs. [16, 17]. The same weakly informative priors as in CM19 are used for the other highest-level parameters ($\alpha$, $Q$, and $F_0$).

## 4 Results

We apply our Bayesian hierarchical model for the PAO14 and TA15 datasets independently and evaluate the association probabilities as in CM19 using the SBG catalogue. In Fig. 2, we show the posterior distribution for the source association fraction of UHECRs with the PAO14 and TA15 dataset as mentioned above, considering four different configurations:

($a$) A pure proton composition, assuming a joint GMF + EGMF magnetic field, following the configuration used in CM19

($b$) A pure proton composition with the GMF deflections as done in this work

($c$) Same as (a) but for a pure nitrogen composition

($d$) Same as (b) but for a pure nitrogen composition

We note that the obtained nitrogen results are preliminary at this stage of our project, as we have not thoroughly considered for further physical effects such as the evolving composition throughout propagation. Here, the source association fraction is defined as the fraction of the flux of UHECRs that are associated to a particular source catalogue such that

$$f = \frac{\sum_k F_k}{F_0 + \sum_k F_k}. \quad (6)$$

In Fig. 3, we show the skymap presenting the individual source-UHECR association probabilities for each dataset, assuming a pure nitrogen composition and including the effects of the GMF deflections from this work (as in case ($d$) from above). The dominant sources, i.e. those with the highest source association probabilities, are NGC4945, M83, and NGC253 for the Southern sky and M82, IC342, and NGC6946 for the Northern sky.

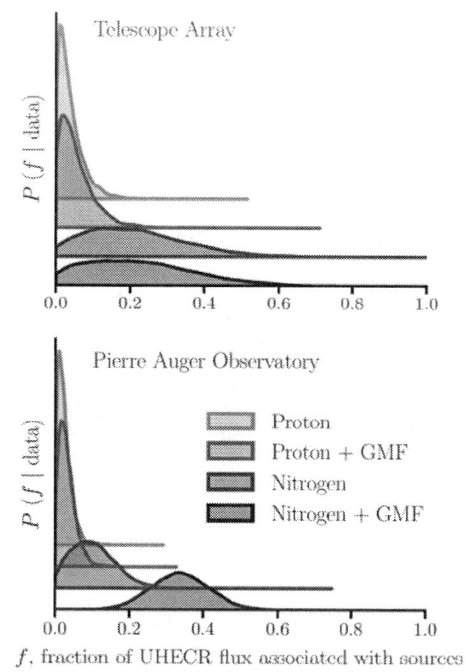


**Figure 2.** Posterior distribution of the source association fraction for TA15 (top) and for PAO14 (bottom). We evaluate the associations with the SBG catalogue for both datasets. The source fractions are determined assuming proton and nitrogen compositions and also with and without the effect of the GMF deflections mentioned in this work, as shown in the legend.

## 5 Discussion

We observe that, with a pure proton composition, the posterior distributions in Fig. 2 are peaked at source fractions close to zero, contrary to the results of CM19. This implies that the narrow priors on the EGMF significantly affect the source-UHECR associations, possibly due to the smaller width of the vMF distribution described by $p(\hat{\omega} \mid \bar{\omega}_k, \bar{\theta})$. We also observe that, with protons, the effects of the GMF deflections do not impact the source association probabilities significantly for both datasets. This is due to the small nuclear charge of protons, which experience fewer deflections from both the EGMF (as seen in Eq. 3) and the GMF due to their $Z$-dependence.

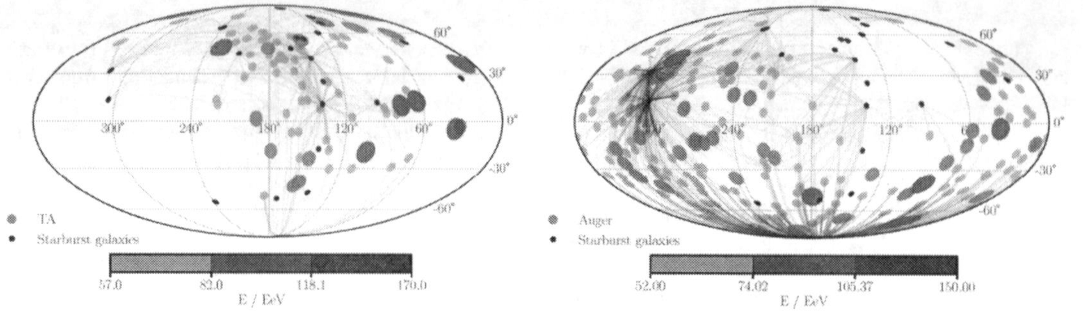


**Figure 3.** Skymaps of individual source-UHECR association probabilities (represented as weighted lines) for the TA15 (left) and for the PAO14 data (right), scaled by their energies (blue). We consider the SBG catalogue and a pure nitrogen composition and further consider the effects of the GMF deflections implemented in this work for both datasets. Coordinates are given in Galactic coordinates with the Galactic anti-center at the origin.

In general, we observe that the source-UHECR associations increase by including the effects of the GMF deflections with a larger nuclear charge. Fig. 2 shows that by considering a pure nitrogen composition, the posterior distributions for the source fraction are small near zero and average at $f \sim 0.2$. This is most likely due to large deflections from both the GMF and EGMF. As both deflections depend on $Z$, a higher nuclear charge will deflect the UHECRs more strongly, increasing the width of the vMF distribution and subsequently allowing for more sources to be associated. This effect is also observed in the skymaps in Fig. 3, as sources outside the field of view of each observatory are associated with the UHECRs in each dataset. Such results may change if we instead replace the vMF distribution by a new distribution at the galactic boundary, which is currently under investigation.

## 6 Conclusion

In our work, we aim to quantify the impact of GMF deflections and UHECR compositions on source-UHECR associations. We perform a preliminary implementation of the GMF deflections and UHECR composition into the joint Bayesian hierarchical model proposed by Capel & Mortlock 2019 [6] with CRPropa3. We evaluate the source-UHECR associations for the PAO14 and TA15 datasets using the SBG catalogue. For a pure proton composition, the source fraction distribution show that the effects of GMF deflections do not impact the source associations strongly. However, the source-UHECR associations, in general, increase when we consider the GMF deflections from this work with a higher nuclear charge, as observed by the source fraction distribution and the skymaps showing the source association probabilities. We plan to more accurately model the UHECR propagation for nuclei by considering the photodisintegration of the nuclei throughout their propagation.

## References

[1] L.A. Anchordoqui, Phys. Rept. **801**, 1 (2019), 1807.09645
[2] R. Alves Batista et al., Front. Astron. Space Sci. **6**, 23 (2019), 1903.06714
[3] L.J. Watson, D.J. Mortlock, A.H. Jaffe, Mon. Not. Roy. Astron. Soc. **418**, 206 (2011), 1010.0911
[4] K. Soiaporn, D. Chernoff, T. Loredo, D. Ruppert, I. Wasserman, Ann. Appl. Stat. **7**, 1249 (2013), 1206.4569
[5] A. Khanin, D.J. Mortlock, Mon. Not. Roy. Astron. Soc. **460**, 2765 (2016), 1601.02305
[6] F. Capel, D.J. Mortlock, Mon. Not. Roy. Astron. Soc. **484**, 2324 (2019), 1811.06464
[7] A. Aab et al. (Pierre Auger), Astrophys. J. Lett. **853**, L29 (2018), 1801.06160
[8] A. Aab et al. (Pierre Auger), Nucl. Instrum. Meth. A **798**, 172 (2015), 1502.01323
[9] A. Aab et al. (Pierre Auger), The Astrophysical Journal **804**, 15 (2015)
[10] T. Abu-Zayyad et al. (Telescope Array), Nucl. Instrum. Meth. A **689**, 87 (2013), 1201.4964
[11] R.U. Abbasi et al. (Telescope Array), Astrophys. J. Lett. **790**, L21 (2014), 1404.5890
[12] R. Alves Batista et al., JCAP **09**, 035 (2022), 2208.00107
[13] R. Fisher, Proceedings of the Royal Society of London Series A **217**, 295 (1953)
[14] R. Jansson, G.R. Farrar, Astrophys. J. **757**, 14 (2012), 1204.3662
[15] Stan Development Team, *Stan Modeling Language Users Guide and Reference Manual*, Online (2022), https://mc-stan.org
[16] R. Durrer, A. Neronov, Astron. Astrophys. Rev. **21**, 62 (2013), 1303.7121
[17] J.D. Bray, A.M.M. Scaife, Astrophys. J. **861**, 3 (2018), 1805.07995

https://doi.org/10.1051/epjconf/202328303010

# CRPropa high statistics simulations for UHECR anisotropy studies

*Simone* Rossoni[1,*] and *Günter* Sigl[1,**]

[1]II. Institute for Theoretical Physics, Hamburg University, Luruper Chaussee 149, 22761, Hamburg, Germany

**Abstract.** Despite the great progress made by modern cosmic ray observatories, the origin and acceleration mechanism of ultra-high-energy cosmic rays (UHECRs) remains unsolved problems to this day. However, there is experimental evidence for an anisotropic component in the UHECR arrival direction at energies greater than few EeV. The search for UHECR sources is further complicated by two main factors: during extragalactic propagation UHECRs interact with background photon fields (like the CMB and the EBL) and, since they are electrically charged, they are deflected by extragalactic and galactic magnetic fields (EGMF and GMF). Moreover, the strength, structure and origin of the EGMF are still not well known, causing reconstructing the deflection of UHECRs to be a non-trivial task. In this work we consider several EGMF models obtained from constrained MagnetoHydroDynamics (MHD) simulations of our local Universe to study the propagation of UHECRs through such a structured environment. We simulate propagation, interactions and observation of UHECRs by using the Monte Carlo code CRPropa3. We also take into account the effect of the GMF by adopting a lensing procedure of the arrival UHECR sky map. We explore several combination of source distribution, EGMF structure and mass composition. As a reference, we also simulate scenarios without the EGMF and with a statistically homogeneous field. We use our simulation results to compute UHECR observables, such as the energy spectrum, the angular power spectrum and the arrival direction map (before and after the GMF) in order to obtain constraints on possible combinations of source distributions and EGMF models.

## 1 Introduction

The origin of the most energetic particles ever observed in nature still remains an unsolved problem in astrophysics. These particles are called ultra-high-energy cosmic rays (UHECRs) and they consist of ionized nuclei of energy higher then 1 EeV ($=10^{18}$ eV). The discovery of UHECR sources would offer the possibility of studying environments capable of reaching such high energies, which are not possible even in modern particle accelerators.

In recent years, cosmic ray observatories have reported several observations showing that the UHECR sky does not appear isotropic above the EeV scale. The Pierre Auger Observatory [1] has reported a dipolar anisotropy signal in the arrival directions of UHECRs with energies above 8 EeV which points $\sim 120°$ away from the galactic center [2]. This observation supports the hypothesis of an extragalactic origin of these incredibly energetic particles.

However, the information we can obtain from the observation of UHECRs must be processed in order to investigate physics at the sources. This is because the propagation in the extragalactic space is perturbed by particle interactions and magnetic deflections. At these high energies, charged nuclei are able to trigger photo-hadronic interactions with the cosmic photon fields, like the CMB and the EBL [3, 4]: these interactions are the electron-positron pair production ($E_{\mathrm{th}} \simeq 2.5$ EeV for protons with CMB photons), the photo-pion production ($E_{\mathrm{th}} \simeq 70$ EeV for protons with CMB photons) and the photodisintegration of heavy nuclei. The presence of magnetic fields in the extragalactic space must also be taken into account. Since UHECRs are electrically charged and they propagate over cosmological distances, a non-negligible deflection can be induced by the Lorentz force (expecially in the case of heavy nuclei) hiding the real position of the source in the sky. Since cosmic magnetic fields are poorly known, numerical simulations represent a fundamental tool to study the role that different magnetic field models play on UHECR observables.

In this work we simulate the propagation of UHECRs in a numerical replica of our local Universe that includes both the structured cosmic magnetic field and the baryon density distribution. We take into account all the relevant interactions and energy loss processes during the extragalactic propagation. Simulation results are then used to compute anisotropy observables to study the resulting UHECR multipole signal when extragalactic and galactic magnetic deflections are considered.

## 2 Simulations and results

In order to reproduce all the relevant physical processes for the extragalactic propagation of UHECRs a numerical framework is necessary. In this work we consider the Monte Carlo code `CRPropa 3` [5], a publicly available simulation code for UHECR propagation and interac-

*e-mail: simone.rossoni@desy.de
**e-mail: guenter.sigl@desy.de

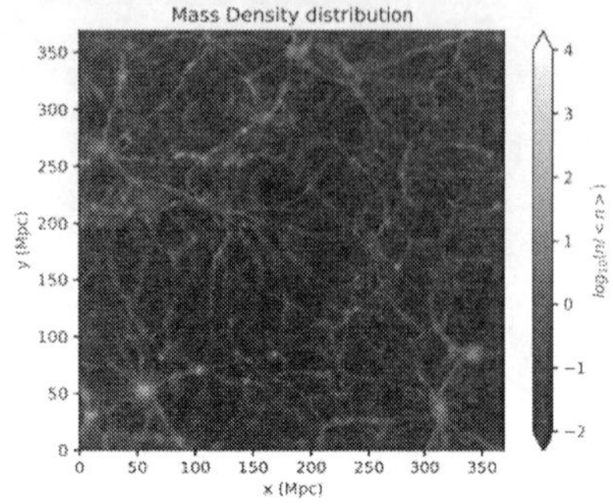


**Figure 1.** Mass density distribution in super-galactic coordinates at $z = 0$. The mass density $n$ is normalized to the average density in the simulation volume $\langle n \rangle$. The plot shows the super-galactic plane orthogonal to the $Z$-axis including the Milky Way at the centre.

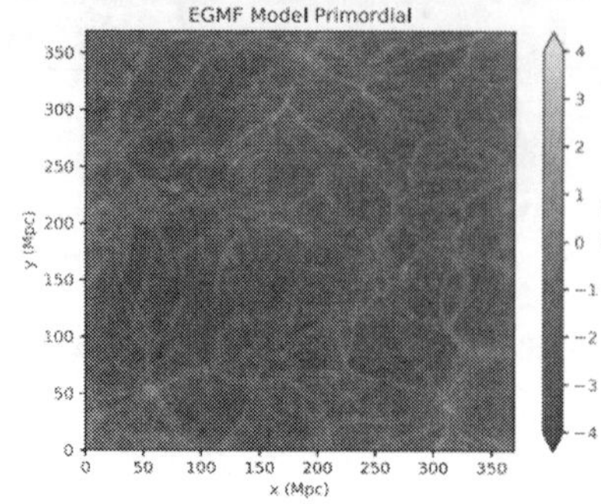


**Figure 2.** EGMF strength in nG for the *primordial* configuration in super-galactic coordinates at $z = 0$. Same coordinate system as Figure 1.

tions[1]. With CRPropa 3 the Lorentz equation is numerically integrated taking into account all energy loss mechanisms, photodisintegration and nuclear decay of heavy nuclei in a 3D volume. In this work we simulate the propagation of UHECRs within a cubic volume of 250 Mpc/h each side. We define a periodic boundary condition: if a particle leaves the volume, its trajectory is continued on the opposite side. An event is observed when it intersects the observer, represented by sphere of radius 1 Mpc located in the centre of the simulation volume.

To study the effect induced by the distribution of cosmic structures in the local Universe, we consider constrained MagnetoHydroDynamics (MHD) cosmological simulations to account the evolution of both the matter distribution and the cosmic magnetic field. The MHD simulations used in this work were obtained with the cosmological grid code ENZO [6]. The numerical replica of our local Universe at redshift $z = 0$ is obtained from MHD simulations started at $z = 60$ within a volume of $(500\,\mathrm{Mpc/h})^3$ with the Milky Way at the centre, as extensively described in [7]. In Figure 1 the normalized mass density distribution at $z = 0$ in the super-galactic plane is shown. Figure 1 shows how the matter distribution of the local Universe is complex and characterized by various types of structures (e.g. clusters, filaments and voids).

The evolution of the extragalactic magnetic field (EGMF) is simulated together with the matter evolution for different assumptions on the origin of the field. Three primordial models for the EGMF origin were considered. In the *primordial* model a uniform magnetic field of strength 0.1 nG along each axis at $z = 60$ is assumed. In both the *primordial2R* and *primordial3R* models the magnetic field is obtained from a power-law distribution with two different slopes for the power spectrum, $n_B = -3$ and $n_B = -4$, respectively. The normalisation of the spectra was chosen such that the root-mean-square field is the same of the *primordial* scenario. The EGMF strength for the *primordial* scenario at $z = 0$ in the super-galactic plane is shown in Figure 2, where the strong correlation between the mass distribution and the EGMF structure is visible. Three cases of astrophysical origin of the magnetic field were also considered. These configurations are called *astrophysical*, *astrophysicalR* and *astrophysical1R* and in each of them the magnetic energy is assumed to be the 50% of the injected thermal energy and it is released as dipoles around the injection region. Further details on the EGMF seeds and the corresponding volume filling factors can be found in [7]. All of these EGMF configurations are considered in our UHECR simulations.

We define three source distribution models: in the *homogeneous* model the injection positions is chosen within the simulated volume with the same injection rate everywhere, in the *mass density* model the injection position is chosen with a probability density function proportional to the mass density distribution shown in Figure 1 and, lastly, in the *mass halos* model the injection is chosen with the same probability from several point-like sources located at the center of halos identified by the gas number density. In the latter case, the number of halos is such that the average number density is $\sim 2 \cdot 10^{-4}\,\mathrm{Mpc}^{-3}$.

Together with the EGMF models described above, we also consider the case without any magnetic field (*NoEGMF*) and with a statistically uniform EGMF with a Kolmogorov-like power law spectrum and $B_{\mathrm{rms}} = 1$ nG (*statistical*). The combination of the various models for the source distribution and the EGMF gives us the opportunity to explore the effects induced on the arrival directions of UHECRs by the presence of structures in both the matter and the magnetic content of the local Universe.

We simulate the propagation of two pure compositions at the injection, one of protons and one of helium nuclei. The initial energy is extracted between 1 EeV and 1000 EeV following a power law spectrum of $E^{-1}$ and the initial momentum is chosen isotropically. We run 10 different simulations for each configuration described above, requesting the observation of 2500 UHECRs with $E \geq 1$ EeV.

[1]Version 3.1.7 of CRPropa 3 is used in this work.

 https://doi.org/10.1051/epjconf/202328303010

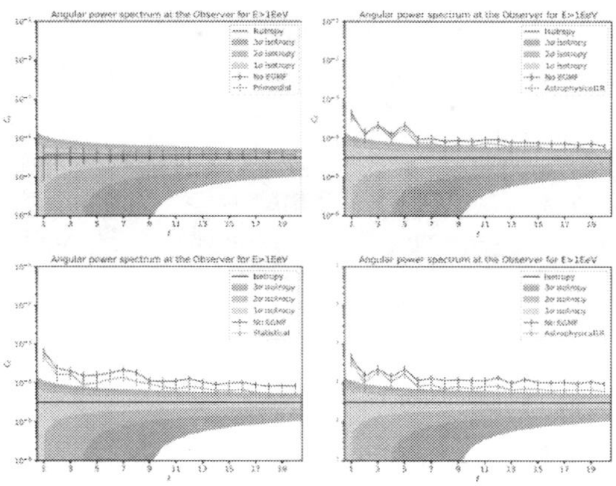

**Figure 3.** APSs at the observer for 2500 UHECRs with $E \geq 1$ EeV. Top left: proton injection and *homogeneous* source distribution. Top right: proton injection and *mass density* source distribution. Bottom left: helium injection and *mass halos* source distribution. Bottom right: helium injection and *mass halos* source distribution. In each panel the isotropic prediction given by (2) (black line and coloured areas), the APS without EGMF (blue line) and the APS with one EGMF model (orange line) are shown.

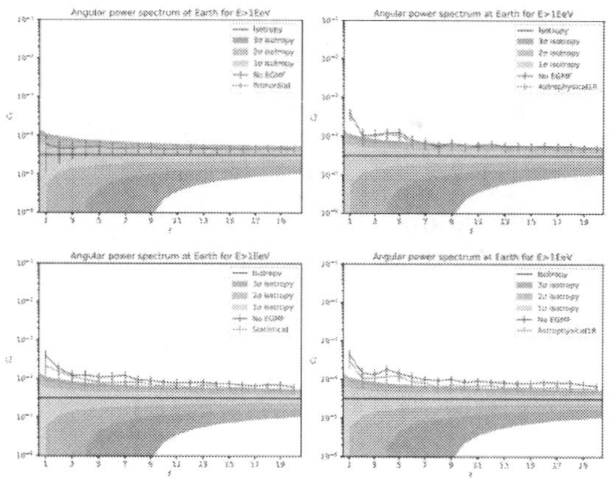

**Figure 4.** Same as Figure 3 after the galactic lensing.

The detected events are used to construct the sky map distribution of the UHECR arrival directions. We use the python package HEALPY, based on the Hierarchical Equal Area isoLatitude Pixelization (HEALPix) scheme [8][2], to handle pixelated data on a sphere. Starting from the sky maps, the Angular Power Spectrum (APS) $C_l$ is defined as

$$C_l = \frac{1}{2l+1}\sum_{m=-l}^{l} |a_{lm}|^2 , \tag{1}$$

where $a_{lm}$ are the spherical coefficients of the sky map spherical decomposition. We compute (1) by using the developed tools in HEALPY for the multipoles $l \leq 20$. We compare these APSs with the expected isotropic prediction given by the same number of observed particles

$$C_l^{\text{iso}} = \frac{1}{4\pi N} \quad , \quad \sigma_l^{\text{iso}} = \sqrt{\frac{2}{2l+1}} C_l^{\text{iso}} . \tag{2}$$

The APSs for several configurations are shown in Figure 3. The blue and orange lines represent the mean value of $C_l$ obtained from the average of the 10 simulations and the error bars represent the corresponding standard deviations. We also show the prediction (2) with the $1\sigma_l^{\text{iso}}$, $2\sigma_l^{\text{iso}}$ and $3\sigma_l^{\text{iso}}$ areas (cyan, blue and green respectively).

From Figure 3 we can see that the combination of *homogeneous* source distribution and *NoEGMF* reproduces the isotropic prediction. The main deviation from the isotropic prediction is introduced by the presence of a structured source distribution regardless of the presence or the model for the EGMF. The low-multipole components decrease towards the isotropic prediction, however the high-multipole components ($l \lesssim 5$) strongly deviate from the isotropy, in particular the dipole component ($l = 1$). The presence of a generic EGMF model, when the source distribution is structured, introduces an additional, but minor, deviation between the corresponding configuration without any magnetic field. This result tells us that the effect of the magnetic deflection is to reduce the anisotropies purely induced by the morphology of the source distribution. This latter effect increases with the increasing of the electric charge at the injection (as expected from the fact that the magnetic deflection depends linearly on the atomic number $Z$). However, this suppression is not strong enough to make the APS compatible with the isotropic prediction.

The results described in the previous paragraphs were obtained by assuming the observation of UHECRs on the observer's surface (a sphere of radius 1 Mpc located in the centre of the simulation volume). If we assume that this surface represents the edge of the Milky Way, then we must take into account the propagation through the galactic environment to compare the angular spectra with experimental data. The galactic propagation can induce further effects on the distribution of the arrival directions because, even if photo-hadronic interactions are negligible on galactic scales, the galactic magnetic field (GMF) strength is $\sim 1 - 10\,\mu$G, and thus the magnetic deflection can be of several degrees. The effect of the galactic propagation on the arrival directions can be reproduced by applying a specific magnetic lens to the sky map obtained at the edge of the Galaxy. Thus, the lensing technique maps a given bin on the surface at the edge of the Galaxy into several bins in the sky (further details on the magnetic lensing technique can be found in [9]). Several magnetic lenses are implemented in CRPropa 3; in this work we use the implementation of the JF12 GMF model [10].

In Figure 4 we report the APSs after the galactic lensing for the same configurations of Figure 3. The effect of the GMF is to suppress the deviation for the isotropy for the high-multipole components ($l \gtrsim 5$). However, the deviations of the low-multipole components are still observ-

[2]http://healpix.sourceforge.net

able. As before the galactic lensing, the main source of anisotropies still remains the structured distribution of the sources. The presence of an EGMF model induces only a weak separation with the ballistic propagation cases. Moreover, this separation is slightly reduced by the galactic deflection.

## 3 Conclusions

In this work we studied the propagation of ultra-high-energy protons and helium nuclei with the Monte Carlo code `CRPropa 3`. We considered all the relevant interaction processes with the background photo fields and the adiabatic energy loss due to the expansion of the Universe. We reproduced the matter and magnetic field distributions of the local Universe by using MHD simulation results. The evolution of the EGMF has been simulated under several assumptions on the origin of the field. We considered six EGMF models, one statistically homogeneous field and the case without magnetic field together with structured and homogeneous source distributions. We also introduced the impact of the GMF using the magnetic lensing technique.

In Figure 3 we can see that deviations from isotropy are mainly due to the structure of the source distribution, which has a greater effect on the low-multipole components. The effect of the EGMF is the introduction of a small deviation from the same configuration obtained without any magnetic field. This effect is more evident in case of heavier nuclei at the injection. After the galactic propagation (Figure 4) the previously mentioned results are still valid. However, the separation between the APSs obtained with and without the EGMF is slightly suppressed.

In the next developments of this work we will be focused on the anisotropic signal due to observed UHECRs with energy above 8 EeV. The propagation of nuclei heavier than helium nuclei (e.g. N, Si and Fe) will be included in order to be able to compare simulated spectra with real anisotropy data. To improve the efficiency of our `CRPropa 3` simulations the recently developed targeted injection technique will be used [11, 12]. This will also give us the possibility to simulate propagation in a larger simulation volume without having to use periodic boundary conditions. The study of the energy spectrum, the mass composition and the dipole direction will allow us to further investigate the impact of different assumptions of the EGMF origin on the UHECR observables.

## References

[1] A. Aab *et al.* [Pierre Auger], "The Pierre Auger Cosmic Ray Observatory," Nucl. Instrum. Meth. A **798** (2015), 172-213 doi:10.1016/j.nima.2015.06.058 [arXiv:1502.01323 [astro-ph.IM]].

[2] A. Aab *et al.* [Pierre Auger], "Observation of a Large-scale Anisotropy in the Arrival Directions of Cosmic Rays above $8 \times 10^{18}$ eV," Science **357** (2017) no.6537, 1266-1270 doi:10.1126/science.aan4338 [arXiv:1709.07321 [astro-ph.HE]].

[3] K. Greisen, "End to the cosmic ray spectrum?," Phys. Rev. Lett. **16** (1966), 748-750 doi:10.1103/PhysRevLett.16.748

[4] G. T. Zatsepin and V. A. Kuzmin, "Upper limit of the spectrum of cosmic rays," JETP Lett. **4** (1966), 78-80

[5] R. Alves Batista, A. Dundovic, M. Erdmann, K. H. Kampert, D. Kuempel, G. Müller, G. Sigl, A. van Vliet, D. Walz and T. Winchen, "CRPropa 3 - a Public Astrophysical Simulation Framework for Propagating Extraterrestrial Ultra-High Energy Particles," JCAP **05** (2016), 038 doi:10.1088/1475-7516/2016/05/038 [arXiv:1603.07142 [astro-ph.IM]].

[6] G. L. Bryan *et al.* [ENZO], "Enzo: An Adaptive Mesh Refinement Code for Astrophysics," Astrophys. J. Suppl. **211** (2014), 19 doi:10.1088/0067-0049/211/2/19 [arXiv:1307.2265 [astro-ph.IM]].

[7] S. Hackstein, F. Vazza, M. Brüggen, J. G. Sorce and S. Gottlöber, "Simulations of ultra-high Energy Cosmic Rays in the local Universe and the origin of Cosmic Magnetic Fields," Mon. Not. Roy. Astron. Soc. **475** (2018) no.2, 2519-2529 doi:10.1093/mnras/stx3354 [arXiv:1710.01353 [astro-ph.CO]].

[8] K. M. Górski, E. Hivon, A. J. Banday, B. D. Wandelt, F. K. Hansen, M. Reinecke and M. Bartelman, "HEALPix - A Framework for high resolution discretization, and fast analysis of data distributed on the sphere," Astrophys. J. **622** (2005), 759-771 doi:10.1086/427976 [arXiv:astro-ph/0409513 [astro-ph]].

[9] H. P. Bretz, M. Erdmann, P. Schiffer, D. Walz and T. Winchen, "PARSEC: A Parametrized Simulation Engine for Ultra-High Energy Cosmic Ray Protons," Astropart. Phys. **54** (2014), 110-117 doi:10.1016/j.astropartphys.2013.12.002 [arXiv:1302.3761 [astro-ph.HE]].

[10] R. Jansson and G. R. Farrar, "The Galactic Magnetic Field," Astrophys. J. Lett. **761** (2012), L11 doi:10.1088/2041-8205/761/1/L11 [arXiv:1210.7820 [astro-ph.GA]].

[11] J. Jasche, A. van Vliet and J. P. Rachen, "Targeting Earth: CRPropa learns to aim," PoS **ICRC2019** (2020), 447 doi:10.22323/1.358.0447 [arXiv:1911.05048 [astro-ph.HE]].

[12] R. Alves Batista, J. Becker Tjus, J. Dörner, A. Dundovic, B. Eichmann, A. Frie, C. Heiter, M. R. Hoerbe, K. H. Kampert and L. Merten, *et al.* "CRPropa 3.2 — an advanced framework for high-energy particle propagation in extragalactic and galactic spaces," JCAP **09** (2022), 035 doi:10.1088/1475-7516/2022/09/035 [arXiv:2208.00107 [astro-ph.HE]].

# Estimation and reduction of the biases by the galactic magnetic field on the UHECR correlation studies

*Ryo* Higuchi[1,*], *Takashi* Sako[2], *Toshihiro* Fujii[3], *Kazumasa* Kawata[2], and *Eiji* Kido[1] for the Telescope Array collaboration

[1]Astrophysical Big Bang Laboratory, RIKEN, 2-1 Hirosawa, Wako, Saitama 351-0198, Japan
[2]Institute for Cosmic Ray Research, The University of Tokyo, 5-1-5 Kashiwanoha, Kashiwa, Chiba 277-8582, Japan
[3]Osaka Metropolitan University, 3-3-138 Sugimoto Sumiyoshi-ku, Osaka-shi, 558-8585, Japan

**Abstract.** The correlation studies between ultrahigh-energy cosmic ray (UHECR) anisotropy and source candidates are keys to understand the origin of UHECRs. Especially, studies with consideration of source models, magnetic fields, and mass compositions are necessary. We estimated the biases caused by the coherent deflection due to the galactic magnetic field (GMF) in the previous maximum-likelihood analysis for searching the UHECR sources (Aab et al. 2018, Abbasi et al. 2018). In our work with simulated mock datasets, we find that the anisotropic fraction $f_{\rm ani}$ is estimated systematically lower than the true value when we ignore the effect caused by the GMF. We also develop the maximum-likelihood method which includes the GMF model and confirm that the estimated parameters would be improved. We apply the method to the observational datasets obtained from the Telescope Array and Auger experiments.

## 1 Introduction

Understanding the origin of ultrahigh-energy cosmic rays (UHECRs) is one of the important problems in physics today. Due to large statistics of the Telescope Array experiment [1] and Auger experiment [2], intermediate-scale anisotropies of UHECRs are found. These UHECR anisotropies are suggested to correlate with cosmic ray sources. Many studies have investigated the correlation between the UHECR anisotropy and source candidates [3–6]. In [5], the Auger collaboration reported the maximum likelihood analysis between UHECRs in south-sky and cosmic ray (CR) flux nearby starburst galaxies (SBGs). The current dataset obtained by the TA experiment does not contradict to the Auger results [6]. On the other hand, however, the UHECR anisotropy is determined by the combination of the source distribution, mass composition in the source, and magnetic fields. It is important to conduct a correlation study with taking into account of magnetic fields and mass compositions.

In [7], we investigated the biases caused by the Galactic magnetic field (GMF) with simulated UHECR datasets (mock datasets). We showed the contribution of SBGs is estimated to be lower with ignorance of GMF. We also suggested that we can reduce the bias when we conduct the maximum-likelihood analysis with consideration of the deflections in the GMF model.

In this study, we apply the method we proposed in [7] to the currently accessible datasets from the TA and Auger experiments. We discuss the parameter space of the SBG source model.

*e-mail: ryo.higuchi@riken.jp

## 2 Method

### 2.1 Models

In this study, we assume three models for the source, magnetic fields, and mass composition. For a source model, we adapt the SBG source model suggested by [6]. The SBG source model is based on the 23 nearby SBGs (see Table 1 in [5]). The total CR flux from each SBG $i$ is described as a sum of the von Mises - Fisher function [8] as follows:

$$F_{\rm org}(\mathbf{n}, \theta) = \frac{\sum_i f_i \exp(\mathbf{n}_i \cdot \mathbf{n}/\theta^2)}{\int_{4\pi} \sum_i f_i \exp(\mathbf{n}_i \cdot \mathbf{n}/\theta^2) d\Omega}, \quad (1)$$

where $\mathbf{n}$, $\mathbf{n}_i$, and $\theta$ are direction in the sky, direction of each SBG $i$, and separation angular scale, respectively. The relative flux weight of each SBG $i$ is shown as $f_i$ (see Table 1 in [5] for a value). The separation angular scale $\theta$ presents a Gaussian-like scattering around the SBG. The anisotropic fraction parameter $f_{\rm ani}$ is introduced as a fraction of UHECR events which can be explained by the assumed source model. The other component $1 - f_{\rm ani}$ is assumed as isotropic background:

$$F_{\rm norm}(\mathbf{n}, f_{\rm ani}, \theta) = f_{\rm ani} F_{\rm org}(\mathbf{n}, \theta) + (1 - f_{\rm ani}) F_{\rm iso}, \quad (2)$$

where $F_{\rm iso} = 1/4\pi$ is isotropic CR flux. With the maximum-likelihood analysis in Section 2.2, the parameter set $(f_{\rm ani}, \theta)$ can be estimated.

For a Galactic magnetic field model, we assume the regular component of Jansson & Farrar 2012 (JF12) model [9, 10]. We assume the separation angular scale $\theta$ presents the extragalactic magnetic fields and random magnetic fields. As a mass-composition model, we assume the Heinze & Fedynitch 2019 (HF19) model [11].

### 2.2 Maximum-likelihood analysis

We describe the maximum-likelihood analysis in [5, 6]. The test statistics $TS$ is determined with a likelihood ratio between the SBG source model and the isotropic background:

$$TS = 2\ln(L(F_{\rm norm})/L(F_{\rm iso})). \quad (3)$$

The likelihood of each model is written as:

$$L(F) = \prod_{\rm CR} \frac{F_{\rm norm}(\mathbf{n}_{\rm CR})\omega(\mathbf{n}_{\rm CR})}{\int_{4\pi} F_{\rm norm}(\mathbf{n})\omega(\mathbf{n})d\Omega}, \quad (4)$$

where $\mathbf{n}_{\rm CR}$ and $\omega(\mathbf{n})$ indicate the arrival direction of each CR event, and the sky coverage of the experiments, respectively. For the sky coverage $\omega(\mathbf{n})$, we adapt the equations in [12] (see also [7]).

### 2.3 Maximum-likelihood analysis with GMF model

The CR flux introduced in equations 1–2 does not reflect the coherent CR deflection by the GMF structure. In this study, we calculate the CR flux after passing through the GMF as follows: First, we define the corresponding function $A_{\rm BT}(\mathbf{n}, R)$ between arrival directions on earth and the outside the Galaxy. The function is determined based on the CR trajectory in backtracking calculation with CRPropa3 [13]. Note that each trajectory of CR depends on the arrival direction on the earth and its rigidity $R = E/Ze$. Because the CR flux conserves along the CR trajectory, we can rewrite the CR flux $F_{\rm org}(\mathbf{n}, \theta)$ to that on the earth $F'_{\rm earth}(\mathbf{n}_{\rm CR}, f_{\rm ani}, \theta, R_{\rm CR})$:

$$F'_{\rm earth}(\mathbf{n}_{\rm CR}, f_{\rm ani}, \theta, R_{\rm CR}) = \frac{F_{\rm org}(A_{\rm BT}(\mathbf{n}_{\rm CR}, R_{\rm CR}), \theta)}{\int_{4\pi} F_{\rm org}(A_{\rm BT}(\mathbf{n}, R), \theta)d\Omega}. \quad (5)$$

With a calculation of $L(F)$ from CR flux on the earth $F'_{\rm earth}(\mathbf{n}_{\rm CR}, f_{\rm ani}, \theta, R_{\rm CR})$, one can estimate the parameter set $(f_{\rm ani}, \theta)$ with reduction of GMF bias.

### 2.4 Maximum-likelihood analysis with GMF model for the observational datasets

In the method introduced in Section 2.3, we need to know the mass $A$ of each CR event. It is difficult to analyze with current UHECR datasets knowing all the event-by-event mass $A$. In this study, we assume mass probability function $p^A(E)$ and calculate CR flux $F'_{\rm earth}$ weighted with the probability. The mass probability function $p^A(E)$ is described with an CR energy spectrum of each mass $A$ ($J_A$),

$$p^A(E_{\rm CR}, Z^A_{\rm CR}) = \frac{J_A(E_{\rm CR})}{\sum_A J_A(E_{\rm CR})}. \quad (6)$$

Regarding $J_A$, we follow the formulas and best-fit parameters in [11] (see also [7]). We include $p^A(E)$ into equation 2 and normalize the CR flux as follows:

$$F_{\rm norm}(\mathbf{n}_{\rm CR}, f_{\rm ani}, \theta, R^A_{\rm CR}) = f_{\rm ani} \sum_A p^A(E_{\rm CR}, Z^A_{\rm CR}) F'_{\rm earth}(\mathbf{n}_{\rm CR}, \theta, R^A_{\rm CR}). + (1 - f_{\rm ani})F_{\rm iso}, \quad (7)$$

We substitute $F_{\rm norm}$ in equation 7 into equations 4 and 3, and calculate $TS$.

## 3 Dataset

We apply the maximum-likelihood analysis in Section 2.4 to observational datasets obtained by TA and Auger experiments (TA-11yr dataset and Auger 2015 dataset). For Auger 2015 dataset, we refer [2]. TA-11yr dataset (Auger 2015 dataset) is taken in 2008 May – 2019 May (2004 January– 2014 March) and includes 279 (225) events above 43 (52) EeV. We follow the energy thresholds for the TA-11yr dataset and Auger 2015 dataset in [6] and [2], respectively.

We also apply the mock datasets calculated from the convolution of the SBG source model, JF12 model, and HF19 model. See [7] for details.

## 4 Results and discussions

### 4.1 Estimation and reduction of the GMF bias

In this section, we briefly summarize the results of the simulated mock datasets in [7]. The left panel in Figure 1 shows an example of the best-fit parameters for 1000 mock datasets in the same manner as previous studies [5, 6]. We see the difference between true parameters $(f^{\rm true}_{\rm ani}, \theta^{\rm true})$ (which are used to generate mock datasets) and estimated parameters. Each best-fit parameter set is estimated to maximize $TS$ value with ignorance of coherent deflection by GMF (Section 2.2). In many true parameters $(f^{\rm true}_{\rm ani}, \theta^{\rm true})$, estimated anisotropic fraction $f_{\rm ani}$ in north-/south- sky are smaller than the true one.

The right panel in Figure 1 shows the same ones in the left panel, but for $TS$ value calculated with consideration for GMF (Section 2.3). When the separation angular scale $\theta$ is small, parameter set $(f_{\rm ani}, \theta)$ is well estimated in $1\sigma$ uncertainty.

### 4.2 Application for TA and Auger datasets

We show $TS$ distribution for observational datasets in Figure 2. The best-fit parameters for TA-11yr & Auger 2015 datasets are estimated to be $(f^{\rm TA}_{\rm ani}, \theta^{\rm TA}) = (3\%, 1\ {\rm deg})$ and $(f^{\rm Auger}_{\rm ani}, \theta^{\rm Auger}) = (5\%, 34\ {\rm deg})$, respectively. Although $TS$-value is small to determine the source model, we search the parameter space $(f_{\rm ani}, \theta)$ in which mock datasets can reproduce the $TS$. Regarding the mock datasets, we assume the same number of events as observational datasets for each dataset. The model assumptions for the generation of the mock dataset are the same as in Section 4.1. We show the parameter space $(f_{\rm ani}, \theta)$ which is consistent with $TS$ from observational datasets in Figure 3. Figure 3 suggests that we can exclude the parameter space $\theta < 12$ (37) deg $\theta < 20$ (45) deg at $f_{\rm ani} = 20\,\%$ (100 %) from the TA 11-yr & Auger 2015 datasets with 95% C.L..

## 5 Summary

Deflections of UHECRs by magnetic fields and mass composition of UHECRs affect the correlation study between UHCR anisotropies and source candidates. Especially

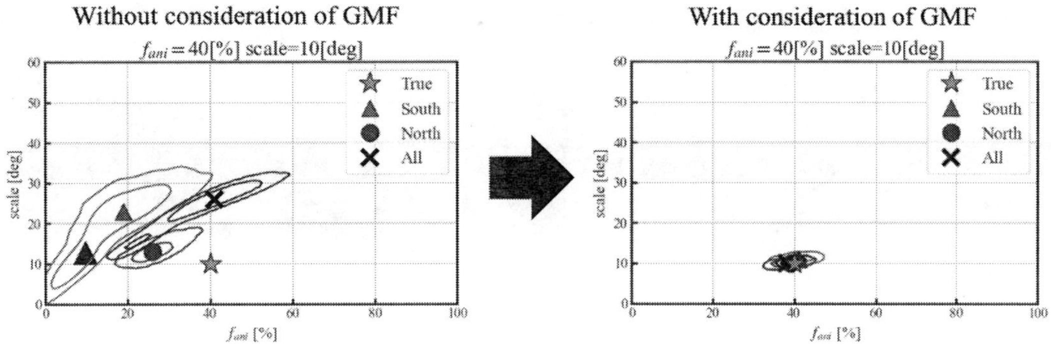


**Figure 1.** Distributions of the best-fit parameters for the 1000 mock event datasets in case of $(f_{\text{ani}}^{\text{true}}, \theta^{\text{true}}) = (40\,\%, 10\text{ deg})$. The left panels show the results with ignorance of coherent deflection by GMF (Section 2.2). The right panels show the same but with consideration for GMF (Section 2.3). In both panels, true parameters $(f_{\text{ani}}^{\text{true}}, \theta^{\text{true}})$ are shown by the grey stars. The black, blue, and red contours indicate the 68 % and 95 % tile containment for all, north and south-sky datasets, respectively. The black cross, blue circle, and red triangle show the most frequent values for all, north and south-sky datasets, respectively. The distributions of best-fit parameters are smoothed with a kernel-Gaussian distribution. In the left panel, best-fit parameter $(f_{\text{ani}}^{\text{Auger}}, \theta^{\text{Auger}}) = (9.7\,\%, 12.9\text{ deg})$ in [5] is shown as a black triangle.

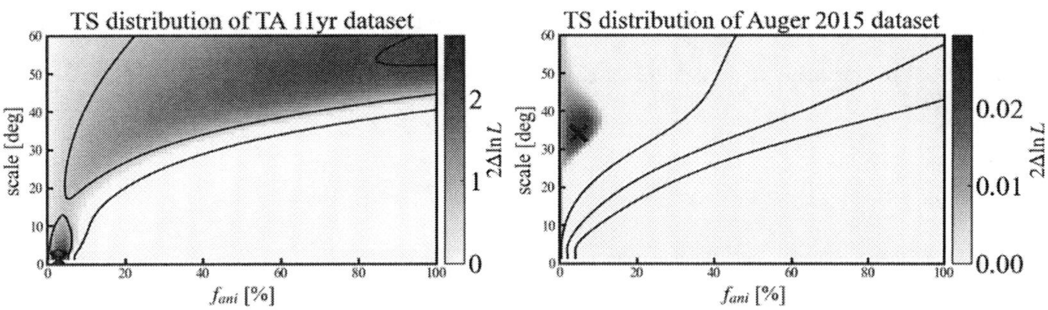


**Figure 2.** The $TS$ distribution (Section 2.4) with reduction of the GMF bias for the TA-11yr dataset (left) and the Auger 2015 dataset (right). The black cross indicates the best-fit parameter set $(f_{\text{ani}}, \theta)$.

when we ignore the coherent deflections by GMF, the contribution from nearby SBGs should be estimated lower [7]. Thus we need the analysis with the assumption of both magnetic fields & mass composition models. In our work [7], we propose the maximum-likelihood analysis technique with consideration of these models. In this study, we apply the technique to current available observational datasets obtained from the TA and Auger experiments. Although the $TS$-value is small to determine the source mode for current observational datasets, we search the parameter space $(f_{\text{ani}}, \theta)$ which mock datasets can reproduce the $TS$. It is suggested that the parameter space $\theta < 12$ (37) deg $\theta < 20$ (45) deg at $f_{\text{ani}} = 20\,\%$ (100 %) are excluded from the TA 11-yr & Auger 2015 datasets.

## References

[1] H. Kawai, S. Yoshida, H. Yoshii, K. Tanaka, F. Cohen, M. Fukushima, N. Hayashida, K. Hiyama, D. Ikeda, E. Kido et al., Nuclear Physics B Proceedings Supplements **175**, 221 (2008)

[2] A. Aab, P. Abreu, M. Aglietta, E.J. Ahn, I.A. Samarai, J.N. Albert, I.F. Albuquerque, I. Allekotte, J. Allen, P. Allison et al., Nuclear Instruments and Methods in Physics Research, Section A: Accelerators, Spectrometers, Detectors and Associated Equipment **798**, 172 (2015)

[3] P. Abreu, M. Aglietta, C. Aguirre, Science **318**, 938 (2007)

[4] P. Abreu, M. Aglietta, E.J. Ahn, D. Allard, I. Allekotte, J. Allen, J.A. Castillo, J. Alvarez-Muñiz, M. Ambrosio, A. Aminaei et al., Astroparticle Physics **34**, 314 (2010)

[5] A. Aab, P. Abreu, M. Aglietta, I.F.M. Albuquerque, I. Allekotte, A. Almela, J.A. Castillo, J. Alvarez-Muñiz, G.A. Anastasi, L. Anchordoqui et al., The Astrophysical Journal **853**, L29 (2018)

[6] R.U. Abbasi, M. Abe, T. Abu-Zayyad, M. Allen, R. Azuma, E. Barcikowski, J.W. Belz, D.R. Bergman, S.A. Blake, R. Cady et al., The Astrophysical Journal **867**, L27 (2018)

https://doi.org/10.1051/epjconf/202328303011

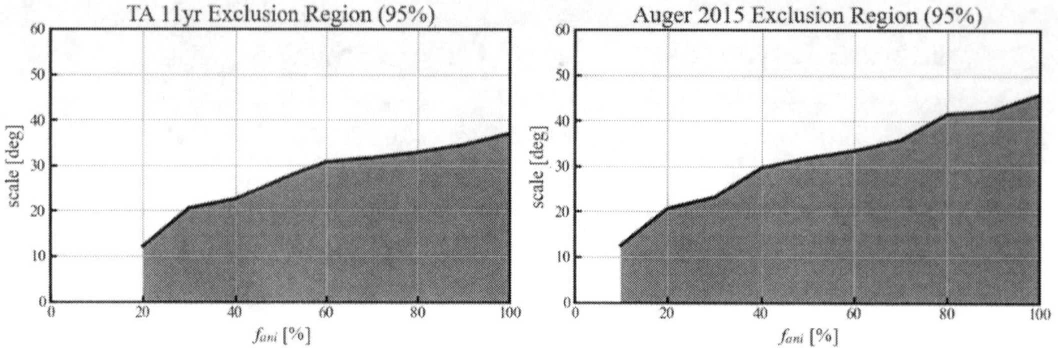


**Figure 3.** Exclusion area of the TA-11yr dataset (left) and the Auger 2015 dataset (right). The gray areas indicate the exclusion region with 95% C.L.

[7] R. Higuchi, T. Sako, T. Fujii, K. Kawata, E. Kido, arXiv e-prints arXiv:2209.00305 (2022), 2209.00305

[8] R. Fisher, *Dispersion on a sphere* (1953), https://www.jstor.org/stable/99186

[9] R. Jansson, G.R. Farrar, Astrophysical Journal **757** (2012)

[10] R. Jansson, G.R. Farrar, Astrophysical Journal Letters **761**, 1 (2012)

[11] J. Heinze, A. Fedynitch, The Astrophysical Journal **873**, 88 (2019)

[12] P. Sommers, Astroparticle Physics **14**, 271 (2001)

[13] R.A. Batista, A. Dundovic, M. Erdmann, K.H. Kampert, D. Kuempel, G. Müller, G. Sigl, A.V. Vliet, D. Walz, T. Winchen, Journal of Cosmology and Astroparticle Physics **2016** (2016)

# Examination of Xmax anisotropy for the next generation Ultra-high energy cosmic rays observations

*Ryosuke* Saito[1,*], *Takayuki* Tomida[1], and *Daisuke* Ikeda[2]

[1]Shinshu University, Nagano, Nagano, Japan
[2]Kanagawa University, Yokohama, Kanagawa, Japan

**Abstract.** The estimation of mass composition ultra-high energy cosmic rays is essential to understand their origin and generation mechanism. Recent experiments are expected to discover anisotropy of mass composition in ultra-high energy cosmic rays. Anisotropy of mass composition using Xmax is currently being performed, but the problem is that the statistics of cosmic rays with measured Xmax is very small. For the next generation experiments, it is important to know how much statistics can be accumulated to find Xmax anisotropy. Therefore, in this analysis, we examine Xmax anisotropic search in ultra-high energy cosmic rays by simulating a cosmic ray air showers under various conditions such as energy spectrum and mass composition.

## 1 introduction

Ultra-high energy cosmic rays(UHECR) have been observed since the Akeno Air Shower Observatory[1] started operation in 1979, and the Fly's Eye experiment[2] in 1981, which established a method for measuring air showers with an atmospheric fluorescence telescope. Since then, various experiments such as the Telescope Array (TA)[3] and the Pierre Auger Observatory[4] have been conducted to observe and study cosmic rays. Cosmic rays have been observed over a wide energy range from $10^8$eV to $10^{20}$eV. In particular, cosmic rays with energies higher than $10^{18}$eV are called ultra-high energy cosmic rays. The question of where and how the ultra-high energy cosmic rays with energies as high as $10^{20}$eV are produced is one of the problems in cosmic ray physics. To solve this question, observational experiments on cosmic rays have been conducted to measure the flux, mass composition, and arrival direction of cosmic rays. Recent experiments have shown signs of anisotropy in the vicinity of HotSpot, WarmSpot, and Supergalactic plane[5]. In addition to energy and arrival direction, mass composition anisotropy analyses using Xmax is now being performed and is expected to be a promising method. However, in the ultra-high energy region, there is a problem that the statistics are very small due to the small arrival frequency and limited duty cycle of the fluorescence technique that can directly measure Xmax. To increase the statistics, it is expected that the next generation of cosmic ray observational experiments will be performed a much larger scale. Therefore, it is important for the planning of the next generation of experiments to investigate what we can learn about the future physics of cosmic rays with a certain amount of statistics.Here, we have focused on the anisotropy analysis of mass composition. In this work, we investigate how much statistics is required for Xmax anisotropy to appear in simulations under different conditions of mass composition and energy spectrum.

*e-mail: 21w2032k@shinshu-u.ac.jp

## 2 Data Set

The following is a detailed description of the data set used for the simulations. The longitudinal evolution of air showers was generated by a Monte Carlo simulation using Corsika.[6] A detailed description of the data obtained from the simulations is given as follows.

- Primary composition : Proton, He, CNO(1:1:1), Si, Iron
- Interaction model : EPOS-LHC
- Log(E/eV) : 19.7-19.8, 19.8-19.9, 19.9-20.0
  (Generated continuously inside each energy bin)
- Zenith angle : 0 - 65 degrees
- Thinning factor : -4

For each combination of energy and mass composition, 1000 air shower events were generated using EPOS-LHC as hadronic interaction model.

## 3 Analysis

In this work, we developed a new analytical method. The outline of this analysis is to compare the Xmax distributions of two regions by randomly selecting assumed regions, energies, and mass compositions according to probabilities, and to verify how much statistics is required for the Xmax distributions of the two regions to be significantly different.Here, two regions are assumed : one is the TA HotSpot region[? ][? ] and the other is the Background region. Including assumed energy and mass compositions, the analysis procedure is described as follows:

 https://doi.org/10.1051/epjconf/202328303012

1. Select two regions (HotSpot, Background) to compare with probability as described in Subsection 3.2
2. Randomly select to follow the assumed energy spectrum and mass composition ratios in the selected region
3. Perform KS test on the two regions for a total of every 10 events.
4. Check the p-value obtained by the KS test and the change in the number of events.

### 3.1 assumption of energy and composition

The parameters assumed in the simulation are the region, energy, and mass composition. Since we have already assumed the region, we must now assume the energy spectrum and mass composition ratios. In the following, we will explain the details of the assumptions for the energy spectrum and the mass composition ratios.

#### *3.1.1 energy*

Figure 1[7] is the result of the energy spectrum obtained by TA experiment. In this analysis, the energy spectrum in the HotSpot region is the same as the energy spectrum in the Background region.Then, the event flux ratio for each energy is calculated based on this spectrum. As seen in Figure1, the spectrum consists of power spectra for each energy region, and for energies above $3 \times 10^{17} eV$, the function is expressed as in Equation1.

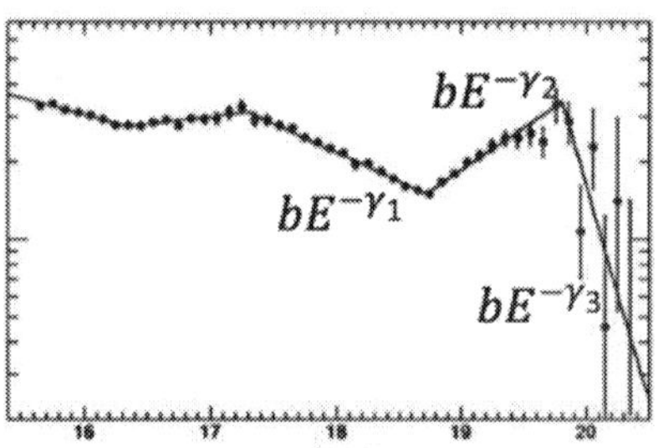


Figure 1: The energy spectrum obtained from the TA experiment[7]

$$
J(E) = \begin{cases} E^{-\gamma_1}(E < E_{ankle}) \\ E^{-\gamma_2}(E_{ankle} < E < E_{break}) \\ E^{-\gamma_3}(E < E_{ankle}) \end{cases} \quad (1)
$$

Based on the above equation and parameters, the energy region from $10^{19.7}$ eV to $10^{20.0}$ eV is divided into three parts and the flux ratio in each region is calculated (Table 2).

Table 1: Values of the broken power-law fit parameters for the TA energy spectrum above $3 \times 10^{17}$eV [7]

| TA spectrum parameters | | | | |
|---|---|---|---|---|
| $\gamma_1$ | $\gamma_2$ | $\gamma_3$ | $E_{ankle}$[EeV] | $E_{break}$[EeV] |
| $3.226 \pm 0.007$ | $2.66 \pm 0.02$ | $4.7 \pm 0.6$ | $5.2 \pm 0.2$ | $60 \pm 7$ |

Table 2: Ratios of fluxes in three energy regions

| | Ratios of fluxes |
|---|---|
| $10^{19.7\sim19.8}$[eV] | 0.65 |
| $10^{19.8\sim19.9}$[eV] | 0.26 |
| $10^{19.9\sim20.0}$[eV] | 0.09 |

#### *3.1.2 compositon*

Figure2 shows the energy densities together with the fit for various mass compositions obtained from the Auger experiment. In the present analysis, we calculate the mass composition ratios in each energy region using Figure 2[8]. The Background mass composition ratios are taken from Figure 2. Table 3 lists the hotspot events for the first five-year period. The number of observed is 19, and the number of expected is 4.49, meaning that 14.51 more events are observed than expected. In other words, more events were observed in the HotSpot region than expected, so excess events must be taken into account.When calculating composition of HotSpot region, 4.49 events are calculated with mixed composition 14.51 events are calculated with pure proton.

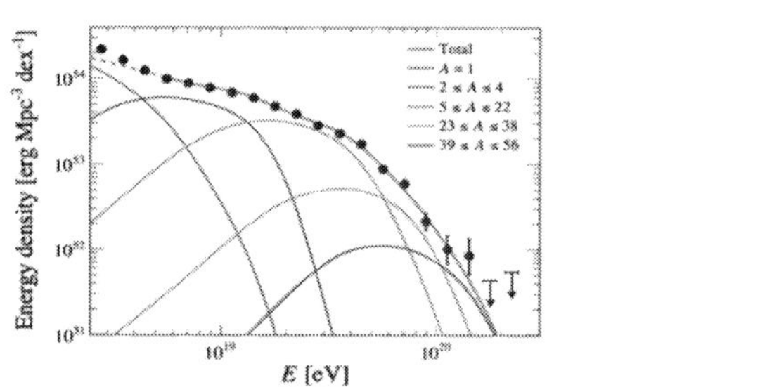


Figure 2: Energy densities with the fit for various mass compositions obtained from the Auger experiment[8]

The calculated mass composition ratios are shown in Table 4.

### 3.2 Comparison of Xmax distributions between Background and HotSpot

Assuming the energy spectra and composition ratios inside and outside the HotSpot, we randomly extract Xmax

Table 3: Numbers of events in the Background and HotSpot regions in the first 5 years of the TA data.

| Number of observed | number of expectation as background | number of excess |
|---|---|---|
| 19 | 4.49 | 14.51 |

Table 4: Composition ratios in three energy regions for the Background and HotSpot

| | Proton : CNO : Si : Iron (Background) |
|---|---|
| $10^{19.7\sim19.8}[eV]$ | 0.0 : 4.4 : 4.4 : 1.2 |
| $10^{19.8\sim19.9}[eV]$ | 0.0 : 2.6 : 5.4 : 2.0 |
| $10^{19.9\sim20.0}[eV]$ | 0.0 : 1.6 : 5.2 : 3.2 |
| | Proton : CNO : Si : Iron (HotSpot) |
| $10^{19.7\sim19.8}[eV]$ | 7.6 : 1.1 : 1.1 : 0.2 |
| $10^{19.8\sim19.9}[eV]$ | 7.6 : 0.6 : 1.3 : 0.5 |
| $10^{19.9\sim20.0}[eV]$ | 7.6 : 0.4 : 1.2 : 0.8 |

from the data set according to the determined probabilities. In this analysis, we focus on the number of events where Xmax anisotropy appears. To examine how much the Xmax distributions in the two regions are statistically significantly separated, we need to compare the Xmax distributions between the HotSpot region and the Background region for each event obtained according to the assumed energy spectrum and mass composition mixing ratios. The observation of the TA experiment shows that there are 72 events above 57 EeV in the last 5 years(2008 2013), of which 19 events are in the HotSpot and 53 events are in the Background. In the following we explain the process of retrieving events and comparing their distributions.

First, two regions (HotSpot or Background), three energy regions, and five mass compositions are randomly selected with probability according to the assumption, and Xmax is randomly obtained from the data set. Figure3 shows the Xmax distributions when a total of 50 events are obtained in the two regions, and the p-value is calculated using KS test every 10 events in total in the Xmax distributions in the HotSpot and Background regions. The test is continued by obtaining events until the p-value obtained from the test exceeds 3 σ. Figure4 is the result. As the number of events increases, the p-value of the test result becomes smaller, and when the number of events exceeds a certain number, the p-value falls below the confidence level, indicating that the Xmax distibutions of the two regions is significantly separated. In other words, the statistic at the point where $3\sigma$ is exceeded is the statistic necessary for the distributions in the two regions to be significantly separated. However, it is also possible that the results shown in Figure4 are obtained by chance, since the events in this simulation are obtained randomly according to the probability. Therefore, it is necessary to consider a new evaluation method. Next, we describe the new method.

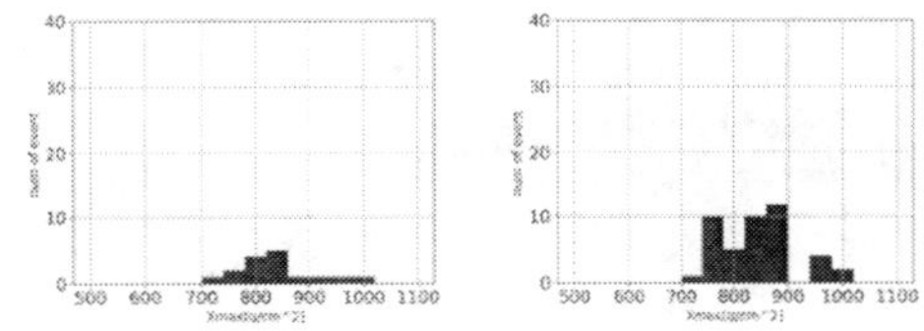

Figure 3: Xmax distributions at a total of 50 events. Left panel shows a distribution of HotSpot .Right panel shows a distribution of Background

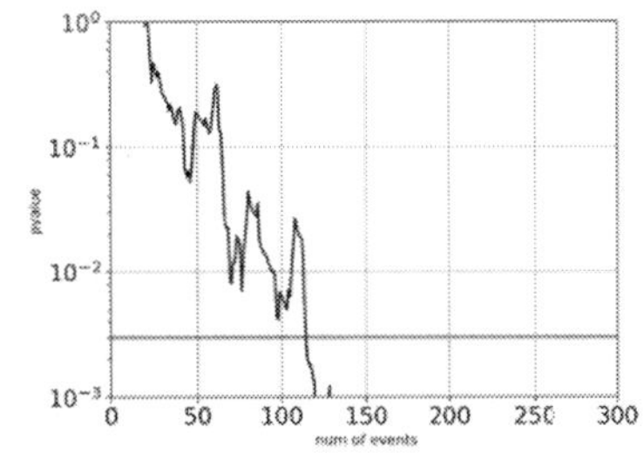


Figure 4: P-value from the comparison between the HotSpot and Background Xmax distributions every 10 events

### 3.3 Number of events required to significantly separate the Xmax distributions of the two regions

In this section, to account for the possibility that the results in Figure4 could have occurred by chance, the sequence of acquisition events is repeated 10,000 times until the p-value obtained by the test exceeds $3\sigma$. This is equivalent to finding the number of Xmax observations required for a sufficient difference in the Xmax distribution to appear in 10,000 experiments. Then, from the results of 10,000 times, we estimate the number of observations required to detect a sufficient difference in the Xmax distribution of $3\sigma$(99.7%).

Figure5 shows the results of 10,000 times, and all 10,000 results are different because the events are randomly obtained and compared by probability. These results are used for evaluation.

As the number of events increases, the p-value decreases and exceeds $3\sigma$ at a certain number of events. The number of experiments with p-values exceeding 3 σ as the number of events also increases. Figure 6 shows the distribution of p-value at 50 events obtained from the simulation. Figure 7 show the distributions for 100 and 150 events, respectively.

Since the number of experiments exceeding $3\sigma$ is expected to increase as the number of events increases, the

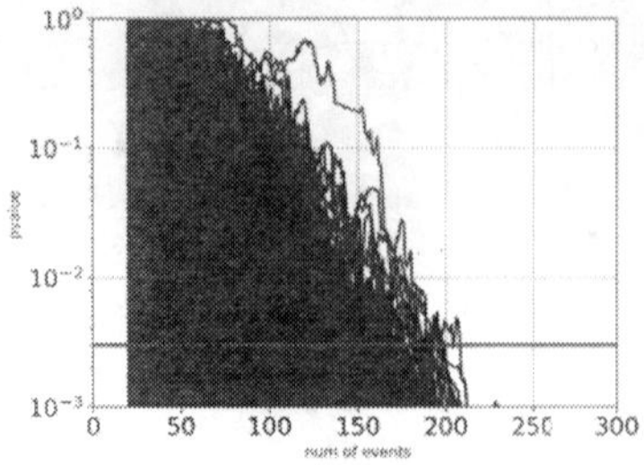


Figure 5: p-value obtained by comparing the HotSpot and Background Xmax distributions for every 10 events for 10,000 results

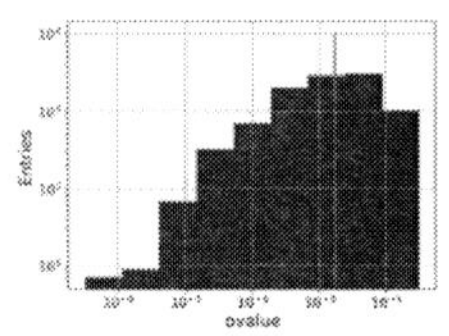


Figure 6: The distributions of p-value at 50 events

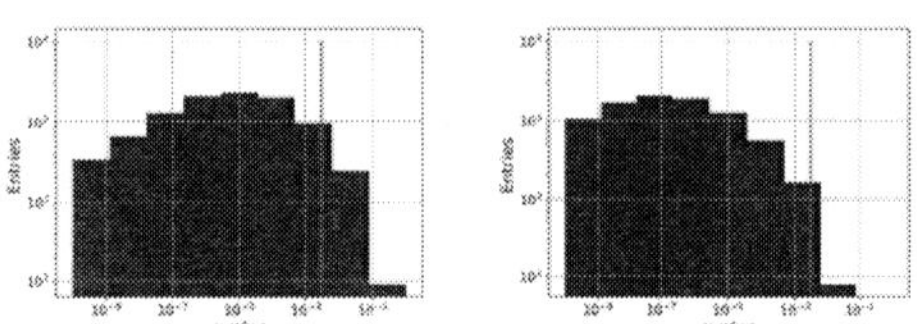


Figure 7: The distributions of p-value. Left figure is for 120 events. Right figure is for 150 event.

number of experiments exceeding $3\sigma$ for a given number of events is 9970 (99.7 %) out of 10000. In this case, the number of events is assumed to be the statistics required to find the Xmax anisotropy. Figure 8 counts the number of experiments exceeding $3\sigma$ for each number of events. Under the conditions of EPOS-LHC for the hadron interactions model and proton for excess particles in the HotSpot, the Xmax distributions of the HotSpot and Background are significantly separated at about 165 events.

In the previous analysis, the excess particles in the HotSpot were protons. Moreover, we analyzed when the excess particles were He, CNO, Si and Iron particles, and the energy range was changed. we calculated some energy range with Figure1 and equation1. Figure 9 shows the results. Table 5 summarizes the results from Figure 9. The statistics required to observe the Xmax anisotropy are, in decreasing order, Proton, He, Fe, CNO, and Si. Since there is no proton in the Background mass composition ratio, the excess particle in the HotSpot region are considered to separate the fastest in the case of proton.

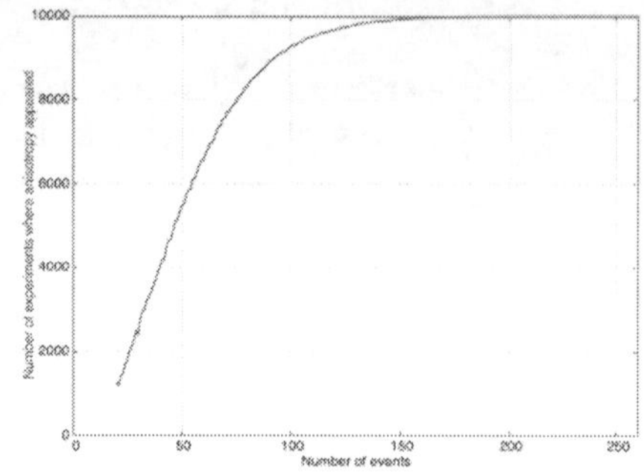


Figure 8: Result of the number of experiments when anisotropy appears as a function of the number of events

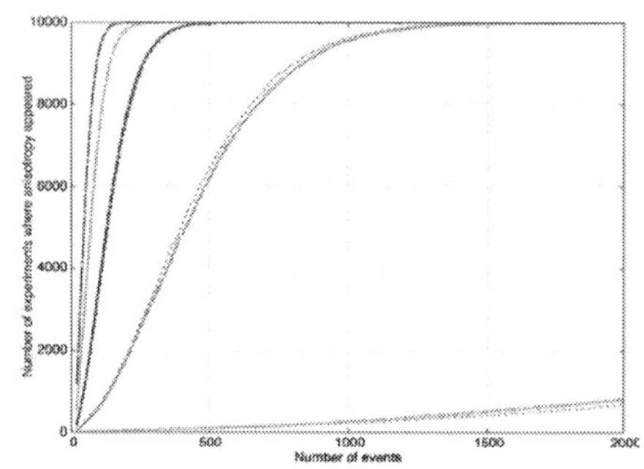


Figure 9: The number of experiments when anisotropy appears as a function of the number of events.The solid line shows the median value of the energy spectrum parameters, and plotted line shows Upper and lower limits of energy spectrum error bars.Proton, He, CNO, Si, Iron are indicated as Red, Orange, Purple, Cyan, Blue.

Table 5: The number of event which required to find Xmax anisotropy.The solid line shows the median value of the energy spectrum parameters, and plotted line shows Upper and lower limits of energy spectrum error bars.Proton, He, CNO, Si, Iron are indicated as Red, Orange, Purple, Cyan, Blue.

| | EPOS-LHC |
|---|---|
| Proton | $166^{+1}_{-5}$ |
| He | $251^{+5}_{-6}$ |
| CNO | $1469^{+20}_{-40}$ |
| Si | $10000_{andmore}$ |
| Iron | $481^{+3}_{-11}$ |

## 4 Conclusion

Recent observations of cosmic rays with TA experiment show signs of anisotropy due to energy in the HotSpot region, WarmSpot, and supergalactic plane. Along with energy and arrival direction analysis, anisotropy analysis of mass composition using Xmax is also expected, but the statistics are very limited due to the duty cycle of the telescopes that can detect Xmax. In this examine Xmax anisotropic search in ultra-high energy cosmic rays by simulating cosmic ray under various conditions, energy spectrum and mass composition. And we estimate statistics required to find out the difference on the Xmax distributions are evaluated by the simulation with some assumptions.

## Acknowledgements

The Telescope Array experiment is supported by the Japan Society for the Promotion of Science(JSPS) through Grants-in-Aid for Priority Area 431, for Specially Promoted Research JP21000002, for Scientific Research (S) JP19104006, for Specially Promoted Research JP15H05693, for Scientific Research (S) JP15H05741 and for Young Scientists (A) JPH26707011; by the joint research program of the Institute for Cosmic Ray Research (ICRR), The University of Tokyo; by the U.S. National Science Foundation awards PHY-0601915, PHY-1404495, PHY-1404502, and PHY-1607727; by the National Research Foundation of Korea (2016R1A2B4014967, 2016R1A5A1013277, 2017K1A4A3015188, 2017R1A2A1A05071429) ; by the Russian Academy of Sciences, RFBR grant 16-02-00962a (INR), IISN project No. 4.4502.13, and Belgian Science Policy under IUAP VII/37 (ULB). The foundations of Dr. Ezekiel R. and Edna Wattis Dumke, Willard L. Eccles, and George S. and Dolores Doré Eccles all helped with generous donations. The State of Utah supported the project through its Economic Development Board, and the University of Utah through the Office of the Vice President for Research. The experimental site became available through the cooperation of the Utah School and Institutional Trust Lands Administration (SITLA), U.S. Bureau of Land Management (BLM), and the U.S. Air Force. We appreciate the assistance of the State of Utah and Fillmore offices of the BLM in crafting the Plan of Development for the site. Patrick Shea assisted the collaboration with valuable advice on a variety of topics. The people and the officials of Millard County, Utah have been a source of steadfast and warm support for our work which we greatly appreciate. We are indebted to the Millard County Road Department for their efforts to maintain and clear the roads which get us to our sites. We gratefully acknowledge the contribution from the technical staffs of our home institutions. An allocation of computer time from the Center for High Performance Computing at the University of Utah is gratefully acknowledged.

## References

[1] N. Chiba et al., Nuclear Instruments and Methods in Physics Research
[2] D. J. Bird et al., The Astrophysical Journal, 424:491-502(1994)
[3] H. Kawai et al., Nuclear Physics B, 175-176(2008)
[4] A. Aab el al. Nuclear Instruments and Methods in Physics Research A(2015)
[5] A. Aab et al., ApJ 790, L21
[6] D.Heck et al., Forschungszentrum Karlsruhe GmbH, Karlsruhe(1998)
[7] R. U. Abbasi et al, 2014, ApJL, 790, L21
[8] R. U. Abbasi et al, 2018, ApJ, 862, 91
[9] D. Ivanov, Proc. 34th ICRC 2015 (The Hague, The Netherlands), PoS (ICRC2015) 349.
[10] A. Aab el al. (Pierre Auger Collaboration),, Phys. Rev. Lett., **125**, 121106(2020).

# Ultra High Energy Cosmic Ray Source Models: Successes, Challenges and General Predictions

*Noémie* Globus[1,*] and *Roger* Blandford[2,**]

[1]Department of Astronomy and Astrophysics, University of California, Santa Cruz, CA 95064, USA
[2]Kavli Institute for Particle Astrophysics and Cosmology (KIPAC), Stanford University, Stanford, CA 94305, USA

**Abstract.** Understanding the acceleration of Ultra High Energy Cosmic Rays is one of the great challenges of contemporary astrophysics. In this short review, we summarize the general observational constraints on their composition, spectrum and isotropy which indicate that nuclei heavier than single protons dominate their spectra above $\sim 5$ EeV, that they are strongly suppressed above energies $\sim 50$ EeV, and that the only significant departure from isotropy is a dipole. Constraints based upon photopion and photodisintegration losses allow their ranges and luminosity density to be estimated. Three general classes of source model are discussed - magnetospheric models (including neutron stars and black holes), jet models (including Gamma Ray Bursts, Active Galactic Nuclei and Tidal Disruption Events) and Diffusive Shock Acceleration models (involving large accretion shocks around rich clusters of galaxies). The value of constructing larger and more capable arrays to measure individual masses at the highest energies and probably identifying their sources is emphasized.

## 1 Introduction

Understanding the acceleration of atomic nuclei to macroscopic energies $\gtrsim 50$ J is one of the great challenges of modern astrophysics. These Ultra High Energy Cosmic Ray (UHECR) accelerators operate undercover because the Universe is filled with electromagnetic and hydromagnetic waves that scatter and attenuate the cosmic radiation making direct source identification nearly impossible. On the other hand, these interactions lie at the heart of cosmic-ray science because they are intimately involved in most acceleration models. They also lead to the production of secondary messengers – $\gamma$-rays and neutrinos – signatures of leptonic and hadronic processes at play in cosmic accelerators. The modeling of these interaction processes connects particle physics to multimessenger astronomy which can reveal the UHECR sources indirectly. $\gamma$-rays and X-rays can be produced by both hadronic and leptonic processes, and a multimessenger effort is made to detect high energy neutrinos as messengers of hadronic accelerators.

In this brief review, we summarize direct and general observational constraints on UHECR sources (Sec. 2). In Sec. 3, we discuss three general UHECR acceleration sites: magnetospheres, jets and non-relativistic shocks. Magnetospheres are mostly associated with compact objects – spinning neutron stars and black holes – and can plausibly accelerate cosmic rays to the observed energies. The challenge is to explain how they can escape without loss. Jet models for UHECR acceleration are mostly associated with black holes jets associated with Gamma Ray Bursts (GRB), and Active Galactic Nuclei (AGN). Here the challenge is to identify sources within the UHECR horizon, as these sources are rare. Shock models are most challenged by the need to accelerate cosmic rays to the highest energies measured. A hierarchical approach in which the entire observed spectrum originates successively at SuperNova Remnant (SNR) shocks, Galactic Wind Termination Shocks (GWTS), and at Cluster Accretion Shocks (CAS), is outlined. The opportunity for deciding between these, other, possible acceleration sites in the near future is discussed in Sec. 4.

## 2 Summary of Observational Constraints on the origin of UHECR

Using observations made with the Telescope Array (TA) [1], and the Pierre Auger Observatory (PAO) [2], over the past decades, it has been established that nuclei are dominating the spectrum already at energies $\sim 5$ EeV. Both experiments confirm the existence of an "ankle" in the cosmic-ray spectrum at $\sim 5$ EeV. Both experiments agree that the observed spectrum starts to be strongly suppressed at energy $\gtrsim 50$ EeV. This could be due to photopion production and photodisintegration by microwave background photons — the GZK effect [3, 4] — or may represent the maximum energy that can be accelerated in any cosmic source.

Despite many claims of association of UHECR with specific sources or source classes, the only significant departure from complete isotropy of UHECR is a $6.6\sigma$ dipole reported by PAO at energies above 8 EeV[5]. It has been shown that the evolution of the dipole amplitude with energy is expected if the sources are extragalactic, as a result of the evolution of the GZK horizon with energy [6].

*e-mail: noglobus@ucsc.edu
**e-mail: rdb3@stanford.edu

This feature could reflect an anisotropy in the distribution of many sources following the large scale structure [7], local sources such as nearby radio galaxies [8] or the presence of a single dominant source, such as a Galactic transient [9] – which may require strong Galactic magnetic turbulence to not upset the anisotropy limits but cannot be totally ruled out. However, the source location could well be significantly shifted across the sky due to cosmic ray deflection by the poorly understood strong Galactic and extragalactic magnetic fields. Given the uncertainty between the magnetic field models, one cannot reach strong conclusions yet regarding the nature of the sources with the dipole anisotropy [10].

It should be noted that while the dipole appears above 8 EeV, the sky in the 4-8 EeV energy range is compatible with isotropy [5]. This is certainly due to the combined effect of a large GZK horizon and the presence of accelerators operating up to lower maximum rigidity. The "light ankle" reported by the KASCADE Grande experiment at $\sim 0.1$ EeV [11] may either be a distinct proton component from more distant sources, or the accumulation of secondaries from the primary UHECR accelerators [12].

Another significant and generally accepted feature of the observations is the change in the cosmic ray composition with energy (see Figs 2.10-2.12 in [13]). Such a statement can only be made on statistical grounds because we cannot determine the composition of individual primary cosmic rays. The nuclei become progressively heavier above the ankle at $\sim 5$ EeV. The cosmic rays are mostly heavy nuclei (CNO or heavier) at energies $\gtrsim 40$ EeV [13]. There are too few showers with larger primary energy to be definitive, but it is generally supposed that the composition remains heavy up to the highest energies.

This change in mindset by the UHECR community has a simple interpretation. Most source and propagation models depend upon the particle's rigidity, $R$, its momentum per unit charge, not its total energy. If we use rigidity instead of energy, the observed cosmic ray spectrum can be accounted for with a maximum rigidity at the accelerator $R_{\rm max} \sim 8$ EV. To account for the whole extragalactic spectrum with only a single extragalactic component, one needs a very hard spectrum at the sources $R^{-1}$ to reproduce the evolution of the composition above the ankle, and a soft proton component ($\sim R^{-2.5}$ for sources avolving as the SFR [14]) to account for the light ankle seen by KASCADE-Grande. Such a single-component model gives a coherent picture of the Galactic-to-extragalactic cosmic ray transition [15] (in this model the protons are the product of photodissociation processes at the source). Other models based on more than a single extragalactic component remain of course possible, and the sub-ankle protons could be the result of another type of accelerator operating in this energy range. The contemporary luminosity density for UHECR acceleration is $\mathcal{L}_{\rm UHECR} \sim 6 \times 10^{44}$ erg Mpc$^{-3}$ yr$^{-1}$ above 5 EeV [16]. For the purposes of astronomical comparison, this is very roughly $3 \sim 10^{-5}$ times the local galaxy luminosity density. More quantitative comparisons require specific source and propagation models.

Regarding the cosmogenic gamma-ray diffuse flux (the by-product of the GZK effect), the mixed-composition model appear to be less constrained by the Fermi-LAT than the electron-positron dip (pure proton) scenario [17–19] that rules out SFR-like and stronger cosmological evolutions. For a mixed composition, however, only very strong evolutions (e.g., FR-II radio galaxies) are excluded by the current observations [20]. The cosmogenic neutrino flux is well below the IceCube sensitivity for a mixed composition [20]. The IceCube collaboration reported a correlation between a PeV neutrino and a blazar [21] implying that blazars are able to accelerate protons to $\sim 0.1$ EeV.

In summary, it is reasonable to assume that any source model needs to be able to accelerate cosmic rays up to a minimum rigidity ~8 EV (equivalently iron at ~200 EeV) to account for the observed spectrum and composition; therefore, in order to be considered as a possible source candidate, several conditions must be fulfilled by a cosmic accelerator:

- Confinement: the acceleration time should be smaller than the escape time; this is related to both the strength of the magnetic field and the geometry, size and age of the accelerator – this is an expression of the famous Hillas criterion [22];
- Power: the central engine must be able to provide the necessary energy (either electromagnetic or kinetic, depending how the energy transfer is mediated);
- Radiation losses at the accelerator: within the accelerating field the acceleration time should be smaller than the radiation loss time;
- Interaction losses (photopion and photodisintegration) at the accelerator: the acceleration time should be smaller than the interaction loss time;
- Emissivity: the density of sources must be enough to account for the observed UHECR flux ($\mathcal{L}_{\rm UHECR} \sim 6 \times 10^{44}$ erg Mpc$^{-3}$ yr$^{-1}$ above 5 EeV) [16];
- Secondary radiation: the accompanying photon and neutrino flux should not be greater than the observed fluxes. This constraint must be satisfied by the secondary gamma-ray and neutrino flux at the sources and also by the cosmogenic, diffuse flux.
- Anisotropy: the source model needs to account for the observed anisotropies. This is the weakest constraint due to the uncertainty regarding the magnetic fields [23, 24].

## 3 UHECR Source Models

### 3.1 Magnetospheric Models

#### *3.1.1 General Considerations*

In order to accelerate a proton to an energy $\sim 8$ EeV, it must follow a trajectory where $\int d\mathbf{r} \cdot \mathbf{E}$ exceeds $\sim 8$ EV. The simplest way to imagine this happening is in the context of a large scale magnetic field. Faraday's law guarantees that if the magnetic field varies with time, then there will be an EMF and an opportunity for particle acceleration. However, even a stationary flow can produce large, accelerating, electric field when there are appropriate boundary conditions. One such configuration is the unipolar inductor [25]. A spinning, conducting body is envisaged to be endowed with an axisymmetric magnetic field distribution. In the simplest case, the body rotates with angular velocity $\Omega$.

If sufficient plasma is continuously produced to sustain an electrical current, without having dynamical relevance, the electromagnetic field in the "magnetosphere" will be force-free. Imposing a boundary condition at the neutron star surface leads to an electric field $\mathbf{E} = -(\mathbf{\Omega}\times\mathbf{r})\times\mathbf{B}$ which implies that $\mathbf{E}\cdot\mathbf{B} = 0$. The magnetic flux surfaces will be equipotential and isorotational and the potential difference between two adjacent flux surfaces containing magnetic flux $d\Phi$ is $dV = \Omega d\Phi/2\pi$. There will also be an electrical current $I(\Phi)$ flowing within a flux surface labeled by $\Phi$ which is associated with a toroidal component of magnetic field and an outward Poynting flow of electromagnetic energy, $IdV$. This Poynting flux extracts rotational energy efficiently from the neutron star with very little dissipation in the neutron star. It can be radiated at great distance, for example in a Pulsar Wind Nebula such as the Crab Nebula.

The relationship between $I$ and $V$ depends upon the detailed boundary conditions but, if the electromagnetic field remains force-free, then $V \sim IZ_0/2\pi$, where $Z_0 \equiv$ 377 Ohm is the impedance of free space. To order of magnitude, a potential difference of, say, 10 EV is then associated with an electromagnetic power $\sim 10^{36}$ W. Whether or not this potential difference is used for efficient UHECR particle acceleration depends upon circumstance but, if it is, then the associated power is inescapable.

### 3.1.2 Neutron Stars

Now consider the application of these general ideas to neutron stars. Spinning magnetized neutron stars are, manifestly, not axisymmetric but quite similar considerations apply. If the magnetic moment is that of a regular pulsar, then the spin period must be close to the maximal allowed value $\sim$ 1.5 ms and the source lifetime is only years [26]. Magnetars, with surface magnetic field $\sim$ 100 GT are more promising [27]. A spin period $\sim$ 30 ms suffices but the source lifetime would be less than a day. It is hard to see how UHECR could escape the environs of a neutron star so soon after its birth given the high density of matter and radiation which should cause catastrophic losses.

However, there is another possibility and this is that the power derives from the magnetic, not the rotational energy. This can be as much as $10^{41}$ J and it could be released intermittently, following magnetic flares, where $\Omega$ would be replaced by the reciprocal of the time it would take light to cross a fraction of the magnetic surface, a few $\mu s$. Furthermore, the magnetic energy stored below the surface might be hundreds of times larger than that within the magnetosphere and accessible for much longer. This is important because the magnetar birth rate should be less than $\sim 10^{-6}\,\mathrm{Mpc}^{-3}\,\mathrm{yr}^{-1}$. requiring an energy per magnetar of $> 10^{43}$ J to account for the UHECR luminosity density.

Despite this energetic discouragement, it is instructive to consider the general principles of particle acceleration exhibited by an axisymmetric pulsar magnetosphere. In this case, the magnetic field is stationary and the EMF formally vanishes. The magnetic surfaces are electrostatically equipotential. However, a current must flow which may require the continuous formation of electron-positron pairs in the magnetosphere. It has long been supposed that this can happen at a "gap", a possibly transient potential drop along the magnetic field where $\mathbf{E}\cdot\mathbf{B} \neq 0$ and single charged particles can be accelerated to sufficient energy to produce $\gamma$-rays which can form pairs and breakdown the vacuum. There has been much modeling of gaps and generally it has been found that the associated potential drops across a gap are order of magnitude smaller than what would be needed to create UHECR directly at gaps. They could, of course be associated with the emission of much less energetic $\gamma$-rays and cosmic rays.

If there is minimal potential drop along the flux surfaces, then the power must associated with electrical current crossing between the flux surfaces. There are many different ways that this has been modeled. For example, there could be enough plasma present that there is a Lorentz force density, $\mathbf{j}\times\mathbf{B}$, that gradually converts the electromagnetic energy into kinetic energy flux with minimal dissipation. This outflow might then be an efficient particle accelerator in a jet or wind (see below). Alternatively, the current may complete dissipatively as particle acceleration. At one extreme, all of the power may heat the thermal plasma and cause it to radiate. An intermediate possibility is that much of this power creates a suprathermal distribution of intermediate energy cosmic rays.

A third possibility is that it is the highest energy particles that carry most of the electrical current across magnetic surfaces. The manner in which this may happen is interesting. A modest energy particle will gyrate around and move along the magnetic field in the local frame in which the electric field vanishes. However, for larger particle energies the gyro radius, $r_L$, increases and the collisionless particle motions also include drift velocities of their guiding centers with speeds $\sim cr_L/L$, where $L$ is the scale length. It is possible that despite the rarity of the highest energy particles, their greater mobility can allow them to carry most of the electrical current and, consequently, be accelerated to energies as large as those for which $r_L$ approaches $L$, specifically, those with energy $\sim ZecBL$. This basic mechanism has been invoked to account for intermediate energy particle acceleration at a pulsar wind termination shock. For this to be possible, in principle, requires that the electrical current flows along and not across magnetic surfaces all the way from the pulsar magnetosphere to the surrounding nebula.

### 3.1.3 Black Holes

Many of these basic physics ideas are translatable to black hole magnetospheres with some important differences. We suppose that magnetic flux is trapped by the inertia and pressure of orbiting gas and some of this flux passes in and out of the event horizon. The black hole is not a perfect conductor. In fact, a formal surface conductivity can be associated with the horizon associating a resistance of $\sim Z_0/2\pi$ with the black hole. This has two consequences. The magnetic field lines rotate with an angular velocity roughly half that of the black hole. There is as much power dissipated behind the horizon as there is extracted outside the black hole.

For the case of massive black holes, the power associated with UHECR acceleration is then, at least, comparable with that radiated by a moderate AGN such as a FRI radio source or a Seyfert galaxy. There is the same need to supply electrical charge continuously and gaps have been invoked as sources of $\gamma$-rays (e.g. [28]). As with pulsars, the potential differences that need to be applied to keep the current flowing are generally tiny compared with those needed to account for UHECR. To drive home this point, if the current carrying pair plasma had the minimum density and energy $\sim 1\,\mathrm{MeV}$ required, then its energy density would be only a fraction $\sim$ the ratio of this energy to the total potential difference — $\sim 1\,\mathrm{MeV}$ to $\sim 10\,\mathrm{EV}$ or $10^{-13}$. (This is also the ratio of the minimum to the maximum gyro radii.) In fact gaps will not even be required if either the magnetic field lines are able to undergo interchange instability to keep the magnetosphere supplied with electron-ion plasma or externally produced $\gamma$-rays can produce pairs directly. For all of these reasons, gaps are not seen as UHECR sources.

As with pulsars, it also possible to imagine particle acceleration occurring when current crosses the toroidal magnetic field lines associated with the electromagnetic outflow produced by a magnetized, spinning black hole. This takes place in the "Espresso" model of UHECR acceleration [29] where lower energy cosmic rays are injected into the outflow and emerge with rigidity comparable with the total potential difference available — an expression of the famous Hillas criterion [22] (see section 3.2.1 below). Assuming $\Gamma \sim 30$ (as in powerful blazars), a one-shot boost of a factor of $\sim \Gamma^2$ in energy, can transform the highest-energy galactic CRs at 100 PeV in the highest-energy UHECRs at 100 EeV. If this mechamism can account for the whole UHECR data is still an open question.

There is a second, quite general constraint on locating UHECR acceleration close to an AGN black hole. A $\sim 100\,\mu$ far infrared photon will be able to create pions in the Coulomb field of a $\sim 10\,\mathrm{EeV}$ cosmic ray. The UHECR acceleration site must be sufficiently far away from the AGN that the photon density does not prevent energy loss or photodisintegration. Similar remarks apply to the pair production threshold which involves $\sim 30\,\mathrm{GHz}$ radio photons. As discussed below, even more stringent constraints must be satisfied to avoid photodisintegration of the heaviest nuclei. The end result is that the only way that processes directly connected to an AGN magnetosphere could account for UHECR is if they involve low luminosity AGN and operate at some remove from the actual black hole magnetosphere.

Similar arguments can be used to rule out UHECR production around or directly associated with the magnetospheres of stellar mass black holes either close to formation in GRB or later on in binary X-ray sources. We now turn to the viability of UHECR acceleration further away from massive or stellar black holes, within relativistic jets.

### 3.2 Relativistic Plasma Jet Models

Relativistic jets are ubiquitous in the universe and their acceleration, collimation and deceleration processes provide a variety of acceleration sites for UHECRs. The central engine of relativistic jets is still under debate. Two driving mechanisms are commonly invoked, heating mediated by accretion ("nurture"), which leads to pressure-driven jets, and the Blandford-Znajek mechanism where the jet is powered by a rapidly spinning black hole ("nature") which leads to magnetically-driven jets [30]. The magnetization factor has important implications for modeling the acceleration of UHECRs and the production of secondary $\gamma$-rays and neutrinos.

Because the jets are structured both laterally but also temporally, the ejected plasma layers have different velocities and hence shear acceleration or diffusive shock acceleration can take place at internal shocks in the jets, lobes, or at the oblique shocks in the jet's boundary layer, or far beyond the Bondi radius at external shocks when the jet decelerates in the ambient medium. These mechanisms allow the transfer of the kinetic energy of the relativistic outflow to cosmic rays. This is mediated by magnetic turbulence and the micro-physics at play in generating the magnetohydrodynamic waves is important and yet poorly constraint in relativistic jets.

Efficient dissipation of magnetic energy can happen through turbulence or reconnection and leads to the emission of high energy photons that can interact with the cosmic rays via photointeractions (pion production for protons or Giant Dipole Resonance, GDR, for nuclei). How this operates in relativistic outflows is still unsettled and depends on the configuration of the magnetic fields. The two configurations commonly invoked are: a large scale, ordered field of a single polarity, leading to a highly magnetized relativistic jet; or a magnetic field of alternating polarity leading to a "striped" jet made of regions with toroidal field of alternating polarity. Instabilities such as the kink can also drive magnetic dissipation. Therefore, for relativistic jets, the following questions remain: What is the best sites for UHECR acceleration in relativistic jets? Are the heavy nuclei able to survive to the high energy photon fields present around relativistic sources? Let us review below some of the most promising jetted source candidates for UHECR sources.

#### 3.2.1 Jet Power requirement for UHECR

Let us recall the jet power requirement for UHECR production. In the following we consider a relativistic plasma jet with half opening angle is $\theta$, $r$ is the jet height above the launching region. We use masses in units of solar mass $\bar{m} = M/M_{\odot}$, luminosities in units of $L_{Edd} = 1.26\,10^{38}\bar{m}$ erg/s and radii in units of gravitational radius $\bar{r} = r/r_g$ with $r_g = 3 \cdot 10^5 \bar{m}$ cm. Quantities in the jet comoving frame are denoted with a prime symbol. The jet expansion time in the comoving frame is $t'_{\rm exp} = r/(\Gamma\beta c)$, where $\Gamma$ is the bulk Lorentz factor of the jet $\Gamma = (1-\beta^2)^{-1/2}$. The condition for jet stability implies that the comoving signal crossing timescale is smaller than $t_{\rm exp}$, which gives the condition

$\Gamma\theta \lesssim \beta_s/\beta$ where $\beta_s c$ is the Alfvén velocity. In relativistic magnetized jets $\beta \approx \beta_s \approx 1$. The acceleration region cannot extend further than the transverse distance $r\theta$. The escape time sideways is thus $t'_{\rm esc} = (r\theta)/(\Gamma c)$.

The jet luminosity $L_j = \pi(\theta r)^2\beta\Gamma^2 c u'_j$ where $u'_j$ is the comoving energy density. The magnetic field of a relativistic jet with magnetic luminosity $L_B \equiv \xi_B L_j$, bulk Lorentz factor $\Gamma$ and half opening angle $\theta$ is

$$B' = \frac{2\left(\xi_B L_j/c\right)^{1/2}}{\theta\Gamma r} \approx 4.3\cdot 10^8(\xi_B\bar{L}_j/\bar{m})^{1/2}\bar{r}^{-1}(\theta\Gamma)^{-1}\mathrm{G}\,. \quad (1)$$

The magnetic field $B'$ is related to the total energy density $u'_j$ by: $B' = \sqrt{4\pi\xi_B u'_j}$.

The acceleration time $t'_{acc} = t'_L/\xi_{acc} = E'/(ZeB'c\xi_{acc})$ with $\xi_{acc} \lesssim 1$ (in numerical experiences $\xi_{acc} \lesssim 0.1$ making the usually assumed Bohm diffusion overoptimistic). The condition $t'_{acc} = t'_{esc}$ (following [22]) gives:

$$E'_{\rm max} = \frac{\xi_{acc} r\theta Z e B'}{\Gamma} \approx 300\,\xi_{acc}Z[B'_{\rm G}r_{\rm cm}](\Gamma\theta)\Gamma^{-2}\,\mathrm{eV} \approx 3.\cdot 10^{16}\xi_{acc}Z\left(\xi_B\bar{L}_j\bar{m}\right)^{1/2}\Gamma^{-2}\mathrm{eV}\,, \quad (2)$$

which gives the condition on the observed jet luminosity:

$$L_j > 8.4\cdot 10^{44}\xi_B^{-1}\Gamma^2\xi_{acc}^{-2}\left(\frac{E_{\rm max}}{Z\,10^{20}\mathrm{eV}}\right)^2 \mathrm{erg\,s^{-1}}\,, \quad (3)$$

where $E_{\rm max}/Z$ is the maximum rigidity reachable in the observer's frame.

It is important to note that here only the mean intensity of the magnetic field $B'$ is considered. However in the usual model of diffusive shock acceleration (DSA), the largest turbulence scale of the magnetic field plays an important role, as the relevant escape time is the escape time *upstream* of the shock front. As the shocks microphysics is poorly understood (especially in the mildly-relativistic regime), the usual prescription for the particles confinement is that a cosmic ray remains accelerated at the shock if the Larmor radius of the particle is smaller than the largest turbulence scale of the magnetic field in the shocked region, i.e. $r_L \lesssim \lambda_{\rm max}$. It is important to keep in mind that $\lambda_{\rm max}$ can be a small fraction of the size of the system. Therefore, in the case of DSA, $t'_{\rm esc} = \xi_{esc}(r\theta)/(\Gamma c)$ with $\xi_{esc} \lesssim 1$.

Also, the above estimate of the maximum rigidity reachable is based on a geometrical argument. However there are several energy losses that will limit the acceleration process. The maximum comoving energy $E'_{\rm max}$ of the accelerated particles is estimated by equating their acceleration time $t'_{acc}$ with the relevant loss time $t'_{loss}$, and the related "opacity" is defined as $\tau_{loss} \equiv t'_{esc}/t'_{loss}$.

The synchrotron opacity is given by

$$\tau_{syn} = \frac{(r\theta)Z^4\sigma_T B^2 E}{6\pi m_e^2 c^4\Gamma}\left(\frac{m_e}{m_A}\right)^4 . \quad (4)$$

Concerning photodissociation, the lowest energy and highest cross-section process is the giant dipole resonance. The energy loss rate due to GDR interaction with the photons of energy $\epsilon'$ is $(t'_{\rm GDR})^{-1} \simeq \sigma_{GDR}n'(\epsilon')c\Delta\epsilon_{\rm GDR}/\epsilon_{\rm GDR}$ s$^{-1}$ with $\sigma_{GDR} = 1.45\,10^{-27}A$ cm$^2$, $\Delta\epsilon_{\rm GDR} = 8$ MeV and $\epsilon_{\rm GDR} \approx 42.65A^{-0.21}$ MeV. The photon energy $\epsilon'$ is given by the threshold condition for GDR interaction: $E'_{\rm N}(A)\epsilon' \approx 2.5\,10^{17}(A/56)$ eV$^2$. The GDR opacity is $\tau_{GDR} \equiv t'_{esc}/t'_{GDR}$. With an opacity close to 1, we are in the optimum working condition for the energy loss process considered. Only when the shock becomes transparent the maximum rigidity is limited by the escape from the magnetized region.

Particle escape acts as a high pass filter: only particles close to the maximum rigidity $\simeq E'_{\rm max}/Z$ can escape from the magnetized region upstream of the shock wave. We expect that the spectrum for the escaping particles is thus very hard, almost a delta function around $\simeq E'_{\rm max}/Z$.

#### 3.2.2 Gamma-Ray Bursts Jets

Gamma-ray bursts (GRBs) are associated with relativistic jets from newly born stellar-mass black holes. There are three different types of GRBs, which are associated with different progenitors. Long GRBs (lGRBs) are associated with the core-collapse of massive stars (more precisely, to SN Ib/Ic). They are the most powerful type of GRBs, with a jet luminosity able to reach $\sim 10^{52}$ erg/s. They are also the rarest, with a rate of $\sim 1$ Gpc$^{-3}$yr$^{-1}$ and a $\gamma$-ray luminosity density $\sim 6\,10^{42}$ erg Mpc$^{-3}$yr$^{-1}$. Short GRBs (sGRBs) are the aftermath of binary neutron star and black hole neutron star mergers and have jet luminosities of typically $\lesssim 10^{49}$ erg s$^{-1}$. Finally, there is another class of GRBs, the low luminosity GRBs (LLGRBs) which have lower luminosities (typically $L_{\rm jet} \lesssim 10^{50}$ erg s$^{-1}$, $\sim 100$ times the rate of lGRBs [31].

In the case of GRBs jets, there are two types of shocks that have been widely considered for diffusive shock acceleration: external shocks and internal shocks [32, 33]. External shocks are caused by the jet deceleration in the ambient medium (these are responsible for the GRB afterglow) and are superluminal. Many authors pointed out the inefficiency of Fermi acceleration at ultra-relativistic shocks, unless a strong small-scale turbulence is present in the downstream medium [34, 35]. Acceleration in ultra-relativistic external shocks can produce cosmic rays with energies above a few $Z \times 10^{15}$ eV (PeVatrons) see e.g., [36, 37]- but not the highest energy cosmic rays.

Internal shocks are expected to form inside the jet once the fast layers of the plasma jet catch up with the slower parts. In that case, the shock in the comoving frame is mildly relativistic (with Lorentz factor of $\sim$1.1 to $\sim$2) so efficient diffusive shock acceleration can take place. It has been shown that the spectrum of UHECR emitted by GRB internal shocks can reach maximum rigidities of 8 EV (iron at $\sim$100 EeV) for the highest luminosities ($L_{\rm jet} \gtrsim 10^{52}$ erg/s), making diffusive shocks acceleration at internal shocks a promising candidate for UHECR production [38]. However, the caveat for the long GRB model is that they fall short to account for the required UHECR luminosity density (typically by a factor $\sim 100$) unless they have a poor $\gamma$-ray efficiency [38, 39]; or, it has been proposed that one single Galactic event can be responsible for most of the UHECR flux if the local extragalactic magnetic field is large and can trap efficiently $\lesssim$8 EV rigidity particles [9].

#### 3.2.3 Active Galactic Nucleus Jets

The jets associated with massive black holes in active galactic nuclei, where the outflow is magnetically independent of a central black hole/accretion disk source, have also been proposed as UHECR accelerators. (By magnetically independent, we mean that the large-scale electrical current that flows along the inner jet has mostly left the jet, to be replaced by small scale current supporting small scale magnetic field.) Here two main mechanisms have been proposed, relativistic shock fonts and magnetic reconnection in a relativistic boundary layer. As just pointed out, relativistic shocks are qualitatively different from their nonrelativistic counterparts and they may be much less efficient in accelerating the highest energy particles. Partly for this reason, it has been suggested that it is the nonrelativistic backflow after a relativistic jet passes through a relativistic termination shock that is where more efficient UHECR acceleration can occur [40]. As the number of suitable sources within the UHECR horizon is small, there is the future possibility of making tentative associations with the highest energy particles may be possible [41].

There is good observational evidence that relativistic jets decelerate, dissipate, accelerate and radiate through their surfaces, especially close to their massive black hole sources. Presumably gas is entrained into the electromagnetic jet from a much more slowly moving external medium. These can be characterized as boundary layers from a gas dynamical perspective and as a cylindrical, relativistic current sheet from an electromagnetic perspective. It is presumably the second description that is most promising as an accelerator. Either way, this is a site where magnetic reconnection will take place. Non-relativistic reconnection is a process which has been extensively studies and there is much space physics data to compare with simulations. Overall it is not very efficent in creating very high energy particles. However, relativistic acceleration has recently been studied quite extensively and it involves novel features, especially the launching of high speed plasmoids [42]. There does not seem to be any natural way to accelerate particles to the very highest energy but this is certainly worth further consideration.

#### 3.2.4 Tidal Disruption Events Jets

There has been much recent observational attention paid to Tidal Disruption Events (TDE), where individual stars orbit massive ($M \lesssim 10^8\,\mathrm{M_\odot}$) black holes in otherwise dormant galaxies passing by so close that they are they are ripped apart by tidal forces with the debris falling back onto the black hole on timescales of months to years and radiating in the optical and X-ray bands especially. A minority of TDEs are found to be jetted and there is evidence that the jets have Lorentz factors $\gtrsim$ 10, like blazars [43]. They have also been proposed as sources of UHECR [44].

The estimated rate of jetted TDEs beamed towards us is $2 \times 10^{-11}\,\mathrm{Mpc^{-3}\,yr^{-1}}$ [45]. A conventional estimate of the total rate of jetted TDE is perhaps a hundred times larger. On this basis, we can estimate the energy required per jetted TDE to account for $\mathcal{L}_{\rm UHECR}$ (Sec. 2) as $\sim 0.03\,\mathrm{M_\odot}c^2$. This is not impossible but it seems implausible given the very much lower radiative efficiencies typically inferred.

### 3.3 Hierarchical Acceleration Model

#### 3.3.1 Diffusive Shock Acceleration

An alternative class of UHECR acceleration models invokes Diffusive Shock Acceleration (DSA, see [46] for a review) at giant intergalactic shock fronts, most notably those surrounding rich clusters of galaxies but also at shock fronts surrounding the filaments which connect these clusters. There are many possibilities. Here, we will strict attention to the more limited but more prescriptive and, consequently, more refutable idea that the entire cosmic ray spectrum observed at Earth is the result of successive DSA from distinct source classes.

Specifically, we consider a hierarchical model in which the observed $\sim 1\,\mathrm{GeV}- \sim 3\,\mathrm{PeV}$ cosmic rays originate in nearby supernova remnants. The pressure from these cosmic rays drives a magnetocentrifugal Galactic wind which passes through a Galactic Wind Termination Shock (GWTS) where re-acceleration takes place. Some of the highest energy particles from the GWTS escape upstream and contributes to the shin part of the spectrum; the remainder are transmitted into the circumgalactic and intergalactic media. GWTS acceleration is even more powerful around galaxies associated with AGN and starbursts.

These intermediate energy cosmic rays permeate the intergalactic medium and are input to DSA at the large intergalactic shocks. This "holistic" interpretation features propagation as well as acceleration. Of course, there are many alternative acceleration sites that could dominate different parts of the spectrum in a hierarchical scheme, as discussed above, or, alternatively, the intergalactic acceleration might operate mainly on mildly relativistic particles injected directly at the shock front.

DSA is a specific mechanism for converting a significant fraction of the kinetic energy flux, measured in the frame of the shock, of the gas upstream of a strong shock front into high energy particles that are transmitted downstream. It relies upon scattering by hydromagnetic disturbances/waves. Individual particles of rigidity (momentum per unit charge) $R$, with mean free paths $\ell(R)$, will diffuse against the flow with an upstream scale height $L \sim \ell c/u$, where $u$ is the speed of the shock relative to the upstream gas. In the simplest, test particle version of this mechanism at a planar shock front, relativistic particles will increase their rigidity by a fractional amount $\Delta R/R \sim 4(1-1/r)u/3c$, where $r$ is the shock compression ratio, every time they make a double crossing of the shock front. On kinematic grounds, the probability that an individual cosmic ray not return upstream is $4u/rc$.

The probability that a relativistic particle incident upon the shock front with initial rigidity $R$ be transmitted downstream with rigidity greater than $xR$ is

$$P(>x) = x^{\frac{3}{r-1}}; \quad x >\geq 1. \tag{5}$$

In this limit the shock behaves as a linear system, convolving the input upstream momentum space particle distribution function with a power law, Greens' function to give the transmitted downstream distribution function. The slope of the power law does not depend upon the details of the scattering and no particles escape upstream. The maximum rigidity, $R_{\rm max}$ that can be accelerated by a given shock is usually determined by the condition that $L$ must be less than the size of the shock, typically its radius of curvature. Note that this condition is, at least, $c/u$ times more stringent that the Hillas criterion. Alternatively, it takes $\sim (c/u)^2$ scatterings, not one orbit, to double a particle's energy. DSA is inherently slower than the more violent mechanisms discussed above. The question is "Can it still account for the highest energy particles we actually observe?".

Realistically, the test particle approximation fails in many ways. The acceleration is influenced by the cosmic-ray pressure gradient decelerating the gas ahead of the shock and changing the kinematics. In addition the cosmic rays can account for a significant fraction of the total energy of the flow and, on account of their different effective specific heat ratio, change the determination of $r$ through the conservation laws. The accelerated cosmic rays also create and sustain the wave turbulence that scatters them [47]. The most direct, and, arguably, the most important, form of scattering involves waves with wavelengths resonant with the particle gyro radii $r_{\rm L}(R)$. If the rms magnetic field amplitude of these waves is $\delta B(r_{\rm L}) = \delta B(R)$, then $\ell \sim (B/\delta B)^2 r_{\rm L}$, assuming random phases and Gaussian statistics (both questionable). If we assume that the waves are nonlinear over some range of $R$, then $\ell \sim r_{\rm L}$ and we call this Bohm diffusion. The efficiency of cosmic ray acceleration, $\epsilon_{\rm CR}$, which can be measured by the ratio of the downstream cosmic ray pressure to the upstream momentum flux, is then determined by the processes that determine the rms magnetic field strength.

There are several possible sources of this wave turbulence. If cosmic rays stream with mean speed in excess of the Alfvén speed in a uniform magnetic field, then they will excite the growth of resonant Alfvén waves propagating along the direction of their mean velocity. These waves will then scatter the cosmic rays in pitch angle and reduce their mean motion so that it is not much more than the Alfvén speed. The linear growth rate is roughly $(n_{\rm CR}/n_i)c/r_{\rm L}$, where $n_{\rm CR}$ measures the number of resonant particles and $n_i$ is the background ion density. An alternative possibility is that waves are created hydromagnetically with wavelength much larger than $r_{\rm L}$ which initiates a turbulence spectrum where the waves with shorter wavelengths will have linear amplitudes. A third possibility is that the waves are generated with wavelengths shorter than $r_{\rm L}$ by the motion of electrons carrying the return current that must balance the current associated with the accelerated cosmic rays. Extensive "PIC" and related simulations have vindicated the generation of wave turbulence and elucidated how these and other processes interplay in DSA.

### 3.3.2 Supernova Remnants

Although we are mostly concerned with questions of UHECR acceleration, it is instructive to contrast intergalactic shocks with precursor acceleration of lower energy particles at smaller shock fronts. It is a reasonable conjecture that analogous physical processes operate at all high Mach number shocks, scaled by $L$ and the momentum flux. While there has long been strong evidence associating supernova remnants with the majority of Galactic cosmic rays up to PeV energy, the argument for the reminder of the spectrum being due to DSA, is no better than circumstantial. What is clear is that if DSA accounts for most of the total spectrum, the acceleration must occur with near maximal efficiency with respect to $R_{\rm max}$ and $\epsilon_{\rm CR}$ in all three sites. For this reason, it is interesting to look at this problem inductively — to ask what would be required from the complex nonlinear plasma processes in the vicinity of the shock to account for the observations— rather than deductively, to attempt to calculate these processes from first principles.

Despite this commonality, there are important distinctions to be made. The first involves the kinematics. Most supernova remnants exhibit a quasi-spherical shock, convex on the upstream side, expanding into a near uniform interstellar medium, though stellar winds and molecular clouds can certainly introduce observable differences in many examples. This geometry facilitates the escape of the highest energy particle upstream as opposed to their transmission downstream, as assumed under the test particle approximation. Over half of the very highest energy particles can leave the remnant in this fashion. The original model of DSA essentially ignored these particles, supposing, instead, that the cosmic rays that we observed were transmitted downstream and were then released, following adiabatic decompression at the end of the remnant's lifetime. By contrast, it could be that most of the observed cosmic rays escape upstream with $R \sim R_{\rm max}$ which decreases as the remnant ages.

To be quantitative, the cosmic ray luminosity, per unit area, of the Galactic disk is generally estimated to be at least ten percent of the local supernova remnant luminosity per unit disk area. ($E_{\rm SNR} \sim 10^{51}$ erg released every $\sim$ 30 yr throughout the Galaxy.) Now, in order to accelerate protons to $\sim$ 3 PeV energy at a SNR expanding with speed $u_{\rm SNR}$ and radius $R_{\rm SNR}$, requires that the gyro radius be $\lesssim u_{\rm SNR}R_{\rm SNR}/c$ or $B \gtrsim 100(u_{\rm SNR}/10{,}000\,{\rm km\,s^{-1}})^{-1}(R_{\rm SNR}/1\,{\rm pc})^{-1}\,\mu$G, perhaps two orders of magnitude large than the rms Galactic field. Furthermore, the magnetic field strength is strongly bounded above by requiring that its contribution to the momentum flux, $B^2/4\pi$ be less than that of the gas, which is $\sim 0.3E_{\rm SNR}R_{\rm SNR}^{-3}$, adopting the Sedov solution, or $B \lesssim 10(R_{\rm SNR}/1{\rm pc})^{-3/2}$ mG. This interval allows efficient cosmic ray acceleration by young SNR up to the knee in the spectrum.

#### *3.3.3 Galactic Wind Termination Shocks*

The next level in this proposed hierarchy involves galactic winds. A large motivation for proposing a wind in our Galaxy comes from cosmic-ray observations. The measured ratio of light to medium nuclides at $\sim$ GeV energy allows us to infer that these cosmic rays escape the disk after traversing a grammage $\sim 6\,\mathrm{g\,cm^{-2}}$. This implies a residence time $\sim 30-100\,\mathrm{Myr}$, much shorter than a disk rotation period. Cosmic rays leave the disk continuously and it is reasonable to suppose that they carry gas with them. The wind provides a chimney or exhaust for the cosmic rays (as well as the hot gas which is unable to cool radiatively) created by the SNR. The existence of an outflowing wind, occupying most of the volume of the Galactic halo, is not inconsistent with the presence of denser gas clouds falling inward, as observed.

Various models of this outflow have been entertained. One recent model posits the presence of a vertical magnetic field at the disk and that $\sim$ GeV cosmic rays stream along it just faster than the Alfvén speed so that Alfvén waves are excited and couple the cosmic rays to the gas, allowing it to achieve the escape speed from the solar radius. The cosmic rays alone drive the wind [48].

A more powerful wind is possible if the vertical magnetic field corotates with the disk. In this case the cosmic rays just elevate the gas to modest altitude $\sim$ a few kpc, at which point the field develops a large radial component and the gas is flung out magnetocentrifugally and it is the orbiting interstellar medium that provides most of the power for the wind. Under this scenario, the interstellar medium should be seen as an open system losing a significant fraction of its mass, and angular momentum as well as energy over a Galactic lifetime. This possibility is not generally entertained in models of the interstellar medium.

The outflow will pass through three magnetohydrodynamic (Alfvénic and magnetosonic) critical points before crossing a termination shock, typically at a radius $R_{\mathrm{GWTS}} \sim 200\,\mathrm{kpc}$, with speed $u_{\mathrm{GWTS}} \sim 1000\,\mathrm{km\,s^{-1}}$, before joining the circumgalactic medium. The flow expands with density $\rho$ decreasing by roughly three orders of magnitude. The cosmic ray rigidities decrease $\propto \rho^{1/3}$, roughly one order of magnitude. The poloidal component of the magnetic field decreases from $\sim 2\,\mu\mathrm{G}$ at the disk to $B_p \sim 3\,\mathrm{nG}$ at the shock, while the toroidal field is roughly ten times larger, $B_\phi \sim 30\,\mathrm{nG}$.

The termination shock can reaccelerate $\sim$ GeV – PeV cosmic rays to fill out the $\sim$ PeV – EeV shin in the spectrum. The shock surface is likely stationary and concave on the upstream side with radius. This changes DSA in significant and observable ways. For modest cosmic-ray rigidity, say below the reduced peak of the cosmic-ray spectrum $R \lesssim 3\,\mathrm{PV}$, the shock will be effectively planar and a reaccelerated spectrum will be transmitted downstream. However, when $\ell \gtrsim u_{\mathrm{GWTS}} R_{\mathrm{GWTS}}/c$, the high energy cosmic rays are able to cross the Galactic halo and continue their acceleration at a quite different part of the shock. This acceleration can continue up to a rigidity where $\ell \sim R_{\mathrm{GWTS}}$, perhaps $\sim 100\,\mathrm{PV}$. When, as is usually the case, cosmic ray energy rather than rigidity spectra are displayed, the composition of the shin particles becomes progressively heavier, as is observed.

Again, to be quantitative, the wind may account for as much as $\sim 1\,M_\odot \mathrm{yr}^{-1}$mass loss from the galaxy and a corresponding power $\sim 5\times 10^{41}\,\mathrm{erg\,s^{-1}}$, more than the cosmic ray power. A turbulent magnetic field strength $B_{\mathrm{RMS}}(100\,\mathrm{PV}) \sim 0.3 B_\phi(R_{\mathrm{GWTS}})$ suffices to sustain acceleration within a closed shock surface throughout the shin observed at Earth.

#### *3.3.4 Intergalactic Shock Fronts*

We now turn to the final and most relevant level in this simple hierarchy when the $\sim$ PeV – EeV cosmic rays accelerated by galaxies, especially those that are more active than our Galaxy, are incident upon intergalactic shock fronts (e.g. [49, 50], Simeon et al. in preparation) . The most powerful shocks, prominent in cosmological simulations, surround rich clusters of galaxies which show Mach numbers that can exceed $\sim 100$. The assembly of clusters, over cosmological time, is complex but we will idealize the shock as a stationary spherical surface surrounding most of the cluster. These shocks have not been observed directly as yet. Deep searches at low radio frequencies seems well-motivated and would be very prescriptive, if successful. Simulations also show quasi-cylindrical filaments that may connect clusters and which may be surrounded by quasi-cylindrical infall shocks. The Milky Way may lie within such a filament.

The kinematics of a stationary cluster accretion shock, CAS, is different from that at SNR and GWTS. The convergence of gas ahead of the shock results in adiabatic heating of lower energy particles. This acceleration, which supplements DSA, will be less efficient for higher energy particles, which diffuse outward, relative to the gas flow. We continue to view the problem inductively and ask what scattering would have to be present in order for nearby cluster shocks to account for the observed UHECR spectrum. If we take an accretion shock surrounding the Virgo cluster then we observation combined with cosmological simulations suggest a shock radius $R_{\mathrm{CAS}} \sim 2\,\mathrm{Mpc}$ an infall speed and density ahead of the shock $u_- \sim 1000\,\mathrm{km\,s^{-1}}$ and $\rho_- \sim 10^{-29}\,\mathrm{g\,cm^{-3}}$. The shock is strong and the rate at which gas kinetic energy crosses it is $\sim 10^{45}\,\mathrm{erg\,s^{-1}}$. If we adopt a rich cluster density of $\sim 10^{-5}\,\mathrm{Mpc^{-3}}$, then the cluster gas kinetic energy luminosity density is $\sim 30\mathcal{L}_{\mathrm{UHECR}}$. There seems to be sufficient power to account for the observed UHECR flux. The larger challenge is to accelerate heavy nuclei, optimally, to $R_{\max} \sim 10\,\mathrm{EV}$.

In this model, we detect the UHECR that escape upstream with a spectrum peaking at a rigidity $\sim 1-8\,\mathrm{EV}$ which will vary from cluster to cluster. The cosmic rays which are transmitted downstream will never be directly observable, though they might be seen at much lower energy through their $\gamma$-ray emission. In order to accelerate cosmic rays to $R_{\max} \sim 10\,\mathrm{EV}$, we require that $R_{\mathrm{CAS}} > (c/3u_-)\ell(R_{\max}$. If we adopt Bohm diffusion, then the rms, resonant field ahead of the shock on these scales is $B_{\mathrm{rms}} \sim 1\,\mu\mathrm{G}$. This contrasts with a general field in the IGM that is generally expected to be $\lesssim 1\,\mathrm{nG}$. The isotropic magnetic pressure associated with this field is roughly an order of magnitude smaller than the nominal momentum flux carried by the background gas $\sim 10^{-13}\,\mathrm{dyne\,cm^{-2}}$. Provided that the turbulence does not dissipate and pre-heat the gas, thereby reducing the shock Mach number and the efficacy of DSA, a maximum rigidity $R_{\max} \sim 10\,\mathrm{EV}$ seems attainable. However, if, for example, protons with significantly larger rigidity than this are identified, it seems hard to see how accretion shocks around relatively well-observed and simulated, local clusters of galaxies could account for their acceleration.

Now turn to the input. Of course, we do not know what is the intergalactic spectrum produced by all the varied types of galaxy that surround the clusters. There are many possible contributors that we we may not be observing at Earth, for example starbursts and the winds they drive, pulsar wind nebulae, jets associated with spinning black holes and a network of weaker intergalactic shock fronts. (It is possible to associate the observed light ankle with filament shock acceleration in which case this comprises the spectrum transmitted downstream.) If we just confine our attention to the SNR spectrum produced by our local Galactic disk, then we expect a source spectrum $S(R) \propto R^{-2.2}$, as produced by all interstellar shocks over their lifetimes. This should extend up to a rigidity of order a few PV. At this point the rigidity spectrum should steepen as these additional sources fill out the shin part of the spectrum. (An energy spectrum will exhibit progressively heavier nuclei as observed.) Incident cosmic rays with rigidity well below the UHECR range will be convected adiabatically into the shock and see it as an essentially planar shock accelerator. Provided that the compression ratio of the subshock is neither too large, specifically $2.5 \lesssim r \lesssim 3.5$, then it will be the cosmic rays near the knee in the spectrum at few $PV$ that will dominate the UHECR spectrum that is reflected upstream back into the IGM. This will be reflected in the UHECR composition. In this model, the GeV particles that dominate the GCR spectrum and freshly injected particles are minor contributors to the output UHECR spectrum.

How reasonable is it for a CAS to behave in this extreme fashion? And, by extension, how reasonable is it for SNR and GWTS to accelerate cosmic rays to energies $\sim 3$ and $\sim 300\,\mathrm{PeV}$ respectively? Here, a rather different approach from that followed in most of the impressive simulations that have been completed may be helpful. Instead of starting from a quiescent initial state and monitoring the progressive acceleration of cosmic rays, and their associated local turbulence, to ever greater rigidity, limited by dynamic range and runtime, it might be interesting to start with the near-maximally turbulent state described and see if it can be self-sustaining. The dynamics is dominated by the highest energy particles and the background gas. PeV cosmic rays can be injected at the shock front. A key feature of this approach is that, as the rigidity increases, the particles diffuse further ahead of the shock front until they either escape upstream or are transmitted downstream with probabilities $\sim$ a half, ending their acceleration. It will also be necessary to include the curvature of the shock which facilitates this escape. These particles, with $R \sim R_{\max}$, will provide the first encounter that the inflowing intergalactic gas has with the shock. The gas itself is, presumably, still quite weakly magnetised with a pressure much smaller than the (anisotropic) pressure of the UHECR. Indeed, the pressure of the cosmic rays is likely to dominate the thermal pressure of the IGM. There is the possibility of hydromagnetic instability, similar to the firehose and mirror instabilities but also including the mean velocity of the UHECR fluid through the gas. This turbulence should grow as the gas falls further into the cluster with wavelengths longer than the local UHECR gyro radius ($\sim 1 - 10B_{\mu\mathrm{G}}^{-1}\,\mathrm{kpc}$), but shorter than the distance from the shock. The turbulence, which should be quasi-stationary in space, will have decreasing outer scale as the shock is approached, and should evolve to shorter wavelengths, progressively scattering lower rigidity particles through strongly nonlinear, resonant interaction as the shock is approached.

This "bootstrap" mechanism (adopting the physics as opposed to the statistics usage [51]) provides a different approach to exploring the complex physics of DSA. In the context of cluster shocks, it needs to be augmented by the inclusion of photopion and photodisintegration loss which are important because, unlike with the SNS, GWTS cases, the acceleration timescale is not short compared with the propagation timescale. Indeed, intergalactic and Galactic diffusion needs to be included in the problem alongside the background velocity field which transitions from inflow to cosmological outflow about 7 Mpc away from the cluster for Virgo [52]. The complex thermal and multiphase nature of the IGM, which is likely to be influenced by the hypothesized magnetic turbulence, is also a large part of the problem. Finally, the downstream boundary condition on particle acceleration, which is normally idealized as strong scattering in a uniform flow could turn out to be important.

## 4 Discussion and Future Goals

Cosmic-rays have been known for almost one century and gave birth to particle physics, but we still don't know where they come from. The question of what else in the Galaxy besides supernovae makes cosmic rays remains an open question and the origin of UHECR (cosmic rays with energy above 0.1 EeV) is still unknown, with both extragalactic and Galactic origin scenarios constrained but not ruled out.

So far, only electromagnetic, gravitational, neutrino signals trace high energy sources. The possible candidate sources of UHECRs – relativistic jets associated with binary neutron star mergers, core-collapse supernovae or active galactic nuclei; magnetars; accretion shocks around clusters of galaxies and filaments; wind bubbles associated with ultra fast outflows; tidal disruption events – still remain candidates and no direct evidence implicating them has been found so far. The question of which of these sources can be the accelerators of the ultra-high energy cosmic rays we detect up to $\sim 200$ EeV still remains unanswered. Although much debated, the common picture is that a transition from a Galactic to an extragalactic origin for cosmic radiation occurs somewhere between the knee and the ankle, and that the cosmic-ray composition becomes heavier as the energy increases. The interpretation of the anisotropies that emerge at the highest energies needs a better understanding of the UHECR composition.

The status is that the lack of observational understanding of the UHECR composition leaves freedom for speculation about their origin. This is also an interesting problem in the understanding of the physics of the extensive air shower development; the observed muon number, which is is a key observable to infer the mass composition of UHECR, shows an excess compared to air shower simulations with state-of-the-art QCD models. This is known as the "Muon Puzzle" ([53] for a recent account). Future data from the LHC, in particular oxygen beam collision would be needed to better address this issue. There is currently an effort from the UHECR community to better probe the cosmic ray composition at the highest energies. AugerPrime, the ongoing upgrade of the PAO has been designed to enhance the sensitivity of composition analyses by adding scintillator surface detectors and radio antennae to the existing Cherenkov detectors [54]. TA×4, the ongoing upgrade of TA (an additional 500 scintillator surface detectors designed to provide a fourfold increase in effective area) is being deployed and is partly in operation [55].

A few energy events have already been detected by Auger and TA above 150 EeV. With better statistics at the highest energies, it may be possible to unveil the sources [56]. The basic reason is that the background of distant sources, which dominates at lower energies, is less important at high energies as a consequence of the GZK cutoff.For example, at 300 EeV, only protons have a horizon $\gtrsim$ 30 Mpc, including the Virgo cluster and M87 at $\sim$ 16 Mpc. Any detection of intermediate or heavy mass nuclei at these energies would have to originate from sources at $\lesssim$ 3 Mpc, again severely limiting the options for sources, probably pointing to transients in the local group.

The detection of doublets or multiplets of "extreme energy" cosmic ray events (i.e. UHECR with energy above 150 EeV) with a composition-sensitive detector could rule out many currently viable source models [56]. Using recent Galactic magnetic field models [57, 58], we calculated "treasure" sky maps to identify the most promising directions for detecting extreme energy cosmic-ray doublets, events that are close in arrival time and direction and predicted the incidence of doublets as a function of the nature of the source host galaxy. Based on the asymmetry in the distribution of time delays, we showed that observation of doublets might distinguish source models. This is again why larger exposures and a better approach to mass composition is needed.

Identifying the sources of UHECR, ushering in the dawn of cosmic ray astronomy and exploring the otherwise inaccessible fundamental physics involved constitute three strong reasons for taking cosmic-ray studies to the next level so as to measure individual cosmic-ray masses at the highest energy. Larger and more capable cosmic-ray detector arrays, such as the Global Cosmic Ray Observatory [59] will be needed to accomplish this.

## Acknowledgments

We thank our colleagues that helped so much in shaping our understanding of UHECR physics and phenomenology: Denis Allard, Chen Ding, Glennys Farrar, Anatoli Fedynitch, Payel Mukhopadhyay, Etienne Parizot, Enrico Peretti, Paul Simeon, Alan Watson. N.G.'s research is supported by the Simons Foundation, the Chancellor Fellowship at UCSC and the Vera Rubin Presidential Chair.

## References

[1] H. Kawai, S. Yoshida, H. Yoshii, K. Tanaka, F. Cohen, M. Fukushima, N. Hayashida, K. Hiyama, D. Ikeda, E. Kido et al., Nuclear Physics B Proceedings Supplements **175**, 221 (2008)

[2] J. Abraham et al. (Pierre Auger), Nucl. Instrum. Meth. A **523**, 50 (2004)

[3] K. Greisen, Phys. Rev. Lett. **16**, 748 (1966)

[4] G.T. Zatsepin, V.A. Kuzmin, JETP Lett. **4**, 78 (1966)

[5] Pierre Auger Collaboration, A. Aab, P. Abreu, M. Aglietta, I.A. Samarai, I.F.M. Albuquerque, I. Allekotte, A. Almela, J. Alvarez Castillo, J. Alvarez-Muñiz et al., Science **357**, 1266 (2017), `1709.07321`

[6] N. Globus, T. Piran, Astrophys. J. Lett. **850**, L25 (2017), `1709.10110`

[7] C. Ding, N. Globus, G.R. Farrar, The Astrophysical Journal Letters **913**, L13 (2021), `2101.04564`

[8] B. Eichmann, M. Kachelrieß, F. Oikonomou, Journal of Cosmology and Astroparticle Physics **2022**, 006 (2022)

[9] D. Eichler, N. Globus, R. Kumar, E. Gavish, Astrophys. J. Lett. **821**, L24 (2016), `1604.05721`

[10] D. Allard, J. Aublin, B. Baret, E. Parizot, Astron. Astrophys. **664**, A120 (2022), `2110.10761`

[11] S. Schoo et al., JPS Conf. Proc. **9**, 010010 (2016)

[12] N. Globus, D. Allard, E. Parizot, Phys. Rev. D **92**, 021302 (2015)

[13] A. Coleman, J. Eser, E. Mayotte, F. Sarazin, F. Schröder, D. Soldin, T. Venters, R. Aloisio, J. Alvarez-Muñiz, R.A. Batista et al., Astroparticle Physics **147**, 102794 (2023)

[14] N. Globus, D. Allard, E. Parizot, T. Piran, The Astrophysical Journal **839**, L22 (2017)

[15] N. Globus, D. Allard, E. Parizot, Phys. Rev. D **92**, 021302 (2015)

[16] A. Aab, P. Abreu, M. Aglietta, J.M. Albury, I. Allekotte, A. Almela, J. Alvarez Castillo, J. Alvarez-Muñiz, R. Alves Batista, G.A. Anastasi et al. (The Pierre Auger Collaboration), Phys. Rev. D **102**, 062005 (2020)

[17] V. Berezinsky, A. Gazizov, M. Kachelrieß, S. Ostapchenko, Physics Letters B **695**, 13 (2011), `1003.1496`

[18] E. Gavish, D. Eichler, The Astrophysical Journal **822**, 56 (2016)

[19] A.D. Supanitsky, Phys. Rev. D **94**, 063002 (2016)

[20] N. Globus, D. Allard, E. Parizot, T. Piran, Astrophysical Journal Letters **839**, L22 (2017), 1703.04158
[21] M.G. Aartsen et al. (IceCube), Science **361**, 147 (2018), 1807.08794
[22] A.M. Hillas, "Annual Review of Astronomy and Astrophysics" **22**, 425 (1984)
[23] M. Unger, G.R. Farrar, *Uncertainties in the Magnetic Field of the Milky Way* (2017), 1707.02339
[24] Unger, Michael, Farrar, Glennys, EPJ Web Conf. **210**, 04005 (2019)
[25] W. Weber, Resultate aus der Beobachtungen des magnetishes Vereins im Jahre 1839 (1940)
[26] K. Fang, K. Kotera, A.V. Olinto, Astrophys. J. **750**, 118 (2012), 1201.5197
[27] J. Arons, Astrophys. J. **589**, 871 (2003), astro-ph/0208444
[28] A. Levinson, F. Rieger, Astrophys. J. **730**, 123 (2011), 1011.5319
[29] D. Caprioli, The Astrophysical Journal Letters **811**, L38 (2015)
[30] R. Blandford, N. Globus, Galaxies **10**, 89 (2022), 2207.05839
[31] E. Liang, B. Zhang, F. Virgili, Z.G. Dai, The Astrophysical Journal **662**, 1111 (2007), astro-ph/0605200
[32] E. Waxman, Physical Review Letters **75**, 386 (1995), astro-ph/9505082
[33] M. Vietri, The Astrophysical Journal **453**, 883 (1995), astro-ph/9506081
[34] J. Niemiec, M. Ostrowski, The Astrophysical Journal **641**, 984 (2006), astro-ph/0510811
[35] M. Lemoine, G. Pelletier, B. Revenu, The Astrophysical Journal Letters **645**, L129 (2006), astro-ph/0606005
[36] Y.A. Gallant, A. Achterberg, Monthly Notices of the Royal Astronomical Society **305**, L6 (1999), astro-ph/9812316
[37] O. Gottlieb, N. Globus, The Astrophysical Journal Letters **915**, L4 (2021), 2105.01076
[38] N. Globus, D. Allard, R. Mochkovitch, E. Parizot, Monthly Notices of the Royal Astronomical Society **451**, 751 (2015)
[39] D. Eichler, D. Guetta, M. Pohl, The Astrophysical Journal **722**, 543 (2010), 1007.3742
[40] J.H. Matthews, A.R. Bell, K.M. Blundell, A.T. Araudo, Monthly Notices of the Royal Astronomical Society **482**, 4303 (2018)
[41] A.R. Bell, J.H. Matthews, Mon. Not. Roy. Astron. Soc. **511**, 448 (2022), 2108.08879
[42] L. Sironi, M.E. Rowan, R. Narayan, Astrophys. J. Lett. **907**, L44 (2021), 2009.11877
[43] L. Rhodes, J.S. Bright, R. Fender, I. Sfaradi, D.A. Green, A. Horesh, K. Mooley, D. Pasham, S. Smartt, D.J. Titterington et al., Monthly Notices of the Royal Astronomical Society (2023)
[44] G.R. Farrar, T. Piran (2014), 1411.0704
[45] I. Andreoni et al., Nature **612**, 430 (2022), [Erratum: Nature 613, E6 (2023)], 2211.16530
[46] R. Blandford, D. Eichler, Physics Reports **154**, 1 (1987)
[47] J. Skilling, Monthly Notices of the Royal Astronomical Society **172**, 557 (1975)
[48] P. Mukhopadhyay, E. Peretti, N. Globus, P. Simeon, R. Blandford (2023), 2301.08902
[49] C.A. Norman, D.B. Melrose, A. Achterberg, The Astrophysical Journal **454**, 60 (1995)
[50] H. Kang, J.P. Rachen, P.L. Biermann, Monthly Notices of the Royal Astronomical Society **286**, 257 (1997), astro-ph/9608071
[51] R. Blandford, S. Funk, AIP Conf. Proc. **921**, 62 (2007)
[52] E.J. Shaya, R.B. Tully, Y. Hoffman, D. Pomarède, Astrophys. J. **850**, 207 (2017), 1710.08935
[53] J. Albrecht et al., Astrophys. Space Sci. **367**, 27 (2022), 2105.06148
[54] J. Stasielak (Pierre Auger), Int. J. Mod. Phys. A **37**, 2240012 (2022), 2110.09487
[55] R. Abbasi et al., Nucl. Instrum. Meth. **A**, 165726 (2021), 2103.01086
[56] N. Globus, A. Fedynitch, R.D. Blandford (2022), 2210.15885
[57] R. Jansson, G.R. Farrar, Astrophys. J. **757**, 14 (2012), 1204.3662
[58] Terral, Philippe, Ferrière, Katia, A&A **600**, A29 (2017)
[59] J.R. Hörandel, arXiv preprint arXiv:2203.01127 (2022)

# The transition from Galactic to extragalactic cosmic rays: the high–energy end of the Galactic spectrum

*Pierre* Cristofari[1,*]

[1]Laboratoire Univers et Théories, Observatoire de Paris, Université PSL, Université Paris Cité, CNRS, F-92190 Meudon, France

**Abstract.** Understanding the transition from Galactic to extragalactic cosmic rays (CRs) is essential to make sense of the Local cosmic ray spectrum. Several models have been proposed to account for this transition in the 0.1 - 10 $\times 10^{18}$ eV range. For instance: ankle models, where the change from a steep Galactic component to a hard extragalactic spectrum occurs in the $4 - 10 \times 10^{18}$ eV region, dip models, where the interactions of CR protons with the CMB producing electron-positron pairs shape the ankle, or mixed composition models, in which extragalactic CRs are composed of nuclei of various types.

In all these scenarios, the low-energy part of the transition involves the high-energy part of the Galactic component. Therefore, any information on the Galactic component, such as maximum energy, chemical composition, and spectrum after propagation, is crucial to understanding the Galactic-extragalactic transition. We briefly review the high-energy part of the CR spectrum expected from the best potential sources of Galactic CRs.

## 1 Introduction

The local interstellar spectrum (LIS) of cosmic rays (CRs), compiling the measurements of dozens of experiments, represents the local flux of CRs at the of Earth's atmosphere as a function of the particle's kinetic energy. The most remarkable features in the LIS include at least the *knee* and the *second knee*, associated to a spectral steepening, and the *ankle*, associated to a spectral hardening. For protons, accounting for about $\sim$ 90 % of CRs, these features are located at energy of $\sim 3 - 4$ PeV($=10^{15}$ eV) (*knee*), $\sim 1 - 5 \times 10^{17}$ PeV (*second knee*) , and $\sim 5$ EeV($=10^{18}$ eV) (*ankle*) (Fig. 1). It is often claimed that at least below the *knee*, CRs must be of Galactic origin, and above the *ankle*, the origin must be extragalactic.

A simple argument for the Galactic origin of the bulk of CRs comes from observing gamma rays from the Galactic disk. If CRs fill the Galaxy, they are expected to interact with the ISM material, and an enhanced gamma-ray signal is expected from Galactic regions where the density is higher, such as the Galactic disk [1]. Such a gamma-ray emission from the Galactic disk has been detected [2]. Moreover, when the Larmor radius of CRs becomes larger than the typical size of the Galactic halo $h$, CRs cannot be confined within the Galaxy, i.e., when $E > 10^{18} \left(\frac{h}{\mathrm{kpc}}\right)\left(\frac{B}{\mu\mathrm{G}}\right)$ eV.

Between the *knee* and the *ankle*, the transition from Galactic to extragalactic CRs has been the subject of numerous discussions, where various scenarios have been explored over the years, e.g., the 1) dip models, where the ankle is the result of pair-production losses of extragalactic CR protons on the CMB photons [3]; 2) mixed composition models, where various nuclei shape the spectrum, and the ankle appears to be a transition energy, and a light extragalactic component below the ankle is usually required [4–6]. 3) ankle models, where the transition at the ankle originates from the intersection of a flat extragalactic component and a very steep Galactic spectrum. The extragalactic component is usually assumed to be pure protons [7–10]. These various models are discussed in several excellent reviews [11–13], and here we focus on several open questions on the Galactic spectrum around the protons knee.

## 2 Galactic cosmic rays around the knee

The minimum requirement for the sources of Galactic CRs (and thus of Galactic protons) is that they must: 1) sustain the measured CR power; 2) inject a CR spectrum compatible with the measured LIS; 3) reach the *knee*, i.e., at least the PeV range, and thus be *pevatrons* for a least some time of their evolution; hence the hunt for pevatrons [14].

### 2.1 The knee

The actual value for the proton knee is still open to discussion, for example, with ARGO-YBL finding a value around $\sim$ 700 TeV [15] and KASCADE a value of $\sim 3 - 4$ PeV [16]. Moreover, measurements have shown that the value of the knee increases with Z and that therefore, the knees for different nuclei should shape the LIS above the knee for protons.

A Z-dependent knee is in agreement with what is expected from diffusive shock acceleration (DSA), the prominent first-order Fermi mechanism responsible for

*e-mail: pierre.cristofari@obspm.fr

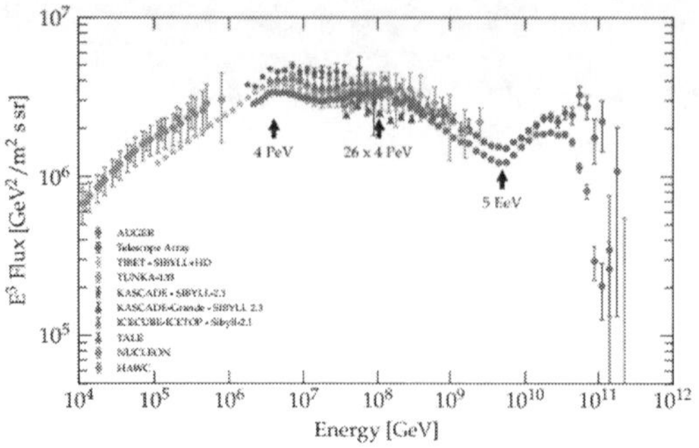


**Figure 1.** The local interstellar proton spectra. Data compiled by C. Evoli https://github.com/carmeloevoli

particle acceleration at strong shocks of SNRs, for which the maximum energy of the accelerated particle is expected to be rigidity dependent. In addition, it has been argued that very sharp cut-offs for the spectra of the different nuclei fail to produce a smooth LIS above the proton knee and that steep spectra above the various knees naturally lead to a smooth LIS above the proton knee [13].

### 2.2 Gamma rays

The search, identification, and characterization of pevatrons are one of the key science projects of most of the major gamma-ray facilities, such as CTA [17], LHAASO [18], or SWGO [19]. In the very-high-energy (VHE) domain (i.e., TeV=$10^{12}$ eV domain), two main mechanisms can efficiently produce gamma rays at Galactic sources: 1) a leptonic mechanism, the inverse Compton diffusion of accelerated electrons; and 2) a hadronic mechanism, the pion production/decay from accelerated nuclei interacting with the ISM. (The bremsstrahlung of accelerated electrons can also potentially produce gamma rays but is often subdominant).

The shape of the hadronic spectrum is roughly mimicking the spectrum of parent protons, and can roughly be approximated by a power-law in energy, exponentially suppressed at a maximum energy $E_\gamma \sim E_p/10$ where $E_p$ is the maximum energy of the parent protons so that the detection of $\sim$ 100 TeV gamma rays can help probe $\sim$ PeV protons. The shape of the leptonic gamma-ray spectrum is expected to be slightly more curved, with a maximum energy limited by the Klein-Nishina suppression in the cross-section of interactions, drastically impinging the efficient production of gamma–rays above $\gtrsim$ 50 TeV.

Most of the current VHE observatories, especially Imaging Atmospheric Cherenkov Telescopes (IACTs), such as H.E.S.S., MAGIC, and VERITAS are typically optimized in the $\sim$ 1-10 TeV range [20], a range in which differentiating between the two mechanisms is highly uneasy. For long, it was, however claimed that instruments optimized in the 100 TeV range could help directly differentiate between a hadronic and leptonic origin for the gamma rays since only the hadronic mechanism could efficiently produce gamma rays. And, at the same time, since $E_\gamma \sim E_p/10$, the detection of 100 TeV would testify of the presence of $\sim$ PeV protons. It is now clear that things are slightly more complicated and that this former claim must be revised. For instance, the detection of the first dozen of Galactic pevatrons by LHAASO [21] showed that for most of these sources, it is not clear which astrophysical object, what acceleration mechanism(s) and what emission mechanism(s) drive the gamma-ray emission. Several of these pevatrons are indeed expected to be leptonic pevatrons [22] – and at least the Crab is a commonly accepted example of a leptonic pevatron. The current difficulty of interpreting the origin of the gamma rays is especially due to the current performances of LHAASO: the exceptional sensitivity of the instrument in the 100 TeV range, which integrates the spectrum of all gamma-ray photons from the whole sky at all times, is counterbalanced by a limited angular and spectral resolution, which will be eventually complemented for forthcoming instruments.

Additionally, it has also been shown theoretically that hard electron spectra, in an environment allowing for fast inverse Compton cooling, can lead to hard gamma-ray spectra in the $\sim$ 100 TeV range so that even the detection of a hard spectrum in the 100 TeV range could not firmly testify of a hadronic origin [23, 24].

### 2.3 The hunt for pevatrons

Therefore, it seems that from the question "Which source(s) can be pevatrons?", we are now moving towards more specific questions:

- Which source(s) can produce a hard-enough spectra above $10^{15}$ eV?
- Which source(s) can produce a sufficient amount of PeV protons?
- Which Galactic source(s) can be *super*pevatrons[1], i.e., accelerate particles up to rigidities $10^{16}$ V? $10^{17}$ V?

## 3 The case of supernova remnants

Supernova remnants (SNRs) have for long been pointed as the most probable sources of Galactic CRs: because they can sustain the power of Galactic CRs; because they can roughly account for the injection spectrum $\propto E^{-2.1...-2.4}$ needed to explain the LIS, thanks to diffusive shock acceleration; and because they can theoretically be pevatrons. The maximum energy of particles accelerated at SNRs can be estimated with the Hillas criterion [25]: $E_{\rm max} \approx \xi \left(\frac{R_{\rm sh}}{\rm pc}\right)\left(\frac{u_{\rm sh}}{1000\ {\rm km/s}}\right)\left(\frac{B}{\mu{\rm G}}\right)$TeV. With typical values of the shock speed and radius of an SNR, the magnetic field needs to reach values significantly larger the one of the ISM in order to reach the PeV range. Such values have been inferred from observations in the X-ray range, up to 2-3 orders of magnitude above the ISM [26]. Theoretically, several mechanisms can lead to the amplification of the magnetic field at SNR shocks [27]. The mechanism

[1] I heard F. Aharonian use this term in a very nice talk in 2021, but not sure it is yet widely used

expected to dominate for young SNRs is the *non-resonant streaming of CRs* [28] (also called the *Bell* mechanism), due to the streaming of accelerated particles upstream of the shock exciting instabilities in the plasma on scales smaller than the Larmor radius of the streaming particles. The maximum energy of accelerated particles assuming magnetic field amplification alla Bell [29, 30] is typically: $E_{\max} \approx \frac{R_{\rm sh}}{10}\frac{\xi c e \sqrt{4\pi\rho(t)}}{\Lambda}\left(\frac{u_{\rm sh}}{c}\right)^2$, where $\Lambda = \ln(p_{\max}/p_{\min})$, and can a priori for some remnants of unusual and rare core-collapse supernovae reach the PeV range. Moreover, it appears that if some SNRs can be pevatrons for some time of their evolution, then in order to not overproduce the total proton spectrum, the rate of these SNe has to be substantially lower than the typical Galactic rate, i.e. of the order $\sim 1-5$ % $\times 2-3$/century [31].

It seems therefore possible that SNRs produce the bulk of CRs, with only a few of them contributing up to the knee. If only focussing on protons, several questions remain:

1. The spectra of particles escaping from SNRs is expected to ber exponentially suppressed at high-energy: how can such suppression produce the transition from $\propto E^{-2.7}$ to $\propto E^{-3}$ at the knee?

2. How do the different types of SNRs overlap to produce the very neat spectrum observed up to the knee?

3. How can the deviation from the perfect power-law, such as the bump at $\sim$ 10 TeV measured by DAMPE, be explained?

For question 1), as for any change of spectral index, the most probable hypotheses are: a break due to transport properties; two (or several) classes of sources overlapping; a steepening/break in the injection term from sources. The latter is probably the most reasonable hypothesis, as it has been shown that several effects, for instance, losses suffered by particles trapped inside an SNR, the leakage of particles in the ISM, could help diverge from a pure exponential suppression at high energy, and thus produce a steepening in the injected spectrum in the VHE range. For alternative candidates, clusters of massive stars, it has been shown that several effects, such as losses, escaping particles, or different diffusion regimes can substantially affect the slope at VHE [32].

For question 2), the question remains very open: several quantities, such as the efficiency of particle acceleration, the duration of the active phase of acceleration, the level of magnetic field downstream, can significantly affect the amount of particle and the spectra released by different SNRs in the ISM. It had already been pointed out that considering several populations of SNRs can naturally lead to bumps and features in the overall LIS proton spectrum [33]. It was additionally shown that in order to reach the PeV range, very peculiar SNRs had to be considered, with, for instance, high mass-loss rates $\dot{M} \gtrsim 10^{-4}$ M⊙/yr, high total explosion energy $E_{\rm SN} \gtrsim 5\times10^{51}$ erg, and low-mass ejecta, typically $M_{\rm ej} \lesssim 2-3$ M⊙. For such objects,

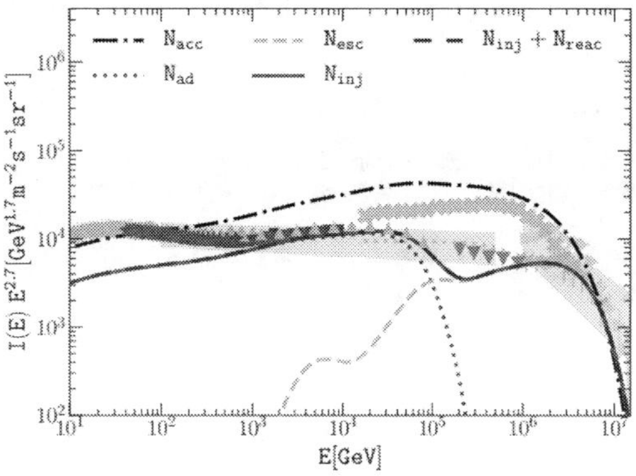


**Figure 2.** Proton spectrum from a core-collapse SN with total explosion energy $E_{\rm SN} = 6\times10^{51}$ erg, Mass-loss rate $\dot{M} = 10^{-5}$ $M_\odot$/yr and efficiency $\xi = 5$ %, the SN rate is $\nu_{\rm SN} = 0.06$/century. The black dot-dashed line corresponds to a naive estimate of all accelerated protons. The dotted green lines correspond to particles trapped inside the SNR and suffering adiabatic losses and released at the end of the adiabatic phase. The orange dashed line corresponds to particles constantly escaping in the ISM. The solid red line corresponds to the sum of trapped and escaping particles. The dashed blue line indicates that diffusive shock re-acceleration does not play a significant role. The data points and yellow area correspond to data from various experiments as in [31].

the total level of protons can naturally come close to the LIS.

When exploring the parameter space accessible to the various SNe, one can then wonder if it may be possible to account for the entire LIS with only type of rare SNe/SNRs. We illustrate the idea of such possibility in Fig. 2. One major issue with this idea is that it relies on a very unusual and rare type of SNe to account for the entire LIS - which is not a satisfying solution - that would leave almost no room for the contribution of the rest of the other SNRs, that are clearly known to accelerate high energy particles (if only because of gamma-ray observations) and are expected to contribute to the LIS.

In short, to the question: "which sources could be pevatrons?" The answer is: 1. Probably SNRs (few and rare); 2. Many other sources could also be involved (massive stars, superbubbles, etc.). To the slightly more precise question "How to produce enough PeV protons, and a hard-enough spectrum above PeV?", we need to specify the role of various SNRs, and the parameter governing particle acceleration (efficiency, duration of the acceleration); the role played by other accelerators, and the origin of fine features (e.g., $\sim$ 10 TeV bump).

Other important issues have been also mentioned for SNRs, such as our inability to understand: a) what is the spectrum of accelerated particles? (i.e. the role of non-linear effects and deviation from the canonical test-particle spectrum $\propto p^{-4}$) b) What is the spectrum of protons/electrons *released* in the ISM? [14, 34, 35].

## 4 Other candidates

Aside from the proton spectrum, the problem of $^{22}Ne/^{20}Ne$ ratio, found to be a factor of $\approx$ 5 larger in CRs than in the local ISM [36], has been interpreted as an indication that stellar winds termination shocks, and especially Wolf-Rayet winds, which are enriched in $^{22}Ne$ must be involved [37–39].

Next to that, the detection of gamma rays from massive stars, such as the Cygnus cocoon, the central molecular zone, or Westerlund 1, with an extended profile scaling as the inverse of the radial distance from the center $\propto 1/R$ [40], suggests a continuous injection of particles from massive stars and has helped dust off the idea that massive stars could be involved in the origin of CRs [41–43]. The role of stellar clusters and interstellar bubbles, whether they are young and compact, with massive stars centered, or older with multiple supernova remnants complexifying the bubble environment has been actively studied [44–48].

If one considers SNRs, superbubbles, and wind termination shocks (WTS) as potential Galactic CR sources, it seems that alone, none of these candidates can by itself easily solve all the issues discussed above - energetics, injected spectrum, acceleration up and beyond PeV with the correct spectrum, 10 TeV bump, $^{22}Ne/^{20}Ne$ ratio. The possibility for a mixed contribution from these candidates has been investigated, and some authors have for instance found, confronting available CR abundance to various benchmark scenarios, that a mixed scenario where most of CRs come from SNRs in superbubbles and a $\sim 5\%$ contribution from WTS of young stellar clusters could decently work [49].

## 5 Open questions

In addition to the major issues discussed so far, several intriguing results on CRs have yet to be explained [see e.g. detailed discussions and references in 50]:

1. the spectral hardening towards the Galactic center
2. the small spatial gradients of CRs
3. the very low level of anisotropy and phase pointing away from the Galactic center (<100 TeV)
4. the origin of small-scale anisotropies
5. the He spectra differing from the proton spectra

Some of these issues might be directly connected to the problem of the interpretation of the Galactic CR spectrum around the knee and to the problem of the identification of Galactic sources. For instance, the wild hypothesis was discussed during UHECR2022, that if the trend of He and proton spectra are extrapolated, what is now understood to be the proton knee at $\sim 3-4$ PeV could in fact be Helium dominated. If this were in fact the case, all the above discussions should need revision, especially the discussions on the nature of the proton pevatrons.

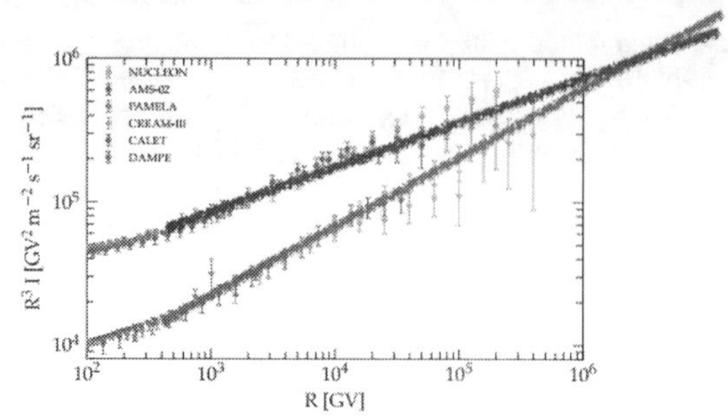


**Figure 3.** The He and proton spectra measured by various experiments. The blue and red lines are very wild extrapolations of the proton and Helium trends to higher energies. Data compiled by C. Evoli https://github.com/carmeloevoli

The next years will be absolutely determinant for the search of the Galactic CR sources. Instruments capable of characterizing the gamma-ray spectra in the 100 TeV range, e.g., CTA, will be of special interest. It will not only help to clarify the role of SNRs, superbubbles, and WTS in the origin of Galactic CRs but also answer essential questions on shock acceleration (efficiency, spectra of accelerated particles, proton/electron ratio accelerated).

## 6 Acknowledgement

PC warmly thanks the organizing committee of UHECR2022, and C. Evoli for the plots.

## References

[1] Hayakawa S., 1952, PThPh, 8, 571. doi:10.1143/ptp/8.5.571

[2] Abdo A. A., Ackermann M., Ajello M., Allafort A., Antolini E., Atwood W. B., Axelsson M., et al., 2010, ApJS, 188, 405. doi:10.1088/0067-0049/188/2/405

[3] Berezinsky V., Gazizov A., Kachelrieß M., 2006, PhRvL, 97, 231101. doi:10.1103/PhysRevLett.97.231101

[4] Allard D., Parizot E., Olinto A. V., Khan E., Goriely S., 2005, A&A, 443, L29. doi:10.1051/0004-6361:200500199

[5] Hooper D., Sarkar S., Taylor A. M., 2007, APh, 27, 199. doi:10.1016/j.astropartphys.2006.10.008

[6] Aloisio R., Berezinsky V., Gazizov A., 2011, APh, 34, 620. doi:10.1016/j.astropartphys.2010.12.008

[7] Unger M., Farrar G. R., Anchordoqui L. A., 2015, PhRvD, 92, 123001. doi:10.1103/PhysRevD.92.123001

[8] Globus N., Allard D., Parizot E., 2015, PhRvD, 92, 021302. doi:10.1103/PhysRevD.92.021302

[9] de Marco D., Stanev T., 2005, PhRvD, 72, 081301. doi:10.1103/PhysRevD.72.081301

[10] Waxman E., 2000, PhST, 85, 117. doi:10.1238/Physica.Topical.085a00117

[11] Kachelrieß M., 2019, EPJWC, 210, 04003. doi:10.1051/epjconf/201921004003
[12] Aloisio R., Berezinsky V., Gazizov A., 2012, APh, 39, 129. doi:10.1016/j.astropartphys.2012.09.007
[13] Parizot E., 2014, NuPhS, 256, 197. doi:10.1016/j.nuclphysbps.2014.10.023
[14] Cristofari P., 2021, Univ, 7, 324. doi:10.3390/universe7090324
[15] Bartoli B., Bernardini P., Bi X. J., Cao Z., Catalanotti S., Chen S. Z., Chen T. L., et al., 2015, PhRvD, 92, 092005. doi:10.1103/PhysRevD.92.092005
[16] Antoni T., Apel W. D., Badea A. F., Bekk K., Bercuci A., Blümer J., Bozdog H., et al., 2005, APh, 24, 1. doi:10.1016/j.astropartphys.2005.04.001
[17] Cherenkov Telescope Array Consortium, Acharya B. S., Agudo I., Al Samarai I., Alfaro R., Alfaro J., Alispach C., et al., 2019, scta.book. doi:10.1142/10986
[18] Cao Z., della Volpe D., Liu S., Editors, :, Bi X., Chen Y., et al., 2019, arXiv, arXiv:1905.02773
[19] Hinton J., SWGO Collaboration, 2022, icrc.conf, 23. doi:10.22323/1.395.0023
[20] Chadwick P., 2021, Univ, 7, 432. doi:10.3390/universe7110432
[21] Cao Z., Aharonian F. A., An Q., Axikegu B., Bai Y. X., Bao Y. W., Bastieri D., et al., 2021, Natur, 594, 33. doi:10.1038/s41586-021-03498-z
[22] de Oña Wilhelmi E., López-Coto R., Amato E., Aharonian F., 2022, ApJL, 930, L2. doi:10.3847/2041-8213/ac66cf
[23] Vannoni G., Gabici S., Aharonian F. A., 2009, A&A, 497, 17. doi:10.1051/0004-6361/200809744
[24] Breuhaus M., Hahn J., Romoli C., Reville B., Giacinti G., Tuffs R., Hinton J. A., 2021, ApJL, 908, L49. doi:10.3847/2041-8213/abe41a
[25] Hillas A. M., 2005, JPhG, 31, R95. doi:10.1088/0954-3899/31/5/R02
[26] Vink J., 2012, A&ARv, 20, 49. doi:10.1007/s00159-011-0049-110.48550/arXiv.1112.0576
[27] Gabici S., Gaggero D., Zandanel F., 2016, arXiv, arXiv:1610.07638
[28] Bell A. R., 2004, MNRAS, 353, 550. doi:10.1111/j.1365-2966.2004.08097.x
[29] Schure K. M., Bell A. R., 2013, MNRAS, 435, 1174. doi:10.1093/mnras/stt137110.48550/arXiv.1307.6575
[30] Schure K. M., Bell A. R., 2014, MNRAS, 437, 2802. doi:10.1093/mnras/stt208910.48550/arXiv.1310.7027
[31] Cristofari P., Blasi P., Amato E., 2020, APh, 123, 102492. doi:10.1016/j.astropartphys.2020.102492
[32] Morlino G., Blasi P., Peretti E., Cristofari P., 2021, MNRAS, 504, 6096. doi:10.1093/mnras/stab690
[33] Ptuskin V., Zirakashvili V., Seo E.-S., 2010, ApJ, 718, 31. doi:10.1088/0004-637X/718/1/3110.48550/arXiv.1006.0034
[34] Cristofari P., Blasi P., Caprioli D., 2021, A&A, 650, A62. doi:10.1051/0004-6361/202140448
[35] Cristofari P., Blasi P., Caprioli D., 2022, ApJ, 930, 28. doi:10.3847/1538-4357/ac6182
[36] Binns W. R., Wiedenbeck M. E., Arnould M., Cummings A. C., George J. S., Goriely S., Israel M. H., et al., 2005, ApJ, 634, 351. doi:10.1086/49695910.48550/arXiv.astro-ph/0508398
[37] Prantzos N., 2012, A&A, 538, A80. doi:10.1051/0004-6361/20111744810.48550/arXiv.1112.4343
[38] Tatischeff V., Gabici S., 2018, ARNPS, 68, 377. doi:10.1146/annurev-nucl-101917-021151
[39] Casse M., Paul J. A., 1982, ApJ, 258, 860. doi:10.1086/160132
[40] Aharonian F., Yang R., de Oña Wilhelmi E., 2019, NatAs, 3, 561. doi:10.1038/s41550-019-0724-0
[41] Casse M., Paul J. A., 1980, ApJ, 237, 236. doi:10.1086/157863
[42] Cesarsky C. J., Montmerle T., 1983, SSRv, 36, 173. doi:10.1007/BF00167503
[43] Voelk H. J., Forman M., 1982, ApJ, 253, 188. doi:10.1086/159623
[44] Vieu T., Gabici S., Tatischeff V., Ravikularaman S., 2022, MNRAS, 512, 1275. doi:10.1093/mnras/stac543
[45] Bykov A. M., 2001, SSRv, 99, 317. doi:10.1023/A:1013817721725
[46] Ferrand G., Marcowith A., 2010, A&A, 510, A101. doi:10.1051/0004-6361/20091352010.48550/arXiv.0911.4457
[47] Parizot E., Marcowith A., van der Swaluw E., Bykov A. M., Tatischeff V., 2004, A&A, 424, 747. doi:10.1051/0004-6361:2004126910.48550/arXiv.astro-ph/0405531
[48] Gupta S., Nath B. B., Sharma P., Eichler D., 2020, MNRAS, 493, 3159. doi:10.1093/mnras/staa286
[49] Tatischeff V., Raymond J. C., Duprat J., Gabici S., Recchia S., 2021, MNRAS, 508, 1321. doi:10.1093/mnras/stab2533
[50] Gabici S., Evoli C., Gaggero D., Lipari P., Mertsch P., Orlando E., Strong A., et al., 2019, IJMPD, 28, 1930022-339. doi:10.1142/S0218271819300222

https://doi.org/10.1051/epjconf/202328304003

# Searching for neutral particles at the highest energies at the Pierre Auger Observatory

*Marcus* Niechciol[1,*] for the Pierre Auger Collaboration[2,**,***]

[1]Center for Particle Physics Siegen, University of Siegen, Walter-Flex-Str. 3, 57072 Siegen, Germany
[2]Observatorio Pierre Auger, Av. San Martín Norte 304, 5613 Malargüe, Argentina

**Abstract.** The Pierre Auger Observatory, being the largest air-shower experiment in the world, offers an unprecedented exposure to neutral particles at the highest energies. Since the beginning of data collection more than 18 years ago, several searches for ultra-high-energy (UHE, $E > 10^{17}$ eV) photons and neutrinos have been performed. The upper limits on the diffuse flux of UHE photons and neutrinos derived from Auger data are among the most stringent in the world, severely constraining models for the origin of UHE cosmic rays. In addition, the Pierre Auger Observatory contributes to current efforts in multimessenger astronomy through follow-up searches for UHE photons and neutrinos in association with transient events, such as gravitational wave events. The various activities concerning searches for UHE photons and neutrinos in the data from the Pierre Auger Observatory are presented and the current results are summarized. In addition, future perspectives will be discussed.

## 1 Introduction

Neutral particles, in particular photons and neutrinos, play a crucial role for current efforts in multimessenger astronomy, where one of the main goals is understanding where and how ultra-high-energy (UHE, commonly referring to the energy range above ~$10^{17}$ eV) cosmic rays are produced. These particles can even exceed energies of $10^{20}$ eV, far beyond what can be reached with man-made particle accelerators. One can exploit the intimate connection between UHE cosmic rays and UHE photons and neutrinos, which appear as "by-products" in almost every scenario aiming at explaining UHE cosmic rays, originating either directly at the sources ("astrophysical" photons and neutrinos) or during the propagation of the UHE cosmic rays through the Universe ("cosmogenic" photons and neutrinos). Photons and neutrinos themselves are complementary messengers: Photons trace the local Universe up to Mpc scales, due to interactions with the cosmic background fields, while neutrinos can traverse the whole Universe. In addition, neutral particles at the highest energies can also serve as cosmic probes of fundamental physics and open a window to new physics, for example in the context of super-heavy dark matter, violations of Lorentz invariance, or axions.

Just as for UHE cosmic rays, also the fluxes of UHE neutral particles are far too small for direct detection using, e.g., satellite experiments. However, photons and neutrinos entering the Earth's atmosphere can initiate extensive air showers like a charged cosmic ray, making indirect detection possible. Therefore, a cosmic-ray observatory measuring air showers can, by construction, also be a photon observatory and even a neutrino observatory, complementing specialized instruments and adding valuable pieces of information to current multimessenger studies.

In these proceedings, current searches for UHE photons and neutrinos with data collected by the Pierre Auger Observatory are summarized. The Observatory itself is briefly introduced in Sec. 2. Sec. 3 deals with searches for UHE photons, while Sec. 4 focusses on UHE neutrinos, including an update of the limits on the diffuse flux of neutrinos with data until the end of 2021. A brief summary is given in Sec. 5, including a discussion of future prospects with the on-going detector upgrade of the Observatory, dubbed AugerPrime.

## 2 The Pierre Auger Observatory

The Pierre Auger Observatory [1], located near the town of Malargüe in the Argentinian *Pampa Amarilla*, is the largest cosmic-ray observatory to date. A key feature of the Pierre Auger Observatory is the hybrid concept, combining a Surface Detector array (SD) with a Fluorescence Detector (FD). The SD consists of about ~1600 water-Cherenkov detectors arranged on a triangular grid with a spacing of 1500 m, covering a total area of about 3000 km$^2$. The SD is overlooked by 24 fluorescence telescopes, located at four sites at the border of the array. The SD samples the lateral shower profile at ground level, i.e., the distribution of particles as a function of the distance from the shower axis, with a duty cycle of ~100 %, while the FD records the longitudinal shower development in

*e-mail: niechciol@physik.uni-siegen.de
**e-mail: spokespersons@auger.org
***Full author list: https://www.auger.org/archive/authors_2022_10.html

the atmosphere above the SD. The FD can only be operated in clear, moonless nights, reducing the duty cycle to ~15 %. In the western part of the SD array, ~50 additional SD stations have been placed between the existing SD stations, forming a sub-array with a spacing of 750 m and covering a total area of about 27.5 km$^2$. With this sub-array, air showers of lower primary energy (below $10^{18}$ eV) with a smaller footprint on the ground can be measured. To allow also for hybrid measurements in this energy range, where air showers develop above the field of view of the standard FD telescopes, three additional High-Elevation Auger Telescopes (HEAT) have been installed at the FD site Coihueco, overlooking the 750 m SD array. The HEAT telescopes operate in the range of elevation angles from 30° to 60°, complementing the Coihueco telescopes operating in the 0° to 30° range. The combination of the data from both HEAT and Coihueco ("HeCo" data) enables fluorescence measurements of air showers over a large range of elevation angles.

## 3 Searches for UHE photons

The biggest challenge when searching for UHE neutral particles is to distinguish air showers initiated by these primaries from those induced by the vast background of primary protons and heavier nuclei in cosmic rays. For the identification of primary photons, one can exploit the fact that the development of photon-induced air showers in the atmosphere is dominated by electromagnetic interactions, which leads to two key differences with respect to air showers initiated by protons and nuclei (for a review, see [2]). Firstly, the atmospheric depth of the shower maximum $X_{max}$ lies deeper in the atmosphere, due to the lower multiplicity of electromagnetic interactions compared to hadronic interactions. Secondly, fewer secondary muons are produced, as only a small fraction of the total energy is transferred to the hadronic (and subsequent muonic) channel through photo-nuclear interactions. One consequence of this is a steeper lateral distribution of secondary particles on the ground and a smaller overall footprint of the shower.

At the Pierre Auger Observatory, these key differences are exploited in various searches for photons. We will focus here on searches for a diffuse—i.e., direction-independent, unresolved—flux of photons, which are performed in three different energy ranges: between $2\times10^{17}$ and $10^{18}$ eV, using HeCo data combined with data from the 750 m SD array [3]; between $10^{18}$ and $10^{19}$ eV using data from the main fluorescence telescopes and the 1500 m SD array [4]; and above $10^{19}$ eV using only data from the 1500 m SD array [5]. In addition, two searches for point sources of UHE photons have been performed: a blind search covering the full field of view of the Observatory [6] and a targeted search involving different classes of potential sources [7]. Both of these analyses are not discussed here further and are only mentioned for completeness. Also dedicated multimessenger studies involving UHE photons have been published, such as follow-up searches for photons in coincidence with transient events like gravitational wave events [8] or the anomalous blazar TXS 0506+056 [9]. These multimessenger analyses are the subject of another contribution [10]. A more detailed summary of all searches for UHE photons at the Pierre Auger Observatory can be found in [11].

In the following, the three searches for a diffuse flux of photons mentioned before are briefly described. In the energy range between $2\times10^{17}$ and $10^{18}$ eV, the analysis is based on three observables: $X_{max}$, $S_b$ and $N_{stations}$. $X_{max}$ can be measured directly with the fluorescence telescopes, while $S_b$ is calculated from the SD measurement through

$$S_b = \sum_i S_i \times \left(\frac{R_i}{1000\,\mathrm{m}}\right)^b ,$$

where $S_i$ and $R_i$ denote the measured signal and the perpendicular distance to the shower axis of the $i$-th triggered station in a given SD event and the parameter $b$ was fixed to 4 [12]. $N_{stations}$ is the number of triggered stations in the event. Both $S_b$ and $N_{stations}$ can be seen as a measure of the steepness of the lateral distribution and hence add complementary information to $X_{max}$, which depends on the longitudinal development of the air shower. These three observables are combined in a multivariate analysis. The output of the boosted decision tree (BDT) analysis, $\beta$, is plotted in Fig. 1(a). The BDT is trained using large samples of photon- and proton-induced air showers. The final value of the photon candidate cut was chosen to ensure a signal efficiency of 50 %. At this signal efficiency, the background rejection (determined from proton simulations) is about 99.9 %. Eventually, the analysis was applied to data collected between 1 June 2010 and 31 December 2015. No events from the data sample pass the candidate cut, hence, upper limits on the incoming photon flux are determined, based on an exposure to photons of about 2.5 km$^2$ sr yr in the given energy range and for the given time period.

The search for photons between $10^{18}$ and $10^{19}$ eV is also based on $X_{max}$, but complemented by the observable $F_\mu$, which is a proxy for the muon content of the air shower [13]. $F_\mu$ is calculated using a model based on air-shower universality, taking the measured signals in the SD stations as input as well as the shower geometry and $X_{max}$ reconstruced from the FD. The combination of both observables in a Fisher discriminant analysis (see Fig. 1(b)) yields an overall background rejection (at a signal efficiency of 50 %) of about 99.9 %. The analysis is then applied to data from the period 1 January 2005 to 31 December 2017, which is equivalent to an exposure to UHE photons of about 1000 km$^2$ sr yr. A total of 22 events from the data sample passed the candidate cut. However, this number is fully compatible with the expectation, determined *a priori* from a sub-sample of the data set which was not used in the final analysis, of 30 ± 15 events. These candidate events were studied in detail, yet none of them could be identified unambiguously as being initiated by a primary photon.

In the energy range above $10^{19}$ eV, the statistics of the data sample collected by the FD is not sufficient anymore to efficiently search for UHE photons. The SD, with a duty cycle of nearly 100 %, can provide much better statistics at the highest energies. The lack of a corresponding FD

https://doi.org/10.1051/epjconf/202328304003

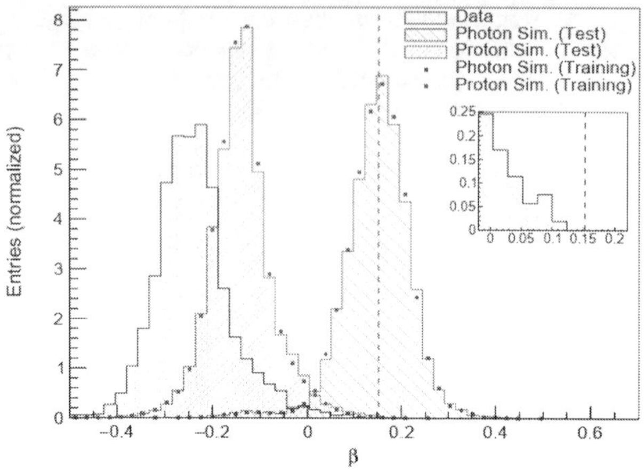


(a)

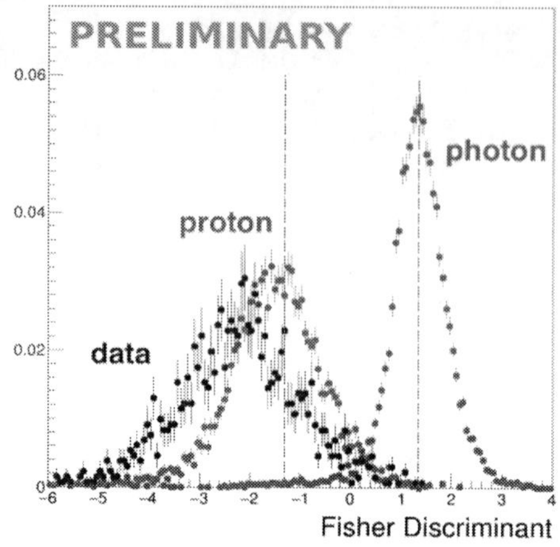


(b)

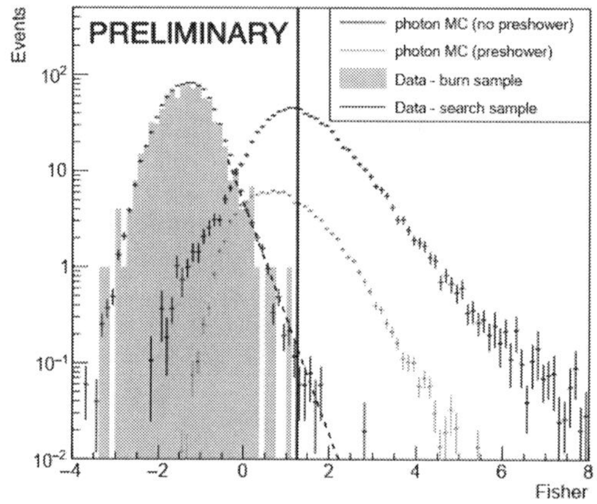


(c)

**Figure 1.** Results of the multivariate analyses used to search for primary photons in the data collected by the Pierre Auger Observatory; (a) energy range between $2{\times}10^{17}$ and $10^{18}$ eV, using HeCo data combined with data from the 750 m SD array [3]; (b) energy range between $10^{18}$ and $10^{19}$ eV using data from the main fluorescence telescopes and the 1500 m SD array [4]; (c) energy range above $10^{19}$ eV using only data from the 1500 m SD array [5].

for the bulk of the data, however, poses some challenges. For example, there is no measurement of $X_{\max}$ available. Also, the energy of the primary particle can only be determined indirectly. Here, we use a look-up table constructed with a large sample of photon-induced air showers. Two observables, $L_{\rm LDF}$ and $\Delta$, are used in this analysis. $L_{\rm LDF}$ quantifies the departure of the observed lateral distribution function (LDF) from the average of all SD data:

$$L_{\rm LDF} = \log_{10}\left(\frac{1}{N}\sum_{i=1}^{N}\frac{S_i}{f_{\rm LDF}(r_i)}\right),$$

where $S_i$ is the total signal measured in the $i$-th SD station and $f_{\rm LDF}(r_i)$ gives the average signal, obtained from all SD data, for a station at the same distance $r_i$ from the shower axis. $\Delta$ is defined through:

$$\Delta = \frac{1}{N}\sum_i \frac{(t^i_{1/2} - t^{\rm bench}_{1/2})}{\sigma_{t_{1/2}}},$$

which can be taken as the average deviation of the measured rise-times from a *data benchmark* $t^{\rm bench}_{1/2}$, describing the average risetime of all of the SD data (assumed to be overwhelmingly constituted by primary nuclei), in units of sampling fluctuations $\sigma_{t_{1/2}}$. The use of the average behavior of all SD data in the two quantities removes the need for assumptions on the composition of the background, which is not known in detail at the highest energies. The two observables are combined in a Fisher discriminant analysis, with the burnt sample—about 2 %

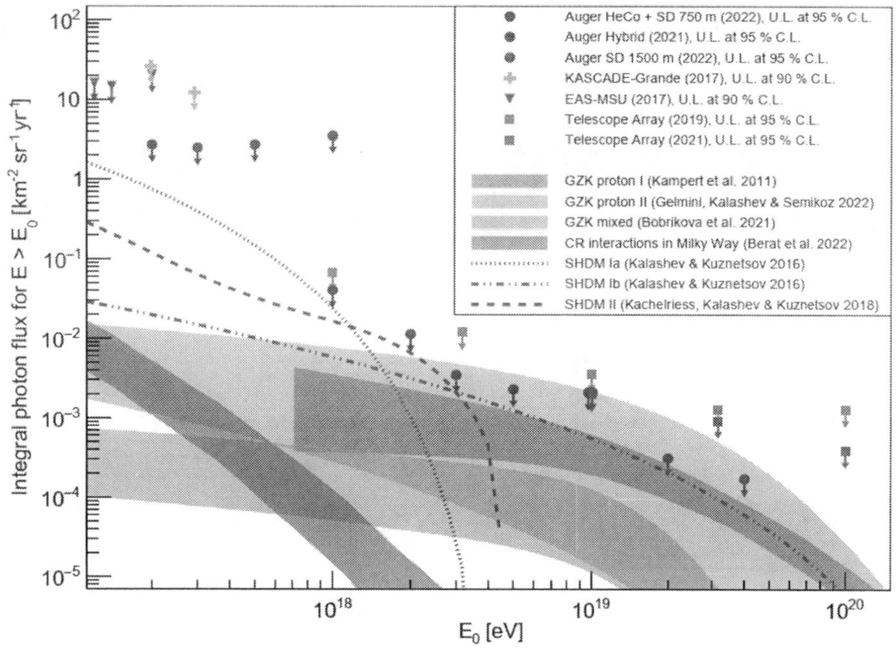


**Figure 2.** Current upper limits on the integral photon flux determined from data collected by the Pierre Auger Observatory (red, blue and gray circles); shown are also upper limits published by other experiments as well as the expected photon fluxes under different theoretical assumptions and scenarios; for a full list of references, see [11].

of the full data sample which are not used in the final analysis—representing the background and photon simulations the signal (see also Fig. 1(c)). The analysis is finally applied to SD data collected between 1 January 2004 and 30 June 2020, which corresponds to an exposure of about 17000 $km^2$ sr yr. Overall, 16 events from the data sample pass the photon candidate cut, which is consistent with the expectation. Again, no primary photon could be unambigously identified.

The upper limits on the integral photon flux determined through the three analyses discussed before are shown in Fig. 2, together with upper limits determined from other experiments as well as the expected photon fluxes under different theoretical assumptions and scenarios. The upper limits determined by the Pierre Auger Observatory are the most stringent ones to date, over a wide energy range spanning from $2\times10^{17}$ eV to the highest energies. More "exotic", top-down models have been strongly constrained—a number of such models were, in fact, already ruled out by previous results published by the Pierre Auger Collaboration. On the other hand, the predictions from some cosmogenic models, for example involving interactions of UHE cosmic rays with the cosmic microwave background, are within reach. For a more detailed discussion of the astrophysical significance of these limits, see [11]. However, we note here that upper limits on the diffuse flux of photons are especially useful to constrain theoretical models involving super-heavy dark matter (see, e.g.,[14, 15]).

## 4 Searches for UHE neutrinos

In a nutshell, searching for UHE neutrinos in air-shower data means searching for inclined showers with an electromagnetic component. Protons and nuclei initiate air showers high in the atmosphere. At large zenith angles, the electromagnetic component is fully absorbed within the atmosphere, and only muons arrive on the ground. Neutrinos, on the other hand, have a much smaller interaction cross section and can thus also initiate showers much deeper in the atmosphere, close to the ground, so that the electromagnetic component of the shower can still be measured even at large zenith angles. This leads to a unique signature that can be efficiently detected with a ground array. Another search channel comes from Earth-skimming $\tau$ neutrinos interacting in the Earth's crust, producing a $\tau$ lepton that enters the atmosphere from below and initiates a slightly upgoing air shower close to the ground—again, a unique signature.

All searches for UHE neutrinos at the Pierre Auger Observatory are based on these signatures. The analyses are categorized into three search channels, based on the zenith angle: Earth-skimming (ES, zenith angles between 90° and 95°; down-going high angle (DGH, zenith angles between 75° and 90°) and down-going low-angle (DGL, zenith angles between 60° and 75°). The down-going channels are further subdivided into a number of sub-channels based on the number of triggered SD stations in the event (for the DGH channel) or based on the zenith angle (for the DGL channel). Using the combined results

https://doi.org/10.1051/epjconf/202328304003

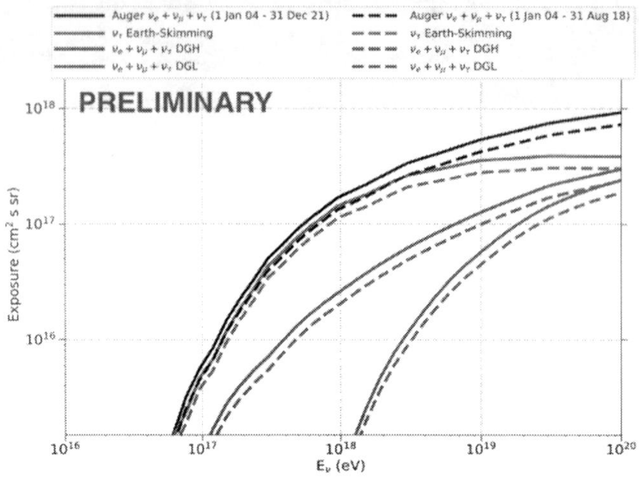


**Figure 3.** Exposure of the 1500 m SD array to UHE neutrinos as a function of energy, for the three search channels (ES, DGH, DGL) and the data period 1 January 2004 to 31 December 2021 (solid lines); for comparison, also the corresponding exposure for the data period from [16] is included (dashed lines).

of all three search channels, upper limits on the diffuse flux of UHE neutrinos have been obtained using data from the 1500 m SD array collected between 1 January 2004 and 31 August 2018 [16]. Here, we update these results with data until 31 December 2021. In addition, limits on point-like sources of UHE neutrinos have been determined [17], which are not discussed further here. Follow-up searches for UHE neutrinos from transient events constitute an important contribution of the Pierre Auger Observatory to current multimessenger studies, in particular for gravitational wave events (see, e.g., [18]). We note again that these multimessenger analyses are summarized in [10].

As mentioned before, we update here our results from [16] with about 3.5 more years of data, which constitutes approximately a 30 % increase in exposure. The exposure to neutrinos, as a function of energy and determined from simulations, is shown in Fig. 3, divided into the three search channels and compared to the exposure from [16]. The exposure is dominated by the ES channel up to about $4\times10^{19}$ eV, then the down-going channels take over.

As examples, we show the results of the data unblinding in two search channels in Fig. 4. In the ES channel (Fig. 4(a)), the average area-over-peak ⟨AoP⟩, defined as the ratio of the integral of the time trace in a SD station to its peak value, averaged over all triggered stations in an SD event, is used as the discriminating observable. The neutrino candidate cut is chosen based on a subset of the data sample, such that a false-positive rate of one event per 50 years is expected. At this cut value, the neutrino selection efficiency is about 95 %. In Fig. 4(b), the results of the unblinding in the DGH channel, specifically the sub-channel with $N_{\text{stations}} \in [7, 11]$, is shown. In the DGL and DGH channels, the analysis is again based on the AoP values, but in this case, several observables that carry information on the time spread of the signals in the SD stations are constructed from the individual AoP values and subsequently combined in a Fisher discriminant analysis. The neutrino candidate cut is independently optimized for each sub-channel, again ensuring a false-positive rate of one event in 50 years. In the specific sub-channel shown in Fig. 4(b), the neutrino selection efficiency is about 85 %. No neutrino candidate event has been identified in any of the search channels.

From the non-observation of any neutrino event, upper limits on the diffuse flux of neutrinos are determined. Assuming a (single flavor) neutrino flux of the form $\phi = k \times E_\nu^{-2}$, an upper limit on the normalization $k$, for the energy range between $10^{17}$ and $2.5\times10^{19}$ eV, was obtained at $3.5\times10^{-9}$ GeV cm$^{-2}$ s$^{-1}$ sr$^{-1}$ or, equivalently, 1.1 EeV km$^{-2}$ yr$^{-1}$ sr$^{-1}$. Since no neutrino candidate events have been identified, the increased exposure from the larger data set directly translate into an improvement of the upper limit on $k$. Upper limits can also be determined for energy bins of width 0.5 in $\log_{10}(E_\nu\,/\,\text{eV})$. These (differential) limits are an effective way of characterizing the energy dependence of the sensitivity of a neutrino experiment. For the Pierre Auger Observatory, the best sensitivity is achieved at energies around $10^{18}$ eV, where it becomes comparable to that of IceCube. Both the integral and differential upper limits are shown in Fig. 5, together with upper limits determined by the IceCube and ANITA experiments, as well as the expected neutrino fluxes under different theoretical assumptions and scenarios. Using these results, we can constrain several classes of models of neutrino production, both cosmogenic and astrophysical. In particular, cosmogenic models involving a pure-proton composition and a strong evolution of the sources with redshift are strongly constrained and even excluded due to the non-observation of neutrinos so far.

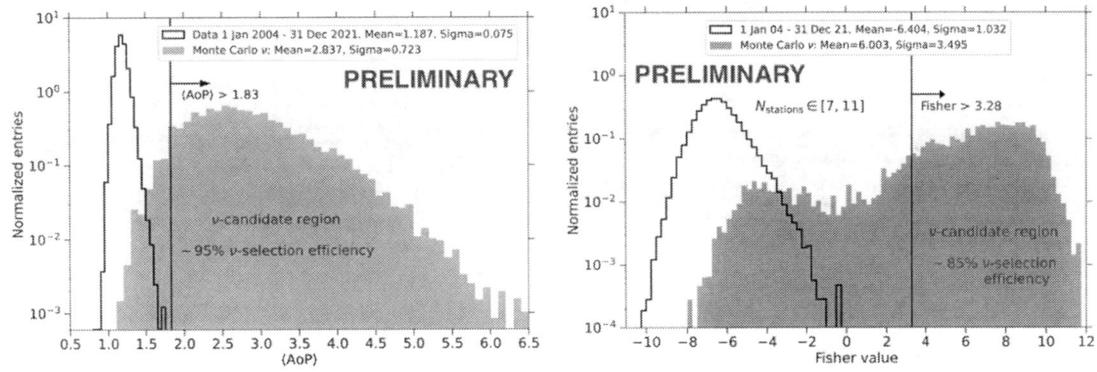


(a) (b)

**Figure 4.** Results of the data unblinding shown exemplarily for two search channels; (a) ES channel, using the average area-over-peak in an SD event as discriminating observable; (b) DGH channel, in the sub-channel with $N_{\mathrm{stations}} \in [7, 11]$, where a Fisher discriminant analysis involving several observables constructed from the individual values of area-over-peak in an SD event.

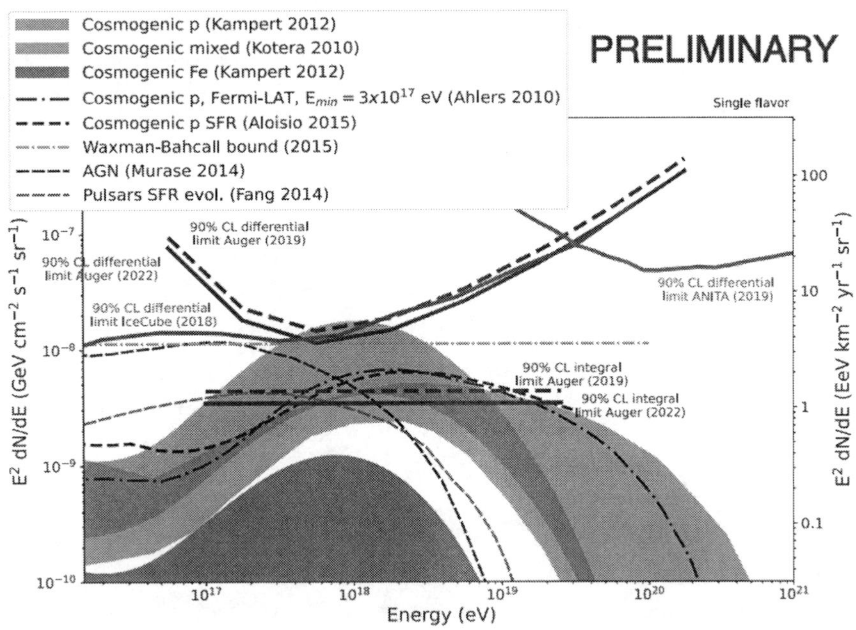


**Figure 5.** Updated upper limits on the diffuse flux of neutrinos, both the integral (straight solid line) as well as differential (curved solid line); for comparison, also the corresponding upper limits from [16] are shown; included are also upper limits from IceCube and ANITA and the expected neutrino fluxes under different theoretical assumptions and scenarios; for a full list of references, see [16].

However, we note that a pure-proton composition in UHE cosmic rays is already disfavored by other measurements from the Pierre Auger Observatory (see, e.g., [19]).

## 5 Summary and Outlook

In summary, the Pierre Auger Observatory offers an unprecedented exposure not only to UHE cosmic rays, but also to UHE photons and neutrinos. Stringent upper limits on the diffuse flux of UHE photons and neutrinos have been obtained from the data collected by the Observatory. Here, we summarized these results and updated the upper limits on the diffuse flux of neutrinos with 3.5 more years of data, leading to an improvement of the upper limits by about 30 %. Overall, these results underline that the Pierre Auger Observatory is a key actor in multimessenger astronomy at the highest energies. The ongoing upgrade of the detector systems, AugerPrime [20–22], will only strengthen this role. A key part of this upgrade is the installation of scintillation detectors on top of the detector stations of the SD. The different detector responses to the electromagnetic and muonic shower components can be used to disentangle the components, from which searches for UHE photons, for example, can directly profit. In addition, the detector stations will be equipped with radio antennas to measure the radio signals emitted by an air shower, and the electronics of the SD stations will be replaced to achieve a better time resolution and a larger dynamic range. All of these efforts combined will significantly improve the upper limits or, in the best case, lead to the first unambiguous detection of a neutral particle at ultra-high energies.

## References

[1] A. Aab et al. (Pierre Auger Collaboration), Nucl. Instrum. Meth. A **798**, 172 (2015), 1502.01323
[2] M. Risse, P. Homola, Mod. Phys. Lett. A **22**, 749 (2007), astro-ph/0702632
[3] P. Abreu et al. (Pierre Auger Collaboration), Astrophys. J. **933**, 125 (2022), 2205.14864
[4] P. Savina for the Pierre Auger Collaboration, PoS **ICRC2021**, 373 (2021)
[5] P. Abreu et al. (Pierre Auger Collaboration), Article submitted to J. Cosmol. Astropart. Phys. (2022), 2209.05926
[6] A. Aab et al. (Pierre Auger Collaboration), Astrophys. J. **789**, 160 (2014), 1406.2912
[7] A. Aab et al. (Pierre Auger Collaboration), Astrophys. J. Lett. **837**, L25 (2017), 1612.04155
[8] P. Ruehl for the Pierre Auger Collaboration, PoS **ICRC2021**, 973 (2021)
[9] A. Aab et al. (Pierre Auger Collaboration), Astrophys. J. **902**, 105 (2020), 2010.10953
[10] L. Perrone for the Pierre Auger Collaboration, these proceedings (2023)
[11] P. Abreu et al. (Pierre Auger Collaboration), Universe **8**, 579 (2022), 2210.12959
[12] G. Ros, A.D. Supanitsky, G.A. Medina-Tanco, L. del Peral, J.C. D'Olivo, M.D. Rodriguez-Frias, Astropart. Phys. **35**, 140 (2011), 1104.3399
[13] P. Savina, C. Bleve, L. Perrone, Proc. 36th International Cosmic Ray Conference (Madison, USA) **PoS (ICRC2019)**, 414 (2019)
[14] P. Abreu et al. (Pierre Auger Collaboration) (2022), 2203.08854
[15] P. Abreu et al. (Pierre Auger Collaboration) (2022), 2208.02353
[16] A. Aab et al. (Pierre Auger Collaboration), JCAP **10**, 022 (2019), 1906.07422
[17] A. Aab et al. (Pierre Auger Collaboration), JCAP **11**, 004 (2019), 1906.07419
[18] A. Albert et al. (ANTARES, IceCube, Pierre Auger, LIGO Scientific, Virgo Collaborations), Astrophys. J. Lett. **850**, L35 (2017), 1710.05839
[19] A. Aab et al. (Pierre Auger Collaboration), Phys. Rev. D **90**, 122005 (2014), 1409.4809
[20] A. Castellina, Pierre Auger Collaboration (Pierre Auger Collaboration), EPJ Web Conf. **210**, 06002 (2019), 1905.04472
[21] A. Aab et al. (Pierre Auger Collaboration) (2016), 1604.03637
[22] C. Bérat for the Pierre Auger Collaboration, these proceedings (2023)

EPJ Web of Conferences **283**, 04004 (2023) https://doi.org/10.1051/epjconf/202328304004
*UHECR 2022*

# Multi-messenger studies with the Pierre Auger Observatory

Lorenzo Perrone[1,*] for the Pierre Auger Collaboration[2,**]

[1]Universtità del Salento and INFN Lecce (Italy)
[2]Observatorio Pierre Auger, Av. San Martín Norte 304, 5613 Malargüe, Argentina
Full author list: http://www.auger.org/archive/authors_2022_10.html

**Abstract.** The combination of data from observatories measuring ultra-high energy cosmic rays, photons, neutrinos and gravitational waves has provided new insights into the most extreme phenomena in the Universe. Sharing information within a broad community is the foundation of the multi-messenger approach. The Pierre Auger Observatory, the world's largest cosmic ray detector, provides sensitivity to photons and neutrinos above $10^{17}$ eV, thus contributing significantly to this joint effort. The latest results from diffuse and targeted searches will be reviewed here, along with results from follow-up analyses and future perspectives. In particular, preliminary limits on photon fluence from a selection of gravitational wave sources detected by LIGO/Virgo and results of the search for ultra-high energy neutrinos from binary black hole mergers will be presented.

## 1 Introduction

In the era of multi-messenger astronomy, ultra-high energy cosmic rays (UHECRs) offer the unique opportunity to investigate the nature of astrophysical sources, and particle interactions, in an energy range far beyond that covered by current particle accelerators. The Pierre Auger Observatory [1], the world's largest cosmic ray detector, combines in a hybrid design the information from fluorescence telescopes, observing the longitudinal profile of extensive air showers, with a surface array, measuring the lateral distributions of secondary particles at the ground.

The Observatory consists of a surface detector (SD) array of about 3000 km$^2$ surrounded by 27 air fluorescence detector (FD) telescopes grouped in four sites, providing a powerful instrument for extensive air shower detection. The SD comprises ~ 1600 water-Cherenkov detectors separated by 1500 m (SD-1500) in a triangular grid, plus a smaller nested array of 61 additional detectors spaced by 750 m (SD-750) covering an area of 23.5 km$^2$. The FD provides a nearly calorimetric estimate of the primary energy, almost independent of the assumptions on hadronic interaction models at the highest energies. The operation of the FD is however restricted to a duty cycle of about 15% because it can only operate during clear, moonless nights and it is limited by the atmospheric conditions while the SD measurements are made for about 100% of the time. The hybrid paradigm relies on the fact that it is possible to calibrate the SD signal using the events simultaneously recorded by the FD, the so-called hybrid events, avoiding to a large extent the use of Monte Carlo simulations and allowing us to achieve good control of the systematic uncertainties in the energy scale [2].

## 2 Multi-messenger astronomy

UHECRs are mostly protons and other atomic nuclei; a fraction of photons would be an ideal probe for fundamental questions in particle physics and cosmology, such as the nature of dark matter and the possibility of Lorentz invariance violation (LIV). Indeed, the decay of superheavy DM particles could result in large fluxes of UHE Standard Model particles [3], with photons and neutrinos dominating the final state. Also, UHE photons could result from the decay of hadrons produced when protons with ~ 10 times the final photon energy interact with microwave/IR/visible/UV background photons, either within the source environment or during their propagation through the intergalactic medium. In the Standard Model, such UHE photons are expected to be quickly absorbed by background photons, but LIV phenomena could inhibit such interactions, resulting in a significantly higher fraction of photons in the UHECR flux.

Only UHECR observatories have capabilities to detect photons at these energies, through the extensive air showers they produce in the atmosphere. In this context, the Pierre Auger Observatory has the highest sensitivity to photon primaries. Due to the lower multiplicity of electromagnetic interactions (compared to hadronic), photons induced showers develop deeper in the atmosphere and have fewer muons thus yielding a larger depth at shower maximum, $X_{max}$, and a steeper lateral distribution function. These distinctive features are also reflected in a longer rise time for the signal shape observed with SD. Since the start of data collection, various searches for ultra-high-energy with energy above $10^{17}$ eV photons have been performed, either for a diffuse flux, for point sources or for photons associated with transient events.

*e-mail: lorenzo.perrone@lc.infn.it
**e-mail: spokespersons@auger.org

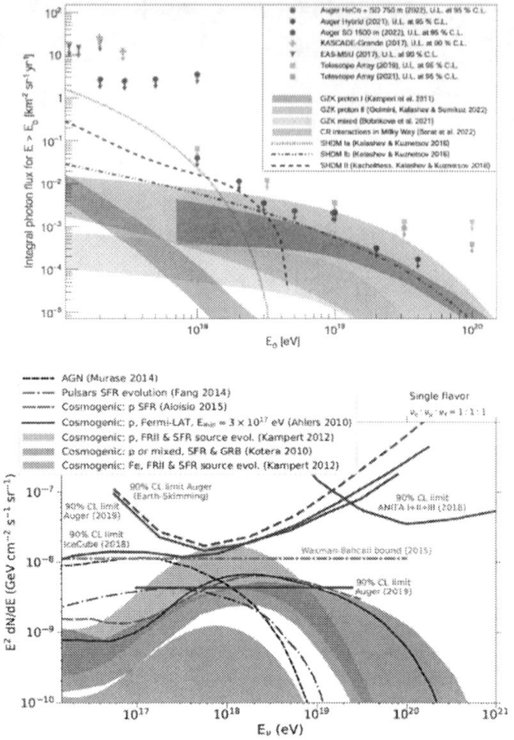


**Figure 1.** Upper limits on the cosmogenic photons (top) and neutrinos (bottom) fluxes compared to limits from other experiments and to model predictions ( [5, 6] and references therein).

A very recent review of current results can be found here [4] and details about the detection techniques for specific photon and neutrino searches have been presented at this conference [7]. As an example, the current status of the search for a diffuse flux of photons is shown in Fig. 1, top panel. Upper limits set by the fluorescence detector operating in hybrid mode play an important role at the lowest and intermediate energies, up the EeV range, those set by the surface detector dominate at the highest energies, above $10^{19}$ eV.

Several model predictions in the context of dark matter scenarions are severely constrained by the Auger results and also the expectation of photons originating from the interaction of UHECRs with the interstallar medium start to be in tension with a scenario of a proton-dominated chemical composition. Thus, the search for photons can provide strong bounds on fundamental physics and, at the same time, valuable indications on UHECRs mass composition.

While photons have a limited horizon due to their interaction with the interstellar radiation field (few Mpc at $10^{18}$ eV), neutrinos can travel substantially unabsorbed over cosmological distances. Their fluxes depend on the properties of the sources and on the composition of the primary beam. Neutrinos can be searched for by selecting "horizontal showers" with a large electromagnetic component close to the detector. This provides a clear identifier of a neutrino-induced shower as, for hadronic cascades, the electromagnetic component would be almost completely absorbed and only muons could be detected in the SD. Two main categories of events are considered: Earth-skimming, induced by $\nu_\tau$ travelling upwards from below the Earth crust in directions just below the horizon and producing a $\tau$ lepton which can then generate a shower above the SD, and downward-going events due to neutrinos of any flavour. Fig. 1, bottom panel, shows the most recent upper limits on cosmogenic neutrinos [6], along with model predictions and limits from other experiments.

Given the very good angular resolution of the Observatory ($\lesssim 1°$) specific follow-up searches were performed with the SD detector looking for point-like sources in spacial and time coincidence with gravitational waves (GW) events. These topics will be reviewed in the following sections.

## 3 Search for photons from GW events

A subset of the most interesting GW events from the LIGO/Virgo collaborations was selected based on their localization quality and distance [9]. Time periods of 1000 s around and 1 day after the GW events are analyzed. To maximize the photon-hadron separation power, the observables are combined using a Fisher discriminant analysis taking advantage from the deviation of the expected photon lateral distribution function and of the observed signal risetime from a benchmark reference value measured in data. The distributions of the Fisher discriminant of data events and a set of simulated photon events are shown in Fig. 2. A typical photon-induced air shower is expected to have a significantly larger Fisher discriminant value than the average event found in data. Then, a threshold value may be placed to define which events will be accepted as "photon candidate events". The photon dis-

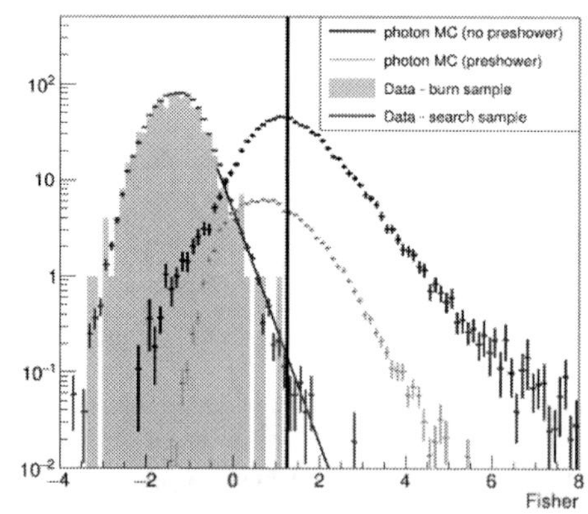


**Figure 2.** Distributions of the Fisher discriminant values of a set of simulated photon events (blue) and data events (red) with photon energies above $10^{19}$ eV recorded by the SD. See text and [8] for further details.

https://doi.org/10.1051/epjconf/202328304004

crimination method is optimized for air showers with incident zenith angles between 30° and 60° and photon energies above $10^{19}$ eV. Following the approach described in [8], out of all air shower events recorded during a period of 16.5 yr (from January 2004 to June 2020), 16 events passed the photon candidate cut which was placed at the median of the distribution of photon simulations (vertical line in Fig. 2.). This number was found to be consistent with the expected hadronic background.

The GW events considered in the analysis described in [9] were recorded by Advanced LIGO and Virgo during their first three observation runs and published in three gravitational wave transient catalogs: GWTC-1 [10], GWTC-2 [11] with its second revision GWTC-2.1 [12], and GWTC-3 [13]. While the first catalog covers the observations of the first two runs O1 (from 2015 September 12 to 2016 January 19) and O2 (from 2016 November 30 to 2017 August 25), the third observation run has been split in two parts, O3a (from 2019 April 1 to 2019 October 1) and O3b (from 2019 November 1 to 2020 March 27), with a maintenance break of 1 month in between. In order to keep the sensitivity to a possible photon signal as high as possible, GW events are additionally selected by their localization quality and distance. Close and well localized sources are preferred over distant and poorly localized ones. Thus, optimal results can be obtained while keeping the expected background at a reasonable level. Four classes of accepted GW events are defined here for which the 50% localization region is analyzed for coincident air shower events (see Fig. 3). These selection criteria can be summarized as: $(D_L < \infty$ and $\Omega_{50\%} < 100\ \mathrm{deg}^2)_{\mathrm{short}}$ (class I), $(D_L < \infty$ and $\Omega_{50\%} < 20\ \mathrm{deg}^2)_{\mathrm{long}}$ (class II), $(D_L < 180$ Mpc and $\Omega_{50\%} < 100\ \mathrm{deg}^2)_{\mathrm{long}}$ (class III), $(D_L < 40$ Mpc and $\Omega_{50\%} < 720\ \mathrm{deg}^2)_{\mathrm{long,short}}$ (class IV), where "long" ("short") stands for a time window of 1 day (1000 s) starting after (before) 500 s from the time of a GW event. Since classically no photon signal is expected from very distant sources, the fact of considering large distances (as for class I and II) keeps a window open for potential discoveries of new physics, while the maximum distance chosen for class III corresponds to the most optimistic horizon for photon primaries at the highest energies. A final class of accepted GW events (class IV) allows well identified sources within a distance providing a realistic chance of observing a potential UHE photon flux or at least placing strong physical constraints on the fraction of energy transferred into UHE photons. The value of 50 Mpc is defined by the maximum distance a source like GW170817 may have so that the fraction of energy transferred into UHE photons could still be constrained by a non-observation of photons at the SD array. In total, 23 GW events are classified in at least one of the classes described above. Out of these, 7 of their localization contours were at least partly covered by the Auger SD field of view in the long time window and 3 in the short time window. All 7 events in the long time window are found in class II while one of those, GW170817, also passes the selection criteria for classes I, III and IV. The 3 events in the short time window belong exclusively to class I, i.e. no event falls in both time windows.

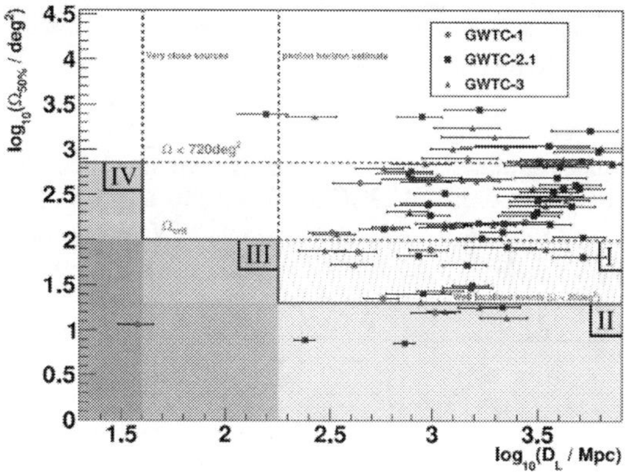


**Figure 3.** All GW events from GWTC-1 (green dots), GWTC-2.1 (blue squares) and GWTC-3 (red triangles) in the space of source distance DL and localization $\Omega_{50\%}$. Events which are not within the field of view in the 1 day time window are drawn with empty markers, while events which do at least partially overlap have solid markers. Three red crosses mark the events which pass the selection criteria for the short time window and also have an overlap with the field of fiew during that time. The shaded regions define the set of accepted events according to the selection citeria described in the text. The hatched region marks class I which is solely relevant for the short analysis time window and the solid regions mark classes II, III and IV [9].

No coincident air showers with energy above $10^{19}$ eV occured for any source in either of the time windows. This is well in agreement with the expected amount of chance coincidences. Consequently, also none of the 16 photon candidate events from [8] was found to be coincident with any of the selected GWs. Following this non-observation of coincident events, for each GW source an upper limit on the number of photons can be placed using the Feldman and Cousins approach [14]. It can then be converted into photon flux upper limits using the directional exposure calculated in [9] assuming for example a power-law shape with spectral index $\alpha$ for the primary photon flux

$$\frac{d\Phi_\gamma^{GW}}{dE_\gamma}(E_\gamma) = k_\gamma E_\gamma^\alpha \tag{1}$$

In addition, an upper limit on the spectral fluence $F_\gamma^{UL}$ of UHE photons arriving from a given source at the Earth can be derived from the flux upper limit by integrating over the chosen time window $[t_0, t_1]$ and energy range $[E_0, E_1]$:

$$F_\gamma^{UL} = \int_{t_0}^{t_1} \int_{E_0}^{E_1} dt\, dE_\gamma\, E_\gamma \frac{d\Phi_\gamma^{GW}}{dE_\gamma}(E_\gamma) \tag{2}$$

While no assumption on the time dependence of the flux is made, the extrapolation of the flux limits (derived for the time of transit of the source through the field of view) to the full time window is done under the assumption that the average flux during the period for which the source has been in the field of view is representative for the whole time window. The limits on the spectral fluence depend on the exact direction of the GW source and on the assumed spectral shape of the UHE photon flux. Hence, in Fig. 4 the results for $F_{\gamma}^{UL}$ are shown for all possible source directions within each localization contour and a variation of the spectral index $\alpha$ between -2.3 and -1.7, for both time windows. In the long time window, all localization regions have been fully covered by the field of view except for GW170818. Hence, this event could not be constrained for all source directions within the $\Omega_{50\%}$-contour. All three GW sources in the short time window have contours which partly leak out of the field of view. No coincidences are observed. Upper limits on the UHE

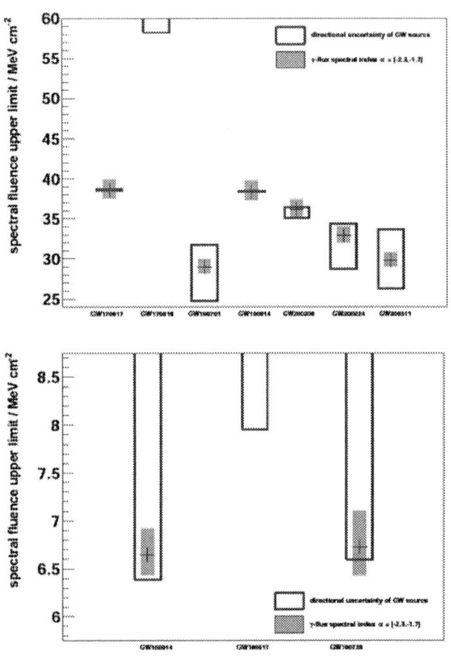


**Figure 4.** Upper limits (at 90% CL) on the spectral fluence of UHE photons from the selected GW sources for the searches in the long (top) and in the short-time window (bottom). The limits for the most likely direction and a spectral index of $\alpha$ = -2 are marked by the cross. The blue (empty boxes) error bars correspond to the variation of the upper limits due to the directional uncertainty of the source. Red (shaded boxes) error bars show the impact of a variation of the spectral index. Further details in [9].

photon fluence from a GW event are typically at the level of ~ 7 MeV cm$^{-2}$ for the time period 1000 s and of ~ 35 MeV cm$^{-2}$ for the time period of 1 day. Due to the proximity of the binary neutron star merger GW170817, the energy of the source transferred into UHE photons above 40 EeV is constrained to be less than 20% of its total gravitational wave energy. These are the first limits on UHE photons from GW sources.

## 4 Search for neutrinos from GW events

The capability of the Pierre Auger Observatory to detect and identify UHE neutrinos has been shown and discussed in section 2 for the diffuse search. The sensitivity to UHE neutrinos covers a large volume from well above 90% of the Sky, in a declination range from -90° to +60° and an accurate knowledge of the Observatory directionaly sensitivity is required for targeted neutrino search. Fig. 5 shows the upper limits derived for point sources as a function of declination [15], under the assumption of a neutrino flux following a power law $\propto E^{-2}$. These limits are derived at

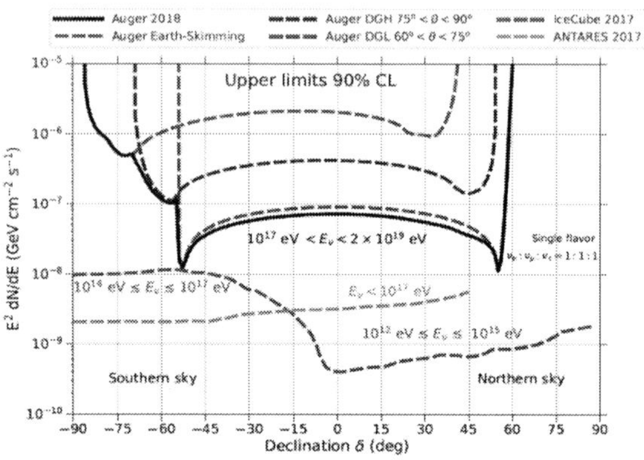


**Figure 5.** Upper limits on the neutrino flux normalization from steady sources [15]. Limits are calculated for a point source at given declination and single flavour neutrinos.

energies higher than the ones covered by ANTARES and IceCube (see the corresponding upper limits also shown in Fig. 5), thus demonstrating the complementarity potential of the Pierre Auger Observatory as neutrino detector in the multi-messenger astronomy joint enterprise.

The UHE neutrino detection capability of the Pierre Auger Observatory makes it also a promising instrument for searches of transient sources such as UHE neutrinos from binary black hole (BBH) mergers that have been observed through the emission of gravitational waves. A dedicated search has been performed for two different durations after each merger: 24-hour and 60-days [16]. By accounting for the time dependent exposure to neutrinos emitted from 3D localization probability distributions provided by LIGO/Virgo, the UHE neutrino luminosity is derived as a function of time after the merger, assuming a universal emission model.

After the first detection of a direct GW signal from a BBH merger, the LIGO and Virgo Collaborations carried out a total of three observational runs (O1, O2, and O3), as reported in section 3. In O2, the first GW signal from a binary neutron star merger, GW170817, was detected, accompanied by an unprecedented multi-wavelength and multi-messenger observation and search campaign [17] featuring a rich variety of observations that remain unique to date. Part of this was also the follow-up search for high and ultra-high energy neutrinos [18]. No neutrino candidates with significant temporal and directional correlation have been observed, leading to constraints in the astrophysical parameter space of the source. In addition, the detection of GW signals from 83 BBH mergers, the most prevalent type of GW events, have been reported and published [11–13] by LIGO/Virgo.

The Pierre Auger Observatory sensitivity across the field of view is not uniform (it is much enhanced close to the horizon), and it depends on the transit and residence time through the field of view of a specific detection channel. An example is again the follow-up search of GW170817, which was located in the most sensitive region directly below the horizon at the time of the merger, allowing to set very stringent UHE neutrino limits in particular for short-term emission. Further objects with follow-up searches for UHE neutrinos include the neutrino-emitting blazar TXS 0506+056 [19].

To quantify the direction-dependent sensitivity to UHE neutrinos for the case of the Pierre Auger Observatory, the effective area $A_{eff}$ is defined as a function of the local zenith angle and neutrino energy, through the relation between the rate of detected neutrino events and the neutrino energy spectral flux $\Phi_\nu$:

$$\frac{dN_\nu(\theta,t)}{d\Omega dt} = \int_0^\infty \phi_\nu(E_\nu) A_{eff}(E_\nu,\theta,t) dE_\nu \tag{3}$$

The average of $A_{eff}$ over the year 2016 is shown in Fig. 6 top panel, as a function of neutrino energy for various zenith angles. It is overall largest for $\tau$ neutrinos arriving from directly below the horizon, peaking at a zenith angle of 90.8°. At this angle, the extensive air showers induced by UHE neutrinos are very well distinguishable from those induced by UHECRs, while the Earth is more transparent to UHE neutrinos than at higher zenith angles, especially at the highest energies. This is the reason why the effective area generally decreases quickly for zenith angles above 91°. Here, UHE neutrino energy spectra are assumed to be $\propto E^{-2}$ which is a common benchmark assumption for neutrino analyses at the highest energies. The effective area can then be folded with the assumed spectrum, yielding to an effective area per energy $\mathcal{A}$

$$\mathcal{A}(\theta,t) = \int_0^\infty E_\nu^{-2} A_{eff}(E_\nu,\theta,t) dE_\nu \tag{4}$$

$\mathcal{A}$ is proportional to the expected rate of detected neutrinos as it combines the efficiency and the effective geometric area of the SD. $\mathcal{A}$ is shown in Fig. 6, bottom as a function of the zenith angle, using the benchmark $A_{eff}$ from the top panel of the same figure. The non-observation of neutrino candidates with a background low enough to be compatible with zero allows to set an upper limit on the expected number of induced neutrinos that can be converted in terms of the UHE neutrino luminosity of the sources. This can be considered as an astrophysical interpretation of the flux limit at Earth that one could also deduce from the non-observation of neutrino candidates at the given search times and directions. The Pierre Auger Observatory approach is based on the assumption of a time-dependent universal UHE neutrino luminosity $L_{up}$(t) with which each BBH merger emits neutrinos isotropically, starting at the time of the merger and lasting for the duration of the respective search period [16].

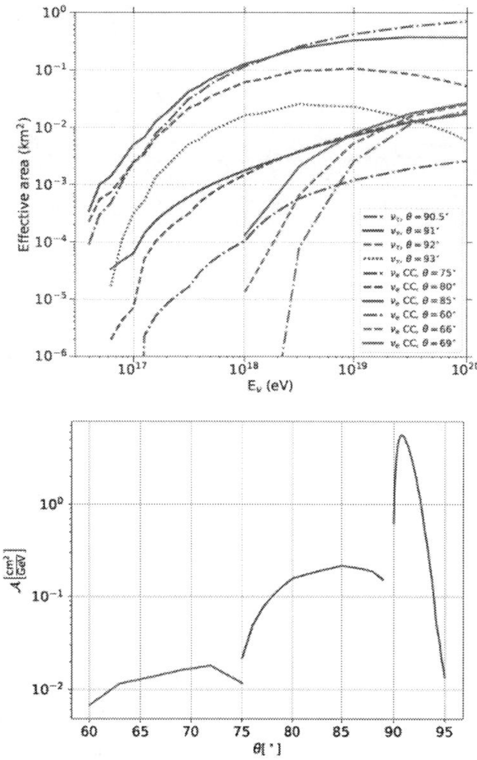


**Figure 6.** Top: the neutrino effective area of the Pierre Auger Observatory for various zenith angles, averaged over the year 2016. Bottom: The 2016 average of the neutrino effective area folded with the assumed $\propto E^{-2}$ neutrino energy spectrum. Colors code as in the top panel [15].

The universal isotropic luminosity upper limit obtained by combining all runs exhibits smaller variations than individual sources or single runs, as the variations in the visibilities of all sources are averaged out. The combination of several sources and different runs substantially improves the sensitivity to the source class of BBH mergers as compared to single sources or runs. The result for the universal isotropic luminosity for the 24-hour (top) and 60-days (bottom) follow-up period is shown in Fig. 7, along with the contributions from the individual LIGO/Virgo observational runs. The variations in $L_{up}$ are caused by the changing visibilities of the sources during a sidereal day. As an example, the most prominent minima in the curve representing the contribution of O2 occur due to the fact that the localizations of the relatively close-by sources GW170608 and GW170814 coincide with the Earth-skimming band at the corresponding times after the merger. The corresponding limit on the total energy emitted in UHE neutrinos per source is ~ 5,6 $10^{51}$ ergs. This bound does not depend on the search duration.

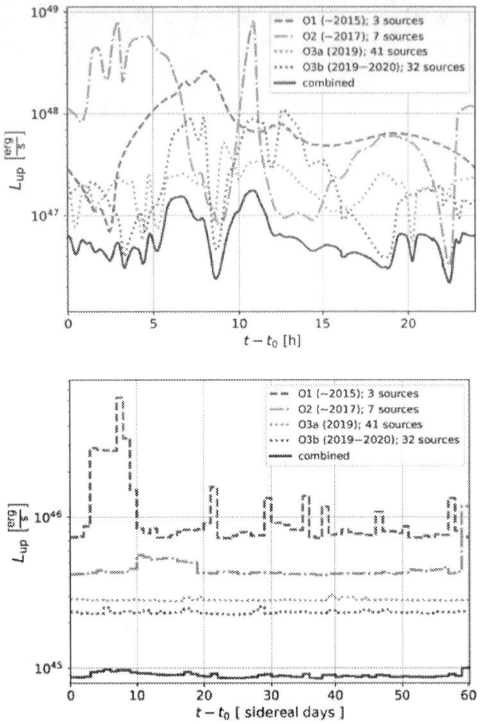


**Figure 7.** Solid line: upper limit (90% C.L.) on the time-dependent universal isotropic neutrino luminosity for the 24 hour (top) or the 60-day (bottom) follow-up period; other lines: contributions from individual LIGO/Virgo observational runs [16].

## 5 Search for upward-going extensive air showers with the FD

The FD is sensitive to upward-going events and can be used to search for upward-going showers without limitation on the zenith angle of the emerging induced showers. The first results from this study, triggered in the context of what reported by the ANITA Collaboration in the energy range above $10^{17}$ eV, have been reported in [20].
Given its operation time and wide field of view, the FD has the potential to support or constrain the "anomalous" observations by the ANITA detector [21], interpreted as upward-going air showers of unexplained nature. We have used 14 years of data collected by the FD to search for upward-going showers using a set of quality selection criteria defined using 10% of the full data sample. To distinguish candidates from false positives, calculate the exposure and obtain the expected background, dedicated simulations for signal (upward-going events) and background (downward-going events) have been performed. Results of the analysis after unblinding the data set are presented.

Upward-going protons have been simulated with energies between $10^{16.5}$ and $10^{18.5}$ eV, zenith angle between 90° and 180° and height of first interaction between 4 and 9 km. Protons have been chosen because they can be easily adapted to fit any interesting scenarios such as neutrinos or Beyond Standard Models (BSM) particles. Inclined downward-going events landing behind an FD site can also mimic an upward-going track in the fluorescence telescope so events with zenith angle between 0° and 90° have also been simulated as background.
When simulating an upward-going air shower the height of the first interaction point $H_{fi}$ can significantly change the trigger efficiency of the FD. For this reason a double differential exposure has been calculated as

$$\frac{d\varepsilon}{dH_{fi}}(E_{cal}, H_{fi}) \simeq 2\pi \cdot S_{gen} \cdot \Delta T \cdot$$
$$\sum_i \eta(E_{cal}, \cos\theta_i, H_{fi}) \cdot \frac{1}{\Delta H_{fi}} \cdot \cos\theta_i \cdot \Delta\cos\theta_i \,. \quad (5)$$

where $E_{cal}$ is the energy released by the shower in the air, $S_{gen}$ is the surface area of generation (a square of $100 \times 100$ km$^2$), $\Delta T$ is the 14 years of operation of the FD, $\eta$ is the fraction of events passing the selection and $\theta$ is the zenith angle. Fig. 8 shows this exposure based on upward-going proton simulations as a function of $E_{cal}$ and $H_{fi}$. As expected the FD exposure increases with energy being null for energies below $10^{17}$ eV. A blind analysis has been per-

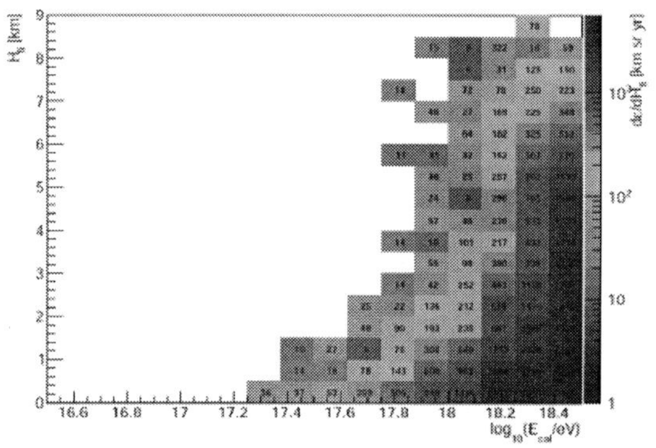


**Figure 8.** Double differential exposure with log10($E_{cal}$/eV) on the x-axis and the height of first interaction on the y-axis for upward-going events [20].

formed using 10% of FD data to study the background due to misreconstructed downward-going events [20]. Moreover upward-going lasers, used by the Collaboration for atmospheric monitoring, represent another possible source of background. They are mainly shot by two facilities located in the middle of the array and by 4 LIDARs located at each FD site. Lasers are mostly fired vertically and the large majority of them is rejected based on their known time stamp. However it may happen that some lasers leak into the data sample as genuinely upward-going events. An algorithm has been developed to identify and reject the remaining lasers by exploiting the time of each event and its position inside the array. A profile constrained geometry fit has been applied to the selected sample testing if any possible upward-going geometry can explain the event. In that case, downward geometries have been tested too. A variable

$$X = \text{atan}(-2\log(\frac{L_{down}}{L_{all}})) \cdot 2/\pi$$

has been defined to compare the likelihood of the two reconstructions so that an event with $X = 0$ is more likely a downward-going event, while if $X = 1$ the upward reconstruction is favoured. Fig. 9 shows the distribution of the $X$ variable for the events in the signal and background simulations as well as for the 10% data sample. According to this distribution a cut value of $X = 0.55$ has been set with an expected background of $\sim 0.5$ events in the full data sample.

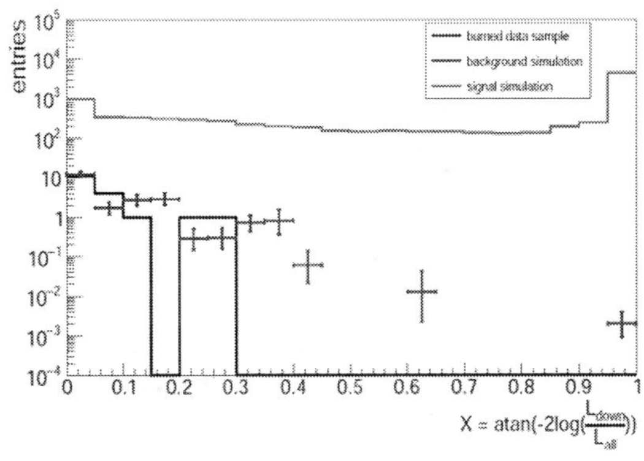


**Figure 9.** Distribution of the X variable (see text for details) for the events from 10% of data (black), background simulations (red), signal simulations (green) [20].

The unblinding procedure leads to only one event passing all the quality cuts, including the final one on the X variable. This result is compatible with the expected background and an integral upper limit to the flux of upward-going air showers has been set at 3.6 $10^{-20}$ $cm^{-2}$ $sr^{-1}$ $s^{-1}$ and 8.5 $10^{-20}$ $cm^{-2}$ $sr^{-1}$ $s^{-1}$ after weightening the exposure with $E_{cal}^{-1}$ and $E_{cal}^{-2}$, respectively.
A specific scenario assuming upward-going air showers are initiated by $\tau$ leptons has been also considered and the corresponding sensitivity of the FD detector calculated in detail [22]. Results have been presented also at this conference [23].

## References

[1] The Pierre Auger Collaboration, Nucl. Instrum. Meth. A **798**, 172 (2015).
[2] B. Dawson (Pierre Auger Collaboration), PoS(ICRC2019)231.
[3] The Pierre Auger Collaboration, accepted for publication on PRL, arXiv.2203.08854
[4] The Pierre Auger Collaboration, Universe **8** (2022) 579.
[5] The Pierre Auger Collaboration, Astrop.J. **933** (2022) 125.
[6] The Pierre Auger Collaboration, JCAP **10** (2019) 022.
[7] M. Niechciol (Pierre Auger Collaboration), at this Conference.
[8] The Pierre Auger Collaboration, submitted to JCAP, http://arxiv.org/abs/2209.05926
[9] The Pierre Auger Collaboration, Submitted to ApJ (2022), Philip Ruehl for the Pierre Auger Collaboration, PoS(ICRC2021)973
[10] LIGO and Virgo Collaborations, Phys. Rev. X **9** (2019) 031040
[11] LIGO and Virgo Collaborations, Phys. Rev. X **11** (2021) 021053
[12] LIGO and Virgo Collaborations, arXiv.2108.01045
[13] LIGO and Virgo Collaborations, arXiv.2111.03606
[14] G.J. Feldman, R.D. Cousins, Phys.Rev.D **57** (1998) 3873-3889
[15] The Pierre Auger Collaboration, JCAP **11** (2019) 004
[16] M. Schimp (Pierre Auger Collaboration), PoS(ICRC2021)968, paper in preparation.
[17] LIGO and Virgo Collaborations, Astrophys. J. Lett. **848** (2017) L12
[18] LIGO and Virgo Collaborations, Astrophys. J. Lett. **850** (2017) L35
[19] The Pierre Auger Collaboration, Astrophys. J. **902** (2020) 105
[20] M. Mastrodicasa (Pierre Auger Collaboration), PoS(ICRC2021)1140
[21] W. Gorham et al. Phys. Rev. Lett. **121** (2018) 161102
[22] I. Caracas (Pierre Auger Collaboration), PoS(ICRC2021)1145
[23] V. Novotny (Pierre Auger Collaboration) at this conference

# Cosmic ray acceleration by multiple shocks in the jets of Active Galactic Nuclei

*Ana Laura* Müller[1,*] and *Anabella* Araudo[1,2,**]

[1]Extreme Light Infrastructure ERIC,ELI Beamlines Facility, Za Radnicí 835, CZ-25241 Dolní Břežany, Czech Republic
[2]Laboratoire Univers et Particules de Montpellier (LUPM) Université Montpellier, CNRS/IN2P3, CC72, place Eugène Bataillon, 34095, Montpellier Cedex 5, France

**Abstract.** Active galactic nuclei are one of the most promising sources for accelerating particles up to the highest energies. In this contribution, we present a scenario in which cosmic rays are accelerated in multiple shocks created by the interaction of relativistic AGN jets with the winds of embedded massive stars. We solve the Fokker-Planck equation considering escape and radiative losses as well as the collective effect of the shocks and the reacceleration of the particles. Finally, we calculate the maximum energies that the particles can achieve and discuss the possibility of producing ultra-high energy cosmic rays in this astrophysical situation.

## 1 Introduction

Starburst galaxies and active galactic nuclei (AGNs) have been in the center of the discussion about the origin of ultra-high-energy cosmic rays (UHECRs) since the last years. Arguments in favour and against each of these astrophysical objects have been exposed by several authors based on theoretical and observational studies [1–3].

When proposing cosmic-ray sources, it is not only necessary to explain the high energies, but also the mass composition inferred from observations, which agrees with intermediate mass nuclei dominating at the higher energies [4, 5]. In the case of starburst galaxies, there is a large amount of massive stars and supernova explosions injecting heavy nuclei in the galactic disk, however, their strong photon fields enhance the photo-disintegration of these particles, and therefore, the superwind region in the halo was proposed as the most suitable place in these galaxies to accelerate particles up to the highest energies [1, 3]. Nevertheless, from the theoretical point of view to achieve energies of $10^{19} - 10^{20}$ eV has been demonstrated to be very unlikely there [see e.g., 6–9]. Besides that, starbursts do not satisfied the minimum energetic requirement to be sources of UHECRs [10, 11]. Jets from powerful radio-galaxies, on the other hand, fulfil the energetic requirements, but justifying the presence of atomic nuclei inside them is not trivial [see e.g., 12–18]. Moreover, the mechanism and efficiency of relativistic jets to accelerate particles are continuously a matter of debate.

The situation of a jet colliding with stellar winds was proposed a long time ago as a way to mass-loading jets with baryons [19]. Star-jet interactions were also suggested as a possible scenario to explain the bright X-ray knots observed in Centaurus A jet and as sources of gamma-rays in AGNs [20–24]. Additionally, the wind of massive stars is very rich in intermediate-mass nuclei. In particular, the presence of oxygen, nitrogen, and carbon makes the atmosphere of Wolf-Rayet (WR) stars to have a composition similar to the one observed in cosmic rays [25, 26]. Motivated by these ideas, we propose in this work a scenario where cosmic rays are accelerated by multiple shocks. Those shocks are produced by the interaction of the relativistic jet with massive stars, as it is sketched in Fig. 1. In Section 2 we describe our scenario. Section 3 contains the description of the procedure that we follow to calculate the particle distribution of the reaccelerated particles. In Section 4 we present our results and finally, Section 5 is dedicated to our conclusions.

## 2 Model

When the jet collides with the wind of a massive star a pair of shock waves is created (see Fig. 1). A non-relativistic shock propagates in the wind with velocity $\sim v_{\rm wind}$ and a relativistic shock propagates backwards into the jet with speed $\sim v_{\rm jet}$. A fraction of the kinetic energy of the jet is dissipated in each interaction and, therefore, the jet slowly decelerates. We assume that its power evolves as $L_{\rm jet}(z) = L^0_{\rm jet} e^{-\tau}$, where $L^0_{\rm jet}$ is the initial jet kinetic luminosity and

$$\tau = \int_{1\ \rm pc}^{z} \left(\frac{\sigma_{\rm sp}}{\sigma_{\rm j}}\right) n_\star(z')\,\sigma_{\rm j}\,{\rm d}z' \qquad (1)$$

where $n_\star(z)$ is the density of stars, and $\sigma_{\rm sp}$ and $\sigma_{\rm j}$ are the cross sections of the bow shock and the jet, respectively [22]. The lower limit of the integral is given by the minimum distance at which the first star can be located, we adopt $z = 1$ pc for this value. The size of the bow shock around the star is determined by the position of the stagna-

*e-mail: analaura.muller@eli-beams.eu
**e-mail: anabella.araudo@eli-beams.eu

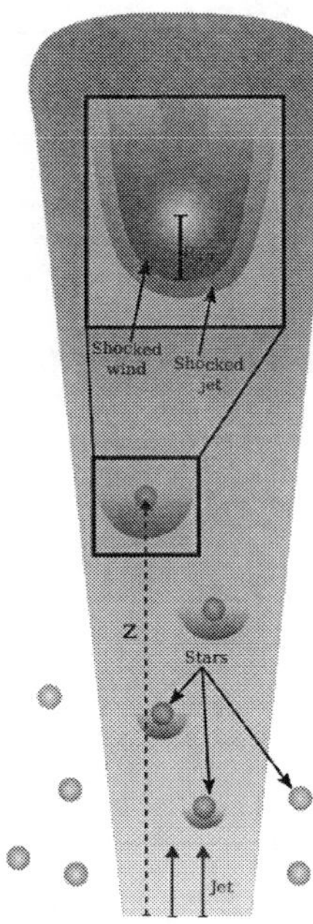


**Figure 1.** Graphic representation of the scenario studied in this work. Several stars are considered inside the AGN jet, represented by the orange spheres. The interaction of the relativistic jet with the stellar winds produces a pair of bow shocks consisting of a pair of shocks, one propagating through the stellar wind and one in the jet. The characteristic size of the bow shocks is given by stagnation point located at a distance $R_{sp}(z)$, which increases with the distance $z$ as indicated in Eq. (3). Particles accelerated in a bow shock can be reaccelerated in another bow shocks at larger $z$.

tion point $R_{sp}$. By equating the jet and wind kinetic pressures we obtain [20]

$$R_{sp}(z) = \sqrt{\frac{\dot{M}_\star\, v_{wind}\, v_{jet}\, R^2_{jet}(z)}{4\, L_{jet}}}, \quad (2)$$

where $\dot{M}_\star$ is the mass-loss rate of the star, $R_{jet}(z) = z \tan\theta$ is the jet radius, and $\theta$ is the jet opening angle. Taking into account the effect of the deceleration, $R_{sp}(z)$ can be written in first approximation as

$$R_{sp}(z) = \frac{R^0_{sp}(z)}{\sqrt{1 - \left(\frac{R^0_{sp}(z)}{R_{jet}(z)}\right)^2 N_\star(z)}}, \quad (3)$$

with $R^0_{sp}$ the the location of the stagnation point when the jet does not decelerate and $N_\star(z) = \int_{1\,pc}^{z} n_\star(z')\,\pi\, R_{jet}(z')^2 dz'$ is the cumulative number of stars into the jet. In Figure 2 we plot $N_\star$ using the stellar distribution from [22]

$$\frac{n_\star}{pc^{-3}} \sim 4.6 \times 10^{-4}\, (497)^y\, \eta^{0.89}_{accr} \left(\frac{M_{BH}}{10^7 M_\odot}\right)^{0.89} \left(\frac{z}{pc}\right)^{-y} \quad (4)$$

with $M_{BH} = 10^8$ $M_\odot$ the mass of the central black hole, $\eta_{accr} = 0.1$ the AGN accretion rate, and $y = 1$ and 2. The opening angle of the jet is assumed to be $\theta = 5°$. We can see in Fig 2 that for $y = 1$ one star is expected in the first 50 pc of the jet, ten stars at 100 pc, and fifty stars at 300 pc.

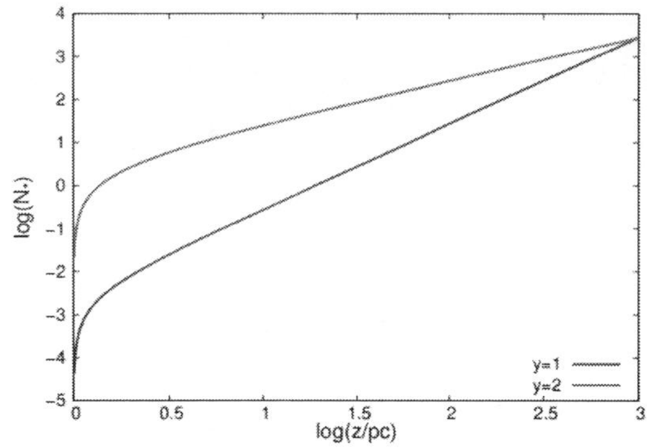


**Figure 2.** Cumulative number of stars inside a jet with opening angle $\theta = 5°$ for a black hole of $M_{BH} = 10^8$ $M_\odot$ and an accretion efficiency $\eta_{accr} = 0.1$. The $0.1 N_\star$ are considered to be WR stars.

The stellar distributions in Eq. (4) correspond to the total distribution of massive stars. In the present contribution we are interested only in WR stars given that they have more powerful winds and a chemical composition rich in intermediate-mass nuclei, although the number of WR stars is expected to be lower than the total number of massive stars. We assume that the number of WR stars is $n_{WR} = 0.1 n_\star$. We adopt standard parameters for a WR star: $\dot{M}_\star = 10^{-4}$ $M_\odot$ yr$^{-1}$, luminosity $L_\star = 10^{38}$ erg s$^{-1}$, temperature $T_\star = 3 \times 10^4$ K, and $v_{wind} = 3000$ km s$^{-1}$. For the jet we adopt $L^0_{jet} = 10^{42}$ erg s$^{-1}$, Lorentz factor $\Gamma_0 = 5$, and $\theta = 5°$. For calculating the deceleration of the jet we consider all the other stars to be OB stars with typical parameters $v_{wind} = 2000$ km s$^{-1}$ and $\dot{M}_\star = 10^{-6}$ $M_\odot$ yr$^{-1}$. In Fig. 3 we plot $R_{sp}(z)$ (and $R_{jet}$ for comparison). Note that $R_{sp}$ increases significantly when the number of WR stars into the jet is much larger than 10 ($z > 180$ pc for $y = 1$ and $z > 63$ pc for $y = 2$). Additionally, the deceleration of the jet makes also the Lorentz factor to be lower than $\Gamma_0$. Its change can be found by using the definition $L_{jet}(z) = [\Gamma(z) - 1]\rho^0_{jet}\, c^2\, \pi\, R^2_{jet}(z)\, v_{jet}(z)$, with $\rho^0_{jet}$ the jet density, and solving the transcendental equation

$$(\Gamma_0 - 1)\sqrt{1 - \frac{1}{\Gamma_0^2}e^{-\tau}} = (\Gamma - 1)\sqrt{1 - \frac{1}{\Gamma^2}}. \quad (5)$$

In Fig. 4 we plot $\Gamma$ for the two stellar distributions considered in the present study. We can see that $\Gamma \sim \Gamma_0$ until $z \sim 60$ pc for $y = 2$ and $z \sim 200$ pc for $y = 1$, where $N_{WR} = 10$ is large enough to affect the dynamics of the jet and also $R_{sp}(z)$ (See Fig. 3).

The magnetic field in the jet is assumed to be $B_{jet}(z) = B_0\, z^{-1}$, where $B_0 = \sqrt{4\, L_{jet}/(R^2_{jet}(z_0)\, c)}$, with $z_0 = 5 \times 10^{-5}(M_{BH}/10^7 M_\odot)$ pc the position where the jet is launched [22, 27, 28]. The magnetic field of the stellar wind is described as in [29]. Using these expressions,

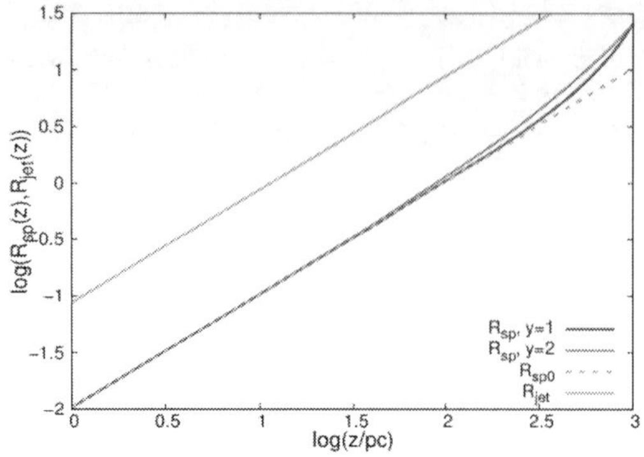


**Figure 3.** Comparison of $R_{sp}(z)$, $R^0_{sp}(z)$, and $R_{jet}(z)$. For the distribution of stars we adopt the expression given in Eq. 4 for $M_{BH} = 10^8\ M_\odot$ and $\eta_{accr} = 0.1$

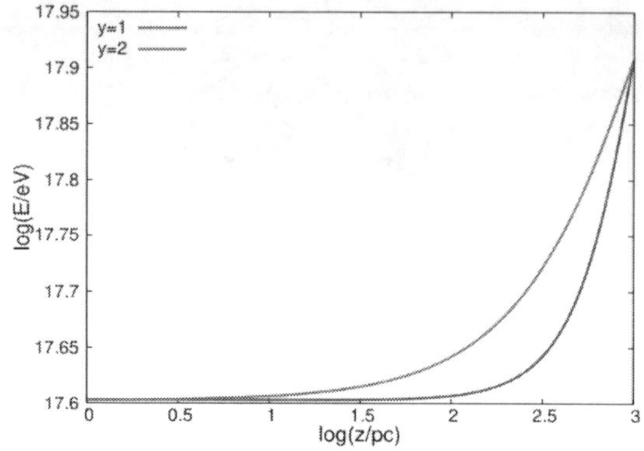


**Figure 5.** Hillas energy at the shocks along the jet considering the stellar population given by Eq. 4 and $Z = 1$ (protons).

along with $R_{sp}(z)$ and $\Gamma(z)$ we estimate the Hillas energy at the shock in the jet as

$$\frac{E_{Hillas}(z)}{\mathrm{eV}} = 5.1 \times 10^{17}\, Z \left(\frac{v_{jet}(z)}{c}\right)\left(\frac{R_{sp}(z)}{5\ \mathrm{pc}}\right)\left(\frac{B_{jet}(z)}{125\ \mu\mathrm{G}}\right), \tag{6}$$

where $Z$ is the atomic number [30]. Note that $R_{sp} = 5$ pc and $B_{jet} = 125\ \mu$G at $z = 338$ pc for $y = 2$. In Fig. 5 we show that $E_{Hillas}$ increases as a result of the faster increment of $R_{sp}(z)$ with respect to the decrease of $B_{jet}$ and $\Gamma$. At $z \sim 1$ kpc, $E_{Hillas} = 10^{18}$ eV and $2.6 \times 10^{19}$ eV, for protons and iron nuclei, respectively. Nevertheless, $E_{Hillas}$ is an upper limit. The actual maximum energies expected at each shock will be determined by the radiative and escape losses, as we discuss in the next section.

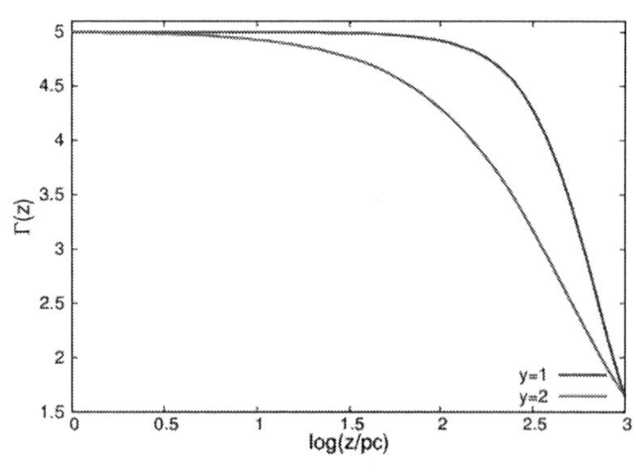


**Figure 4.** Evolution of the Lorentz factor because of the deceleration of the jet for the stellar distribution given in Eq. 5 with $y = 1$ and $y = 2$.

## 3 Multiple shock acceleration

Particles can be accelerated in the bow shocks created by the interaction of the jet with the winds of WR stars. Acceleration at single bow shocks has been considered before in order to study the gamma-ray emission from misaligned AGN jets [20–22]. Because a large number of stars can be simultaneously into the jet, particles accelerated in a shock can be reaccelerated at another shock at larger $z$. We follow the next steps to obtain the particle distributions after multiple-shock interactions

1. As we are interested in accelerating nuclei, first we consider the injection of particles by the shock into the wind, because heavy particles are present there. This shock is non-relativistic and strong, with a velocity $v_{sh} \sim v_{wind}$. We assume that particles are accelerated by the Fermi I mechanism whose typical injection is $S(p) = S_0 p^{-\alpha} e^{-p/p_{max}}$, with $\alpha = 4$ for strong shocks, and the maximum momentum $p_{max}$ determined by equalling the acceleration and losses timescales. The only relevant radiative process for the particles are proton-proton inelastic collisions, and we consider also diffusion and convection escape losses. The normalisation is

$$S_0 = \frac{L_{nt}}{\int_{p_{min}}^{p_{max}} 4\pi p^2 p^{-\alpha} e^{-p/p_{max}}\, p\, c\, dp} \tag{7}$$

with $L_{nt}$ the luminosity injected into non-thermal particles. We adopt $L_{nt} = 0.1\,\dot{M}_\star\, v^2_{wind}/2$, i.e., 10% of the kinetic energy of the stellar wind. The particle distribution $f(p)$ is obtained by solving the Fokker-Planck equation as in [31].

2. The particles distribution $f(p)$ produced at the first shock propagates to the next one. During propagation, we only consider that particles can lose energy because of the radiative processes or escape the system. To obtain the particle distribution reaching the next shock, we solve the Fokker-Planck equation with temporal and energy dependency and assume that the particles move along the jet convected by the outflow material, thus time and position inside the jet relates as $t = \int_{z_0}^{z} [v_{jet}(z')]^{-1}\, dz'$.

3. The amount of accelerated particles encountering the next bow shock is $f_{prop} \propto (R_{sp}(z)/R_{jet})^2$. These

cosmic rays will be reaccelerated at the relativistic shocks in the jet. We estimate the distribution after reacceleration in one shock using the following expression [32–34]

$$f_{\rm reacc}(p) = \alpha_{\rm rel} \int_{p_0}^{p} \frac{dp'}{p'} f_{\rm prop}(p') \left(\frac{p}{p'}\right)^{-\alpha_{\rm rel}}, \quad (8)$$

where $\alpha_{\rm rel} = 3/(1 - \xi^{-1}) = 4.5$ when the compression factor is $\xi = 3$. This expression is formally valid for the case of reacceleration in non-relativistic shocks. We consider it as good approximation in our scenario since $\Gamma_0 = 5$ and the shocks are mildly relativistic [35]. We calculate $p_{\rm max}$ at each shock and introduce a cut-off $e^{-p/p_{\rm max}}$ to the distribution.

4. All these steps are repeated until there are not more bow shocks inside the jet.

## 4 Preliminary results

As a preliminary calculation, we consider the shocks produced by three WR stars arbitrarily located at $z_1 = 100$, $z_2 = 150$, and $z_3 = 200$ pc. At $z_1$ we consider only acceleration in the wind bow shock, because we are interested in accelerating heavy nuclei. At $z_2$ and $z_3$ we consider reacceleration at the shocks in the jet. For the preliminary results presented in this contribution, we do not consider acceleration of particles in the wind shocks at $z_1$ nor at $z_3$.

### 4.1 Maximum energies

Fig. 6 shows the timescales related to the parameters mentioned before for a wind shock at $z_1$ (top panel), and jet shocks at $z_2$ (middle panel) and $z_3$ (bottom panel), respectively. Adopting the Bohm diffusion regime and the escape distance $0.5R_{\rm sp}(z)$, the maximum energy of the particles ($E_{\rm max}$) is determined by the particle escape out of the acceleration region. In the case of the wind shock, it can be seen that $E_{\rm max}$ is much smaller than in the case of the shocks in the jet, mostly because $B_{\rm jet} > B_{\rm wind}$. As mentioned in Section 2, $R_{\rm sp}(z)$ increases faster with $z$ than the decay of the magnetic field of the jet, and therefore $E_{\rm max}$ slowly increases with $z$, obtaining $E_{\rm max}(150\ {\rm pc}) = 5 \times 10^{16}$ eV and $E_{\rm max}(200\ {\rm pc}) = 6 \times 10^{16}$ eV for protons.

### 4.2 Particle distributions

We calculate the particle distributions following the procedure described above. In Fig. 7 we exhibit the proton distribution after each step. The distribution injected in one shock arrives to the next one after propagation almost without being modified. This is a consequence of the low jet matter density, which prevents the cooling through proton-proton collisions.After reacceleration, the distribution becomes slightly harder ($\alpha = -3.8$).

Figure 8 displays the injected distribution at $z_1$ and the result after the interaction with two successive shocks at $z_2$, and $z_3$. The distribution extends from shock to shock up to higher energies because of the higher $E_{\rm max}$ at each shock. After the interaction with two shocks, the spectral index of the particle distribution becomes $\alpha = -3.6$.

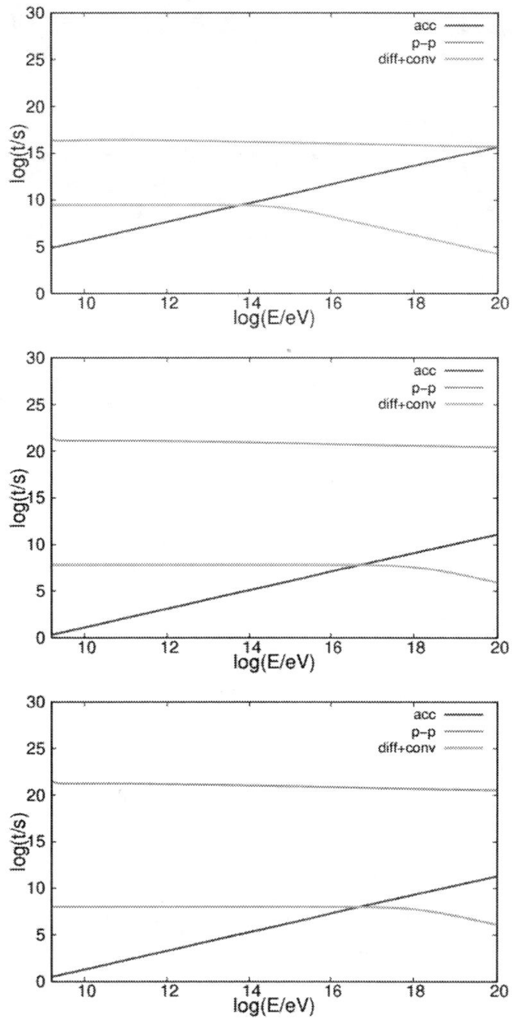


**Figure 6.** These plots show the acceleration, cooling by proton-proton inelastic collisions, and escape timescales for a shock into the wind of a star located at $z = 100$ pc (top panel), for shocks in the jet of a system at $z = 150$ pc (middle panel), and at $z = 200$ pc (bottom panel). The maximum energies obtained are $10^{14}$ eV, $5 \times 10^{16}$ eV, and $6 \times 10^{16}$ eV, respectively.

## 5 Conclusions

We investigate the acceleration of particles at multiple shocks inside AGN jets. These shocks are formed by the interaction of the jet with the winds of WR stars. The jet is charged with protons and heavier nuclei which are removed from the stellar winds. In particular, WR stars con-

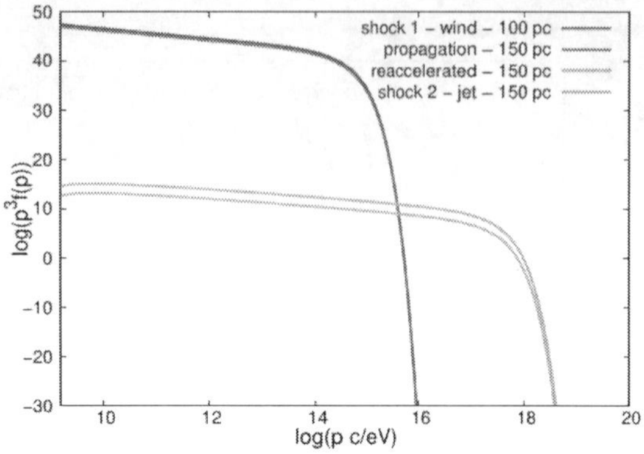


**Figure 7.** Evolution of the particle distribution between two shocks. Protons are injected at the shock into the wind at $z = 100$ pc, propagate up to the next shock located at $z = 150$ pc, where they are reaccelerated.

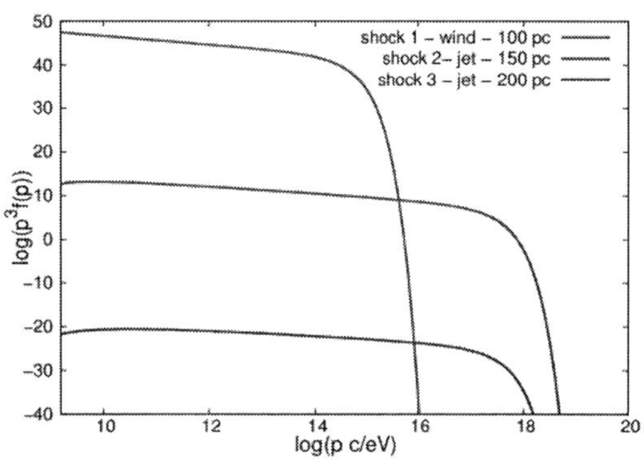


**Figure 8.** Particle distribution injected by the wind shock at $z = 100$ pc, and the distribution after reacceleration in the jet shocks at 150 pc and 200 pc.

tribute with oxygen, nitrogen, and carbon, alike the mass composition inferred from the observation of UHECRs.

Particles are accelerated and reaccelerated in the bow shocks by the Fermi I mechanism. We find that the spectrum of the cosmic rays is just slightly modified by propagation between the shocks because of the low jet matter and photon densities. Accounting for the jet deceleration after each interaction, we find that the Hillas energy at the shocks increases with $z$ allowing to reach higher energies. The energy dissipated in the collisions also reduces significantly the Lorentz factor of the jet after the interaction with tens of stars, causing the shocks into the jet to become mildly relativistic. Reacceleration by multiple shocks, on the other hand, causes the spectrum to become flatter than the typical injection slope through first order Fermi acceleration. Overall, we show the potential of this scenario to produce UHECR in AGN jets. Nevertheless, better estimations require to improve the characterisation of the stellar populations and their spatial distribution in the surroundings of AGN jets.

The authors thank the Czech Science Foundation under the grant GAČR 20-19854S. A.T.A thanks the Marie Skłodowska-Curie fellowship.

## References

[1] L.A. Anchordoqui, G.E. Romero, J.A. Combi, Phys. Rev. D **60**, 103001 (1999), astro-ph/9903145
[2] A. Aab, P. Abreu, M. Aglietta, I.F.M. Albuquerque, I. Allekotte, A. Almela, J. Alvarez Castillo, J. Alvarez-Muñiz, G.A. Anastasi, L. Anchordoqui et al., ApJL **853**, L29 (2018), 1801.06160
[3] L.A. Anchordoqui, Phys. Rev. D **97**, 063010 (2018), 1801.07170
[4] The Pierre Auger Collaboration, A. Aab, P. Abreu, M. Aglietta, I.F.M. Albuquerque, J.M. Albury, I. Allekotte, A. Almela, J. Alvarez Castillo, J. Alvarez-Muñiz et al., arXiv e-prints arXiv:1909.09073 (2019), 1909.09073
[5] The Pierre Auger Collaboration, A.A. Halim, P. Abreu, M. Aglietta, I. Allekotte, K. Almeida Cheminant, A. Almela, J. Alvarez-Muñiz, J. Ammerman Yebra, G.A. Anastasi et al., arXiv e-prints arXiv:2211.02857 (2022), 2211.02857
[6] G.E. Romero, A.L. Müller, M. Roth, A&A **616**, A57 (2018), 1801.06483
[7] J.H. Matthews, A.R. Bell, K.M. Blundell, A.T. Araudo, MNRAS **479**, L76 (2018), 1805.01902
[8] A.L. Müller, G.E. Romero, M. Roth, MNRAS **496**, 2474 (2020), 2006.12259
[9] E. Peretti, G. Morlino, P. Blasi, P. Cristofari, MNRAS **511**, 1336 (2022), 2104.10978
[10] M. Kachelriess, Extragalactic cosmic rays, in 37th International Cosmic Ray Conference (2022), p. 18, 2201.04535
[11] A.R. Bell, J.H. Matthews, MNRAS **511**, 448 (2022), 2108.08879
[12] S.S. Kimura, K. Murase, B.T. Zhang, Phys. Rev. D **97**, 023026 (2018), 1705.05027
[13] B. Eichmann, J.P. Rachen, L. Merten, A. van Vliet, J. Becker Tjus, JCAP **2018**, 036 (2018), 1701.06792
[14] J.H. Matthews, A.R. Bell, K.M. Blundell, A.T. Araudo, MNRAS **482**, 4303 (2019), 1810.12350
[15] F.M. Rieger, P. Duffy, ApJL **907**, L2 (2021), 2101.06014
[16] L. Merten, M. Boughelilba, A. Reimer, P. Da Vela, S. Vorobiov, F. Tavecchio, G. Bonnoli, J.P. Lundquist, C. Righi, Astroparticle Physics **128**, 102564 (2021), 2102.01087
[17] R. Mbarek, D. Caprioli, ApJ **921**, 85 (2021), 2105.05262
[18] J.H. Matthews, A.M. Taylor, arXiv e-prints arXiv:2301.02682 (2023), 2301.02682
[19] S.S. Komissarov, MNRAS **269**, 394 (1994)
[20] W. Bednarek, R.J. Protheroe, MNRAS **287**, L9 (1997), astro-ph/9612073

[21] M.V. Barkov, F.A. Aharonian, V. Bosch-Ramon, ApJ **724**, 1517 (2010), 1005.5252

[22] A.T. Araudo, V. Bosch-Ramon, G.E. Romero, MNRAS **436**, 3626 (2013), 1309.7114

[23] S. Wykes, M.J. Hardcastle, A.I. Karakas, J.S. Vink, MNRAS **447**, 1001 (2015), 1409.5785

[24] B. Snios, S. Wykes, P.E.J. Nulsen, R.P. Kraft, E.T. Meyer, M. Birkinshaw, D.M. Worrall, M.J. Hardcastle, E. Roediger, W.R. Forman et al., ApJ **871**, 248 (2019), 1901.00509

[25] S. Thoudam, J.P. Rachen, A. van Vliet, A. Achterberg, S. Buitink, H. Falcke, J.R. Hörandel, ApJ **595**, A33 (2016), 1605.03111

[26] S. Wykes, B.T. Snios, P.E.J. Nulsen, R.P. Kraft, M. Birkinshaw, M.J. Hardcastle, D.M. Worrall, I. McDonald, M. Rejkuba, T.W. Jones et al., MNRAS **485**, 872 (2019), 1812.04587

[27] J.H. Krolik, Active Galactic Nuclei (Princeton University Press, Princeton, 1999), ISBN 9780691227474, https://doi.org/10.1515/9780691227474

[28] N. Torres-Albà, V. Bosch-Ramon, A&A **623**, A91 (2019), 1902.05008

[29] D. Eichler, V. Usov, ApJ **402**, 271 (1993)

[30] A.M. Hillas, ARA&A **22**, 425 (1984)

[31] B.T. Park, V. Petrosian, ApJS **103**, 255 (1996)

[32] R.D. Blandford, J.P. Ostriker, ApJ **237**, 793 (1980)

[33] P. Schneider, A&A **278**, 315 (1993)

[34] M. Cardillo, E. Amato, P. Blasi, A&A **595**, A58 (2016), 1604.02321

[35] A. Zech, M. Lemoine, A&A **654**, A96 (2021), 2108.12271

# Production of high-energy neutrinos in binary-neutron-star merger events

*Simone* Rossoni[1,*], *Denise* Boncioli[2,3,**], and *Günter* Sigl[1,***]

[1] II. Institute for Theoretical Physics, Hamburg University, Luruper Chaussee 149, 22761, Hamburg, Germany
[2] Università degli Studi dell'Aquila, Dipartimento di Scienze Fisiche e Chimiche, Via Vetoio, 67100, L'Aquila, Italy
[3] INFN Laboratori Nazionali del Gran Sasso, Assergi (L'Aquila), Italy

**Abstract.** High-energy neutral astrophysical messengers, such as neutrinos and photons, can be produced by the interaction of ultra-high-energy cosmic rays (UHECRs) with radiation fields, either during extragalactic propagation or within source environments. Neutrinos and gamma-rays can play a crucial role in the study of acceleration mechanisms of cosmic rays. In particular, after being produced, neutrinos leave the source environment and propagate to the Earth without further interactions. They are only subject to energy redshift and flavour oscillation, which makes them bearers of otherwise inaccessible information about their sources. We study high-energy environments of the type that are likely to be the end states of a binary-neutron-star (BNS) merger, and we model their local photon field as a black body at a given temperature. Using a modified version of the Monte Carlo code *SimProp* v2r4 we simulate the propagation and interaction of UHECRs through these environments. We consider several combinations for the spectral index and high-energy cutoff of the UHECR protons, in order to obtain the escaped neutrino flux. We propagate these fluxes to the Earth and compare to the astrophysical IceCube neutrino flux to obtain constraints on the BNS merger spectra properties, emissivity and density rate.

## 1 Introduction

Ultra-high-energy cosmic rays (UHECRs) are ionized nuclei with energy $E \gtrsim 10^{17}$ eV and represent the most energetic class of particles ever observed. Acceleration mechanisms and source classes which enable them to reach these extreme energies are still an open issue in modern astroparticle physics. Magnetic deflection due to galactic and extragalactic magnetic fields and interactions with background photon fields of cosmic origin make it difficult to use UHECRs as astrophysical probes to reveal their sources. On the other hand, neutral and stable particles could represent an important class of new astrophysical messengers: in particular neutrinos propagate almost undisturbed through the Universe being electrically neutral and only weakly interacting particles.

In the last few years, the IceCube Neutrino Observatory reported the evidence of a diffusive neutrino flux in the TeV - PeV energy range consistent with an astrophysical origin [1]. Extremely energetic astrophysical environments (e.g. SNRs, AGNs and GRBs) might be able to accelerate cosmic rays to the production of high-energy neutrinos. However, the coincidence analysis between neutrino events and plausible source catalogues did not lead to a definitive answer on the origin of the observed neutrino events [2, 3].

In this work we consider as a neutrino source class the final state of binary-neutron-star (BNS) mergers. As suggested in [4], we assume the presence of some acceleration mechanisms focusing on the study of photo-hadronic interactions of confined UHECRs with the local photon fields generated by the surrounding material ejected by the merger (more details can be found in [4]). From these interactions charged pions can be produced, and subsequently neutrinos show up in their decay chains. Due to the low cross-section, neutrinos can easily leave the source environment providing information not accessible through charged particles and photons.

After interactions within the source environment, cosmic rays (CRs) and neutrinos propagation to the Earth is considered. Due to the very efficient relativistic boost of the energy of cosmic photons in the CR rest frame, photo-hadronic interactions with the CMB and the EBL (i.e. the Cosmic Microwave Background and the Extragalactic Background Light) are possible [5, 6]. Therefore photo-hadronic interactions like electron-positron pair production, photo-pion production and photo-disintegration and adiabatic energy loss due to the expansion of the universe ($-1/E \cdot dE/dt = H_0$, where $H_0$ is the Hubble constant at the present time) have been taken into consideration in order to predict the observed neutrinos flux from BNS mergers.

## 2 UHECR interactions

Monte Carlo simulation codes are essential tools to produce statistically significant ensembles of data to unveil possible characteristics of UHECR and neutrino sources. In order to take into account all the relevant processes

*e-mail: simone.rossoni@desy.de
**e-mail: denise.boncioli@univaq.it
***e-mail: guenter.sigl@desy.de

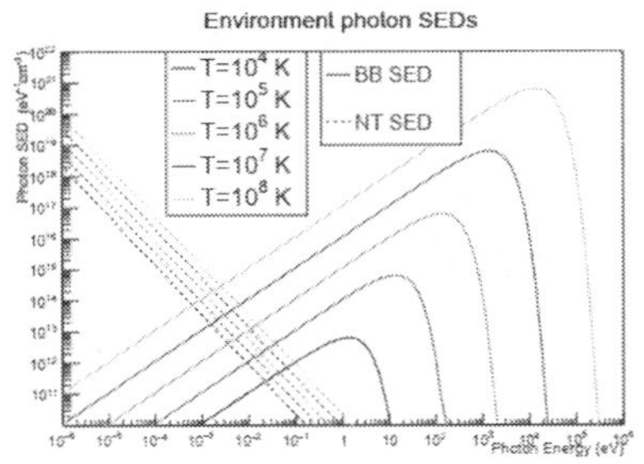


**Figure 1.** Thermal (continuous lines) and non-thermal (dashed lines) photon SEDs considered in this work at different black body temperatures (from yellow $10^8$ K to black $10^4$ K).

during the propagation of UHECRs over extragalactic distances, the Monte Carlo code *SimProp* v2r4 [7] has been used in this work. The energy loss mechanisms mentioned above and the decay of unstable particles are implemented in *SimProp* v2r4. A generic run of *SimProp* consists of $N$ nucl injected into extragalactic space at a given redshift $z_{\rm inj}$ with specific energy $E_{\rm inj}$ and mass composition $(A_{\rm inj}, Z_{\rm inj})$. After the injection, the nucleus and its interaction products are followed step by step to the Earth at $z = 0$ representing the end-condition of the simulation.

The interaction rate $\tau_{ij}^{-1}$ for the process $i$ due to the interaction between the nucleus and the photon field $j$ is computed as

$$\frac{1}{\tau_{ij}} = \frac{c}{2\Gamma^2} \int_{\epsilon_{\rm th}}^{\infty} d\epsilon\, \sigma_i(\epsilon)\, \epsilon \int_{\epsilon/2\Gamma}^{\infty} dx\, n_j(x)\, x^{-2}, \qquad (1)$$

where $x$ is the photon energy in the laboratory rest frame, $n_j(x)$ is the photon spectral energy density (SED, i.e. the number of photons per unit of volume and unit of energy) in the laboratory rest frame, $\sigma_i(\epsilon)$ is the total cross-section expressed as a function of photon energy in the cosmic-ray rest frame and $\epsilon_{\rm th}$ is the photon energy threshold in the same frame. Since the interaction rate in (1) is linear in both the cross-section and the photon field, the total interaction rate of a series of different stochastic processes is given by

$$\tau_{\rm tot}^{-1} = \sum_{i,j} \tau_{ij}^{-1}, \qquad (2)$$

and the corresponding interaction probability is $p_{ij} = \tau_{ij}^{-1}/\tau_{\rm tot}^{-1}$. With regards to leptons produced in photo-hadronic interactions, the propagation of electromagnetic particles is not implemented in *SimProp* v2r4, and neutrinos produced at the redshift $z_{\nu,\rm prod}$ with energy $E_{\nu,\rm prod}$ are immediately propagated to $z = 0$ with energy

$$E_{\nu,\rm Earth} = \frac{E_{\nu,\rm prod}}{1 + z_{\nu,\rm prod}}. \qquad (3)$$

### 2.1 Simulations within the source environment

In this work the original code *SimProp* v2r4 has been modified in order to study the interactions between confined UHECRs and local photon fields present in the source environment. Particles escaping from the interaction region will represent the injection configuration for the extragalactic propagation that is computed by the original *SimProp* code.

The first modification to *SimProp* v2r4 concerns the target photon fields for UHECR interactions. Cosmic photon fields (CMB and EBL) have been replaced by local photon fields within the source environment. We report the SEDs of these photon fields in Figure 1 in which continuous lines correspond to the black body (BB) SEDs, while the dashed lines correspond to the non-thermal (NT) SEDs obtained from [8]. After the merger, the ejected material induces the synthesis of heavy nuclei, responsible for the thermal photon field powered by the nuclear decay of heavy elements. This thermal emission subsequently decreases until the non-thermal domination phase. At this point the source environment is mostly characterised by synchrotron emission.

As shown in [8], the non-thermal emission can be modeled by the flux density function $\phi(\epsilon) \propto \epsilon^{-\beta}$, where $\epsilon$ is the photon energy. The spectral index $\beta$ is given by observations and its values is $\sim 0.6$; time evolution of $\phi(\epsilon)$ can then be included in the normalization factor. Starting from the flux density in [8], we obtain the non-thermal photon SED evolution shown in Figure 1.

The black body temperature evolution has been obtained from the so called "Optimistic" scenario reported in [4] (this scenario has been designed to optimise the production of high-energy neutrinos within the source environment). Following the approach in [4], the source temperature can be semi-analytically computed obtaining the time evolution

$$\log\left(\frac{t}{1\,\rm s}\right) \simeq -\frac{1}{2}\log\left(\frac{T}{1\,\rm K}\right) + 6. \qquad (4)$$

Different temperatures are shown with different colors in Figure 1.

In order to reproduce the propagation within the source, we define a new end-condition for our simulations to introduce the escape from the interaction region. We define the typical radius of the source as

$$\lambda_{\rm esc}(t) = \beta_{\rm ej} c t, \qquad (5)$$

where $\beta_{\rm ej}$ represents the speed of the ejected material in units of speed of light by the merger (hereinafter this parameter is fixed to $\beta_{\rm ej} = 0.3$, as done in [4]). The escape condition is obtained assuming the leaky box model, parametrised by the escape rate $\tau_{\rm esc}^{-1} = c\,\lambda_{\rm esc}(t)^{-1}$, that can be compared to the other interaction rates $\tau_{ij}^{-1}$. We want to remind that, in this simple scheme, the confinement of a charged particle is only given by the dimension of the source itself and does not depend on the particle rigidity.

We evaluate the efficiency of the photo-hadronic interactions in these environments by defining the total opacity of the source $\eta_j$ with respect to a specific photon field $j$ as the ratio between the typical escape length and the total interaction length of a nucleus with the photon field $n_j(\epsilon)$:

$$\eta_j = \frac{\lambda_{\rm esc}}{\lambda_j} = \frac{\sum_i \tau_{ij}^{-1}}{\tau_{\rm esc}^{-1}}. \qquad (6)$$

 https://doi.org/10.1051/epjconf/202328304006

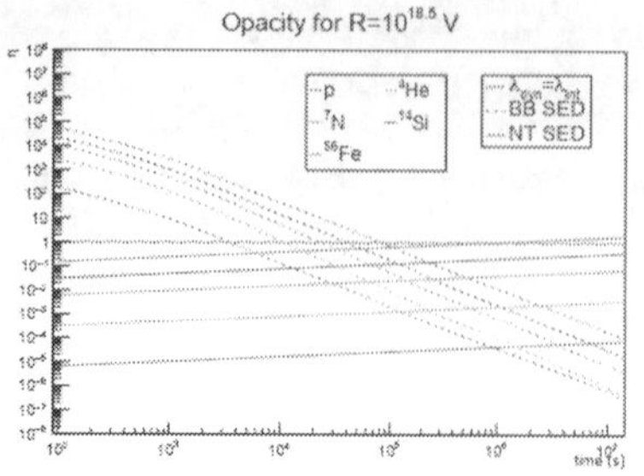


**Figure 2.** Source opacity for several nuclei with rigidity $R = 10^{18.5}$ V obtained with the photon SEDs in Figure 1. The escape condition is given by the black dashed line.

The opacity is thus a time-dependent quantity; we can see that for $\eta_j > 1$ ($\eta_j < 1$) interactions are more (less) efficient than escape and thus the source environment is opaque (transparent) for a given nucleus. In Figure 2 we show the opacity for several nuclei with rigidity $10^{18.5}$ V where all the interactions are considered. Interactions of UHECRs with the thermal photon field are very efficient right after the merger, but this contribution decreases rapidly with the cooling down of the source. The non-thermal contribution increases very slowly and it is negligible almost in every time. Similar results were obtained in [9].

In this work we considered an injection scenario of protons only. Based on our previous results, we simulate a source environment only with the thermal photon field. Therefore, the maximum relevant time after the merger for neutrino production is $\sim 10^4$ s (or the minimal relevant temperature $\sim 10^4$ K). We would like to point out that a non-thermal contribution would become relevant in the case of a heavy composition confined within the source environment for times $t \gtrsim 10$ days after the merger.

## 3 Results

### 3.1 Source escape

The modified version of the simulation code *SimProp* v2r4 has been used to perform in-source simulations as described in Section 2.1. We consider the five temperatures reported in Figure 1 and an injected UHECR spectrum of the form

$$J_{\rm inj}(E, Z) \propto E^{-\gamma} \exp\left(-\frac{E/Z}{R_{\rm max}}\right), \quad (7)$$

where $E$ and $Z$ are the energy and the charge number of the injected nucleus, $\gamma$ is the spectral index and $R_{\rm max}$ is the high-rigidity cutoff. In this work we only consider protons at the injection ($Z = 1$).The choice of different values for $\gamma$ and $R_{\rm max}$ allows the study of the contribution of different energies at the injection.

We simulate the propagation of a pure proton flux with energy uniformly distributed in $\log E$ between $10^{14}$ eV and $10^{20}$ eV. As mentioned above, the non-thermal contribution is negligible for all nuclei and therefore we only considered the black body photon field in this study.

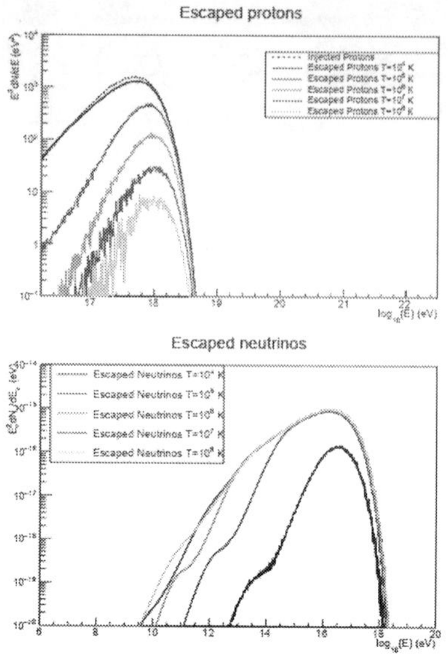


**Figure 3.** CR (top) and neutrino (bottom) spectra escaped from the source environment; CR fluxes are multiplied by $E^3$ while the neutrino ones by $E^2$. The parameters at injection are $\gamma = 1.5$ and $R_{\rm max} = 10^{17.5}$ V and the mass composition is pure protons.

In Figure 3 we show the escaped UHECR (top) and neutrino (bottom) spectra form the source environment for the following injection parameters: $\gamma = 1.5$ and $R_{\rm max} = 10^{17.5}$ V. The color scheme is the same of Figure 1 and the black dotted line in the top panel represent the injected spectrum. In the top panel of Figure 3 we can see that the UHECR emission is larger for low temperatures (black line) than for high temperatures (yellow line) of the thermal black body. The interpretation of this result is that as the temperature decreases, the source gradually becomes larger, but with the condition $\eta_{\rm BB} > 1$. The latter condition implies that the radius of the source $\lambda_{\rm esc}$ increases slower than the photo-pion interaction length. On the other hand, the production of neutrinos (Figure 3, bottom panel) is maximal right after the merger, when the temperature is the highest possible and it decreases with time. We obtain that all these considerations are independent of the injection CR parameters, which only affect the slope and the maximum energy of the escape spectra.

### 3.2 Extragalactic propagation

UHECR and neutrino spectra obtained above have been used as inputs for the simulations of the extragalactic propagation. This second phase requires further assumptions about the source distribution and cosmological evolution. A uniform distribution of identical sources is assumed. Since in-source simulations do not take into account the source evolution, we can introduce the redshift dependence with the following factorisation

$$J_{\rm inj}(E, Z, z) = (1 + z)^m J_{\rm inj}(E, Z), \quad (8)$$

https://doi.org/10.1051/epjconf/202328304006

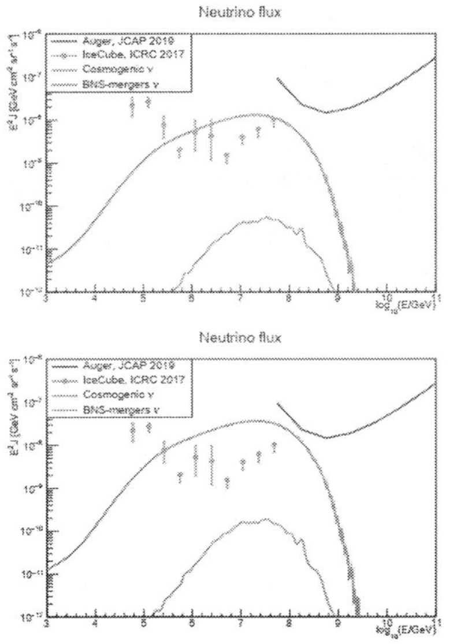


**Figure 4.** Expected BNS mergers and cosmogenic neutrino fluxes, with the measured flux by IceCube [12] and the limit for cosmogenic neutrinos by the Pierre Auger Observatory [13]. The expected neutrino fluxes refer to the following parameters of a pure-proton spectrum at injection and for a GW-like source class: $\gamma = 1.5$, $R_{max} = 10^{18}$ V and $m = 0$ (top panel); $\gamma = 1.5$, $R_{max} = 10^{18}$ V and $m = 3.4$ (bottom panel).

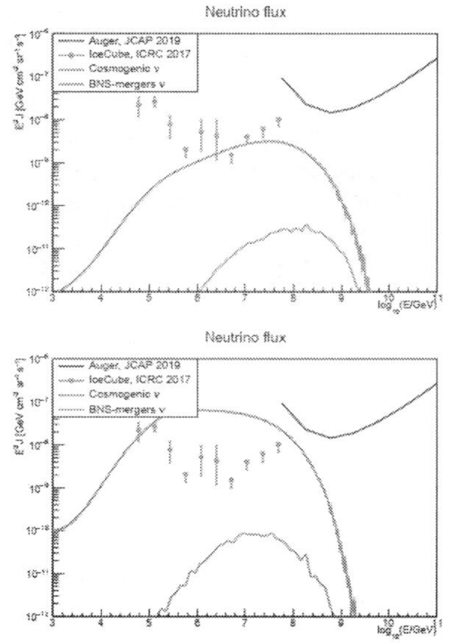


**Figure 5.** The same of Figure 4, but for the following parameters: $\gamma = 1.5$, $R_{max} = 10^{18.5}$ V and $m = 0$ (top panel); $\gamma = 2.1$, $R_{max} = 10^{18}$ V and $m = 0$ (bottom panel).

where $z$ is the redshift and the parameter $m$ defines the cosmological evolution of the source class. In this work different cosmological evolutions are taken into account for the distribution of sources, such as the case of non-evolution ($m = 0$) and the one following the star-forming rate (SFR). The diffuse fluxes of CRs and neutrinos at Earth were computed using *SimProp* v2r4 for the propagation in the extragalactic space.

Since the injection of particles within the source environment occurs over time after the merger, it is reasonable to expect that the production of high-energy neutrinos is a time dependent quantity. Then the total injected spectrum will be given by the integral over time since the merger event. We take into account the time evolution of the merger by fixing the initial black body temperature and summing the lower temperature simulation results. Then we defined two classes of sources: the GW-like class in which the initial temperature is $10^6$ K and the Optimistic class for $10^8$ K (as inspired by [4]). We report only the results for the GW-like class since the Optimistic class gives rise to neutrino fluxes far above the experimental limits.

In order to compare our results with experimental results we chose to normalize the propagated cosmic ray spectrum to the observed flux by the Pierre Auger Observatory below the ankle [11]. This therefore also fixes the normalization for the corresponding neutrino fluxes, that include both the neutrinos produced in the source environment and the ones produced in the extragalactic propagation (the so called cosmogenic neutrinos).

In Figures 4 and 5 we show the neutrinos fluxes for several choices of injection and propagation parameters (details in figure captions). In Figure 4 we can see the effect of different cosmological evolution. When a SFR evolution is assumed ($m = 3.4$) the expected neutrino flux is increased with respect to the non-evolution case due to the larger contribution of high redshift sources. This is visible both for the source neutrinos (blue lines) and for the cosmogenic ones (gray lines). In Figure 5 the effect of different injection scenarios is shown: in the top panel a higher value for the maximum rigidity is used, while in the bottom panel the spectral index is changed. When increasing the maximum rigidity the fluxes can reach higher cut-off energies, but the source neutrinos result in a slightly smaller flux, mainly due to the normalization of the corresponding cosmic ray flux. The effect of a softer spectral index for the injected protons is given by an observed neutrinos flux which is increased in the low energy range.

## 4 Conclusions

We have studied the production of high-energy neutrinos within the environment produced by BNS mergers with a modified version of the Monte Carlo code *SimProp* v2r4. We assumed a pure proton composition and an energy range of the UHECR spectrum, given in (7), injected into the source environment leaving the other parameters free in order to investigate the source phase space of BNS mergers, that could explain the observed astrophysical neutrino flux.

The first result, valid in all the configuration studied in this work, is that cosmogenic neutrinos produced during the extragalactic propagation cannot be responsible of the observed diffuse neutrino flux. For any source configuration we found that the cosmogenic neutrino flux is a subdominant component.

The neutrino flux produced in the sources is strongly connected to the time after the merger, being this linked to the temperature of the black-body photon field and to the dimension of the source. In this scenario, colder sources seem to be the most favoured since scenarios where $T \gtrsim 10^7$ K give rise to neutrino fluxes far above the experimental limits. Different spectral parameters also affected the expected flux; we obtained that a hard spectral index and a large maximum rigidity of the protons at the source could better reproduce the experimental results. Moreover, source classes that evolve weakly with redshift seems to be favoured.

The normalization procedure assumed in this work allows us to compute the emissivity at the escape from the source environment. Considering the parameter configuration shown in Figure 5 top panel, we obtain $\mathcal{E}_{esc} \simeq 2.5 \cdot 10^{45}\,\mathrm{erg\,Mpc^{-3}\,yr^{-1}}$. With this quantity we can estimate the BNS merger density rate $\dot{n}$ considering the relation $\mathcal{E}_{esc} = \mathcal{L}_{esc} \cdot n = E_{esc} \cdot \dot{n}$, where n is the BNS merger density, $\mathcal{L}_{esc}$ is the luminosity at the escape and $E_{esc}$ is the total energy at the escape. We compute the energy at the escape by considering the fallback luminosity reported in [4]. The corresponding BNS merger density rate obtained for the above configuration corresponds to $\dot{n} \simeq 5 \cdot 10^5\,\mathrm{Gpc^{-3}\,yr^{-1}}$. The experimentally estimated BNS merger rate (reported in [4]) is $\dot{n} \sim 600\,\mathrm{Gpc^{-3}\,yr^{-1}}$; we can see that our choice of normalization lead us to a merger rate greater than the experimental range.

Future outcomes of this work will take into account heavier nuclear species injected into the source region: this new scenario will require the study of the propagation through both the thermal and the non-thermal SEDs. It will be also relevant to extrapolate the required efficiency for powering the cosmic-ray flux below the ankle, taking into account the luminosity of the SED in a BNS merger, given the current estimates of the rate of BNS mergers.

## References

[1] M. G. Aartsen *et al.* [The IceCube Collaboration], Evidence for high-energy extraterrestrial neutrinos at the IceCube detector. Science. 2013 Nov 22;342(6161):1242856. doi: 10.1126/science.1242856. PMID: 24264993.

[2] M. G. Aartsen *et al.* [IceCube], "The contribution of Fermi-2LAC blazars to the diffuse TeV-PeV neutrino flux," Astrophys. J. **835** (2017) no.1, 45 doi:10.3847/1538-4357/835/1/45 [arXiv:1611.03874 [astro-ph.HE]].

[3] M. G. Aartsen *et al.* [IceCube], "Constraints on Galactic Neutrino Emission with Seven Years of IceCube Data," Astrophys. J. **849** (2017) no.1, 67 doi:10.3847/1538-4357/aa8dfb [arXiv:1707.03416 [astro-ph.HE]].

[4] V. Decoene, C. Guépin, K. Fang, K. Kotera and B. D. Metzger, "High-energy neutrinos from fallback accretion of binary neutron star merger remnants," JCAP **04** (2020), 045 doi:10.1088/1475-7516/2020/04/045 [arXiv:1910.06578 [astro-ph.HE]].

[5] K. Greisen, "End to the cosmic ray spectrum?," Phys. Rev. Lett. **16** (1966), 748-750 doi:10.1103/PhysRevLett.16.748

[6] G. T. Zatsepin and V. A. Kuzmin, "Upper limit of the spectrum of cosmic rays," JETP Lett. **4** (1966), 78-80

[7] R. Aloisio, D. Boncioli, A. Di Matteo, A. F. Grillo, S. Petrera and F. Salamida, "SimProp v2r4: Monte Carlo simulation code for UHECR propagation," JCAP **11** (2017), 009 doi:10.1088/1475-7516/2017/11/009 [arXiv:1705.03729 [astro-ph.HE]].

[8] R. Margutti, K. D. Alexander, X. Xie, L. Sironi, B. D. Metzger, A. Kathirgamaraju, W. Fong, P. K. Blanchard, E. Berger and A. MacFadyen, *et al.* "The Binary Neutron Star Event LIGO/Virgo GW170817 160 Days after Merger: Synchrotron Emission across the Electromagnetic Spectrum," Astrophys. J. Lett. **856** (2018) no.1, L18 doi:10.3847/2041-8213/aab2ad [arXiv:1801.03531 [astro-ph.HE]].

[9] X. Rodrigues, D. Biehl, D. Boncioli and A. M. Taylor, "Binary neutron star merger remnants as sources of cosmic rays below the "Ankle"," Astropart. Phys. **106** (2019), 10-17 doi:10.1016/j.astropartphys.2018.10.007 [arXiv:1806.01624 [astro-ph.HE]].

[10] A. Mucke, R. Engel, J. P. Rachen, R. J. Protheroe and T. Stanev, "SOPHIA: Monte Carlo simulations of photohadronic processes in astrophysics," Comput. Phys. Commun. **124** (2000), 290-314 doi:10.1016/S0010-4655(99)00446-4 [arXiv:astro-ph/9903478 [astro-ph]].

[11] A. Aab *et al.* [Pierre Auger], "The Pierre Auger Observatory: Contributions to the 36th International Cosmic Ray Conference (ICRC 2019): Madison, Wisconsin, USA, July 24- August 1, 2019," [arXiv:1909.09073 [astro-ph.HE]].

[12] M. G. Aartsen *et al.* [IceCube], "The IceCube Neutrino Observatory - Contributions to ICRC 2017 Part II: Properties of the Atmospheric and Astrophysical Neutrino Flux," [arXiv:1710.01191 [astro-ph.HE]].

[13] A. Aab *et al.* [Pierre Auger], "Probing the origin of ultra-high-energy cosmic rays with neutrinos in the EeV energy range using the Pierre Auger Observatory," JCAP **10** (2019), 022 doi:10.1088/1475-7516/2019/10/022 [arXiv:1906.07422 [astro-ph.HE]].

# Starburst Nuclei and Cosmic-Rays Transport Mechanisms: future Opportunities for Neutrino Astronomy and Beyond Standard Model Studies

*Antonio* Ambrosone[1,2,*], *Marco* Chianese[1,2], *Damiano F.G.* Fiorillo[5], *Antonio* Marinelli[1,2,4], and *Gennaro* Miele[1,2,3]

[1]Dipartimento di Fisica "Ettore Pancini", Università degli studi di Napoli "Federico II", Complesso Univ. Monte S. Angelo, I-80126 Napoli, Italy
[2]INFN - Sezione di Napoli, Complesso Univ. Monte S. Angelo, I-80126 Napoli, Italy
[3]Scuola Superiore Meridionale, Università degli studi di Napoli "Federico II", Largo San Marcellino 10, 80138 Napoli, Italy
[4]INAF-Osservatorio Astronomico di Capodimonte, Salita Moiariello 16, I-80131 Naples, Italy
[5]Niels Bohr International Academy, Niels Bohr Institute, University of Copenhagen, Copenhagen, Denmark

**Abstract.** Experimental observations have demonstrated a strong correlation between star-forming processes and gamma-ray luminosities, giving strong hints about the nature of the Cosmic-Rays (CRs) transport mechanisms inside Starburst Nuclei. In this contribution, we discuss the imprints on nearby Starburst galaxies (SBGs) gamma-ray spectra left by different CR transport models, quantifying the potentiality of future telescopes to distinguish between them. We also investigate the possibility of constraining the properties of light Dark Matter (DM) particles exploiting the peculiar nature of CR transport inside SBNs. We show that the property of scattering between high-energy CRs and DM particles leads to observable features, thereby posing stringent constraints on the DM parameter space.

## 1 Introduction

Starforming and Starburst Galaxies (SFGs and SBGs) are peculiar astrophysical sources which are expected to inject and accelerate high-energy CRs in their cores (see [1–6] and Refs. therein). Indeed, some nearby SBGs have been discovered as a GeV gamma-ray emitters [7]. Furthermore, M82 and NGC 253 have also been detected in the TeV domain by respectively VERITAS [9] and HESS detectors [10]. The Auger collaboration has also found a correlation of about $\sim 4\,\sigma$ with some of these sources and ultra high-energy cosmic-rays [11]. All of these demonstrate, to some extent, that SBGs produce high-energy CRs (at least up to $\sim$ 1 PeV energy) and are able to confine them for long enough to produce secondary particles such gamma-rays and probably neutrinos. Indeed, high-energy cosmic rays are expected to be accelerated by supernovae explosions (SNe) and they should collide with interstellar medium particles producing secondary particles through inelastic p-p collisions. The spectra of secondary particles strictly depend on the parent high-energy protons distribution in starburst nuclei (SBNi)[5]. As a result, the cosmic-ray transport mechanisms are expected to leave indelible signatures on the gamma-ray and neutrino spectra which are observables. Therefore, we can use these as indirect tools to study the properties of the transport conditions inside SBGs, which are still under debate [5, 6, 16]. In fact, while electrons are predominatly confined inside the starburst nuclei, the high-energy protons are only partially confined and it is still unclear which is the main escape phenomenon out of SBNi: advection or diffusion [5] (see also [16]). Furthermore, it is still not completely clear which is the source of the diffusions mechanisms. For instance, Refs. [1–4] indicate external turbulence as diffusion mechanism, while Refs. [13–15] have proposed diffusion driven by self-induce interactions with Alfen waves generated though streaming instability. In this proceeding, we analyze the imprints left by different cosmic-ray transport mechanisms inside SBGs on gamma-ray observations (see [5] for more details). In particular, we demonstrate how the next-generation CTA telescope and SWGO are going to discriminate between different cosmic ray transport mechamisms. Furthermore, our findings suggest that advective phemonena seem to be preferable as escape mechanism until $E \sim 10 - 100\,\mathrm{TeV}$ for protons [5]. We also exploit the nature of our results to perform a beyond standard model study [6]. In fact, a potential interaction between a dark-matter particle $\chi$ and protons would modify the gamma-ray spectra leading to a depletion of photons, which is not currently observed. Therefore, this leads to strict bounds on $\sigma_{\chi,p} \sim 10^{-32}\,\mathrm{cm}^2$ for $m_\chi \leq 10^{-6}\,\mathrm{GeV}$ [6]. The proceeding is structured as follows: in Sec. 2, we discuss the CR transport models considered in our analysis and how we evaluated the secondary particle production rates. In Sec. 3, we discuss the different gamma-ray phenomenology arising from the corresponding CR transport models and in Sec. 4 we discuss possible incoming constraints from neutrino telescopes. On the other hand, In Sec. 5, we discuss the implications of scattering between sub GeV DM particles and cosmic rays inside SBNi. Fi-

*e-mail: antonio.ambrosone@unina.it

nally, in Sec. 6 we discuss all the results and derive our conclusions.

## 2 Cosmic-ray Mechanims Models and Secondary Particles

Many authors have described the cosmic-ray transport mechanisms inside SBGs [1, 2, 13, 16–18, 22](see also [4, 5] and Refs therein for more details ). In particular, we work in a steady-state assumptions, leading to a balance between the high-energy cosmic rays injected by SNe and all the mechanisms such as p-p interaction, advection and diffusion occuring inside Starburst Nuclei (SBNi) [5]. Therefore, the proton distribution $f(p)$ inside the nuclei can be written as [5]

$$f(p) = Q(p) \cdot \tau_{pp} \cdot F_{\rm cal}(p) \tag{1}$$

where $Q(p) \propto (p/mp)^{-\alpha} \times e^{-p/p_{max}}$ is the injected amount of high-energy cosmic rays from SNe with $p_{max} \sim 1-10$ PeV [1–6]. $\tau_{pp}$ is the proton-proton interaction timescale:

$$\tau_{pp} = \frac{1}{k \cdot n_{\rm ISM} \cdot \sigma_{pp} \cdot c} \tag{2}$$

where $k = 0.5$ is the inelasticity of the process. $c$ is the light velocity. $n_{\rm ISM}$ is the interstellar medium density. $\sigma_{pp}$ is the inelastic cross section [20]. $F_{\rm cal}(p)$ is the calorimetric fraction, namely the fraction of high-energy protons which effectively lose energy inside the nucleus producing gamma-rays and neutrinos rather than escape via advection or diffusion. $F_{\rm cal}(p)$ theoretically depends on the geometrical assumptions of the system (e.g. spherical, cilindrical or even toroidal). This is because under a different geometries, protons can escape differently from the nucleus. Secondly, different assumptions on the escape mechanisms can affect the energy-dependency of $F_{\rm cal}$. In this proceeding, along with Ref. [5], we quantify the differences between the CR model developed by Ref. [1, 2] which was also used by Ref. [3–6] and the model put forward by [13] which was also used by Ref. [14, 15]. The former, which we call model A (see also [5] for more details), assumes that the nucleus is spherical and the diffusion is described by a Kolmogorov-like scenario in a regime of strong turbulence with an external magnetic field $B$. Given $R$ as the dimension of the nucleus, the diffusion timescale $\tau_{\rm diff} = R^2/D$ where $D$ is the diffusion coefficient (see [1] for more details). The model takes also into account the advection phenomen which consist in cosmirays driven by the motion of the interstellar medium. Indeed, given $v$ the wind velocity, the advection timescale $\tau_{\rm adv} = R/v$. In this model, the escape mechanims is given by the competition between $\tau_{\rm adv}$ and $\tau_{\rm diff}$. In fact, the escape timescale is $(\tau_{\rm adv}^{-1} + \tau_{\rm diff}^{-1})^{-1}$ [5]. For this model the calorimetric fraction can be express as [5]:

$$F_{\rm cal} = \frac{\tau_{esc}/\tau_{pp}}{\tau_{esc}/\tau_{pp} + 1} \tag{3}$$

where $\tau_{\rm eff} = \tau_{esc}/\tau_{pp}$. In this model, $\tau_{\rm diff} > \tau_{\rm adv}$, hence the escape timescale is driven by the winds. The latter, namely model B, on the other hand, assumes a cilindrical geometry and, consequently, the calorimetric fraction is given by [5, 13–15]:

$$F_{\rm cal} = 1 - \left[{}_0F_1\left(\frac{1}{5}, \frac{16}{25}\tau_{\rm eff}\right) + \frac{3\tau_{\rm eff}}{4M_A^3}\,{}_0F_1\left(\frac{9}{5}, \frac{16}{25}\tau_{\rm eff}\right)\right]^{-1} \tag{4}$$

where $M_A \simeq 2$ is the mach number and ${}_0F_1$ is the generalized hypergeometric function [13–15]. The cosmic ray transport description, in this model, is different since neglects advection phenomena arguing that they could only affect the ionised gas and it is still not clear whether they could also affect the cold interstellar medium gas or not. Furthermore, it considers diffusion driven by the self-turbulence by the alven waves generated by CRs themselves. In this framework the diffusion coefficient is proportional to the streaming velocity of cosmic rays [5, 13–15]:

$$D \propto V_s = {\rm Min}\Big[c, {\rm V_{al}}(1 + 2.3 \cdot 10^{-3} \cdot \left(\frac{\rm E}{\rm m_p}\right)^{\alpha-3} \times \\ \times \left(\frac{n_{\rm ISM}}{10^3}\right)^{3/2}\left(\frac{\chi}{10^{-4}}\right)\left(\frac{\sigma_g}{10\sqrt{2}\,{\rm Kms^{-1}}}\right)\Big] \tag{5}$$

where $\sigma_g$ is the dispersion velocity, $\chi$ is the ionisation fraction. $E$ and $m_p$ are respectively proton energy and mass. $\tau_{diff}$ in model B is defined as $h^2/D$ where $h$ is the height of the galatic disk. One of the differences between model A and model B is that in the former the calorimetric fraction is practically energy independent up to $\sim 1$ PeV, while for the latter it has a strong trend with energy. Indeed, above $\sim 100\,{\rm GeV} - 1\,{\rm TeV}$ SBGs stop being calorimetric leading to a suppresion to the consequent gamma-ray and neutrino production.

In both models, the main contribution to the gamma-ray and neutrino production is given by hadronic production, which we evaluate using the delta-function approximation following [20]. We also take into account bremsstrahlung and Inverse Compton contributons following [4, 5]. We finally emphasize that we neglect gamma-ray and neutrino production due to photo hadronic ($p\gamma$) collisions due to the fact that the energy threshold for such process requires CR energy to be greater than $\simeq 150$ PeV which is a scenario which we do not consider. Furthermore, Refs. [18, 19] have demonstrated that their contribution is completely negligible with respect to the p-p collision component.

## 3 Gamma-ray Phenomenology

In order to analyze the different phenomenology arising from model A and model B, we fit gamma-ray data of known SFGs and SBGs (see [5] for more details). The gamma-ray data analized are the *Fermi*-LAT data from [7, 21]. For M82 and NGC 253, we also take into account VERITAS [9] and HESS [10] data. We consider as free parameters the star formation rate $\dot{M}$ of the source as well as the spectral index $\alpha$ (see [5] for details about the values of the other structural parameters used). We stress that we impose the star formation rate of the source to be consistent with the one inferred by experimental data [16] within a factor 3 [5].

Fig. 1 shows the comparison between the best-fit obtained from model A and B and its current gamma-ray data [7, 9].

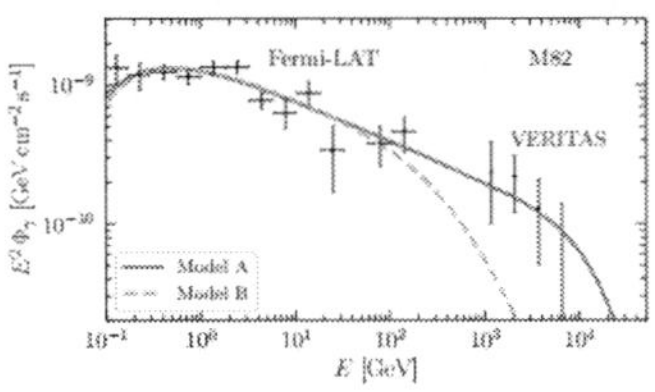


**Figure 1.** Comparison between the gamma-ray best fit from model A and model B for M82 and its gamma-ray data [7, 9]. Figure taken from Ref. [5].

Interestingly, model A seems to slighly fit better the gamma-ray data. Indeed, model B seems to be disfavoured at $2\sigma$ level [5]. This seems to slightly point towards the conclusion that phenomenlogically advection phenomena might be important for describing the escape mechanisms inside starburst nuclei (SBNi). Indeed, the fact that gamma-ray data are compatible with a simple power-law spectrum might be indicative of CRs dominated by energy-independent timescales [22]. Future experimental campaigns will be fundamental in order to get more and more precise data. In particular, the CTA telescope [26] will provide measurements increasing the statistical power in order to distinguish between model A and model B [5] (take a look also at Refs. [27, 28] the potentiality of the ASTRI-mini array to fully understand the potentiality of Imagine Atmosphere Cherenkov Telescopes (IACTs). We end this section by saying that also other experiments could be able to constrain phenomenological properties of SBNi such as SWGO [29].

## 4 (Potential) Future Neutrino Constraints

Different CR transport descriptions lead to different neutrino spectra expectations as well. In this section, focusing on model A, we show the prospects for future constrains from current and upcoming high-energy neutrino telescopes (following Ref. [4]). Fig. 2 shows the comparison between the expected 1 TeV neutrino normalizations from nearby SFGs and SBGs with the point-like IceCube sensitivity [23] and the expected sensitivity of the KM3NeT/ARCA [24] detector and ICeCube gen 2.

Interestingly, KM3NeT/ARCA [25] could constrain the hadronic emissions of the cores of Small Magellanic Cloud (SMC) and Circinus Galaxy in 6 years of data taking. It is important to say that the sensitivities refer to $E^{-2}$ spectra while the neutrino expectations of the sources might be slightly softer (see [4] for more details). Nonetheless, the emission of the core of SMC could be constrained in 6 years of data taking. The next decade will be fundamental for constraining the property of SFGs

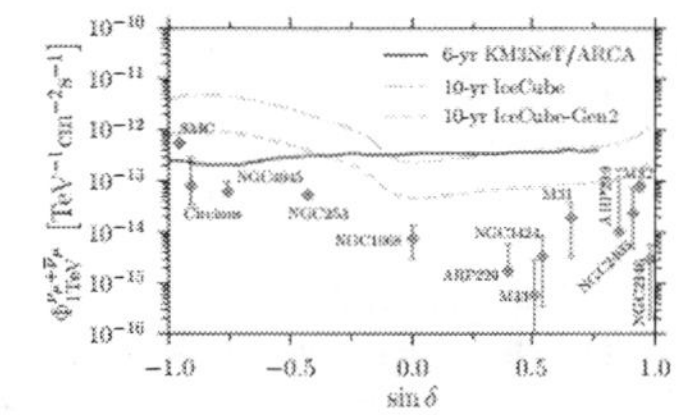


**Figure 2.** Comparison between the 1 TeV neutrino normalizations according to model A (see [4] for more details) and the IceCube, KM3NeT and IceCube Gen 2 sensitivities for $E^{-2}$ point-like sources. Figure taken from Ref. [4].

and SBGs. Indeed, a potential neutrino emission measurement from these sources not only would demonstrate that star formation processes can emit neutrinos but also they will allow incredible step forward for understanding the CR transport inside these sources.

## 5 DM-p Interactions

In the previous sections, we have analized the CR transport mechanisms inside SBGs. It seems that current data slightly favour model A. In this section, we exploit this fact in order to search for new physics. In particular, astrophysical and cosmological observations show that galaxies have a halo of dark matter (DM) [34–41]. Therefore, a potential interaction between a DM particle ($\chi$) and protons could modify the CR transport inside the SBGs leading to a suppression of the secondary particles fluxes. For instance, a further a loss timescale $\tau_{\rm loss}$ modifies $\tau_{\rm esc}$ in Eq. 3 leading to a different $F_{\rm cal}$. The new timescale is

$$\tau^{\rm el}_{\chi p} = \left[\frac{-1}{E}\left(\frac{dE}{dt}\right)_{\chi p}\right]^{-1} \tag{6}$$

which represent the elastic timescale between $\chi$ and protons. $dE/dt$ is the time rate with which CR exchange energy with dark matter particles. It strictly depends on the DM density $\rho_\chi$ as well as on the elastic cross section $\sigma_{\rm el}$ [6]:

$$\left(\frac{dE}{dt}\right)_{\chi p} = \frac{\rho_\chi}{m_\chi}\int_0^{T^{\max}} dT_\chi T_\chi \frac{d\sigma_{\rm el}}{dT_{\rm el}} \tag{7}$$

where $T^{\max}$ is the maximal kinetic energy allowed by DM particles in a collision with a proton with energy $E$. The expression of $T^{\max}$ can be found in Ref. [6] and in Refs. therein. For this contribution (see Ref. [6] for more details), we consider the DM particle to be a fermion interacting with protons through a high massive mediator. In this context, the differential elastic cross section is [6, 30]:

$$\frac{d\sigma_{\rm el}}{dT_\chi} = \begin{cases} \frac{\sigma_{\chi p}}{T^{\max}}\frac{F^2(q^2)}{16\mu^2_{\chi p}s}(q^2+4m_p^2)(q^2+4m_\chi^2) & \text{if } T_\chi < T^{\max} \\ 0 & \text{if } T_\chi > T^{\max} \end{cases} \tag{8}$$

where $\sigma_{\chi p}$ is the DM-p at zero CM momentum [6]. $\mu_{\chi p}$ is reduced mass between $\chi$ and protons. $s = m_\chi^2 + m_p^2 + 2Em_\chi$

is center of mass energy. $F(q) = (\frac{1}{1+q^2/\Lambda^2})^2$ (with $q^2 = 2m_\chi T_\chi$) is the proton form factor. This factor practically implies that at high-energy the elastic collision becomes unlikely with respect to the inelastic one. Indeed, for energies $E > m_p^2/(2m_\chi)$ the proton start being broken, producing additional gamma-rays through pion decay. We parametrize the inelastic collisons between $\chi$ and protons, using a simple semy-analytical approximation (similarly to [31–33]) assuming the $\sigma_{\text{inel}\,\chi p}$ scales as the $\nu p$ (neutrino nucleon) cross section and rescaling it to match the elastic $\chi p$ cross section [6]. This allows us to define the inelastic cross section as a function of $\sigma_{\chi p}$ defined in Eq. 8 (see [6] for more details). The inelastic collisions timescales can be defined analogously to the proton proton collisions as $\tau_{\text{inel}} = \left(k\sigma_{\text{inel}} \cdot c\rho_\chi/m_\chi\right)^{-1}$, where k is the inelasticity of the collision and $\rho_\chi$ is the DM energy density. We consider $k = 0.5$ analogously to p-p collisions [6].

### 5.1 Dark Matter Density inside Starburst Nuclei

Each galaxy is expected to be embedded in a Dark Matter halo [34–41]. The *correct* DM profile is still uncertain because of the uncertainty in the DM distribution in the core of the galaxies [42] (see also [6] and Refs. therein for further details). One of the benchmarks which can be used are the NFW profile [43]:

$$\rho_{DM}(r) = \rho_s \frac{1}{r/r_s(1 + r/r_s)^2} \tag{9}$$

Another benchmark DM density profile is the Burkert profile [44]:

$$\rho_{DM}(r) = \frac{\rho_0 r_s^3}{(r + r_s)(r^2 + r_s^2)} \tag{10}$$

where $r$ represents the distance between the DM halo. We consider it to be also coincident with the position of the Starbust Nucleus of the SBG considered. In order to parametrize $\rho_s$, $r_s$ and $\rho_0$ for the two cases, we follow Refs. [37–39, 41]. In particular, we do it by means of $c_{200}$ and $M_{200}$. $c_{200} = r_{200}/r_s$, where $r_{200}$ is the radius corresponding to $M_{200}$ where the mean density is 200 times the critical density of the Universe [6]. These two parameters are not measured, but they are constrained by cosmological simulations to be $7 \leq c_{200} \leq 12$ and $10^{10} \leq M_{200}/M_\odot \leq 10^{12}$ [37–39, 41]. Fig. 3 shows the DM density distribution for both benchmark cases. In particular, the blue band corresponds to the NFW profile and the orange one corresponds to the Burkert one. The main difference between the NFW and Burkert profile resides in the fact that the former is divergent for $r \to 0$. The figure also shows the value corresponding to $r = R_{\text{SBN}} = 200$ pc which corresponds to a benchmark value of a prototypical SBN (see also [1, 2, 4, 5]).

### 5.2 Signatures on Gamma-ray Spectra: M82 Example

In this section, we analyse which are phenomenological signatures of DM-p interactions. Fig. 4 shows the case of M82.

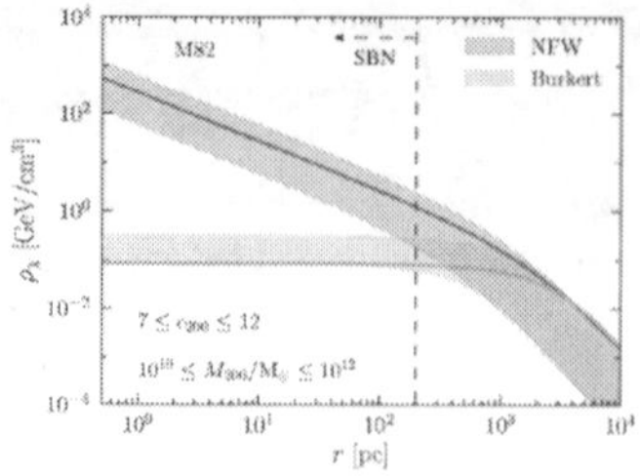


**Figure 3.** DM energy density as a function of $r$. The Blue band corresponds to the NFW profile, while the orange one corresponds to the Burkert profile. The vertical line correspond to the dimension R of a typical starburst nucleus $R = 200$ pc.

In particular, on the left panel, we show the comparison between the standard timescales (losses, advection and diffusion) as black lines with the elastic and the inelastic DM-p interaction timescales for three different values of $(m_\chi, \sigma_{\chi p})$. $\tau_{el}^{\chi p}$, for $E_p \leq E_{\min} = m_p^2/2m_\chi$, scales as $T^{-3}$, where $T$ is the proton kinetic energy. By constrast, for $E > E_{\min}$, $\tau_{el}^{\chi p} \sim T^3 \times \text{Log(T)}$ (see [6] for more details). As a result, at high energies ($E > E_{\min}$), the elastic collisions become unlikely leading to a strongly energy- increasing timescale. The right panel of Fig. 4 shows the corresponding the gamma-ray spectrum predicted. The black line corresponds to the standard scenario (model A), while the coloured lines correspond to the gamma-rays with DM-p interactions. The elastic collisions clearly reduce the spectrum leading to a dip in the gamma-ray spectrum. This is due to the fact that protons interacting with $\chi$ reduce their energy without producing gamma-rays. In practice this strongly reduces $F_{\text{cal}}$. For $E_\gamma \gtrsim 0.1E_{\min}$ (the $\sim 0.1$ factor is due to the fact that gamma-rays produced through pion decays carry about 1/10 of the energy of the parent high-energy proton), the gamma-rays get replenished by the photons produced by inelastic DM-p collisions. Indeed, in this regime, SBGs start being calorimetric again leading to a copious gamma-rays production. It is important to highlight that for energies greater than $E_{\min}$, $F_{\text{cal}} \to 100\%$. On the contrary, in the standard scenario $F_{\text{cal}} \sim 40$–$50\%$ given the competition between advection and p-p timescales. All of this leads to a higher normalization after the dip.

### 5.3 Phenomenological Constraints On $\sigma_{\chi p}$

The signature of DM-p interaction might significantly modify the gamma-ray spectrum of the SBG. Therefore, current data can be used to put strict constraints on $\sigma_{\chi p}$. For this purpose, following Ref. [6], we analyze the gamma-ray data of two nearby SBNi: M82 and NGC 253. In particular we analyze 10-year data of Fermi-LAT data [7] as well as VERITAS [9] and HESS [10] data respectively for M82 and NGC 253. We perform the statistical analysis following Refs. [4–6]; indeed, the likelihood is defined by $\mathcal{L} = e^{-\frac{1}{2}\chi^2}$, where $\chi^2 = \sum_i (SED_i - E_i^2\phi(E_i, m_\chi, \sigma_{\chi p}|\theta))^2/\sigma_i^2)$, where $i$ runs over all the measurments [4–6] and $\theta$ represents all the astrophysical parameters treated as nuisance parameters (see

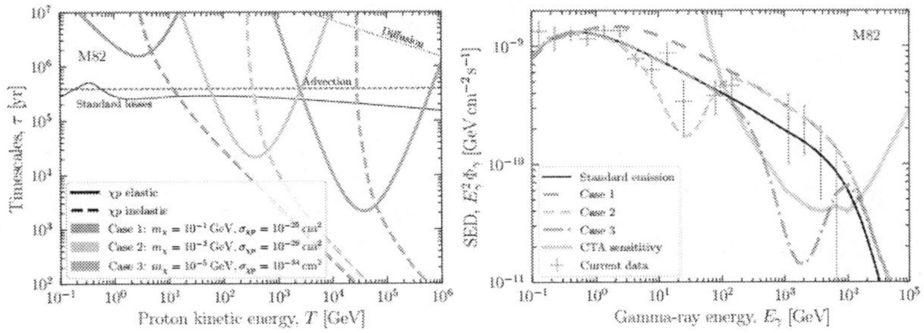


**Figure 4.** **Left Panel** The comparison between the *standard* timescales (losses, advection and diffusion) in black lines and three different cases regarding $m_\chi$ and $\sigma_{\chi p}$. Please be careful that in this figure the losses timescales are divided by $\alpha - 3$ [6]. **Right Panel** The corresponding theoretical expected gamma-ray fluxes for the source compared with the experimental Fermi-LAT and VERITAS data [7, 9]. Image taken from Ref. [6].

[6]). The bounds on $\sigma_{\chi p}$ by means of the test statistic: $\Delta\chi^2 = \chi^2(m_\chi, \sigma_{\chi p}) - \chi^2(m_\chi, 0)$ [6], where the chi squared is marginalised over the astrophysical parameters. The bounds are set with $\Delta\chi^2 = 23.6$, namely at $5\sigma$ level [6].

Fig. 5 shows the bounds imposed on $\sigma_{\chi p}$, on the left regarding M82, and on the right NGC 253. The blue band corresponds to the NFW profile and the orange band corresponds to the Burkert one. Indeed, the band mirrors the uncertainty on the cosmological parameters $c_{200}$ and $M_{200}$ of the DM halo. The shape of the bounds do not depend on the DM profile since the gamma-ray spectrum is dependent on the mean density in the SBN $\Phi_\gamma \propto \int \rho_\chi^{-1}(r)/V dV$ [6]. Indeed, the blue and the orange bands scale exaclty with the difference of the average DM density inside SBN. The results of the two sources slightly different because of the difference between the two datasets. For instance, at high energies NGC 253 has slighlty more and more precise data than M82. For $m_\chi \lessapprox 1\,\mathrm{KeV}$ the bounds flatten because the $E_{\min} > 100\,\mathrm{TeV}$, corresponding to a dip in the gamma-ray spectrum at energies $E_\gamma > 10\,\mathrm{TeV}$ which cannot be probed for these two extragalactic sources due to EBL absorption (see for instance [45]). The higher $m_\chi$, the bounds weaken because of the proton form factor which make the elastic collision unlikely.

## 6 Discussion and Conclusions

In this contribution, we have showed several updates about the CR transport description inside SBNi. In particular, we have studied the gamma-ray and neutrino phenomenology of different CR transport description. Indeed, following Ref. [5], we have studied two different CR models: model A [1], in which the main escape CR phenomenon is advection, leading to a energy-independent calorimetric fraction. Model B [13], on the contrary, neglecting advection phenomena and considering diffusion caused by self-turbulence, leads to a energy-depedent calorimetric fraction above $\sim 100\,\mathrm{GeV}$, which makes SBNi totally non calorimetric. We have demonstrated that current gamma-ray seem prefering model A, pointing to a simple power-law spectrum. However, given the current astrophysical uncertainty, it is not possible to statistically distinguish between the models at more than $1 - 2\,\sigma$ [5]. Future experiments such as CTA [26], SWGO [29] and ASTRI mini array [27, 28] will be able to provide us with more pieces of information regarding these sources, solving the mystery of their CR transport mechanisms. We have also used model A in order to search for new physics. Indeed, a new interaction between a DM particle and protons strongly modifies the proton distribution inside the nucleus. Indeed, this would create a dip in the gamma-ray spectrum, namely modyfing the standard power-law spectrum. Since in the experimental data, there is no dip, we set strong constrains on the cross section $\sigma_{\chi p}$ analysing M82 and NGC 253 data. The SBG bounds on light dark matter particle pose strict bounds on $\sigma_{\chi p}$ at level of $10^{-31} - 10^{-34}\,\mathrm{cm}^2$ depending on the source.

## References

[1] E. Peretti, P. Blasi, F. Aharonian and G. Morlino, Mon. Not. Roy. Astron. Soc. **487** (2019) no.1, 168-180 doi:10.1093/mnras/stz1161 [arXiv:1812.01996 [astro-ph.HE]].

[2] E. Peretti, P. Blasi, F. Aharonian, G. Morlino and P. Cristofari, Mon. Not. Roy. Astron. Soc. **493** (2020) no.4, 5880-5891 doi:10.1093/mnras/staa698 [arXiv:1911.06163 [astro-ph.HE]].

[3] A. Ambrosone, M. Chianese, D. F. G. Fiorillo, A. Marinelli, G. Miele and O. Pisanti, Mon. Not. Roy. Astron. Soc. **503** (2021) no.3, 4032-4049 doi:10.1093/mnras/stab659 [arXiv:2011.02483 [astro-ph.HE]].

[4] A. Ambrosone, M. Chianese, D. F. G. Fiorillo, A. Marinelli and G. Miele, Astrophys. J. Lett. **919** (2021) no.2, L32 doi:10.3847/2041-8213/ac25ff [arXiv:2106.13248 [astro-ph.HE]].

[5] A. Ambrosone, M. Chianese, D. F. G. Fiorillo, A. Marinelli and G. Miele, Mon. Not. Roy. Astron. Soc. **515** (2022) no.4, 5389-5399 doi:10.1093/mnras/stac2133 [arXiv:2203.03642 [astro-ph.HE]].

[6] A. Ambrosone, M. Chianese, D. F. G. Fiorillo, A. Marinelli and G. Miele, "Starburst Nuclei as Light

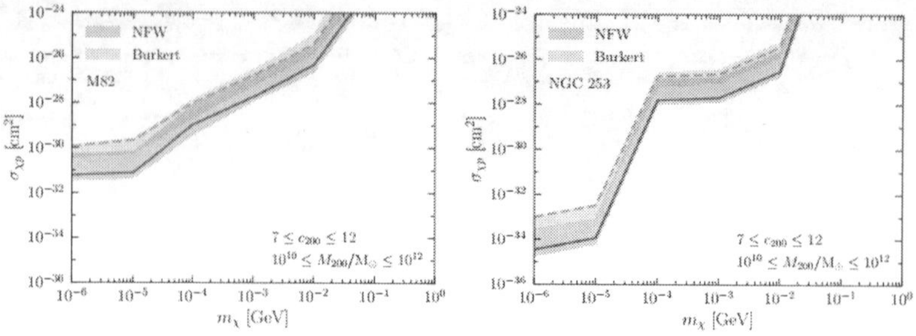


**Figure 5. Left Panel** $\sigma_{\chi p}$ constrains as a function of $m_\chi$ for M82. The blue band corresponds to the NFW profile, while the orange one corresponds to the Burkert one. **Right Panel** The corresponding constraints for NGC 253. Image taken from the supplementary material of Ref. [6].

Dark Matter Laboratories," [arXiv:2210.05685 [astro-ph.HE]].

[7] M. Ajello, M. Di Mauro, V. S. Paliya and S. Garrappa, Astrophys. J. **894** (2020) no.2, 88 doi:10.3847/1538-4357/ab86a6 [arXiv:2003.05493 [astro-ph.GA]].

[8] T. A. Thompson, E. Quataert, E. Waxman, N. Murray and C. L. Martin, Astrophys. J. **645** (2006), 186-198 doi:10.1086/504035 [arXiv:astro-ph/0601626 [astro-ph]].

[9] 2009Natur.462..770V VERITAS Collaboration and 92 colleagues 2009. A connection between star formation activity and cosmic rays in the starburst galaxy M82. Nature 462, 770–772. doi:10.1038/nature08557

[10] H. Abdalla *et al.* [H.E.S.S.], Astron. Astrophys. **617** (2018), A73 doi:10.1051/0004-6361/201833202 [arXiv:1806.03866 [astro-ph.HE]].

[11] P. Abreu *et al.* [Pierre Auger], Astrophys. J. **935**, no.2, 170 (2022) doi:10.3847/1538-4357/ac7d4e [arXiv:2206.13492 [astro-ph.HE]].

[12] A. Abdul Halim *et al.* [Pierre Auger], 'Constraining the sources of ultra-high-energy cosmic rays across and above the ankle with the spectrum and composition data measured at the Pierre Auger Observatory," [arXiv:2211.02857 [astro-ph.HE]].

[13] M. R. Krumholz, R. M. Crocker, S. Xu, A. Lazarian, M. T. Rosevear and J. Bedwell-Wilson, Mon. Not. Roy. Astron. Soc. **493**, no.2, 2817-2833 (2020) doi:10.1093/mnras/staa493 [arXiv:1911.09774 [astro-ph.HE]].

[14] M. A. Roth, M. R. Krumholz, R. M. Crocker and S. Celli, Nature **597**, no.7876, 341-344 (2021) doi:10.1038/s41586-021-03802-x [arXiv:2109.07598 [astro-ph.HE]].

[15] M. A. Roth, M. R. Krumholz, R. M. Crocker and T. A. Thompson, "CONGRuENTS (COsmic-ray, Neutrino, Gamma-ray and Radio Non-Thermal Spectra). I. A predictive model for galactic non-thermal emission," [arXiv:2212.09428 [astro-ph.HE]].

[16] P. Kornecki, E. Peretti, S. del Palacio, P. Benaglia and L. J. Pellizza, Astron. Astrophys. **657**, A49 (2022) doi:10.1051/0004-6361/202141295 [arXiv:2107.00823 [astro-ph.HE]].

[17] P. Kornecki, L. J. Pellizza, S. del Palacio, A. L. Müller, J. F. Albacete-Colombo and G. E. Romero, Astron. Astrophys. **641**, A147 (2020) doi:10.1051/0004-6361/202038428 [arXiv:2007.07430 [astro-ph.HE]].

[18] A. Condorelli, D. Boncioli, E. Peretti and S. Petrera, [arXiv:2209.08593 [astro-ph.HE]].

[19] E. Peretti, G. Morlino, P. Blasi and P. Cristofari, Mon. Not. Roy. Astron. Soc. **511**, no.1, 1336-1348 (2022) doi:10.1093/mnras/stac084 [arXiv:2104.10978 [astro-ph.HE]].

[20] S. R. Kelner, F. A. Aharonian and V. V. Bugayov, Phys. Rev. D **74**, 034018 (2006) [erratum: Phys. Rev. D **79**, 039901 (2009)] doi:10.1103/PhysRevD.74.034018 [arXiv:astro-ph/0606058 [astro-ph]].

[21] M. Hayashida, Ł. Stawarz, C. C. Cheung, K. Bechtol, G. M. Madejski, M. Ajello, F. Massaro, I. V. Moskalenko, A. Strong and L. Tibaldo, Astrophys. J. **779** (2013), 131 doi:10.1088/0004-637X/779/2/131 [arXiv:1310.1913 [astro-ph.HE]].

[22] B. C. Lacki and R. Beck, Mon. Not. Roy. Astron. Soc. **430**, 3171 (2013) doi:10.1093/mnras/stt122 [arXiv:1301.5391 [astro-ph.CO]].

[23] M. G. Aartsen *et al.* [IceCube], Phys. Rev. Lett. **124**, no.5, 051103 (2020) doi:10.1103/PhysRevLett.124.051103 [arXiv:1910.08488 [astro-ph.HE]].

[24] S. Aiello *et al.* [KM3NeT], Astropart. Phys. **111**, 100-110 (2019) doi:10.1016/j.astropartphys.2019.04.002 [arXiv:1810.08499 [astro-ph.HE]].

[25] S. Adrian-Martinez *et al.* [KM3Net], J. Phys. G **43**, no.8, 084001 (2016) doi:10.1088/0954-3899/43/8/084001 [arXiv:1601.07459 [astro-ph.IM]].

[26] B. S. Acharya *et al.* [CTA Consortium], WSP, 2018, ISBN 978-981-327-008-4 doi:10.1142/10986 [arXiv:1709.07997 [astro-ph.IM]].

[27] F. G. Saturni, C. H. E. .Arcaro, B. .Balmaverde, J. B. González, A. Caccianiga, M. Capalbi, A. .Lamastra, S. Lombardi, F. Lucarelli and R. Alves Batista, *et al.* JHEAp **35**, 91-111 (2022) doi:10.1016/j.jheap.2022.06.004 [arXiv:2208.03176

[astro-ph.HE]].

[28] S. Scuderi, A. Giuliani, G. Pareschi, G. Tosti, O. Catalano, E. Amato, A. L. A., J. Becerra Gonzàles, G. Bellassai and C. Bigongiari, *et al.* JHEAp **35**, 52-68 (2022) doi:10.1016/j.jheap.2022.05.001 [arXiv:2208.04571 [astro-ph.IM]].

[29] A. Albert, R. Alfaro, H. Ashkar, C. Alvarez, J. Alvarez, J. C. Arteaga-Velázquez, H. A. Ayala Solares, R. Arceo, J. A. Bellido and S. BenZvi, *et al.* [arXiv:1902.08429 [astro-ph.HE]].

[30] Y. Ema, F. Sala and R. Sato, SciPost Phys. **10**, no.3, 072 (2021) doi:10.21468/SciPostPhys.10.3.072 [arXiv:2011.01939 [hep-ph]].

[31] G. Guo, Y. L. S. Tsai, M. R. Wu and Q. Yuan, Phys. Rev. D **102**, no.10, 103004 (2020) doi:10.1103/PhysRevD.102.103004 [arXiv:2008.12137 [astro-ph.HE]].

[32] R. H. Cyburt, B. D. Fields, V. Pavlidou and B. D. Wandelt, Phys. Rev. D **65**, 123503 (2002) doi:10.1103/PhysRevD.65.123503 [arXiv:astro-ph/0203240 [astro-ph]].

[33] D. Hooper and S. D. McDermott, Phys. Rev. D **97**, no.11, 115006 (2018) doi:10.1103/PhysRevD.97.115006 [arXiv:1802.03025 [hep-ph]].

[34] F. Iocco, M. Pato and G. Bertone, Nature Phys. **11**, 245-248 (2015) doi:10.1038/nphys3237 [arXiv:1502.03821 [astro-ph.GA]].

[35] G. Bertone and D. Hooper, Rev. Mod. Phys. **90**, no.4, 045002 (2018) doi:10.1103/RevModPhys.90.045002 [arXiv:1605.04909 [astro-ph.CO]].

[36] P. Salucci, "The distribution of dark matter in galaxies," Astron. Astrophys. Rev. **27**, no.1, 2 (2019) doi:10.1007/s00159-018-0113-1 [arXiv:1811.08843 [astro-ph.GA]].

[37] M. Werhahn, C. Pfrommer, P. Girichidis and G. Winner, Mon. Not. Roy. Astron. Soc. **505**, no.3, 3295-3313 (2021) doi:10.1093/mnras/stab1325 [arXiv:2105.11463 [astro-ph.HE]].

[38] M. Werhahn, C. Pfrommer, P. Girichidis, E. Puchwein and R. Pakmor, Mon. Not. Roy. Astron. Soc. **505**, no.3, 3273-3294 (2021) doi:10.1093/mnras/stab1324 [arXiv:2105.10509 [astro-ph.HE]].

[39] M. Werhahn, C. Pfrommer and P. Girichidis, Mon. Not. Roy. Astron. Soc. **508**, no.3, 4072-4095 (2021) doi:10.1093/mnras/stab2535 [arXiv:2105.12134 [astro-ph.GA]].

[40] C. Pfrommer, M. Werhahn, R. Pakmor, P. Girichidis and C. M. Simpson, Mon. Not. Roy. Astron. Soc. **515**, no.3, 4229-4264 (2022) doi:10.1093/mnras/stac1808 [arXiv:2105.12132 [astro-ph.GA]].

[41] M. Werhahn, P. Girichidis, C. Pfrommer and J. Whittingham, [arXiv:2301.04163 [astro-ph.HE]].

[42] M. Benito, N. Bernal, N. Bozorgnia, F. Calore and F. Iocco, JCAP **02**, 007 (2017) [erratum: JCAP **06**, E01 (2018)] doi:10.1088/1475-7516/2017/02/007 [arXiv:1612.02010 [hep-ph]].

[43] J. F. Navarro, C. S. Frenk and S. D. M. White, Astrophys. J. **462**, 563-575 (1996) doi:10.1086/177173 [arXiv:astro-ph/9508025 [astro-ph]].

[44] A. Burkert, Astrophys. J. Lett. **447**, L25 (1995) doi:10.1086/309560 [arXiv:astro-ph/9504041 [astro-ph]].

[45] A. Franceschini and G. Rodighiero, Astron. Astrophys. **603**, A34 (2017) doi:10.1051/0004-6361/201629684 [arXiv:1705.10256 [astro-ph.HE]].

# Testing hadronic and photo-hadronic interactions as responsible for UHECR and neutrino fluxes from Starburst Galaxies

*Antonio* Condorelli[1,2,*], *Denise* Boncioli[3,4], *Enrico* Peretti[5], and *Sergio* Petrera[1,4]

[1]Université Paris-Saclay, CNRS/IN2P3, IJCLab, Orsay, France.

[2]Gran Sasso Science Institute, Via F. Crispi 7, 67100, L'Aquila, Italy.

[3]Dipartimento di Scienze Fisiche e Chimiche, Università degli Studi dell'Aquila, via Vetoio, 67100, L'Aquila, Italy.

[4]INFN/Laboratori Nazionali del Gran Sasso, via G. Acitelli 22, 67100, Assergi (AQ), Italy.

[5]Niels Bohr International Academy, Niels Bohr Institute, University of Copenhagen, Blegdamsvej 17, DK-2100 Copenhagen, Denmark.

**Abstract.** We test the hypothesis that starburst galaxies are the sources of ultra-high energy cosmic rays and high-energy neutrinos. The computation of interactions of ultra-high energy cosmic rays in the starburst environment as well as in the propagation to the Earth is made using a modified version of the Monte Carlo code SimProp, where hadronic processes are implemented for the first time. Taking into account a star-formation-rate distribution of sources, the fluxes of ultra-high energy cosmic rays and high-energy neutrinos are computed and compared with observations, and the explored parameter space for the source characteristics is discussed. We find that, depending on the density of the gas in the source environment, spallation reactions could hide the outcome in neutrinos from photo-hadronic interactions in the source environment and in extra-galactic space. We confirm that source-propagation models constitute a promising way to improve the discrimination power of models considering only ultra-high energy cosmic rays, on the way to unveiling the source class responsible for ultra-high energy cosmic rays and high-energy neutrinos.

## 1 Introduction

The discovery of a diffuse stream of cosmic particles, that extends in energy up to $\sim 10^{20}$ eV, an energy range that significantly exceeds every Earth-based accelerator, is one of the most intriguing astrophysics discoveries of the past century. We have been able to investigate its spectral behavior and composition in terms of atomic nuclei thanks to decades of observations. Despite all, one of the most intriguing open problems in contemporary astrophysics is the mystery concerning ultra-high energy cosmic rays (UHECRs), whose nature and origin are still unknown. In order to provide an answer to such a question, an analysis of measurements by the Pierre Auger Observatory [1] was recently updated in [2], where the correlation between source catalogs and UHECRs at the highest energies is investigated. In particular, a strong correlation has been found ($4.2\sigma$, predicted to increase to $5\sigma$ in 2026) between the arrival directions of UHECRs and the coordinates of the starburst galaxies (SBGs) in the catalogue [3], taking into account the 23 brightest nearby objects with a radio flux greater than 0.3 Jy, even if the contribution from SBGs is summed to an isotropic background; these studies support the hypothesis that SBGs could be a potentially interesting class of sources for UHECRs.

One of the most important aspects to investigate would be how the starburst environment affects the UHECR interactions if UHECRs were created in the most active regions of SBGs, known as starburst nuclei (SBNi). As a matter of fact, in various studies [see e.g. 4–8] it was shown that the post-processing of UHECRs through photo-disintegration of CR nuclei in the environment surrounding a hypothetical source can qualitatively explain the UHECR spectrum and composition across the so-called "ankle," the feature corresponding to the flattening of the spectrum near $5 \times 10^{18}$ eV [10].

The photo-disintegration process in these models, which will be referred to as "source-propagation models," functions as a high-pass filter, allowing the highest energy cosmic ray nuclei to escape unscattered while the lowest energy ones are disintegrated inside the source region, resulting in a pile-up of nucleons with energy scaling as $1/A$, where $A$ is the mass of the nucleus injected in the acceleration region. Particles escaping the source environment are then propagated through the intergalactic medium and finally the obtained diffuse fluxes are compared to the experimental data at Earth.

*e-mail:condorelli@ijclab.in2p3.fr

Hadronic interactions with the source environment have recently been taken into account for a generic source, demonstrating that they can potentially contribute to the same escape effect [8].

Diffuse fluxes of gamma rays (up to TeV energy) and HE neutrinos (up to PeV energy) have been observed respectively by Fermi-LAT [see 11] and IceCube [see 12, and references therein]. Starburst galaxies have already been suggested as potential candidates for such fluxes, however a comprehensive modeling of UHECR interactions in such environment has not yet been thoroughly explored, primarily because there is currently no acceleration model that can inject particles at the highest energies. In fact, while cosmic accelerators such as powerful supernova remnants [see e.g. 13, and references therein] may be able to accelerate particles up to PeV, it is unclear whether or not the EeV range of acceleration may occur in a starburst environment or in the surrounding wind bubbles [see e.g. 14–17]. Because there is currently no comprehensive theory for particle acceleration, it is important and relevant to conduct a phenomenological study of UHECRs in starburst environments in order to comprehend the fundamental characteristics of this particle population.

In this work, we assume that SBNi are capable of powering UHECRs and, for the first time, we use a source-propagation model to derive the UHECR and high-energy neutrino fluxes from these sources. We develop an extension of the Monte Carlo code *SimProp* [see 18] to simulate the behavior of UHECRs and study the multimessenger implications in terms of associated High Energy (HE) neutrino flux, focusing on the role of the hadronic and photo-hadronic interactions in these environments.

## 2 UHECR interactions in Starburst Nuclei

Starburst galaxies are peculiar astrophysical structures characterized by an extreme star forming activity that can be as high as $\sim 10 - 10^3\,M_\odot yr^{-1}$ [see 19]. A corresponding increase in the rate of supernovae (SNe) ($\mathcal{R}_{SN} \sim 0.1 \div 1\,yr^{-1}$) is frequently seen as the star formation rate (SFR) increases along with the infrared luminosity. SBGs are particularly effective cosmic ray (CR) factories thanks to their enhanced SNe rate, and connections between the non-thermal brightness and the SFR have been observed [see e.g. 20, and references therein].

The SBNi environment exhibits extreme conditions such as a gas density as high as (or higher than) $n_{ISM} \sim 10^2\,cm^{-3}$, magnetic field ($B$) at the level of $\sim 0.1 - 1\,mG$ and infrared photon density ($U_{RAD}$) often higher than $10^3\,eV\,cm^{-3}$. Additionally, a strong wind might be launched under favorable conditions due to the superposition of many SNe and the intensive star formation activity with an estimated speed: ($v_W$) of about $\sim 10^2 - 10^3\,km\,s^{-1}$.

PeV and sub-PeV CR protons may lose a constant fraction of their energy through proton-proton interactions before being able to escape, primarily because of the advection in the wind, according to the high level of turbulence expected for the SBN environment [21]. Recent investigations proposed that CRs can be additionally accelerated up to $\sim 10^2$ PeV at shocks in the wind bubbles [14, 17], whereas in [16] it is argued that in the same conditions energies up to $\sim 10^2$ EeV could be achieved. The proton-proton ($pp$) and proton-gamma ($p\gamma$) interactions on the SBN diluted photon field will still cause some of these HE particles to lose some of their energy, but their energy will effectively allow them to diffuse away from the starburst environment.

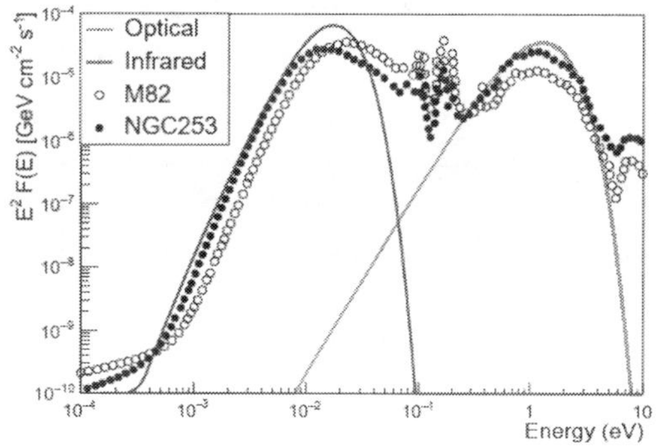


**Figure 1.** Photon spectrum of the prototype SBG, as inspired by [21]: thermal dust modified black bodies (red line) and optical star black body (green line). The black points refer to the measurements from [22] for two different SBG: M82 and NGC253.

In this study, we focus on the multi-messenger implications of such particle population in terms of UHECRs and HE neutrinos, assuming that CR nuclei are accelerated in the SBN environment up to the highest energies.

### 2.1 Interactions and escape from starburst environment

The low energy photon field of SBGs is complex and is composed of the superposition of a number of thermal components of various wavelengths, from the far infrared (FIR) to the optical (OPT) and ultraviolet (UV) [22]. In particular, the most prominent spectral components are 1) a blackbody associated to the starlight peaking at $\varepsilon_{opt} \simeq 1$ eV and 2) a second thermal component peaking at $\varepsilon_{IR} \simeq 10$ meV resulting from the reprocessing of the UV light by interstellar dust. We assume a stereotypical low-energy spectral energy distribution (SED) approximated by these two thermal components. In our investigation, it was decided to adopt a prototype SBG, that is, an SBG with the specifications specified in in Tab. 1 of [23], in order to analyze the environment surrounding the SBN and how it effects UHECRs. The photon spectrum for our prototype SBG is shown in Fig. 1 where it is compared with the spectra of two nearby SBGs: M82 and NGC253 [22].
Using a modified version of *SimProp*, typical timescales for photo-hadronic and hadronic interactions of CR particles in the SBN are presented in the following.

Under the assumption of a monochromatic photon field of photon density $n_\gamma$, the typical interaction rate between a relativistic atomic nucleus ($A$) and a low energy

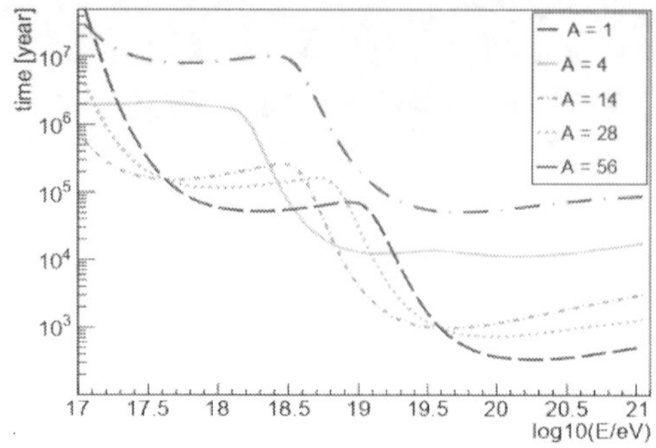


**Figure 2.** Timescales for photo-interactions in our prototype SBG for five different injected CR nuclear species as indicated in the legenda (see [23] for details).

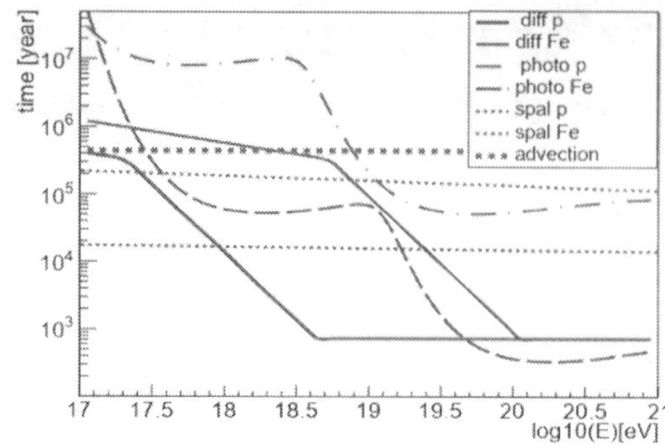


**Figure 3.** Interaction and escaping timescales for our prototype SBG: photo-hadronic interaction times (dashed-dot lines), spallation times (dashed lines) and diffusion times (solid lines) for protons (red) and Iron nuclei (blue). The green dashed line is the advection time.

photon is approximately $\tau_{A\gamma}^{-1} \simeq c\sigma_{A\gamma}n_\gamma$, where $\sigma_{A\gamma}$ represents the cross section of the process. The interaction rate reads as follows when a more realistic photon density is taken into account along with the cross section's dependence on energy:

$$\frac{dN_{\rm int}}{dt} = \frac{c}{2\Gamma}\int_{\epsilon'_{\rm th}}^{\infty} \sigma_{A\gamma}(\epsilon')\epsilon' \int_{\epsilon'/2\Gamma}^{\infty} \frac{n_\gamma(\epsilon)}{\epsilon^2}\, d\epsilon\, d\epsilon' \quad (1)$$

where $\Gamma$ is the interacting nucleus's Lorentz factor. The primed symbols (such $\epsilon'$) relate to quantities in the nucleus rest frame, whereas the unmarked symbols refer to quantities in the laboratory frame. The interaction timescales, which are the inverse of Eq. (1), are displayed in Fig. 2 photo time for various nuclear species, taking into account the reactions of electron-positron pair photo-production, nuclear photo-disintegration, and photo-pion production. The photo-disintegration on the OPT component causes the drop at low energy, while the FIR peak dominates the interaction timescale at higher energies. The other reactions indicated above have a smaller influence than photo-disintegration.

Despite having little effect on the extra-galactic medium, spallation processes between CR nuclei and their surrounding gas play a remarkable role in the ISM of SBNi due to the typical densities encountered in active star-forming regions. The timescale of the spallation process is given as follows::

$$\tau_{\rm spal} = \frac{1}{n_{\rm ISM}\,\sigma\, c}, \quad (2)$$

where $n_{\rm ISM}$ is the ISM gas density in the SBN environment and $\sigma$ refers here to the spallation cross section. This process has been implemented in *SimProp* adopting the most recent hadronic model, Sibyll 2.3d [24], an event generator designed for Monte Carlo simulations of atmospheric cascades at ultra-high energies. The cross section for proton-nucleus $pA$ and $pp$ interactions can be calculated using Sibyll 2.3d, which then enables the computation of the typical timescale for the spallation process. Additionally, Sibyll 2.3d allows for the estimation of hadronic interactions while accounting for nuclear fragmentation and the rapidity of secondary particles produced at each interaction. The computation of the longitudinal momentum distribution, in particular, is critical for determining secondary particle fluxes.

High energy particles usually remain confined in an astrophysical environment for a short amount of time before escaping. In reality, particles might escape the system through diffusion or advection in a wind. The advection timescale can be expressed specifically as $t_{\rm adv} = R/v_{\rm W}$, where $R$ denotes the size of the source and $v_{\rm W}$ denotes the wind speed. The diffusion timescale reads: $t_{\rm D} = R^2/D$, where $D$ is the CR diffusion coefficient computed in the context of quasi-linear theory and assuming a coherence length $l_c \sim 1$ pc for the magnetic field. Such an assumption for $l_c$ is motivated by the typical scale at which the turbulence is expected to be injected in the SBN [see e.g. 21]. In reality, supernovae are thought to be the source of the turbulence, and $\sim$ 1 pc is in line with the typical size of a young supernova remnant. The expression of the diffusion coefficient is: $D \simeq c r_L^{2-\delta} l_c^{\delta-1}/3$, where $r_L = E/qB$ is the particle Larmor radius and $\delta$ is the spectral slope of the turbulence, $E$ is the energy and $q$ is the charge of the particle while $B$ is the strength of the magnetic field. In particular, we assume $\delta = 5/3$ as prescribed for a Kolmogorov turbulence cascade. Following [25], we also take into account the change in the diffusion regime occurring when $r_L \gtrsim l_c$. In this energy range the diffusion coefficient is estimated as $D = D_0(r_L/l_c)^2$, where $D_0$ is the value of the diffusion coefficient computed at the energy $E_0$ such that $r_L(E_0) = l_c$. We finally estimate the escape timescale, shown in Fig. 3, as the minimum between the advection and the diffusion time, namely $t_{\rm esc} = \min[t_{\rm adv}, t_{\rm D}]$.

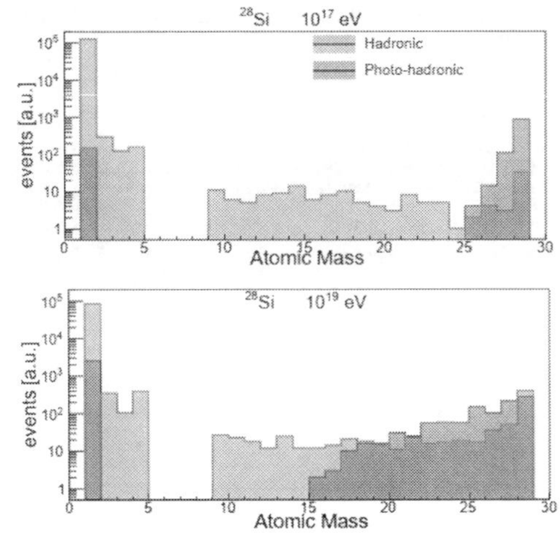


**Figure 4.** Mass distributions of nuclei escaping from the prototype SBG, for primary Silicon nuclei with energies $10^{17}$ (top) and $10^{19}$ eV (bottom panel). The blue (red) histogram refers to the propagation in the source when only photo-hadronic (hadronic) interactions are permitted.

The typical timelines for interactions and escape in the source environment for our prototype SBG are summarized in Fig. 3. Depending on the source characteristics and the CR spectrum at the acceleration site, the interplay between timescales affects the shape of the CR fluxes to be released in extra-galactic space as well as the mass composition. We note that the spallation has the shortest timescale and so dominates the transport in the lowest energy range ($E \lesssim 10^{18}$ eV). The photo-interaction with infrared photons is the dominating mechanism at higher energies $E \gtrsim 10^{18}$ eV). Take into account that the diffusion is ballistic in this energy range, and these two processes are in competition.

Spallation produces more secondary particles than photo-hadronic processes, which leads to more photons, neutrinos, and lighter nuclear fragments. The evolution of the nuclear cascade inside the source environment is changed as a result of the mixed composition of propagating nuclei. The mass distributions of nuclei escaping our prototype SBG are shown in Fig. 4 when the injection is assumed to consist entirely of silicon nuclei with energies of $10^{17}$ and $10^{19}$ eV. We particularly concentrate on differentiating the transport effects when there is only one interaction mechanism occurring inside the source: the photo-hadronic-only scenario (blue) is displayed separately from the spallation-only scenario (red). It can be seen that the photo-hadronic scenario does not efficiently produce intermediate mass (C, N, and O) nuclei, but the spallation scenario easily produces all lighter nuclei.

### 2.2 Implementation of source-interactions in *SimProp*

We created an extension of an existing Monte Carlo code called *SimProp* [18] in order to estimate the escaping flux from an SBG. . This software has so far been developed and used in relation to the extra-galactic propagation of UHECRs, such as in [26, 27], while in this work, it has been modified to represent both the transport inside the source . All important quantities are considered to be constant within the SBN, where they are propagated, for the purpose of this analysis. UHECRs are considered to be unaffected by the SBG environment once they have left this area and have started to propagate through the extra-galactic medium. According to the following assumptions, the propagation inside the source is carried out in the context of a leaky-box model: Particles are introduced into the SBN in the following ways: 1) they are injected; 2) all typical timescales are independent of position; 3) they escape if the interaction probability is smaller than the escape probability; otherwise, they lose energy and all of their byproducts are taken into account in the following step of propagation; and 4) they are no longer propagated if their energy is outside of our range of interest.

*SimProp* simulates UHECR propagation in the extra-galactic medium under the assumption of a predetermined spectrum of injected particles. It should be noted that the propagation in the source depends on the source parameters but not on the spectral parameters. Due to this, a single flux ($\propto E^{-1}$) is used for each set of source parameters when performing the in-source propagation. Re-weighting the elements spectra will provide the corresponding ejected spectra when spectral parameters (spectral index and rigidity cut) are modified. The benefit of this method is that it drastically cuts down on the amount of time needed to compute to explore the parameter space. For the final stage, the propagation from the SBN to the Earth, we use the same steps outlined in appendix A of [26].

## 3 Comparison to experimental data

The purpose of this study is to investigate the claim that SBGs are UHECR's sources. This is done by comparing experimental data to the CR flux as it has been affected by interactions in the SBNi and extra-galactic propagation. We adopt a measurement of the energy spectrum in $\log_{10}(E/\mathrm{eV})$ bins of 0.1 width from 17.8 eV to 20.2 eV, obtained with the data collected over 15 years with the surface detector array of the Pierre Auger Observatory [28]. As for the depth of the shower maxima $X_{\max}$ distributions [29], we consider $\log_{10}(E/\mathrm{eV})$ bins of 0.1 from 17.8 eV to 19.6 eV, plus one additional larger bin containing events with energies above $10^{19.6}$ eV; each $X_{\max}$ distribution is binned in intervals of 20 g cm$^{-2}$. Following [26], we use the deviance $D = -2\ln(\mathcal{L}/\mathcal{L}_{\rm sat})$ as estimator of the agreement of our parametric model to data, where $\mathcal{L}$ is our model and $\mathcal{L}_{\rm sat}$ is a model that perfectly describes the data. The total deviance consists of two terms, $D_J$ and $D_{X_{\max}}$. The former, which deals with the energy spectrum, is a product of Gaussian distributions,

https://doi.org/10.1051/epjconf/202328304008

whereas the latter, which is used to fit the $X_{max}$, is a result of multinomial distributions. Gumbel distribution functions with hadronic interaction model-dependent parameters are used to model them. For the current analysis we adopt EPOS-LHC [30] as the hadronic interaction model. It has been demonstrated in [27] that utilizing Sibyll 2.3d as the hadronic interaction model in the atmosphere has a minimal influence on the deviance value but has no significant negative effects on the main features of the propagated spectrum and composition, leading to the same general scenario.

### 3.1 Characterization of the parameter space

In what follows, we provide the set of free parameters and assumptions used in the current study to characterize the source environment, the injection parameters of the accelerated CRs, and the details of extra-galactic propagation, as well as the additional parameters required for the low-energy region of the measured spectrum and composition.

*Source parameters:*

The total infrared luminosity of the SBG, $L_{IR}$ ,and the radius of the SBN region, $R$, are the free parameters related to the source. The former is allowed to range from the value $10^{44}$ erg s$^{-1}$ up to $10^{46}$ erg s$^{-1}$, featured by the most powerful ultra-luminous infrared galaxies (ULIRGs), such as Arp 220 [22]. The latter is assumed to vary from a minimum value of 150 pc, as inferred for NGC253 [see e.g. 21], up to 250 pc as a standard value for the scale-height of thin disks in spiral galaxies like the Milky Way. The size and luminosity both have a significant impact on the transport of UHECRs. Energy losses are specifically impacted by $L_{IR}$, whereas $R$ has an impact on both interaction and escape time. The Kennicutt-Schmidt scaling [31] relates the target density $n_{ISM}$ to the star formation rate and, in turn, to the IR luminosity. Finally, it is assumed that the magnetic field $B$ in the SBN is a fixed parameter with a representative value of 200 $\mu$G for the SBNi. The magnetic field's coherence length is set to 1 pc [see e.g. 21].

*Injection parameters:*

We assume that CRs are injected as a power-law spectrum of index $\gamma$, with maximum rigidity $R_{cut}$, and that the injected flux is proportional to $E^{-\gamma}$. The range of $\gamma$ is specifically taken into account, with a minimum value of $\gamma = 1$ and a maximum value of $\gamma = 2$ as anticipated by the diffusive shock acceleration. Hard spectra from a variety of conceivable circumstances have previously been proposed in the literature. According to recent findings on the combined fit performed by the Pierre Auger Collaboration [26], we assume the range $10^{18} - 10^{19}$ eV for the rigidity threshold. For the sake of simplicity, we operate under the presumption that a single heavy nuclear mass, $A$, was injected into the SBN environment. The exploration of heavy nuclei fragmentation and the resulting formation of lighter byproducts is made possible by such an assumption. In agreement with [5] and [26] we fix $A$ to the atomic mass value of Silicon-28.

*Extra-galactic propagation:*

Two photo-disintegration cross section models are implemented by *SimProp*: TALYS [32] and PSB [33] as well as two potential EBL models: Gilmore [34] and Dominguez [35]. In this study, we used the TALYS and Gilmore photo-disintegration cross sections and the EBL model, respectively, to compute the interactions in the source environment and in extra-galactic space. Finally, we make the assumption that the redshift distribution of our UHECR sources follows the evolution of the star formation rate (SFR) up to $z_{max} = 6$.

*Low-energy component:*

We also add a significant CR flux below the ankle, similar to [5]. This is necessary because the measured mass composition does not match what is expected by the disintegration of nuclei in the source environment, which would result in only light nuclear fragments in the low energy range. Such a spectral component might belong to a distinct category of extra-galactic sources. [see e.g. 17] or the re-acceleration of Galactic CRs at the Galactic wind termination shock are examples of such phenomena. In our work, we take this new component to have a fixed spectral index $\gamma = 4.2$, with the nitrogen mass group as its dominant mass group [27], while we allow for a free normalization.

## 4 Results

We run a parameter space scan to determine the best configuration, starting from the set of parameters of our prototype SBG (see Tab. 1 of [23]), which is indicative of the most prevalent class of mild starbursts (including nearby sources like M82 and NGC253). The shape of the star-formation-rate function, which shows that M82-like galaxies are the most prevalent and plentiful in the local Universe, led us to choose mild starbursts as the central focus of our investigation. The best source parameters found correspond to a luminosity ~ 5 times higher than our starting prototype; such a luminosity is typical of a more active class of galaxies known as luminous infrared galaxies (LIRGs). Given the properties of the luminosity functions [36], SBGs with these characteristics are somehow less frequent than the prototype model. On the other hand, ULIRGs like Arp220 can be discovered at the highest end of the luminosity function, which is not where LIRGs may be found. As a result, it is possible to infer that galaxies with infrared luminosities above a particular threshold are more likely to have UHECR acceleration sites.

The spectrum and mass composition of UHECRs at Earth are shown in relation to the best fit parameters in Fig. 5. While our calculation produces a complex evolution of the mass composition with energy that closely resembles the ankle feature, the hierarchical order of the par-

 https://doi.org/10.1051/epjconf/202328304008

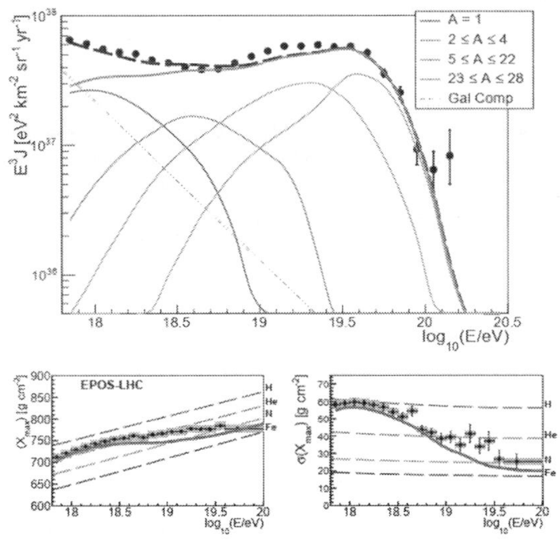


**Figure 5.** Top: All-particle best-fit scenario and partial spectra related to different detected mass groups compared to the measured spectrum [37]. Bottom: Average (left panel) and standard deviation (right panel) of the experimental (black dots as reported in [38]) and expected (lines) $X_{\max}$ distribution.

tial fluxes follows the first two moments of the $X_{\max}$ distributions. This is despite the simplicity of our model, which allows us to qualitatively reproduce the ankle feature with a precision of the order of $\sim$ 10% (at the energy of the ankle). Particularly, at low energies, light secondary nucleons predominate while at high energies, the composition becomes heavier as a result of the dominant escape rate over the interaction rate. Only below $\sim 10^{18}$ eV the additional low-energy component is relevant; at higher energies, the contribution from SBGs dominates. We emphasize that, rather than using a more appropriate luminosity function, our results were obtained under the hypothesis of a single injected nuclear species and a single prototypical SBG representative for the entire population. The effects of such assumptions are discussed in the section 5.

The number of SBGs required to describe the data should also not be greater than the number of such galaxies detected in the nearby universe. It is possible to estimate the necessary emissivity $\varepsilon$ needed to power the UHECRs at redshift $z = 0$ by comparing the model prediction with data as follows:

$$\varepsilon = \int_{E_{\min}}^{\infty} J_{\text{inj}}(E)\, E\, dE \qquad (3)$$

where $J_{\text{inj}}$ is the spectrum injected at the source, before considering the interactions, and $E_{\min}$ is an arbitrarily low energy value (here $E_{\min} = 10^{17}$ eV). As already discussed in [23], we define $\alpha$ as the ratio between the CR luminosity ($L_{\text{CR}}$) and IR luminosity obtained from the best fit ($L_{\text{IR}} = 1.2 \cdot 10^{45}$ erg/s). The number density of sources, $n_{\text{SBG}}$, can be estimated as

$$n_{\text{SBG}} = \frac{\varepsilon_{\text{CR}}}{\alpha \cdot L_{\text{IR}}} = 5.1 \cdot 10^{-5} \left[\frac{\alpha}{0.1}\right]^{-1} \text{Mpc}^{-3}, \qquad (4)$$

where $\alpha$ is normalized to 0.1 under the assumption of a sub-equipartition of the UHECRs compared to the background photon fields.

In [36] the luminosity density is computed as a function of the redshift using the luminosity functions. By integrating these functions for luminosity above the best fit one , we find:

$$n_{\text{SBG}} \simeq 3.3 \cdot 10^{-4}\ \text{Mpc}^{-3}. \qquad (5)$$

It can be seen that, assuming a fiducial value of $\alpha = 10\%$, the number density of sources derived from the integral of the luminosity function (eq. 5) is an order of magnitude greater than the one produced using our model (eq. 4). This discovery is encouraging since it can be explained by the fact that UHECRs in the starburst environment requires a lower energy budget.

### 4.1 Constraining the neutrino flux

We track the hadronic and photo-hadronic byproducts such gamma rays and neutrinos in addition to the escaping flux of UHECRs.

Due to the strong magnetic fields characteristic of the SBN environment, which is highly opaque to gamma rays with energy $\gtrsim 10$ TeV, HE electron-positron pairs are predicted to cool via synchrotron rather than starting cascades that could result in spectral features in the TeV range [21]. There is hence no reason to expect a meaningful gamma-ray counterpart to the presence of UHECRs in the SBN environment. Due to this, we postpone to future works the exploration of the multi-wavelength effects of UHECRs in the SBN environment. Different from gamma rays, neutrinos travel almost undisturbed once they are produced. Both within the SBG environment and during the propagation of UHECRs from their sources to the Earth, we calculate the production of HE neutrinos. We simply take into consideration the adiabatic energy loss effect of the Universe's expansion on the neutrino flux due to their low interaction cross section. After oscillating through cosmological distances, we finally assume an average flavour composition for Earth (1:1:1).

In Fig. 6 we show the diffuse neutrino fluxes associated to the different contributions considered in this work: cosmogenic neutrinos (grey lines), namely those neutrinos produced by the interaction of UHECRs with the cosmic microwave background (CMB) and the extra-galactic background light (EBL), the neutrinos produced by photo-hadronic interactions of UHECRs in the source (blue lines) as well as the neutrinos produced by hadronic interactions of UHECRs in the source (magenta lines). The neutrino flux resulting from our calculation is compared with the Auger limit, the limit expected by GRAND [40] after three years of operation and the flux observed by IceCube [41, 42]. Two different model predictions, solid and dashed lines, are shown and compared in Fig. 6. They

https://doi.org/10.1051/epjconf/202328304008

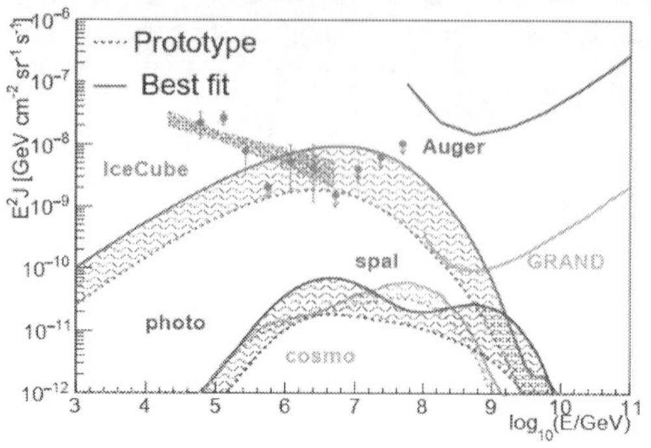


**Figure 6.** Single-flavour neutrino fluxes for the prototype case (dashed) and best case (solid), compared to the IceCube neutrino flux (both High Energy Starting Events [41] (points) and single power-law flux from muon-neutrinos [42] (band)), the expected sensitivity of GRAND after three years of operation [40] and to the limits for the cosmogenic neutrinos by the Pierre Auger Collaboration [39]. The shaded area is drawn in order to highlight the difference.

refer respectively to our best fit SBG and the prototype SBG (M82-like assumption). It's interesting to note that in the cases presented, the source-neutrinos produced in $p\gamma$ interactions are almost equivalent to the expected cosmogenic fluxes, whereas the source-neutrinos produced in $pp$ and $pA$ interactions predominate the neutrino flux, which is of the same order of magnitude as the observed one. This finding suggests that the neutrino flux measured by IceCube above $\sim$ 100 TeV may be a direct outcome of UHECR accelerators found in SBGs. This multi-messenger relationship suggests that we may be able to look at the galaxies that are emitting neutrinos with energies between 1 and 100 PeV in order to learn more about the origins of UHECRs.

## 5 Discussion and conclusions

In this work, we develop a source-propagation model to quantify the interactions occurring in the SBG environment and in the propagation to the Earth in order to investigate whether SBGs can be the sources of UHECRs. We precisely compute proton-proton and proton-nucleus interactions for the first time in the SBG environment and analyze the effects on the UHECR and neutrino fluxes in addition to photo-hadronic interactions. We compare our model prediction with the energy spectrum and mass composition observed by the Pierre Auger Observatory, working under the assumption that the sources of UHECRs are all characterized by some representative properties. We demonstrate how SBGs can qualitatively properly represent the composition and spectrum of the measured UHECR. Additionally, we compute the neutrino flux related to the transport of UHECRs and demonstrate how this enhances the constraining power of our model. This could enable us to consider whether a set of source parameters can adequately describe the UHECR data without exceeding the measured neutrino fluxes. In particular, we show that the UHECRs accelerators would contribute significantly to the neutrino flux measured by IceCube at energy $\gtrsim 10^2$ TeV if SBGs were hosting them. We observe that the expected neutrino flux from sources dominates the cosmogenic one. Additionally, we demonstrate that hadronic interactions might be essential for explaining the observed neutrino flux and deserve further research in source environments , since they could be able to hide the outcome in neutrinos from photo-hadronic interactions in the source environment and in the extra-galactic space. Therefore, in the near future, a complementary search for the origins of UHECRs could be guided by the identification of steady PeV neutrinos emitters.

We find that the data may be precisely described within the investigated parameter space if the UHECR nuclei are injected with a hard spectrum. Here, we emphasize that even if the conventional injection $\sim E^{-2}$, typical of the diffusive shock acceleration, is not preferred, it is not necessarily impossible for such a mechanism to accelerate UHECRs. In fact, it is possible to obtain hard spectra in the context of diffusive shock acceleration under a number of core SBG conditions, such as multiple shocks, converging flows, particle re-acceleration, or transport conditions in the acceleration region that are different from those in the rest of the SBG.

The results that are being given here were obtained under the assumption that just one heavy nuclear species had been injected. The analysis of scenarios in which many masses are injected, each with a distinct relative composition that might also soften the spectral index discovered at the injection, goes beyond the scope of this work and is left for further research. Nonetheless, when the injected masses are heavier than silicon nuclei, the expected scenario fails. We specifically tested the iron nuclei at injection scenario and found that the data description significantly worsens. This is due to two factors: 1) The expected composition at Earth is too heavy compared to the observational data from the Pierre Auger Observatory; 2) The maximum rigidity of the iron nuclei is determined by comparing the expected and measured UHECR spectra; and 3) The maximum energy of the nucleons from the disintegration ($1/A$ of the maximum energy of nuclei) is too low to adequately describe the data at the energy of the ankle.

Different luminosities of the prototype SBG and various particle spectra are just two of the hypotheses that have been investigated in relation to the source characteristics. A clear limitation of this research is that it only considers sources with standard characteristics. On the other hand, this approach can highlight the possibility of some average properties characterizing a class of UHECR sources. Instead of relying on a single prototype, an intriguing upgrade of the current study might make use of a catalog of the luminosity functions of galaxies. In fact, it is reasonable to assume that SBGs with varying luminosities might contribute to the energy spectrum at varying levels and probably better represent the Auger data. The sources of UHECRs could be constrained by many messengers using the HE neutrino flux. The ability of UHECR data to constrain increases as the measured neutrino flow is exceeded. The assumption that all similar sources are distributed throughout the universe has, however, simply been taken into consideration up to a redshift of $z = 6$ in this work. Instead of the UHECR fluxes, which are predicted to originate close to $z = 1$, this hypothesis has a greater impact on the predicted neutrino fluxes. Additionally, it is anticipated that using luminosity functions as opposed to a single prototype will reduce and broaden neutrino fluxes.

Future developments of this work will involve the production and propagation of photons within SBGs, as additional multi-wavelength restrictions may be discovered, particularly in hard X-rays, where electrons and positron pairs are predicted to emit through synchrotron on the strong magnetic fields typical of SBGs. The expected photon fluxes can be compared to the actual data as the neutrino fluxes, strengthening the constraining power of our model. Finally, further advancements in source-propagation analyses can be expected due to the Pierre Auger Observatory upgrade [43] and its increased precision in determining the mass composition at the highest energies.

## References

[1] Aab, A. & Others, *NIM A.* **798** pp. 172-213 (2015)

[2] Abreu, P. & Others, [arXiv:2206.13492 [astro-ph.HE]].

[3] Lunardini, C., Vance, G., Emig, K. & Windhorst, R., *JCAP.* **10** pp. 073 (2019)

[4] Allard, D. & Protheroe, R., *A & A.* **502**, 803 - 815 (2009)

[5] Unger, M., Farrar, G. & Anchordoqui, L., *Phys. Rev..* **D92**, 123001 (2015)

[6] Supanitsky, A., Cobos, A. & Etchegoyen, A., *Phys. Rev. D.* **98** pp. 103016 (2018)

[7] Muzio, M., Unger, M. & Farrar, G., *Phys. Rev. D* **100**, 103008 (2019).

[8] Muzio, M., Farrar, G. & Unger, M., *Phys. Rev. D.* **105**, 023022 (2022)

[9] Globus, N., Allard, D. & Parizot, E., *Phys. Rev. D.* **92**, 021302 (2015)

[10] Aab, A. & Others, *Phys. Rev. D.* **102**, 062005 (2020)

[11] Ackermann, M. & Others *Astrophys. J..* **799** pp. 86 (2015)

[12] Abbasi, R. & Others, *Phys. Rev. D.* **104**, e022002 (2021,7)

[13] Cristofari, P., *Universe.* **7**, 324 (2021,8)

[14] Romero, G., Müller, A. & Roth, M., *Astron. Astrophys..* **616** pp. A57 (2018)

[15] Müller, A., Romero, G. & Roth, M., *MNRAS.* **496**, 2474-2481 (2020,6)

[16] Anchordoqui, L., *Physical Review D.* **97**, e063010 (2018,3)

[17] Peretti, E., Morlino, G., Blasi, P. & Cristofari, P., *MNRAS.* **511**, 1336-1348 (2022,3)

[18] Aloisio, R., Boncioli, D., Grillo, A., Petrera, S. & Salamida, F., *Journal Of Cosmology And Astroparticle Physics.* **2012**, 007-007 (2012,10), http://dx.doi.org/10.1088/1475-7516/2012/10/007

[19] Gao, Y. & Solomon, P. *The Astrophysical Journal.* **606** pp. 271-290 (2004,5)

[20] Kornecki, P., Peretti, E., Palacio, S., Benaglia, P. & Pellizza, L.,*Astronomy & Astrophysics.* **657** pp. A49 (2022,1), https://doi.org/10.1051%5C%252F0004-6361%5C%252F202141295

[21] Peretti, E., Blasi, P., Aharonian, F. & Morlino, G., *Mon. Not. Roy. Astron. Soc..* **487**, 168-180 (2019)

[22] Galliano, F., Dwek, E. & Chanial, P.. *The Astrophysical Journal.* **672** pp. 214-243 (2007)

[23] Condorelli, A., Boncioli, D., Peretti, E. & Petrera, S., (arXiv,2022), https://arxiv.org/abs/2209.08593

[24] Riehn, F., Engel, R., Fedynitch, A., Gaisser, T. & Stanev, T., *Physical Review D.* **102** (2020,9), http://dx.doi.org/10.1103/PhysRevD.102.063002

[25] Subedi, P., **837** pp. 140-150 (2017)

[26] Aab, A. & Others, **2017**, 038-038 (2017,4), http://dx.doi.org/10.1088/1475-7516/2017/04/038

[27] Abdul Halim, A. & Others, [arXiv:2211.02857 [astro-ph.HE]].

[28] Aab, A. & Others, *Phys. Rev. Lett..* **125**, 121106 (2020,9)

[29] Aab, A. & Others, *Physical Review D.* **90** (2014,12), http://dx.doi.org/10.1103/PhysRevD.90.122005

[30] Pierog, T., Karpenko, I., Katzy, J., Yatsenko, E. & Werner, K., *Physical Review C.* **92** (2015,9), http://dx.doi.org/10.1103/PhysRevC.92.034906

[31] Kennicutt, R., *Astrophys. J..* **498** pp. 541 (1998)

[32] Koning, A., Hilaire, S. & Duijvestijn, M., *International Conference On Nuclear Data For Science And Technology.* **769** pp. 1154-1159 (2005,5), https://ui.adsabs.harvard.edu/abs/2005AIPC..769.1154K

[33] J. Puget, F. & J. Bredekamp, *Astrophys.J..* **205** pp. 638-654 (1976)

[34] Stecker, F. & Salamon, M., *The Astrophysical Journal.* **512**, 521-526 (1999,2), http://dx.doi.org/10.1086/306816

[35] Dominguez, A. & Primack, J., *Mon. Not. Roy. Astron. Soc..* **410** pp. 2556 (2011)

[36] Gruppioni, C. & Others *MNRAS.* **432**, 23-52 (2013,6)

[37] V. Verzi, *Proc. 36rd ICRC.*, https://pos.sissa.it/358/450/

[38] A. Yushkov *Proc. 36rd ICRC.*, https://pos.sissa.it/358/482/

[39] Aab, A. & Others, *Journal Of Cosmology And Astroparticle Physics.* **2019**, 022-022 (2019,10), https://doi.org/10.1088/1475-7516/2019/10/022

[40] Álvarez-Muñiz, J. & Others, *Sci. China Phys. Mech. Astron..* **63**, 219501 (2020)

[41] Kopper, C., *PoS.* **ICRC2017** pp. 981 (2018)

[42] Abbasi, R. & Others, *Astrophys. J..* **928**, 50 (2022)

[43] Cataldi, G. & Others, *PoS.* **ICRC2021** pp. 251 (2021)

# First scientific results of the KM3NeT neutrino telescope

*Francesco* Benfenati[1,2,*], *Francesco* Filippini[1,2,**], and *Tommaso* Chiarusi[2,***] on behalf of the KM3NeT Collaboration

[1]INFN - Sezione di Bologna, Viale Berti-Pichat 6/2, 40127 Bologna, Italy
[2]Dipartimento di Fisica e Astronomia dell'Università, Viale Berti Pichat 6/2, 40127

**Abstract.** KM3NeT is a multidisciplinary observatory, for the detection and study of cosmic neutrinos and their sources in the Universe, as well as the measurement of neutrino properties such as the mass hierarchy and oscillation parameters. Two underwater detectors are under construction in the Mediterranean Sea. The configuration of the ARCA detector, located off-shore Sicily, Italy, is optimized for the detection of neutrinos in the energy range of 1 TeV-100 PeV. The ORCA detector off-shore Toulon, France is configured for the measurement of neutrinos with energy from about 100 MeV up to the sub-TeV region. At present, 21 and 14 detection units are taking data at the ARCA and the ORCA sites, respectively. Installation of additional detection units is foreseen in the next few years. In this contribution an overview of the expected performances of the full detectors will be reviewed and the main physics results obtained with ARCA and ORCA, still in their partial configuration, will be reported. Finally, in the context of the multi-messenger scenario the KM3NeT online alert system will be presented.

## 1 Introduction

Neutrino Astrophysics is a branch of Astroparticle Physics that mainly focuses on the detection of high-energy neutrinos in order to determine the sources and accelerators of primary cosmic-rays, constraining their propagation through the Cosmos. Due to the extremely small cross-section in their weak interaction with matter, neutrinos are the optimal messengers to investigate far and violent phenomena in the Universe, complementary to the charged nuclei and photons which can be diverted or absorbed during their long travel from the sources to Earth. The detection of cosmic neutrinos coming from highly relativistic jets out of Active Galactic Nuclei, micro-quasar systems, Gamma Ray Bursts and other similar objects, represents a clear signature of the occurrence of hadronic interactions in such extremely energetic engines. The measurement of the neutrino fluxes can be used to constrain the composition of the matter surrounding the astrophysical sources, even of those hidden behind optical thick shells, and to disentangle the underlying acceleration mechanisms against the thermal / not thermal processes (e.g. leptonic models). Also neutron star mergers are expected to be intense sources of high-energy neutrinos, whose detection can contribute to a more precise localization of the event in the sky and help constraining the parameter space of the coalescent pair, completing what can be inferred with the detection of the preceding gravitational wave signals, in an effective multi-messenger approach. Finally, other targets of neutrino Astrophysics are supernova recognition, indirect detection of dark matter annihilation in dense matter environments and searches for magnetic monopoles, nuclearities and other exotic-matter candidates.

Neutrino telescopes are optimized for the detection of high-energy extraterrestrial neutrinos. They must first remove the background coming from atmospheric muons and neutrinos generated in cosmic-ray induced air showers in the Earth atmosphere. Generally, atmospheric muons can be removed by using the Earth as a shield when looking for upward-going particles. Moreover, the irreducible background of atmospheric neutrinos becomes less relevant as energy increases, enhancing the probability of observing high-energy cosmic candidates. At the same time, interesting sources might be located in the sky above the neutrino telescope; in this case, if the detector is large enough, some parts of it (generally the peripheral ones) can be exploited as vetoing elements to select only neutrino induced events originating inside the detector volume.

The implementation of a neutrino telescope consists of a large-volume 3D array of photodetectors placed at large depths in a transparent medium, such as water or ice. Cherenkov light, induced by relativistic charged particles passing through the medium, is detected in the array and the recorded information can be used to reconstruct the direction and energy of the incoming neutrino, by which those charged particles were produced. Neutrino telescopes can detect all-flavour neutrino interactions [1]. Together with the pure Astroparticle goals mentioned above, other appealing Physics objectives can be achieved with neutrino telescopes, like the study of the fundamental properties of neutrinos within the Standard Model.

Currently, the largest and most sensitive neutrino telescope on Earth is the IceCube detector, placed at a depth

*e-mail: francesco.benfenati@bo.infn.it
**e-mail: francesco.filippini@bo.infn.it
***e-mail: tommaso.chiarusi@bo.infn.it

between 1.4 and 2.4 km below the ice surface at the South Pole. Since 2010, IceCube has been taking data with 1 km$^3$ size detector. It also has a DeepCore infill, mainly used to study the atmospheric neutrinos below 100 GeV. Exploiting a complementary surface array for a more efficient vetoing approach, IceCube was capable to obtain the first evidence of the presence of cosmic neutrinos [2]. More recent analyses have improved the significance of such observations. This signal is compatible with an isotropic diffuse flux of cosmic neutrinos, at present with only a few associations to possible sources [3].

Along with IceCube, other under-water telescopes are participating to the quest for cosmic neutrinos. Taking the legacy of a precursor installation, and aiming at the km$^3$ scale, the GVD telescope is under construction in the Baikal lake in Russia. GVD first results are reported in [4], showing a small preliminary excess in the search for diffuse flux of extra-terrestrial neutrinos. In the Mediterranean Sea, the ANTARES neutrino telescope was located at a depth between 2 and 2.45 km below sea level in front of the French coast, 40 km off-shore Toulon. ANTARES operated for 15 years from 2007. Although much smaller than IceCube, ANTARES had been able to obtain significant results in various analyses concerning both neutrino Physics and Astronomy, especially when searching for neutrinos below 100 TeV from the Southern Sky. Dismantled in spring 2022, ANTARES has passed the baton to the next-generation neutrino telescope in the Mediterranean Sea, the KM3NeT observatory, whose first results are described in this contribution.

In Section 2 the two KM3NeT telescopes ARCA and ORCA are introduced. In Section 3 the measurement strategy and the detection performance for both ARCA and ORCA is reviewed. In Section 4 a selection of the most recent scientific results are presented and discussed. Finally, conclusions and outlooks are summarized in Section 5.

## 2 The KM3NeT ARCA and ORCA detectors

The KM3NeT observatory consists of two neutrino telescopes currently under construction at the bottom of the Mediterranean Sea. ARCA (Astroparticle Research with Cosmics in the Abyss) is located off-shore Portopalo di Capo Passero, Sicily, Italy, at a depth of ~3500 m and ORCA (Oscillation Research with Cosmics in the Abyss) off-shore Toulon, France, at a depth of ~2450 m. ARCA is designed to study cosmic neutrinos and their possible astrophysical sources, as described above in the introductory section. The focus of ORCA is the study of the intrinsic properties of neutrinos, such as the oscillation parameters and the neutrino mass-hierarchy [5].

### 2.1 The dectection elements and the sea-floor network

Both ARCA and ORCA are made of vertically aligned Detection Units (DUs), each hosting 18 Digital Optical Modules (DOMs) mounted on Dyneema® ropes. Each DOM contains 31 3-inch photomultiplier tubes (PMTs), calibration and positioning instrumentation and readout electronic boards [6]. Each DU is pulled vertical by a buoy on top and it is anchored to the sea-floor by a special heavy anchor, which hosts a further active element, the DU-Base Module (DU-BM). Together with hosting dedicated instrumentation like hydrophones and other calibration devices, such as lasers and acoustic beacons, the DU-BM concentrates the optical links for the DOM communication, and also acts as the power hub for all the devices of the DU. Groups of DUs are connected to submarine Junction Boxes (in ARCA) or Nodes (in ORCA), which act as electric power hubs for the connected DUs, and concentrate their optical signals via the main electrical-optical cables (MEOCs) up to the shore stations. Although sharing the same detection elements, ARCA and ORCA have different layouts, according to their scientific targets. For accomplishing the detection of very high energy neutrinos (up to ~100 PeV), ARCA must have a large extension, aiming at the size of at least 1 km$^3$ instrumented volume. For this reason, in ARCA the spacing between the DUs is ~90 m and between the DOMs on each DU is ~36 m. The final configuration of the ARCA installation will be divided in two building blocks of 115 DUs each, connected to shore by two 100 km MEOCs (see Figure 1a). The ORCA instrumented volume is about 7 MTon, ~1% of the ARCA extension. In ORCA the inter-distance between the DUs is 20 m and the DOMs in a DU are spaced by 9 m, allowing for an optimized detection of neutrinos with energies spanning from 100 MeV up to 100 GeV. ORCA is made of one building block of 115 DUs, and it is also connected by two 50 km MEOCs to the shore-station (see Figure 1b).

### 2.2 The data acquisition model

ARCA and ORCA both implement a trigger-less streaming readout paradigm for data-taking. All the DOMs and DU-BMs independently send continuous streams of data from the PMTs and the acoustic sensors to shore. Such data streams are processed (aggregated and filtered) by a totally online system of software processes running in the computing facilities of the shore station. This approach, also called *all-data-to-shore*, allows to deploy the simplest hardware in the abyssal sites. On shore, the networking infrastructure is modular and the computing resources dynamically scale to cope with a larger and larger incoming throughput from the growing detectors. The details of the KM3NeT data acquisition (DAQ) system are reported in [7]. Although running with two independent DAQ installations, the data taking from the two telescopes is combined at higher level by the KM3NeT online multi-messenger system [8].

### 2.3 Detector monitoring

KM3NeT detectors are equipped with a live monitoring system, which allows various online checks for determining the detector status and the quality of the data taking with short latencies ranging from few seconds to ~ 1

https://doi.org/10.1051/epjconf/202328304009

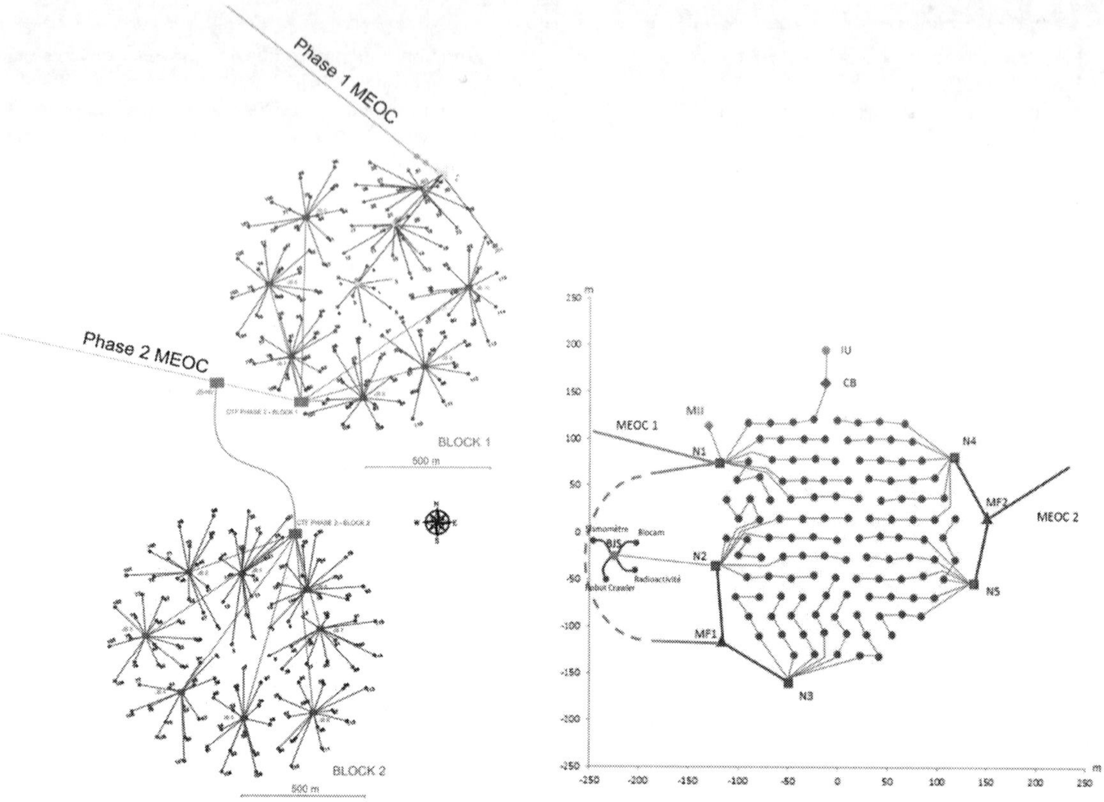


(a) (b)

Figure 1: The ARCA (a) and ORCA (b) sea-floor maps. Each dot stands for a DU, while the straight lines indicate the inter-link cables connecting the DUs to intermediate stages of aggregation devices, up to the main electrical-optical cables (MEOCs). The aggregation designs of the DUs in the detector modules are different: ARCA follows a star-center approach, with groups of up to 14 DUs concentrated on devices called Junction Boxes (red squares); in ORCA, six daisy-chains made of 4 DUs each are concentrated on a device called Node (brown squares).

minute. Examples of such monitoring consists of time-plots of low-level information, such as the PMT single rates and environmental parameters like temperature, humidity and orientation of the DOMs, together with more complex representation of derived quantities, obtained after the data processing, like the per-event hit maps distributed over the active DUs versus the hit-time, or the time-plots of the triggered event rates. All the monitoring plots or complementary information are published on web pages that are used by the Collaboration during the data-taking shifts.

### 2.4 ARCA and ORCA evolution

Thanks to the scalable features of the KM3NeT DAQ system, the data taking started, for both ARCA and ORCA, immediately after the very first lines were deployed. The data sets used in the analyses presented in this contribution (see Sec.4) are referred to particular layouts of the telescopes, reporting the name of the detector used for collecting the data together with the number of active DUs. ARCA1 and ARCA2 data sets correspond to a period of almost 5 years starting in 2015, during which the KM3NeT Collaboration was engaged to optimize the off-shore detector, the deployment technique and to set up the mass-production stages for DOMs, DU-BMs and other devices of the sea-floor infrastructure. ORCA1 was deployed in early 2019, and the ORCA6 layout was completed in mid-2020. Despite the harsh working conditions occurred during the COVID pandemic, the KM3NeT Collaboration started to boost both the detector mass production and the sea campaigns, reaching the current situation of ARCA21 and ORCA14 in September 2022. During 2023, ARCA32 (see the black sector of Fig.1a) and ORCA24 (see the DUs connected to N1 in Fig.1b) will be completed. Thanks to dedicated funds for enhancing the scientific installations of Italian Southern Regions (PON and PNRR calls [9, 10]), additional ~100 ARCA DUs will be built in the next 5 years, completing the first ARCA building block and part of the second one. At the same time, thanks to French funds (NEUMED project), also the ORCA detector will increase its instrumented volume up to 48 DUs, fully populating also Node N2, together with the expansion of the sea-floor network by 2 more additional Nodes (N3 and N4).

In addition to the above-mentioned main scientific objective, the measurements of the instruments hosted by the KM3NeT infrastructure and the measurements of the environmental parameters which are taken for calibration purposes (see Sec.2.1) [11] can be exploited for Earth and Marine Sciences studies, offering the opportunity to online monitor the marine ecosystem for long periods of time, and to carry out researches on the deep marine environment and on possible signatures of the climate change.

## 3 Neutrino measurements and detector performance

### 3.1 Event signatures

The events detected in KM3NeT can be separated in two distinct classes: track-like events and shower-like events. Track-like events are mostly due to muons originated in charged current $\nu_\mu$ interactions and they are characterized by long tracks since muons have a long path length in water, while shower-like events are produced by the charged current interactions of $\nu_e$, $\nu_\tau$ and neutral current interactions of all neutrino flavors, and they are characterized by the presence of electromagnetic or hadronic (or both) showers, as the produced particles propagate only for a few tens of meters. The signature of a charged current interaction of $\nu_\tau$ with energy larger than ~1 PeV is referred to as a *double-bang* event, since the track of the neutrino induced $\tau$ can be long enough to allow a separation between the light yield of the first shower, resulting from the original neutrino interaction, and the light yield of the second shower caused by the $\tau$ decay. Background events consist of downward-going muons, which can be produced in the decays of unstable mesons present in the cosmic-ray induced atmospheric showers. At the considered depth, the flux of direct atmospheric muons exceeds by several orders of magnitude the flux of muons created in charge-current interactions of neutrinos. As a consequence, the expected background is almost exclusively constituted of downward-going muons and therefore it can be rejected by restricting signal searches with upward-going events. As mentioned above in Sec.1, atmospheric neutrinos represent background only for Astrophysics searches, while their measurement is the main channel to infer the neutrino oscillation parameters and constrain the mass-hierarchy problem.

### 3.2 Detector performance

The effective geometrical acceptance of the neutrino telescope scales with the detector size and with the energy of the impinging neutrinos. Useful figures of merit are the effective area and the effective volume, computed thanks to Monte Carlo simulations. Being proportional to the expected rate of reconstructed events, such quantities are proxies of the detection performance, and can be used to benchmark different neutrino telescopes. In Fig.2, ARCA115 and ORCA115 effective volumes are compared to the effective volume of ANTARES: over a wide energy range, the KM3NeT detectors are expected to outperform, by almost an order of magnitude, the ANTARES telescopes. Recent computations show that, already in their current building stage, ARCA21 and ORCA 14 have exceeded the ANTARES effective area all over the respective energy range.

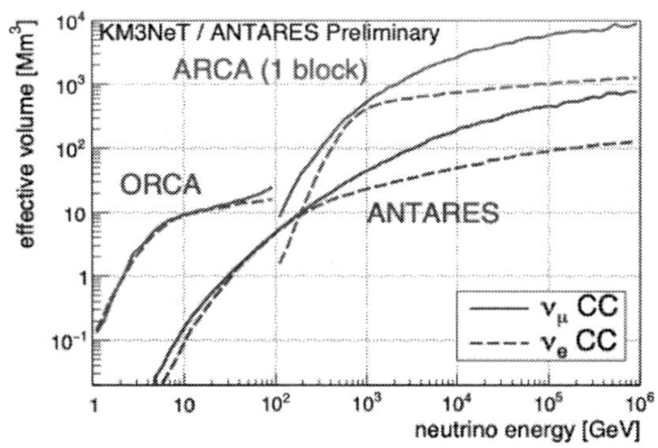


Figure 2: Detector effective volumes as a function of neutrino energy: comparison between ANTARES detector, ORCA and ARCA (1 building block). Solid lines refer to track event signatures ($\nu_\mu$ charged coupled interactions) while dashed ones to shower signatures ($\nu_e$ charged coupled interactions).

The choice to install the KM3NeT telescopes in abyssal sites presents clear complications due to the difficulties of arranging the deployments in such extreme conditions. Moreover the continuous optical background due to $^{40}$K decays and the bursts due to bio-luminescence imply a high background level in the acquired data, with respect to the rare neutrino induced signals. On the other hand, the light yield of $^{40}$K decays is exploited to optimise the measurement of the relative time-offsets between the PMTs in a DOM [12] and to perform long-term monitoring of the PMT photon detection efficiency.

Another advantage of the sea water is that it is far more homogeneous than ice, with a larger scattering length [1]. This results in a better angular resolution of the undersea detectors with respect to the under-ice ones. Thanks to Monte Carlo simulations it was possible to compute the angular deviation between the reconstructed neutrino tracks and their true directions as a function of energy. The figure of merit of the ARCA angular resolution, computed for the 2 building blocks configuration, is shown in Fig.3. Due to kinematical effects and for the topology of the event itself, a smaller angular resolution is obtained for track-type events than for shower-type events. For neutrino energies larger than few hundreds of TeV, an angular resolution below 0.1° is obtained.

Thanks to the optimal optical properties of the sea-water, especially at the depth of the ARCA installation, and thanks to a new reconstruction algorithm [13], it has been possible to obtain a sub-degree median angular resolution even for shower-like events from neutrinos with energies larger than 100 TeV. The energy resolution for track-like events in the range 10 TeV$\leq E_\mu \leq$10 PeV is, for the case of ARCA, ~0.27 expressed in $\log_{10}(E_\mu/E_{reco})$ units, where $E_\mu$ is the true energy of the simulated muon and $E_{reco}$ is the reconstructed one. For shower-like contained events, and for $E_\nu$ >50 TeV, the relative energy resolution gets smaller than 5%.

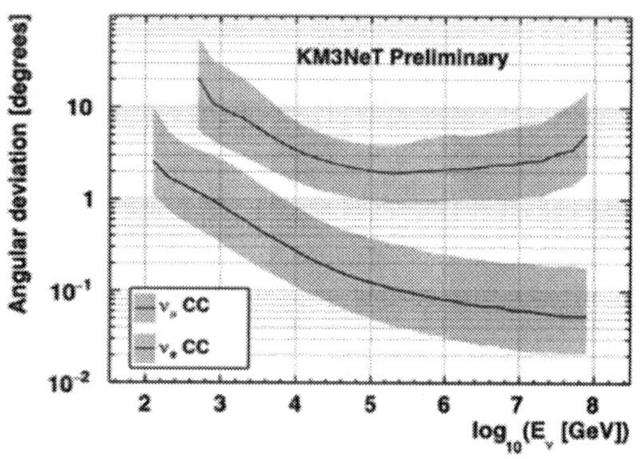


Figure 3: Angular resolution for ARCA track-like events (from $\nu_\mu$ charged current interactions) and shower-like events ($\nu_e$ charged current interactions), as a function of the neutrino energy.

## 4 Review of selected Physics results

### 4.1 Atmospheric muon flux depth-intensity relation

Data collected from the first three detection units of KM3NeT, ARCA2 and ORCA1, have been exploited to measure the atmospheric muon flux between 2232 and 3386 meters. This flux is shown to be compatible with the Bugaev flux calculation [14]. In this analysis, the capability of the multi-PMTs layout of the KM3NeT DOM allowed to gather information concerning the number of photons, as well as their arrival times and directions. Coincident photon-hits on different PMTs, here defined to be within a time window of 15 ns, can be exploited in order to discriminate between signals from atmospheric muons, bioluminescence and radioactive $^{40}$K decays. The result is a high-purity sample of atmospheric muons. After background rejection, efficiency-corrected muon rates are computed. This information, together with the effective area of a single optical module, coming from ad-hoc simulations, allows to determine the muon flux. The flux itself is then estimated along the detector depth, expressed in meters of

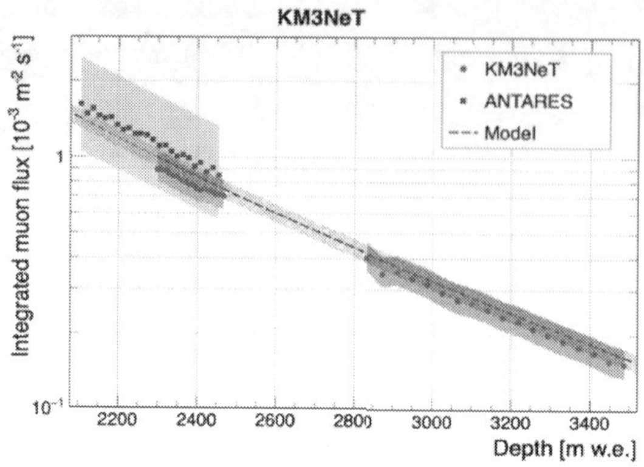


Figure 4: Integrated atmospheric muon flux as a function of depth below the sea level (expressed in meter of water equivalent) measured with ORCA1 and ARCA2 detectors, red points. Bugaev model is reported with black dashed line. ANTARES data are also included as blue point for comparison. Systematic uncertainties are reported as shadowed area.

water equivalent, as reported in Fig.6. The obtained measurement is in very good agreement with theoretical models, and have shown the synergy and complementarity of the two sites of the KM3NeT experiment. For major details, see [15].

### 4.2 Point source studies with ARCA6

One of the main goals in the neutrino astronomy is the identification of the sources of cosmic neutrinos. Therefore a time-integrated binned likelihood analysis was applied on the data acquired with ARCA6, for a total live-time of 92 days. After the event selection, the surviving events were binned in a two-dimensional space, determined by the angular distance of the events to a given source and their energy. A likelihood-ratio test statistic was then computed. A catalogue of 46 candidate sources was selected, and for each of them an $E^{-2}$ spectrum was tested: 6 of them are known to be spatially extended in the sky, and therefore the detector point spread function was modified with a Gaussian or a disk-like smearing. The smallest p-value obtained (~ 0.020), compatible with the background-only hypothesis, was found in correspondence of the radio galaxy Centaurus A (RA,DEC) = (201.36°,-43.02°). No strong neutrino emitters were found in the period of data analyzed (May 2021 - September 2021), and for this reason the 90% upper limit has been calculated for each candidate, as reported in Fig.5. While the effective area of the detector is comparable with the size of ANTARES, the main limiting factor for placing more stringent limits is the short live-time of the used data set.

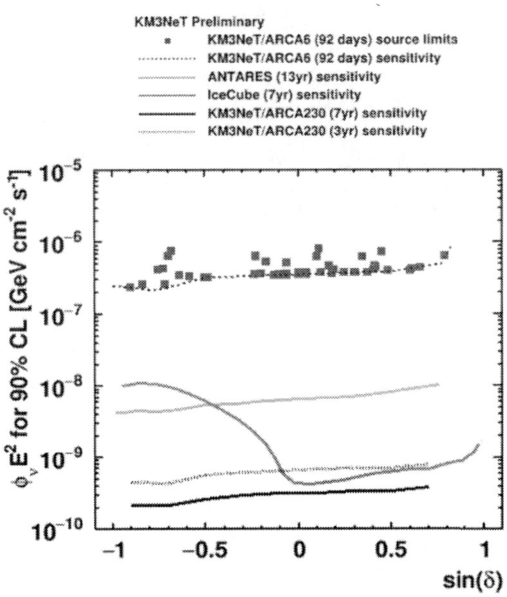


Figure 5: 90% flux limit as a function of declination reported for each candidate source in the catalogue. Sensitivity of ARCA6 (92 days, dashed blue line), ANTARES (13 yr, solid green line) and IceCube (7 yr, solid red line) are shown. The estimated sensitivity for ARCA in its final configuration is also shown, for respectively 3 and 7 years of data-taking (dashed and solid black lines).

### 4.3 KM3NeT potential for the next core-collapse supernovae

In the final configuration, the KM3NeT detectors will comprise a total of ~200,000 PMTs (grouped in ~6200 Optical modules). Despite the distance of optical modules within a DU, the neutrino telescopes are sensitive to 10 MeV neutrinos from Galactic and near-Galactic core-collapse supernovae. The detection technique exploits the possible excess of photon-hit coincidences on PMTs belonging to the same DOM, over the background. Data from the first six lines of ARCA and ORCA detectors have been analyzed to characterize the background expectation. Signal expectation is instead derived thanks to specific simulations, based on a custom software developed by the KM3NeT Collaboration. Searches for signal over background events is carried out considering time windows of 500 ms: such duration was chosen in order to cover, on average, the majority of the neutrino emission in the supernova accretion phase. In this way the sensitivity of the KM3NeT detectors to the neutrino burst is evaluated [16]. For a supernova at 10 kpc, KM3NeT will be able to estimate the mean energy of the core-collapse supernova neutrinos with sub-MeV accuracy. KM3NeT will be also able to share data of neutrino light curve in quasi-real time with millisecond time resolution. This is crucial for multi-messenger networks such as SNEWS2.0 [17] to confirm the detection and to be able to provide a precise localisation of the source to the astronomy community.

### 4.4 Diffuse and Galactic ridge neutrino emission

The IceCube Collaboration has reported evidence of a diffuse flux of astrophysical neutrinos in several detection channels [2, 18, 19]. This detection is important in order to complement direct searches for neutrino sources and to shed light on the mechanisms at the origin of acceleration and propagation of high-energy cosmic-rays. Several analyses by the IceCube Collaboration were performed on different data samples: in the single power-law hypothesis, best fit parameters and contour uncertainties showed a mild tension. In this scenario, KM3NeT can represent a complementary observer and provide an independent measurement to fully constrain the diffuse emission. For this reason, the data coming from ARCA6 have been analyzed, searching for an excess of high-energy events, over the atmospheric background. A total livetime of 101 days have been used. The analysis selected tracks with zenith angle larger than 90°, and to further reduce the remaining background, coming from wrongly reconstructed atmospheric muons, a Boosted Decision Tree has been built. No excess was found, and therefore the 90% upper limit has been reported, assuming the best-fit flux reported from the IceCube Collaboration [20]. A possible explanation

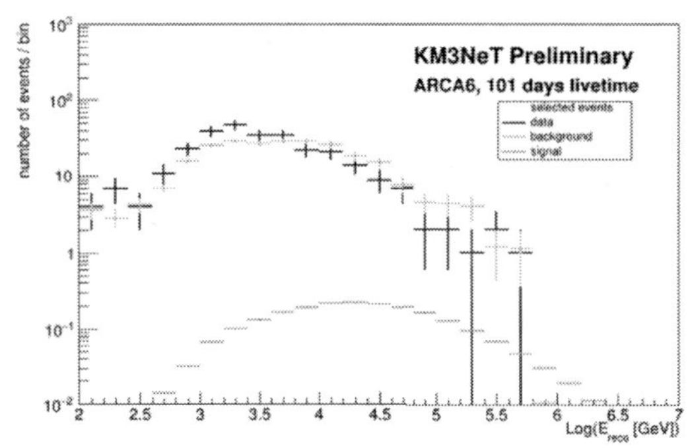


Figure 6: Energy distribution of selected data events (black points). The superimposed energy distributions of atmospheric background (light blue points) and of cosmic neutrino events (green points) estimated from MonteCarlo simulations are also shown.

for the tension observed and the declination distribution of the IceCube events, pointing at a North/South asymmetry of the neutrino flux, can be the presence of a bulk of neutrinos generated inside our Galaxy. Both ANTARES and KM3NeT, being placed in the northern hemisphere, are privileged neutrino observatories for the detection of a possible emission coming from the Galactic center. At the same time, multi-messenger measurements of neutrino and $\gamma$-ray spectra from the Galactic center provide a unique possibility to indirectly measure the cosmic-ray spectrum. For these reasons an ON-OFF analysis, based on the work reported in [21], considering an extension for the Galactic center neutrino emission of $|b_{gal}| < 3°$ and $|l_{gal}| < 40°$ , was performed on the data collected from ARCA6. Even in this case no excess was found, and the 90% upper limit has been reported. However, the recent analysis performed by the ANTARES Collaboration, with an extended data set and reported in [22], is pointing to a deviation from the background-only hypothesis above $2\sigma$: future analysis comprising an increased livetime and effective area of the KM3NeT detectors will allow to put even more stringent limits on the neutrino emission originated in the center of our Galaxy.

### 4.5 IceCube alert follow-up

An analysis pipeline has been developed and put in place in order to follow-up alerts coming from external observatories. In particular, several recent IceCube real-time alerts of potential astrophysical neutrinos (IC211208A, IC220205B, IC220225A, IC220304A), using the complementarity of the ORCA and ARCA data, have been inspected. At the time of the alerts both ARCA and ORCA detectors were operational with respectively 8 and 10 DUs. Each follow-up has been performed searching for a signal excess around the best Active Galactic Nuclei counterpart position. For the analysis, a default ±1 day time window around the alert time has been taken into account. Only for the alert IC211208A, in coincidence with the blazar PKS0735+17 found in a flaring state, an extended time window of 1 month (full December 2021) has also been analyzed. No neutrino candidate has been detected in the standard time window follow-up for all the considered alerts. Only for the search in the extended time window of 1 month, a candidate neutrino event has been detected, with an estimated energy ~18 TeV and fully compatible with the background expectation. Future alerts will be followed-up using a procedure similar but fully integrated in the online framework illustrated in Sec.4.6.

### 4.6 Online analysis framework

Online multi-messenger searches have already demonstrated the capability to answer to many open questions in Astroparticle Physics. A tight collaboration of neutrino with multi-wavelength (radio to $\gamma$-ray) and gravitational wave astronomers could in the near future shed light on the origin of the most energetic particles and the acceleration mechanisms in the Universe. The main requirement for these multi-messenger studies is the fast communication of potentially interesting observations through circulars ("alerts") sent to partner observatories, with latency of few minutes. With the scientific return of experience from the ANTARES case, the KM3NeT online analysis platform has been designed and implemented: it comprises track and shower reconstruction algorithms from ORCA and ARCA detectors, selection in real time of neutrino sample, receivers and filters of alerts from the external community, analysis pipelines searching for coincidences in time/space around these alerts, selection of a sample of interesting events to send alerts and core-collapse supernovae analysis. Considering the exploited data taking approach used by KM3NeT, i.e. the trigger-less streaming readout, the most consuming processing, in term of computing-resources, is performed directly at the ARCA and ORCA shore stations. The common centralized part contains some services like tools to make event display, event storage, monitoring and the internal/external reporting, the SN final processing, the neutrino alert sending module and the online analysis module.

### 4.7 Neutrino oscillation studies

From the measurement of the atmospheric neutrino flux with the ORCA detector, it is possible to infer the neutrino oscillation parameters $\Delta m^2_{31}$ and $\theta_{23}$ through $\nu_\mu$ disappearance, exploiting baselines spanning from tens of meters for downward-going events to ~13000 km for upward-going events. The first results for the oscillation parameters were obtained from a data set consisting in 355 days of livetime with ORCA6 and are reported in [23]. The first measurement shows that oscillations are preferred with a significance of $5.9\,\sigma$ over the "no oscillation" scenario. The contour plot at 90% CL with the sensitivity of ORCA6 to $\Delta m^2_{31}$ and $\theta_{23}$ shows that with almost one year of data for ORCA6 the result is of the same order of magnitude as other competing experiments. This shows a promising scenario when a larger detector is available.

### 4.8 Moon and Sun shadows

The analysis of the ORCA6 data, corresponding to 499 days live-time, allowed the first observation of the Moon and the Sun cosmic-ray shadows in the downward going atmospheric muon flux [24]. Both the one-dimensional and the two-dimensional analyses in the phase space of the angular distance between the reconstructed event coordinates in the sky and the position of the two celestial objects have been performed. The significance of the observed shadows has been computed according to an ON/OFF approach. The cosmic-ray shadows induced by the Moon and the Sun were detected with a statistical significance of $4.1\,\sigma$ and $6.2\,\sigma$, respectively. By assuming the Moon and Sun nominal coordinates, it was possible to verify the positioning accuracy of the detector array together with additional checks of other possible systematic effects, for example possible time calibration biases. The ORCA detector angular resolution for downward-going muons was found to be $0.8° \pm 0.14°$ for the Sun analysis and $0.49° \pm 0.15°$ for the Moon. Even from these early results it is possible to confirm the high potential of the KM3NeT technology.

## 5 Conclusions

In this contribution the status and the first results of the KM3NeT neutrino telescopes, ARCA and ORCA, located in two abyssal sites of the Mediterranean Sea, have been reported. The KM3NeT Collaboration has entered in the massive production stage of its detector parts: in the next few years additional DUs will be deployed, aiming to realize 0.7 $km^3$ size for ARCA and almost half of the ORCA building block. At the time of writing, 21 ARCA DUs and 14 ORCA DUs have been succesfully deployed and are continuously taking data. Selected results from the analyses of the data collected by the ARCA and ORCA detectors since 2015 have been presented. Although still at early stages, the current ARCA and ORCA configurations allow to match, if not exceed, the detection potential of ANTARES. Measurements of the downward-going flux have been exploited to check the detector performance and alignment (e.g. with the Moon and Sun cosmic-ray shadows). Searches for diffuse fluxes and point-like sources, as well as the measurement of the neutrino oscillation parameters have been started, confirming the exceptional potential discovery that can be reached in few years, given the scheduled enlargement of the size of the telescopes. Complementary to the standard data taking chain, an efficient online reconstruction and analysis framework has been optimized to exchange quick alerts with other space and ground-based experiments, such as optical telescopes, X-ray, $\gamma$-ray and gravitational wave observatories, promptly contributing to a rich multi-messenger program.

## References

[1] T. Chiarusi, M. Spurio, Eur. Phys. J. C **65**, 649 (2010);

[2] M. G. Aartsen *et al.* [IceCube], Phys. Rev. Lett. **111**, 2, 021103 (2013);

[3] M. Aartsen *et al.* [IceCube], Science **361**, 6398, 147 (2018);

[4] V. A. Allakhverdyan *et al.* [Baikal], [arXiv:2211.09447 [astro-ph.HE]];

[5] S. Adrián-Martínez *et al.* [KM3NeT], Journal of Physics G: Nuclear and Particle Physics **43**, 8, 084001 (2016);

[6] E. Leonora *et al.* [KM3NeT], Journal of Physics: Conference Series **1056**, 1, 012031 (2018);

[7] T. Chiarusi *et al.* [KM3NeT], EPJ Web of Conferences, in press. (2023);

[8] W. Assal *et al.* [KM3NeT], J. of Instrumentations **16**, C09034 (2021);

[9] http://www.ponricerca.gov.it;

[10] https://www.mur.gov.it/it/pnrr/missione-istruzione-e-ricerca;

[11] R. Le Breton *et al.* [KM3NeT], JINST **16**, 9, C09004 (2021);

[12] K. Melis *et al.*[KM3NeT], PoS ICRC2017, 1059 (2017);

[13] T. Van Eeden *et al.* [KM3NeT], Zenodo (2022) https://doi.org/10.5281/zenodo.6804829;

[14] E. V. Bugaev, A. Misaki, A. Naumov Vadim, T. S. Sinegovskaya, S. I. Sinegovsky, N. Takahashi, Phys. Rev. D **58**, 054001 (1998);

[15] M. Ageron *et al.* [KM3NeT], EPJC **80**, 2, 99 (2020);

[16] S. Aiello *et al.* [KM3NeT], EPJC **81**, 5, 445 (2021);

[17] The Supernovae Early Warning System, https://snews.bnl.gov/;

[18] M. G. Aartsen *et al.* [IceCube], AAS **883**, 1, 3 (2016);

[19] M. G. Aartsen *et al.* [IceCube], Phys. Rev. Lett. **125**, 12, 121104 (2020);

[20] J. Stettner *et al.* [IceCube], PoS ICRC2019 1017 (2019);

[21] S. Adrian-Martinez *et al.* [ANTARES], Phys. Lett. B **760**, 143 (2016);

[22] A. Albert *et al.* [ANTARES], arXiv:2212.11876 (2022);

[23] L. Nauta *et al.* [KM3NeT], PoS ICRC2021, 1123 (2021);

[24] KM3NeT Collaboration, [arXiv:2211.08977 [astro-ph.IM]].

# UHECR Signatures and Sources

*Daniele* Fargion[1,*], *Pier Giorgio* De Sanctis Lucentini[2], and *Maxim Y.* Khlopov[3,4]

[1]Rome University "La Sapienza" and MIFP, Rome, Italy.
[2]National University of Oil and Gas «Gubkin University», Moscow, Russia.
[3]National Research Nuclear University "MEPHI", Moscow, Russia.
[4]Institute of Physics, Southern Federal University, Rostov on Don, Russia.

**Abstract.** We discuss recent results on the clustering, composition and distribution of Ultra-High Energy Cosmic Rays (UHECR) in the sky; from the energy of several tens of EeV in the dipole anisotropy, up to the highest energy of a few narrow clusters, those of Hot Spots. Following the early UHECR composition records deviations from proton we noted that the UHECR events above 40 EeV can be made not just by any light or heavy nuclei, but mainly by the lightest ones as He,D, Li,Be. The remarkable Virgo absence and the few localized nearby extragalactic sources, such as CenA, NGC 253 and M82, are naturally understood: lightest UHECR nuclei cannot reach us from the Virgo distance of twenty Mpc, due to their nuclei fragility above a few Mpc distances. Their deflection and smearing in wide hot spots is better tuned to the lighter nuclei than to the preferred proton or heavy nuclei candidate courier. We note that these lightest nuclei still suffer of a partial photodistruction even from such close sources. Therefore, their distruption in fragments, within few tens EeV multiplet chain of events, have been expected and later on observed by Auger collaboration, nearly a decade ago. These multiplet presences, strongly correlate with the same CenA, NGC253 sources. The statistical weight of such correlation is reminded. We conclude that the same role of NGC 253 clustering at lower energies could also feed the Auger dipole anisotropy at lower energy ranges. Such lower energy anisotropy could be fed and integrated by nearest Vela, Crab, LMC and Cas A contributes. In our present UHECR model, based on lightest nuclei in local volumes of a few Mpcs, closest AGN, Star-Burst or very close SNR are superimposing their signals, frozen in different epochs, distances and directions, feeding small and wide anisotropy. Possible tests to confirm, or untangle the current model from alternative ones, are suggested and updated.

## 1 Introduction

Cosmic rays (CR) mainly contain charged particles. As charges they are bent, deflected and smeared in their flight through cosmic and galactic magnetic fields losing most of the astrophysical source imprint. Thus for a century the CR origin remained a mystery and a compelling question continually addressed by high-energy astrophysics[1]. Indeed the Ultra High Energy CR (UHECR) nature itself connects the most violent sources of the Universe with the deep inner secrets of nuclear and particle physics.

The extreme boundaries of micro and macro physics find a rare connection by the UHECR-source correlation. We offer here our reading key based on last decades discovered maps and composition data. UHECR proton above EeV ($10^{18}$ eV) are expected to be less deflected because of their harder rigidity. They may more closely represent the location of their source leading to connections on an astrophysical or cosmic scale.
With increasing energy the UHECR protons are deflected less and less, but above nearly 60 EeV where they should be somewhat beamed within a few degrees, they also become opaque and constrained by the Cosmic Black Body, due to the photopion production. That effect is known as the GKZ cutoff [2, 3].
Such UHECR nucleons are then enclosed within one to two hundred Mpc leading to a small volume bound within our much larger Universe (4Gpc). This proton scenario, if real, could facilitate the identification of the source associated with such an inhomogeneous scale mass distribution.

UHECRs could also be nuclei, but these are affected by increased deflection and smearing as well as partial opacity of the photopions. In addition, the lighter UHECR nuclei undergo additional photon-nuclear destruction, reducing the UHECR propagation distance to a few Mpc. Their fragments at tens of EeV, must be lighter and more diffused. Therefore, the path of an incoming UHECR depends not only on the energy but also on the mass and composition of the charge (such as p,He,N,Fe,Ni..), on the intensity of the magnetic fields encountered and on the timing of any in-flight decays.

Indeed, for a more in-depth understanding of the phenomenon we have to consider not only the distribution of the possible sources, but also the *history* of the CR, the timing of the detected signals, the delay due to the random-walk [4], each of them setting specific limits to the allowed energy windows.

Candidate sources, whether they are small as SN, GRB, Star-Burst, or larger as AGN Jet, can be found both

*e-mail: daniele.fargion@fondazione.uniroma1.it

far from us and close –even in our galaxy– in a complex inhomogeneous cosmic structure. We show how this difficult source correlation search can be approached and partially solved by constrains based on last two decades of observational discoveries [1] on UHECR composition, their clustering and anisotropies at different energy windows.

## 2 UHECR composition and Virgo absence

In the early 2000s most of the theoretical models were based on a proton composition for UHECR at the GZK energy limits: $6 \cdot 10^{19}$-$10^{20}$ eV. In fact, the GZK cutoff was the main focus looked for in the UHECR spectra. The proton at such energies should cluster within tight angles, 2° for coherent bending, or near 8° for incoherent random bending [5]; it must also be contained in volumes of $1 - 2$ hundred Mpc, the GZK volume. Thus they were expected to rise and shine from the nearest and more mass populated regions, similar to how infrared galaxies are painting the infrared sky: the Virgo cluster, at a distance of about 20 Mpc, is the dominant mass in the Auger southern emisphere sky. Surprisingly, at those energies the UHECR signals in Auger, and later also in Telescope Array (TA) in the northern hemisphere, did not show a sharp signature confirming this expectation: the *surprising Virgo absence*.

Rather both Auger and TA discovered some wider ($8° - 20°$) clustering elsewhere, uncorrelated with Virgo. A first Hot Spot in southern sky was pointing to Cen-A, the nearest AGN in Auger sky. A second Hot Spot in northern sky was found by TA pointing toward a nearest AGN M82. A third clustering, or Hot Spot, was later recognized [6], pointing to NGC 253 a micro AGN or star-burst galaxy still in the Auger sky.

A combined map of the Auger and TA signals is presented in equatorial coordinates in Fig.1. In the center the area with the most abundant dark dots, thousands of galaxies of the Virgo Cluster, at a distance of 18 Mpc well within the GZK volume [7].

Assuming UHECR protons, the Virgo absence was (and still is) puzzling and inexplicable. Since 2007 a harder spectra in Auger composition studies [8] based on the recently observed most energetic UHECR airshower profile, disfavored protons above 10 EeV in favour of light and lightest nuclei. Protons have actually only been found to be dominant at sub-EeV energy ranges. This led us to suggest that the absence of Virgo could be explained if UHECRs consist mainly of lighter nuclei, such as D, He, Li, Be [9–12].

The lightest nuclei are indeed too fragile to survive the 20 Mpc flight from Virgo, due to photon-nuclear distruption. On the other hand, Cen-A, M82 and NGC 253 are just few Mpc away, much closer than Virgo. These smaller distances allow most UHECRs to survive photon nuclear distruption and opacity. These sources are among the ideal candidates star-burst or active AGN around us, whose signals are well tuned and observable in the Auger-TA data. As the data of the last decade had shown. Only a decade later (2017), the UHECR signature of the light nuclei was largely confirmed (see note on page 26 of [13]) by their detailed slant depth averaged models of the UHECR shower

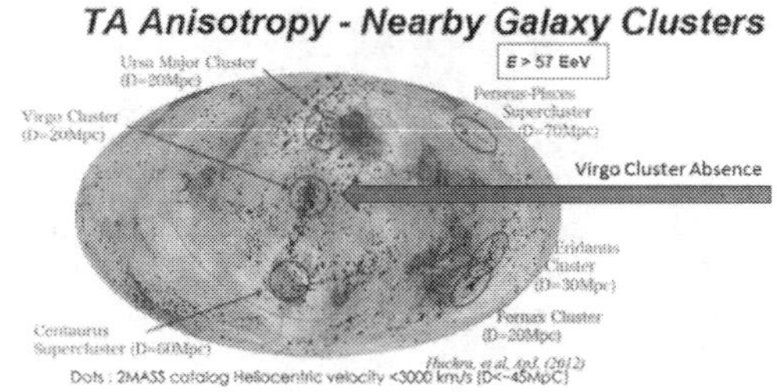


Figure 1: The combined map, in equatorial coordinates, based on the Auger and TA data. Note in red two Hot Spot UHECR clustering signals. The highest UHECR density event rates are north around M82 AGN for TA and south around Cen-A for Auger. The dark dots circled in red in the center [7] correspond to the closest potential sources included in the GZK volume, pointing to the Virgo cluster. Those sources were ideal for any UHECR proton carrier. The lighter nuclei are much more limited, a few Mpc, by photonuclear destruction. These are the ideal candidates who can explain the absence of Virgo in the Auger map.

maximum, which show the key role of light and even the lightest nuclei.

Therefore, around $3 \cdot 10^{19}$ eV, UHECRs mainly He like nuclei may be cosmic but, as already mentioned, extremely local just within a few Mpc, almost all in our Local Group [6].

Other near and far cosmic UHECR sources could further add dilute homogeneous and isotropic noises. Rarer and heavier nuclei, such as Ni and Fe, may also be present at higher energies (around or higher than $7 \cdot 10^{19}$ eV), but so bent and smeared that they are often confined within our own galaxy [11]. Ultimately these near candidates, Cen-A, M82, NGC 253, are contributing the most to the local anisotropy [6].

### 2.1 Hot Spot and correlated multiplet fragments

A lightest nuclei whose charge is two or four times that of the proton undergoes a somewhat larger deflection than the protons, a smearing comparable to those observed in the Southern and Northern skies. The charge $Z = 2 - 4$ of lightest nuclei explains the observed sizes of the Hot Spots due to their random incoherent deflection at about $3 \cdot 10^{19}$ eV, assuming He-like nuclei are bent within angles of $16° - 32°$ [5].

We note that other light but heavier nuclei, like N,O, suffer no photo-nuclear distruption from Virgo, but are more smeared in a wider solid angle. These nuclei would cause a signal from Virgo diluted in a large angle of the sky, $56° - 64°$, with a possible negligible role in the observed Hot Spot, but with a potential role in the dipole anisotropy. Such a dipole signature toward Virgo is again absent: it points elsewhere, almost half the sky away. Therefore, this is a compelling argument for considering UHECR nuclei consisting mostly of the lightest ones, unable to reach us from Virgo.

https://doi.org/10.1051/epjconf/202328304010

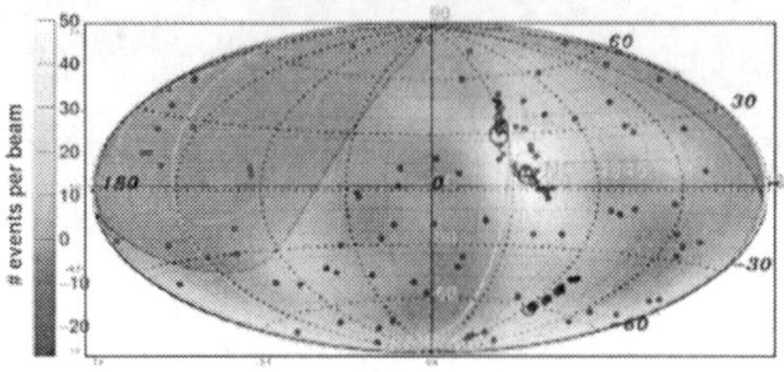


Figure 2: The galactic UHECR map overlaid on the recent 39 EeV anisotropy map, shows the earliest (2011) UHECR events and the associated 20 EeV smeared multiplet, up and down respect to the Cen A position. We interpret these nearly vertical spread of events, followed by the fragment signals, as due to the horizontal planar galactic spiral magnetic fields, whose spin flip up and (later) down during the UHECR flight, is leading to an up-down deflection of the charged He,as well as D, Li, Be, lightest UHECR fragments. Let us note, here for the first time, also the presence of a train of UHECR fragments located around the SMC and LMC galactic direction, pointing to NGC 253 source in the South galactic Pole [14]. The probability that these fragments are correlated by chance with both to the two main candidate of UHECR in Auger sky, Cen-A and NGC 253, is quite negligible, below 0.001%.

The same photo-nuclear distruption can occur for D, He,Li, Be nuclei even from a nearby Cen-A, or NGC 253; in part it had to happen, and as we have noted it happened. These fragments were to be centered or distributed along the main Hot Spot sources in the Auger sky: Cen-A, as well as in the additional clustering around NGC 253. M82 is not visible from Auger sky, but is within the TA sky.

We had foreseen [5, 11], such a tail of fragments around Cen-A, a couple of years before its observation by Auger [14] in 2012, see Figure 2. In the figure three multiplets of events are marked with black dots. Their extrapolated endpoints are signed with a blue cross. We note that two of them are within 8° of Cen-A. And the third one is within 8°, to the other relevant source, NGC 253.

One may inquire how often such a correlation may occur by chance. The solid angle distance of each source from its fragment tail cross center is about $7° - 8°$, which represents less than 1% of the AUGER sky.

The probability that two such related events occurred within 3 trials, each of which point to the same Cen-A area within 1% of the sky, is less than $2.97 \cdot 10^{-4}$.

Furthermore, the probability that three signals appearing above the two selected sources, in an area that is 2% of the Auger sky, and occurring by chance is about $8 \cdot 10^{-6}$. We recall that Cen-A and NGC 253 are the only two in the Auger sky.

Thus, the directional clustering of the fragments towards our a priori selected sources offers another strong argument in favor of the current model of lighter nuclei.

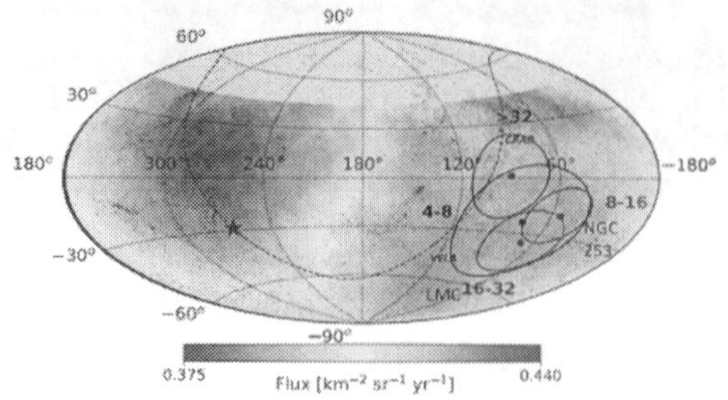


Figure 3: The UHECR dipole clustering in equatorial coordinates. Blue, green, red, and violet points and curves indicate different energy ranges (4-8, 8-16, 16-32 and >32). Few suggested galactic or nearby extragalactic source candidate are labeled. The main extragalactic source, the star-burst NGC 253 role, is well correlated to the $8-16; 16-32$ EeV anisotropy energy ranges. An additional minor influence of galactic sources, as Crab, Vela and LMC, could explain the lower energy, 4 EeV, anisotropy and its directional variability with the different energy increase.

### 2.2 UHECR at AGN 3C 454: a Z resonance?

Few unexplained UHECR events correlated with very distant AGN (such as 3C 454), if confirmed, may require a different mechanism to exceed the GZK cut-off. A ZeV neutrino that scatters on the relic neutrinos, forming the final UHECR and effectively exceeding any GZK cut-off. The calorimeter for ZeV neutrinos scattering is in the huge light (even sterile) relic cosmic neutrinos halo spread around origin galaxy or local group. They are converting the ultrarelativistic energy of the Z boson, by its decay, into secondary nucleons and antinucleons. These nucleons can reach us, overcoming any GZK distance. These UHE ZeV neutrinos [16], complementary to the GZK neutrinos at tens of EeV, could rise as ascending or horizontal airshowers in deep valleys [17], induced by a tau decay skimming and escaping from the Earth [18].

## 3 Conclusion

Many scenario inspired by many spare parameter are confusing the UHECR connection to their sources. The UHECR anisotropy [19] and asymmetry [20] show that their main origin is within a very local Universe, not a hundred Mpc away.

UHECR are very probably formed in jets both in AGN quasars and in micro-quasars (star forming regions, accretion disk and ejecta), powered by black Hole or neutron star binary tidal disruptions. These jets [21] may arise in both AGN and in smaller size GRB, micro quasars and SGR, shining at different epochs and locations. The charged hadronic nuclei UHECR flying from those jets are possibly bent, spread by random walk, suffering large time delay. This explain why they are not expected to be correlated with present optical active AGN at distances of

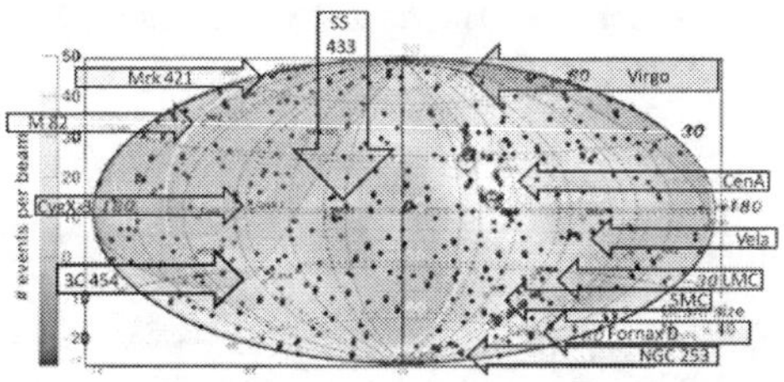


Figure 4: UHECR map in galactic coordinate with marked the main probable identified sources and the missing ones as Virgo. The role of 3C 454 may be relevant only if the narrow clustering will be more enhanced [15]. A 39 EeV anisotropy and a 20 EeV multiplet map is also overlaid. These candidate sources are the same ones discussed since 2015 [6]. In particular we note the growing signals around the Vela, LMC, SMC and Crab in nearby sky, possibly feeding the low energy dipole Auger anisotropy. The extreme powerful gamma source, 3C 454, shows some minor clustering of UHECR. The 3C 454 is too far in GZK distance unity to be explained via any directional nuclear event. It can be an UHECR secondary source: an eventual UHE ZeV neutrino courier, whose hitting the relic cosmic neutrino background, in a hot wide neutrino halo [16] is leading to the Z boson capable to decay also into UHECR nucleon or anti-nucleon. The relic neutrinos are possibly, also, at $\simeq 1.6eV$ mass, as the recent candidate sterile ones. The ZeV scatterinng could be leading to Z resonant bosons. The secondary decay of such ultrarelativistic Z boson, may shine as a final UHECR nucleon and antinucleon, observed finally on Earth as a minor UHECR clustering along the 3C 454 source [4].

hundred Mpc[4]. There is a possibility that the heavier UHECRs such as NI, Co, Fe are mostly bent, twisted and even contained in spiral paths within halos of their own origin. The lightest nuclei *Helium-like*, possibly secondary fragments of heavier ones, are quite directional; they are able to escape from those AGN galaxy, star-burst sources as nearby Cen-A, NGC 253, M82, reaching us with some memory of their origin in the observed few Hot Spots.

These lightest UHECR nuclei at tens of EeV have the virtue of flying only within a space of very few Mpc, hiding the presence of Virgo and offering only a local group detection view. Protons only play a role at EeV or lower, escaping hundreds of Mpcs rather smeared in their arrival directions. Their bending and overlapping should lead to a homogeneous isotropic sky, like the one observed at the EeVs energy windows. Lastly, the very few and rare heaviest and more energetic Ni-Co-Fe UHECRs can also arise around our galactic halo, perhaps also polluted by nearest galactic sources.

In summary lightest nuclei in UHECR are tagging just a few sources in the very local universe Cen-A, M82, NGC 253; AGN or starburst or microquasars. Also at lower energy, in the 4 – 30 EeV range, UHECR are feeding an Auger dipole anisotropy, mostly linked to NGC 253, see Fig.3, and partially to Vela, Crab, LMC, SMC. Other minor, unexplained UHECR cluster can, if confirmed, revive earliest Z boson resonance models, based on UHE ZeV neutrinos scattering on relic cosmic ones, with mass, see Fig.4. The eventual multi-peak energy signature [22] in their spectra, may in principle in a far future, test the different neutrino mass splitting or even the recent sterile neutrino masses candidature. Their final nucleon anti-nucleon shower composition signature (not the lightest nuclei one) is a key test, to verify and disentangle their Z boson nature, see Fig. 4, [4].

In conclusion, early popular proton courier at the GZK energy of 40 – 60 EeV, is not longer a viable UHECR candidate [8]. The recent, simplest models based on starburst versus AGN nearby sources nature, while ignoring the UHECR nuclei composition role, cannot explain the puzzling absence of Virgo. In a sentence, for the moment, lightest nuclei [9] and their clustering along a few correlated local sources [6] are the first and main guaranteed messages from last decade of Auger and TA data understanding.

The three tens of EeV events in the Auger sky clustered in multiplets pointing two towards Cen-A and one towards NGC 253, as predicted [10, 11, 23] and observed [14] (see Fig. 2), are providing key support for the lightest nuclei model.

**Acknowledgement**

The research by M.K. was financially supported by Southern Federal University in the framework of the State contract with the Ministry of Science and Education of Russian Federation.

## References

[1] R. Aloisio, *Ultra High Energy Cosmic Rays an overview*, in *Journal of Physics: Conference Series* (IOP Publishing, 2023), Vol. 2429, p. 012008

[2] K. Greisen, Physical Review Letters **16**, 748 (1966)

[3] G.T. Zatsepin, V.A. Kuz'min, Soviet Journal of Experimental and Theoretical Physics Letters **4**, 78 (1966)

[4] D. Fargion, P. De Sanctis Lucentini, M.Yu. Khlopov, Proceedings of Science **306**, 15 (2018)

[5] D. Fargion, Nuclear Instruments and Methods in Physics Research Section A: Accelerators, Spectrometers, Detectors and Associated Equipment **630**, 111 (2011)

[6] D. Fargion, G. Ucci, P. Oliva, P. De Sanctis Lucentini, *The meaning of the UHECR Hot Spots-A light nuclei nearby astronomy*, in *EPJ Web of Conferences* (EDP Sciences, 2015), Vol. 99, p. 08002

[7] J.P. Huchra, L.M. Macri, K.L. Masters, T.H. Jarrett, P. Berlind, M. Calkins, A.C. Crook, R. Cutri, P. Erdoğdu, E. Falco et al., The Astrophysical Journal Supplement Series **199**, 26 (2012)

[8] A. Watson, arXiv preprint arXiv:2301.07392 (2023)

[9] D. Fargion, Physica Scripta **78**, 045901 (2008)

[10] D. Fargion, Progress in Particle and Nuclear Physics **67**, 569 (2012)

[11] D. Fargion, Progress in Particle and Nuclear Physics **64**, 363 (2010)

[12] D. Fargion, D. D'Armiento, P. Paggi, S. Patri', Nuclear Physics B - Proceedings Supplements **190**, 162 (2009), proceedings of the Cosmic Ray International Seminars

[13] A. Aab, P. Abreu, M. Aglietta, I. Al Samarai, I. Albuquerque, I. Allekotte, A. Almela, J.A. Castillo, J. Alvarez-Muñiz, G. Anastasi et al., Journal of Cosmology and Astroparticle Physics **2017**, 038 (2017)

[14] P. Abreu, M. Aglietta, E. Ahn, I.F.d.M. Albuquerque, D. Allard, I. Allekotte, J. Allen, P. Allison, J.A. Castillo, J. Alvarez-Muñiz et al., Astroparticle Physics **35**, 354 (2012)

[15] D. Fargion, G. Ucci, P. Oliva, P. De Sanctis Lucentini, in [6], p. 08002

[16] D. Fargion, B. Mele, A. Salis, The Astrophysical Journal **517**, 725 (1999)

[17] D. Fargion, The Astrophysical Journal **570**, 909 (2002)

[18] D. Fargion, P. De Sanctis Lucentini, M. De Santis, M. Grossi, The Astrophysical Journal **613**, 1285 (2004)

[19] E. Mayotte, T. Fitoussi, *Update on the indication of a mass-dependent anisotropy above 1e18.7ev in the hybrid data of the pierre auger observatory*, in 6th International Symposium on UHECR, L'Aquila, Italy (2018)

[20] P. Plotko, A. van Vliet, X. Rodrigues, W. Winter, arXiv preprint arXiv:2208.12274 (2022)

[21] D. Fargion, Astronomy and Astrophysics Supplement Series **138**, 507 (1999)

[22] D. Fargion, D. D'Armiento, O. Lanciano, P. Oliva, M. Iacobelli, P. De Sanctis Lucentini, M. Grossi, M. De Santis, Nuclear Physics B - Proceedings Supplements **168**, 292 (2007)

[23] D. Fargion, Nuclear Instruments and Methods in Physics Research Section A: Accelerators, Spectrometers, Detectors and Associated Equipment **692**, 174 (2012)

 https://doi.org/10.1051/epjconf/202328304011

# Multl-messenger probe of Cosmic Ray Origins: MICRO project

*Antonio* Condorelli[1,*] on behalf of MICRO *project*[**]

[1]Université Paris-Saclay, CNRS/IN2P3, IJCLab, Orsay, France

**Abstract.** We present the milestones achieved by the MICRO (MultI-messenger probe of Cosmic Ray Origins) project. This multi-institute project consists of a study of bursting astrophysical sources as candidate sources for ultra-high-energy cosmic rays (UHECRs). We aim at identifying source classes that correlate best with existing observational data (arrival direction distribution, energy spectrum and primary mass composition) and to constrain the origin of UHECRs on the basis of the largest datasets acquired to this day. The study of the secondary fluxes of neutrinos and gamma rays will provide a powerful test of the most suitable astrophysical scenarios. For these achievements, a public software enabling a joint fit of both composition and flux data as a function of energy and direction is being developed. In addition, a modelling of transient sources is foreseen, including the treatment of hadronic interactions within the sources, which represents a novelty for UHECR simulation codes. This contribution will highlight the first results of the project and detail upcoming software releases, which will benefit the astroparticle community at large.

## 1 Introduction

Identifying the sources of Ultra-High Energy Cosmic Rays (UHECR) is one of the most pressing questions in high-energy astrophysics. Despite the advent of high-statistics and high-quality data, most prominently obtained by the Pierre Auger Observatory, providing an astrophysical explanation to the observed features at Earth is not trivial. To shed light on the origin of UHECRs, many studies, such as for instance refs [1–3] and references therein, have considered a simple cosmological model, assuming all the sources identical, pointlike and steady in time.

MICRO (MultI-messenger probe of Cosmic Ray Origins) is a project which aims to go beyond the conventional modelling of steady sources of UHECRs. A novelty of this approach is a complete study of bursting sources starting from the modelling of selected source classes, including hadronic interactions within the source and over the propagation down to Earth, to predict the UHECR sky as a function of energy and mass composition. The objectives specifically addressed in the MICRO project will provide important answers to the leading questions of identifying the sources of UHECRs: (i) how do burst-like signatures such as Gamma Rays Bursts (GRBs) or Active Galactic Nuclei (AGN) flares fit the cosmic-ray data, (ii) how can we constrain the 3D distribution of sources from available UHECR observables, and (iii) could astrophysical high-energy neutrinos, some high-energy gamma rays, and UHECRs come from the same bursting sources.

Two different subjects are addressed within the MICRO project: first the development of a public software, enabling a joint fit of both UHECRs composition and flux data as functions of energy and direction; second the modelling the source environment, studying how this can affect UHECR ejection.

## 2 Towards a public software for UHECR observables

One of the milestones of the project is to release a public software capable of providing acceptable solutions to the astrophysical origin of UHECRs. The idea is to define it an astrophysical scenario, assuming bursting sources, following a certain distribution in redshift. The injected flux for each mass at the source can be written as:

$$J_j(E_i) = \varepsilon_j^{\mathrm{tot}} \cdot k \cdot \left(\frac{E_i}{E_0}\right)^{-\gamma} \cdot f_{\mathrm{cut}}(E_i, Z \cdot R_{\mathrm{cut}}) \tag{1}$$

where $E_0$ is a reference energy, $\varepsilon_j^{\mathrm{tot}} \cdot k$ governs the normalization for the mass $j$, which represents the injected energy, e.g. per stellar mass per rate , $R_{\mathrm{cut}}$ is a cutoff rigidity, $\gamma$ is the spectral index at escape from the sources and $f_{\mathrm{cut}}(E_i, Z \cdot R_{\mathrm{cut}})$ is the cutoff function.

In order to connect properties at the sources with observables at Earth, the extra-galactic propagation has to be performed. This is done considering Monte Carlo codes, such as *SimProp* [4] or *CRPropa* [5]. All the events generated are stored in a normalized 5D tensor $T\left(E_{\mathrm{det}}, A_{\mathrm{det}}, z \middle| E_{\mathrm{inj}}, A_{\mathrm{inj}}\right)$. In this way, the probability that a particle with energy $E_{\mathrm{inj}}$ and mass $A_{\mathrm{inj}}$ at a redshift $z$ would be detected with an energy $E_{\mathrm{det}}$ and a mass $A_{\mathrm{det}}$ is performed. After properly taking into account the possible interactions of UHECRs with the cosmic microwave background (CMB) and the extra-galactic background light (EBL) during the extra-galactic propagation, the propagated flux at Earth can be calculated and then compared

*e-mail:condorelli@ijclab.in2p3.fr
**e-mail:MICRO-L@in2p3.fr
A list of contributors can be found here: https://micro-uhecr.github.io

 https://doi.org/10.1051/epjconf/202328304011

to the experimental data. In Fig. 1 the expected and the experimental spectra are shown, as well as the first two moments of the depth of the shower maxima $X_{\max}$ distributions.

As in similar works [1, 2], it was found that it is pos-

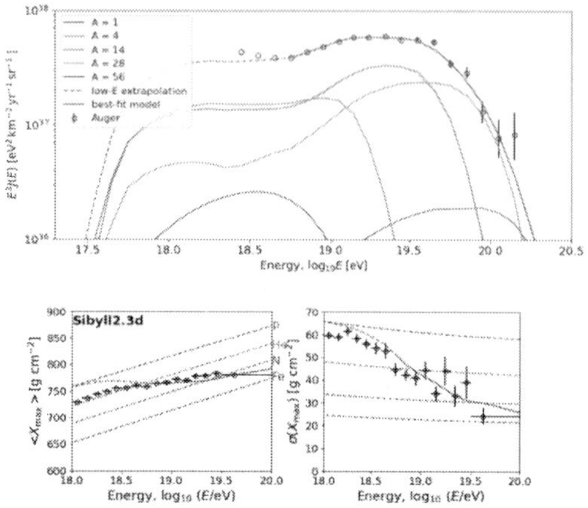


**Figure 1.** Top: Fluxes of nuclei with primary injected particles as labelled. Curves with different colours show the sum of the flux of primaries with a given mass number and all secondaries produced by the same nuclear species. The fit is performed only at energies above the ankle ($10^{18.7}$ eV). Experimental data are the flux measurements by the Pierre Auger Collaboration [6].
Bottom: black points are the average and standard deviation of the $X_{\max}$ distribution at each energy, as measured by the Pierre Auger collaboration ; coloured lines are predicted by Sibyll2.3d [7] for various pure compositions (thin lines) and for the composition predicted by our model at the best fit (thick brown line).

sible to have a good description of both UHECR energy spectrum and mass composition in terms of a hard spectral index $\gamma < 2$, and a low maximum energy of protons $\simeq 2 \cdot 10^{18}$ eV.

An advantage of using a Monte Carlo simulations for computing the extra-galactic propagation is that the secondary productions can be easily computed. Neutrinos propagate through the inter-galactic space undergoing only adiabatic energy losses due to the expansion of the Universe, hence their energy spectrum at Earth, unlike that of UHECRs, is expected to be also affected by the presence of sources at large redshifts and by their cosmological evolution. For this reason, the flux of predicted cosmogenic neutrinos associated with the best fit results of each chosen scenario can be compared with the measured fluxes (or, at higher energies, with the experimental upper bounds) and such comparison could possibly constrain the cosmological evolution of sources. This is done in Fig. 2, where the expected neutrino flux is compared to the IceCube flux [8] and different computed limits for cosmogenic neutrinos.

The expected neutrino flux could serve as multi-messenger constraint for the sources of UHECRs. A neutrino flux exceeding the measured one improves the constraining capability with respect to UHECR data only and then it can improve the search for the best-fit

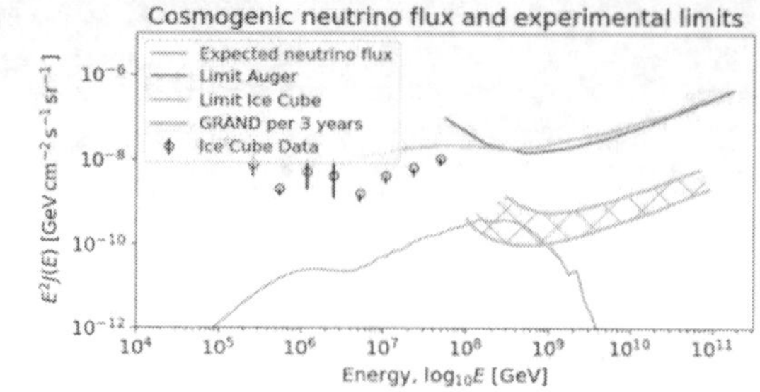


**Figure 2.** The predicted flux of cosmogenic neutrinos for a combined fit above the ankle, compared to the IceCube neutrino flux [8], the GRAND sensitivity after three years of operation [9] and to the limits for the cosmogenic neutrinos by the Pierre Auger [10] and IceCube [11] collaborations.

parameters at the source.

### 2.1 Organization of the code

The code has been divided in several different modules in Python, which separate the different steps of the fit.

- **Tensor**: This module handles the reading of the tensor, stored in a `numpy`'s compress array format for each injected mass. This module allows us to take into account the extra-galactic propagation: UHECRs travel through the extra-galactic environment and interact with photon backgrounds changing their energy and, in the case of nuclei, their species too. This module computes the weights that will be used to transform the spectra at the source to the expected spectra at Earth.
- **Spectrum**: This module contains methods to simulate the emission of UHECR nuclei from sources. Nuclei of mass $A$ and charge $Z$ are injected in the extra-galactic space with a power-law energy spectrum with spectral index $\gamma$ and rigidity cutoff $R_{\mathrm{cut}}$, as shown in eq. (1). Starting from the injected spectra at the source and considering a defined evolution of the source to a certain $z_{\max}$, methods within this class can be used to compute the expected energy spectrum at Earth up to the atmosphere. In addition, this module reads in the experimental spectrum, to provide both the model and the data to the module `minimizer`, which deals with the minimization of the deviance.
- **Mass** and **Gumbel**: These classes deal with the effects of the cosmic-ray cascade production in the atmosphere which, due to stochastic effects, introduce shower to shower fluctuations in particular for the slant depth of the shower maximum ($X_{\max}$). The expected $X_{\max}$ distributions are calculated making use of the generalized Gumbel probability distribution functions [12], and then corrected for the detector effects. A great advantage of using this parametric function is that it allows us to evaluate the model $X_{\max}$ distribution for any mixture of nuclei without the need to simulate showers for each

primary. In this analysis all the nuclei from $A = 1$ up to $A = 56$ are taken into account. The spectra are unaffected by atmosphere, so they are not managed in this class. Alternatively, it is possible to read and compute the mean mass $\langle \ln A \rangle$ and its variance at Earth, or the first two moments of the $X_{\text{max}}$ distributions, using parametrizations obtained by CONEX showers [13]. As in the module `spectrum`, in `mass` the reading of the experimental mass composition indicator, as well as the computation of the expected one, is performed.

- **Minimizer**: This class implements the methods for calculating the deviances of the spectrum and mass composition. The deviances are calculated comparing experimental data with model expectations, where the latter depend on fit parameters. The $X_{\text{max}}$ deviance calculation is performed at each energy bin of the corresponding $X_{\text{max}}$ distribution.

Others modules used in the code are: **Draw**, which contains all the utilities to produce the plots, **Xmax_tools**, which is a library containing all the parametrizations for computing $X_{\text{max}}$ and $\sigma(X_{\text{max}})$. A module called **Skymaps** for taking into account the arrival directions in the combined fit is under development.

## 3 Including the hadronic interaction in UHECRs code

A key part of the MICRO project regards the inclusion of the hadronic interaction, i.e. interactions with of UHECRs with other hadrons, in the UHECR propagation codes. Hadronic interactions are negligible for the extra-galactic propagation, due to the low density of the gas in the extra-galactic medium; however, for propagation inside the source environment, they are a crucial quantity to be taken into account. The computation of hadronic process is a new entry in UHECR propagation codes. In fact, as shown in [14], where hadronic interactions have been implemented in SimProp, hadronic interactions generate more secondary particles than photo-hadronic process, consequently more photons, neutrinos and a larger number of light nuclear fragments. This different composition of propagated nuclei produces, in turn, a change in the evolution of the nuclear cascade inside the source environment. Fig. 3 illustrates the mass distributions of nuclei escaping from a prototype starburst galaxy. In particular, nuclear fragments are shown considering a single type of target: the photo-hadronic-only scenario (blue) is shown separately from the hadronic-only one (red). It can be seen that, while the hadronic scenario produces efficiently all lighter nuclei, the photo-hadronic scenario does not produce efficiently intermediate mass (C, N, O) nuclei.

An effort to include hadronic interaction in CRPropa is ongoing. This is aimed to be done using `impy` [15]. A good advantage of calling impy consists in directly comparing different hadronic interaction models and their predictions, for example, the secondary production, as shown in Fig. 4; in this way it is possible to study the systematic uncertanties related to the choice of one model. An ongoing work inside the CRPropa team concerns the extension of this module for nuclei interactions. Finally, a comparison between the implementation of the hadronic interactions between SimProp and CRPropa is foreseen, which will benefit the astroparticle community at large.

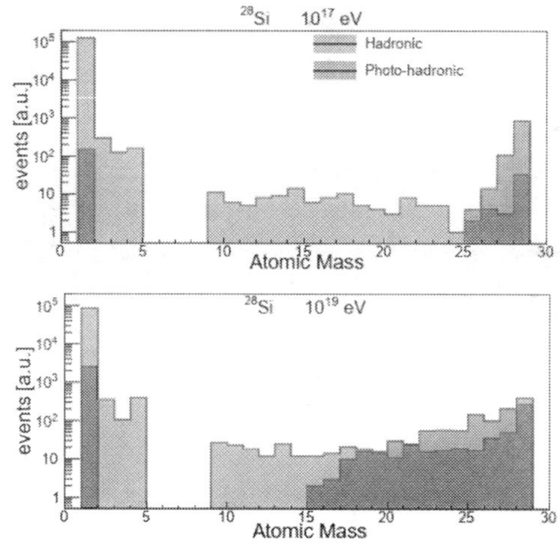


**Figure 3.** Mass distributions of nuclei escaping from a prototype starburst galaxy, for primary silicon nuclei with energies $10^{17}$ (top) and $10^{19}$ eV (bottom panel). The blue (red) histogram refers to the propagation in the source when only photo-hadronic (hadronic) interactions are permitted. From [14].

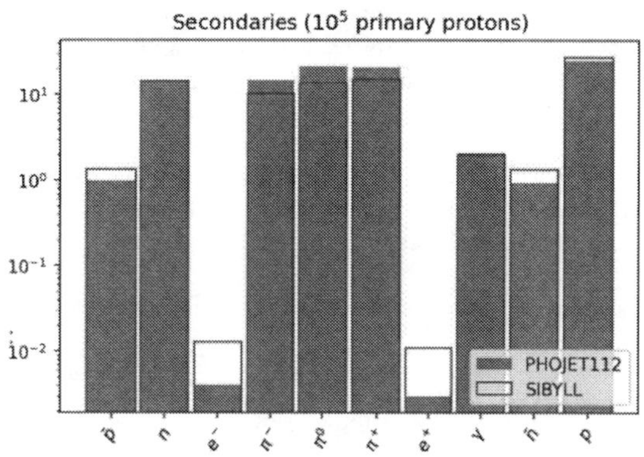


**Figure 4.** Percentage of particles produced in p-p interactions according to two different hadronic interaction model: PHOJET [16] and SIBYLL [7]. Credits: L. Morejon.

## 4 Modelling the source

A goal of the project consists in modelling the source environment of potentially interesting class of sources. Within this spirit, many collaborators are involved in modelling of multimessenger flares from blazars [17]. The idea is to understand the light curves of gamma and radio fluxes and their temporal correlations to putative neutrino emission from the blazar PKS 1502+106 . This requires detailed nu-

https://doi.org/10.1051/epjconf/202328304011

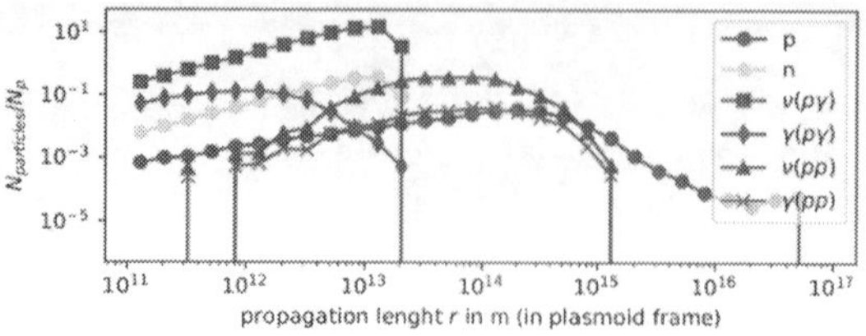


**Figure 5.** Percentage of escaping particles produced in $10^8$ GeV proton flare as a function of the distance to the central supermassive black hole. Note that it is possible to distinguish the secondary production from different interactions. Credits: L. Schlegel.

merical modelling of the local environment of AGN jets, covering the relevant leptonic and hadronic interactions. To properly take into account the physics on sub-kpc scales, several ingredients are required: 1) The possibility to select custom photon fields for the chosen source, e.g. the thermal field emitted by the accretion disc or the synchrotron-radiation field emitted by relativistic electrons; 2) To have time-dependent interaction rates realizing the temporal evolution during propagation along the jet axis; 3) To include hadronic interactions; 4) To tag the secondary emission, in order to resolve origin of secondaries by processes; 5) the spatial resolution of (turbulent and ordered) magnetic fields in sources are important quantities to determine. It was chosen to use CRPropa, using the same framework of the model exploited in [18]; in this way it was possible to have time resolved multimessenger signatures of relativistic plasmoids, i.e. coherent structure of plasma and magnetic fields, as shown in Fig. 5.

## 5 Future perspectives

The MICRO project is reaching its prefixed milestones and aims to have a useful impact in the scientific community, especially releasing a public software for UHECR analysis and including hadronic interactions in UHECR propagation codes. The great novelties consists in developing a cosmological scenario for describing the UHECR observables using transient sources. Injecting bursts of cosmic rays will be important, since it will provide insights on the UHECR spectrum prior to interactions within the cosmic environment and thus it will constrain the sources. In addition, we aim at accounting for the in-source propagation scenario in the individual treatment of AGN/GRBs, with a broad scope of the possible emission scenarios. As for the emission of secondaries, the different photon fields, magnetic fields, and gas distributions obviously play an important role as well.
As another important product of the MICRO project, we aim at predicting the flux of neutral secondaries produced within the sources: this will be made possible by implementing the in-source propagation scenario, by determining whether the flaring scenarios being explored are consistent with current limits from Fermi-LAT, IceCube, and Auger for diffuse emissions [19], and whether neutrino and $\gamma$-ray point-like emission could arise from the propagation of nuclei within the source environment.

## 6 Aknowledgement

This project is jointly supported by the Agence Nationale de la Recherche (ANR) via the grant ANR-20-CE92-0052 and the Deutsche Forschungsgemeinschaft (DFG).

## References

[1] R. Aloisio, V. Berezinsky and P. Blasi, JCAP **10** 020 (2014)
[2] Aab, A.*et al.* [Pierre Auger] , JCAP **4** 038 (2017), [Erratum: JCAP **03**, E02 (2018)]
[3] Adila, A. H. *et al.* [Pierre Auger] , [arXiv:2211.02857 [astro-ph.HE]].
[4] R. Aloisio *et al.* JCAP **11** 009 (2017).
[5] R. Alves Batista *et al.* , JCAP **09** 035 (2022)
[6] P. Abreu *et al.* [Pierre Auger], Eur. Phys. J. C **81** (2021), 966
[7] F. Riehn *et al.* Phys. Rev. D **102** (2020) no.6, 063002
[8] M. G. Aartsen *et al.* [IceCube],
[9] J. Álvarez-Muñiz *et al.* [GRAND], Sci. China Phys. Mech. Astron. **63** (2020) no.1, 219501
[10] A. Aab *et al.* [Pierre Auger], JCAP **10** (2019), 022
[11] M. G. Aartsen *et al.* [IceCube], APS **98** (2019), 6
[12] M. De Domenico et al., JCAP **07** (2013) 050
[13] T. Bergmann et al., Astroparticle Physics **26** (2007), 420
[14] A. Condorelli et al., [arXiv:2209.08593 [astro-ph.HE]].
[15] A.Fedynitch, https://github.com/impy-project/impy
[16] F. W Bopp et al., [arXiv:hep-ph/9803437 [hep-ph]].
[17] Emma Kun et al., ApJ **940** (2022), 163
[18] M. R. Hoerbe et al., MNRAS **496** (2020), 2885
[19] A. Aab *et al.* [Pierre Auger], Front. Astron. Space Sci. (2019)

# Ultra-High-Energy Cosmic Rays from a Population of Non-identical Sources

*Domenik* Ehlert[1,*], *Foteini* Oikonomou[1], and *Michael* Unger[2,1]

[1]Institutt for fysikk, Norwegian University of Science and Technology (NTNU), Trondheim, Norway
[2]Institute for Astroparticle Physics, Karlsruhe Institute of Technology (KIT), Karlsruhe, Germany

**Abstract.** Astrophysical candidates for the sources of ultra-high-energy cosmic rays (UHECRs) exhibit a large diversity in terms of their properties relevant for the acceleration of charged particles, such as luminosity, Lorentz factor, size and magnetic field. Yet, fits of the observed UHECR spectrum and composition often assume identical sources. Here we investigate a population of sources with a power-law distribution of maximum energies. We show that the allowed source-to-source variance of the maximum energy must be small to describe the UHECR data. Even in the most extreme scenario, with a very sharp cutoff of individual source spectra and negative redshift evolution of the accelerators, the maximum energies of 90% of sources must be identical within a factor of three – in contrast to the variance expected for astrophysical sources. However, the overall population variance can be large when maximum rigidities are distributed as a broken power law, with a steep decline above the break and with hard source spectra. In this scenario, most of the observed UHECR flux is produced by sources near the break.

## 1 Introduction

With energies up to and above $10^{20}$ eV, ultra-high-energy cosmic rays (UHECRs) provide exciting prospects for identifying and understanding the most extreme environments in the Universe. However, although tremendous progress has been made since their discovery more than a century ago – and accelerated in the last decades by the commissioning of the latest generation of large-scale UHECR observatories, the Pierre Auger Observatory (Auger) and Telescope Array (TA) – central questions about their origin and acceleration mechanism remain to be answered, and individual sources have so far avoided detection at appreciable significance.

A popular approach for constraining the properties of the UHECR sources is based on the combined fit of the observed spectrum and composition at Earth and cross-reference with the predictions made by "effective" models that attempt to describe the source population with a limited number of free parameters (e.g. [1–7]). While these studies have identified some common features present over most analyses, significant assumptions are made in the process. Arguably the strongest of these is that all sources are identical in terms of the parameters relevant for cosmic rays acceleration, such as maximum rigidity [1], spectral index and emitted cosmic-ray composition.

It is standard to assume an acceleration mechanism universal in rigidity up to some maximum rigidity $R_{\max}$. In this "Peters Cycle" [8] scenario, a remarkably good fit to the observed spectrum and composition can be achieved even with these simplified models if one accepts injection spectra harder than $\mathrm{d}N/\mathrm{d}E \sim E^{-1}$ and heavy source compositions dominated by CNO-Fe nuclei at the highest energies [3–7]. However, the assumption of identical sources remains questionable.

Motivated by the observational fact that the most likely astrophysical objects where acceleration is predicted to occur, such as active galactic nuclei (AGN), gamma-ray bursts (GRBs) and tidal disruption events (TDEs), exhibit significant diversity in important parameters such as luminosity, magnetic field strength and Lorentz factor, we extend the simplistic model of identical sources by allowing for source diversity in terms of maximum CR rigidities. In particular, we investigate the case of a source population with a power-law distribution of maximum rigidities.

In these proceedings, we summarise our investigation of such non-identical sources and the constraints on source diversity imposed by current UHECR observational data. The extended version of our study and conclusions is available at [9]. We find that observations of UHECR spectrum and composition require their sources to be nearly identical for realistic source parameters.

## 2 Model

### 2.1 Population Spectrum of non-identical Sources

Following the Peters cycle prescription, we assume that the cosmic-ray spectrum of an individual source and injected nuclear species $i$ can be described by a power law with cutoff at high rigidities

$$\phi_{\mathrm{src}} = \frac{\mathrm{d}^2 N}{\mathrm{d}R\,\mathrm{d}t} = \sum_i \phi_0(Z_i)\, R^{-\gamma_{\mathrm{src}}}\, f(R, R_{\max}). \quad (1)$$

*e-mail: domenik.ehlert@ntnu.no

[1]The rigidity of a particle in the relativistic limit, with charge Z and using natural units, is $R = E/Z$.

https://doi.org/10.1051/epjconf/202328304012

The flux normalisation factors $\phi_0(Z_i)$ are the averaged normalisation of the entire population and can already include a hidden variance of the per-source normalisations. They are the same for all sources.

The population spectrum, i.e. the cosmic-ray flux produced by the entire population of, possibly non-identical, sources can be obtained from the individual source spectra $\phi_{\mathrm{src}}$ and the distribution of maximum rigidities $p(R_{\max})$ as

$$\phi_{\mathrm{pop}}(R) = \int_0^{\infty} \phi_{\mathrm{src}}(R, R_{\max})\, p(R_{\max})\, \mathrm{d}R_{\max} \,. \quad (2)$$

The sum over all injection elements is assumed implicitly. If sources are identical, then $p(R_{\max}) = \delta(R_0 - R_{\max})$ and the population spectrum will exhibit the same shape as the individual source spectra.

Several different choices can be made for the high-rigidity cutoff. We adopt as fiducial scenario a commonly-used exponential cutoff that is expected in some astrophysical contexts [10, 11]. It is obtained from a more general super-exponential solution

$$\phi_{\mathrm{src}}^{\text{s-exp}} = \phi_0\, R^{-\gamma_{\mathrm{src}}} \exp\left(-\frac{R}{R_{\max}}\right)^{\lambda_{\mathrm{cut}}}, \quad \lambda_{\mathrm{cut}} > 0 \,. \quad (3)$$

in the limit of $\lambda_{\mathrm{cut}} = 1$. We also consider other values that will result in less ($\lambda_{\mathrm{cut}} < 1$) or more ($\lambda_{\mathrm{cut}} > 1$) intrinsic variance of the maximum rigidity. Assuming a power-law distribution of maximum rigidities

$$p(R_{\max}) = \begin{cases} 0 & R_{\max} < R_0 \\ \frac{\beta_{\mathrm{pop}}-1}{R_0} \left(\frac{R_{\max}}{R_0}\right)^{-\beta_{\mathrm{pop}}} & \text{otherwise,} \end{cases} \quad (4)$$

and source spectra with super-exponential cutoff, Eq. (2) can be evaluated analytically and the predicted population spectrum reads

$$\begin{aligned} \phi_{\mathrm{pop}}^{\text{s-exp}} &= \phi_0\, R^{-\gamma_{\mathrm{src}}} \left(\frac{R}{R_0}\right)^{-\beta_{\mathrm{pop}}+1} \frac{\beta_{\mathrm{pop}}-1}{\lambda_{\mathrm{cut}}} \\ &\cdot \gamma\left(\frac{\beta_{\mathrm{pop}}-1}{\lambda_{\mathrm{cut}}}, \left(\frac{R}{R_0}\right)^{\lambda_{\mathrm{cut}}}\right), \end{aligned} \quad (5)$$

where $\gamma(\dots)$ is the lower incomplete gamma function. The source spectra and resulting population spectra for a range of different cutoff choices are shown in Fig. 1. As evident from the figure and provable analytically, the asymptotic behaviour is entirely independent of the cutoff choice, and we always retrieve

$$\lim_{R/R_0 \to 0} \phi_{\mathrm{pop}}(R) \propto R^{-\gamma_{\mathrm{src}}} \quad (6)$$

$$\lim_{R/R_0 \to \infty} \phi_{\mathrm{pop}}(R) \propto R^{-\gamma_{\mathrm{src}}-\beta_{\mathrm{pop}}+1} \,. \quad (7)$$

### 2.2 Astrophysical Connection

The simple parameterisation given in the previous section can be connected to observable distributions of specific properties of astrophysical objects to gain further insights into the nature of UHECR sources. In the following, we propose two such connections, one with the Lorentz factor of jetted acceleration environments and one with the observed source luminosity.

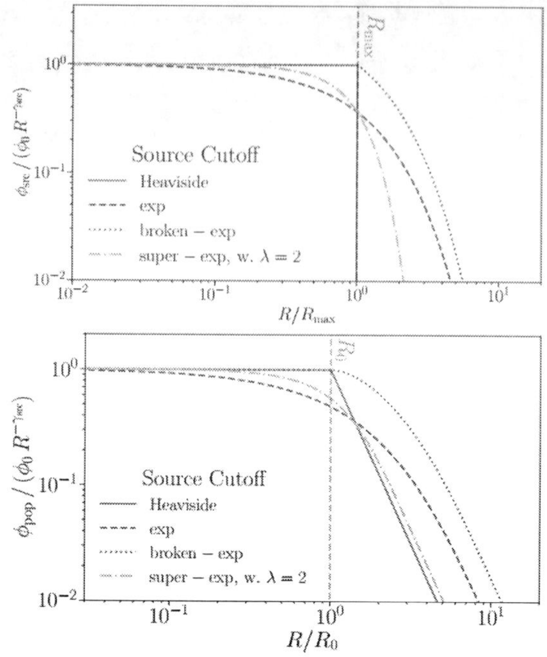


**Figure 1. Top:** Cosmic-ray source spectra for different high-rigidity cutoffs. $R_{\max}$ denotes the maximum rigidity, and the $y$-axis is scaled to show the ratio to an unmodified power law with spectral index $\gamma_{\mathrm{src}}$. **Bottom:** Population spectra resulting from the convolution of a power-law distribution of maximum rigidities above rigidity $R_0$ (for $\beta_{\mathrm{pop}} = 4$) and the source spectra displayed in the previous panel.

#### *2.2.1 Lorentz Factor*

In the simplified picture where acceleration is limited only by the size of the source – the Hillas criterion [12] – for UHECR production in relativistic jets, the maximum rigidity is connected to the bulk Lorentz factor as $R_{\max} = R_0\, \Gamma_{\mathrm{jet}}$, with $R_0 \propto l B$ related to the size $l$ of the source region and the magnetic field $B$. More generally, this scenario can be expressed as

$$R_{\max} = R_0\, \Gamma_{\mathrm{jet}}^{\alpha} \,, \quad (8)$$

where the Hillas scenario is captured by $\alpha = 1$ and *Espresso* acceleration [13–15] imposes $\alpha \leq 2$.

Assuming a power-law distribution of Lorentz factors

$$\frac{\mathrm{d}p}{\mathrm{d}\Gamma_{\mathrm{jet}}} = (\eta - 1)\, \Gamma_{\mathrm{jet}}^{-\eta} \,, \quad (9)$$

in agreement with observations of jetted AGN [16, 17], this results in the distribution of maximum rigidities

$$\begin{aligned} p(R_{\max}) &= \frac{\mathrm{d}p}{\mathrm{d}\Gamma_{\mathrm{jet}}} \left| \frac{\mathrm{d}\Gamma_{\mathrm{jet}}}{\mathrm{d}R_{\max}} \right| \\ &= \frac{\eta - 1}{\alpha}\, R_0^{-1} \left(\frac{R_{\max}}{R_0}\right)^{\frac{1-\eta}{\alpha}-1} \theta(R_{\max} - R_0). \end{aligned} \quad (10)$$

Including the effect of energy-boosting on the total output flux of a particular source, the above expression becomes the same as our simple ansatz Eq. (5) after identifying the slope of the $R_{max}$-distribution as

$$\beta_{pop} = \frac{\eta - 1}{\alpha} + 2 - \gamma_{src} + \xi/\alpha\,, \tag{11}$$

with a time dilation factor $\xi = 1$ for UHECR production and acceleration inside the jet and $\xi = 0$ for *espresso*-style re-acceleration.

### 2.2.2 Luminosity

Alternatively, we connect the maximum rigidity to the source luminosity by using the Lovelace-Waxman-Blandford argument for the minimum luminosity required for cosmic ray acceleration in expanding flows [18–22]

$$L_0 \approx 10^{45.5}\,\frac{1}{\beta}\left(\frac{R_0}{10^{20}\,\mathrm{V}}\right)^2 \mathrm{erg\ s^{-1}}, \tag{12}$$

with $\beta$ the bulk velocity of the outflow. The maximum rigidity as a function of luminosity is then

$$R_{max} \sim R_0\,\beta^{1/2}\left(\frac{L}{L_0}\right)^{1/2}. \tag{13}$$

The additional source diversity introduced by non-identical outflow speeds is expected to be small and can be neglected.

For a power-law luminosity function with slope $y_2$, as observed for a large range of potential source classes (e.g. AGN [23, 24], TDEs [25, 26]),

$$\frac{\mathrm{d}p}{\mathrm{d}L} = \frac{y_2 - 1}{L_0}\left(\frac{L}{L_0}\right)^{-y_2}, \tag{14}$$

our original power-law description of $p(R_{max})$ is reproduced exactly, except for an additional normalisation factor $\kappa$, for

$$\beta_{pop} = 2y_2 - 3 \quad \text{and} \quad \kappa = \frac{\beta_{pop} + 1}{\beta_{pop} - 1}. \tag{15}$$

## 3 Methods

We simulate the injection and propagation of UHECRs numerically with `CRPropa3` [27] and compare the predictions with the latest available spectrum [28] and composition data [29, 30] from the Pierre Auger Observatory. For our fiducial model the free parameters are the minimum maximum rigidity of sources $R_0$, spectral index $\gamma_{src}$ and slope $\beta_{pop}$ of the $R_{max}$-distribution. Other parameters, such as the redshift evolution of the source emissivity $n(z)$ and of $R_0$, and the shape of the UHE cutoff $\lambda_{cut}$ are also investigated and mentioned briefly. For a full discussion of these variations, we refer the reader to the primary paper [9].

Sources are assumed to follow a continuous distribution with number density constant in redshift up to a maximum redshift $z_{max} = 4$. Except for the case of $R_0$ evolving strongly with redshift, existing multimessenger limits and observations do not constrain the source parameters in our model, and we omit their discussion for brevity.

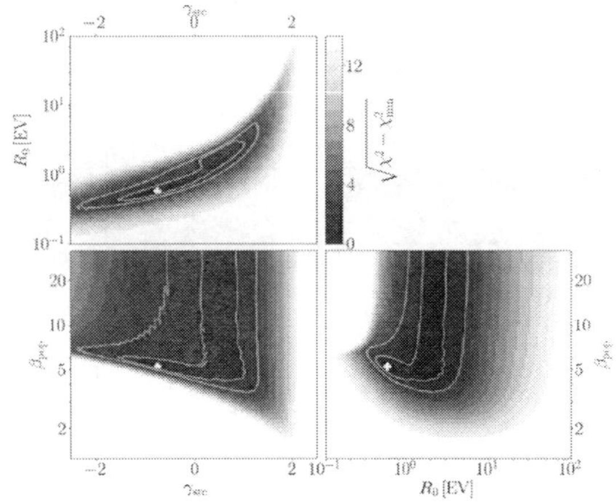


**Figure 2.** Source parameter scan for the fiducial model marginalised along all but two axes respectively. The surface plot shows the agreement between prediction and Auger observations in terms of the $\chi^2$ estimator, and the contour lines indicate the one (green) and three (red) sigma confidence intervals for two degrees of freedom. The best fit is marked with a white cross.

The propagation is treated ballistically, and magnetic fields are not included. For the extragalactic background light, we use the model of Gilmore et al. [31]. To limit the complexity of the simulation and the number of adjustable injection fractions, we choose five representative elements ($^1H$, $^4He$, $^{14}N$, $^{28}Si$ and $^{56}Fe$), which are commonly used for similar fits and provide adequate coverage over the entire mass range.

Auger observations indicate a small variance of the mean shower depth at UHE, implying a relatively pure composition. A spread in maximum rigidity within the source population will increase the mixing between different elements even if the composition at the sources is identical. To establish an upper limit on the allowed population variance, it is necessary to minimise the intrinsic shower variance, which can be achieved by assuming the heaviest UHECR composition compatible with observations since heavy CRs produce less variance in the air showers. To that end, we select Sibyll2.3c [32] as hadronic interaction model for our fiducial scenario and shift the mean and variance of the shower depth to their heaviest realisation, i.e. $-1\sigma_{syst}$ & $+1\sigma_{syst}$ respectively.

## 4 Results and Discussion

In contrast to the sizeable diversity of sources expected from the astrophysical connections, we find that sources with power-law distribution in $R_{max}$ must be effectively identical in terms of maximum rigidity if the un-shifted Auger data is considered. Even after including the fiducial shifts of the mass scales, the best-fit population variance is limited, and the softness of the $R_{max}$-distribution is

 https://doi.org/10.1051/epjconf/202328304012

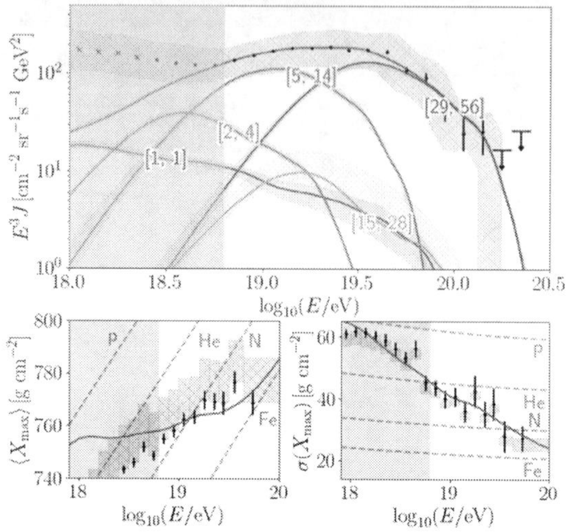


**Figure 3.** Predicted spectrum and composition at Earth for the best-fit scenario of the fiducial model (Sibyll2.3c, $\langle X_{max}\rangle - \sigma_{syst}$, $\sigma(X_{max}) + \sigma_{syst}$). The coloured bands indicate the contributions of the separate mass groups with $[A_{min}, A_{max}]$, including the 68% uncertainties (1 dof). Hatched areas indicate systematic uncertainties of the data. Only points above $10^{18.8}$ eV are fitted.

bounded as (see Fig. 2)

$$\beta_{pop} \gtrsim -\gamma_{src} + 4\,. \tag{16}$$

This is a consequence of the high-rigidity limiting behaviour of the population spectrum Eq. (6). If source spectra are soft and the distribution of maximum rigidities is relatively flat then $\phi_{pop}$ is characterised by a large tail toward the highest CR energies even beyond the GZK limit. These extremely-UHE cosmic rays will experience strong interactions and rapid disintegration, resulting in a very mixed cosmic-ray flux at Earth. This is in disagreement with the observed pureness of the UHE composition, and the corresponding source configurations are consequently disfavoured. From the above condition, significant source diversity is only possible for $\gamma_{src} \gtrsim 1$, which is again disfavoured because of the associated, relatively mixed CR flux. At the best fit (Fig. 3), where $\beta_{pop} \approx 5.2$, sources are functionally identical, and population variance exceeding half a decade for 90% of sources

$$(R^{0.90}_{max}/R_0 > 5) \Longleftrightarrow (\beta_{pop} \lesssim 2.4) \tag{17}$$

is excluded at more than $6\sigma$. Quasi-identical "standard candle" sources ($\beta_{pop} \gtrsim 10$) cannot be rejected.

Varying the redshift evolution of the source number density, $n(z) \propto (1+z)^m$, we find that the allowed population variance is increased marginally for highly local distributions while sources are required to be even more similar if they are further away on average. This can be explained intuitively by the increased interaction and fragmentation of cosmic rays from more distant sources, which results in more intrinsic mixing between the different mass groups, reducing the permitted inter-source variance. More variance is also possible if the UHE cutoff of the source spectra is steeper than exponential, which again reduces the intrinsic mixing of different nuclei and leaves more room for source-to-source variance. At $\lambda_{cut} = 5.4^{+1.7}_{-2.3}$, the best fit is close to a Heaviside-like spectral cutoff.

Even in the most extreme scenario, with Heaviside source spectrum cutoff and negative density evolution ($m = -3$), the distribution of maximum rigidities cannot be flatter than $\beta_{pop} \approx 3-4$; corresponding to a factor of only two to three difference in the maximum rigidity of 90% of sources.

A larger overall population variance is possible if a broken power-law distribution of maximum rigidities is assumed instead. In this case, good agreement with observations can be achieved if the individual source spectra are hard and the $R_{max}$ distribution decays rapidly above the break. The observed UHECR flux is then dominated by sources around the break while the population variance is driven by sources far below the break.

## 5 Conclusions

We have performed an analysis of the dispersion in maximum rigidity of a population of cosmic ray sources allowed by current observations of the UHECR spectrum and composition. We have derived analytic expressions for the expected population spectrum for a range of high-energy cutoff functions, assuming a power-law distribution of maximum rigidities.

With these parametrisations as input to our numerical simulation of UHECR propagation, we determine that the sources responsible for the observed UHECR flux must be effectively identical in their maximum rigidity. If the most optimistic, i.e. 'heaviest', mass scale of the observational data is applied, the allowed population variance is increased slightly. However, even in the most extreme scenario of sudden spectral cutoff and predominantly nearby sources, we find that the majority of sources cannot differ by more than a factor of three in maximum rigidity. This is in stark contrast to the population diversity expected for the most probable source classes if a connection between maximum rigidity and jet Lorentz factor or source luminosity is considered.

Only if a broken power-law distribution in maximum rigidity is considered can the total population variance become large, provided the distribution decays rapidly above the break and source spectra are hard.

## References

[1] D. Allard, E. Parizot, A.V. Olinto, Astropart. Phys. **27**, 61 (2007), astro-ph/0512345

[2] A.M. Taylor, Astropart. Phys. **54**, 48 (2014), 1401.0199

[3] M. Unger, G.R. Farrar, L.A. Anchordoqui, Phys. Rev. D **92**, 123001 (2015), 1505.02153

[4] A. Aab et al. (Pierre Auger), JCAP **04**, 038 (2017), [Erratum: JCAP 03, E02 (2018)], 1612.07155

[5] J. Heinze, A. Fedynitch, D. Boncioli, W. Winter, Astrophys. J. **873**, 88 (2019), 1901.03338

[6] R. Alves Batista, R.M. de Almeida, B. Lago, K. Kotera, JCAP **01**, 002 (2019), 1806.10879

[7] A. Abdul Halim et al. (Pierre Auger) (2022), 2211.02857

[8] B. Peters, Il Nuovo Cimento **22**, 800 (1961)

[9] D. Ehlert, F. Oikonomou, M. Unger (2022), 2207.10691

[10] R.J. Protheroe, T. Stanev, Astropart. Phys. **10**, 185 (1999), astro-ph/9808129

[11] V.N. Zirakashvili, F. Aharonian, Astron. Astrophys. **465**, 695 (2007), astro-ph/0612717

[12] A.M. Hillas, Ann. Rev. Astron. Astrophys. **22**, 425 (1984)

[13] D. Caprioli, Astrophys. J. Lett. **811**, L38 (2015), 1505.06739

[14] R. Mbarek, D. Caprioli, Astrophys. J. **886**, 8 (2019), 1904.02720

[15] R. Mbarek, D. Caprioli, Astrophys. J. **921**, 85 (2021), 2105.05262

[16] M.L. Lister, A.P. Marscher, The Astrophysical Journal **476**, 572 (1997)

[17] M.L. Lister et al., apj **874**, 43 (2019), 1902.09591

[18] R.V.E. Lovelace, Nature **262**, 649–652 (1976)

[19] E. Waxman, Phys. Rev. Lett. **75**, 386 (1995), astro-ph/9505082

[20] E. Waxman, ICTP Lect. Notes Ser. **4**, 309 (2001), astro-ph/0103186

[21] R.D. Blandford, Physica Scripta **T85**, 191 (2000)

[22] M. Lemoine, E. Waxman, JCAP **11**, 009 (2009), 0907.1354

[23] M. Ajello et al., Astrophys. J. **699**, 603 (2009), 0905.0472

[24] L. Marcotulli et al. (BASS), Astrophys. J. **940**, 77 (2022), 2209.09929

[25] S. van Velzen, Astrophys. J. **852**, 72 (2018), 1707.03458

[26] Z. Lin, N. Jiang, X. Kong, S. Huang, Z. Lin, J. Zhu, Y. Wang, Astrophys. J. Lett. **939**, L33 (2022), 2210.14950

[27] R. Alves Batista et al., JCAP **05**, 038 (2016), 1603.07142

[28] A. Aab et al. (Pierre Auger), Phys. Rev. D **102**, 062005 (2020), 2008.06486

[29] A. Aab et al. (Pierre Auger), Phys. Rev. D **90**, 122005 (2014), 1409.4809

[30] A. Yushkov (Auger), PoS **ICRC2019**, 482 (2020)

[31] R.C. Gilmore, R.S. Somerville, J.R. Primack, A. Domínguez, Mon. Not. Roy. Astron. Soc. **422**, 3189 (2012)

[32] F. Riehn, H.P. Dembinski, R. Engel, A. Fedynitch, T.K. Gaisser, T. Stanev, PoS **ICRC2017**, 301 (2018), 1709.07227

# Hadronic Interactions Studies at the LHC

*Tanguy* Pierog[1,*], *Matias* Perlin[2,3], and *Klaus* Werner[4]

[1]KIT, IAP, Postfach 3640, D-76021 Karlsruhe, Germany
[2]Instituto de Tecnologías en Detección y Astropartículas (CNEA, CONICET, UNSAM), Buenos Aires, Argentina
[3]Departamento de Física, FCEN, Universidad de Buenos Aires and IFIBA, CONICET, Buenos Aires, Argentina
[4]SUBATECH, University of Nantes – IN2P3/CNRS – IMT Atlantique, Nantes, France

**Abstract.** The deficit of muons in the simulation of extensive air showers is a long-standing problem and the origin of large uncertainties in the reconstruction of the mass of the high-energy primary cosmic rays. Hadronic interaction models re-tuned after early LHC data have a more consistent description of the muon content among them but still disagree with the data. Furthermore, some inconsistencies are also visible in the electromagnetic component. Ten years after the first LHC-tuned model release, much more detailed data are available both from LHC, SPS and hybrid air shower measurements, allowing for the understanding of some deficiencies in the model which could lead to a change in both $X_{max}$ and the muon production by air showers. A better treatment of the core-corona approach according to LHC data has important consequences on the muon production while an update of diffraction and nuclear fragmentation is changing the $X_{max}$ distribution. An updated version of EPOS LHC will be presented addressing the main issues of this model.

## 1 Introduction

Despite all the efforts made to take into account the first results of proton-proton collisions at the LHC in hadronic interaction models used for air-shower simulations, the observed number of muons, their height of production, or even the depth of shower maximum are still not reproduced consistently by the models [1]. Furthermore, the differences among model predictions introduce uncertainties in cosmic-ray data analysis, which are currently smaller than in the past but still exceed the experimental uncertainties in certain cases [2]. Nevertheless before claiming the need for "new physics", it is important to guarantee that all the standard QCD physics is properly taken into account in these models. For that, it is necessary to go beyond the simplest observables which are usually used to test them. After ten years of running, the various LHC experiments provided a large amount of complex data to analyze and understand, in particular, thanks to the correlations between different observables, which are not yet fully investigated.

Among the hadronic interaction models used for air-shower analysis, only Epos LHC [3–6] includes all the features needed to have a detailed description of the correlation between various observables [1]. Indeed, the core-corona approach in this model, which allows the production of a collective hadronization phase, appears to be a key element to reproduce LHC data. Before LHC, it was usually accepted that hydrodynamical phase expansion, for instance due to the formation of a quark-gluon plasma, was possible only in central heavy-ion collisions. Proton-nucleus (pA) collisions were then used as a reference to probe the effect of such collective behavior (final state) but with some nuclear effect at the initial state level, while proton-proton (pp) interactions were free of any nuclear effects. With the LHC operated in pp, pPb and PbPb mode, it is now possible to compare high-multiplicity pp or pPb events with low-multiplicity PbPb events (which correspond to the same number of particles measured at mid-rapidity) and surprisingly, the very same phenomena are observed [7, 8] concerning the soft-particle production.

One of the most striking features observed in all systems is the long-range two-particle correlation and the evolution of the particle flow as described in Ref. [9]. In Ref. [10], the authors demonstrate how these data from the CMS Collaboration can be reproduced and explained using an approach combining standard perturbative calculations for initial conditions and hydrodynamical calculations for the final state interactions.

At the same time, the recent results compiled by the Working group on Hadronic Interactions and Shower Physics (WHISP) [11] clearly indicate that the discrepancy between the muon production in simulations and data gradually increases with energy. It is a strong indication of a different hadronization than the one used in the current hadronic models [12–17], including Epos LHC, which does not have enough core contribution according to data [7] published after the release of the model in 2012.

Furthermore, other studies showed some deficiencies of Epos LHC and other models not only for the number of muons but also for $X_{max}$ measurements [18]. Since this is the most important observable for the mass composition of cosmic rays, one should check the model against the

*e-mail: tanguy.pierog@kit.edu

latest LHC data in terms of cross-section and multiplicity distributions which are key factors for the shower development [19]. This will be done in Section 2.

In Ref. [20], we presented a modified version of Epos LHC [4] based on Epos3 [3] to study the consequence of the extended range of collective hadronization on air-shower physics in this particular model with positive results on the number of muons. To have a more general approach, new results based on modified secondary particle spectra following the core-corona approach applicable to any model are now presented before a full implementation in a new version of the model. In Section 3, we will discuss the impact of collective hadronization on the total number of muons produced by air showers. In Section 4 the basic principles the core-corona approach will be presented and we will discuss the changes made in CONEX [21] to take this modifications into account and check the results.

Finally, in Section 5, we will discuss future developments.

## 2 Latest LHC data and $X_{\max}$

Using the simple Heitler-Matthews model [22] or studying Monte-Carlo simulations [19], it is well known that the inelastic cross-section, the elasticity and the multiplicity of hadronic interactions, together with the mass and energy of the primary cosmic rays are the most important factors which determine the position of the maximum air shower development $X_{\max}$. For Epos LHC, this was fixed using the early data of different LHC experiments at 7 TeV center-of-mass energy. Since then, new and more precise data have been released at 13 TeV or using lead as a projectile giving new constraints to the models which should be taken into account to give proper $X_{\max}$ predictions. A new preliminary version of the model, called Epos LHC-R, is presented here.

### 2.1 pp cross-sections

The various pp cross-sections (inelastic, elastic and total) are key ingredients to tune the model parameters. As described in [4], the amplitude of the elementary scattering in Epos LHC can be used directly to compute these cross-sections via the profile function. Really recently, the ATLAS Collaboration, using its ALFA detector, published a very precise measurement at 13 TeV [23], which is a few millibarn lower than the TOTEM measurements published early on and on which the models were tuned [24]. These new data are chosen to define the cross-sections in the new Epos LHC-R.

In Fig. 1, the elastic and total cross-sections are shown. The various lines correspond to the model Epos LHC-R (dotted line), Epos LHC (full line), QGSJetII.04 [27, 28] (dashed line), and Sibyll 2.3d [29] (dash-dotted line) predictions. The data points are taken from [23, 25, 26]. The first generation of hadronic models tuned to TOTEM data predict larger cross-sections than Epos LHC-R tuned to ALFA data by about 10 to 15% at the highest energy. These reduced cross-sections in the new model will have a direct impact on the depth of the shower maximum $X_{\max}$.

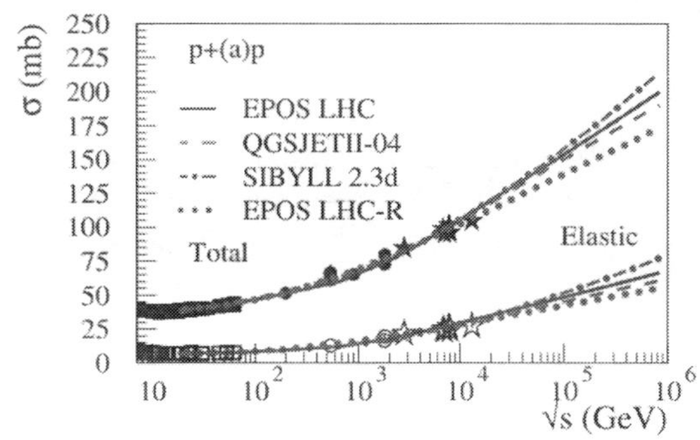


Figure 1. Total and elastic p-p cross sections as a function of center-of-mass energy calculated with Epos LHC-R (dotted line), Epos LHC (full line), QGSJetII.04 (dashed line), and Sibyll 2.3d (dash-dotted line) predictions. Points are data from [25], and the stars are the LHC measurements [26] including the latest ALFA measurements [23].

### 2.2 Multiplicity and elasticity in pA

The elasticity is important for the shower development but is complicated to measure at LHC. There is no direct measurement, but some other distributions could be sensitive to the energy fraction of the leading particle. The average multiplicity of pp and pPb interactions, another important measurement for $X_{\max}$, is well described by the hadronic models. In the meantime, the fluctuations of the multiplicity in pPb collision have been published and is, in fact, not so well reproduced by the two models which can run with lead projectile Epos LHC and QGSJetII.04. In fact, the latter is linked to the elasticity because the frequency of low multiplicity events is related to highly elastic events while the high multiplicity events are linked to low elasticity. So, a better description of the fluctuations of the multiplicity should improve the elasticity prediction.

In Fig. 2, the predictions for the multiplicity distribution of Pb-p interactions at 5.02 TeV from Epos LHC-R (dotted line), Epos LHC (full line) and QGSJetII.04 (dashed line) are compared to CMS data [30]. On the left-hand side, a linear scale for the y-axis is used to focus on the low multiplicity (high elasticity) events. Epos LHC clearly underestimates the very low multiplicity peak (related to diffraction, so with high elasticity), while QGSJetII.04 underestimates the intermediate range. On the right-hand side, a logarithmic scale is used to observe the tale at high multiplicity where the multiplicity is underestimated by Epos LHC and overestimated by QGSJetII.04. To correct Epos LHC, it is not possible to simply change some parameters. The lack of fluctuations at low and high multiplicity is due to the usage of fixed screening parameters which can reproduce the average behavior correctly but underestimate the natural fluctuations that are due to different configurations of elementary interaction reinteractions which are not taken into account explicitly. A contrario, in QGSJetII.04 these Pomeron-

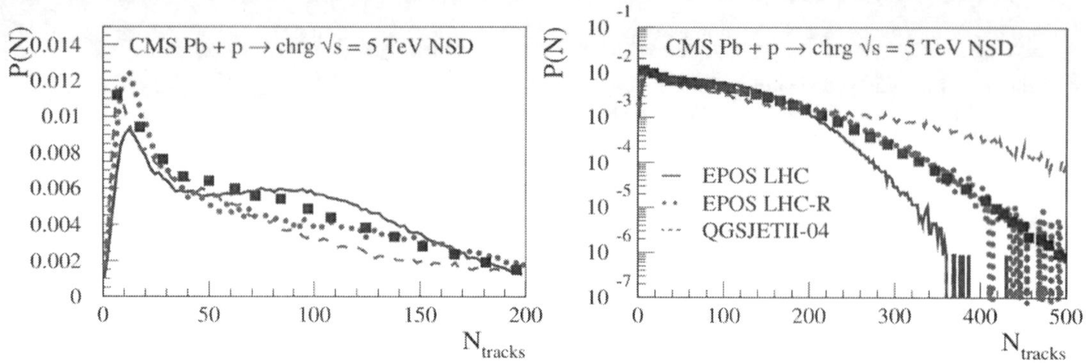


**Figure 2.** Multiplicity distribution of Pb-p collisions at 5 TeV with Epos LHC-R (dotted line), Epos LHC (full line), QGSJetII.04 (dashed line) simulations using a Non-Single-Diffractive (NSD) experimental trigger. Data points are from [30] corrected for acceptance and efficiency. On the left-hand side, with a linear scale for the y-axis and with a logarithmic scale on the right-hand side.

Pomeron interactions are taken into account but neglecting the energy conservation at the amplitude level, leading to the overestimation of the high multiplicity events. In Epos LHC-R, a distribution of screening parameters has been introduced to take into account these fluctuations in an effective way with only one parameter to be tuned to these data, leading to a wider range of fluctuations. One important consequence is that more events have a low multiplicity because less nucleons are interacting leading to a larger elasticity.

### 2.3 Nuclear fragmentation

As stated in [31], the fluctuations of $X_{max}$ for nuclear primaries are underestimated in Epos LHC because of an error in the generation of wounded nucleons which were produced first by the model itself, which is based on the list of nucleons, and then again by the subroutine creating the nuclear fragment from the list of the remaining nucleons, which includes some evaporation which was not necessary. Hence, the produced fragments are too light with too many free nucleons. This is now corrected in Epos LHC-R but the RMS of $X_{max}$ for iron showers is increased only from 20 g/cm$^2$ to 22 g/cm$^2$ while other models have 25 g/cm$^2$. The remaining difference is coming from the generation of nuclei with large neutron imbalance compared to protons which cannot be considered as stable. This can still be improved by not using a fully random choice between protons and neutrons interacting during the collision, but this will probably not cover the full difference with QGSJetII.04 and Sibyll 2.3d which do not take this into account at all and generally underestimate the number of free nucleons after a nuclear collision.

### 2.4 Hadronic interactions in Air and $X_{max}$

In order to understand the impact of these modifications on the $X_{max}$ predictions by Epos LHC-R we can first check the p-air cross-section and elasticity. In Fig. 3, the inelastic p-air cross-section is shown on the left-hand side and the elasticity is on the right-hand side. The lines are for Epos LHC-R (dotted line), Epos LHC (full line), QGSJetII.04 (dashed line), and Sibyll 2.3d (dash-dotted line), and the points are data from [25]. As expected, Epos LHC-R has a cross-section about 10% lower than Epos LHC at high energy and an elasticity which is increased by around 15%.

These are not huge differences, and as shown in Fig. 4, it is only a 2% increase of $X_{max}$ for proton-induced showers. Nevertheless this represents already around 15 g/cm$^2$ which was the difference between Epos LHC (line with full stars) and Sibyll 2.3d (line with full triangles). So, in fact, Epos LHC-R (line with open circles) predicts around the same $X_{max}$ values than Sibyll 2.3d, much deeper than the predictions from QGSJetII.04 (line with open squares).

In terms of mass composition, it means that the mean logarithmic mass deduced from $X_{max}$ measurements using Epos LHC-R will also be around 15% larger, hence reducing the muon deficit in simulation compared to data. It will probably be very similar to Sibyll 2.3d, but these are still preliminary results.

## 3 Collective hadronization and air-shower physics

The dominant mechanism for the production of muons in air showers is via the decay of light charged mesons. The vast majority of mesons are produced at the end of the hadron cascade, after typically five to ten generations of hadronic interactions (depending on the energy and zenith angle of the cosmic ray). The energy carried by neutral pions, however, is directly fed into the electromagnetic shower component and is not available for further production of more mesons and subsequently muons. Thus, the energy carried by hadrons other than neutral pions is typically available to produce more hadrons and ultimately

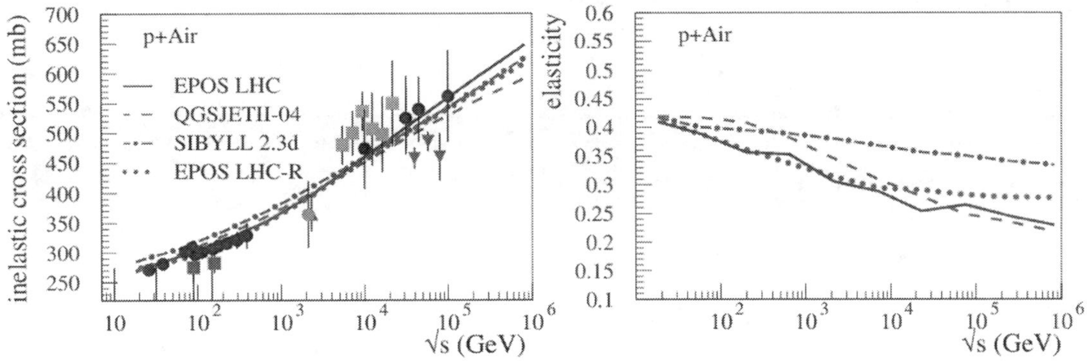


**Figure 3.** Inelastic cross-section (left-hand side) and elasticity (energy fraction of the leading particle) (right-hand side) for p-air interactions as a function of center of mass energy. Simulations are done with Epos LHC-R (dotted line), Epos LHC (full line), QGSJetII.04 (dashed line), and Sibyll 2.3d (dash-dotted line). Points are data from [25].

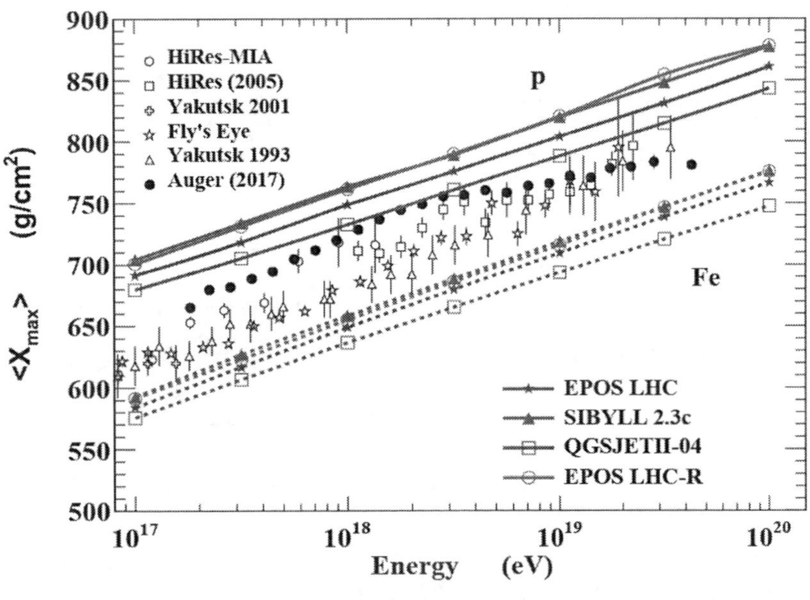


**Figure 4.** $\langle X_{\max} \rangle$ for proton- and iron-induced showers as a function of the primary energy. Predictions of different high-energy hadronic interaction models are presented with full lines for proton and dashed lines for iron with full triangles for Sibyll 2.3d, open circles for Epos LHC-R, open squares for QGSJetII.04, full stars for Epos LHC. Refs. to the data can be found in [32] and [33].

muons in following interactions and decays. As explained in Ref. [12, 34], the ratio of the average electromagnetic to average hadronic energy, called $R$, and its dependence on center-of-mass energy, are thus related to the muon abundance in air showers: if this energy ratio is smaller (larger), more (less) energy is available for the production of muons at the end of the hadronic cascade and ultimately more (fewer) muons are produced. In a simplified model where all particles have the same energy, we can define $R = c/(1 - c)$ where $c = N_{\pi^0}/N_{\text{all}}$ and $N_{\text{all}}$ is the total multiplicity. Then it can even be demonstrated using the simple Heitler-Matthews model [22] that the exponent $\beta$ of the energy dependence of the muon production is directly related to $c$ and thus $R$ as

$$\beta = 1 + \ln(1 - c)/\ln(N_{\text{all}}). \qquad (1)$$

Since in a collective hadronization (or statistical model) the production of particles with higher mass (in particular, strange quarks) [35] is not suppressed as in a string hadronization, the fraction of secondary pions in the dense core is reduced because many other more massive hadrons and resonances are produced. This leads to

 https://doi.org/10.1051/epjconf/202328305001

a lower ratio of the electromagnetic to hadronic energy density in particles produced from the core. As a consequence, such reduction of $R$ due to more collectivity in secondary particle production should increase the slope $\beta$ of the energy dependence of the muon production in air showers when compared to traditional hadronic models where only the string hadronization is taken into account.

## 4 Core-corona effect

The discussion in the previous section suggests that a change in $R$ is a potential way to reduce the discrepancy between measurements and air shower simulations. Nevertheless, $R$ is quite well constrained by theory as well as laboratory measurements and, thus, cannot be changed entirely arbitrarily as studied in another analysis [19]. In a naive model like Ref. [22], where only pions are considered as secondary particles, $R = 0.5$. In a more realistic approach based on string fragmentation, we have $R \approx 0.41$ to 0.45.

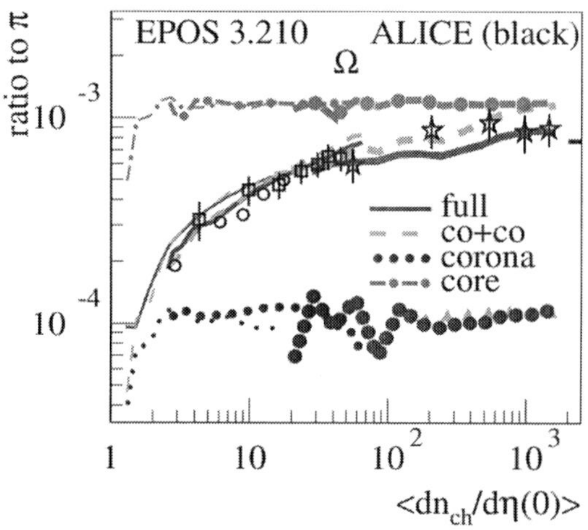


**Figure 5.** Particle to pion ratio for the Ω baryon versus multiplicity at mid-rapidity, for different contributions (core (dash-dotted), corona (dotted), core+corona (dashed) and all (core+corona+hadronic gas) (full)) from the Epos simulations, for different systems (pp (thin), pPb (normal), PbPb (bold)). We also plot ALICE data from [7].

But as shown in Ref. [7], particle ratios such as $K/\pi$, $p/\pi$ or $\Lambda/\pi$ change with increasing secondary-particle density, saturating to the value given by a thermal/statistical model with a freeze-out temperature of 156.5 MeV [36] yielding $R \approx 0.34$. Such a behavior can be explained in terms of a core-corona picture [35]. This approach has been used in the framework of realistic simulations [37], but also in simple model calculations [38–41]. The basic idea is that some fraction of the volume of an event (or even a fraction of events) behaves as a QGP and decays according to statistical hadronization (core), whereas the other part produces particles via string fragmentation (corona). The particle yield $N_i$ for particle

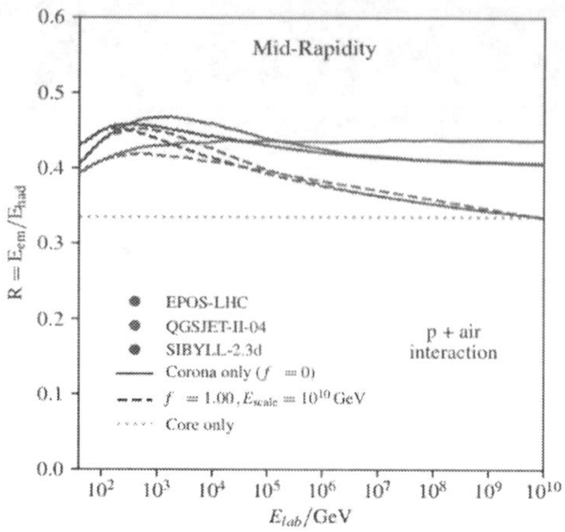


**Figure 6.** Evolution of the ratio $R$ from pure corona to pure core in the case of a linear increase of core fraction as a function of the logarithm of the energy for three different models in proton-air interactions at mid-rapidity.

species $i$ is then a sum of two contributions

$$N_i = \omega_{\mathrm{core}}\, N_i^{\mathrm{core}} + (1 - \omega_{\mathrm{core}})\, N_i^{\mathrm{corona}}, \tag{2}$$

where $N_i^{\mathrm{core}}$ represents statistical (grand canonical) particle production, and $N_i^{\mathrm{corona}}$ is the yield from string decay. Crucial is the core weight $\omega_{\mathrm{core}}$. In order to explain LHC data [7], the weight $\omega_{\mathrm{core}}$ needs to increase monotonically with the multiplicity, starting from zero for low-multiplicity pp scattering, up to 0.5 or more for very high multiplicity pp, reaching unity for central heavy-ion collisions (PbPb), as illustrated in Fig. 5 using the Epos3 model [3]. This process was already present in Epos LHC but in a very early stage without data to constrain the evolution of the core and corona ratio leading to a very small effect on the muon production. The development done for Epos3 can be used to improve Epos LHC but this requires non trivial changes which are not yet available. In this paper, we will still present the results of a more general study.

In order to know whether such a constrained value of $R$ could be low enough to increase the number of muons in air-shower simulations such that the data could be reproduced, a simplified core-corona approach can be used to at least set some realistic upper limit under the following assumptions:

- the fraction of core effectively increases with energy: the core-corona approach is originally a function of the multiplicity and independent of the collision energy for a given multiplicity. Since the average multiplicity increases with the energy, the average core fraction must increase with energy.
- only the change in hadronization is taken into account: collective effects in the core in principle include particle correlations and flow, but since the longitudinal particle

momentum is dominating over the transverse momentum down to relatively low energy, these effects can be neglected in first order.

- no nuclear effect is introduced: the multiplicity increases with the mass of the projectile, so the core fraction would increase for higher primary mass. This is not taken into account due to technical limitations (minimizing the effect for higher primary mass but only for the first few hadronic generations).
- core-corona effect is applied on full phase space: core hadronization has been observed at mid-rapidity, but in order to see the maximal effect on air shower to set an upper limit, the modification should apply at larger rapidities.

So, in the following, we are going to employ a straightforward core-corona approach, based on Eq. (2), for any hadronic interaction model in CONEX air-shower simulations. The particle yield from the chosen interaction model is by definition considered to be the corona yield, whereas we use the standard statistical hadronization (also referred to as resonance gas) for the core part. So $\omega_{\rm core} = 0$ would be the "normal" simulation with the default interaction model. Choosing $\omega_{\rm core} > 0$, amounts to mixing the yields from the interaction model according to the core-corona superposition shown in Eq. (2) and changes the value of $R$ depicted in Fig. 6 for proton-air interactions for the three different hadronic interaction models EPOS LHC, QGSJETII.04 and SIBYLL 2.3d.

Technically, we directly modify individual particle ratios of the secondary particle spectra $dN_i/dE_j$, for particle species $i$ and energy bins $dE_j$, of hadronic interactions with air nuclei used by CONEX for numerical air-shower simulations based on cascade equations. Knowing the initial ratios $\pi^0/\pi^\pm$, $p/\pi^\pm$, $K^\pm/\pi^\pm$, $p/n$, $K^0/K^\pm$ (taking into account strange baryon decays) from a corona-type model and the value of the same ratios from the core model, we compute new spectra in which the particle yields include both, core and corona according to $\omega_{\rm core}$. Since the hadronization mechanism can affect only newly produced particles, the properties of the leading particle should be preserved. To achieve that, the new particle yields are computed for all secondaries, but excluding the one corresponding to the respective projectile type, i.e. protons in proton-air, kaons in kaon-air interactions, and so on. The yield of the projectile-type particles is determined subsequently by exploiting energy conservation in all energy bins $dE_j$ summed over all secondary particle species $i$: the sum $\sum_i E_j\, dN_i/dE_j$ must be conserved. Since at high $x_{\rm F} = E_j/E_{\rm lab}$ only the projectile-type particles will have $dN_i/dE_j$ significantly different from zero (aka leading-particle effect), the resulting modified leading-particle type spectra at high $x_{\rm F}$ follow the original distribution, and are only affected by the scaling procedure at lower values of $x_{\rm F}$. Together, this assures that energy conservation as well as the total multiplicity are not affected, but only the particle ratios. More details will be given in a future publication.

We expect the core weight $\omega_{\rm core}$ to increase with energy in a logarithmic way like the total multiplicity. Thus,

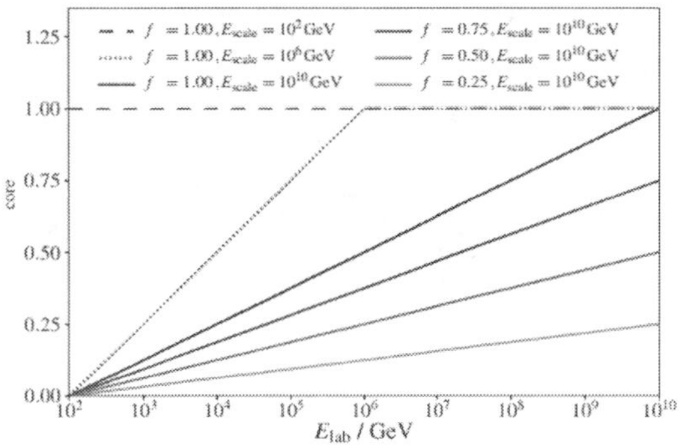


**Figure 7.** Different energy evolution probed for $\omega_{\rm core}$. The solid lines represent changing the scale $f_\omega$ of the effect, while the dashed lines also indicate the effect of changing $E_{\rm scale}$.

we use

$$\omega_{\rm core}(E_{\rm lab}) = f_\omega\, F(E_{\rm lab}; E_{\rm th}, E_{\rm scale}) \tag{3}$$

with

$$F(E_{\rm lab}; E_{\rm th}, E_{\rm scale}) = \frac{\log_{10}(E_{\rm lab}/E_{\rm th})}{\log_{10}(E_{\rm scale}/E_{\rm th})} \text{ for } E_{\rm lab} > E_{\rm th}, \tag{4}$$

to model this, starting already at fixed-target energies, $E_{\rm th} = 100\,\text{GeV}$. Different energy dependencies are explored by changing $E_{\rm scale}$ from 100 GeV (corresponding to a step function), to $10^6$ GeV, and $10^{10}$ GeV. The $f_\omega$ scale is varied from 0.25, 0.5, 0.75 to 1.0; in addition, we enforce $F(E_{\rm lab}; E_{\rm th}, E_{\rm scale}) \stackrel{!}{=} 1$ for all $E_{\rm lab} \geq E_{\rm scale}$. This yields the $\omega_{\rm core}$ energy dependencies as depicted in Fig. 7.

All these scenarios have been used to simulate full air showers with CONEX, using cascade equations from the first interaction to the ground, for proton and iron primary particles at $E_0 = 10^{19}$ eV. In Fig. 8 ,the results are shown in the $X_{\rm max}$-ln$N_\mu$ plane for QGSJETII.04. This typical example illustrates that it is well possible with modified hadronization in air-shower cascades to describe the data of the Pierre Auger Observatory. As expected, more core-like contributions are needed compared to what is currently provided by the models. This means, QGP-like effects also in light colliding systems and starting in central collisions at much lower center-of-mass energies may play a decisive role. It is worth noticing that if $X_{\rm max}$ is shifted to the right in this plot, a smaller value of core fraction is needed to reduce the difference between the predicted and the measured number of muons.

## 5 Summary

The better description of the collective hadronization and, in particular, the fact that the core is produced earlier than predicted by EPOS LHC can have very important consequences for the muon production in air showers. The effect of QGP using the standard EPOS LHC was shown not to be significant. Indeed in this model, the QGP was produced only for very high-multiplicity events and at mid-rapidity

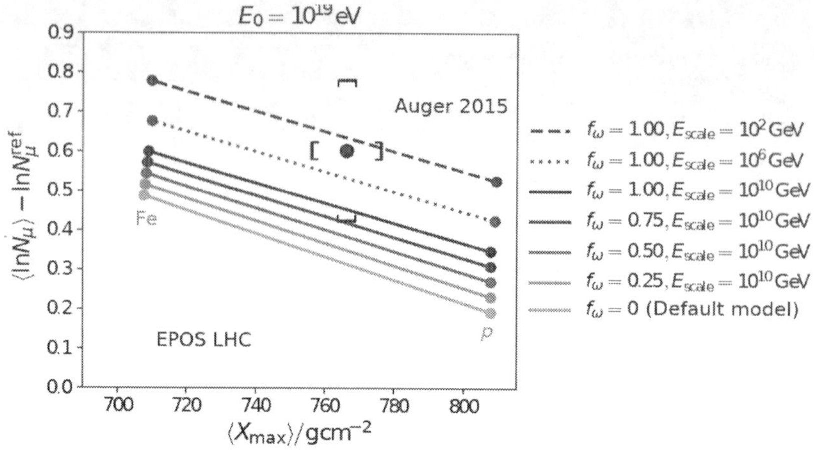


**Figure 8.** Comparison of different core-corona mixing scenarios, as described in the text, on air shower simulations at $10^{19}$ eV QGSJETII.04 in the $X_{max}$-ln$N_\mu$ plane. The solid lines represent changing the scale $f_\omega$, while the dashed lines also indicate the effect of changing $E_{scale}$. The *default* model corresponds to the corona-only simulations. The data point is from the Pierre Auger Observatory [42]. Each model line represents all values that can be obtained for any mixture of cosmic nuclei from proton (bottom right) to iron (top left).

which are both rare and not so important for air-shower development. Other studies using a QGP or alternative hadronization as a possible new source of muons were all based on changes under extreme conditions too [13, 43] or with extreme consequences not observed at the LHC [14]. As shown here and in Ref. [17, 44, 45], according to the most recent LHC results, the collective hadronization happens at a much lower multiplicity and, consequence, with effects at larger rapidities (lower particle densities than foreseen). In that case, the fact that much more particles coming from the hadronization of a QGP play a significant role in the air-shower development. A stronger effect could be observed in case of a nuclear projectile that could create a collective phase with a non-zero chemical potential, which could lead to an even stronger increase of the non-electromagnetic secondary particles [16]. The combination of a mild increase like observed in this study with a proton primary with an even stronger effect for a heavier projectile associated with a deeper $X_{max}$ (thus a larger mean mass) could be the complete solution of the muon-deficit problem in air-shower simulations. The new EPOS LHC-R taking into account the core-corona effect in a detailed enough way to reproduce all major effects observed at the LHC is still under development, but the preliminary results on $X_{max}$ using up-to-date cross-section and multiplicity distributions have been presented and lead indeed to larger values of $X_{max}$.

## References

[1] T. Pierog, PoS **ICRC2017**, 1100 (2018)
[2] K.H. Kampert, M. Unger, Astropart. Phys. **35**, 660 (2012), `1201.0018`
[3] K. Werner, B. Guiot, I. Karpenko, T. Pierog, Phys. Rev. C **89**, 064903 (2014), `1312.1233`
[4] T. Pierog, I. Karpenko, J.M. Katzy, E. Yatsenko, K. Werner, Phys. Rev. C **92**, 034906 (2015), `1306.0121`
[5] K. Werner, I. Karpenko, T. Pierog, M. Bleicher, K. Mikhailov, Phys. Rev. C **82**, 044904 (2010), `1004.0805`
[6] K. Werner, F.M. Liu, T. Pierog, Phys. Rev. C **74**, 044902 (2006), `hep-ph/0506232`
[7] J. Adam et al. (ALICE), Nature Phys. **13**, 535 (2017), `1606.07424`
[8] J. Pan (ALICE), J. Phys. Conf. Ser. **832**, 012044 (2017)
[9] V. Khachatryan et al. (CMS), Phys. Lett. B **742**, 200 (2015), `1409.3392`
[10] K. Werner, J. Phys. Conf. Ser. **636**, 012006 (2015)
[11] H.P. Dembinski, others (EAS-MSU, IceCube, KASCADE-Grande, NEVOD-DECOR, Pierre Auger, SUGAR, Telescope Array, Yakutsk EAS Array), *Report on Tests and Measurements of Hadronic Interaction Properties with Air Showers* (2019), `1902.08124`
[12] S. Baur, H. Dembinski, M. Perlin, T. Pierog, R. Ulrich, K. Werner (2019), `1902.09265`
[13] J.F. Soriano, L.A. Anchordoqui, T.C. Paul, T.J. Weiler, PoS **ICRC2017**, 342 (2018), `1811.07728`
[14] J. Alvarez-Muniz, L. Cazon, R. Conceição, J.D. de Deus, C. Pajares, M. Pimenta (2012), `1209.6474`
[15] G.R. Farrar, J.D. Allen, EPJ Web Conf. **53**, 07007 (2013), `1307.2322`
[16] L.A. Anchordoqui, H. Goldberg, T.J. Weiler, Phys. Rev. D **95**, 063005 (2017), `1612.07328`
[17] L.A. Anchordoqui, C.G. Canal, F. Kling, S.J. Sciutto, J.F. Soriano, JHEAp **34**, 19 (2022), `2202.03095`

[18] J. Vicha et al. (Pierre Auger), PoS **ICRC2021**, 310 (2021)

[19] R. Ulrich, R. Engel, M. Unger, Phys. Rev. D **83**, 054026 (2011), 1010.4310

[20] T. Pierog, S. Baur, H.P. Dembinski, R. Ulrich, K. Werner, PoS **ICRC2019**, 387 (2020)

[21] T. Bergmann, R. Engel, D. Heck, N.N. Kalmykov, S. Ostapchenko, T. Pierog, T. Thouw, K. Werner, Astropart. Phys. **26**, 420 (2007), astro-ph/0606564

[22] J. Matthews, Astropart. Phys. **22**, 387 (2005)

[23] (2022), 2207.12246

[24] G. Latino et al. (TOTEM), *First Results from the TOTEM Experiment*, in *2nd International Workshop on Multiple Partonic Interactions at the LHC* (2011), 1110.1008

[25] C. Caso et al. (Particle Data Group), Eur. Phys. J. **C3**, 1 (1998)

[26] M. Aaboud et al. (ATLAS), Phys. Rev. Lett. **117**, 182002 (2016), 1606.02625

[27] S. Ostapchenko, Phys. Lett. B **636**, 40 (2006), hep-ph/0602139

[28] S. Ostapchenko, Phys. Rev. D **81**, 114028 (2010), 1003.0196

[29] F. Riehn, R. Engel, A. Fedynitch, T.K. Gaisser, T. Stanev, Phys. Rev. D **102**, 063002 (2020), 1912.03300

[30] S. Chatrchyan et al. (CMS), Phys. Lett. B **724**, 213 (2013), 1305.0609

[31] S. Ostapchenko, *LHC results and hadronic interaction models*, in *25th European Cosmic Ray Symposium* (2016), 1612.09461

[32] J. Blumer, R. Engel, J.R. Horandel, Prog. Part. Nucl. Phys. **63**, 293 (2009), 0904.0725

[33] A. Aab, others (Pierre Auger), Phys. Rev. D **90**, 122005 (2014), 1409.4809

[34] T. Pierog, K. Werner, Phys. Rev. Lett. **101**, 171101 (2008), astro-ph/0611311

[35] K. Werner, A.G. Knospe, C. Markert, B. Guiot, I. Karpenko, T. Pierog, G. Sophys, M. Stefaniak, M. Bleicher, J. Steinheimer, EPJ Web Conf. **17**, 10900 (2018), 1812.06330

[36] A. Andronic, P. Braun-Munzinger, K. Redlich, J. Stachel, J. Phys. Conf. Ser. **779**, 012012 (2017), 1611.01347

[37] K. Werner, Phys. Rev. Lett. **98**, 152301 (2007), 0704.1270

[38] J. Manninen, F. Becattini, Phys. Rev. C **78**, 054901 (2008), 0806.4100

[39] F. Becattini, J. Manninen, J. Phys. G **35**, 104013 (2008), 0805.0098

[40] J. Aichelin, K. Werner, Phys. Rev. C **82**, 034906 (2010), 1001.1545

[41] J. Aichelin, K. Werner, Phys. Rev. C **79**, 064907 (2009), [Erratum: Phys. Rev.C81,029902(2010)], 0810.4465

[42] A. Aab et al. (Pierre Auger), Phys. Rev. D **91**, 032003 (2015), [Erratum: Phys.Rev.D 91, 059901 (2015)], 1408.1421

[43] D. LaHurd, C.E. Covault, JCAP **2018**, 007 (2018), 1707.01563

[44] L.A. Anchordoqui, C. Garcia Canal, S.J. Sciutto, J.F. Soriano (2019), 1907.09816

[45] J. Manshanden, G. Sigl, M.V. Garzelli (2022), 2208.04266

# Status of the LHCf experiment

*Ken* Ohashi[1,*], *Oscar* Adriani[2,3], *Eugenio* Berti[2], *Pietro* Betti[2,3], *Lorenzo* Bonechi[2], *Massimo* Bongi[2,3], *Raffaello* D'Alessandro[2,3], *Sebastiano* Detti[2], *Maurice* Haguenauer[4], *Yoshitaka* Itow[1,5], *Katsuaki* Kasahara[6], *Yuga* Kitagami[1], *Moe* Kondo[1], *Yutaka* Matsubara[1], *Hiroaki* Menjo[1], *Yasushi* Muraki[1], *Paolo* Papini[2], *Giuseppe* Piparo[7,8,9], *Sergio* Ricciarini[2,10], *Takashi* Sako[11], *Nobuyuki* Sakurai[12], *Monica* Scaringella[2], *Yuki* Shimizu[13], *Tadashi* Tamura[13], *Alessio* Tiberio[2,3], *Shoji* Torii[14], *Alessia* Tricomi[7,8,9], *William C.* Turner[15], and *Kenji* Yoshida[6]

[1]Institute for Space-Earth Environmental Research, Nagoya University, Nagoya, Japan
[2]INFN Section of Florence, Florence, Italy
[3]University of Florence, Florence, Italy
[4]Ecole-Polytechnique, Palaiseau, France
[5]Kobayashi-Maskawa Institute for the Origin of Particles and the Universe, Nagoya University, Nagoya, Japan
[6]Faculty of System Engineering, Shibaura Institute of Technology, Japan
[7]INFN Section of Catania, Italy
[8]University of Catania, Catania, Italy
[9]Centro Siciliano di Fisica Nucleare e di Struttura della Materia, CSFNSM, Catania, Italy
[10]IFAC-CNR, Florence, Italy
[11]ICRR, University of Tokyo, Kashiwa, Japan
[12]Tokushima University, Tokushima, Japan
[13]Kanagawa University, Kanagawa, Japan
[14]RISE, Waseda University, Shinjuku, Tokyo, Japan
[15]LBNL, Berkeley, California, USA

**Abstract.** A precise understanding of hadronic interactions is essential to interpreting the mass composition of ultra-high energy cosmic rays from the results of air shower experiments. The Large Hadron Collier forward (LHCf) experiment aims to measure forward neutral particles for validation of hadronic interaction models adopted in air shower simulations. We already published the production cross sections of forward photons and neutrons for proton-proton collisions at $\sqrt{s}$ = 13 TeV. Recently, we showed a preliminary result of the energy spectrum of forward $\eta$ mesons for proton-proton collisions at $\sqrt{s}$ = 13 TeV. Moreover, in September 2022, we had another data-taking for proton-proton collisions at $\sqrt{s}$ =13.6 TeV. In data taking, we planned to obtain a number of $\pi^0$ and $\eta$ candidates ten times larger for precise measurements and to perform the joint operation with ATLAS Roman pots and zero-degree calorimeters. Thanks to the joint operation with the ATLAS Roman pots, we can measure diffractive mass and neutral particles from diffractive dissociation simultaneously. Furthermore, energy resolution for neutrons is expected to be improved from 40% to 20% by combining the LHCf and the ATLAS zero-degree calorimeters. In this work, we report the status and prospects of the LHCf experiment.

## 1 Introduction

The origin of the ultra-high energy cosmic rays is unknown. The acceleration at the source and the propagation from the origin to Earth depends on the charge of these cosmic rays and therefore on the composition of these cosmic rays. These cosmic rays have been measured by air shower experiments such as the Pierre Auger Observatory [1] and the Telescope Array experiment [2]. The depth of maximum of the air shower development, $X_{\max}$, and the number of muons on the ground were widely adopted for the mass-sensitive observables. The interpretation of the composition has been performed by comparing the measurements of these observables and their predictions by Monte Carlo simulation. However, predictions for both $X_{\max}$ and the number of muons show a large uncertainty due to the hadronic interaction models. Large uncertainty makes the interpretation challenging. In particular, the number of muons measured by the experiments was much larger than expected by the simulations [3]. This problem is known as the muon deficit problem. The ratio of pions to kaons in particle production of each hadronic interaction can be a source of the muon deficit problem [4]. Thus, the production of strange mesons is important. Furthermore, the large uncertainty in pion-nucleus collisions affects both $X_{\max}$ and muon components in air showers.

The Large Hadron Collider forward (LHCf) experiment measures forward neutral particle productions at the Large Hadron Collider (LHC). It shares the interaction point with the A Toroidal LHC Apparatus (ATLAS) experiment; thus, a common analysis using the LHCf de-

*e-mail: ohashi.ken@isee.nagoya-u.ac.jp

tectors and the ATLAS detectors can be performed. We completed the data taking of proton-proton collisions at $\sqrt{s} = 0.9, 2.76, 7$, and 13 TeV and proton-lead collisions at $\sqrt{s_{NN}} = 5.02$ and 8 TeV. Photons, $\pi^0$, and neutrons in the forward regions were measured. For proton-proton collisions with $\sqrt{s} = 13$ TeV, for example, measurements of forward photons [5] and neutrons [6] were published and the preliminary result for $\pi^0$ was shown [7]. Moreover, forward photons produced in diffractive dissociation were analyzed by the common analysis of the ATLAS and LHCf experiments [8]. However, several measurements are still missing: measurements of forward strange mesons, measurements of high-energy pion-proton collisions using the one-pion exchange process in the proton-proton collisions, and measurements of proton-light nucleus collisions.

Recently, we focused on the analysis of forward $\eta$ mesons using data taken in 2015 and on the new data-taking in 2022. In this work, we present a preliminary result for the forward $\eta$ mesons, the status of the common analysis, and the data-taking successfully completed in September 2022.

## 2 Status of the analysis of proton-proton collisions of $\sqrt{s} = 13$ TeV

### 2.1 Preliminary result of forward $\eta$ meson

The $\eta$ meson is one of the mesons that consists of strange and other quarks. It decays into two photons with a branching ratio of 39.14% [9]. We can reconstruct the meson from the two photons measured by the LHCf-Arm2 detector. The preliminary result of the $\eta$ meson was shown in [10] and the final result is in preparation.

The analysis was performed using data taken between June 12th and 13th, 2015 (LHC Fill 3855), a special low luminosity run with $\beta^* = 19$ m. The half crossing angle was 145 $\mu$rad. Two data sets were used for the analysis; one is the data taken from 22:30 June 12th to 1:30 June 13th (CEST). The other is the data taken from 1:40 to 12:10 on June 13th. The collisions per bunch crossing ranged from 0.007 to 0.012 for the first data set and from 0.03 to 0.04 for the second data set. The integrated luminosity was 0.194 nb$^{-1}$ and 1.9378 nb$^{-1}$ for the first and second data sets, respectively.

The energy and positions were reconstructed from the energy deposit in each layer and measured by silicon microstrip layers. Several event selection criteria were applied to select events with photon-like signals in the both calorimeter towers of the LHCf-Arm2 detector. Events with a hit in 2mm from the calorimeter edge or with a reconstructed energy less than 200 GeV were removed. Photon-like events were selected using the longitudinal development of signals in the calorimeter. The depth $L_{90\%}$ that 90% of energy deposit was observed in a calorimeter tower was calculated and adopted as a parameter of the longitudinal development. An electromagnetic shower induced by a photon shows a larger energy deposit in the first several layers of the calorimeter tower, thus we can select photons by selecting events with small $L_{90\%}$. Events with more than one particle hit in a calorimeter tower were removed.

The invariant mass of two photons was reconstructed from the energy and positions of two photons by assuming a decay at the interaction point. Using the opening angle of two photons, $\theta$, the invariant mass is calculated as follows;

$$M_{\gamma\gamma} = \sqrt{2E_1E_2(1 - \cos\theta)}, \qquad (1)$$

where $E_1$ and $E_2$ are the energy of two photons. We found that the mass of $\pi^0$ was shifted by -2.57 ± 0.04%. Thus the mass was artificially shifted the energy of single photons by -2.65% to correct the position of the $\pi^0$ mass peak to the reference value. The left plot of Fig. 1[10] shows the invariant mass distributions around the mass of the $\eta$ meson. The distributions were fitted by an asymmetric Gaussian function and the third-order Chebyshev polynomial function for signals and backgrounds, respectively. The region within $3\sigma$ from the peak was considered as signal region, and the region from $4\sigma$ to $7\sigma$ from the peak was considered as background region. The energy spectrum of the backgrounds was estimated by scaling the spectrum of the events in the background regions and then we subtracted them from the events in the signal regions.

Correction factors were applied to the signal distributions; acceptance and selection efficiency was corrected using MC simulations. The branching ratio was corrected by the constant value. The inefficiency due to multi-hit rejections was corrected using the fraction of multi-hit events in each bin. Systematic uncertainties due to energy scale, beam center stability, particle identification, background subtraction, luminosity, and correction factors were considered. The right plot of Fig. 1[10] shows the preliminary result of the production cross sections of $\eta$ for $P_T < 1.10$ GeV. The black points are the results of the LHCf-Arm2 detector. Predictions by hadronic interaction models are shown by the solid color lines. QGSJET II-04 shows the best agreements among the models.

### 2.2 ATLAS-LHCf common analysis

After the preliminary result in [8], the common analysis of the ATLAS and LHCf collaborations is focusing on two directions: the energy spectrum of photons from single diffractive dissociation and the correlation between forward hadrons and the number of particles in central regions for the modeling of the multi-parton interaction. The latter was proposed in Ref. [11]. We are preparing the final result for the former and finalizing the analysis for the latter.

## 3 Data-taking of proton-proton collisions at $\sqrt{s} = 13.6$ TeV

### 3.1 Upgrades from the last data-taking and the common data-taking with ATLAS detectors

Several upgrades were applied for new data-taking in 2022. One is the upgrade to the LHCf-Arm2 detector. The other is the common data-taking with ATLAS detectors.

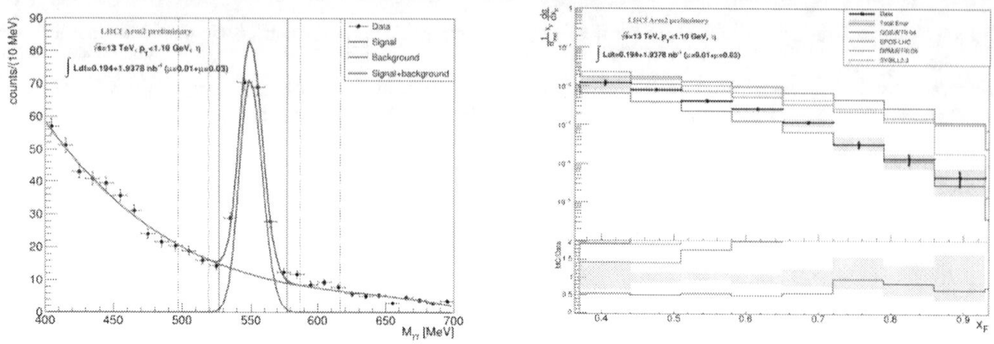


**Figure 1.** (left) Invariant mass distribution reconstructed from two photons [10]. (right) Production cross-sections of $\eta$ meson [10].

The former one is the upgrade of read-out of the silicon microstrip layers of the LHCf-Arm2 detector. It allows us to improve read-out time and therefore the trigger rate. The latter one is the common data-taking with Zero degree calorimeters (ZDC) and Roman pots detectors of the ATLAS experiment. Figure 2 shows a schematic view of the common data-taking. Three modules of the ATLAS-ZDC detector were installed behind each LHCf detector. The thickness of the LHCf detectors to hadrons is 1.6 interaction length, therefore, not enough to measure hadrons precisely. By installing ATLAS-ZDC detectors, we can improve the energy resolutions for hadrons from 40% to approximately 20% [13]. Moreover, the common data-taking with the Roman pots detectors and the LHCf detectors allows us to measure both protons and particles in the dissociation system in single diffractive dissociation.

### 3.2 Physics targets

We have four main physics targets; a) precise measurements of $\pi^0$, $\eta$, and $K^0$ mesons, b) the measurements of pion-proton collisions using the one-pion exchange process, c) measurements of diffractive dissociation using the LHCf detectors and the Roman-pot detectors, d) measurements of $\Delta^+$.

a Precise measurements of the $\pi^0$, $\eta$, and $K^0$ mesons

Measurements of $\pi^0$ and $\eta$ mesons were performed using the data taken in 2015, while statistics of these mesons were limited. In particular, $\eta$ meson candidates were approximately 1500 events and not enough to analyze $p_T$ dependencies. Thanks to the upgrade in the read-out of the LHCf-Arm2 detector, we can measure ten times larger statistics of $\pi^0$ and $\eta$ candidates than the previous data-taking. Moreover, we expect a few hundred $K_s^0$ candidates.

b The measurements of pion-proton collisions using the one-pion exchange process

Measurements of the one-pion exchange process in proton-proton collisions allow us to measure cross sections and particle productions in high-energy pion-proton collisions [14]. For measurements of the one-pion exchange process, we need to precisely measure the energy and positions of each neutron. By combining the ATLAS-ZDC and LHCf detectors, we can measure neutrons with good precisions: approximately 20% energy resolutions and better than 0.3 mm for position resolutions. By combining information from the ATLAS inner detector, we can reject contributions from the other processes and measure the one-pion exchange process [15].

c Measurements of diffractive dissociation using the LHCf detectors and the Roman-pot detectors

In the single diffractive dissociation, one proton is produced and particles are produced in the forward regions on the opposite side to the proton. The proton and the particles on the opposite side can be measured by the ATLAS Roman pots detectors and the LHCf detector, respectively. In this way, we can measure the diffractive mass and particles produced in dissociation simultaneously. It helps us to understand particle productions in diffractive dissociation.

d Measurements of $\Delta^+$

By measuring a proton and a $\pi^0$ produced on the same side, we can measure $\Delta^+$ produced in the forward regions if exists. The predictions of forward $\Delta^+$ vary among hadronic interaction models, thus we can constrain predictions of hadronic interaction models.

### 3.3 Data-taking

The data-taking was successfully operated from 23rd to 26th September 2022. It was a special low-luminosity run with $\beta^* = 19$ m and the number of collisions per bunch crossing $\mu \sim 0.04$. Figure 3 shows the statics over time. We had approximately 48 hours of data-taking during this operation. We performed the data-taking with two detector positions to increase the detector coverage; the beam center position and the 5 mm higher position. The total statistics taken during this operation is approximately 300 million events.

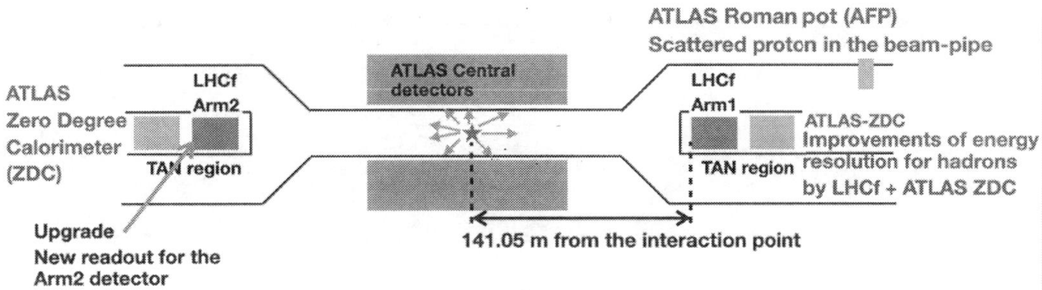


**Figure 2.** A schematic view of detector configurations of the data-taking in Sept. 2022.

The two-photon invariant mass distribution was calculated using the quarter data set of the LHCf-Arm2 beam-center position. Events with a photon in each calorimeter tower were selected. Figure 4 shows the preliminary two-photon invariant mass distributions. We found two peaks corresponding to $\pi^0$ and $\eta$ mesons. Although a quarter of the one data set was analyzed, we have large statistics for $\pi^0$ and $\eta$ candidates. The position of the peak is lower than expected. The detector is needed to be calibrated. We had a beam test of detectors a the Super Proton Synchrotron in October 2022 and will calibrate the energy scale later.

## 4 Summary

The LHCf experiment measures forward neutral particles to improve hadronic interaction models. Recently, we showed the preliminary result of forward $\eta$ meson productions in proton-proton collisions at $\sqrt{s}$ = 13 TeV. QGSJET II-04 shows the best agreements among models. Moreover, we successfully completed the data-taking of proton-proton collisions at $\sqrt{s}$ = 13.6 TeV in September 2022. We took the larger number of $\pi^0$ and $\eta$ candidates. It was confirmed using the invariant mass distributions of two photons. We found the shift of the peak position corresponding to $\pi^0$ and $\eta$ mesons. Thus, we will perform the detector calibration using the test-beam data taken at the Super Proton Synchrotron in October 2022. In parallel, we will finalize the analysis of data taken in 2015 and start the analysis of data taken in 2022.

## References

[1] A. Aab et al., The Pierre Auger Cosmic Ray Observatory, Nuclear Instruments and Methods in Physics Research, Section A: Accelerators, Spectrometers, Detectors and Associated Equipment 798, 172 (2015), doi:10.1016/j.nima.2015.06.058, 1502.01323.

[2] R. Abbasi et al., Highlights from the Telescope Array experiment, In Proceedings of 37th International Cosmic Ray Conference — PoS(ICRC2021), vol. 395, p. 012, doi:10.22323/1.395.0012 (2021).

[3] L. Cazon et al., "Working Group Report on the Combined Analysis of Muon Density Measurements from Eight Air Shower Experiments", In Proceedings of 36th International Cosmic Ray Conference — PoS(ICRC2019), vol. 358, p. 012, doi:10.22323/1.358.0214 (2021).

[4] J. Albrecht, L. Cazon, H. Dembinski et al., The Muon Puzzle in cosmic-ray induced air showers and its connection to the Large Hadron Collider, Astrophys.Space Sci. 367 (2022) 3, 27

[5] O. Adriani et al., the LHCf Collaboration, Phys. Lett. B 780 (2018) 233–239.

[6] O. Adriani et al., the LHCf Collaboration, JHEP07(2020)016.

[7] A. Tiberio et al., PoS (ICRC2021) 386, https://pos.sissa.it/395/386/.

[8] The ATLAS and LHCf Collaborations, ATLAS-CONF-2017-075, https://cds.cern.ch/record/2291387.

[9] P.D.G., R. Workman, V. Burkert, V. Crede, E. Klempt, U. Thoma, L. Tiator, K. Agashe, G. Aielli, B. Allanach, et al., Review of particle physics, Progress of theoretical and experimental physics 2022, 083C01 (2022).

[10] G. Piparo et al., To be published in the Proceedings of The Tenth Annual Conference on Large Hadron Collider Physics PoS(LHCP2022). The paper with the final result is in preparation.

[11] S. Ostapchenko, M. Bleicher, T. Pierog, and K. Werner, Phys. Rev. D 94 (2016) 114026.968 https://link.aps.org/doi/10.1103/PhysRevD.94.114026.

[12] LHCf Collaboration, "LHCf— Technical Proposal for the LHC Run 3." https://cds.cern.ch/record/2679323.

[13] M. Kondo et al., in this proceedings

[14] V.A. Petrov et al., Eur. Phys. J. C (2010) 65: 637—647.

[15] K. Ohashi et al., PoS (ICRC2021) 190, https://pos.sissa.it/395/190/.

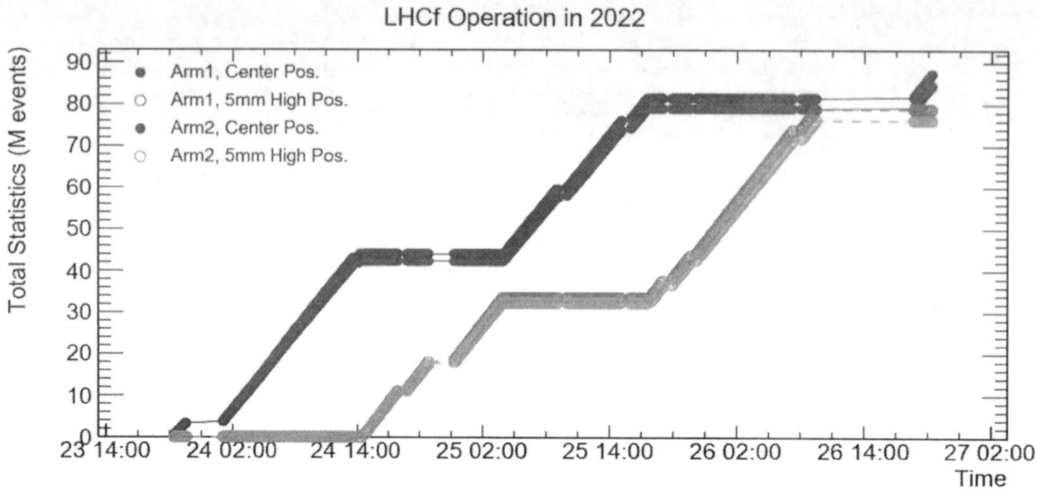


**Figure 3.** Total statistics in the data-taking.

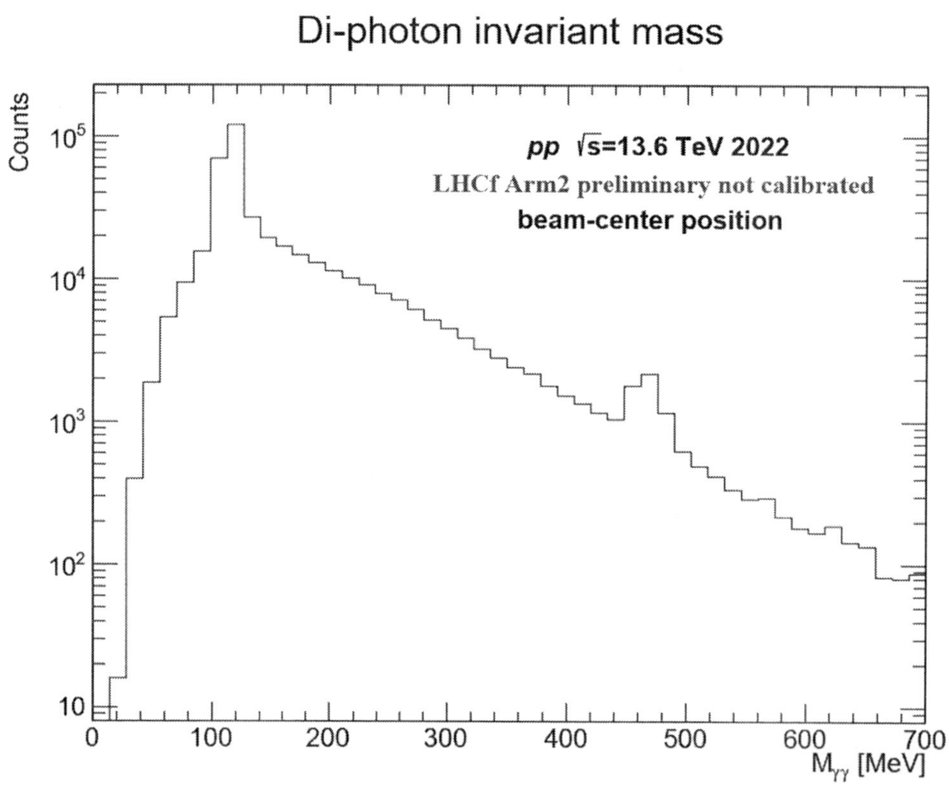


**Figure 4.** Preliminary invariant mass distribution reconstructed from the two photons for the LHCf-Arm2 detector. A quarter of the total statistics for the beam-center positions were analyzed.

# Searches for Lorentz Invariance Violation at the Pierre Auger Observatory

*Caterina* Trimarelli[1,2,*] for the Pierre Auger Collaboration[3,**]

[1]Università dell'Aquila
[2]INFN Laboratori Nazionali del Gran Sasso, Assergi (L'Aquila), Italy
[3]Observatorio Pierre Auger, Av. San Martín Norte 304, 5613 Malargüe, Argentina

**Abstract.** Lorentz Invariance (LI) requires the space-time structure to be the same for all observers, but various quantum gravity theories suggest that it may be violated when approaching the Planck scale. Even a small violation of LI could easily affect the Ultra-High-Energy Cosmic Ray (UHECRs) propagation on a cosmological scale. Moreover, at the extreme energies, like those available in the collisions of UHECRs in the atmosphere, one should also expect a change in the development of Extensive Air-Showers (EAS). In this contribution, Lorentz Invariance Violation (LIV) has been introduced as a perturbation term in the single particle dispersion relation considering a phenomenological approach. As a result, the kinematics of the interactions in both the extragalactic propagation and in the shower development in the atmosphere is affected. The comparison between the model predictions and the data measured by the Pierre Auger Observatory are considered to constrain the LIV scenarios. The impact of LIV has been tested considering the measured energy spectrum and the composition and the resulting upper limits on the photon flux. Finally, the effects on the shower cascade in the atmosphere are studied considering the muon content distribution.

## 1 Introduction

Violations of Lorentz symmetry could change the energy threshold of photo-hadronic interactions; in particular, depending on the composition of the UHECRs at the highest energies, the attenuation length of photo-meson production or photo-disintegration may become extremely large and suppress particle interaction during propagation in the extragalactic space [1–4]. As a consequence, the existing evidence of the suppression of the flux at the highest energies [5] can be used to put a limit on LIV. However, data from the Pierre Auger Observatory can be interpreted in a scenario in which the origin of the suppression is due to the maximum acceleration energy at the sources rather than to interactions with the background radiation. In this scenario, UHECR data can no longer yield bounds on LIV. On the other hand, violation of Lorentz invariance can be tested by deriving limits on violation parameters from UHECR phenomena other than propagation. In particular, LIV can be tested by searching the best description of the UHECR observables, under LIV assumptions, as already done in [6–9]. Moreover, to set limits in the electromagnetic sector the upper limits on the UHE photon flux found by the Pierre Auger Observatory is compared with the result produced in the presence of LIV.

An alternative approach to constrain LIV models is that, depending on the strength of the violation, the high energy available in the collision of cosmic rays with the atmosphere can lead to modifications of the shower development with respect to the standard LI case [10].

## 2 The Pierre Auger Observatory and datasets

The Pierre Auger Observatory [11], located on a vast plain in Argentina, just northeast of the town of Malargüe, in the Province of Mendoza, 1440 m above the sea level, is the largest observatory to detect UHECRs ever built and it has been in operation since 2004. It covers an area of 3000 km$^2$ with a Surface Detector array (SD) overlooked by a Fluorescence Detector (FD). The SD consists of 1660 water-Cherenkov detectors arranged in a triangular grid operating with a nearly 100% duty cycle. Each SD station detects at ground level the secondary particles of the EAS produced by the primary UHECR interacting in the atmosphere. The FD consists of a set of telescopes that measure the UV fluorescence light from nitrogen molecules excited by the EAS particles along their path in the atmosphere. FD operations are limited to clear moonless nights, resulting in a duty cycle of about 15%. This hybrid detection technique combines the calorimetric measurement of the shower energy through fluorescent light with the high-statistics data of the surface array. The combination of the information from both techniques results in a quasi-calorimetric determination of the energy scale, a geometric direction reconstruction and an estimator of the primary particle mass.

In this paper, I summarize some recent results about the searches of Lorentz Invariance Violation using the data

*e-mail: caterina.trimarelli@aquila.infn.it
**e-mail: spokespersons@auger.org
Full author list: https://www.auger.org/archive/authors_2022_10.html

measured by the Pierre Auger Observatory described in [9, 10].

## 3 Lorentz Invariance Violation framework

A well established phenomenological approach to introduce LIV effects consists of adding effective terms in the dispersion relation of particles [12]. The resultant dispersion relation can be expressed as:

$$E_i^2 - p_i^2 = m_i^2 + \epsilon_i s_i^2, \tag{1}$$

where $i$ defines the particle, $E$, $m$ and $p$ correspond, respectively, to the energy, mass and momentum of the particle, and $s$ can be chosen to be either the energy or the momentum (for ultra-relativistic particles). In the following we consider $s = p$. At $p \ll M_{\rm Pl}$, the factor $\epsilon$ can be expanded as a polynomial series as a function of $p$, so that:

$$E_i^2 - p_i^2 = m_i^2 + \sum_{n=0}^{N} \delta_i^{(n)} p_i^{2+n}, \tag{2}$$

where $n$ is the LIV order. The parameters $\delta_i^{(n)}$ are independent for each particle species and define the energy scale associated with the violation. It can be noticed that the LIV parameters is sometimes introduced as:

$$\delta_i^{(n)} = \frac{\eta_i^{(n)}}{M_{\rm Pl}^n} \tag{3}$$

where $M_{\rm Pl} \approx 1.22 \cdot 10^{22}$ GeV/$c^2$ is the Planck mass. If we assume that only the lowest-order non-vanishing term has a non-negligible effect it is possible to set individual limits on $\delta_i^{(n)}$.

Considering Eq. 3, the modified dispersion relation (for only the leading term) can be expressed in terms of the dimensionless constant parameter $\eta_i^{(n)}$ as follows:

$$E_i^2 - p_i^2 = m_i^2 + \eta_i^{(n)} \frac{p_i^{n+2}}{M_{\rm Pl}^n} \tag{4}$$

## 4 LIV searches in the extragalactic propagation of UHECRs

In this section, the searches for LIV signatures reported in [9] in the cosmic-ray interactions during their propagation through the universe are described. In that work, we consider the modified dispersion relation in terms of $\delta_i^{(n)}$ (see Eq. 2). In particular, new bounds for the LIV parameter $\delta_i^{(n)}$ are found using the data collected by the Pierre Auger Observatory. The kinematics and the energy thresholds of interactions are modified once the LIV terms become comparable to the squared masses involved. Two independent LIV sectors are tested. Limits on the electromagnetic sector are set considering that, at sufficiently high energies, the cosmogenic photons produced by cosmic ray interactions with the background radiation fields could be subluminal and be attenuated much less than in the LI case over cosmological distances. On the other hand, for the hadronic sector, how the modifications on the cosmic rays interactions affect the energy spectrum and the mass composition at Earth is considered.

Signatures of LIV are studied considering a specified UHECR scenarios. We consider isotropically distributed sources with $(1 - z)^m$ cosmological evolution emitting an energy spectrum given by:

$$\frac{dN_A}{dE} = J_0 f_A \left(\frac{E}{10^{18}\text{eV}}\right)^{-\Gamma} \times \begin{cases} 1, & R < R_{\rm cut} \\ \exp(1 - R/R_{\rm cut}), & R \geq R_{\rm cut} \end{cases} \tag{5}$$

where $z$ is the redshift, $\Gamma$ is the spectral index at the injection, $R_{\rm cut}$ is the cutoff rigidity, $f_A$ is the fraction of nuclei with mass $A$, and $J_0$ is the normalization factor of the flux. UHECRs interact with photon backgrounds and at these energies, the most important photon backgrounds are the cosmic microwave background (CMB) and the extra-galactic background light (EBL). Given the uncertainties in the models, two EBL distributions and two photo-nuclear cross section models among the UHECR propagation models are used in this study [13]. To compare the events at Earth with the $X_{\rm max}$ data, the mass composition is transformed into an $X_{\rm max}$ distribution with the use of EAS simulations according to different hadronic interaction models. The $X_{\rm max}$ distributions are parameterized using the Gumbel functions [14]. Finally, a UHECR scenario is here defined by the set of parameters from the source, the propagation model and the hadronic interaction model.

### 4.1 LIV limits on the electromagnetic sector

In this section, the LIV signatures in the electromagnetic sector are briefly described. The interaction of UHECRs with the background radiation produces pions as studied by Greisen [15], Zatepin and Kuzmin (GZK) [16]. The neutral pions decay into UHE photons, hereinafter call GZK photons, and are absorbed during propagation due to electron-positron pair production with background photon fields [17]. In presence of LIV in the photon sector, the kinematics of this pair production is changed. Considering Eq. 2 for subluminal LIV ($\delta_\gamma < 0$), the modified mean free path can be retrieved. Figure 1 shows the ratio between the LIV and LI mean free paths for different LIV parameters corresponding to the LIV order $n = 0$. The main effect consists on a significant increase of the mean free path above a critical energy. The same result is found for LIV at first and second orders. As a consequence, for negative values of the LIV parameters, fewer UHE photons will be absorbed and, the arriving flux of these particles will be larger. To quantify the LIV effects, the pair production of UHE photons is simulated in the extragalactic space under LIV hypotheses and the UHE photon flux are used to test LIV in the electromagnetic sector. In particular, the LIV-modified mean free path is implemented in the software packages CR-Propa3/EleCa [18, 19] in order to obtain the arriving GZK photon flux. Further simulation details can be found in [9]. Comparing the resulting photon flux in presence of LIV with the upper limits imposed by the Pierre Auger Observatory a new bound for

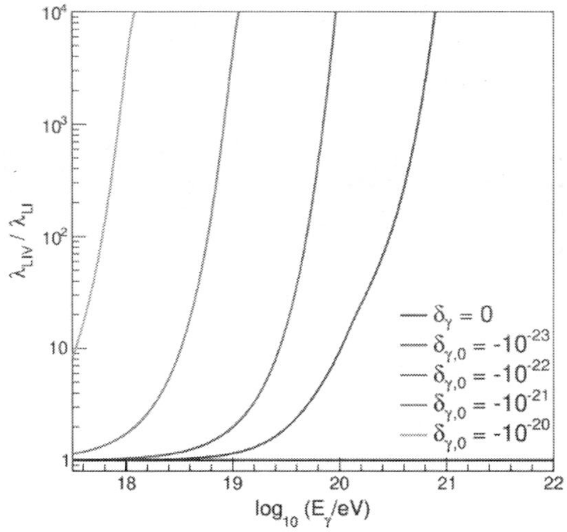


**Figure 1.** Ratio of the mean free path of pair production as a function of energy when LIV and LI are considered. The black line represents the LI case, while the shades of purple represent different LIV coefficients. Figure from [9].

the LIV parameters for different LIV orders is found. Different UHECR scenarios are taken into account. In Figures 2, the results are shown for photons produced in extragalctic propagation by UHECRs with an additional proton component at high energies. The Gilmore EBL model [20], Talys [21] cross-section and the EPOS LHC interaction model [22] are used. In particular, the Figures show the simulated GZK photon flux arriving at Earth under this scenario for LIV orders $n = 1, 2$ compared to the measured upper limits on the photon flux from the Pierre Auger Observatory. For some coefficients, the simulated LIV GZK photon flux is higher than the upper limits. The direct comparison of the simulation to the data give new limits for the LIV parameters: $\delta_\gamma^{(1)} > -10^{-40}$ eV$^{-1}$, $\delta_\gamma^{(2)} > -10^{-58}$ eV$^{-2}$.

### 4.2 LIV limits on the hadronic sector

In this section, the LIV signatures in the hadronic sector are described. During the propagation, the UHECRs interact with the background radiation and the kinematics of the dominant energy losses, photo-pion production and photodisintegration are modified under LIV assumptions. Only positive values of the LIV parameter $\delta_{\rm had}^{(n)}$ are taken into account. In the superluminal case, above a critical energy the photopion production for protons and the photodisintegration for nuclei change. The number of interactions is reduced and, consequently, the cosmic rays can travel farther than they would do under the LI hypotheses. To quantify the LIV effects both the modified attenuation length for pion production and the LIV modified energy threshold for photodisintegration were implemented in SimProp v2r4 [23]. Simulation details can be found in [9]. In this case instead of considering a fixed scenario, the source parameters and the LIV coefficients are fitted to the energy spectrum and composition from the Pierre Auger Observatory. Spectrum and composition are given by the LIV modified $10^5$ events produced with SimProp v2r4 (simulations performed for 5 nuclei). The simulated energy spectrum and composition arriving at Earth are used to fit to the data of the Pierre Auger Observatory for energies above $10^{18.7}$ eV. The fit procedure follows the explanation in [24]. Within each UHECR scenario, the free parameters of the fit are: a) the nuclei fractions, $f_A$, b) the index of the energy spectrum, $\Gamma$, c) the maximum rigidity, $R_{\rm cut}$, d) the normalization factor of the flux, $J_0$ and e) the LIV coefficient, $\delta_{\rm had}^{(0)}$. The cosmological evolution of the sources was fixed to $m = 0$. For each value of $\delta_{\rm had}^{(0)}$ ranging from $10^{-24}$ to $10^{-18}$ in $\log_{10}$ steps of 1 and for the LI case ($\delta_{\rm had}^{(0)}$) a log-likelihood fit was done searching for the combination of the parameters which best describes the data. Comparing the deviances obtained in LIV and LI scenarios limits on $\delta_{\rm had}^{(0)}$ have been imposed. In Figure 3 the energy spectrum and the first two moments of the $X_{\rm max}$ distributions in comparison with these corresponding to the LI case are shown. The comparison between the two panels with LI (left panel) and LIV (right panel) shows the correlation between mass composition and LIV in the fit. Lorentz violation suppresses the interactions during propagation which is compensated with a lighter composition at the sources in order to obtain the same composition on Earth.

## 5 LIV searches in the air-shower development

In this section, the approach considered to understand and determine the modification that the air-shower development undergoes in presence of LIV is described.

Interpreting the right-hand side of Eq. 4 as an energy dependent mass, $m_{\rm LIV}^2 = m^2 + \eta^{(n)} p^{n+2} / M_{\rm Pl}^n$, the Lorentz factor for a LI violating particle at energy $E$ can be defined as:

$$\gamma_{\rm LIV} = E / m_{\rm LIV} \tag{6}$$

Depending on the value assumed by $\eta^{(n)}$, the LIV effects can be easly analyzed considering the lifetime of the considered particle $\tau = \gamma_{\rm LIV}\tau_0$, where $\tau_0$ is the lifetime at rest, that will change accordingly. For negative/positive values of $\eta^{(n)}$ the lifetime of the particle should increase/decrease with respect to the LI case producing modifications in the EAS development which depends both on the energy and the strength of the violation. To understand the expectations, the $\pi^0$ decay can be taken into account. The $\pi^0$ lifetime as a function of the energy for the standard case and for different values of the LIV parameters is shown in Fig. 4. The energy at which the lifetime evolution deviates from the standard LI case depends both on the order and the strength of the violation. To have a qualitative idea of what one should expect, let us consider the simple model [25] where a primary hadron interacting in the atmosphere produces 2/3 of charged pions $\pi^\pm$ and 1/3 of $\pi^0 s$. In the standard case, charged pions decay producing muons and neutrinos while the neutral ones quickly decay

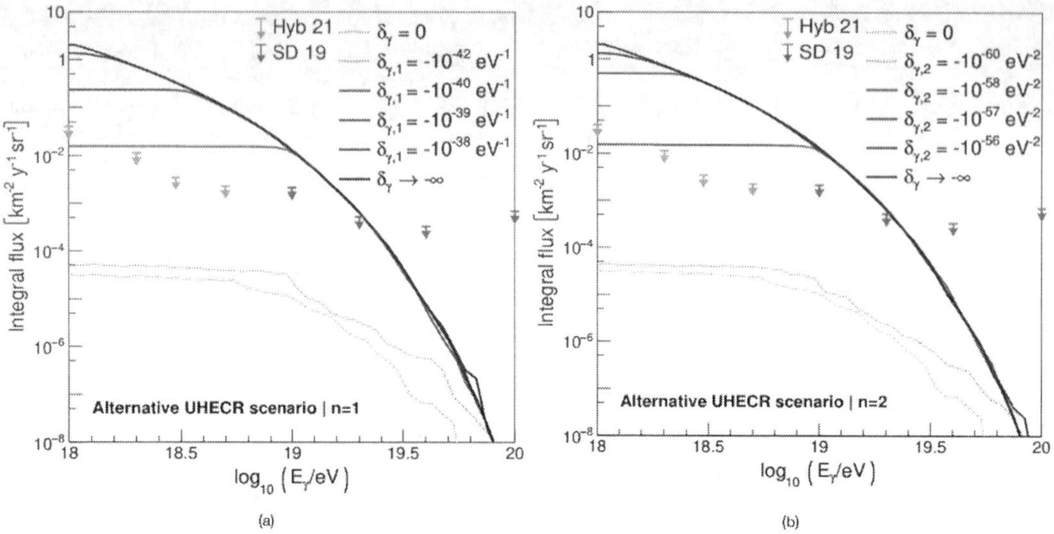


**Figure 2.** Simulated integral flux of GZK photons as a function of the energy for an alternative scenario with subdominant proton component. Continuous lines show the rejected LIV scenarios. The arrows show the flux determined by analysis of the Pierre Auger Observatory data. LIV at first order (a) and LIV at second order (b). Figures from [9].

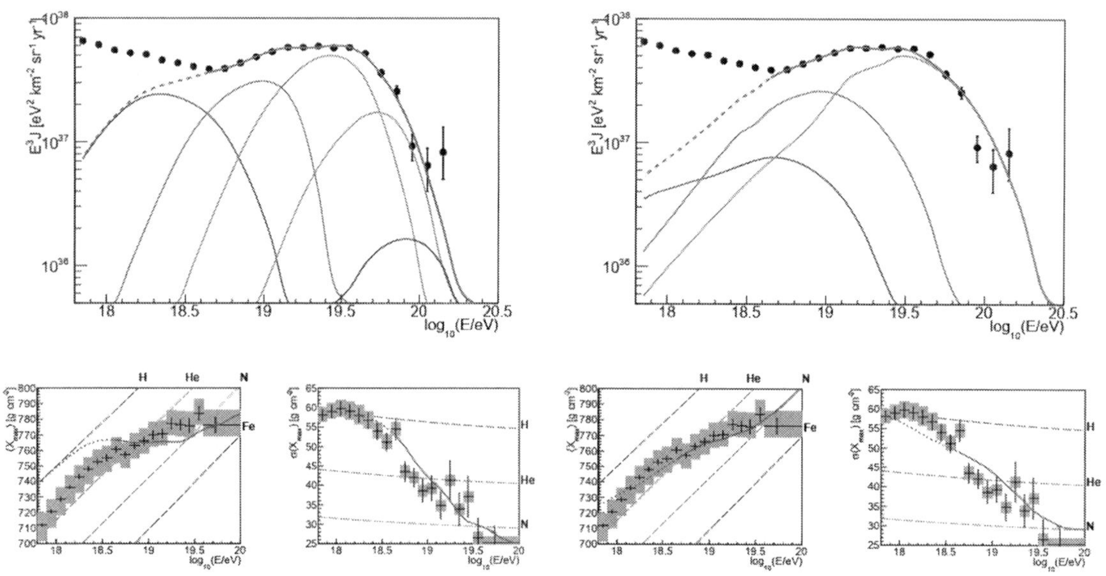


**Figure 3.** Energy spectrum (top) and first two moments of the $X_{\max}$ distributions (bottom), for the LI (left panels) and LIV $\delta^{(0)}_{\text{had}} = 10^{-20}$ cases (right panels) for the STGE propagation model compared to the Pierre Auger Observatory data. Partial distributions are grouped according to the mass number as follows: $A = 1$ (red), $2 \leq A \leq 4$ (grey), $5 \leq A \leq 22$ (green), $23 \leq A \leq 38$ (cyan), $39 \leq A \leq 56$ (blue), total (brown). Dashed brown lines show the energy region not used for the fit. Dashed lines in the bottom panes show simulations predictions for each element. Figures from [9].

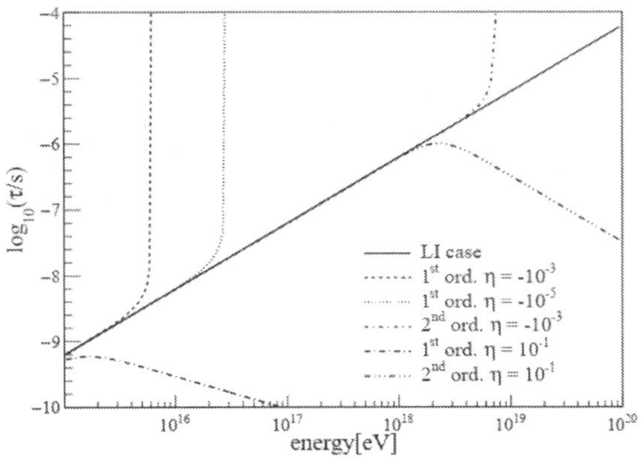


**Figure 4.** Neutral pion mean lifetime as a function of energy for the Lorentz invariant case and for different strengths of LIV. Figure from [10].

in two photons producing an electromagnetic sub-shower. Otherwise, in the presence of LIV and for negative values of $\eta^{(n)}$, the $\pi^0$ lifetime grows and the probability to interact before decaying increases. The re-interacting $\pi^0$s will behave as the source of a hadronic sub-showers like those initiated by charged pions. As the energy decreases in the further shower generations, $\pi^0$s will start again to produce a standard electromagnetic sub-shower. The consequence is a modification of the shower development in the atmosphere. The amount of energy deposited in the atmosphere will be reduced (i.e. invisible energy going to neutrinos will grow) leading to an underestimation of the primary energy if the event is treated as a standard physics one. Moreover, the position of the shower maximum ($X_{\text{max}}$) [26] will be slightly modified with respect to the standard LI case. In addition, as the muon content correlates with the energy of the hadronic component of the shower, we can expect that the number of muons produced in the EAS will increase and the physical fluctuations will decrease, as almost all the energy is kept into the hadronic component after the first stages of the shower developement, with little room for stochastic leakage to the electromagnetic component [27]. In this work only the effects due to negative values of the $\eta$ parameter are considered. Our purpose is to find only the lower limit of the bound for LIV at first order.

To quantify the effect of LIV on the shower development, we have performed a library of simulated showers by using CONEX software [28, 29] in both LI and LIV cases. For the violated scenario, the software has been modified by changing the lifetimes of all the unstable particles according to Eq. 6. In particular, in the velocity definition $\beta = \frac{|\mathbf{P}|}{m\gamma}$, the Lorentz factor $\gamma$ has been replaced with the LIV expression $\gamma_{\text{LIV}} = E/m_{\text{LIV}}$. The values of $\eta^{(n)}$ considered[1] for this study are $-10^{-1}$, $-10^{-2}$, $-10^{-3}$, $-10^{-4}$, $-10^{-5}$, $-10^{-6}$, $-10^{-7}$, $-10^{-8}$ and order of the violation $n = 1, 2$. For each value of $\eta$, 5000 primary cosmic ray particles have been produced in the energy range between $10^{16}$ eV and $10^{21}$ eV, using EPOS LHC [22] and QGSJETII-04 [30] hadronic interaction models and for different primary particle types i.e. proton, helium, nitrogen, silicon and iron.

### 5.1 The muon content distribution

As a result of the modified air shower simulation in the presence of LIV at first order, we have considered the mean longitudinal profile $dE/dx$. For larger values of $|\eta|$ (only negative LIV parameters are taken into account) the effects due to the violation increase. In fact, we have found a shift of $X_{\text{max}}$ , and a reduction in the height of the maximum energy deposit in the atmosphere.

First of all, the displacement in the position of the maximum of the longitudinal profile leads to different values of $X_{\text{max}}$. The change in the energy threshold of particle decays (mainly neutral pions), consuming the electromagnetic part of the shower faster, generates the effect to move the shower maximum to higher altitudes. This is due to the fact that a proton from the point of view of the shower development is behaving like a heavier nucleus. In fact, in the presence of LIV, the measured mass composition corresponds to a greater fraction of protons at the highest energies. This result has been already used in order to constrain LIV models in a previous work [31] and it is not considered in this contribution.

On the other hand, the reduction in the normalization of the longitudinal profile is linked to a change in the number of muons at the ground. In particular, the calorimetric energy deposited in the atmosphere in the presence of LIV is lower than the standard one. The modification of the energy-momentum relation allows hadronic interactions of neutral pions that contribute to the growth of the hadronic cascade producing an increase in the number of muons, as shown in Fig. 5, where the average number of muons at ground as a function of the primary energy in LI and LIV cases are shown. For proton-induced air showers, the number of muons is considerably increased, in any of the LIV cases considered here, while for iron primaries the effect is in general milder.

These two main effects, caused by the LIV framework, affect the fluctuations of the number of muons. In fact, the ratio of the fluctuations to the average number of muons (hereinafter referred to as relative fluctuations), dominated mostly by the first interaction [27], considerably decreases in the presence of LIV, as shown in Fig. 5, where the relative fluctuations of the number of muons as a function of the primary energy are reported.

Limits on LIV parameter $\eta$ can be derived comparing the observed strong decrease of the relative fluctuations with the muon fluctuation measurement [32] from the Pierre Auger Observatory. Considering the dependence of the decrease of the relative fluctuations on the different LIV strengths, a new bound for the LIV parameter

[1]Positive values of LIV parameter have been also considered but no effect on the shower development has been found because in these cases, the lifetime of neutral pions, already negligible in the LI case, decreases above the critical energy and the pions decay faster than in the standard one.

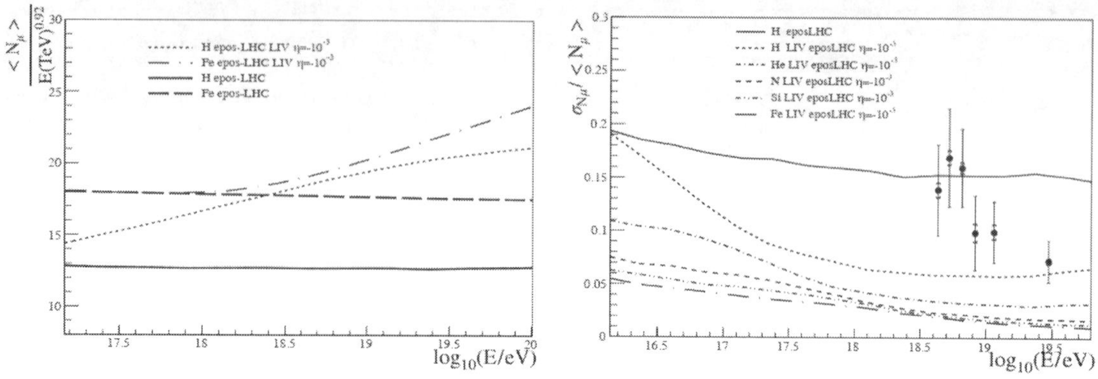


(a) (b)

**Figure 5.** (a) Average number of muons at ground vs primary energy in LI and LIV cases; (b) Data (black points with error bars) compared to LI and LIV models for the relative fluctuation of the number of muons vs primary energy. The LIV case corresponds to $\eta = -10^{-3}$. The statistical uncertainty is indicated by the error bars. The total systematic uncertainty is indicated by the square brackets. Figures from [10].

$\eta^{(1)}$ has been obtained. To find this, the most conservative LIV model with respect to data, depending on the particular mixture of primary particles, should be considered. In particular, the effects of different composition scenarios on both fluctuations and average number of muons have been taken into account. It can be noticed, referring to Fig. 3 in [32], that the mixture of the two components, p and Fe, gives the maximum value of relative fluctuations. Therefore, the most conservative LIV model corresponds to the ratio of the fluctuations to the average number of muons for a mixture of proton and iron.

We have defined the average number of muons $\langle N_\mu \rangle_{\rm mix}$ and the fluctuations $\mathrm{RMSD}(N_\mu)_{\rm mix}$ for a mixture of p and Fe using the following expressions:

$$\begin{aligned}\langle N_\mu \rangle_{\rm mix}(\alpha;\eta) &= (1-\alpha)\langle N_\mu \rangle_p + \alpha \langle N_\mu \rangle_{Fe} \\ \mathrm{RMSD}^2(N_\mu)_{\rm mix}(\alpha;\eta) &= (1-\alpha)\,\mathrm{RMSD}^2(N_\mu)_p + \\ &\alpha\,\mathrm{RMSD}^2(N_\mu)_{Fe} + \alpha(1-\alpha)(\langle N_\mu \rangle_p - \langle N_\mu \rangle_{Fe})^2 \quad (7)\end{aligned}$$

where $\alpha$, which depends on the energy, is the relative abundance of iron nuclei and the subscripts p and Fe label the averages and RMSD of pure proton and iron primaries respectively that are retrieved from the air shower simulations. [2] These values have been used to parametrize $\langle N_\mu \rangle$ and the RMSD between $\log_{10}(E/\mathrm{eV}) = 16.5$ and $\log_{10}(E/\mathrm{eV}) = 20$, for $-10^{-3} < \eta < -10^{-15}$ and for different masses. For each value of $\eta$, the parameterizations can be used to calculate the expected relative fluctuations for a certain mixture of proton and iron nuclei (hereinafter referred to as mixed relative fluctuations) as:

$$\frac{\sigma_\mu}{\langle N_\mu \rangle}(\alpha;\eta) = \frac{\sqrt{\mathrm{RMSD}^2(N_\mu)_{\rm mix}(\alpha;\eta)}}{\langle N_\mu \rangle_{\rm mix}(\alpha;\eta)} \quad (8)$$

To determine the most conservative LIV model the result of the mixed relative fluctuations for different mixtures of proton-iron composition are considered as a function of the energy. In particular, it has been found that the maximum with respect to $\alpha$ is above all the mixed relative fluctuations for a scan of the relative abundance $\alpha$ between 0 and 1. This effect has been observed for all the violation strengths. Therefore, only if all the curves are below the data points, the maximum with respect to $\alpha(E)$ of Eq. 8 in each energy bin corresponds to the most conservative LIV model. For any LIV parameter value, the most conservative LIV relative fluctuations as a function of the energy can be found without repeating any shower simulation. In Fig. 6, the coloured thin curves represent the maxima with respect to $\alpha$ for the relative fluctuations obtained from the parameterizations considering the $\eta$ parameter in the range $[-10^{-3}, -10^{-15}]$. Considering only the curves that are below the data, the $\chi^2$ can be calculated as a function of $\eta$ using each resulting mixed fluctuation and all the experimental data points. In this way, a continuous confidence level to exclude the LIV model has been found. This continuous result allows to determine the strictest lower $\eta$ parameter bound. As a consequence, the new bound for $\eta^{(1)}$ is $[-5.95 \cdot 10^{-6}, 10^{-1}]$ at 90.5% of CL.

In conclusion, we have found a new lower bound of the $\eta$ parameter range of values using the maximum relative fluctuation for a mixed initial proton-iron composition for LIV at first order. A similar approach using the minimum of the relative fluctuation with respect to $\alpha$ could lead to the definition of a negative upper bound of the LIV parameter.

[2] $\langle N_\mu \rangle_p$, $\langle N_\mu \rangle_{Fe}$, $\mathrm{RMSD}^2_{\rm p}$ and $\mathrm{RMSD}^2_{\rm Fe}$ depend on energy and on LIV parameter $\eta$

## References

[1] R. Aloisio et al. *Phys. Rev. D* **62** (2000) 053010.
[2] L. Maccione et al. *JCAP* **04** (2009) 022.
[3] A. Saveliev, L. Maccione and G. Sigl, *JCAP* **03** (2011) 046.

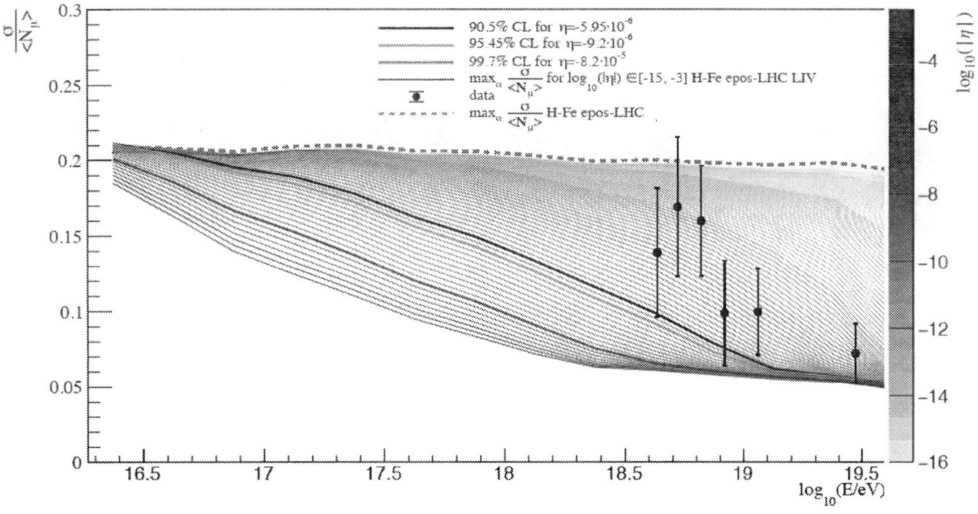


**Figure 6.** Maximum with respect to $\alpha$ of the mixed relative fluctuations obtained using the parameterizations in the standard case (dashed curve) and in the presence of LIV considering $\eta$ in the range $[-10^{-3}, -10^{-15}]$ (coloured curves) as a function of the primary energy. Each color corresponds to a different violation strength (right axis). The black points with error bars (statistical uncertainties) represent the measured relative fluctuations in the number of muons. Figure from [10].

[4] L. Maccione, A. Saveliev and G. Sigl, *Proc. 12th Int. Conf. Astroparticle and Underground Physics, Munich, Germany (2011)*, PoS(TAUP2011)375.

[5] A. Aab et al. [Pierre Auger Coll.], *The Pierre Auger Observatory: Contributions to the 36th International Cosmic Ray Conference: Madison, Wisconsin, USA(2019)*, Compendium of PoS(ICRC2019).

[6] D. Boncioli et al. [for the Pierre Auger Coll.], *Proc. 34th Int. Cosmic Ray Conf., The Hague, The Netherlands (2015)*, PoS(ICRC2015)521.

[7] D. Boncioli [for the Pierre Auger Coll.], *Proc. 35th Int. Cosmic Ray Conf., Busan, Korea (2017)*, PoS(ICRC2017)561.

[8] R. G. Lang [for the Pierre Auger Coll.], *Proc. 36th Int. Cosmic Ray Conf., Madison, Wisconsin, USA (2019)*, PoS(ICRC2019)327.

[9] P. Abreu *et al.* [Pierre Auger Coll.], *JCAP* **01** (2022) no.01, 023.

[10] C. Trimarelli et al. [Pierre Auger Coll.], *Proc. 37th ICRC, Berlin, Germany (2021)*, PoS(ICRC2021)340.

[11] A. Aab et al. [Pierre Auger Coll.], *Nucl. Instrum. Meth. A* **798** (2015) 172-213.

[12] S. R. Coleman and S. L. Glashow, *Phys. Rev. D* **59** (1999) 116008.

[13] R. Alves Batista, D. Boncioli, A. di Matteo, A. van Vliet and D. Walz, JCAP **10** (2015), 063 doi:10.1088/1475-7516/2015/10/063 [arXiv:1508.01824 [astro-ph.HE]].

[14] M. De Domenico, M. Settimo, S. Riggi and E. Bertin, JCAP **07** (2013), 050 doi:10.1088/1475-7516/2013/07/050 [arXiv:1305.2331 [hep-ph]].

[15] K. Greisen, Phys. Rev. Lett. **16** (1966), 748-750 doi:10.1103/PhysRevLett.16.748

[16] G. T. Zatsepin and V. A. Kuzmin, JETP Lett. **4** (1966), 78-80

[17] A. De Angelis, G. Galanti and M. Roncadelli, Mon. Not. Roy. Astron. Soc. **432** (2013), 3245-3249 doi:10.1093/mnras/stt684 [arXiv:1302.6460 [astro-ph.HE]].

[18] R. Alves Batista, A. Dundovic, M. Erdmann, K. H. Kampert, D. Kuempel, G. Müller, G. Sigl, A. van Vliet, D. Walz and T. Winchen, JCAP **05** (2016), 038 doi:10.1088/1475-7516/2016/05/038 [arXiv:1603.07142 [astro-ph.IM]].

[19] Settimo, M. & De Domenico, M. 2013, International Cosmic Ray Conference, 33, 2488

[20] Gilmore, R. C., Somerville, R. S., Primack, J. R., et al. 2012, 422, 3189.

[21] A. J. Koning et al. TALYS 1.0. Int. Conf. on Nuclear Data for Science and Tecnology, pages 211–214, 2007.

[22] T. Pierog, I. Karpenko, J. M. Katzy, E. Yatsenko and K. Werner, *Phys. Rev. C* **92** (2015) 034906, **1306.0121**].

[23] R. Aloisio, D. Boncioli, A. Di Matteo, A. F. Grillo, S. Petrera and F. Salamida, JCAP **11** (2017), 009 doi:10.1088/1475-7516/2017/11/009 [arXiv:1705.03729 [astro-ph.HE]].

[24] A. Aab *et al.* [Pierre Auger], JCAP **04** (2017), 038 [erratum: JCAP **03** (2018), E02] doi:10.1088/1475-7516/2017/04/038 [arXiv:1612.07155 [astro-ph.HE]].

[25] J. Matthews, *Astropart. Phys.* **22** (2005) 387-397.

[26] J. Linsley and A. A. Watson, *Phys. Rev. Lett.* **46** (1981) 459-463.

[27] L. Cazon, R. Conceição and F. Riehn, *Phys. Lett. B* **784** (2018) 68-76, 1803.05699].

[28] T. Pierog et al. *Nucl. Phys. B Proc. Suppl.* **151** (2006) 159-162, astro-ph/0411260].

[29] T. Bergmann et al.*Astropart. Phys.* **26** (2007) 420-432, astro-ph/0606564].

[30] S. Ostapchenko, *Phys. Rev. D* **83** (2011) 014018, 1010.1869].

[31] F. R. Klinkhamer, M. Niechciol and M. Risse, *Phys. Rev. D* **96** (2017) 116011, 1710.02507].

[32] A. Aab et al. [Pierre Auger Coll.], *Phys. Rev. Lett.* **126** (2021) 152002, 2102.07797].

# Diffuse flux of ultra-high energy photons from cosmic-ray interactions in the disk of the Galaxy and implications for the search for decaying super-heavy dark matter

*Corinne* Bérat[1], *Carla* Bleve[1], *Olivier* Deligny[2], *Pierpaolo* Savina[2,3], *François* Montanet[1], and *Zoé* Torrès[1,*]

[1]Laboratoire de Physique Subatomique et de Cosmologie, CNRS/IN2P3, Université Grenoble Alpes, Grenoble, France
[2]Laboratoire de Physique des 2 Infinis Irène Joliot-Curie, CNRS/IN2P3, Université Paris-Saclay, Orsay, France
[3]University of Wisconsin-Madison, Department of Physics and WIPAC, Madison, WI, U.S.A

**Abstract.** When propagating to Earth, UHECRs can interact with the gas in our Milky Way and produce secondary particles including photons. This can impact the search for UHE photons as we face a diffuse flux of UHE photons resulting from their propagation. This flux, together with the photon flux expected from the GZK interactions, participates in the diffuse photon flux which is limiting the detection of UHE-gamma sources. We present an estimate of the diffuse flux resulting from the interactions of UHECRs in our galaxy above $10^{17}$ eV, using results from the Pierre Auger Collaboration concerning the flux and its composition, and two different interstellar gas density models. We also discuss the impact of the evaluated diffuse flux of UHE photons on SHDM searches, as the former can be considered as a floor below which other signals would be overwhelmed. Similarly, the neutrino flux produced during the same process can be evaluated, by accounting for the neutrino mixing.

## 1 Introduction

On their way to Earth, UHECRs may interact with the interstellar gas in our galaxy. Above $10^{17}$ eV, the cosmic ray flux composition consists of protons and heavier nuclei. Hence, the cosmic rays interact with the interstellar gas through nucleon-nucleon collisions, whereas at lower energies other interactions can also take place, such as Bremsstrahlung and inverse-Compton. Following these interactions, secondary particles such as UHE photons are produced. This additional flux of UHE photons can be problematic in the search for UHE photons. Indeed, these photons do not originate from the vicinity of potential astrophysical sources of UHECRs, but are the result of their propagation. By combining the knowledge of this flux, with that of GZK photons, one could estimate the diffuse flux limiting the detection of UHE photons coming from actual sources.

In this contribution, we give an estimate of the flux of UHE photons arising from the irradiation of the interstellar gas above $10^{17}$ eV. The full study has been published in [1]. These photons essentially propagate through the galaxy without interacting, as their mean free path is higher than the galactic diameter in this energy range, as showed in Fig. 1 [2]. Therefore, the estimation of the diffuse flux per steradian of UHE photons with an energy $E_\gamma$, $\phi_\gamma(E_\gamma, \mathbf{n})$ can be simply done by integrating the emission rate at position $\mathbf{n}$, $q_\gamma(E_\gamma, \mathbf{n})$, per unit of volume and energy along the line of sight, as follows :

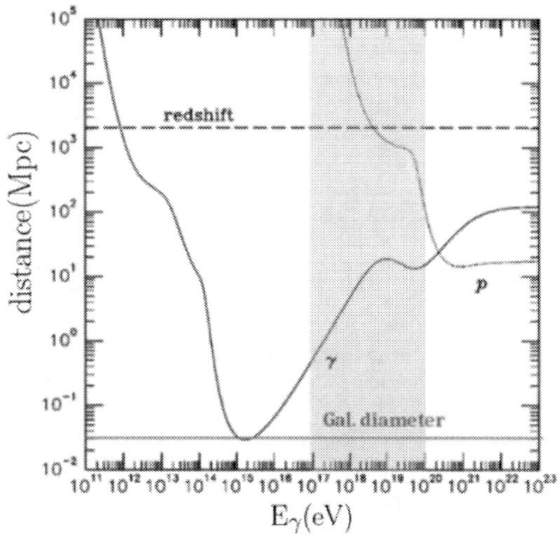


**Figure 1.** The attenuation distance of photons as function of the photon's energy. The energy range considered is showed by the pink area. The galactic diameter is reported by the blue line. Figure taken from [2].

$$\phi_\gamma(E_\gamma, \mathbf{n}) = \frac{1}{4\pi}\int_0^\infty \mathrm{d}s\, q_\gamma(E_\gamma, s\mathbf{n} + \mathbf{x}_\odot) \qquad (1)$$

Here, $\mathbf{x}_\odot$ is the position of the solar system in the galaxy, and $\mathbf{n} \equiv \mathbf{n}(l, b)$ is a unit vector on the sphere

*e-mail: torres@lpsc.in2p3.fr

 https://doi.org/10.1051/epjconf/202328305004

pointing to the longitude $l$ and latitude $b$ in galactic coordinates. It is important to note that above $10^{17}$ eV, the observation of UHECRs shows a high degree of isotropy [3], which leads to a equally isotropic irradiation of the interstellar gas and therefore an isotropic emission of UHE photons, hence the $\frac{1}{4\pi}$ factor in Eq. 1.

To compute the emission rate $q_\gamma(E_\gamma, \mathbf{x})$, corresponding to the number of photons of energy $E_\gamma$ produced by UHECRs with energy above $E_0 = 10^{17}$ eV at position $\mathbf{x}$ in the galaxy, we consider :

- The inelastic cross section of the interaction gas-UHECR : $\sigma(E_{CR})$.
- The UHECR flux $\Phi_{CR}(E_{CR})$ which can be decomposed in four nuclei groups: H, He, CNO and Fe. This is done so that we can use the data from the Pierre Auger Collaboration in section 2.
- The mean number of UHE photons of energy $E_\gamma$ produced during the interaction of an UHECR of energy $E_{CR}$ with the interstellar gas : $\frac{dN}{dE_\gamma}(E_{CR}, E_\gamma)$.
- The Milky Way interstellar gas density in nuclei per cm$^3$ in at position $\mathbf{x}$ : $\rho(\mathbf{x})$.

$$q_\gamma(E_\gamma, \mathbf{x}) = 4\pi \sum_{i,j} \rho_j(\mathbf{x})$$
$$\int_{E_0}^{\infty} dE_{CR}\, \Phi_{CR,i}(E_{CR})\, \sigma_{i,j}(E_{CR})\, \frac{dN_{i,j}}{dE_\gamma}(E_{CR}, E_\gamma) \quad (2)$$

In Eq. 2, the summation over $j$ corresponds to the different interstellar gas elements : molecular and atomic hydrogen, and helium. The one over $i$ accounts for the four groups of UHECRs nuclei defined above.

## 2 Cosmic Ray Flux

Above $10^{17}$ eV, the energy spectrum of cosmic rays and its mass composition are obtained via the indirect measurements of extensive air showers. With measurements of the slant depth corresponding to the maximum of the shower development, a parameter strongly correlated to the mass of the primary cosmic ray, the composition is determined on a statistical basis.

To obtain the cosmic ray flux for a defined group of primaries $\Phi_{CR,i}(E_{CR})$, we combine the all-particle spectrum of UHECRs, $\Phi_{CR}(E_{CR})$, with the energy-dependent abundances of the primaries $f_i$, giving : $\Phi_{CR}(E_{CR}) = \sum_i f_i\, \Phi_{CR,i}(E_{CR})$.

As for the spectrum, we used the one measured at the Pierre Auger Observatory [4] using the surface detector. This choice was made because the observatory owns the largest cumulated exposure. It also uses a single type of detector avoiding the combination of measurements and therefore the addition of systematics to the study. Uncertainties on the spectrum must be propagated when calculating the diffuse flux. Finally, the primary energy is known calorimetrically as it is measured by the fluorescence telescopes of the observatory, allowing no assumption on the mass of the particle.

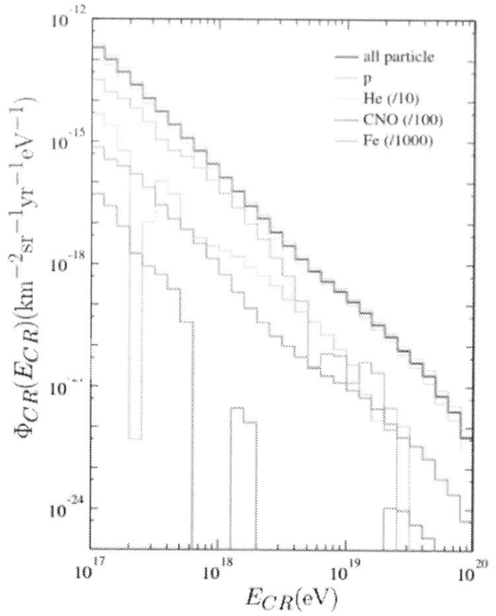


**Figure 2.** The all particle spectrum from [4] (black) with the contributions of H (red), He (orange), CNO (green) and Fe (blue).

Concerning the fraction of the four primary groups, they are also obtained from measurements carried at the Pierre Auger Observatory. The determination of the primary composition is performed by comparing the measured maximum of development of the shower, with expectations according to high energy hadronic interaction models for the four types of nuclei introduced before. We use the results from [5], where the fractions of the different primary groups are reported as a function of the cosmic ray energy, namely EPOS-LHC [6], Sibyll2.3 [7] and QJSJetII-04 [8].

Fig. 2 shows the combination of the all particle spectrum with the abundances of the nuclei types. Except for Hydrogen, the species have been re-scaled for clarity issues (see legend). The systematic uncertainties arising from both the all-particle spectrum of UHECRs and the fractions of mass groups inferred from the three different hadronic interaction models, will be propagated when calculating the estimate of the diffuse flux. As a cross-check, we took the fractions from the Global Spline Fit (GSF) [9], a data-driven fit using different ground based experiments. We regained similar results for the estimate of the flux in section 5.

## 3 Interstellar Gas Density

The interstellar gas in the Milky way is mostly composed of Hydrogen (90%) and Helium (10%), distributed essentially in the disk. Hydrogen, contributing to most of the interstellar gas mass, can be found in three different forms depending on its temperature : atomic $H_I$, molecular $H_{II}$ and ionised. On the other hand, Helium stays neutral as its first ionisation potential is higher. The case of ionised hydrogen will not be considered in this work as its contribution is smaller.

The distribution of the $H_I$ component, representing about two thirds of the Hydrogen mass, is rather constant in the distance range [4-10 pc] from the galactic centre and falls at larger distances. The density decreases exponentially in the perpendicular direction to the galactic plane with a scale length depending on the gas temperature. This distribution is known following observations via the 21-cm radio hyperfine line, whether by absorption or emission. In the case of $H_{II}$, the observation is not direct. Indeed, a homo-nuclear diatomic molecule like $H_{II}$ does not possess a permanent electric dipole moment, and therefore has no simple rotational transition. Thus, the column density of $H_{II}$ is inferred through the observations of carbon dioxide (CO) emission lines. Molecules of CO, the second most abundant molecule in the interstellar medium after $H_{II}$, are excited by their collisions with $H_{II}$ molecules, resulting in a correlation between the integrated intensity of CO lines and the column density of molecular hydrogen. Finally, the distribution of Helium is not really known, being only observable in a specific ionised state. Therefore, it is assumed that it follows the hydrogen distribution but with a factor 10%.

We use two recent models of gas distribution relying on data, and compute the diffuse flux for both. The differences between the results will contribute to the systematics of $\phi_\gamma(E_\gamma, n)$.

The first model considered, labelled as model A in the following and developed by [10], captures the large-scale characteristics of the distribution without attempting to describe finer details. Thus, this distribution is axially and up-down symmetric neglecting features like the spiral arms and the disk wrap of our galaxy. In this model, the gas is concentrated along the galactic plane with a thickness increasing with the distance to the galactic centre $r$. The column density averaged on the galactic longitude $l$, is reported in Fig. 3 by the grey lines for atomic (plain) and molecular (dashed) Hydrogen, as function of the galactic latitude $b$.

The second model of interstellar gas distribution, referred as model B hereafter, was developed by [11]. This is a more detailed description with the modelisation of the spiral arms as well as the wrapping of the disk, which is useful for studying cosmic ray propagation and traces the structure of the Galaxy. The full parametrization of the model can be found in [11]. Moreover, as for model A, the surface density averaged on the galactic longitude $l$ is also shown in Fig. 3 by the blue lines.

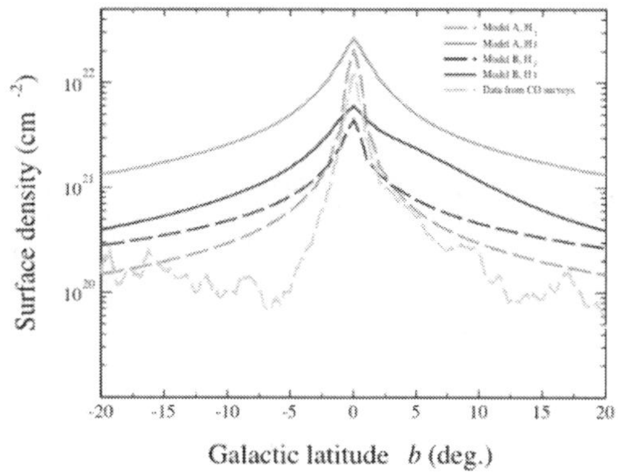


**Figure 3.** The surface density average over Galactic longitude as a function of Galactic latitude for atomic and molecular hydrogen, for model A [10] in grey and in blue for model B [11]. The surface density of molecular Hydrogen as obtained from the composite survey of [12] is shown by the orange line.

When summing the surface density of $H_I$ and $H_{II}$, we observe a difference of the order of a few units of $10^{22}\text{cm}^{-2}$ in the surface density averaged over the galactic longitude. Theses differences will be contributing in the systematics when calculating the photon flux in section 5.

## 4 Production of UHE Photons

Neutral pions $\pi^0$ are among the secondary particles produced following the interaction of the UHECRs with the gas. They play a key role as their most probable decay mode is two photons:

$$\pi^0 \rightarrow 2\gamma$$

The production of secondaries such as $K, \rho, \eta$, decaying into $\pi^0$ also occurs during the interaction, resulting in an additional flux of photons.

To obtain the inelastic cross sections $\sigma_{i,j}(E_{CR})$ along with the energy spectra of UHE photons $\frac{dN_{i,j}}{dE_\gamma}(E_{CR}, E_\gamma)$, we use Cosmic Ray Monte Carlo (CRMC) [13]. This package gives access to several cosmic ray event generators that model the production of secondary particles following a hadronic interaction. It also allows the follow-up of the decays of the secondary particles that are likely to produce photons. In this work, we chose the hadronic interaction model EPOS-LHC [6].

Using CRMC, we simulate 100,000 collisions for each couple $(i, j)$ of UHECR and gas element, for different energies of UHECR ranging from $10^{17}$ to $10^{20}$ eV (in steps of $\Delta\log_{10}(E_{CR}) = 0.5$). The spectra of the photon energies are then produced for each cosmic ray energy,

 https://doi.org/10.1051/epjconf/202328305004

with a bin width of $\Delta\log_{10}(E_\gamma) = 0.05$.

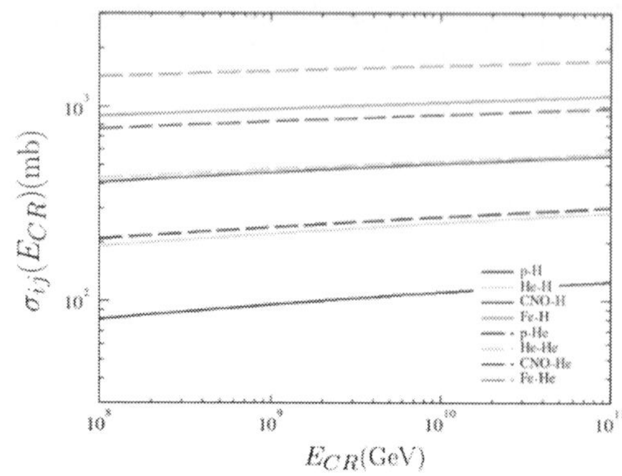


**Figure 4.** The inelastic cross sections $\sigma_{i,j}(E_{CR})$ obtained with CRMC, as function of the cosmic ray energy, for each couple (i,j) of cosmic ray and gas element.

The inelastic cross sections given by CRMC are reported in Fig. 4 as function of the cosmic ray energies. They are of the order of a couple of hundreds of mb, up to thousands for those involving heavier elements. The photon yields $\frac{dN_{i,j}}{dE_\gamma}(E_{CR}, E_\gamma)$ are shown in Fig. 5 for three different comic ray energies ($10^{18}$, $10^{19}$ and $10^{20}$ eV). One can note that, as expected, for a fixed photon energy, the photon yields increase with the cosmic ray energy. On the other hand, for a fixed cosmic ray energy, the yields seem to depend on the mass. Indeed, during the interaction, only a single nucleon intervenes. The energy per nucleon being divided by the atomic number A, the number of photons produced is bigger for higher cosmic ray energies and lighter masses.

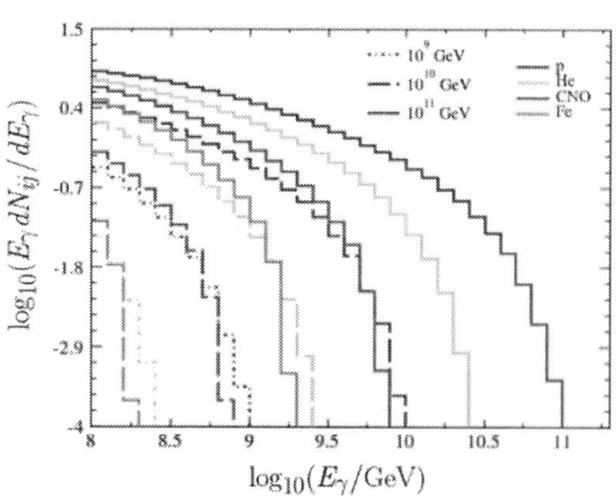


**Figure 5.** The photon yields $\frac{dN_{i,j}}{dE_\gamma}(E_{CR}, E_\gamma)$ obtained with CRMC, for different cosmic ray energies and UHECR masses.

## 5 Diffuse Flux of UHE Photons

By combining the various ingredients introduced in the previous sections in Eq. 2, then Eq. 1, we obtain the diffuse flux of UHE photons. The results for both models of interstellar gas distribution are reported in Fig. 6 for model A and in Fig. 7 for model B. As expected, the flux is concentrated along the galactic plane, reaching $\simeq 8.7.10^{-2}$ km$^{-2}$yr$^{-1}$sr$^{-1}$ when averaged over $|b| < 5°$ in case of model A ($\simeq 3.2.10^{-2}$ km$^{-2}$yr$^{-1}$sr$^{-1}$ for model B).

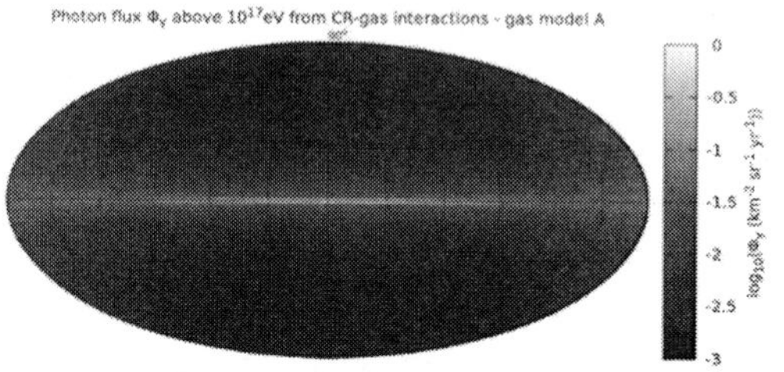


**Figure 6.** The diffuse flux UHE photons with $E_\gamma > 10^{17}$eV in Galactic coordinates, for model A.

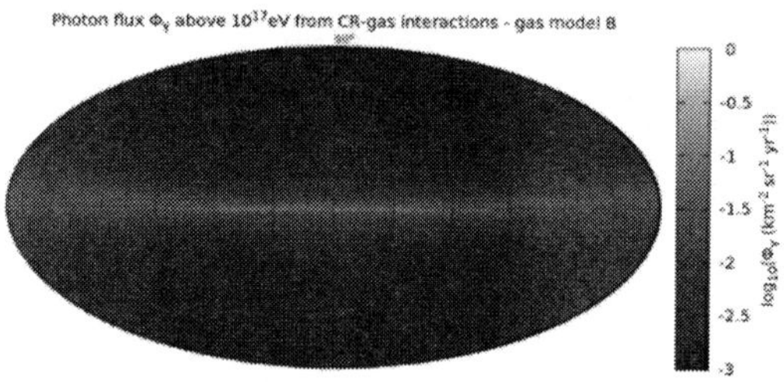


**Figure 7.** The diffuse flux UHE photons with $E_\gamma > 10^{17}$eV in Galactic coordinates, for model B.

The energy spectrum has a similar slope as the UHECR spectrum, but is shifted a decade down in energy. When averaged over a 5° band around the galactic plane, we find that its value is about $10^{-5}$ that of the UHECRs when considering energy thresholds between $10^{17}$ to $10^{18}$ eV. For higher thresholds however, the spectrum of UHE photons becomes steeper, and this ratio decreases to $10^{-6}$ for thresholds above $10^{19}$ eV, before dropping sharply at highest energies.

$$\psi_\gamma(l) = \frac{1}{2.sin(5°)} \int dE_\gamma \int \mathrm{dsin(b)} \int \mathbf{dn'}\, \mathrm{f}_{\theta_0}(\mathbf{n}, \mathbf{n'})\, \phi_\gamma(\mathrm{E}_\gamma, \mathbf{n}) \quad (3)$$

The calculated photon fluxes can be compared to searches for UHE photons fluxes above $10^{17}$ eV. The directional limits reported in [14], for energies ranging from $10^{17.3}$ to $10^{18.5}$ eV, come from a search for point-like sources in the exposed sky of the Pierre Auger Observatory (zenith angles below 60°). These limits averaged over a 5° band around the galactic plane are reported as the red line in Fig. 8, as function of the galactic longitude $l$. To compare our work to these results, we convert our directional fluxes into a collection of point-like sources detected through the point-spread function of the Pierre Auger Observatory $f_{\theta_0}$. This is equivalent to applying a Gaussian filter on the sphere with an angular scale of $\theta_0 = 1°$, as seen in Eq. 3. The integration range of the photon energy is the one mentioned before, and $b$ ranges from -5 to 5°. The limits in this work were computed with the EPOS-LHC hadronic interaction model. However, the choice of the hadronic interaction model when creating the photons energy spectra with CRMC only intervenes at the second order in the computation, thus it does not impact the results. They are reported for model A and B by the grey and blue bands respectively, and are about 3 orders of magnitude below the limits reported in [14]. The bands depict the systematics coming both from those in the UHECRs sprectrum and in the different hadronic interaction models used in the determination of the mass composition. The propagation of uncertainties can be found in [1]. One can conclude easily that these limits cannot be reached with current observation capabilities. Although these limits could be improved by a factor 2 or 3 by collecting more data at the Pierre Auger Observatory, the conclusion would be the same. Moreover, the limits reported here are for a $E_\gamma^{-2}$ photon spectrum, the difference would be higher for steeper spectra.

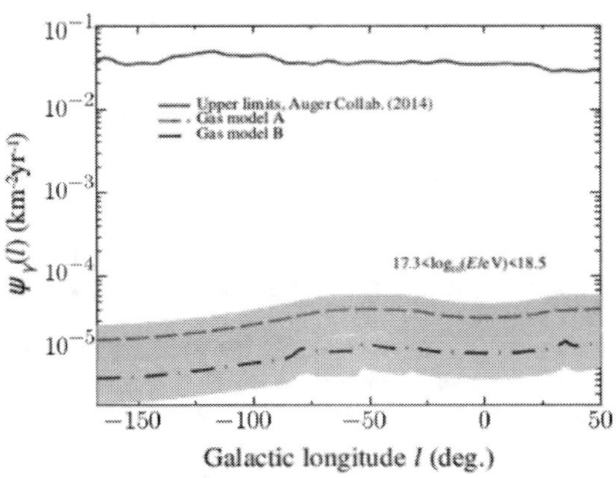


**Figure 8.** The directional photon fluxes averaged over an 5°-band along the galactic plane, when converting the fluxes into a collection of point-like sources detected through the point-spread function of the Auger Observatory. The energy range considered ranges from $10^{17.3}$ to $10^{18.5}$ eV. Dashed: gas model A. Dotted-dashed: gas model B. Continuous red line: upper limits at 95% confidence level obtained at the Pierre Auger Observatory [14].

The UHE photons fluxes obtained can also be compared to other searches for a diffuse flux above $10^{17}$ eV, performed by several experiments like EAS-MSU [15], KASCADE-Grande [16], Telescope Array [17], and the Pierre Auger Collaboration [18]. The upper limits for each of them are reported of Fig. 9 by an orange dot, a black dot, a pink triangle and a blue square respectively, while our results correspond to the blue and grey bands. One must note that, to compare to upper limits on the diffuse fluxes illuminating the whole field of view of the observatory considered, it is needed to calibrate the expected fluxes on the whole sphere, as showed in Eq. 4. The limits obtained by the different observatories appear to be between 2.5 to 3 orders of magnitude above our expectations for energy thresholds between $10^{17}$ and $10^{18}$ eV, and even higher for larger thresholds.

$$\Phi_\gamma(E_\gamma) = \frac{1}{4\pi}\int_{E_\gamma}^{\infty} \mathrm{d}E'_\gamma \int_{4\pi} \mathrm{d}\mathbf{n}\,\phi_\gamma(E'_\gamma, \mathbf{n}) \qquad (4)$$

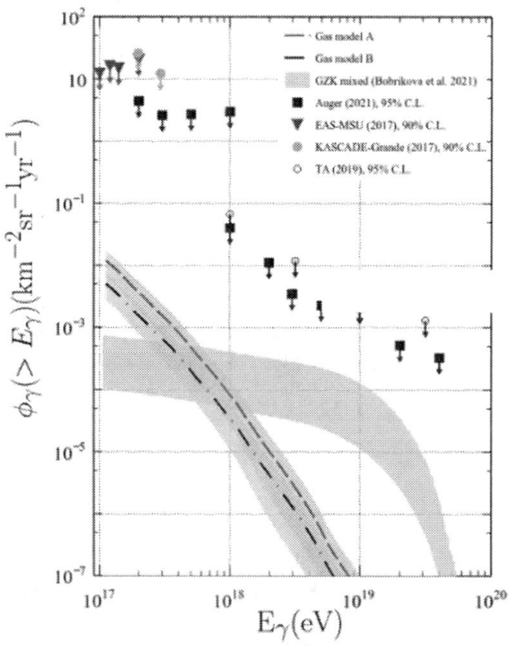


**Figure 9.** Upper limits on diffuse photon fluxes reported by [15] (pink triangles, EAS-MSU), [16](filled orange circles, KASCADE-Grande), [17](open circles, Telescope Array), and [18](dark blue squares, Auger). Expected fluxes from UHECRs interactions with the gas distribution in the Galaxy are shown as the grey dashed line (model A) and grey dashed- dotted line (model B), as well as from UHECRs interac tions with background photon fields (GZK mixed) estimated by [19].

In Fig. 9, the orange band represents the expected GZK photon flux reported in [19], a cosmogenic photon flux arising from the interaction of UHECRs with the photon fields present the universe. This flux depends on the nature of the UHECR and it is reported here for a mixed composition fitting the Pierre Auger Observatory data, namely the measured energy spectrum and mass composition. The systematic uncertainties arising from these measurements lead to the orange band.

For energy thresholds between $10^{17}$ and $10^{18}$ eV, we expect the cosmogenic flux originating from the interactions of UHECRs with the interstellar gas to be the dominant one, while the GZK takes over for higher thresholds.

## 6 Implication on SHDM Searches

The region between $10^{17}$ to $10^{18}$ eV, where the flux computed in this work is dominant, is of interest. Indeed, it can been seen as a floor that could prevent probing sources, in particular in the galactic disk, or indications on the presence of super-heavy darkm matter (SHDM) produced in the early universe and decaying today generating a photon flux $\phi_\gamma^{DM}$ in the direction of the galactic centre. In the field of SHDM searches, a photon flux in this peculiar direction has important implications. Depending on the mass of the super heavy particle $M_X$, the photon flux can be translated into a ceiling for the life-time $\tau_X$ of this particle.

The production of UHE photons following the decay of SHMD particles can occur if these particles have a long enough life-time. This is possible when they are protected from decays in the pertubative domain, but disintegrate by what is called instanton non-pertubative effects [20]. This mechanism provides metastable particles of life-time $\tau_X$ which can produce secondary particles such as photons. The expected photon flux from these decays can be obtained by integrating the position-dependent emission rate per unit volume and energy along the direction **n** as in [21]:

$$\phi_\gamma^{DM}(E_\gamma, \mathbf{n}) = \frac{1}{4\pi M_X \tau_X} \frac{dN}{dE_\gamma} \int_0^\infty \mathrm{d}s\, \rho_{DM}(\mathbf{x} + s\mathbf{n}) \quad (5)$$

In Eq. 5, $\rho_{DM}$ is the energy-density profile of dark matter in the galaxy, and $\frac{dN}{dE_\gamma}$ represents the energy spectrum of UHE photons produced in the decay which depends on the hadronization process. In this work we considered the Navarro-Frenk-White profile found in [22], normalized so that $\rho(\odot) = 0.3$ GeV.$cm^{-3}$. In [23], it is shown that the spectra of the final state particles, varies as $E^{-1.9}$, and we used this energy dependence in the following.

It is possible to infer constraints in the $(M_X, \tau_X)$ plane from the limits on UHE photon fluxes by requiring that $\phi_\gamma^{DM}(E)$, the flux presented in Eq. 5 averaged over all directions **n**, is smaller than the limits. Indeed, for an upper limit at a certain energy threshold, one can scan the value of $M_X$ in order to obtain a lower limit on the life-time $\tau_X$. By repeating the procedure on all energy thresholds, we obtained a collection of excluded regions of $(M_X, \tau_X)$ whose sum is reported as the red area on Fig. 10.

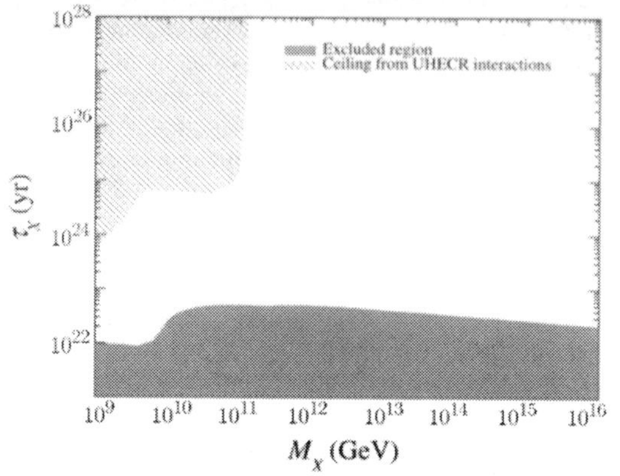


**Figure 10.** Allowed region of mass and life-time of SHDM particles decaying into standard-model ones. The filled red region is excluded from the upper limits in UHE photon fluxes. The hatched one corresponds to the ceiling region discussed in section 5

Finally, by scanning $M_X$ we can also determine the corresponding $\tau_X$ for which $\phi_\gamma^{DM}(E_\gamma) < \phi_\gamma(E_\gamma)$, meaning that the flux originating from dark matter is overwhelmed by the cosmogenic flux. This results in a ceiling region showed in Fig. 10 as the pink hatched area. The ceiling affects masses up to $10^{11}$ GeV, which corresponds to the cut-off in the cosmogenic photon fluxes. While masses below $10^{10}$ GeV are constrained by the photon flux discussed in this work, for masses above this threshold the GZK expected photon flux from [19] takes over. In the end, a large region of the plane remains unaffected by the ceiling.

## 7 Conclusion and Outlooks

The expected flux of UHE photons above $10^{17}$ produced by the interactions of UHECRs with the interstellar gas amounts to $1.1 \times 10^{-2}$ /km$^2$/yr/sr (a few degrees around the galactic plane). It is the dominant cosmogenic one in the energy thresholds ranging from $10^{17}$ to about $10^{18}$ eV. Although its detection remains out of reach for current observatories, it is relevant for searches on SHDM in the direction of the galactic centre, as it sets a floor below which other signals would be overwhelmed. Indeed, the results set a floor for a ceiling region for the life-time of SHDM particles that seems to affect masses up to $10^{11}$ GeV.

A similar interesting study is ongoing at a lower threshold in energy of $10^{15}$eV, with other neutral messengers such as neutrinos produced by charged pions decays. In this case, the abundances of nuclei are taken from the GSF [9] mentioned in section 2. Moreover, when decreasing the energy threshold, one has to account for the anisotropy of UHECRs sources and add the contribution of galactic sources of cosmic rays. As for now, the flux has been computed with the assumtpion of the isotropy of UHECRs and appears to be below current upper limits on neutrino fluxes.

## References

[1] C. Bérat, C. Bleve, O. Deligny, F. Montanet, P. Savina, Z. Torrès, The Astrophysical Journal **929**, 55 (2022)
[2] Sangjin Lee, Physical Review D **58**, 043004 (1998)
[3] Alexander Aab et al., The Astrophysical Journal **891**, 142 (2020)
[4] Alexander Aab et al., The European Physical Journal C **81**, 996 (2021)
[5] J. Bellido, PoS **ICRC2017**, 506 (2018)
[6] T. Pierog, I. Karpenko, J.M. Katzy, E. Yatsenko, K. Werner, Physical Review C **92** 034906 (2015)
[7] F. Riehn, H.P. Dembinski, R. Engel, A. Fedynitch, T. Gaisser, T. Stanev, PoS **ICRC2017**, 301 (2017)
[8] S. Ostapchenko, *EPJ Web of Conferences* (EDP Sciences,2013) 02001
[9] H. Dembinski, R. Engel, A. Fedynitch, T. Gaisser, F. Riehn, T. Stanev, PoS **ICRC 2017**, 533 (2017)
[10] P. Lipari, S. Vernetto, Physical Review D **98** 043003 (2018)
[11] J. Guđlaugur, T.A. Porter, I.V. Moskalenko, The Astrophysical Journal **856**, 45 (2018)
[12] T.M. Dame, D. Hartmann, P. Thaddeus, The Astrophysical Journal **547**, 792 (2001)
[13] R. Ulrich, T. Pierog, C. Baus, *Cosmic Ray Monte Carlo Package, CRMC* (2021) source
[14] A. Aab et al., The Astrophysical Journal Letters **789**, 160 (2014)
[15] Y.A. Fomin, N.N. Kalmykov, I.S. Karpikov, G.V. Kulikov, M.Y. Kuznetsov, G.I. Rubtsov, V.P. Sulakov, S.V. Troitsky, Physical Review D **95**, 123011 (2017)
[16] W.D. Apel et al., The Astrophysical Journal **848**, 1 (2017)
[17] R.U. Abbasi et al., Astroparticle Physics **110**, 8–14 (2019)
[18] P. Savina, PoS **ICRC2021**, 373 (2021)
[19] A. Bobrikova, M. Niechciol, M. Risse, P. Ruehl, PoS **ICRC2021**, 449 (2021)
[20] V.A. Kuzmin, V.A. Rubakov, Physics of Atomic Nuclei **61**, 1028 (1998)
[21] R. Aloisio, V. Berezinsky, M. Kachelriess, Physical Review D **74**, 023516 (2006)
[22] J.F. Navarro, C.S. Frenk, S.D.M. White, The Astrophysical Journal **462**, 563 (1996)
[23] R. Aloisio, V. Berezinsky, M. Kachelriess, Nuclear Physics B Proceedings Supplements **136**, 319 (2004)

# A Study of Modified Characteristics of Hadronic Interactions

*Jiri* Blazek[1,*], *Jan* Ebr[1], *Jakub* Vicha[1], *Tanguy* Pierog[2], and *Petr* Travnicek[1]

[1]FZU Prague, Czech Republic
[2]Karlsruhe Institute of Technology, Germany

**Abstract.** We have implemented ad-hoc modifications to the CORSIKA Monte-Carlo generator which allow us to simultaneously adjust the multiplicity, elasticity and cross-section of hadronic interactions with respect to the predictions of the Sibyll 2.3d interaction model, in order to assess whether a reasonable combination of changes (that is not excluded by current experimental data) could alleviate the observed tension between the model predictions and observed features of extensive air showers induced by ultra-high energy cosmic rays (UHECR). Previously, we have studied the effects of such changes on proton-initiated showers. Because a multitude of experimental data suggest that the primary composition of the UHECR is mixed, we have expanded the modification procedure to include nuclear projectiles in a consistent way based on the superposition model, in a similar manner as was used in the previous studies carried out using one-dimensional simulation methods. As we are using a fully three-dimensional approach, we can quantify the effects of the changes on both longitudinal and lateral features of the showers. With the inclusion of nuclear projectiles, we can study the impact of the changes on observable quantities for realistic primary beams as well as on the determination of the primary composition from data under the assumption of the modified hadronic interactions.

## 1 Introduction - The Muon Puzzle

The muon puzzle in the physics of ultra-high energy of cosmic-rays (UHECR, energy above $10^{18}$ eV) is a long-standing and intriguing problem that has challenged the current understanding of primary particles and their interactions with the Earth's atmosphere [1]. In the UHECR region, there exists a well-established tension between the mass composition derived from fluorescence measurements using the maximum of the shower development $X_{\max}$ and mass composition inferred from ground-based observational techniques. More specifically, measurements at Earth's surface observe a number of muons [2], [3], [4], [5] which is larger than the predictions of various high-energy hadronic interaction models when considering the composition derived using $X_{\max}$, leading to an interpretation of heavier composition compared to fluorescence techniques. This discrepancy is supported by measurements by various other experiments, and there is evidence that it gets more significant with the rise of energy [6].

The predictions of hadronic interaction models, such as Sibyll 2.3d [7], QGSJET II-04 [8] and EPOS-LHC [9], are based on data coming from collider experiments. These measurements of cross sections, hadron production spectra etc. are then extrapolated to the energy and centrality region of ultra-high energy cosmic rays. A thorough overview of the relation of different collider observables to the features of extensive atmospheric air showers is given in [1]. Naturally, there is a systematic uncertainty associated with the extrapolation. In this work, we will investigate whether modifications applied to features of hadronic interactions - cross section, inelasticity and multiplicity, could possibly lead to the reconcilement of the models' predictions and data. It should be stressed that any such introduced modifications are purely ad-hoc, with no basis in hadronization physics, and should serve only as a guidance to the model builders. We follow the approach introduced by Ulrich et al. in [10] and we build upon our previous result [11] where we implemented a set of conservative modifications for a proton primary. In this work, we introduce also a self-consistent approach to modifying interactions of heavier nuclei and we more thoroughly explore the generated phase space of observable features of extensive air showers such as the $X_{\max}$ and the muon signal on ground, $N_{\mu}$.

## 2 Ad-hoc modifications of hadronic interactions

Following [10], we consider energy-dependent changes to the number and properties of particles produced by a hadronic interaction generator, in our case Sibyll 2.3d, and to the cross-section of the interactions. We utilize this approach, rather than changing the intrinsic parameters of the model, in order to more fully explore the available phase space. The choice of Sibyll 2.3d, which uses a superposition model for general nucleon-nucleon interactions, makes this procedure straightforward both on conceptual and technical levels. We make use of the CORSIKA pack-

*e-mail: blazekj@fzu.cz

age, which allows us to obtain also information about the lateral features of an extensive air shower.

Concretely, we recalculate the following set of three parameters of a hadronic interaction:

- The hadronic particle cross-section, $\sigma_{\text{prod}}$.
- The elasticity, defined as $\kappa_{\text{el}} = E_{\text{leading}}/E_{\text{tot}}$, where $E_{\text{leading}}$ is the energy of leading particle in the lab system and $E_{\text{tot}}$ is the energy of the incoming particle in the interaction.
- The secondary multiplicity, defined as the total number of particles escaping the interaction.

The modifications are implemented in such a way that obeys all the expected symmetries - most notably the energy, momentum, charge and isospin are conserved. The parameters are modified concurrently. We have already demonstrated in [11] that applying two simultaneous modifications has a small but non-negligible effect when compared with super-imposing the respective changes.

Recent studies have shown that the muon puzzle could be resolved by various mechanisms in which the probabilities to produce particles of different types are affected – [12], [13], [14]. We could have simulated this effect by varying the $\pi^0/\pi^{\pm}$ ratio. This would however likely cover our entire phase space, and wouldn't allow us to explore the effect of the other modifications in detail. We thus chose a constrained approach, featuring three modifications with constraints from collider measurements.

The modification is applied by calculating an energy dependent factor and multiplying the respective parameter.

$$f(E, f_{19}) = 1 + (f_{19} - 1) \cdot F(E) \tag{1}$$

where $F(E) = 0$ below some threshold energy $E_{\text{thr}}$ and otherwise

$$F(E) = \frac{\log_{10}(E/E_{\text{thr}})}{\log_{10}(10\ \text{EeV}/E_{\text{thr}})}. \tag{2}$$

The values of $E_{\text{thr}}$ can in principle be different for each modified parameter. The functional form of the scaling was chosen such that it represents the increasing uncertainty of the model's prediction in a straightforward linear fashion. Ideally, we would like to start within $\pm 3\ \sigma$ of a known collider measurement at the respective $E_{\text{thr}}$ and extrapolate to UHECR energies. We discuss the choice of the threshold energies more extensively in [11], here we will just state the outcomes:

- For cross section, we set $E_{\text{thr}} = 10^{16}$ eV, corresponding to $f_{19} \subset (0.8, 1.2)$
- For multiplicity, we set $E_{\text{thr}} = 10^{15}$ eV, corresponding to $f_{19} \subset (0.6, 1.7)$
- For elasticity, we set $E_{\text{thr}} = 10^{14}$ eV, corresponding to $f_{19} \subset (0.6, 1.5)$

Note that, especially for the modifications of the cross section, this approach is much more conservative than previous work.

The technical implementation of the respective modifications is described in detail in [10] and we shall not attempt to reproduce it here.

The modifications are considered to apply to a proton-air (or, in general, nucleon-air) interaction. For nuclei, the factor $f$ is thus calculated from the energy of the interacting particle per nucleon. The Sibyll 2.3d model treats the individual nucleon-air sub-interactions in a nucleus-air collision separately and thus the same procedures for resampling of the produced particles used to adjust the resulting multiplicity and/or inelasticity in proton-air collisions are simply applied to each of the nucleon-air sub-interactions. For cross-section modifications, the precise way of calculating would be to adjust the proton-air cross-section using the factor $f$ (for the appropriate per-nucleon energy), convert it into a modified proton-proton cross-section and then calculate the nucleus-air cross-section from the modified value in the same manner as it is calculated in the model. This is however a complex calculation; moreover, in the implementation of the model in Corsika, the nuclear cross-sections are pre-calculated, making the implementation of this approach cumbersome. To make changes to them on the fly, we use a simpler parametrization based on the observation that in Sibyll 2.3d, both proton and nuclear cross-sections at the relevant energies follow very well a simple power law function $\sigma(E) = CE^B$ (with parameters depending on $A$) and the assumption that the nucleus-air cross-section at energy $E$ is a function of just the proton-air cross-section at $E/A$. Then the modified proton-air cross-section is equal to proton-air cross-section at a different energy

$$\begin{aligned}\sigma_{\text{p}}^{\text{mod}}(E/A) &= f\sigma_{\text{p}}(E/A) = fC_{\text{p}}(E/A)^{B_{\text{p}}} \\ &= C_{\text{p}}(Ef^{-B_{\text{p}}}/A)^{B_{\text{p}}} = \sigma_{\text{p}}(Ef^{-B_{\text{p}}}/A).\end{aligned} \tag{3}$$

The nucleus-air cross-section corresponding to this proton-air cross section is then simply obtained by evaluating the formula at the corresponding energy

$$\begin{aligned}\sigma_{\text{A}}^{\text{mod}}(E) &= \sigma_{\text{A}}(Ef^{-B_{\text{p}}}) = C_{\text{A}}(Ef^{-B_{\text{p}}})^{B_{\text{A}}} \\ &= f^{B_{\text{A}}/B_{\text{p}}}\sigma_{\text{A}}(E);\end{aligned} \tag{4}$$

so that the factor $f$ is only modified by the ratio of the two exponents. Compared to just using the factor $f$ calculated at $E/A$ as it is, this correction introduces a shift no larger than 0.6 g cm$^{-2}$ in interaction depth for the nuclei, energies and $f_{19}$ values in question, from which we can conclude that any further improvements in this calculation would have a negligible impact on the results.

## 3 Simulation setup

For reasons described in the previous section we utilize the Sibyll 2.3d model. We have performed tests with the QgsJet II-04 and Epos-LHC with sparser binning of modification parameters and smaller statistics, using only a proton primary. The choice of a model doesn't have a strong effect on the general behavior of e.g. the dependence of $X_{\text{max}}$ change on $f_{19}$ for the case of a proton primary particle.

The simulated primary beam consists of protons and iron nuclei of equal energy, $10^{18.7}$ eV. We simulate the incoming primary particles having two zenith angles, 0 and 60 degrees. Previously, we also simulated intermediate zenith angles with the aim of observing any emerging zenith dependence and eventually constructing a ground based quantity similar to $S_{38}$ utilized by the Pierre Auger Observatory [15], [16]. Since we did not attempt this approach in this work, we restricted ourselves to only the two extreme zenith angles to save computing time.

We make use of the CORSIKA Monte-Carlo generator [17] in order to capture the lateral features of the shower's development. More precisely, we utilize the option CONEX in CORSIKA, with our modifications that allow for resampling of the hadronic interactions and for steering of the simulation. Particles in the shower cascade are initially treated with the one-dimensional Monte-Carlo CONEX code and every interaction is resampled if its energy is above the respective $E_{\text{thr}}$. The particles are then given over to the CORSIKA code at 300 GeV, which performs a full 3D Monte-Carlo simulation and outputs particle densities at ground level. We set this observational level to 1400 meters above the sea level, in correspondence with the characteristics of the site of the Pierre Auger Observatory. Muons produced within the one-dimensional simulation are handed over to the CORSIKA code immediately at all energies.

Overall, we carry out simulations for proton and iron primaries, two values of zenith angle, 0 and 60 degrees, with three modifications of the cross-section $f_{19}^{\sigma} = (0.8, 1.0, 1.2)$, five modifications of the elasticity $f_{19}^{\text{el}} = (0.6, 0.8, 1.0, 1.2, 1.5)$ and five modifications in multiplicity $f_{19}^{\text{mult}} = (0.6, 0.8, 1.0, 1.3, 1.7)$, covering a total phase space of 300 combinations. We simulate 1000 showers for every such bin.

The large generated libraries of 3-dimensional shower simulations allow for the extraction of many observables, related both to the longitudinal development of the showers and to the numbers, types and spectra of particles at the ground. Here we focus on the maximum of the longitudinal development $X_{\text{max}}$ (which is straightfowardly extracted from the longitudinal shower profile saved by CORSIKA) and the number of muons at ground. To extract this number, we sum the number of particles at ground in rings with a given radius, perpendicular to the shower axis. To again facilitate a better comparison with the data from the Pierre Auger Observatory, we set the radius to 1000 meters, which is the distance where Auger reports their results. To maintain independence of any particular experiment, we use simply the number of muons $N_{\mu}^{1000}$; for a typical detector, the measured signal will be proportional to this number. We track muons down to the threshold energy of 10 MeV. We performed extensive tests in order to verify that the simulations obtained in this hybrid approach are compatible within statistical uncertainties with simulations performed with the full 3D simulation using the standalone CORSIKA treatment, in particular that the shapes of the lateral distribution functions exhibit a good match.

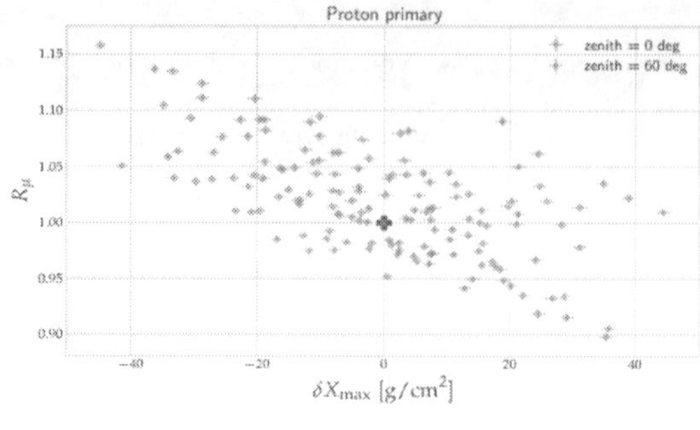


**Figure 1.** Results of simulations in $[\delta X_{\text{max}}, R_{\mu}]$, proton primary.

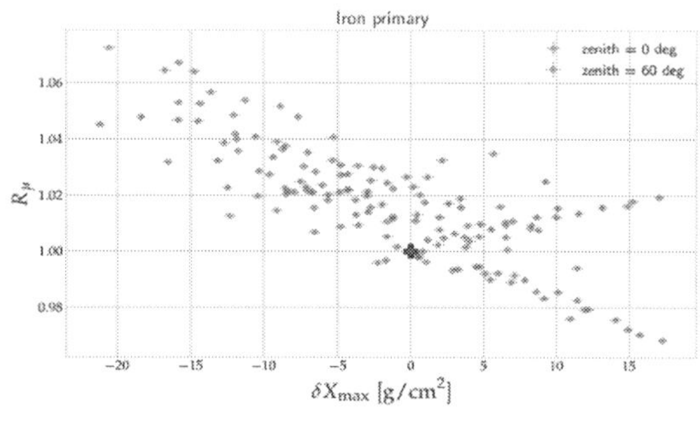


**Figure 2.** Results of simulations in $[\delta X_{\text{max}}, R_{\mu}]$, iron primary.

## 4 Results

Because the exact values of $X_{\text{max}}$ and $N_{\mu}^{1000}$ fluctuate from shower to shower, we consider the following four observables: the mean shift in $X_{\text{max}}$: $\delta X_{\text{max}} = X_{\text{max}} - X_{\text{max}}(\text{ref})$, its standard deviation $\sigma(X_{\text{max}})$, the muon rescaling: $R_{\mu} = N_{\mu}^{1000}/N_{\mu}^{1000}(\text{ref})$ and the change in the standard deviation of the muon number $\sigma(N_{\mu}^{1000})/\sigma(N_{\mu}^{1000}(\text{ref}))$, where the reference values are determined from the unmodified simulations in which $f_{19}^{\sigma} = f_{19}^{\text{el}} = f_{19}^{\text{mult}} = 1$. Note that while absolute $X_{\text{max}}$ and $\sigma(X_{\text{max}})$ values in g/cm$^2$ have a straightforward interpretation, the numbers for $N_{\mu}^{1000}$ are usually not measured directly and thus are best expressed as dimensionless ratio with respect to the reference value.

Each of Figs. 1–8 shows the results of all the simulations, with each point representing one of the possible combinations of $[f_{19}^{\sigma}, f_{19}^{\text{el}}, f_{19}^{\text{mult}}]$ projected on a different plane in the $[\delta X_{\text{max}}, \sigma(X_{\text{max}}), R_{\mu}, \sigma(N_{\mu}^{1000})/\sigma(N_{\mu}^{1000}(\text{ref}))]$ space. In each of the plots, thick crosses indicate the reference values for unmodified simulations $f_{19}^{\sigma} = f_{19}^{\text{el}} = f_{19}^{\text{mult}} = 1$. These plots show both the extent of the possible changes of the observables within the range of the considered modifications and the correlations between different observables. Note that the ranges of the axes differ for proton and iron primaries; the effects of the modifications on iron primaries are universally smaller, as expected from the way the superposition model is applied to generate them.

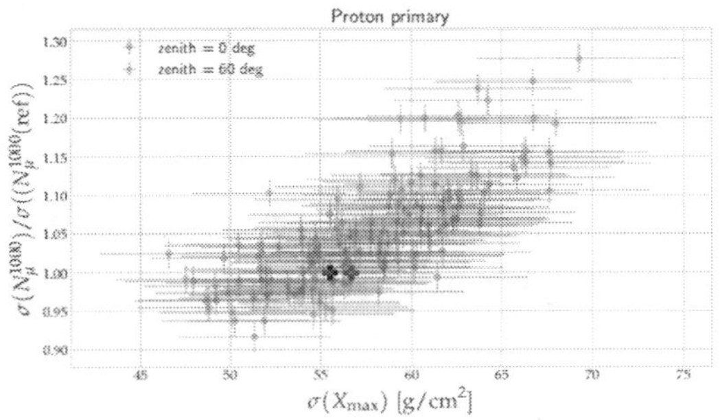


**Figure 3.** Results of simulations in [$\sigma(X_{\max})$, $\sigma(N_\mu^{1000})/\sigma(N_\mu^{1000}(\mathrm{ref}))$], proton primary.

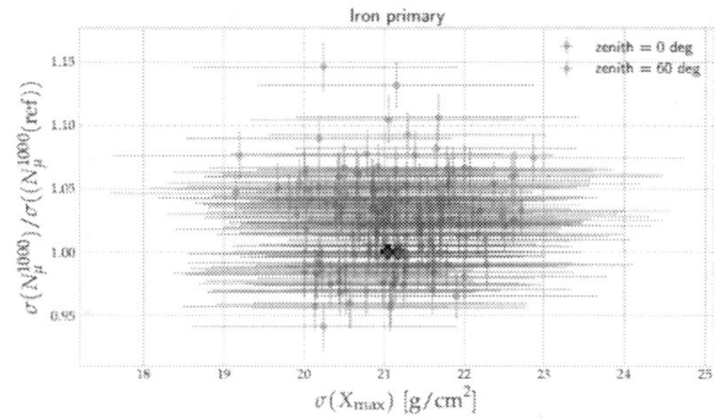


**Figure 6.** Results of simulations in [$\sigma(X_{\max})$, $\sigma(N_\mu^{1000})/\sigma(N_\mu^{1000}(\mathrm{ref}))$], iron primary. Note the difference in scale compared to Fig. 3.

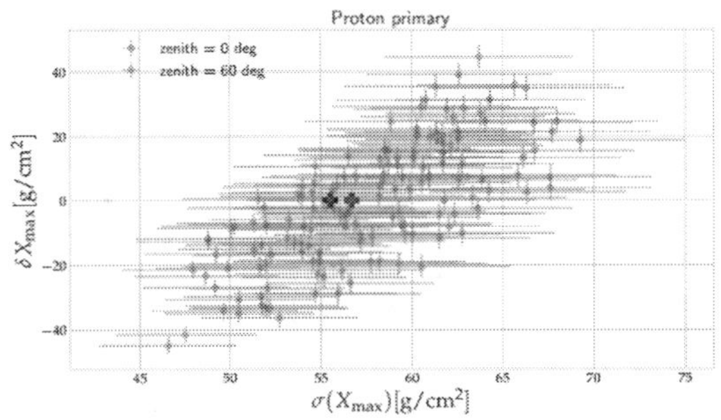


**Figure 4.** Results of simulations in [$\sigma(X_{\max})$, $\delta X_{\max}$], proton primary.

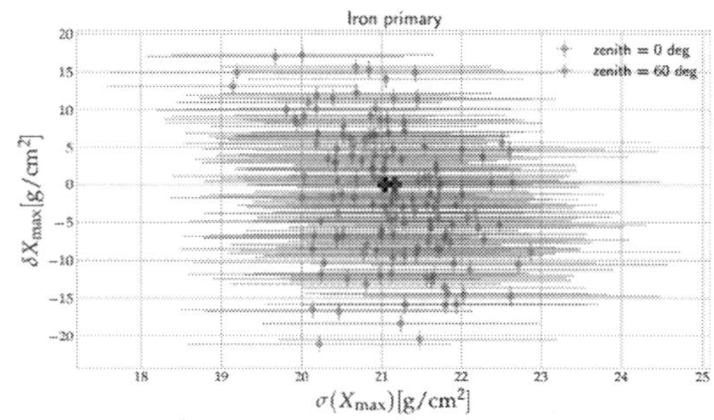


**Figure 7.** Results of simulations in [$\sigma(X_{\max})$, $\delta X_{\max}$], iron primary.

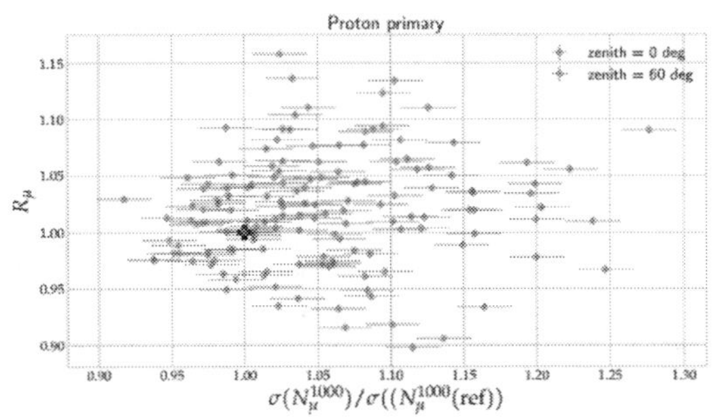


**Figure 5.** Results of simulations in [$\sigma(N_\mu^{1000})/\sigma(N_\mu^{1000}(\mathrm{ref}))$, $R_\mu$], proton primary.

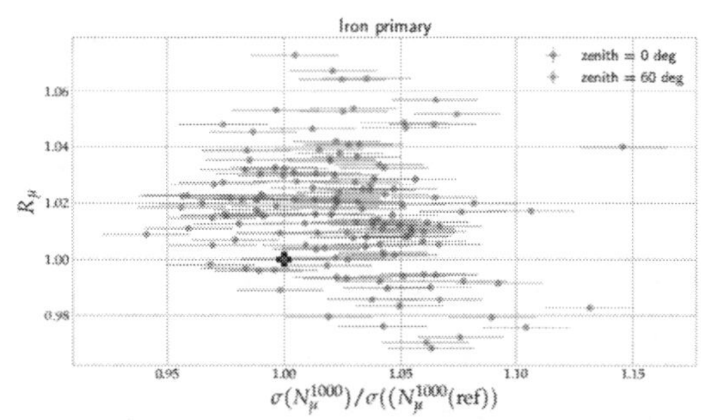


**Figure 8.** Results of simulations in [$\sigma(N_\mu^{1000})/\sigma(N_\mu^{1000}(\mathrm{ref}))$, $R_\mu$], iron primary.

In [11] we showed the effect of each the modifications in detail, here we shall only qualitatively describe their direction in the phase space. Changes in cross-section move the points along the $\delta X_{\max}$ axis, as increasing the cross-section leads to lower $\delta X_{\max}$, i.e. shallower showers, as expected, with only a small effect on the number of muons produced. Changes in multiplicity and elasticity result in anti-correlated shifts in $X_{\max}$ and $R_\mu$ (along the anti-diagonal in out plots), but for multiplicity, the change in $X_{\max}$ is, for the same change in $R_\mu$, much smaller than for elasticity.

Figs. 1 and 2 show that for vertical showers, there is a strong correlation where an increase in the number of muons implies shallower showers. This is particularly interesting in view of the recent results from the Pierre Auger Observatory [18] which indicate that the data is best described if the number of muons and $X_{\max}$ in simulations are increased simultaneously, which is very difficult to achieve with any combination of hereby considered modifications. Similarly interesting is the observation that for inclined showers the effect of the modification on the

https://doi.org/10.1051/epjconf/202328305005

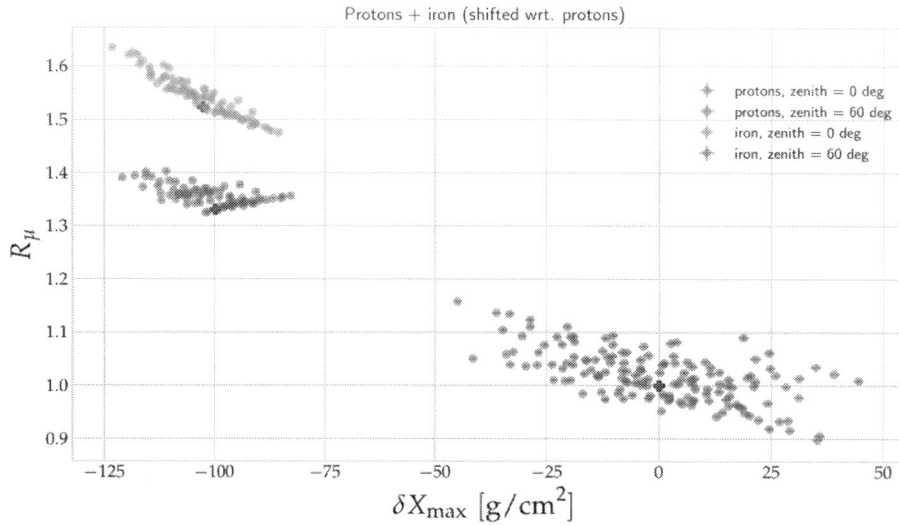


**Figure 9.** Mean muon rescaling $R_\mu$ and mean shift of $X_{max}$ for proton and iron primaries shown in both cases with respect to unmodified simulations for proton. Results are shown for two zenith angles, in each case the reference is taken at the corresponding zenith angle; each point corresponds to a particular combination of $f_{19}^{sigma}$, $f_{19}^{el}$ and $f_{19}^{mult}$.

number of muons is generally smaller. Figs. 3–8 show that the effect of modifications on the variances of the variables is far more pronounced for proton than for iron - in particular $\sigma(X_{max})$ for iron is very difficult to change by any modifications. For protons, the changes in $X_{max}$ itself, $\sigma(X_{max})$ and $\sigma(N_\mu^{1000})$ are well correlated, showing again (similarly to the case of the correlation between $R_\mu$ and $X_{max}$ discussed above) that even making parallel modifications in several significantly different ways still enables the coverage of only a restricted region of the space of possible values of observables. Such correlations may have deep implications when compared with experimental data.

Fig. 9 presents the same data as Figs. 1 and 2 but in a way that allows for additional insight. The values for muon rescaling $R_\mu$ and mean shift of $X_{max}$ for proton and iron primaries for all the applicable combinations of $f_{19}^{sigma}$, $f_{19}^{el}$ and $f_{19}^{mult}$ and two zenith angles are shown with respect to unmodified simulations *for proton* at the corresponding zenith angle. From this plot it is immediately clear that the effects of the modifications on iron showers are significantly smaller as expected. It also shows that even for unmodified simulations (highlighted by thick crosses), the difference in $R_\mu$ between iron and proton primaries is much more pronounced for vertical than for inclined showers; the same is then true for the influence of the modifications on $R_\mu$. Secondly, one can see that for vertical showers, the points are concentrated along a line that almost, but not quite follows the line between proton and iron. Such a trend is expected, as increasing multiplicity and cross-section makes proton act more "iron-like", but at the same time it is clear that there is more to the difference between proton and iron showers than simply the difference in the three macroscopic parameters. This is even more pronounced for inclined showers where the proton points are arranged along an almost horizontal axis. This fact is actually encouraging for the outlooks for determining the properties of hadronic interactions from UHECR data because it shows that there is not a complete degeneracy between the – both unknown – chemical composition and properties of hadronic interactions of the cosmic rays, at least not for the three basic properties of cross-section, multiplicity and elasticity.

## 5 Conclusions and Outlook

We have shown that ad-hoc modifications of macroscopic properties of a model of hadronic interactions, carried out within the currently applicable experimental constraints, allow for substantial changes in air-shower observables, however only for certain combinations of these observables. In particular we have shown that achieving an increase in the number of muons on the ground at 1000 meters from the shower axis simultaneously with an increase in the atmospheric depth of the shower maximum is challenging. We have demonstrated that such ad-hoc changes can be implemented in a consistent way for proton and nuclear primaries without the need to introduce additional parameters, if the superposition model of nuclear interactions is adopted. This opens the door to further studies with mixed primary beams that would provide more realistic description of the real cosmic-ray flux, more suitable for direct comparison with experiments.

In further work, we plan to study the effects of the modifications on the signal at different distances from the shower core and at different energy thresholds for the detection of secondary particles at the ground as well as on muon production depth, frequency of anomalous showers, the correlation between $X_{\max}$ and the ground signal and further observables. We can also determine which observables are the most sensitive to the individual parameters of the hadronic interactions, thus helping to guide the design of future cosmic-ray observatories.

## Acknowledgements

This work is funded by the Czech Science Foundation under the project GACR21 – 02226M.

## References

[1] J. Albrecht et al., Astrophys. Space Sci. **367**, 27 (2022), 2105.06148
[2] A. Aab et al. (Pierre Auger), Phys. Rev. Lett. **117**, 192001 (2016), 1610.08509
[3] R.U. Abbasi et al. (Telescope Array), Phys. Rev. D **98**, 022002 (2018), 1804.03877
[4] A. Aab et al. (Pierre Auger), Phys. Rev. D **91**, 032003 (2015), [Erratum: Phys.Rev.D 91, 059901 (2015)], 1408.1421
[5] A. Aab et al. (Pierre Auger Collaboration), Eur. Phys. J. C **210**, 751 (2020)
[6] D. Soldin (EAS-MSU, IceCube, KASCADE-Grande, NEVOD-DECOR, Pierre Auger, SUGAR, Telescope Array, Yakutsk EAS Array), PoS **ICRC2021**, 349 (2021), 2108.08341
[7] F. Riehn, R. Engel, A. Fedynitch, T.K. Gaisser, T. Stanev, Phys. Rev. D **102**, 063002 (2020), 1912.03300
[8] S. Ostapchenko, Phys. Rev. D **83**, 014018 (2011), 1010.1869
[9] T. Pierog, I. Karpenko, J.M. Katzy, E. Yatsenko, K. Werner, Phys. Rev. C **92**, 034906 (2015), 1306.0121
[10] R. Ulrich, R. Engel, M. Unger, Phys. Rev. D **83**, 054026 (2011), 1010.4310
[11] J. Blazek, J. Vicha, J. Ebr, R. Ulrich, T. Pierog, P. Travnicek, PoS **ICRC2021**, 441 (2021)
[12] S. Baur, H. Dembinski, M. Perlin, T. Pierog, R. Ulrich, K. Werner (2019), 1902.09265
[13] L.A. Anchordoqui, C.G. Canal, F. Kling, S.J. Sciutto, J.F. Soriano, JHEAp **34**, 19 (2022), 2202.03095
[14] J. Manshanden, G. Sigl, M.V. Garzelli, JCAP **02**, 017 (2023), 2208.04266
[15] A. Aab et al. (Pierre Auger), JINST **15**, P10021 (2020), 2007.09035
[16] A. Aab et al. (Pierre Auger), Phys. Rev. D **102**, 062005 (2020), 2008.06486
[17] D. Heck, J. Knapp, J.N. Capdevielle, G. Schatz, T. Thouw (1998)
[18] J. Vícha (Pierre Auger) (2022), 2209.00744

# Neutrons in simulations of extensive air showers

*Ralph* Engel[1,2], *Alfredo* Ferrari[1], *Markus* Roth[1], *Martin* Schimassek[1,3], *David* Schmidt[2,*], and *Darko* Veberic[1]

[1]Institute for Astroparticle Physics, Karlsruhe Institute of Technology, Germany
[2]Institute of Experimental Particle Physics, Karlsruhe Institute of Technology, Germany
[3]CNRS/IN2P3, Université Paris-Saclay, Orsay, France

**Abstract.** We study neutrons produced in simulations of extensive air showers. By using the Monte Carlo simulation package FLUKA, our examination is able to extend from the highest energy neutrons, produced in hadronic interactions, all the way down to thermal energies. The energy spectra, arrival times, and lateral distributions of neutrons at the ground are compared for different primary species, as are the longitudinal profiles of the neutron fluence. Direct comparisons are drawn with the analogous distributions for muons.

## 1 Introduction

Given their electrical neutrality and stability on the time scales in question, neutrons produced in air showers exhibit characteristics distinct from showers' electromagnetic and muonic components. A number of measurements of neutrons in air showers have been performed (for a summary, see [1–3]); however, a lack of reliable predictions as to the properties of the neutron cloud impedes its interpretation.

We use FLUKA [4–6], a Monte Carlo simulation package for particle transport and interaction with matter, to investigate the properties of neutrons in extensive air showers. Air showers induced by proton, iron, and photon primaries ranging from sub-PeV to EeV in energy are covered in these studies. This work is motivated by the planned, ongoing, and existing deployments of scintillator detectors at large-area very high and ultra-high energy cosmic ray air shower arrays.

FLUKA permits detailed simulation of neutron production and propagation in air showers and provides access to the characteristics of the neutron cloud down to even thermal energies, which are inaccessible with other air shower simulation programs. FLUKA has been validated both with high-energy accelerator measurements (for a validation of neutron spectra, see [7]) and cosmic ray measurements at different atmospheric depths and under different conditions (for examples, see [8–12]). For the studies presented here, neutrons were transported down to $10^{-5}$ eV. Earth's atmosphere was described by 100 layers with densities in accordance with the US Standard Atmosphere.

## 2 Energy spectrum and arrival times

The energy spectra of neutrons at sea level for vertical air showers induced by proton, iron, and photon primaries of $5.6 \times 10^{16}$ eV are shown in figure 1. These spectra are normalized such that the integral is equal to the total number of neutrons arriving at the ground.

As evidenced by the nearly identical neutron and antineutron spectra for energies above a few GeV, the highest energy neutrons are produced in baryon, anti-baryon pairs in hadronic interactions. These are the same hadronic interactions as those producing the pions and kaons that go on to decay to muons, and as with muons, the multiplicity of high-energy neutrons is much lower for photon-induced air showers as compared with hadronically-induced showers. The peak in the neutron spectrum between 50 and 150 MeV is the so-called quasi-elastic peak. Below 10 to 20 MeV, neutrons propagate through diffusion. The structures between 0.1 and 10 MeV are due to resonances largely dependent on the target material (in this case, air). The peak at the lowest energies exists for targets with sufficient hydrogen to fully thermalize the neutrons, which was the case for the soil simulated at sea-level.

The extension of the energy spectrum down to low, even thermal, energies stands in sharp contrast to that of muons, which suffer from decay and ionization losses during propagation. At the depth of sea-level, the neutron spectra for proton and iron primaries are strikingly similar, particularly for the highest energy neutrons. The source of this coincidence is further elaborated upon in section 3.

In figure 2, the energy spectrum of neutrons in air showers initiated by protons of $5.6 \times 10^{16}$ eV is broken down into the arrival times at the ground. The arrival time of a particle is defined here as the delay with respect to the time light would require to travel from the point where the primary first interacts to the impact point with the measurement plane. The broad range of neutron energies is reflected by a distribution of arrival times also covering many orders of magnitude. Whereas the bulk of muons arrive at the ground well within 1 μs, the majority of neutrons arrive after 1 μs and as late as milliseconds. This de-

*e-mail: david.schmidt@kit.edu

https://doi.org/10.1051/epjconf/202328305006

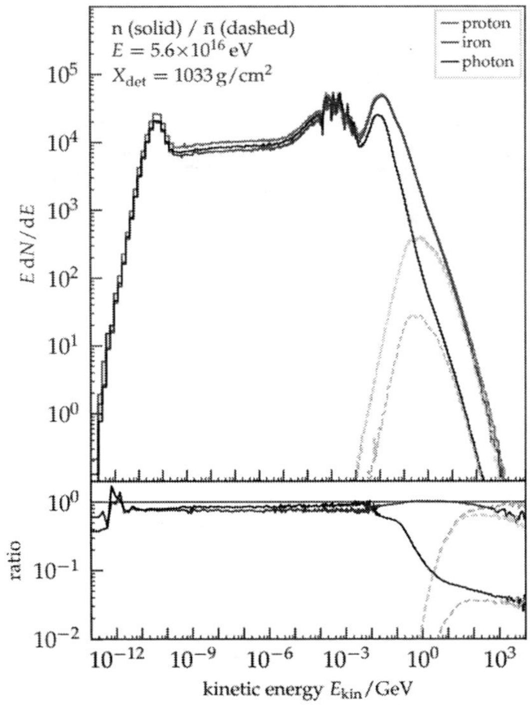


**Figure 1.** Energy spectra of neutrons and antineutrons reaching sea level for vertical air showers induced by proton, iron, and photon primaries of $5.6 \times 10^{16}$ eV. A layer of soil was present at the observation depth in the simulations. The lower panel depicts the ratio of the spectra, where the proton spectra serves as the reference.

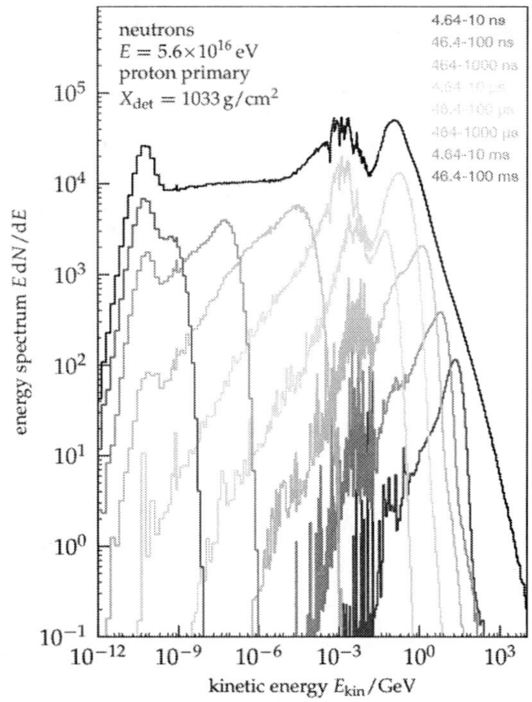


**Figure 2.** Energy spectrum of neutrons for vertical air showers induced by proton primaries of $5.6 \times 10^{16}$ eV. The spectrum is binned in the arrival times of neutrons at the observation depth of sea-level.

layed arrival of the majority of neutrons provides a means of identifying them and differentiating them from muons.

In figure 3, the energy spectra of neutrons and muons arriving at a depth of 675 g/cm$^2$, which is similar to that of the IceTop surface detector array, are juxtaposed for ranges in energy and arrival time representative of those for which surface detector arrays are typically sensitive. Beyond 1 μs, the multiplicity of muons and neutrons are, to first order, comparable. The presence of an ice surface at the observation depth on the neutron energy spectrum appears to have only a very limited impact for the arrival times relevant for detection and only below $\sim$ 1 MeV.

## 3 Two coincidences

When observing the longitudinal profiles of neutrons, two coincidences may be observed that have implications for the scaling of neutron multiplicities with the primary energy and mass. Both coincidences arise due to the interplay between the energy in the hadronic cascade and the degree to which the resulting neutrons are attenuated prior to arrival at the depth of observation.

As a proxy for the longitudinal profile, the fluence for different particle species is calculated. This unitless quantity is defined as the product of the density of a given slice of atmosphere and the sum of the tracklengths of all particles of a given type therein normalized by the column density (i.e. thickness in g/cm$^2$) of the slice.

### 3.1 Energy scaling

The first coincidence is depicted in figure 4, where the longitudinal profiles of the neutron fluence are shown for proton and photon primaries over a range of energies. Each profile is scaled inversely proportional to the primary energy to highlight departure from a linear scaling with energy. Only neutrons with energies greater than 10 MeV were considered when constructing the profiles to narrow the focus on neutrons where detection is possible with typical air shower arrays. The maximum of the longitudinal profile of the neutron fluence scales with energy according to $E^{\beta}$ where $\beta \approx 0.9$, analogous to muons. The increased depth of the maximum for primaries of higher energy results in a smaller degree of attenuation at a fixed observation depth, however. The result is that the number of neutrons reaching ground scales approximately linearly with energy for hadronic primaries at an observation depth of $\sim$ 850g/cm$^2$, which approximately corresponds to the depth of the surface detector arrays of existing ultra-high-energy cosmic ray observatories.

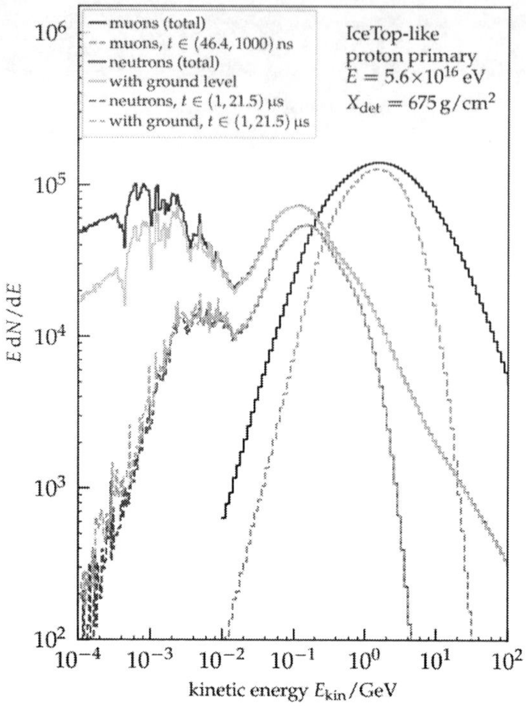


**Figure 3.** Energy spectra of muons and neutrons arriving at the ground for air showers induced by proton primaries of $5.6 \times 10^{16}$ eV. The energy spectra of all particles of each species are shown in addition to the spectra for arrival times representative of those for which surface detector arrays are typically sensitive.

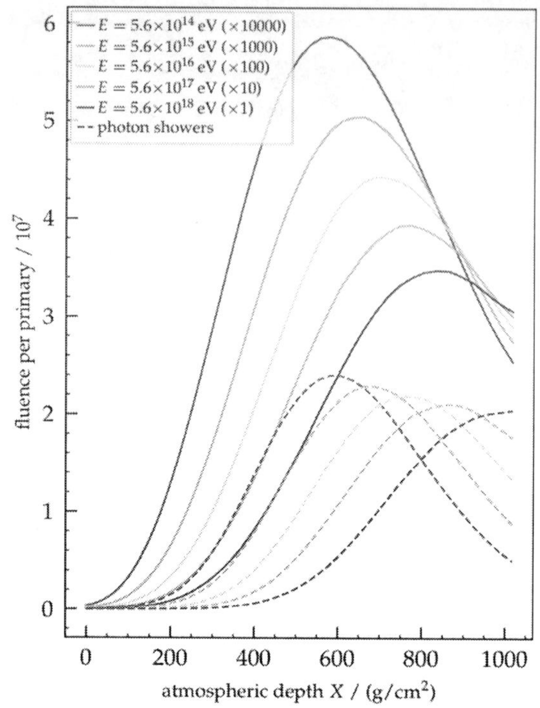


**Figure 4.** Longitudinal profiles of the fluence of neutrons with energies greater than 10 MeV for air showers induced by proton and photon primaries of various energies. Each profile is scaled inversely proportional to the primary energy to highlight departures from a linear scaling with energy.

### 3.2 Mass scaling

The second coincidence is depicted in figure 5, where the evolution of the neutron and muon fluences with atmospheric depth are juxtaposed for different species of primary particle. As before, only neutrons with energies greater than 10 MeV were considered in the calculation of the neutron fluence. The additional energy channeled into the hadronic component for heavier primaries manifests in a larger number of neutrons at the maximum. Due to their shorter removal length (100 g/cm$^2$, as a rule of thumb), however, the number of neutrons is attenuated more quickly than that of muons. While a difference in the muon fluence of $\sim$ 25% remains between proton and iron primaries at an observation depth of $\sim$ 850g/cm$^2$, very little difference in the neutron fluence is observed.

## 4 Lateral distribution

To examine the radial distribution of neutrons, we normalize the integral of the energy spectra above a given energy and normalize the results with the area of the given radial bin. We do the same for muons. The resulting lateral distributions are shown in figure 6 for proton, iron, and photon primaries of $5.6 \times 10^{18}$ eV at an observation depth of 878g/cm$^2$. This primary energy and observation depth are characteristic of ultra-high-energy cosmic ray observatories. The distributions for neutrons are shown with a cut on the energy of 1 MeV and 1 GeV to highlight detectability and crudely differentiate between production mechanisms, respectively.

The lateral distribution of neutrons with energies above 1 MeV appears broader and has a slope less steep than that of muons. Similar to muons, the slope for proton primaries appears slightly steeper than that of iron primaries. An interesting coincidence is that for photon primaries, the total number of neutrons above 1 MeV is approximately equal to the number of muons, whereas for hadronic primaries, it is significantly less.

## 5 Conclusions

The neutrons produced in extensive air showers exhibit energy spectra and arrival time distributions distinct from those of the electromagnetic and muonic shower components. These distinct characteristics and the sheer abundance of neutrons make them of potential interest for air-shower experiments.

Our predictions indicate that the considerable delay in the arrival times of the bulk of neutrons serves as a means of identifying their presence in measurements. We have also shown that an interplay between the energy in the

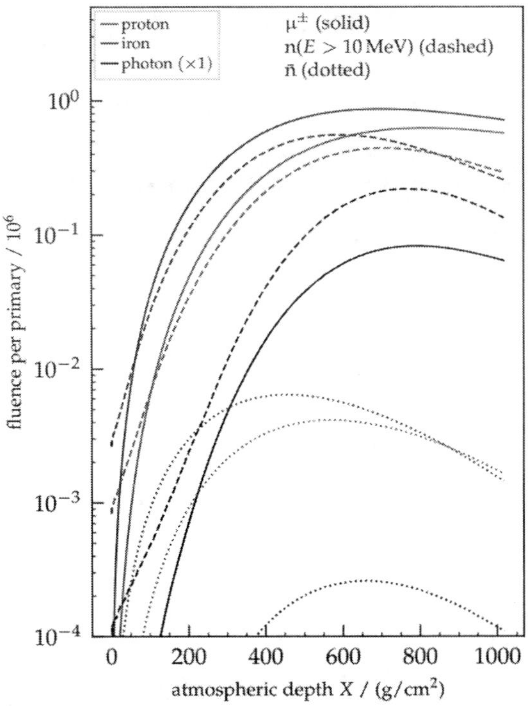


**Figure 5.** Longitudinal profiles of the fluence of muons, neutrons, and antineutrons for air showers induced by proton, iron, and photon primaries of $5.6 \times 10^{16}$ eV. Only neutrons with energies greater than 10 MeV were considered in the calculation of the neutron fluence.

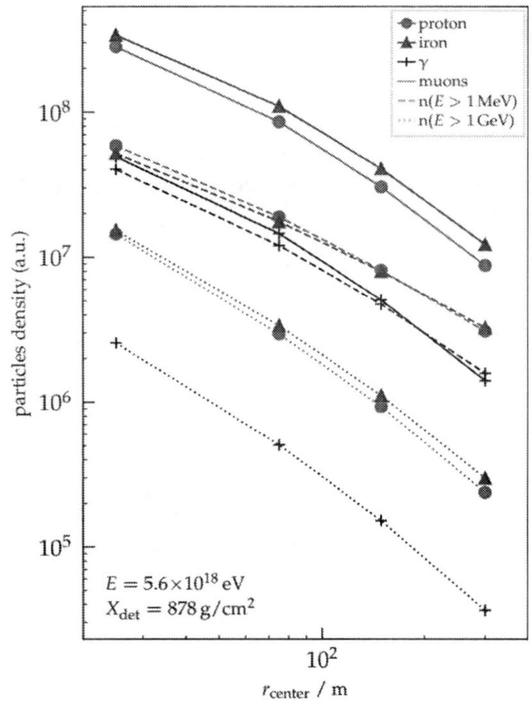


**Figure 6.** Radial dependence of neutron and muon densities at a depth of 878 g/cm$^2$ for vertical air showers initiated by proton, iron, and photon primaries of $5.6 \times 10^{18}$ eV

hadronic component of air showers and attenuative effects results in a linear scaling with the primary energy of the number of potentially detectable neutrons arriving at observation depths typical of ultra-high-energy cosmic ray observatories. An additional coincidence resulting from the interplay between the same two effects also results in a striking similarity of the number of potentially detectable neutrons for different hadronic primaries. A lower abundance of neutrons in photon-induced air showers may provide additional information to differentiate them from hadronically-induced showers.

To verify the predicted characteristics of the neutron cloud with measurements, detailed detector simulations will be necessary.

## Acknowledgments

We would like to acknowledge discussions with our valued colleagues of the Pierre Auger Collaboration and thank Alan Watson, in particular, for bringing the early works of Linsley [13] and Hillas on this subject matter to our attention.

## References

[1] A. Shepetov et al., Eur. Phys. J. Plus **135**, 96 (2020), 1912.13173
[2] Y.V. Stenkin, D.D. Dzhappuev, J.F. Valdes-Galicia, Phys. Atom. Nucl. **70**, 1088 (2007)
[3] A.D. Erlykin, Nucl. Phys. B Proc. Suppl. **175-176**, 330 (2008), hep-ph/0612118
[4] A. Ferrari, P.R. Sala, A. Fasso, J. Ranft, *FLUKA: A multi-particle transport code (Program version 2005)* (2005)
[5] T.T. Böhlen, F. Cerutti, M.P.W. Chin et al., Nucl. Data Sheets **120**, 211 (2014)
[6] www.fluka.org
[7] M. Brugger, P. Cennini, A. Ferrari et al., Nucl. Tech. **168**, 752 (2009)
[8] G. Battistoni, A. Ferrari, T. Montaruli, P.R. Sala, Astropart. Phys. **17**, 477 (2002), hep-ph/0107241
[9] G. Battistoni, S. Muraro, P.R. Sala, A. Ferrari, Nucl. Phys. B Proc. Suppl. **168**, 286 (2007)
[10] P. Zuccon et al., Astropart. Phys. **20**, 221 (2003)
[11] A. Ferrari, M. Pelliccioni, T. Rancati, Radiat. Prot. Dosim **93(2)**, 101 (2001)
[12] N. Combier, A. Claret, P. Laurent et al., IEEE Transactions on Nuclear Science **64**, 614 (2017)
[13] J. Linsley, J. Phys. G **10**, L191 (1984)

https://doi.org/10.1051/epjconf/202328305007

# Probing Lorentz violation at ultra-high energies using air showers

*Marcus* Niechciol[1,*], *Fabian* Duenkel[1], and *Markus* Risse[1]

[1]Center for Particle Physics Siegen, University of Siegen, Walter-Flex-Str. 3, 57072 Siegen, Germany

**Abstract.** In air showers initiated by ultra-high-energy cosmic rays in the Earth's atmosphere, even the secondary particles created in the start-up phase are produced at energies far above those accessible by other means. These high-energy particles can be used to search for New Physics, such as a violation of Lorentz invariance. We focus on isotropic, nonbirefringent Lorentz violation in the photon sector and consider the two cases $\kappa < 0$ and $\kappa > 0$ (i.e., the velocity of photons is larger/smaller than the maximum attainable velocity of standard Dirac fermions). In both cases, processes that are forbidden in the standard, Lorentz-invariant theory ($\kappa = 0$) become allowed, in particular photon decay in the case $\kappa < 0$ and vacuum-Cherenkov radiation for $\kappa > 0$. Implementing these processes into air-shower simulations, we found that the development of an air shower at the highest energies can be significantly impacted, specifically the average atmospheric depth of the shower maximum $\langle X_{\max}\rangle$ and its shower-to-shower fluctuations $\sigma(X_{\max})$. Comparing these simulations to actual measurements, we were able to obtain much stricter bounds on this specific type of LV in the case $\kappa < 0$ than possible with previous methods. We discuss these limits and, in addition, present first results for the case $\kappa > 0$.

## 1 Introduction

In current theories aiming to establish a more fundamental understanding of particle physics beyond the current Standard Model (SM), Lorentz invariance (LI) can be broken (see, e.g., [1, 2]). In the search for the effects of a violation of Lorentz invariance (LV), ultra-high energy (UHE) cosmic rays were used to produce some of the strongest bounds (see [3] for a yearly-updated compilation). In the analyses presented here, isotropic, nonbirefringent LV in the photon sector is implemented through the framework of modified Maxwell theory (Sec. 2). We study the impact of LV on the development of air showers initiated by UHE cosmic rays (Sec. 3). Through the comparison to current air shower measurements, LI can be probed and improved bounds on LV can be set (Sec. 4).

## 2 Theory Background

We use a fairly simple extension of standard quantum electrodynamics (QED). A single term is added to the Langrange density which breaks Lorentz invariance but preserves CPT and gauge invariance [4, 5][1]:

$$\mathcal{L} = \underbrace{-\frac{1}{4}F^{\mu\nu}F_{\mu\nu} + \overline{\psi}\left[\gamma^{\mu}(i\partial_{\mu} - eA_{\mu}) - m\right]\psi}_{\text{standard QED}} \underbrace{-\frac{1}{4}(k_F)_{\mu\nu\rho\sigma}F^{\mu\nu}F^{\rho\sigma}}_{\text{CPT-even LV term}}. \quad (1)$$

*e-mail: niechciol@physik.uni-siegen.de

[1]Note that natural units ($\hbar = c = 1$) and the Minkowski metric $g_{\mu\nu}(x) = \eta_{\mu\nu} = [\text{diag}(+1,-1,-1,-1)]_{\mu\nu}$ are used here.

For isotropic, nonbirefringent LV in the photon sector, the LV coefficient $(k_F)_{\mu\nu\rho\sigma}$ depends only on a single, dimensionless parameter $\kappa \in (-1, 1]$, which is related to $k_F$ by [6]

$$(k_F)^{\lambda}_{\mu\lambda\nu} = \frac{\kappa}{2}\left[\text{diag}(3,1,1,1)\right]_{\mu\nu}. \quad (2)$$

For $\kappa \neq 0$, the LV term allows processes that are kinematically forbidden in the SM. We exploit the fact that in an air shower initiated by a UHE cosmic ray, secondary particles with very high energies are produced, for which these LV processes become relevant. We can then derive bounds on LV using the expected changes in the air-shower development (see Sec. 3).

In the case of $\kappa < 0$, nonstandard photons decay almost immediately into an electron- positron pair if they are above the threshold energy

$$E_{\gamma}^{\text{th}}(\kappa) = 2m_e\sqrt{\frac{1-\kappa}{-2\kappa}} \simeq \frac{2m_e}{\sqrt{-2\kappa}}. \quad (3)$$

The exact decay rate for the process $\tilde{\gamma} \rightarrow e^+ + e^-$ is given by [7]

$$\Gamma_{PhD}(E_\gamma) = \frac{\alpha}{3}\frac{-\kappa}{1-\kappa^2}\sqrt{E_\gamma^2 - \left(E_\gamma^{th}\right)^2}\left(2 + \left(E_\gamma^{th}\right)^2/E_\gamma^2\right) \quad (4)$$

Changes in the decay time of neutral pions also occur [8], but they have a negligible impact on the air-shower observables used here. A previous bound of $\kappa > -9\times10^{-16}$ (98% C.L.) has been derived from observations of TeV $\gamma$-rays [9]. A much stricter bound of $\kappa > -3\times10^{-19}$ (98% C.L.) was obtained using secondary photons from air showers [10].

In the case of $\kappa > 0$, vacuum Cherenkov radiation (VCh) can occur, leading to charged particles with mass $m$ losing their energy quickly by radiating photons if they are above the threshold energy

$$E^{\text{th}}_{\text{VCh}}(\kappa) = m\sqrt{\frac{1+\kappa}{2\kappa}} \simeq \frac{m}{\sqrt{2\kappa}}, \tag{5}$$

The decay rate for this process is given by [9]

$$\Gamma_{VCh}(E) = \alpha E\left(\left(\frac{4}{3}\kappa + \frac{1}{6}\kappa^2\right) + \left(-\frac{3}{2} - \frac{1}{3}\kappa - \frac{23}{24}\kappa^2\right)\frac{m^2}{E^2}\right) \tag{6}$$

VCh photons can inherit a significant fraction of the radiating particle's energy. A previous bound ($\kappa < 6 \times 10^{-20}$ at 98 % C.L.) has been derived from observations of UHE cosmic rays with energies above 100 EeV [9, 11, 12]. In an air shower, mostly electrons and positrons are affected by VCh, as they are the lightest charged particles.

## 3 Simulation Studies

We used the Monte Carlo code CONEX v2r7.50 [13, 14] to simulate extensive air showers. The code was modified to include the LV effects outlined in the preceding section. Large simulation samples have been created for different values of $\kappa$ to analyze the potential impact of LV on the development of air showers. The full MC approach used here is outlined in greater detail in [10, 15]. Simulations were performed using the most up-to-date hadronic interaction model SIBYLL 2.3d [16]. In addition, detailed cross checks with other models (EPOS LHC [17], QGSJET-II-04 [18]) were done. We found that SIBYLL 2.3d gave the most conservative results, hence we focus on this model. An example of the impact of LV on the shower observables $\langle X_{\max}\rangle$ and $\sigma(X_{\max})$ can be seen in Fig. 1. The average depth of the shower maximum $\langle X_{\max}\rangle$ decreases with increasing $|\kappa|$, if the energies are above the respective threshold, while the shower-to-shower fluctuations $\sigma(X_{\max})$ remain largely unaffected. The previous analysis for $\kappa < 0$ [10] only used $\langle X_{\max}\rangle$. This approach was extended to include also information from $\sigma(X_{\max})$ as well as the primary composition, taking into account any allowed combination of protons, helium nuclei, oxygen nuclei and iron nuclei [15]. We apply a similar approach for the first time to the case $\kappa > 0$. [19]. In the following section, we briefly summarize the main results.

## 4 Results

We focus first on the analysis for $\kappa < 0$ [15]. For a given $\kappa$ and primary energy, plotting $\langle X_{\max}\rangle$ against $\sigma(X_{\max})$ for all combinations of the four primary particle types used in this study leads to the "umbrella plots" shown in Fig. 2(a).

We then compare our simulations to measurements of $\langle X_{\max}\rangle$ and $\sigma(X_{\max})$ from the Pierre Auger Observatory [20], taking into account both statistical and systematic uncertainties in the form of two-dimensional confidence intervals. It is checked if there are primary particle compositions that are allowed by both simulations and

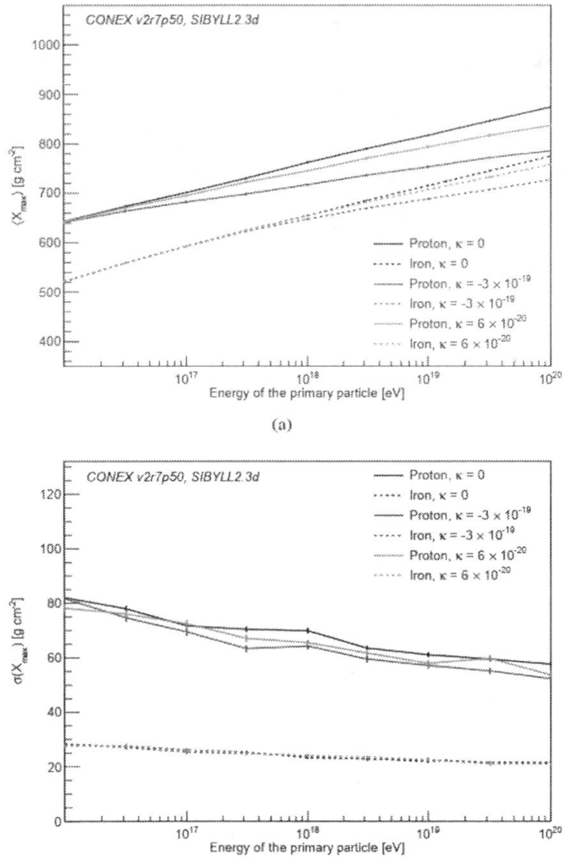


(a)

(b)

**Figure 1.** $\langle X_{\max}\rangle$ and $\sigma(X_{\max})$ as a function of the primary energy for primary protons and iron nuclei for the absence of LV ($\kappa = 0$) and for different vaues of $\kappa$. All lines were obtained from CONEX simulations using the hadronic interaction model SIBYLL 2.3d.

data. If there are no such compositions for a given $\kappa$, then this value of $\kappa$ can be excluded. Using this approach, the bound on $\kappa < 0$ was improved to $\kappa > -6 \times 10^{-21}$ at 98 % C.L. [15].

For $\kappa > 0$, lighter primaries at high energies lose their energy due to VCh even before reaching Earth, restraining possible compositions to higher masses, as visualized in the umbrella plots in Fig. 2(b): with increasing $\kappa$, the umbrellas shrink, since primary particles of increasing mass do not reach Earth anymore. Apart from this major effect, also a shift of the umbrellas can be seen, similar to the shift observed in Fig. 2(a). This is expected, as VCh radiation affecting electrons and positrons—which are the most abundant charged secondary particles in an air shower—is expected to lead to shorter showers in a similar way as photon decay affecting secondary photons leads to shorter showers for the case $\kappa < 0$. We also performed a preliminary comparison to data, taking into account the most common isotopes in primary cosmic rays and following

 https://doi.org/10.1051/epjconf/202328305007

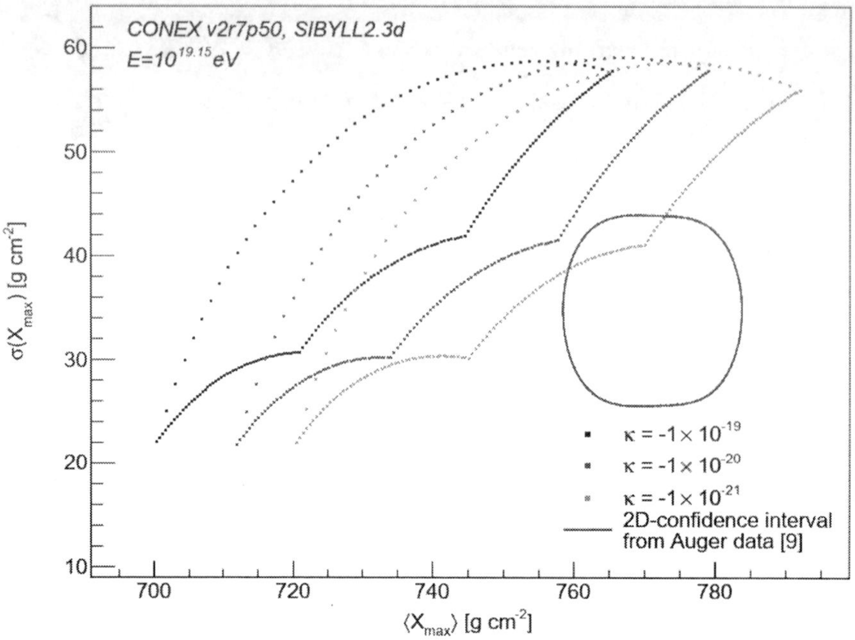


(a)

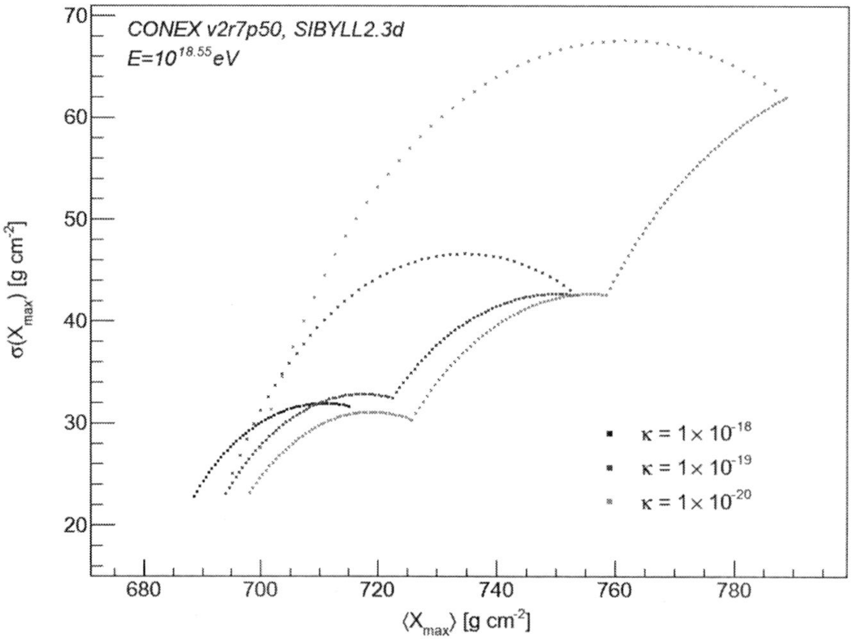


(b)

**Figure 2.** Comparison of combinations of $\langle X_{\max}\rangle$ and $\sigma(X_{\max})$ derived by simulations using CONEX and the hadronic interaction model SIBYLL 2.3d. For a given $\kappa$, the corresponding "umbrella" covers all values allowed by arbitrary combinations of these primaries. In (a), the case $\kappa < 0$ is shown, for different values of $\kappa < 0$ and a primary energy of $10^{19.15}$ eV. In addition, the two-dimensional confidence interval given by measurements from the Pierre Auger Observatory [20] is included. In (b), the case $\kappa > 0$ is shown, for different values of $\kappa > 0$ and a primary energy of $10^{18.55}$ eV. Note that in this case, combinations with particles above the VCh radiation threshold are excluded.

the same logic as for $\kappa < 0$. This comparison yields a preliminary bound of $\kappa < 3 \times 10^{-20}$ at 98 % C.L. confirming and slightly improving the previous bound from [9]. This bound is the first bound on $\kappa > 0$ using air showers and electrons/positrons. Future extensions of this work will also include the number of muons on ground ($N_\mu$) as well as the correlation between $X_{\max}$ and $N_\mu$ and the resulting composition constraints.

## Acknowledgements

The many fruitful discussions with F.R. Klinkhamer are greatly appreciated. We thank T. Pierog for his help in modifying the CONEX source code. This work was supported by the German Research Foundation (DFG Project No. 408049454).

## References

[1] S. Liberati, L. Maccione, Ann. Rev. Nucl. Part. Sci. **59**, 245 (2009), 0906.0681
[2] A. Addazi et al., Prog. Part. Nucl. Phys. **125**, 103948 (2022), 2111.05659
[3] V.A. Kostelecký, N. Russell, Rev. Mod. Phys. **83**, 11 (2011), 0801.0287
[4] S. Chadha, H.B. Nielsen, Nucl. Phys. B **217**, 125 (1983)
[5] V.A. Kostelecký, M. Mewes, Phys. Rev. D **66**, 056005 (2002), hep-ph/0205211
[6] J.S. Díaz, F.R. Klinkhamer, Phys. Rev. D **92**, 025007 (2015), 1504.01324
[7] J.S. Díaz, F.R. Klinkhamer, M. Risse, Phys. Rev. D **94**, 085025 (2016), 1607.02099
[8] F.R. Klinkhamer, Mod. Phys. Lett. A **33**, 1850104 (2018), 1610.03315
[9] F.R. Klinkhamer, M. Schreck, Phys. Rev. D **78**, 085026 (2008), 0809.3217
[10] F.R. Klinkhamer, M. Niechciol, M. Risse, Phys. Rev. D **96**, 116011 (2017), 1710.02507
[11] F.R. Klinkhamer, M. Risse, Phys. Rev. D **77**, 016002 (2008), 0709.2502
[12] F.R. Klinkhamer, M. Risse, Phys. Rev. D **77**, 117901 (2008), 0806.4351
[13] T. Bergmann, R. Engel, D. Heck, N.N. Kalmykov, S. Ostapchenko, T. Pierog, T. Thouw, K. Werner, Astropart. Phys. **26**, 420 (2007), astro-ph/0606564
[14] T. Pierog, M.K. Alekseeva, T. Bergmann, V. Chernatkin, R. Engel, D. Heck, N.N. Kalmykov, J. Moyon, S. Ostapchenko, T. Thouw et al., Nucl. Phys. Proc. Suppl. **151**, 159 (2006), astro-ph/0411260
[15] F. Duenkel, M. Niechciol, M. Risse, Phys. Rev. D **104**, 015010 (2021)
[16] F. Riehn, R. Engel, A. Fedynitch, T.K. Gaisser, T. Stanev, Phys. Rev. D **102**, 063002 (2020), 1912.03300
[17] T. Pierog, I. Karpenko, J.M. Katzy, E. Yatsenko, K. Werner, Phys. Rev. C **92**, 034906 (2015), 1306.0121
[18] S. Ostapchenko, Physical Review D **83**, 014018 (2011), 1010.1869
[19] F. Duenkel, M. Niechciol, M. Risse, publication in preparation (2023)
[20] A. Yushkov for the Pierre Auger Collaboration, PoS **ICRC2019**, 482 (2019), 1909.09073

https://doi.org/10.1051/epjconf/202328305008

# On the mystery of the multi-muon flux at the TeV cosmic-ray energy range

*Jordi* Tuneu[1,*], *Peter* Filip[1,**], and *Eva* Santos[1,***]

[1]FZU - Institute of Physics of the Czech Academy of Sciences, Prague, Czech Republic

**Abstract.** Current Monte Carlo simulations do not provide a good description of the muon component of extensive air showers. Many air shower experiments report discrepancies between their data and Monte Carlo predictions, ranging from the TeV scale up to the highest energies. In these proceedings, we address the seasonal variation of the multi-muon events observed by the NOvA Near Detector (ND). For our studies, we use the general-purpose Monte Carlo code FLUKA to treat the transport and interaction of the air-shower particles in the atmosphere and other media. Our design considers a multilayered atmosphere and a layered underground approximated to match the NOvA ND location and detector geometry. Our atmospheric model uses air densities for winter and summer calculated from the temperature and geopotential information for the pressure levels given by the European Center for Medium-Range Weather Forecasts (ECMWF) datasets in situ. Understanding the multi-muon flux at the cosmic ray high-energy range may lead to a better description of the muon production mechanisms in ultra-high-energy extensive air showers. In addition, it can help to improve future Monte Carlo codes or hint at new physics processes or interactions.

## 1 Introduction

The seasonal variation of the underground muon flux, originating from the interaction of high-energy cosmic rays in the Earth's atmosphere, has been studied by the MINOS and NOvA near detectors (NDs) located 100 m underground [1, 2]. MINOS cosmic data [3] show that the total (single)-muon flux increases in summer and decreases in winter by $\approx$ 1%. In contrast, both experiments find the multi-muon event rates to reach a maximum during winter and a minimum in the summer [1, 2]. The magnitude of the seasonal multi-muon flux oscillation is five times larger than the single-muon flux oscillation and has an opposite phase: the maximum is shifted by approximately six months. In [2], the MINOS Collaboration proposed several mechanisms to explain the winter maximum for the multi-muon flux. However, none of them fully describe the magnitude of the oscillation. A recent work [5] using CORSIKA simulations [6] has also addressed this issue. Still, a quantitative agreement was not obtained: the predicted seasonal multi-muon oscillation magnitude (see Fig. 9, in [5]) is a factor four times smaller than the MINOS and NOvA data, showing that even in the well-known TeV range, discrepancies between data and Monte Carlo simulations exist. A proper understanding of this problem could help to improve future Monte Carlo codes or even hint at new physics processes in hadronic interactions. It could also contribute to the work of the WHISP [7] group reporting on the muon puzzle arising at $\sqrt{s} \sim 10$ TeV.

*e-mail: tuneu@fzu.cz
**e-mail: filip@fzu.cz
***e-mail: esantos@fzu.cz

In this work, we use the general-purpose Monte Carlo code FLUKA [9] to treat the transport and interaction of the shower particles in the atmosphere and other media. In particular, we have based on the FLUKA's 100-layered atmospheric model, where we have implemented the air density for the summer and winter profiles as provided by the ECMWF database [10] for the NOvA/MINOS NDs location. The muon component was analyzed at the surface (226 m above sea level) and the detector plane (99 m underground) for both atmospheric profiles. The atmospheric model and geometry are described in section 2. Preliminary results are discussed in section 3. A summary of our results is given in section 4.

## 2 Atmospheric Model and Detector Geometry

Our atmospheric model is based on the FLUKA's 100-layered model [8], adapted for the NOvA ND geometry and location, considering different atmospheric density profiles for winter and summer. Since our detector area is $\approx 20$ km $\times$ 20 km at the surface and $\approx 1$ km $\times$ 1 km at the detector plane, we can neglect the effects of the Earth's curvature and have adopted a cylindrical geometry instead of the one from the original model. Our layout includes four further underground layers, allowing for more proper treatment of the interaction and propagation of the muons underground.

The description of our atmosphere starts at 72 km altitude, with a density of $8.78 \times 10^{-8}$ g cm$^{-3}$ for both atmospheric profiles. The air density is gradually increasing over the one hundred layers, reaching $1.21 \times 10^{-3}$ and 1.28 $\times$ $10^{-3}$ g cm$^{-3}$ at the surface (226 m a.s.l.) for the sum-

https://doi.org/10.1051/epjconf/202328305008

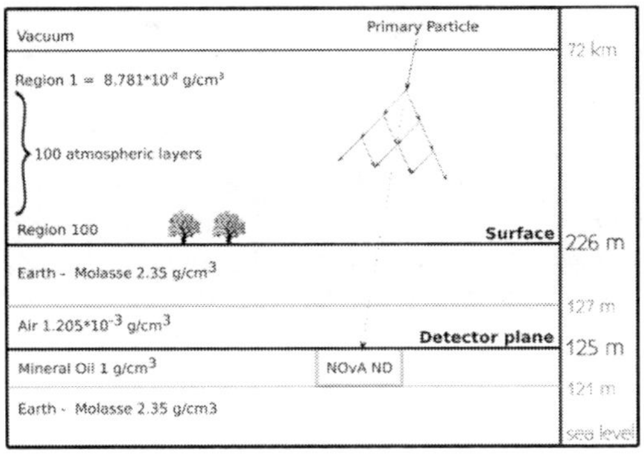


Figure 1: Geometry implemented to study the multi-muon seasonal variation observed by NOvA ND.

mer and winter profiles, respectively. The air densities for winter and summer have been calculated from the temperatures and geopotential information for 37 pressure levels given by the ERA5 ECMWF dataset in situ (averaged data from January and July 2017 periods were used). We have implemented a simple uniform magnetic field permeating the whole volume with components $B_x$ [T] = $1.93 \times 10^{-5}$, $B_y$ [T]= 0, and $B_z$ [T] = $4.97 \times 10^{-5}$.

The first underground layer has a thickness of 99 m, ranging from the surface level down to an altitude of 127 m a.s.l., where the NOvA ND experiment is installed. It has a molasse composition with a density of 2.35 g cm$^{-3}$. At this depth, we implement the second underground layer, which consists of a cavity of 2 m height filled with air, with a density of $1.205 \times 10^{-3}$ g cm$^{-3}$. The end of the second layer coincides with the altitude of the top of the NOvA detector, at 125 m a.s.l., followed by a four meters layer of mineral oil (1 g cm$^{-3}$). Finally, we add a molasse layer that reaches the sea level, which further slows down and absorbs the particles, as illustrated in Figure 1.

### 2.1 FLUKA cosmic ray shower library

For the work presented in this paper, we have used the Monte Carlo code FLUKA [9], version 4-2.2, with the DPMJET package [4] as the hadronic interaction model. We simulate a flux of high-energy cosmic rays arriving from two zenith angles, $\theta = 30°$ and $50°$, at an altitude of $z \approx 72$ km, which coincides with the top of our atmosphere. We adjust the XY initial coordinates of cosmic rays so that the shower axis intersects the center of the surface plane, located at 125 m a.s.l., at $x = y = 0$. For each zenith angle, we simulate six energies: 10, 30, 50, 100, 500, and 1000 TeV. In order to mitigate the statistical uncertainties at the lowest energies, where few muons reach the NOvA detector, we have produced a shower library in which the number of simulations decreases with energy as follows: $10^4$ showers at 10 TeV, $4 \times 10^3$ at 30 and 50 TeV, and $10^3$ at 100, 500, and 1000 TeV. We are assuming a pure proton cosmic ray mass composition since protons are 5 - 6 times more abundant than helium nuclei [11] at this energy range (with reference to energy per nucleon)

## 3 Preliminary Results

In this section, we report on our findings regarding the differences in the muon fluxes for the winter and summer atmospheres. In subsection 3.1, we present our results for the total flux of muons (single-muon flux), and in subsection 3.2, we focus on the multi-muons case, which is a sub-sample of the total muon flux.

We use the following notation for all our figures and tables: winter atmosphere results are denoted by a W and plotted in blue, while for the summer, we will use an S and red color. For our analysis, we used the FLUKA shower library described in subsection 2.1, from which we stored the information regarding the position and momenta of all muons produced along the shower development.

### 3.1 Single-muon flux

We observe that the total muon flux at the surface is always higher in the winter than in the summer. On average, $12 - 22\%$ ($10 - 18\%$) more muons are detected at the surface in winter than in summer for $\theta = 30°$ ($50°$) showers. This difference was found to increase with the cosmic ray energy. We find that our results are in agreement with the observed seasonal muon flux variation reported in [12].

In a second step, we calculated the muon flux summer-to-winter ratio at the detector level, located at 99 m depth. We found that, at the detector plane, the ratio becomes compatible with one within statistical errors, meaning that a seasonal variation is not significant. A summary of the results discussed above for the surface and detector planes is given in Tables 1 and 2.

Table 1: Ratios of the summer and winter muon flux for showers at $\theta = 30°$ and different energies, measured at the detector and the surface planes, respectively.

| E (TeV) | Detector $S/W$ | Surface $S/W$ |
|---|---|---|
| 10 | $1.006 \pm 0.006$ | $0.882 \pm 0.001$ |
| 30 | $1.002 \pm 0.006$ | $0.884 \pm 0.001$ |
| 50 | $1.001 \pm 0.005$ | $0.878 \pm 0.001$ |
| 100 | $1.001 \pm 0.007$ | $0.884 \pm 0.001$ |
| 500 | $1.002 \pm 0.004$ | $0.874 \pm 0.001$ |
| 1000 | $0.889 \pm 0.003$ | $0.775 \pm 0.001$ |

Table 2: Ratios of the summer and winter muon flux for showers at $\theta = 50°$ and different energies, measured at the detector and the surface planes, respectively.

| E (TeV) | Detector $S/W$ | Surface $S/W$ |
|---|---|---|
| 10 | $1.002 \pm 0.006$ | $0.898 \pm 0.001$ |
| 30 | $1.015 \pm 0.006$ | $0.903 \pm 0.001$ |
| 50 | $1.004 \pm 0.005$ | $0.893 \pm 0.001$ |
| 100 | $1.013 \pm 0.008$ | $0.864 \pm 0.001$ |
| 500 | $0.948 \pm 0.004$ | $0.811 \pm 0.001$ |
| 1000 | $0.945 \pm 0.004$ | $0.820 \pm 0.001$ |

We have calculated the fraction of muons reaching the detector plane from the surface to be 3.2 – 6.1% (2.8 – 5.4%) in winter (summer). Assuming that this ratio is mainly related to the decay of low-energy muons propagating in the rock, the energy spectrum of muons should be softer in summer than in winter. One possible explanation is that, in winter, the atmosphere is more compressed due to lower temperatures leading to cosmic rays interacting deeper and producing air showers developing in a denser medium. That leads to an increased probability of pions and kaons interacting more times in the atmosphere before decaying into (lower energy) muons. In Figure 2, we plot the $x$ and $y$ positions of muons (top row) and the muon density, $\rho = \frac{dN}{dR}\frac{1}{2\pi\langle R\rangle}$, as a function of the distance from the shower core (bottom row) at the surface (left), and the detector plane (right).

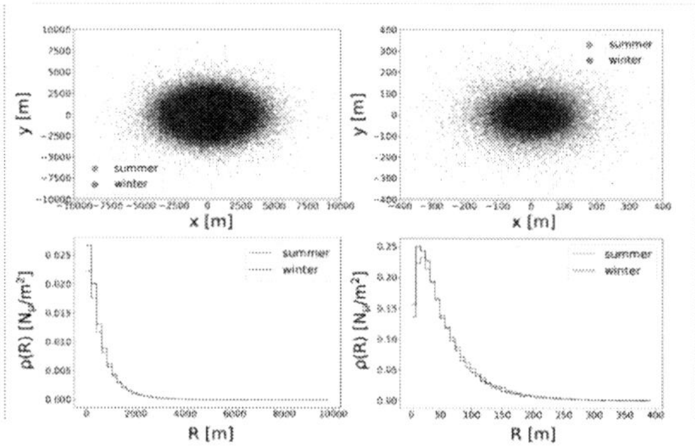


Figure 2: Spatial distribution (first row) and muon density $\rho_\mu$ (second row) of summer muons (red) and winter muons (blue) crossing the surface (first column) and the detector plane (second column) from primary protons with 100 TeV and $\theta = 30$.

We have also calculated the mean distance of muons from the shower axis as $\sigma_r = \langle (x_i^2 + y_i^2)^{1/2} \rangle$ at the detector plane for both profiles. Our results are shown in Table 3.

Table 3: Standard deviation of the distance of muons from the shower axis $r$ arriving at the detector plane, $z = 125$ m, for summer (S) and winter (W) for zenith angles $\theta = 30°$, and $\theta = 50°$.

| E (TeV) | 30° $\sigma_r$ (m) | | 50° $\sigma_r$ (m) | |
|---|---|---|---|---|
| | S | W | S | W |
| 10 | 62.2 ±0.2 | 59.8 ±0.2 | 95.7 ±0.4 | 91.3 ±0.4 |
| 30 | 61.3 ±0.2 | 58.0 ±0.2 | 93.1 ±0.4 | 88.8 ±0.4 |
| 50 | 59.3 ±0.2 | 55.8 ±0.2 | 91.5 ±0.3 | 87.0 ±0.3 |
| 100 | 56.1 ±0.3 | 52.9 ±0.2 | 86.5 ±0.5 | 82.8 ±0.5 |
| 500 | 54.7 ±0.1 | 50.6 ±0.1 | 81.0 ±0.3 | 74.0 ±0.2 |
| 1000 | 53.3 ±0.1 | 47.9 ±0.1 | 75.9 ±0.2 | 73.9 ±0.2 |

From Figure 2 and Table 3, we confirm that the lateral distribution of muons is slightly denser in winter, suggesting that it could play a key role in the seasonal variation detected by the NOvA ND experiment [1].

## 3.2 Multi-muons

To reduce the statistical uncertainties in the measurement of multi-muon events in each shower, we used a grid of identical NOvA-ND detectors (16 × 4 m large) located in $|x| < 400$ m; $|y| < 400$ m area, thus effectively simulating a total of $10^4$ detectors. For each shower, we check for multi-muon events in each one of the detectors. This way, we simulate the randomly shifted position of NOvA-ND relative to the distance from the shower axis.

We have evaluated the average multi-muon flux for the whole simulated shower library. The number of multi-muon events is shown in Tables 4 and 5, where $s = n_S$ and $w = n_W$ are the number of multi-muons detected in summer and winter, $s/w$ is their ratio and $D_{SW}$ quantity is the $S - W$ difference in percentage of $S + W$ (which is two times the average flux), as given by:

$$D_{SW} = \frac{s - w}{s + w}, \tag{1}$$

with statistical error $\delta$ evaluated as:

$$\delta = \pm \frac{2 \cdot s \cdot w \sqrt{\frac{1}{s} + \frac{1}{w}}}{(s + w)^2}, \tag{2}$$

A negative value of $D_{SW}$ means a deficit of multi-muons in summer with respect to winter.

Table 4: Number of multi-muons events as a function of energy for summer ($n_S$) and winter ($n_W$), their ratio ($n_S/n_W$) and percentage difference ($D_{SW}$) between summer and winter multi-muon events for $\theta = 30°$.

| E (TeV) | $n_S$ | $n_W$ | $s/w$ | $D_{SW}$ (%) |
|---|---|---|---|---|
| 10 | 647 | 702 | 0.92 ±0.05 | -4.1 ± 2.7 |
| 30 | 1820 | 2028 | 0.89 ±0.03 | -5.4 ± 1.6 |
| 50 | 4177 | 5032 | 0.83 ±0.02 | -9.3 ± 1.0 |
| 100 | 3263 | 3668 | 0.89 ±0.02 | -5.8 ± 1.2 |
| 500 | 18709 | 19104 | 0.98 ±0.01 | -1.0 ± 0.5 |
| 1000 | 20011 | 20634 | 0.98 ±0.01 | -1.5 ± 0.5 |

Table 5: Number of multi-muons events as a function of energy for summer ($n_S$) and winter ($n_W$), their ratio ($n_S/n_W$) and percentage difference ($D_{SW}$) between summer and winter multi-muon events for $\theta = 50°$.

| E (TeV) | $n_S$ | $n_W$ | $s/w$ | $D_{SW}$ (%) |
|---|---|---|---|---|
| 10 | 153 | 186 | 0.82 ±0.09 | -9.7 ± 5.4 |
| 30 | 490 | 647 | 0.76 ±0.05 | -13.8 ± 2.9 |
| 50 | 1241 | 1457 | 0.85 ±0.03 | -8.0 ± 1.9 |
| 100 | 1003 | 1218 | 0.82 ±0.04 | -9.7 ± 2.1 |
| 500 | 7420 | 8445 | 0.88 ±0.01 | -6.5 ± 0.8 |
| 1000 | 9584 | 9883 | 0.97 ±0.01 | -1.5 ± 0.7 |

Our outcome shows a significant difference between the multiple-muon flux in winter and summer. For $\theta = 30°$, this difference was found to increase for primaries up to 50 TeV and decrease afterward. For $\theta = 50°$, the maximum difference is obtained at 30 TeV, decreasing above 100 TeV, and reaching minimal values at the highest energies. The multiplicity dependence of the simulated winter and summer multi-muon fluxes as a function of energy for $\theta = 30°$ is shown in Figure 3, from which observes a higher flux of single-muon events in summer for $E = 50$ TeV ($0.47 \pm 0.26\%$) and $E = 100$ TeV ($1.33 \pm 0.40\%$).

We find our results compatible with the observed sea-

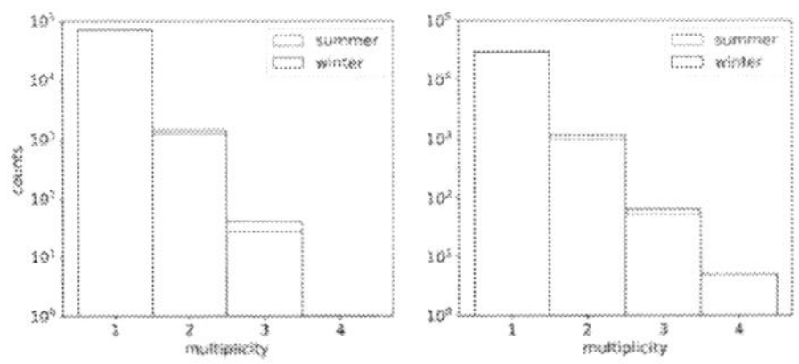


Figure 3: Muon multiplicity distribution from $\theta = 50°$ zenith angle showers and energies of 50 TeV (left) and 100 TeV (right) for summer (red) and winter (blue).

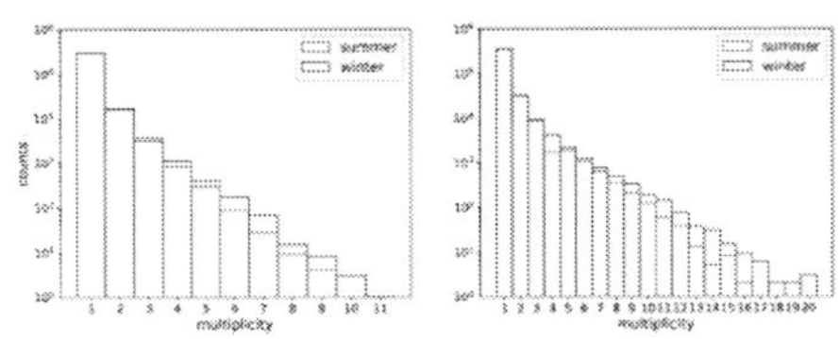

Figure 4: Accumulated multiplicity distribution considering all the energy showers (10, 30, 50, 100, 500, and 1000 TeV) for zenith angle showers of (a) $\theta = 30°$ and (b) $\theta = 50°$) for summer (red) and winter (blue).

sonal oscillation of a summer maximum of the single-muon flux observed by the MINOS Near Detector [3] located at the same depth and location as NOvA-ND. For the multi-muon ($N_\mu > 2$) events, an excess of the flux in winter, relative to summer, is clearly visible. Accumulated distributions for the whole simulation library, shown in Figure 4, suggest a multi-muon flux excess in winter is increasing with $N_\mu$ muon event multiplicity, as observed by the NOvA-ND [1].

## 4 Conclusions

We reported on our attempt to explain the winter multi-muon flux excess observed by the NOvA and MINOS Near Detectors [1, 2] as all previous works based on Monte Carlo simulations proved unsuccessful. In particular, in [5], a study using CORSIKA simulations has predicted an amplitude for the observed seasonal variation by a factor of four smaller than the one observed. That motivated us to use the Monte Carlo FLUKA package [8] since it allows for a more realistic modelling of the atmosphere while adequately treating the muon propagation in various media. We show that even considering the simple case of a pure proton nuclear mass composition for a set of fixed zenith angles and energies, we were able, for the first time, to describe the multi-muon flux excess in winter, as reported in [1, 2]. Our results show that the winter excess seems to increase with the muon number $N_\mu$ of the multi-muon events. It rises from the lowest simulated shower energy of 10 TeV, reaching the maximum between 30 and 50 TeV, and decreases for 500 and 1000 TeV. A direct quantitative comparison of (winter-summer) differences in our calculations with the NOvA-ND result [1] is being evaluated and will be reported in a forthcoming work.

## Acknowledgments

This work is fully funded by the Czech Science Foundation under the project GA ČR 21-02226M and by the Inter-Excellence-Inter-Transfer project LTT1801 financially supported by Ministry of Education, Youth and Sports of the Czech Republic.

## References

[1] Acero, M. A. et al. (NOvA Collaboration), Phys. Rev. D, **99**, 122004 (2019)

[2] Adamson, P. et al. (MINOS Collaboration), Phys. Rev. D, **91**, 112006 (2015)

[3] Adamson P. et al. (MINOS Collaboration), Phys. Rev. D **90**, 012010 (2014)

[4] Kling, A., et al., *Advanced Monte Carlo for Radiation Physics, Particle Transport Simulation and Applications* (Springer, Berlin, Heidelberg, 2001) 1033–1038

[5] Gaisser, T. K. and Verpoest, S., Astropart. Phys., **133**, 102630 (2021)

[6] Heck, D., Knapp, J., Capdevielle, J. N., Schatz, G. and Thouw, T., FZKA-6019.

[7] Albrecht, J. et al., Astrophys. Space Sci. **367**, 27 (2022)

[8] Battistoni, G., Margiotta, A., Muraro, S. and Sioli, M., Nucl. Instrum. Meth. A **626-627**, S191-S192 (2011)

[9] Battistoni, G. et al., Annals Nucl. Energy **82**, 10-18 (2015)

[10] https://www.ecmwf.int/

[11] Yoon, Y. S. et al., Astrophys. J. **839** 1 (2017)

[12] Yurina E.A., Bogdanov A. G., Dmitrieva A. N., Kokoulin R. P., Shutenko V. V., J. Phys. Conf. Ser. **1189**, 012010, (2019)

# Reconstruction of the muon production longitudinal profiles in extensive air showers

*Antonín* Kravka[1,*], *Eva* Santos[2,**], *Maximilian* Stadelmaier[2], and *Alexey* Yushkov[2]

[1]Czech Technical University - Faculty of Nuclear Sciences and Physical Engineering
[2]FZU - Institute of Physics of the Czech Academy of Sciences

**Abstract.**
Muons produced in extensive air showers have large decay lengths and small radiative energy losses, enabling a large fraction of them to reach surface and underground detector arrays while keeping relevant information about the hadronic interactions that occurred high in the atmosphere. We can relate a muon's arrival time and position at the detector to its production depth in the atmosphere. The total delay of muons with respect to the shower plane is primarily due to their geometric path and energy, we call these contributions the geometric and kinematic delays, respectively. We are working on the improvement of the current kinematic delay parameterizations using Deep Neural Networks for muons arriving at surface and underground detector arrays. We aim to reconstruct the longitudinal profile of muons for future arrays of buried scintillator detectors at energies from around the second knee to the ankle of the cosmic ray spectrum, where there is an overlap with the nominal energies at the LHC. Given the low acceptance of scintillator detectors to inclined air showers and the richness of the forward physics near the shower core, we aim at applying a radial cut of 200 m instead of the usual 1000 m used in previous works.

## 1 Introduction

Cosmic rays with energies above $10^{15}$ eV can only be detected indirectly through the secondary particles produced from their interaction in the Earth's atmosphere. Their energy and, particularly, nuclear mass composition are inferred by comparing the air-shower particles' time, lateral, and longitudinal distributions with those of Monte Carlo simulations using hadronic interaction models. However, the most relevant hadronic interactions of the most energetic cosmic rays occur at $\sqrt{s} \sim 100$ TeV, one order of magnitude above the nominal energy at the LHC, at phase-space regions not well-covered by most detectors. Lacking a theory to calculate multi-particle production from first principles, hadronic interaction models are our most significant source of systematic uncertainty [1].

Muons, mostly produced from the decay of charged pions, are the best messengers of the hadronic cascade. After production, they propagate in nearly straight lines to the ground without suffering significant energy losses, keeping relevant information about their mother particle [2–4]. In [5], from a compilation of the eight largest cosmic-ray experiments, a deficit of muons in simulations with respect to data is reported for $E > 10^{17}$ eV with a significance of more than $8\sigma$, right where the center-of-mass energy equals the nominal energy at the LHC. Also, in [6], the Pierre Auger Collaboration observed a tension between the EPOS-LHC model predictions and data for the maximum depth of the muon production longitudinal profile.

In this work, we extend the method of the reconstruction of the muon production depth (MPD) described in [2, 3], used in [6], for an array of underground muon detectors, as is the case of AMIGA [7], currently being deployed at the Pierre Auger Observatory.

## 2 Muon Production Depth

In [2, 3], a method relating the arrival time of muons with their production distance is proposed. It is assumed that muons are produced in the shower axis and propagate to the ground in straight lines, as shown in Figure 1. According to Figure 1, muons deviate from the shower axis

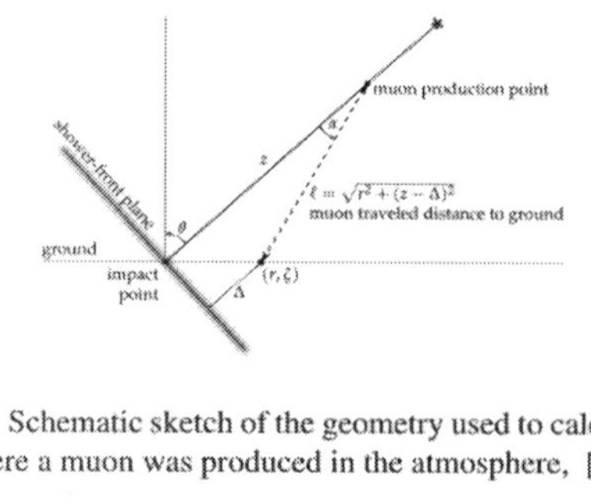


**Figure 1.** Schematic sketch of the geometry used to calculate the point where a muon was produced in the atmosphere, [6].

*e-mail: kravkant@fjfi.cvut.cz
**e-mail: esantos@fzu.cz

by an angle $\alpha$, arriving at the ground with coordinates $(r, \zeta)$ expressed on the shower-front plane coordinate system. Here $r$ is the muon distance from the shower axis, and $\zeta$ is its azimuthal angle. The total distance traveled by the muon is defined as $l = \sqrt{r^2 + (z - \Delta)^2}$, where $z$ is the muon production distance, measured along the shower axis, $\Delta = r \cos\zeta \tan\theta$ is the $z$-coordinate of the muon at the detector, expressed on the shower-front plane coordinate system, and $\theta$ is the shower zenith angle. Given the longer path traveled by muons, they will arrive delayed with respect to the arrival of the shower front. Assuming that muons are massless particles that travel at $c$, using a purely geometric relation, we can calculate the muon production distance, $z$, as

$$z = \frac{1}{2}\left(\frac{r^2}{ct_g} - ct_g\right) + \Delta, \qquad (1)$$

where $t_g$ is the geometric delay of the muon. However, there are additional sources of delay, being the most relevant one the so-called kinematic delay, which is intrinsically associated with the energy of the muon [2–4]. Therefore, Equation (1) is rewritten as $t_g \simeq t - \langle t_\epsilon \rangle$, where $t$ is the total delay, i.e., the one we can measure, and $\langle t_\epsilon \rangle$ is the kinematic delay. The latter is usually calculated from a parametrization assuming the average energy of muons arriving at a given distance from the shower axis, see Appendix B in [3]. Finally, the Muon Production Depth (MPD) is calculated from the muon production distance $z$ as

$$X = \int_z^\infty \rho(z')\, dz', \qquad (2)$$

where $\rho$ is the atmospheric density.

The method proposed in [2, 3] is intended for a detector array, as the one of the Pierre Auger Observatory [8], for $\theta \sim 60°$ showers, applying a radial cut of $r > 1000$ m. The reason for these cuts is two-fold; namely, they ensure negligible electromagnetic contamination in the total signal, while the contribution of the geometric delay, from which the MPD is calculated, is the dominant source of delay. In [6], this method was applied to the Pierre Auger data for events with $E > 10^{19.3}$ eV and $55° \leq \theta \leq 65°$, using detectors at $1700 < r\,[\mathrm{m}] < 4000$. Given the 1500 m spacing of the Auger detectors, the array proved too sparse, limiting the method's applicability to fewer than 500 events.

Our goal is to extend the present method's applicability to an array of buried scintillator detectors, as in the one of AMIGA [7]. Another advantage of AMIGA is its 750 m and 433 m detector spacing, allowing us to encompass the $10^{17}$ eV energy region, where overlap with the LHC energies exists. That enables us to reconstruct the MPD in an energy region where we expect reduced systematic uncertainties from hadronic interaction models. However, since typical scintillator detectors have $\sim$ 5 cm width, their acceptance is limited to $\theta < 55°$, [9]. Also, the shower footprint is smaller at low zenith angles and energies, requiring a significantly lower radial cut of $r \gtrsim 200$ m.

### 2.1 Towards the reconstruction of the MPD

In this subsection, we report on our assessments of the feasibility of the extension of the present method to $\theta < 60°$, applying a radial cut of 200 m, as required for an array of buried scintillator detectors as it the case of AMIGA [7]. The AMIGA detectors are being deployed at 2.3 m depth, resulting on an additional mass overburden of 540 g cm$^{-2}$ for a local soil density is $\rho = 2.4$ g cm$^{-3}$ [7]. For our calculations we assume that all muons behave as minimum ionizing particles, i.e., losing a constant fraction of their energy per unit of crossed superficial density of $a(E) = 1.8$ MeV/g cm$^{-2}$ [10], from which we calculate the minimum kinetic energy for the muons to reach the AMIGA detectors as

$$E^{\mu}_{kin} > a(E)\, \rho\, h / \cos\theta, \qquad (3)$$

where $h = 2.3$ m is the depth at which the detectors are buried, and $\theta$ is the zenith angle of the shower. From Equation (3), we expect that the minimum kinetic energy for a muon to arrive to the underground detectors to increase as a function of the zenith angle, resulting it to be $E^{\mu}_{kin,min} = 1$ GeV (2 GeV) at $\theta = 0°$ ($\theta = 60°$). Throughout our work, we will use the results of Equation (3) to select the muons from which we reconstruct the MPD at 2.3 m depth.

*CORSIKA shower libraries*

For our studies, we produced two CORSIKA [11] shower libraries, one with a standard thinning of $10^{-6}$, combined with a radial thinning of $r < 50$ m, and a second one in which we did not activate the thinning algorithm. We set the ground level at 1452 m, the average height of the Pierre Auger Observatory. The second library was aimed to study the impact of thinning on the muon distributions close to the shower axis. In both cases, we use a fixed energy of $10^{17}$ eV, and several zenith angles, assuming proton and iron-initiated showers. We have simulated 200 showers per nuclear species, zenith angle, and high-energy hadronic interaction model.

The "*thinned*" shower library was produced with CORSIKA 7.7402 for the following zenith angles: $\theta = \{0°, 25°, 35°, 45°, 55°, 60°\}$. We consider the high-energy hadronic interaction models EPOS-LHC [12], QGSJet-II.04 [13] and Sibyll 2.3d [14], and FLUKA 2020.0.6 [15] as the low-energy hadronic model. For the "*unthinned*" shower library, we used CORSIKA 7.7410 with FLUKA 2020.0.10. Only Sibyll 2.3d as the high-energy hadronic model and $\theta = \{0°, 35°, 60°\}$ were considered.

*Testing the original model assumptions*

In [2, 3], it was assumed that muons are produced in the shower axis. In an update in [4], the average distance of the production of muons from the shower axis was estimated to be about a few tens of meters for $10^{19}$ eV showers with $\theta = 60°$, using pre-LHC model predictions. We check if these findings hold for the current models and at lower-energy and zenith angles. In this section, for illustration, we show our results for 20 unthinned Sibyll 2.3d

 https://doi.org/10.1051/epjconf/202328305009

iron-initiated showers at $10^{17}$ eV and $\theta = 0°$. In the text, we compare the results from the figures with the outcome of proton-initiated showers and also at $\theta = 60°$, when pertinent.

In Figure 2, we plot the distance from the shower axis at which the muons are produced. We observe that, for

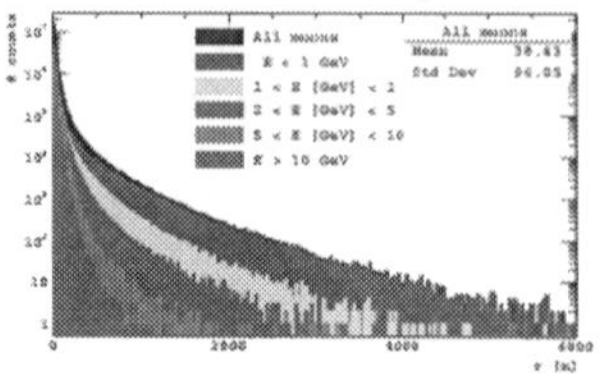

**Figure 2.** Distance from the shower axis at which muons are produced. The colors represent several ranges for the initial energy muons, namely: black – all energies; red – $E_i < 1$ GeV; yellow – $1 < E_i\,[\mathrm{GeV}] < 2$; green – $2 < E_i\,[\mathrm{GeV}] < 5$; blue – $5 < E_i\,[\mathrm{GeV}] < 10$; violet – $E_i > 10$ GeV.

both primary species, at $\theta = 0°$, muons are produced at $\langle r_{prod} \rangle \sim 40$ m, a distance that reduces to $\sim 30$ m at $\theta = 60°$, confirming the findings of [4]. We also see that at the largest distances we are dominated by a population of low-energy muons. Quantitatively, at 0°, muons with $1 < E_i\,[\mathrm{GeV}] < 2$ ($2 < E_i\,[\mathrm{GeV}] < 5$) are produced at an average distance of 45 m (30 m) from the shower axis.

In a second step, we studied the muon spectra and arrival time distributions of muons at the ground level as a function of the distance to the shower axis, as shown in the left and right panels of Figure 3, respectively. The arrival time of muons was calculated as $t' = t - t_0$, where $t$ is the elapsed time since the first interaction, and $t_0 = \left(h_1 - h_g\right) / (c \cos\theta)$ is the time needed for a hypothetical massless particle to travel, parallel to the shower axis, from the first interaction point to the ground. In the equation, $h_1$ is the height of the first interaction, as given by CORSIKA, $h_g$ is the height of the ground level, and $c$ is the speed of light. From the top panel of Figure 3, we observe that, near

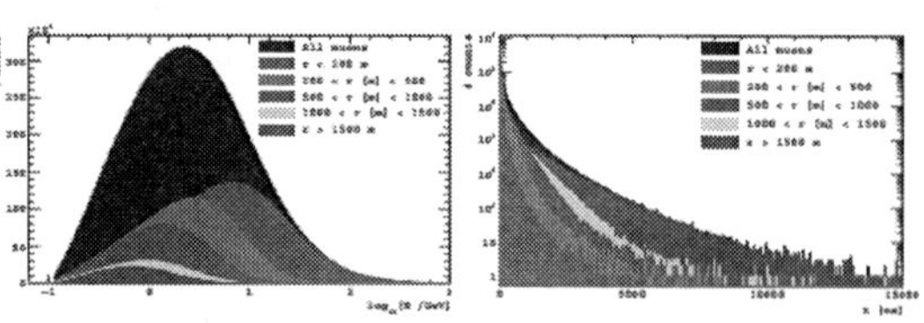

**Figure 3.** Energy spectra (left) and arrival time distribution (right) of muons arriving at the ground plane. We use the same notation for the two plots: black – all muons; violet – $r < 200$ m; blue – $200 < r\,[m] < 500$; green – $500 < r\,[m] < 1000$, yellow – $1000 < r\,[m] < 1500$, and red – $r > 1500$ m.

the core, we have a very rich muon spectrum, ranging from hundreds of GeV to less than 1 GeV. However, as we move further away from the shower axis, we become dominated by a less varied and less energetic muon component. We also find that the muon spectrum hardens with the zenith angle, most likely due to the decay of low-energy muons in the increased air overburden, which grows by a factor of two at $\theta = 60°$. We further observe that for $\theta = 0°$, there is a much richer energy spectrum at $r < 200$ m than the one for 60°. On the bottom panel of Figure 3, we see that the arrival time distribution of muons is narrower closer to the shower axis, broadening as we move further away from it. There is a change in behavior in the arrival time distributions as a function of the zenith angle for the near and far regions from the shower axis. Namely, at 0°, we have a broader arrival time distribution of muons arriving at $r < 200$ m, than at 60°.

Finally, in Figure 4, we study the impact of the energy on the muon arrival time distributions at $200 < r\,[\mathrm{m}] < 500$ (left) and $500 < r\,[\mathrm{m}] < 1000$ (right). From Fig-

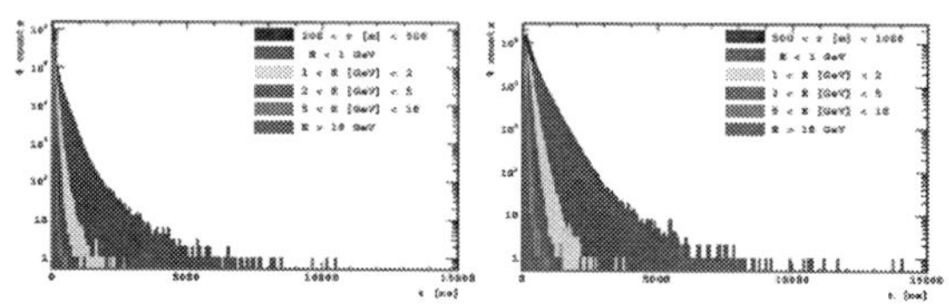

**Figure 4.** Arrival time distribution of muons arriving at $200 < r\,[\mathrm{m}] < 500$ (left) and $500 < r\,[\mathrm{m}] < 1000$ (right). The colors represent several ranges for the initial energy muons, namely: black – all energies; red – $E_i < 1$ GeV; yellow – $1 < E_i\,[\mathrm{GeV}] < 2$; green – $2 < E_i\,[\mathrm{GeV}] < 5$; blue – $5 < E_i\,[\mathrm{GeV}] < 10$; violet – $E_i > 10$ GeV.

ure 4, we observe that the most (least) energetic muons dominate the early (late) part of the distribution. However, at the beginning of the arrival time distribution, the muon energy spectrum is very diverse, ranging from the lowest to the highest energies, making the proper estimation of the kinematic delay challenging.

### *Estimating the kinematic delay*

In [2, 3], the estimated kinematic delay of muons traveling a distance $l$ is estimated calculated as:

$$t_\epsilon \simeq \frac{1}{2} \frac{\left(m_0 c^2\right)^2}{c \rho a} \left[\frac{1}{E_f} - \frac{1}{Ei}\right], \tag{4}$$

where $m_0$ is the muon rest mass, $a = 2\ \mathrm{MeV/g\,cm^{-2}}$, assuming that muons behave as minimum ionizing particles [10]. $E_i$, and $E_f$ are the muon energy at production, and at the detector, respectively.

We will use Equation (4) to explicitly calculate the kinematic delay of muons $t_\epsilon$. In Figure 5, we plot the total (black), geometric (blue), and kinematic (red) delays of muons as a function from the shower axis. The estimate the total delay as the $t' = t - t_0$, as explained above, and the geometric delay as $t_g = t' - t_\epsilon$. From the middle and bottom panels of Figure 5, we confirm that the kinematic delay is intrinsically related with the energy of muons, and it is larger for low-energy muons. We also observe that the kinematic delay is dominant close to the shower axis, prevailing over the geometric delay until $r = 200$ m for $E < 1$ GeV.

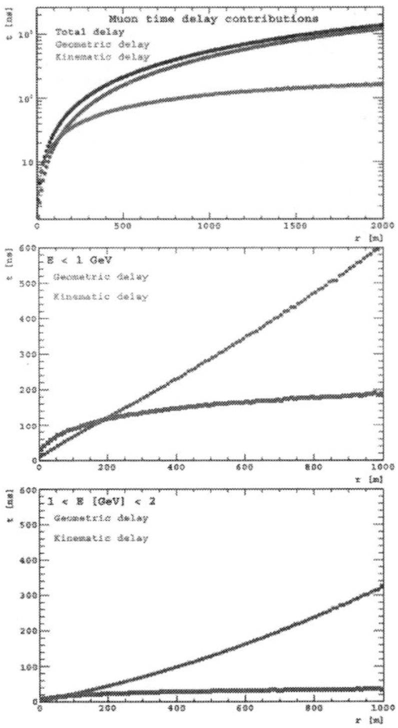


**Figure 5.** Contributions to the total muon delay for an average of 20 iron-initiated showers with $E = 10^{17}$ eV, and $\theta = 0°$. Top panel: Total distributions for the total (black), geometric (blue) and kinematic (red) delay. Middle panel: geometric (blue) and kinematic (red) delays for muons with $E < 1$GeV. Bottom panel: geometric (blue) and kinematic (red) delays for muons with $1 < E\,[\mathrm{GeV}] < 2$. Please, mind the logarithmic scale in the $y$-axis used in the top panel.

## 3 Calculating the MPD using DNNs

For the MPD reconstruction, we used a Fully Connected Neural Network with 6 hidden layers in a pyramid scheme, where the number of nodes decreases from 1024 to 32 from the first to the last hidden layer. We use ReLU as the activation function and the ADAM optimizer. Our input features to the network are the sec $\theta$ of the shower, the ground coordinates of the muon $(r, \zeta)$, and its arrival time $t'$, as explained above. We also add a boolean flag that tells us if the muons had enough energy to reach the underground detectors, according to Equation (3). Our loss function was derived empirically as

$$L = \left(\frac{z_{DNN} - z_{MC}}{z_{MC} + 1}\right)^2 + [(1 - \rho) + \gamma_3]\,c, \tag{5}$$

where $z_{DNN}$ and $z_{MC}$ are the predicted and Monte Carlo values of muon production distance, respectively. $\rho$ and $\gamma_3$ are the Pearson correlation coefficient and the distribution skewness, respectively, and $c$ is a normalization constant. Finally, we convert the estimated muon production distance $z$ into the MPD by applying Equation (2). We used 10% of the showers for the training of the DNN.

Our main results are shown in Figure 6. In the top panel, we show the correlation matrix between the reconstructed and Monte Carlo values of the MPD for all muons reaching the ground independently if they had enough energy to reach 2.3 m depth, from which observe a Pearson correlation of 0.936 between the reconstructed and true MPD. On the bottom panel, we show one MPD profile reconstructed at 2.3 m depth for one $10^{17}$ eV proton-initiated shower at $\theta = 0°$, assuming a radial cut of 200 m. We see

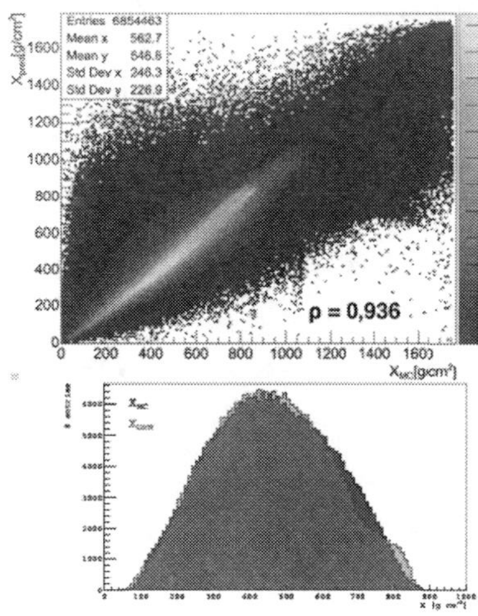


**Figure 6.** Top panel: Correlation matrix between the predicted and Monte Carlo MPD. Bottom panel: Example of one reconstructed *apparent* MPD profile using DNNs (red histogram). For comparison, the Monte Carlo profile is also plotted in the figure as the black histogram.

a good agreement between the reconstructed and Monte Carlo profiles up to a production depth of 500 g cm$^{-2}$. However, the DNN seems to perform worse for deeper MPD, probably due to the estimation of the kinematic delay since we expect that, at the deeper regions, the profile is dominated by low-energy muons.

## 4 Conclusions

We extended the MPD reconstruction method presented in [2, 3] to an array of buried scintillator detectors. The applicability region of our work is intended for $\theta < 60°$, and $r > 200$ m, a fully complementary to the previous method. We were able to successfully reconstruct the MPD both for surface and underground muons using a Fully Connected DNN. As we expected, the estimation of the kinematic delay is the largest source of systematic uncertainty in the model. Our preliminary results show that, for the current DNN architecture, we achieve a muon-by-muon reconstruction of the MPD profiles at 2.3 m depth of $\langle\Delta X\rangle < 10\,\mathrm{g\,cm^{-2}}$, and $\sigma(\Delta X) \sim 80\,\mathrm{g\,cm^{-2}}$ for $E = 10^{17}$ eV proton and iron-initiated showers, for $0° < \theta < 60°$, when a radial cut of $r > 200$ m is applied. These results can be improved if a different type of Neural Network or loss function is used. We plan a journal publication with updated results.

## Acknowledgments

This work is fully funded by the Czech Republic grant of MEYS CR GA ČR 21-02226M.

## References

[1] R. Engel, et al., Ann. Rev. Nucl. Part. Sci. **61**, 467-489 (2011)
[2] L. Cazon, et al., Astropart. Phys. **21**, 71-86 (2004)
[3] L. Cazon, et al., Astropart. Phys. **23**, 393-409 (2005)
[4] L. Cazon, et al., Astropart. Phys. **36**, 211-223 (2012)
[5] J. Albrecht, et al., Astrophys. Space Sci. **367**, 27 (2022)
[6] The Pierre Auger Coll., Phys. Rev. D **90**, 012012 (2014) [erratum: Phys. Rev. D **92**, 019903 (2015)]
[7] The Pierre Auger Coll., JINST **11**, P02012, (2016)
[8] The Pierre Auger Coll., Nucl. Instrum. Meth. A **798**, 172-213 (2015)
[9] The Pierre Auger Coll., Eur. Phys. J. C **80**, 751 (2020)
[10] D. E. Groom, et al., Atom. Data Nucl. Data Tabl. **78**, 183-356 (2001)
[11] D. Heck, et al., FZKA-6019.
[12] T. Pierog, et al. Phys. Rev. C **92**, 034906 (2015)
[13] S. Ostapchenko, Phys. Rev. D **83**, 014018 (2011)
[14] F. Riehn, et al., PoS ICRC2015, **558** (2016)
[15] G. Battistoni, et al., Annals Nucl. Energy **82**, 10-18 (2015)

# Pythia 8 as hadronic interaction model in air shower simulations

*Maximilian* Reininghaus[1,*] for the CORSIKA 8 Collaboration*, *Torbjörn* Sjöstrand[2], and *Marius* Utheim[3]

[1]Karlsruher Institut für Technologie (KIT), Postfach 3640, 76021 Karlsruhe, Germany
[2]Department of Physics, Lund University, Sölvegatan 14A, S-223 62 Lund, Sweden
[3]University of Jyvaskyla, Department of Physics, P.O. Box 35, FI-40014 University of Jyvaskyla, Finland

**Abstract.** Hadronic interaction models are a core ingredient of simulations of extensive air showers and pose the major source of uncertainties of predictions of air shower observables. Recently, Pythia 8, a hadronic interaction model popular in accelerator-based high-energy physics, became usable in air shower simulations as well. We have integrated Pythia 8 with its new capabilities into the air shower simulation framework CORSIKA 8. First results show significantly shallower shower development, which we attribute to higher cross-section predictions by the new simplified nuclear model of Pythia.

## 1 Introduction

Large-scale experiments in modern ultra-high energy cosmic ray research rely heavily on simulations of extensive air showers to link air shower observables to properties of the primary cosmic ray. An important ingredient in these simulations are hadronic interaction models (also known as event generators) that govern the interactions of hadronic shower particles with air nuclei. Due to the nature of the strong interaction, the wealth of hadronic interactions and multiparticle production is difficult or impossible to calculate from first principles alone. Only hard processes, i.e. those involving a large momentum transfer, can be treated within the framework of perturbative quantum chromodynamics (pQCD). The bulk of soft interactions, however, relies mainly on phenomenological modelling in combination with theoretical constraints and pQCD [1]. The most widely used up-to-date models used in air shower simulations are EPOS-LHC [2], QGSJet-II.04 [3] and SIBYLL 2.3d [4]. While EPOS-LHC has its origins in heavy-ion physics, the other two are specifically tailored to the needs of air shower simulations and the features of hadronic interactions important in that context.

In accelerator-based high-energy physics (HEP), a very popular event generator is Pythia [5], currently at version 8.3 [6]. For a long time, Pythia was not suitable for air shower simulations, mainly due to entirely different setups employed in HEP and air shower simulations: While simulations for accelerators typically generate a large number of events with the same settings (beam particle IDs and momenta), air shower simulations require event generation with these settings randomly varying event by event. Recently, progress has been made in making Pythia 8 more suitable for that setup [7]. On the one hand, this pertains to an accelerated context switching between beam parameters. On the other hand, the number of valid beam particles has been extended and a simplified model of hadron-nucleus collisions has been developed, while the fully-fledged Angantyr module [8] for heavy-ion collisions is not yet usable in air shower simulations. Moreover, the range of valid beam energies has been extended down to $\sim 200\,\mathrm{MeV}$ (lab), which means that Pythia can be used without an additional low-energy hadronic interaction model or just as such together with another high-energy model.

In this contribution we describe and analyse first results obtained using Pythia 8 in a realistic air shower simulation. For this, we have integrated Pythia 8.307 into the air shower simulation framework CORSIKA 8 [9] (Note that Pythia 8 already has been used to handle particle decays in CORSIKA 8). This work represents a continuation of the study started in ref. [10].

## 2 Setup

We simulate air showers in a hybrid fashion: Hadronic interactions and propagation of hadron and muons are treated in a Monte Carlo manner in CORSIKA 8. Electromagnetic particles are passed to CONEX [11], which generates longitudinal profiles by solving the cascade equations numerically. Hadrons and muons stemming from photohadronic interactions are therefore missing in this setup. We consider showers at $10^{17.5}\,\mathrm{eV}$ with an inclination of 60° using Linsley's parameterization of the US Standard atmosphere (see e.g. ref. [12]). The observation level is set to sea level. We consider SIBYLL 2.3d, EPOS-LHC and Pythia 8.307 as high-energy interaction model above 63.1 GeV. In each case, Pythia is used as low-energy model down to the hadron/muon cut energy that is set to 1 GeV. For nuclear projectiles that cannot be treated directly in the simplified nuclear model, we employ the semi-superposition model implemented in SIBYLL that

*full author list available at https://s.kit.edu/c8-authorlist-uhecr2022, e-mail: reininghaus@kit.edu

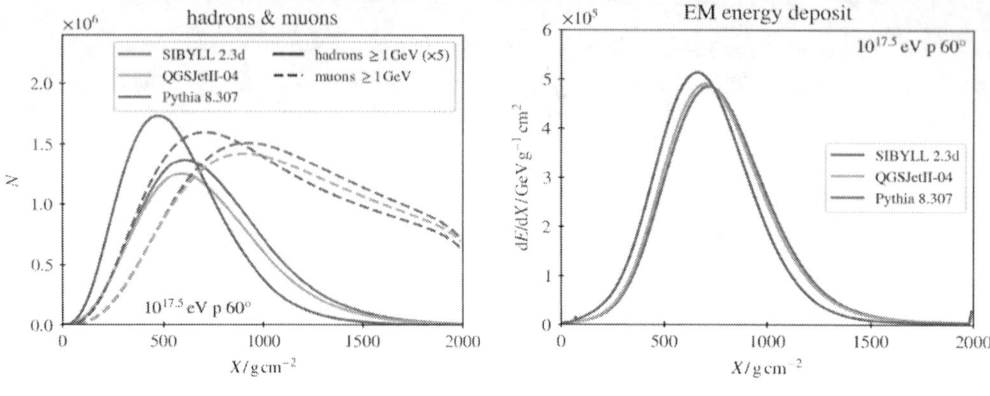


**Figure 1.** Longitudinal profiles. Left: hadrons and muons. Right: electromagnetic energy deposit.

breaks down $A-A$ collisions into multiple $p/n-A$ collisions [13].

## 3 Results

Figure 1 shows the average longitudinal profiles of proton-induced showers. We observe that Pythia produces some 30 % more hadrons than the traditional models in its maximum, which occurs $120\,\mathrm{g\,cm^{-2}}$ to $140\,\mathrm{g\,cm^{-2}}$ earlier than with the other models. Regarding the muon profile, the maximum with Pythia is more than $200\,\mathrm{g\,cm^{-2}}$ shallower and the number of muons at the maximum is $\sim 6\,\%$ higher than with QGSJet-II.04. At ground, however, the number of muons $N_\mu$ is smaller with Pythia due to the longer propagated distance causing higher energy losses and a higher probablity of in-flight decay. The electromagnetic (EM) longitudinal profiles differ less severly as the EM cascade quickly decouples from the hadronic one, so that deviations in the hadronic interactions cannot accumulate that much.

Figure 2 displays the results in the $X_{\mathrm{max}}$-$N_\mu$ plane also for heavier primaries. We note a significant shift of the Pythia line by $\sim 40\,\mathrm{g\,cm^{-2}}$ to $50\,\mathrm{g\,cm^{-2}}$ towards lower $X_{\mathrm{max}}$ values w.r.t. the other models.

It has been shown by Ulrich et al. [14] that a variation of the inelastic hadron-air cross-sections has a large impact on $X_{\mathrm{max}}$ while leaving $N_\mu$ almost unchanged. For that reason, we show the model predictions of the inelastic cross-sections in fig. 3. Due to the wealth of precise data on $pp$ collisions from accelerator measurements up to LHC energies ($\sim 10^{17}$ eV in lab frame) to which the models have been tuned, the predictions differ only slightly. Precise data on $\pi^{\pm}p$ collisions, however, exist only up a few 100 GeV, so that predictions diverge especially above $10^{14}$ eV, with Pythia yielding the lowest values. At present, it is unknown whether the $\pi p$ cross-sections eventually converge to the $pp$ ones, which is expected assuming a universal saturation of low-$x$ gluons, or stay below as expected from a Pomeron-style rise.

When considering oxygen targets, the picture is different. Pythia predicts cross-sections significantly higher

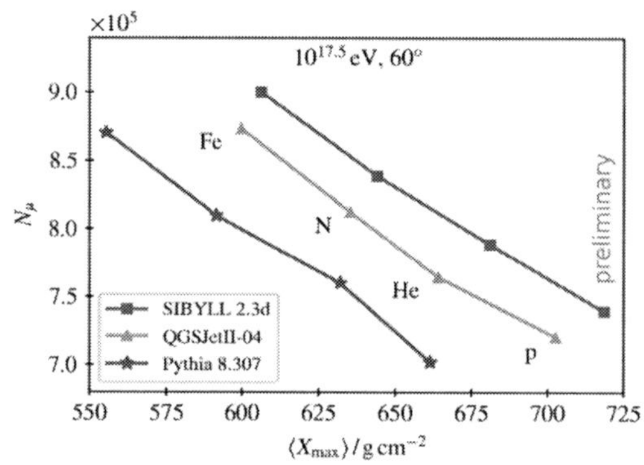


**Figure 2.** Number of muons at ground $N_\mu$ vs. shower maximum $X_{\mathrm{max}}$

than the other models. The simplified nuclear model of ref. [5] considers only total cross-sections $\sigma_{\mathrm{tot}}$ by employing the relation $\sigma_{\mathrm{tot}}^{(hA)} = A\sigma_{\mathrm{tot}}^{(hp)}/\langle n_{\mathrm{subcoll}}\rangle$, with the mean number of subcollisions $\langle n_{\mathrm{subcoll}}\rangle$ parameterized from full Angantyr events. Therefore, we estimate the inelastic cross-section by scaling $\sigma_{\mathrm{tot}}$ with the ratio $f_{\mathrm{inel}}$ of inelastic events, which we determined to be approx. 92 % in case of $\pi O$ and 90 % in case of $pO$ events with negligible energy dependence. It is noteworthy that Pythia yields the smallest cross-sections among the considered models in case of $\pi p$ but the largest in case of $\pi O$ and $pO$.

## 4 Conclusions and outlook

We have integrated the latest version of Pythia 8 into CORSIKA 8 to be used as hadronic interaction model for realistic air shower simulations for the first time. The results presented demonstrate that Pythia is capable to meet the higher demands (more projectile/target configurations, extrapolations to beyond-LHC energies) of such simulations compared to its original use-case in accelerator-

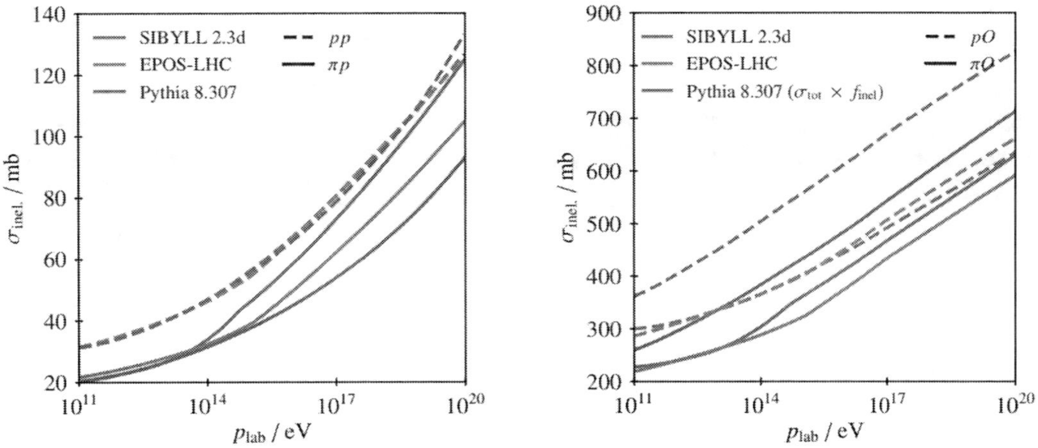


**Figure 3.** Inelastic cross-section predictions. Left: proton target, right: Oxygen target

based high-energy physics. The observed differences in the longitudinal development can be attributed to cross-section predictions significantly higher than those of the other models – an issue that requires further investigation and improved modelling. Further refinements and improvements, also regarding the use of Angantyr directly, are ongoing and expected in upcoming releases.

The availability of Pythia 8 in air shower simulations does not only provide yet another interaction model but also interesting opportunities: The possibility of tuning the model by the users themselves may offer new insights into the production of muons and its uncertainties by systematically studying the impact on air shower observables and accelerator measurements at the same time. Moreover, Pythia 8 is the only model treating the production of all quark flavors. Until now, only SIBYLL models charm production. Finally, the advent of Pythia 8, being an object-oriented C++ code, marks an important step towards enabling parallelization of CORSIKA 8 simulations by multithreading.

## Acknowledgements

The authors acknowledge support by the High Performance and Cloud Computing Group at the Zentrum für Datenverarbeitung of the University of Tübingen, the state of Baden-Württemberg through bwHPC and the German Research Foundation (DFG) through grant no. INST 37/935-1 FUGG.

## References

[1] R. Engel, D. Heck, T. Pierog, Ann. Rev. Nucl. Part. Sci. **61**, 467 (2011)

[2] T. Pierog, I. Karpenko, J.M. Katzy, E. Yatsenko, K. Werner, Phys. Rev. C **92**, 034906 (2015), `1306.0121`

[3] S. Ostapchenko, Phys. Rev. D **83**, 014018 (2011), `1010.1869`

[4] F. Riehn, R. Engel, A. Fedynitch, T.K. Gaisser, T. Stanev, Phys. Rev. D **102**, 063002 (2020), `1912.03300`

[5] T. Sjöstrand, Comput. Phys. Commun. **246**, 106910 (2020), `1907.09874`

[6] C. Bierlich et al., SciPost Phys. Codebases **8** (2022), `2203.11601`

[7] T. Sjöstrand, M. Utheim, Eur. Phys. J. C **82**, 21 (2022), `2108.03481`

[8] C. Bierlich, G. Gustafson, L. Lönnblad, H. Shah, JHEP **10**, 134 (2018), `1806.10820`

[9] R. Engel, D. Heck, T. Huege, T. Pierog, M. Reininghaus, F. Riehn, R. Ulrich, M. Unger, D. Veberič, Comput. Softw. Big Sci. **3**, 2 (2019), `1808.08226`

[10] M. Reininghaus (CORSIKA), *Air showers and hadronic interactions with CORSIKA 8*, in *51st International Symposium on Multiparticle Dynamics* (2022), `2210.07797`

[11] T. Bergmann, R. Engel, D. Heck, N.N. Kalmykov, S. Ostapchenko, T. Pierog, T. Thouw, K. Werner, Astropart. Phys. **26**, 420 (2007), `astro-ph/0606564`

[12] J. Cruz Moreno, S. Sciutto, Eur. Phys. J. Plus **128**, 104 (2013)

[13] R. Engel, T.K. Gaisser, T. Stanev, P. Lipari, Phys. Rev. D **46**, 5013 (1992)

[14] R. Ulrich, R. Engel, M. Unger, Phys. Rev. D **83**, 054026 (2011), `1010.4310`

https://doi.org/10.1051/epjconf/202328305011

# Search for upward-going showers with the Pierre Auger Observatory

*Vladimír* Novotný[1] for the Pierre Auger Collaboration[2,*]

[1]Institute of Particle and Nuclear Physics, Faculty of Mathematics and Physics, Charles University, V Holešovičkách 2, 180 00 Prague 8, Czech Republic

[2]Full author list: http://www.auger.org/archive/authors_2022_10.html

**Abstract.** Motivated by the ANITA report of pulses compatible with upward-going extensive air showers, we present a recent search for such showers using the Pierre Auger Observatory. The dataset, registered using the fluorescence detector of the Observatory, is scanned to identify showers ascending from the ground. Consistent with the exit angles reported from the first and third ANITA flights, we focus on shower geometries that are incompatible with Standard Model interactions of neutrinos in the Earth. We present the effective area of the Observatory to generic upward-going showers as a function of shower energy and altitude. This can be used to constrain predictions based on physics beyond Standard Model and the upgoing-shower interpretation of the ANITA results. To demonstrate the method, we calculate limits to the production rate of tau leptons near the ground emerging under a range of exit angles.

## 1 Introduction

The Pierre Auger Observatory [1], located near Malargüe, Argentina was built to investigate ultra-high energy cosmic rays (UHECRs). Nevertheless, it is also capable of measuring upward-going (UG) extensive air showers (EASs) that could be created by other particles than UHECRs. In this contribution, we search for EASs that would emerge from the ground with zenith angle ($\theta$) above 110°. To this end, we utilize the fluorescence detector (FD) of the Observatory that is composed of 24 telescopes at four sites covering elevations up to 30°, and three High-Elevation Auger Telescopes (HEAT) looking between 30° and 60° in elevation.

Our search is motivated by ANITA report of radio pulses compatible with upward-going extensive air showers [2] and our choice of investigated zenith angles comply with these findings. It is worth noting that presence of UG EASs with that large $\theta$ implies phenomena beyond the Standard Model of particle physics, because such showers would be inconsistent with interactions of neutrinos in the Earth and current limits on neutrino flux provided by the Auger Observatory [3].

## 2 Analysis

Showers ascending from the ground can be distinguished in the FD from downward-going (DG) EASs using the trigger time information from individual FD pixels together with their pointing. On top of that, we utilize an approximately universal longitudinal shape of EASs that must be compatible with calculated profile of the energy deposited in the atmosphere [4]. The later is especially important to assess the geometry of showers detected only in one FD station for which the time and pointing information is not sufficient for short tracks.

After the cleaning of the dataset from lasers used to monitor the atmosphere, Section 2.1, we estimate the number of background events, Section 2.2, calculate the FD exposure to UG events, Section 2.3, and, finally, unblind the data and assess the number of candidate events, Section 2.4.

### 2.1 Laser cleaning

To efficiently filter events coming form monitoring lasers of the Observatory, occurring occasionally in undefined times due to technical malfunctions, a burnt dataset (10 % of the full sample) was unblinded in advance and not used in the final analysis. The distribution of impact points of events in the burnt sample before and after the laser cleaning is shown in Fig. 1. The apparent lines occur due to misreconstruction of laser events that are shot from the two laser facilities in the centre of the array and from LIDARs located at four FD sites. Thus, the residual laser events are rejected based on the azimuth of the direction of normal to the shower-detector plane and on specific frequencies of laser shots.

### 2.2 Background

In the case that a shower is detected only in one FD station, i.e. we use a monocular reconstruction to calculate the shower axis, the shower can be wrongly reconstructed. As a consequence, ordinary downward-going EASs can mimic UG signal. A shower falling behind the FD telescope produces an image on the FD camera that is moving

*e-mail: spokespersons@auger.org

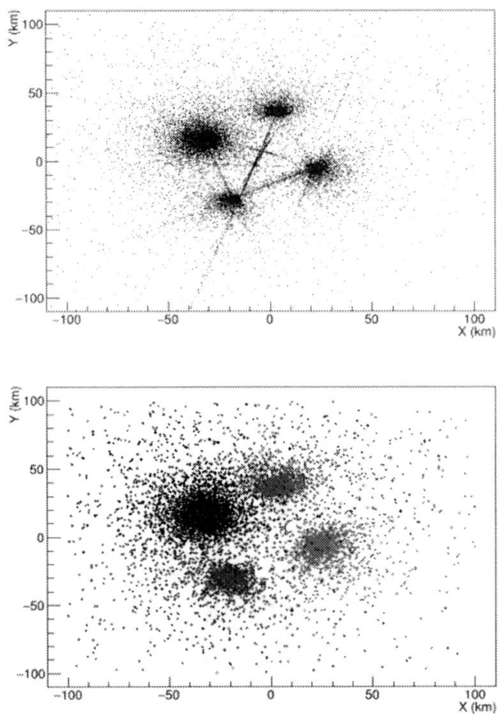


**Figure 1.** Distribution of shower cores in the burnt dataset before and after the laser cleaning process in the top and bottom panel, respectively. Apparent lines in the upper panel correspond to the misreconstructed laser events, and colours in the bottom panel represent FD sites that detected individual showers.

from low to high elevations, which is also consistent with the UG shower emerging in front of the telescope. This is visualised in Fig. 2.

To identify as many of events falling behind the telescope as possible, we apply a specific data cleaning that uses maximum likelihood values $L_{\mathrm{up}}$ and $L_{\mathrm{down}}$ coming from competing DG and UG geometry reconstructions, respectively. The decisive quantity that selects UG candidates is

$$l = \frac{\arctan\left(-2\log\left(L_{\mathrm{down}}/\max\left(L_{\mathrm{up}}, L_{\mathrm{down}}\right)\right)/50\right)}{\pi/2}, \quad (1)$$

which is cut, after optimization using simulations [5], above 0.55 to select signal.

After the cleaning procedure, a background of $0.45 \pm 0.18$ events is expected in the full data sample. The distributions of the $l$ quantity, Eq. (1), are shown in Fig. 3 for the burnt dataset, background simulations, and signal simulations. The correspondence between the burnt sample and the background distributions is a non-trivial result that comes from the precise knowledge of the detector and energy spectrum of UHECRs provided by the Pierre Auger Observatory.

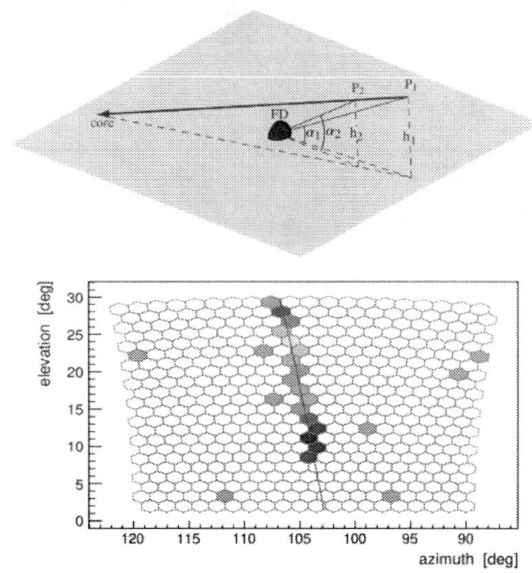


**Figure 2.** Geometry of a downward-going EAS with an impact point behind the field of view of an FD telescope (top panel). An image of such a shower moves in the field of view from bottom (magenta - former) to up (red - later) as marked by colours of triggered pixels (bottom panel). Gray pixels correspond to random triggers. The red line represents a projection of the shower axis.

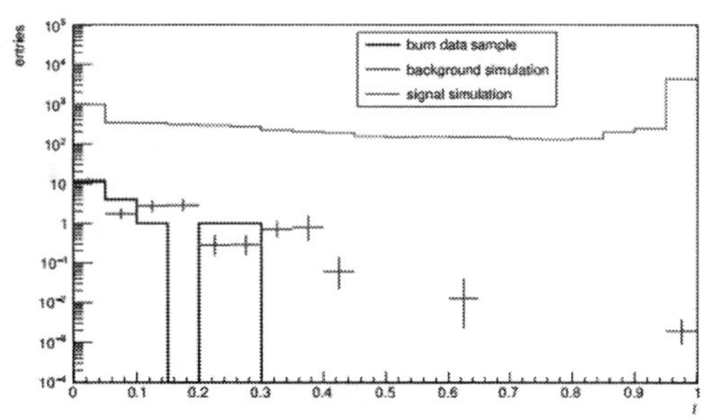


**Figure 3.** Distributions of the $l$ quantity in the burnt dataset (black), background simulations (red), and signal simulations (green). Background simulations are weighted to the UHECR flux measured by the Auger Observatory and normalized to 10 % of the data. The normalization of the signal simulations is arbitrary.

## 2.3 Exposure

Extensive Monte Carlo (MC) simulations were produced to estimate the FD exposure to UG EASs. Protons, chosen as primaries, were thrown in geometries that correspond to an isotropic emergence from the ground with the height of the first interaction ($H_{\mathrm{fi}}$) sampled equally between 0 and 9 km. This range is limited by the field of view (FoV) of the FD telescopes. Quantities used to define the injection of particles into the MC simulations are visualized in Fig. 4.

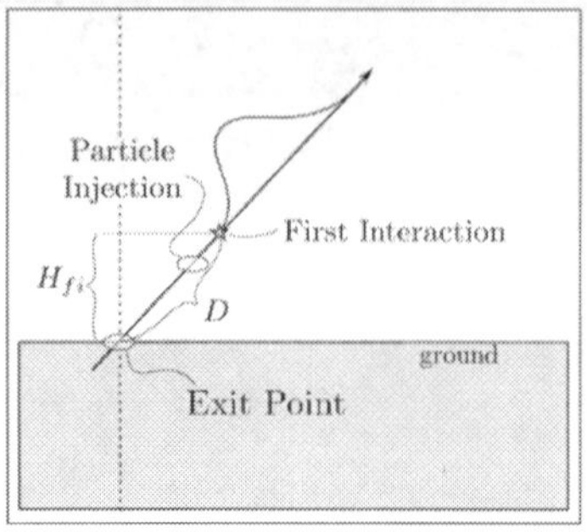


**Figure 4.** Parameters used in the simulation of UG EASs. Height of the first interaction $H_{\rm fi}$, and the distance between the particle exit point from the ground and the first interaction $D$ are shown.

After the application of the selection criteria, briefly described in Section 2.2, to the MC simulated showers, we evaluate selection efficiency and exposure in individual energy–$H_{\rm fi}$ bins. In Fig. 5, we present the double-differential exposure of the FD to UG EASs w.r.t. $H_{\rm fi}$ and calorimetric energy of showers ($E_{\rm cal}$). It allows one to test any type of exotic model of production of EASs, provided that the distribution of emergence of showers with altitude is predicted. Effective areas restricted to zenith angle bins of $\theta \in [110°, 124.2°]$, $\theta \in [124.2°, 141.3°]$, and $\theta \in [141.3°, 180°]$ were also calculated [5].

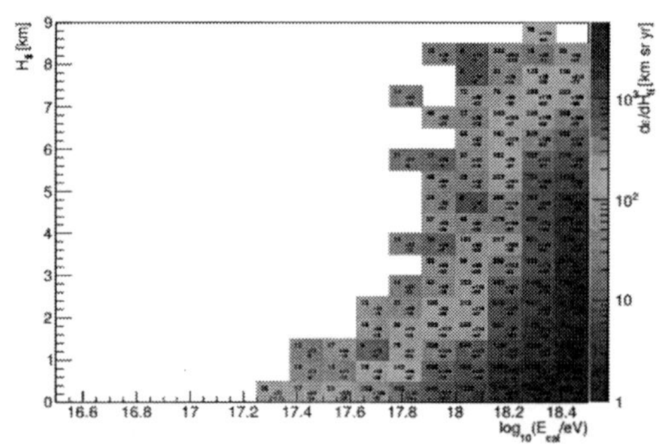

**Figure 5.** The double-differential exposure of the FD to UG EASs. An isotropic emergence of events for $\theta \in [110°, 180°]$ is assumed.

### 2.4 Candidates

After unblinding, 1 event has been observed to pass all the selection criteria in the full data sample. This is in line with the background expectation. Thus, we place upper limits on the integral flux of UG showers at 95 % confidence level, assuming two different spectral indices $\gamma = 1, 2$

$$F^{95\,\%}_{\gamma=1} = 3.6 \cdot 10^{-20}\,\mathrm{cm^{-2}sr^{-1}s^{-1}}, \tag{2}$$

$$F^{95\,\%}_{\gamma=2} = 8.5 \cdot 10^{-20}\,\mathrm{cm^{-2}sr^{-1}s^{-1}}. \tag{3}$$

These limits are valid for any UG EASs with $E_{\rm cal}$ above $10^{17.5}$ eV.

The double-differential exposure table shown in Section 2.3 can be used also to place differential limits in individual energy bins, and even place limits using specific scenario of the production of UG showers. This is illustrated in Section 3.

## 3 Upper limits on the flux of $\tau$ leptons

Having the previous results in hand, upper limits on a particular scenario of the EAS production from $\tau$ leptons can be set [6]. We assume a flux of $\tau$s in the energy range of $E_{0,\tau} \in [10^{16.5}\,\mathrm{eV}, 10^{20}\,\mathrm{eV}]$.

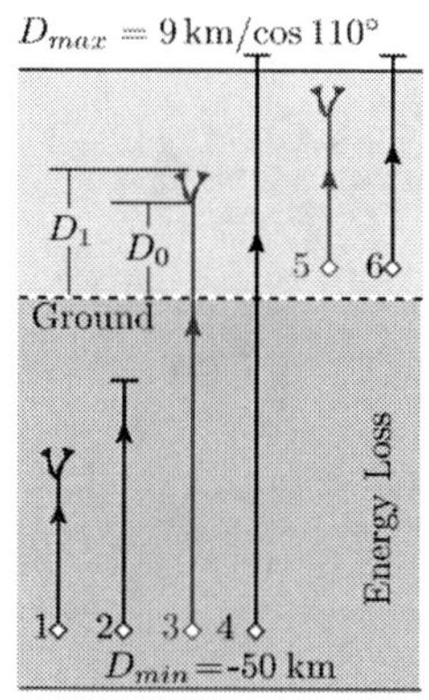


**Figure 6.** Possible outcomes of the $\tau$ simulations in NuTauSim. The height $H_{\rm fi}$ in Fig. 4 corresponds to the distance $D_1$, where the decay products of the $\tau$ interact.

Using the MC simulations in NuTauSim [7], we investigated the $\tau$ propagation in the Earth assuming their isotropic production in the distance range of $D \in [-50\,\mathrm{km}, 26.3\,\mathrm{km}]$ around the ground plane. According to the sketch in Fig. 6, only in cases 3 and 5, for which the $\tau$ decay appears in the FD-observable part of the atmosphere, we simulated the decay of $\tau$ in TAUOLA [8]. We obtained the distribution of $\tau$-decay induced EASs within the FoV of the FD shown in the top panel of Fig. 7.

The distribution of $\tau$ induced EASs is then folded with the double-differential exposure of the FD from Fig. 5 to get the number of events per height bin that would trigger the FD, given in the bottom panel of Fig. 7. It is further integrated in height for each energy bin.

After the calorimetric energy of $\tau$ EASs ($E_{\rm sh}$) is tracked back to $E_{0,\tau}$, we place differential upper limits on the emergence of $\tau$ leptons near the ground shown in Fig. 8. Because each $E_{0,\tau}$ contributes to several $E_{\rm sh}$, due to the energy distribution of $\tau$-daughter particles, we assume two spectral indices of the $\tau$ flux, $\gamma = 1, 2$, to calculate the limits.

These limits hold irrespectively of the underlying scenario of the $\tau$ production.

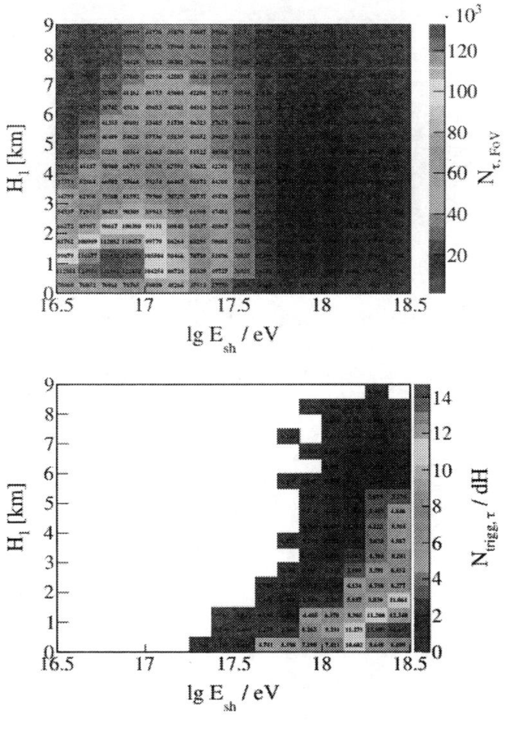


**Figure 7.** Distribution of MC-generated $\tau$-decay induced EASs within the FoV of the FD (top panel) and corresponding number of FD-triggered events per height bin (bottom panel). The height $H_1$ corresponds to the distance $D_1$ in Fig. 6 and coincides with $H_{fi}$.

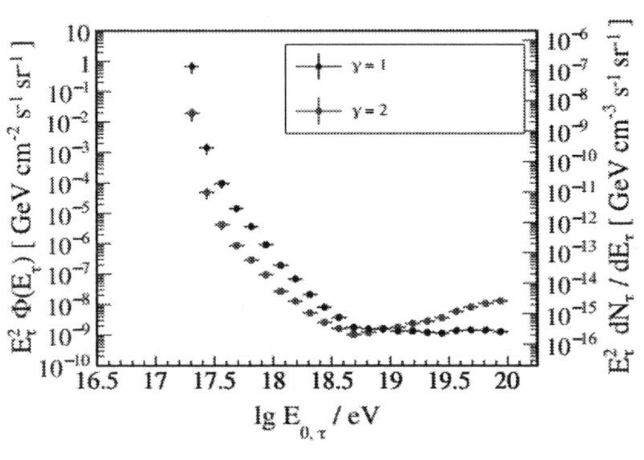


**Figure 8.** Differential upper limits on the isotropic emergence of $\tau$ leptons near the ground at 95 % confidence level. The left axis corresponds to the flux of $\tau$-leptons created within 50 km of path length below the Earth's surface, and the right scale shows the limits on the rate of $\tau$s generated per unit volume, energy, and solid angle. Two indicated spectral indices of the $\tau$ flux are assumed.

## 4 Conclusions

The data of 14 years of operation of the fluorescence detector of the Pierre Auger Observatory have been searched for upward-going extensive air showers with zenith angles between 110° and 180°. One candidate event has been found, which is consistent with the background expectation of ~0.5 events, coming from wrongly reconstructed downward-going cosmic-ray showers. Integral upper limits on the flux of upward-going showers were determined for two spectral indices, $\gamma = 1, 2$, used in the calculation of an average exposure. The double-differential exposure of the fluorescence detector was presented, and was applied to derive upper limits on the $\tau$ lepton emergence near the ground.

## Acknowledgement

This work was supported by the Czech Ministry of Education, Youth and Sports grant LTT18004.

## References

[1] A. Aab et al. (Pierre Auger), Nucl. Instrum. Meth. A **798**, 172 (2015)
[2] P.W. Gorham et al. (ANITA), Phys. Rev. Lett. **121**, 161102 (2018)
[3] L. Perrone et al. (Pierre Auger), these proceedings (2022)
[4] V. Novotny et al. (Pierre Auger), PoS **ICRC2019**, 374 (2021)
[5] M. Mastrodicasa et al. (Pierre Auger), PoS **ICRC2021**, 1140 (2021)
[6] I.A. Caracas et al. (Pierre Auger), PoS **ICRC2021**, 1145 (2021)
[7] J. Alvarez-Muñiz et al., Phys. Rev. D **97**, 023021 (2018), [Erratum: Phys.Rev.D 99, 069902 (2019)]
[8] M. Chrzaszcz, T. Przedzinski, Z. Was, J. Zaremba, Comput. Phys. Commun. **232**, 220 (2018)

# Performance evaluation of LHCf-ATLAS ZDC joint measurement using proton beam

LHCf Collaboration
*Moe* Kondo[1,*], *Oscar* Adriani[2,3], *Eugenio* Berti[2], *Pietro* Betti[2,3], *Lorenzo* Bonechi[2], *Massimo* Bongi[2,3], *Raffaello* D'Alessandro[2,3], *Sebastiano* Detti[2], *Maurice* Haguenauer[4], *Yoshitaka* Itow[1,5], *Katsuaki* Kasahara[6], *Yuga* Kitagami[1], *Yutaka* Matsubara[1], *Hiroaki* Menjo[1], *Yasushi* Muraki[1], *Ken* Ohashi[1], *Paolo* Papini[2], *Giuseppe* Piparo[7,8,15], *Sergio* Ricciarini[2,9], *Takashi* Sako[10], *Nobuyuki* Sakurai[11], *Monica* Scaringella[2], *Yuki* Shimizu[12], *Tadashi* Tamura[12], *Alessio* Tiberio[2,3], *Shoji* Torii[13], *Alessia* Tricomi[7,8,15], *William C.* Turner[14], *Kenji* Yoshida[6],
ATLAS ZDC group
*Zvi* Citron[16], *Brian* Cole[17], *Matthias* Grosse-Perdekamp[18], *Chad* Lantz[18], *Riccardo* Longo[18], *Daniel* MacLean[18], *Yftach* Moyal[16], *Shir* Shenkar[16], *Peter* Steinberg[19], *Lion* Sudit[16], and *Aric* Tate[18]

[1]Institute for Space-Earth Environmental Research, Nagoya University, Nagoya, Japan
[2]INFN Section of Florence, Florence, Italy
[3]University of Florence, Florence, Italy
[4]Ecole-Polytechnique, Palaiseau, France
[5]Kobayashi-Maskawa Institute for the Origin of Particles and the Universe, Nagoya University, Japan
[6]Faculty of System Engineering, Shibaura Institute of Technology, Japan
[7]INFN Section of Catania, Catania, Italy
[8]University of Catania, Catania, Italy
[9]IFAC-CNR, Florence, Italy
[10]ICRR, University of Tokyo, Kashiwa, Japan
[11]Tokushima University, Tokushima, Japan
[12]Kanagawa University, Kanagawa, Japan
[13]RISE, Waseda University, Shinjuku, Tokyo, Japan
[14]LBNL, Berkeley, California, USA
[15]Centro Siciliano di Fisica Nucleare e di Struttura della Materia, CSFNSM, Catania, Italy
[16]Ben-Gurion University of the Negev, Israel
[17]Columbia University, New York, USA
[18]University of Illinois at Urbana-Champaign, USA
[19]Physics Department, Brookhaven National Laboratory, New York, USA

**Abstract.** Measurements of forward neutrons in $pp$ collisions will allow us to investigate $\pi$-$p$ cross-section via one-pion exchange process, which are important for air shower development. However, the precision of these measurements is limited by the energy resolution of the LHCf detectors. To improve it, a joint measurement with the ATLAS ZDC was planned. In 2021, a beam test was conducted to evaluate the performance of the joint measurement of the LHCf-Arm1 and ZDC detectors using proton beams of 350 GeV at SPS. Combining the LHCf data with the ZDC data, we confirmed that the energy resolution improved from about 40% to 21.4%.

## 1 Introduction

The muon excess and model discrepancy of mass composition of ultra-high energy cosmic rays are mainly caused by a poor understanding of high-energy hadronic interaction [1]. Especially for high energies, $\pi$ interactions are hardly understood. To study the hadronic interaction and to contribute to improving the models, LHC-forward (LHCf) experiment [2] measures neutral particles emitted to the very forward region of $pp$ collision at the Large Hadron Collider (LHC). If LHCf detects the neutron energy properly, we can tag the one pion exchange events [3]. From the one pion exchange, we will be able to measure total cross-section and multiplicity of $\pi$-$p$ interactions using a joint operation data of LHCf and ATLAS. For these reasons, the LHCf experiment is expected to contribute more to model improvement due to the improved energy resolution for neutrons. Energy resolution for neutrons will be improved by combining energies deposited in LHCf with that in the ATLAS Zero Degree Calorimeter (ZDC), located immediately downstream of the LHCf detector.

*e-mail: kondo.moe@isee.nagoya-u.ac.jp

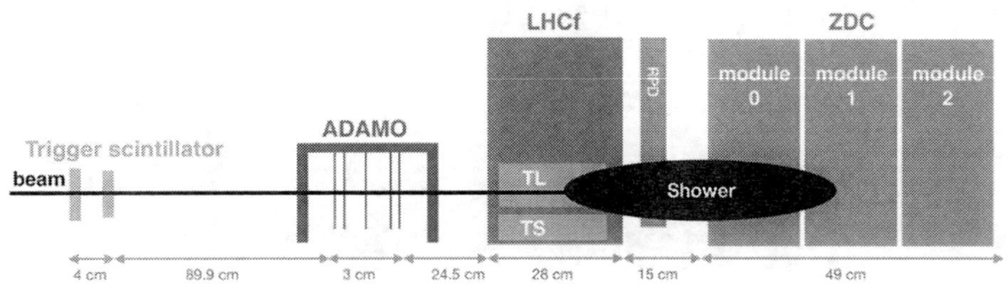


**Figure 1.** Experimental setup of the beam test. Beams were injected from the left-hand side, and they passed through trigger scintillators, ADAMO, LHCf, and ZDC in that order.

## 2 LHCf experiment and Arm1 detector

LHCf detectors are located 140 m apart on either side of the ATLAS Interaction point (IP1). One is called Arm1 and the other Arm2. We measure photons, $\pi^0$s, and neutrons. Arm1 detector consists of two sampling calorimeter towers, the Large Tower (TL) and the Small Tower (TS). Each tower has a sandwich structure of 16 layers of GSO plates, 4 layers of GSO-bar XY hodoscopes, and 16 layers of tungsten plates. The total length is $44X_0$[1] and $1.6\lambda_I$[2]. The energy resolution for neutrons is ~40%. The position resolution is better than 1 mm for neutrons [5].

## 3 Strategy of the operation in 2022

The LHCf had an LHC operation with proton-proton collisions at $\sqrt{s}$ = 13.6 TeV during the low-luminosity special run period scheduled for September 2022. To improve the energy resolution for neutrons, LHCf performs a joint operation with ZDC detector. Three hadronic modules of the ZDC detector [6], called as ZDC-HAD, were installed behind the LHCf detector. Each module consists of tungsten layers and fused-silica fiber layers, and produced Cherenkov lights in the fibers were read by a PMT. The interaction length of the ZDC-HAD modules is $3.4\lambda_I$, and the total interaction length of LHCf and ZDC-HAD detectors is $5.0\lambda_I$. Therefore, most of the hadron shower can be contained and the energy resolution is expected to be improved.

## 4 Beam test at SPS in 2021

### 4.1 Experimental setup

We conducted a beam test at the H4 beam line of SPS at CERN from 24th to 28th September 2021 to verify the performance of the joint measurement and to calibrate the energy. A set of trigger scintillators and a silicon tracker (ADAMO) were installed in front of the LHCf and ZDC detectors as shown in Fig. 1. ADAMO is a silicon strip detector, and it precisely determines the position of incident particles. Proton beams with 350 GeV/c were injected to the detectors with changing the position of detectors. In total, 860 k events were collected.

[1] $X_0$ is the radiation length.
[2] $\lambda_I$ is the interaction length. For reference, $1\lambda_I$ of tungsten is 191.9 g/cm$^{-2}$ [4].

### 4.2 Analysis

#### 4.2.1 Event quality cut and incident position cut

Events with only one track observed in ADAMO were selected to remove the events affected by pile-up and with showering in materials of the beam line. By projecting the track at ADAMO to the LHCf detector, the incident positions of the beam particles were determined. Moreover, a cut to select event with beam position in a 8 mm squared area in the center of the TL was applied.

#### 4.2.2 Energy Reconstruction

Total energy deposit in LHCf and ZDC detector was used as the energy estimator of the hadronic showers. The energy deposit in the LHCf-Arm1 detector, $E_{\mathrm{LHCf}}$, was calculated from the summation of energy deposit in 16 scintillator layers after pedestal subtraction and gain calibration. The energy deposit of ZDC, $E_{\mathrm{ZDC}}$, was obtained as a simple summation of pedestal-subtracted ADC counts from three ZDC modules because of non-availability of gain parameters in this moment. Leaked shower particles from the LHCf detector were measured in the ZDC detector. So, we were able to see the anti-correlation between LHCf energy deposit and ZDC energy deposit (Fig. 2). Considering the difference of the energy scale between $E_{\mathrm{LHCf}}$ and $E_{\mathrm{ZDC}}$, the total energy deposit of the two detector, $E_{\mathrm{LHCf+ZDC}}$, was defined using the function;

$$E_{\mathrm{LHCf+ZDC}} = E_{\mathrm{LHCf}} + a \times E_{\mathrm{ZDC}} \qquad (1)$$

where $a$ is a energy scale factor, which was obtained by fitting the correlation plot of Fig. 2 with a linear function. The value of $a$ is ~ 0.0089 ± 0.0006.

### 4.3 Result

Figure 3 shows the distribution of obtained $E_{\mathrm{LHCf+ZDC}}$ in the selected events. The $E_{\mathrm{LHCf+ZDC}}$ values were scaled as the mean value of the distribution to 350 GeV, and it is called 'corrected energy'. The corrected energy distribution from only the LHCf-measured energies was shown for the comparison. The energy resolution was estimated from the fitting result of the distribution by a gaussian function from 200 GeV to 500 GeV. The obtained energy resolution for LHCf + ZDC was 21.4%, which is considerably better than the 46% achieved by LHCf alone.

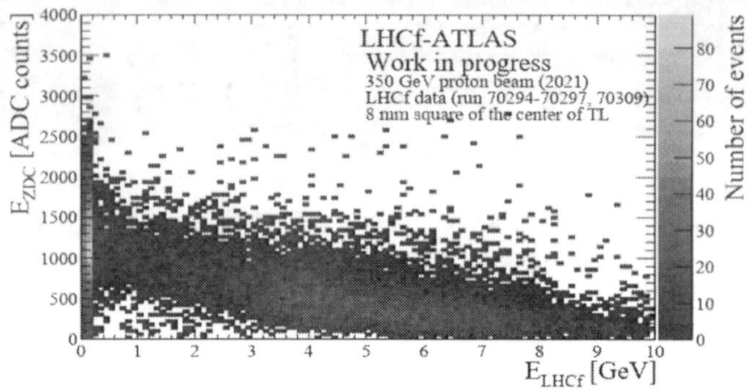


**Figure 2.** Correlation of the energy deposit in LHCf and the ZDC. The horizontal axis shows the total of LHCf energy deposit [GeV]. The vertical axis shows the summation of three ZDC modules' ADC counts.

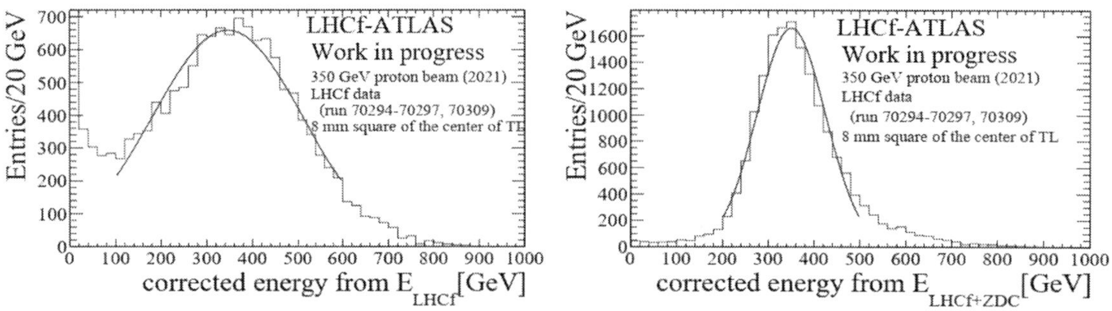


**Figure 3.** The distributions of 'corrected energies'. The red curves show Gaussian fits to the data. (left) Corrected energy from total energy deposit of LHCf only. (right) Corrected energy estimated from the summation of total energies in LHCf and ZDC. $a \sim 0.0089 \pm 0.0006$ was used as the scale factor of Eq.(1).

## 5 Summary

By analyzing data taken with 350 GeV proton beams at the SPS H4 beam line, we characterized the energy resolution resulting from joint operations of the LHCf and ATLAS ZDC detectors. The resolution was measured to be 21.4%. This value represents a strong improvement compared to the 46% resolution achieved by LHCf alone. The combined method will be applied to the data obtained with proton-proton collisions at $\sqrt{s} = 13.6$ TeV at LHC in September 2022 for studying the $\pi$-$p$ interactions.

Finally, we would like to thank the staff of the SPS and the North area test beam facility for making this beam test possible.

## References

[1] Albrecht, J., Cazon, L., Dembinski, H. et al., Astrophys Space Sci **367**, 27 (2022)
[2] O. Adriani et al., JINST **3** S08006 (2008)
[3] K. Ohashi et al., PoS **ICRC2021**, 190 (2022)
[4] Workman, R. L. and others, PTEP **2022**, 083C01 (2022)
[5] O. Adriani, E. Berti, L. Bonechi et al., JHEP **11**, 073 (2018)
[6] S. White, Nucl. Instrum. Meth. A **617**, 126-128 (2010)

# AugerPrime status and prospects

*Corinne* Berat[1,*] *on behalf of the* Pierre Auger Collaboration[2,**]

[1]Univ. Grenoble Alpes, CNRS, Grenoble INP, LPSC-IN2P3, 38000 Grenoble, France
[2]Observatorio Pierre Auger, Av. San Martín Norte 304, 5613 Malargüe, Argentina
Full author list: https://www.auger.org/archive/authors_2022_10.html

**Abstract.** The Pierre Auger Collaboration started a few years ago the AugerPrime project to increase the Surface Detector (SD) performance of the Pierre Auger Observatory. It aims to address the still open questions on the origin and composition of the highest energy cosmic rays by allowing a better identification of the nature of the primaries. The key element of this major upgrade is the capability of measuring the different components of extensive air showers, which will be significantly improved by the addition of a Surface Scintillator Detector (SSD) on each water-Cherenkov detector (WCD) constituting the SD. Moreover, the dynamic range of measurement is extended through an additional small photomultiplier tube inside the WCD. New electronics is processing the signals from the WCD and the SSD with higher sampling frequency and enhanced resolution. The scintillator module deployment started in 2019, and the new electronics in December 2020. The collected data allow for the evaluation of the first performances of the upgraded array and to adapt the whole data acquisition chain necessary for an efficient and sustainable operation of the Observatory. After the recall of the motivations for the upgrade, the main characteristics of the new detection setup are reviewed, as well as the status of its deployment and commissioning. The expected prospects are also discussed.

## 1 Introduction

The Pierre Auger Observatory [1] is the largest cosmic ray observatory, measuring extensive air showers (EAS) produced by very high-energy astroparticles that penetrate the Earth's atmosphere. It was designed to be a hybrid detector: it comprises a Fluorescence Detector (FD) and a Surface Detector (SD). The SD consists in 1600 water-Cherenkov detectors (WCD) distributed over 3000 km$^2$ on a triangular grid with a regular spacing of 1.5 km. It also includes two denser arrays with reduced spacing: 61 additional SD stations are deployed over 27.5 km$^2$ with a 750 m spacing, and 1.95 km$^2$ with a 433 m spacing. In each WCD, filled with ultra pure water, the Cherenkov light produced by EAS particles is viewed by three 9-inch photomultipliers (PMT). To provide a common reference for calibrating signals between all the WCD of the SD, the recorded signals are measured in units of the signal produced by a vertical muon traversing the tank, a vertical equivalent muon (VEM). The conversion of signal from hardware units to VEM is obtained by a local and automatic calibration procedure [2].

The FD is composed by 24 optical telescopes installed in 4 buildings, and overlooking the SD array, having a field of view of 30° in elevation with a minimum of 1.5° above the horizon, plus 3 additional ones, overlooking the denser arrays, with an elevation between 30 – 60° measuring lower energy air showers. To operate the FD, detailed knowledge of the atmospheric conditions is mandatory, therefore measurement and monitoring devices are installed at various locations. In the denser sub-arrays, two Engineering arrays are deployed, one to operate radio antennas (AERA[1]) [3], and the other to operate buried muon detectors (AMIGA[2]) [4].

With ~17 years of data taking during Phase I, the Observatory has accumulated high quality data, and the largest exposure to Ultra-High Energy Cosmic Rays (UHECR) reaching 90000 km$^2$ sr yr for spectrum reconstruction, and 122000 km$^2$ sr yr for anisotropy studies. This has led to a wealth of results providing major advances in our understanding of these highly energetic astroparticles, and notable results concern the energy spectrum [5], the flux composition [6], the searches for neutral particles and multi-messengers astrophysics [7] [8]. Nevertheless, several questions are still open: what are the origins of the spectrum index changes and the suppression observed at very high energy? What is the UHECR flux composition at the highest energies? Thanks to the searches for anisotropies in UHECR arrival directions, the extra-galactic origin of these particles has been proved by the observation of a dipole [9] above ≃ 8 EeV. Strong hints of correlation between UHECR arrival directions and the massive, star-forming, or active galaxies positions within 200 Mpc have been obtained above 40 EeV in intermediate scale anisotropy studies [10]. However, the UHECR sources remain unknown, even if candidates are suspected. Understanding the spectrum features and measuring the

*e-mail: berat@lpsc.in2p3.fr
**e-mail: spokespersons@auger.org

[1]Auger Engineering Radio Array
[2]Auger Muons and Infill for the Ground Array

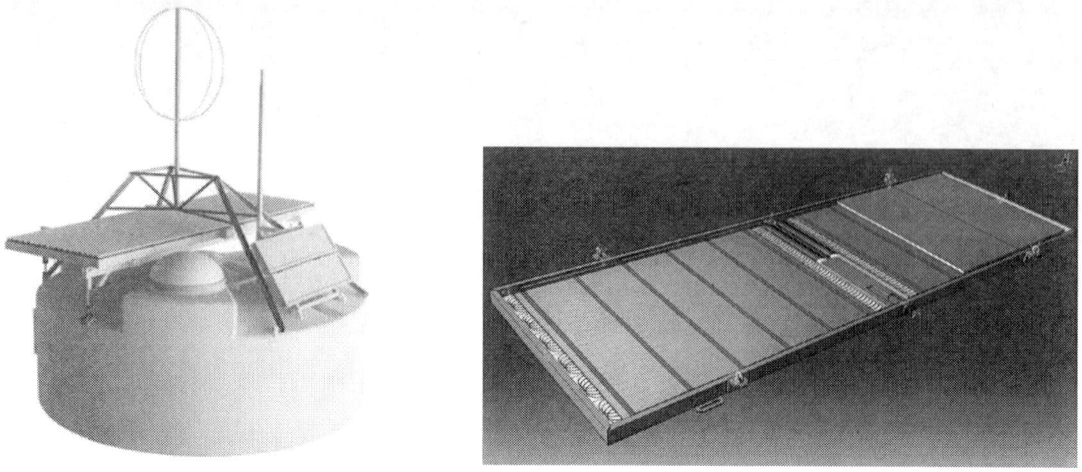

**Figure 1.** Left: drawing of a complete detection station for AugerPrime, with, on the WCD, a SSD module and an antenna. Right: design of one SSD module.

composition on the full energy range could constrain the source characteristics [11]. We are eager to do mass driven anisotropy studies, with the opportunity, in the case of a substantial light component of the flux, to do astronomy using these light nuclei. The various EAS measurements carried out have revealed inconsistencies between the muon content of the showers predicted by the models and that actually measured in the data (see e.g. [12] and references therein). This issue should be addressed by studying hadronic interactions in the kinematic region beyond man-made accelerators reaches.

Based on these findings, the objective of the Observatory upgrade, undertaken by the Pierre Auger collaboration and dubbed *AugerPrime*, is to measure the composition of the cosmic-ray flux at the highest energies, and to obtain an event-by-event composition. The aim is also to enhance the sensitivity to the $\gamma$ and $\nu$ fluxes exploring the potential of future experiments. The upgrade should provide a unique opportunity to study EAS, hadronic interactions and multiparticle production above a 70 TeV centre-of-mass energy, to explore particle physics beyond human accelerators and thus improve constraints on new physics phenomena.

## 2 AugerPrime components

To study the origin of UHECRs, it is necessary to improve the knowledge of the composition of the flux at high energy and on an event-by-event basis. Currently, determinations of the cosmic-ray mass composition are obtained by direct observations with the FD of $X_{max}$, the slant depth of the maximum of the shower development, and indirectly by measuring observables related to $X_{max}$ obtained with SD data. Due to the FD operating time, $X_{max}$ direct measurements can be performed with a duty cycle of only 13% to 15%. Adding mass sensitivity to the SD is thus the key for the upgrade.

As the number of muons present in the shower is strongly correlated with the mass of the primary, it is a valuable indicator for determining the composition. Therefore, disentangling the muonic and electromagnetic components of the EAS at ground provides the required improvement towards the mentioned science goals. This can be achieved either with a muon detector or with a detector having a different sensitivity to both EAS components. The WCD is significantly more sensitive to muons, thus we need to complement the information, delivered by the WCD with other detectors more sensitive to the electromagnetic part of the shower. For AugerPrime, plastic scintillators are installed on each WCD. These surface scintillator detectors (SSD) being mostly sensitive to vertical showers, radio antennas are installed on each WCD to improve the measurements of inclined ones, via the detection of their radio emission in the 30–80 MHz frequency range (Fig. 1, left). To increase the dynamic range of the delivered signals, a small photomultiplier tube (SPMT) is added in the WCD to register large signals near the shower core impact point (the 9-inch PMTs are then named large PMTs (LPMT)). Direct measurements of the shower muon content will be performed for lower energy showers using muon detectors, consisting in scintillators buried in the denser array. To process signals delivered by existing and new detectors, new surface detector electronics is required, with higher performances than the current one. In the following are described the SSD, the SPMT, and the electronics Upgraded Unified Board (UUB). Details concerning the radio detector (RD) and the underground muon detector (UMD) can be found in these proceedings [13] [14].

### 2.1 Surface Scintillator Detector

A SSD unit (Fig. 1, right) consists of a box of 3.8 m x 1.3 m containing two scintillator panels. Each panel is composed of 24 extruded scintillator bars (1600x50x10

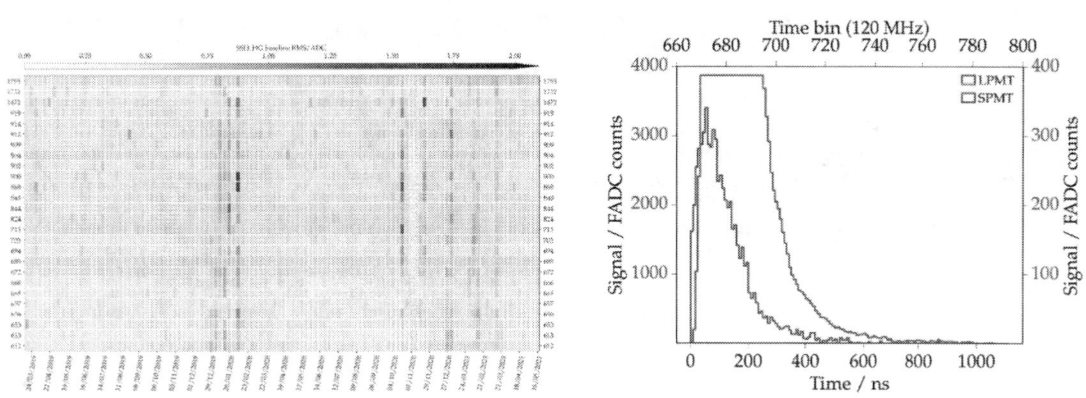


**Figure 2.** Left: evolution of the baseline of the PMTs high gain trace for a subsample of SSD of the preproduction array. Right: large PMT (LPMT) and small PMT (SPMT) signals in a detection station: the anode channel of the LPMT is clearly saturated (red), while the SPMT is providing an unsaturated signal (green).

mm). The scintillation light produced by particles is collected with wavelength-shifting fibres inserted into 2 straight extruded holes in the bars. The 1 mm diameter fibers (Kuraray Y11(300)M S-type) are positioned following the grooves of the routers at both ends in a U-configuration. The fiber ends are all together bundled in a PMMA cylinder and optically coupled to a single 1.5 inch photomultiplier tube (Hamamatsu R9420), with 18% quantum efficiency at 500 nm. The power supply of the PMT is based on a custom-made design manufactured by the ISEG company. Two PMT channels are available: high gain (amplified anode signal) and low gain (attenuated anode signal) with a factor of 32 between them. With these two channels, the dynamic range of the SSD spans from the signal of a single particle, mandatory for calibration purpose, to large signals, up to more than 10000 Minimum Ionizing Particles (MIP)s. The two scintillator panels are housed in a light and waterproof aluminium casing. In addition, a second roof is installed to provide thermal protection. The SSD module is attached to the top of the WCD tank using a support frame designed to withstand environmental conditions, particularly high winds.

A total of 1518 SSD units have been assembled and tested in seven European research institutes [15], members of the Pierre Auger Collaboration. Using atmospheric muons, the light tightness of each module has been checked, and the charge over peak of the MIP signal has been measured. The MIP is the unit for the SSD signal calibration, as the VEM is used for the WCD signal calibration. The number of photo-electrons per MIP was also measured to check the quality of each assembled module, and the obtained mean value is around (30 ± 2) photo-electrons per MIP. The uniformity in the response of the SSD detectors has been studied on a sample of detectors using external trackers on a muon tower setup. The linearity has been checked to be 5% in one scintillator bar, and 10% between bars. All photomultipliers to be installed in the SSD modules for converting the collected light have been tested in two European research institutes. Their characteristics have been validated, in particular their linear response when operated at low gain.

### 2.2 Additional small photomultiplier tube

As previously mentioned, for the purpose of extending the dynamic range of the signal acquisition, the WCD is equipped with an additional small photomultiplier tube, installed on an available view port on the tank liner, located close to one of the three large PMTs (LPMT). The photomultiplier tubes are Hamamatsu R8619 1-inch diameter ones, with passive base, and the power supply in a separate box [16]. Linearity and gain curves of each SPMT were carefully measured [17].

### 2.3 Upgraded Unified Board

The functionalities of the Upgraded Unified Board should at least be the same as the current one, but extended in performances and adapted to new detectors: analog signal processing, GPS time tagging, triggering, calibration, data acquisition, communications via radio transmitter [18]. It has also to provide interfaces with UMD and RD systems. Compared to the Phase I unified board, its capabilities have been improved: FADCs with higher sampling rate (from 40 to 120 MHz), larger dynamic range (from 10 to 12 bits), more accurate time tagging, significantly more powerful FPGA, upgraded CPU (more than 10 times faster). It is designed to ensure mechanical compatibility, by fitting the enclosure of current electronics, and accepting the existing connection cables. Backward compatibility with the current dataset are also maintained; in particular digital filtering and trace down-sampling are provided to emulate current triggers in addition to any new triggers. The design goals also include improved reliability and ease of maintenance.

The boards production started on January 2020, by the A4F (formely SITAEL SpA) company. Boards are tested after production, and environmental stress screening tests

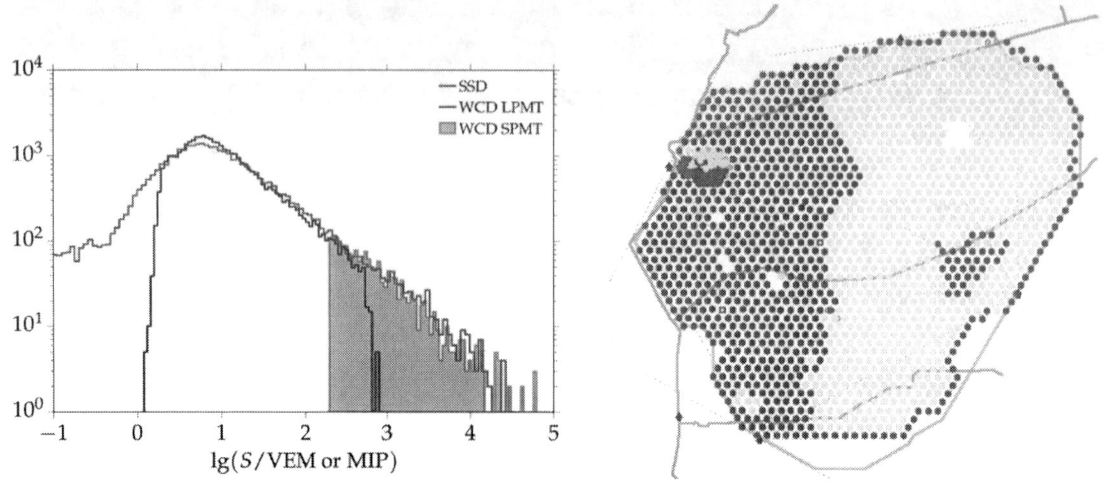


**Figure 3.** Left: The dynamic range of both WCD (blue) and SSD (red). The extension of the WCD dynamic range is obtained thanks to the SPMT. Right: map showing the status of the AugerPrime deployment on 03/10/2022. Red dots: WCD+SSD+UUB stations (except on the border) ; yellow dots: WCD+SSD; purple dots: WCD only.

are done in one European research institute of the Pierre Auger collaboration, before delivery to Argentina.

## 3 Pre-production array

SSD modules, SSD-PMTs and SPMT, once ready and tested, have been shipped to the Pierre Auger Observatory. To check the performances in situ, a pre-production array of 77 SSD was deployed in March 2019. The reading of the SSD-PMTs output was performed using a specific electronic module to adapt the PMT signal to the Phase I electronics (1 LPMT of the WCD had to be disconnected to free one channel and connect the SSD PMT).

The pre-production array has been monitored during several months, the behaviour of the SSD modules has been controlled, and the SSD parameter values checked, as well as the SPMT ones [19]. As an example, the evolution of the baseline of the PMTs trace (high gain channel) for a subsample of SSD of the preproduction array is shown in Fig. 2, left. In the vertical axis the identification numbers of the WCD are reported. The color uniformity along horizontal rows clearly shows the stability of the detectors. The vertical patterns signal the occurrence of thunderstorms or temporary communication default in the SD array. When comparing the amplitude of the baseline variations during days and nights, no difference has been found, confirming the light tightness of the SSD modules.

The SPMT have been also evaluated in the AugerPrime Engineering Array[3] (several upgraded detection stations), using UUB modules being tested in the fields. The SPMT active area being almost 100 times smaller than the LPMT one, with the appropriate gain value, the signal produced in a WCD as close as ~250 m from the shower core is unsaturated (Fig. 2, right). The signals in the WCD are measured up to saturation (around 700 VEM) by the LPMTs in the WCD tanks, and above the saturation region, they are derived from the SPMT: the non saturated range of the WCD extends to more than 20000 VEM (Fig. 3, left).

[3]Started in 2016, an Engineering Array of twelve upgraded stations, with detector and electronic board prototypes has been taking data in the SD array.

## 4 AugerPrime commissionning

### 4.1 Deployment status

The installation of the SSD modules on top of each WCD, over the 3000 km$^2$ in the pampa, started at the end of 2018. In spite of the COVID-19 pandemic, and the sometimes harsh conditions of access to the WCD, the deployment of the SSD was completed (except on the boarder, as planned, and in two particular areas, as it is impossible to access them) in March 2022. The deployment of the new electronics and the installation of the SPMT started in December 2020. In October 2022, thanks to the strong commitment and effort from the staff in Malargüe, more than 30% of the array has been upgraded (see Fig. 3, right). The production of electronics boards should be completed in January 2023, and the deployment will continue up to mid-2023.

### 4.2 Measured performances

Since beginning of 2021, the behaviour of the increasing part of the array which is upgraded is studied, and available data are analyzed. The data acquisition system of the Observatory has been running without major interruption. All triggers run at 40 MHz, to ensure backward compatibility, until the full upgrade of the array. Nevertheless, the traces with the binning corresponding to 120 MHz are recorded to benefit of the high sampling provided by the

https://doi.org/10.1051/epjconf/202328306001

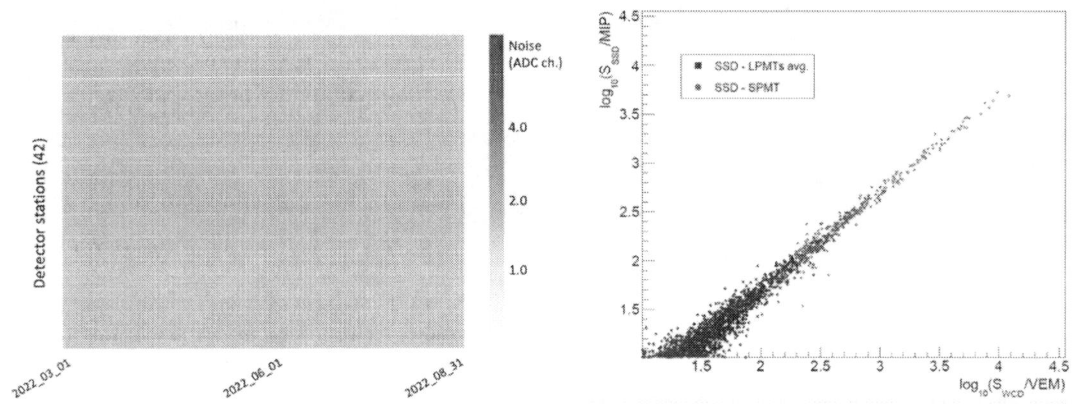


**Figure 4.** Left: RMS(baseline, high gain channel) of large PMTs as a function of time for a sub-sample of upgraded stations running without interruption. Right: Correlation between SSD and WCD signals. For the WCD, the large PMTs are used up to the saturation (blue dots); the measurements are extended further by the small PMT (red dots).

new electronics. Each SSD unit is triggered by the associated WCD, in a slave-mode.

For each upgraded detection station, composed by one WCD plus one SSD unit, the baselines of the 5 PMTs (3 LPMTs, 1 SPMT, 1 SSD-PMT) are constantly monitored. The fluctuations show that the level of noise is in the expected range, being below 2 FADC channels for the high-gain channel of the LPMTs (Fig. 4, left). The time resolution, measured using a set of showers triggering two nearby stations, is found to be in the order of 5 ns. Some issues concerning the GPS time tagging and the trigger rates were observed and are being resolved.

### 4.3 AugerPrime calibration

As explained in 1 and 2.2, the WCD signals are expressed in terms of VEM units, and the SSD signals in MIP. Signal calibration, which converts electronic units in VEM or MIP units, is performed using atmospheric muon signals acquired by dedicated triggers. About 40% of WCD calibration triggers produce a MIP in the SSD.

Due to the small sensitive area of the SPMT, it is not possible to calibrate their signals with the atmospheric muons. Instead, small local showers are selected to cross-calibrate SPMT using the VEM signals of the LPMTs. A very good correlation between the calibrated signals of the WCD (large and small PMTs) and the SSD is obtained (Fig. 4, right), on the full dynamic range, up to 20000 VEM. The calibration parameters (VEM, MIP) are constantly monitored, and a good stability of the corresponding values is observed. The possibility to improve WCD calibration taking advantage of the SSD is currently under study.

### 4.4 Reconstruction of events

The reconstruction of events with the SD data applied during Phase I is described in detail in [20]. Algorithms in the reconstruction process need to be revised to incorporate the information provided by the new detectors, in order to benefit from their valuable inputs.

Once the signals, measured by the individual detectors, including the SSDs, have been properly calibrated, they can be used in the reconstruction of the showers. After correctly defining the time integration range of the SSD signal corresponding to the passage of the shower particles, the uncertainties in the measured signals were quantified and shown to be proportional to the square root of the signal, with a proportionality coefficient depending on the zenith angle [21].

Algorithms have been developed for reconstructing shower-size estimator for SSD as complement to WCD estimator. These two shower size estimates are expected to allow the deconvolution of the relative intensity of the muonic and electromagnetic components of air showers. The signal as a function of the distance to the shower axis is expressed as $S(r) = S(r_{\rm opt}) f_{NKG}(r)$, using a modified NKG-like function, expressed by

$$f_{NKG}(r) = \left(\frac{r}{r_{\rm opt}}\right)^{\beta} \left(\frac{r + r_s}{r_{\rm opt}}\right)^{\beta+\gamma}$$

The optimal distance $r_{\rm opt}$, largely driven by detector spacing and array geometry, has been set to 1000 m for the SD in the event reconstruction [20]. Considering that SSDs and WCDs are co-located, the same value is adopted for the reconstruction of the LDF using the SSD signals. An example of the lateral distributions measured by both SSD and WCD are displayed in Fig. 5, showing different slopes. Nevertheless there is room for improvements, searching for more appropriate LDFs. Up to now, only the shower size at 1000 m is fitted, but for events with high SSD multiplicity, it will be possible to fit the LDF slope. More sophisticated algorithms using also the time structure of traces, and not only the signal magnitudes are in development.

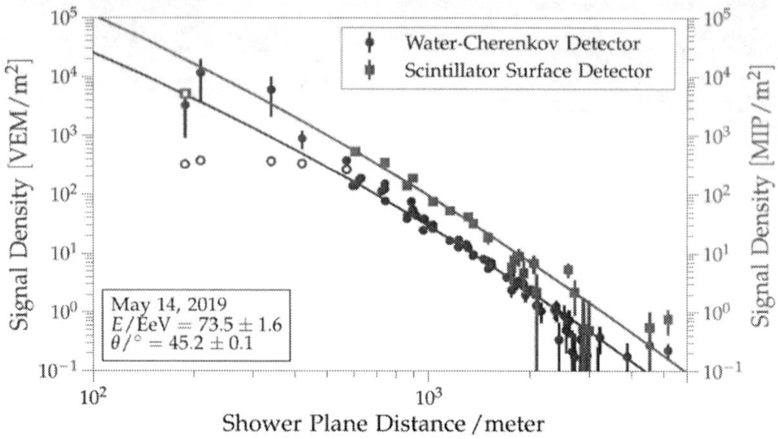


**Figure 5.** Exemple of a high-energy event measured in the pre-production array. The WCD (blue) and SSD (red) signal density are plotted as a function of the shower plane distance. Both corresponding fitted LDFs are shown.

## 5 Outlook

The objective of AugerPrime is to obtain enhanced composition information on the UHECR detected by the Pierre Auger Observatory, to answer still open questions on the origin of these astroparticles. It will provide a multi-hybrid cosmic ray detector, that will allow simultaneous measurement of a shower with water-Cherenkov detectors, surface scintillator detectors, radio detectors, muon counters and fluorescence detectors. Operation of the full upgraded Observatory is expected from 2023 to 2030, the goal being to reach an exposure of $\sim$ 40000km$^2$ sr yr for vertical events, i.e. those with a zenith angle below 60°.

More than 30% of the detection stations of the SD have been equipped with scintillator detectors, one additional small PMT in the water-Cherenkov detector and new electronics, allowing data taking and commissionning procedures. A stable behaviour of the different detectors is observed. Event reconstruction using information from the scintillation is underway, results are promising, and several studies are going on to improve it, based on principles of air shower universality or using machine learning techniques. Development of methods to deconvolve the contributions of the electromagnetic and muonic shower components are on-going, moving towards primary mass identification.

## References

[1] The Pierre Auger Collaboration, Nucl. Instrum. Meth. **A 798** 172-213 (2015)

[2] X. Bertou, et al., Pierre Auger Collaboration, Nucl. Instr. and Meth. **A 568** 839 (2006)

[3] The Pierre Auger Collaboration, Phys. Rev. **D 93** 122005 (2016)

[4] The Pierre Auger Collaboration, Eur. Phys. J. **C 80** 751 (2020)

[5] The Pierre Auger Collaboration, Eur. Phys. J. **C 81** 966 (2021)

[6] A. Yushkov for the Pierre Auger Collaboration, Proceeding of Science, 37th International Cosmic Ray Conference, 358 (2021).

[7] The Pierre Auger Collaboration, Universe **8** 579, (2022)

[8] Karl-Heinz Kampert, Miguel Alejandro Mostafa, Enrique Zas and The Pierre Auger Collaboration, Front. Astron. Space Sci. **6** 24 (2019)

[9] The Pierre Auger Collaboration, ApJ **868** 4 (2018)

[10] The Pierre Auger Collaboration, ApJ **935** 170 (2022).

[11] E. Guido, for the Pierre Auger Collaboration, PoS(ICRC2021)311

[12] D. Soldin, PoS(ICRC2021)349

[13] T. Huege, for the Pierre Auger Collaboration, in these proceedings.

[14] M. Scornavacche , for the Pierre Auger Collaboration, in these proceedings.

[15] J. Pekala, for the Pierre Auger Collaboration, PoS(ICRC2019)380

[16] G.A. Anastasi et al., JINST **17** T04003 (2022)

[17] M. Buscemi et al., JINST **15** P07011) (2020).

[18] G. Marsella for the Pierre Auger Collaboration, PoS(ICRC2021)230.

[19] G. Cataldi for the Pierre Auger Collaboration, PoS(ICRC2021)251.

[20] The Pierre Auger Collaboration, JINST 15 P10021 (2020)

[21] D. Schmidt, for the Pierre Auger Collaboration, PoS(ICRC2021)218

 https://doi.org/10.1051/epjconf/202328306002

# The Radio Detector of the Pierre Auger Observatory – status and expected performance

*Tim* Huege[1,2,*] for the Pierre Auger Collaboration[3,**]

[1]Karlsruhe Institute of Technology, Institute for Astroparticle Physics (IAP), Karlsruhe, Germany
[2]Vrije Universiteit Brussel, Astrophysical Institue, Brussels, Belgium
[3]Observatorio Pierre Auger, Avenida San Martín Norte 304, Malargüe, Argentina

**Abstract.** As part of the ongoing AugerPrime upgrade of the Pierre Auger Observatory, we are deploying short aperiodic loaded loop antennas measuring radio signals from extensive air showers in the 30 – 80 MHz band on each of the 1,660 surface detector stations. This new Radio Detector of the Observatory allows us to measure the energy in the electromagnetic cascade of inclined air showers with zenith angles larger than ~ 65°. The water-Cherenkov detectors, in turn, perform a virtually pure measurement of the muon component of inclined air showers. The combination of both thus extends the mass-composition sensitivity of the upgraded Observatory to high zenith angles and therefore enlarges the sky coverage of mass-sensitive measurements at the highest energies while at the same time allowing us to cross-check the performance of the established detectors with an additional measurement technique. In this contribution, we outline the concept and design of the Radio Detector, report on its current status and initial results from the first deployed stations, and illustrate its expected performance with a detailed, end-to-end simulation study.

## 1 Introduction

The Pierre Auger Observatory is currently undergoing the AugerPrime upgrade [1]. In addition to equipping every water-Cherenkov detector (WCD) with a surface scintillator detector (SSD) [2] and improving the dynamic range and the readout electronics, every WCD will also be equipped with a dual-polarized antenna measuring the radio signals from extensive air showers in the 30 – 80 MHz band. With the 1.5 km spacing of the Auger surface detector (SD) array, this will allow the measurement of inclined air showers, with zenith angles beyond ~ 65°, which have been both predicted by simulations [3] and confirmed with measurements of the Auger Engineering Radio Array (AERA) [4] to illuminate areas of tens of km$^2$ with detectable radio signals. For such geometries, the electromagnetic cascade of extensive air showers will be effectively absorbed in the atmosphere, and the WCDs will measure the virtually pure muon content of the particle showers. The radio antennas, in contrast, are purely sensitive to the electromagnetic cascade of the air showers. The combination of the two provides very good sensitivity to the mass composition of the primary cosmic rays, with different systematics from the combination of WCD and SSD measurements provided by AugerPrime for more vertical air showers.

In this article, we will shortly describe the concept and design of the *Auger Radio Detector* (RD), report on its current status and plans for deployment, and finally present the results of an end-to-end simulation study of its expected performance.

## 2 Concept of the Auger Radio Detector

The Auger RD builds on the available infrastructure of the upgraded SD. An upgraded detector station including a radio antenna is shown in Fig. 1. Two aluminum rings forming the dual-polarized antenna are mounted on a fiberglass mast in which the signal cables are contained. The mast is fixed by an aluminum frame directly connected to the structure of the SD tank (not touching the surface scintillator detector beneath). Guy wires additionally support the mast and reduce vibrations that could otherwise be induced by strong winds.

The antenna is a *Short Aperiodic Loaded Loop Antenna* (SALLA) [6], a revised and improved design based on the one originally developed in the context of AERA [7]. It features a simple mechanical design, minimizing cost and easing handling and maintenance. With its diameter of 122 cm, it is tailored at the frequency range of interest of 30 – 80 MHz, for which it delivers a virtually uniform response with very little dispersion. The antenna features a 392 Ω resistor at the bottom which shapes the antenna's main lobe towards the zenith and suppresses dependence on structures below the antenna, in particular the SSD, the WCD and potentially variable ground conditions. Although originally designed for AERA, the SALLA was not favored at the time because its resistor leads to a decreased sensitivity; only ~ 10% of the captured signal in-

*e-mail: tim.huege@kit.edu
**e-mail: spokespersons@auger.org, full author list available at https://www.auger.org/archive/authors_2022_10.html

**Figure 1.** Photo of a prototype RD station in the field. The dual-polarized SALLA antenna is mounted on a fiberglass mast held in place by an aluminum frame, which is connected directly to the WCD tank. Guy wires help suppress vibrations. The additional RD readout electronics is encapsulated in the dome visible in the foreground.

tensity is available at the input of the low-noise amplifier at the top of the rings. However, this is not a relevant problem for the Auger RD, as it aims at the detection of the highest-energy cosmic rays, where signal strengths are not a limiting factor.

The concept for the readout is illustrated in Fig. 2. Each of the two antenna channels is read out by an analog-to-digital converter with a sampling rate of 250 MHz and a dynamic range of 12 bits. Before digitization, the signals are amplified by a total of 36 dB and filtered by a band-pass filter with a width of 30 – 80 MHz. An FPGA (not shown in the diagram) coordinates data exchange with the *Upgraded Unified Board* (UUB) which queries 2,048 samples of the radio data whenever a trigger was received from the WCD. In the future, we plan to also include information from the radio detector in the trigger decision, which would be useful in particular for the detection of photon-induced air showers.

Finally, the data are sent via the existing wireless communications system as part of the regular data stream. Further monitoring information, such as a regular characterisation of the Galactic radio background, will in addition be transmitted as part of a monitoring data stream.

## 3 Status of the Auger Radio Detector

The design of the RD hardware has undergone several iterations with various prototypes and has been finalized after completion of a critical design review. Since November 2019, a hexagonal configuration of seven prototype RD stations have been operated within an *RD engineering array* (RDEA). The mechanical structures have proven to withstand the harsh conditions of the Argentinian pampa, and also the electronics have performed within specifications. At the site of the RDEA, the background situation is very favorable with very low rates of pulsed radio-frequency interference (RFI) and a comparatively clean measurement band.

In Fig. 3 we show events measured over the course of 12 months, triggered by the WCDs, for which at least one pulsed signal above Galactic background was seen at the expected time. The left panel shows the distribution of arrival directions, and as usual for air-shower radio emission, a north–south asymmetry can be seen due to threshold effects related to the angle the shower makes with the Earth's magnetic field (cf. magnitude of the Lorentz force). The right panel shows the distribution of the cores of the detected air showers.

For the calibration of the detector we perform calculations of the antenna directional gain pattern using the NEC2 code and in-the-lab measurements of the responses of the LNAs, the filteramplifiers and the digitizers. These characteristics and full capability for reading in and processing measured data as well as simulations have already been included in the Offline analysis framework of the Pierre Auger collaboration [8], building on the analysis functionality previously developed for AERA [9]. For an in-situ absolute calibration, we then compare the frequency- and local-sidereal-time-dependent power received from the sky with a model of the Galactic background folded through our above-mentioned detector response. The comparison shows that the two are in agreement within ~ 5% [10], confirming the accuracy of the characterisation of the system and illustrating the feasibility of a continued absolute calibration on Galactic background signals [11].

With the design finalized and proven, we have initiated mass production of the required 1,660 units plus spares in 2022. Except for some cables, the LNAs and the RD digitizer boards, the required hardware is already on site in Malargüe and being assembled for deployment. After a temperature-cycling and in-the-lab characterization of the LNAs, they and the missing cables will be shipped to Argentina shortly. The production of the digitizers currently suffers from delivery chain disruptions resulting in the unavailability of certain components. That said, deployment of a first series of complete detectors is envisaged in the first half of 2023, and the deployment of the full RD array is envisaged to be completed by the end of 2023.

## 4 Expected performance of the Auger Radio Detector

In an earlier publication [12], we had estimated the expected performance of the Auger Radio Detector in terms of achievable event statistics, detection threshold and potential for muon-number measurements on the basis of a signal model and an assumed background. Here, we present a study based on a fully realistic, end-to-end sim-

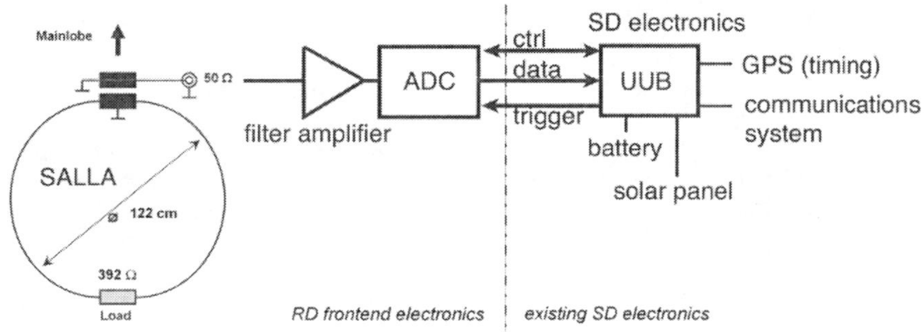


**Figure 2.** Schematic diagram for the RD readout chain and its connection to the existing Surface Detector electronics via the Upgraded Unified Board (UUB). Updated from [5].

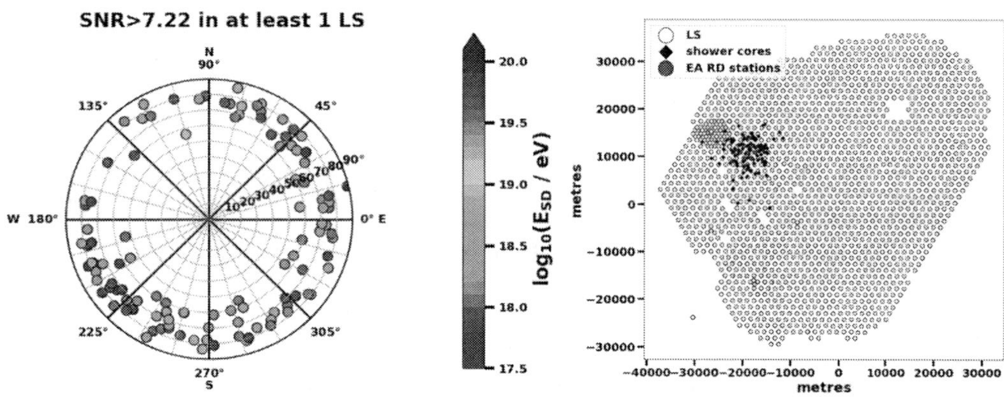


**Figure 3.** Data taken with the RD engineering array over the course of 12 months. *Left:* Angular distribution of air showers for which at least one radio pulse with a signal-to-noise ratio (SNR) exceeding a value of 7.22 (defined as the maximum amplitude squared divided by the RMS power in a noise window) was seen in at least one antenna at the expected pulse arrival time after triggering the readout by the WCDs. *Right:* Distribution of the impact points of the showers for which an air-shower radio signal was detected.

ulation of the actual detector response, Monte Carlo simulations with the CoREAS code [13], and an event reconstruction approach tailored to radio detection of inclined air showers [14], using realistic uncertainties and only information that will also be available in measured data. This allows us to provide a detailed account of the expected performance of the Auger RD in terms of event statistics, detection threshold, achievable energy resolution, potential for muon number measurements, and mass composition sensitivity.

### 4.1 Simulation setup

As a basis for this study, we use CORSIKA [15] simulations, including CoREAS [13] for the simulation of the radio emission, for proton, helium, nitrogen and iron nuclei arriving with zenith angles in the range from 65° to 85° following a $\sin^2\theta$ distribution, energies sampled uniformly in the logarithm of the energy in the range from $10^{18.4}$ eV to $10^{20.2}$ eV, and random, uniformly distributed azimuth angles. The atmosphere and magnetic field configuration were chosen with representative values for the site of the Auger Observatory. A total of ~ 8,000 simulations each were generated with the QGSJETII-04 [16] and Sibyll 2.3d [17] interaction models and used to derive the results presented in this article. An additional 8,000 simulations probing the effects of more conservative particle thinning and the influence of seasonal changes in the atmospheric conditions were also generated and used for further cross-checks.

These simulations undergo a realistic end-to-end detector simulation using the Auger Offline analysis framework. The particle content of the showers is used to simulate the triggering of the WCD stations; only radio signals of particle-triggered stations are used in the following. The radio signals are convolved with the direction- and frequency-dependent antenna gain for the dual-polarized antennas, followed by bandpass filtering, signal amplification and digitization. Actual measured noise from the RDEA stations is added to the simulated ADC traces. A Gaussian smearing with a width of 5% is applied to the ADC traces of individual antennas to mimic uncertainties and variations from antenna to antenna. Finally, the signal timing is smeared out with a Gaussian of width 6 ns to mimic the accuracy reachable by the GPS timing system, yielding "simulated traces like measured data".

### 4.2 Detection efficiency, aperture and event statistics

First, we study the detectability of air showers with the Auger RD. We deconvolve the detector response and reconstruct the electric field traces. On the basis of these we determine a signal-to-noise ratio (SNR), measured as the ratio of the square of the maximum, Hilbert-enveloped electric field, divided by the RMS power determined from a noise window. We consider a signal detectable if it exceeds an SNR of 10, which ensures a signal purity of better than 99%. We count a shower as "detected" if three antennas were triggered to be read out by the associated WCDs and exceeded this SNR-threshold. The detection efficiency we then calculate as the ratio of such-detected showers and the number of showers that triggered the SD. (Calculating the ratio to all simulated showers would underestimate the detection efficiency because of the muon number deficit in simulations; the SD trigger simulation would be too pessimistic.)

The resulting detection efficiency with the Auger Radio Detector as a function of energy and zenith angle is shown in Fig. 4 (left). It is obvious, and expected, that the zenith angle has a much higher influence on detectability than the energy, as it governs the size of the radio-emission footprint and thus is the main factor determining whether a signal can be detected in three antennas coincidentally. The Auger Radio detector will be fully efficient as of $\sim 70^\circ$ zenith angle.

Using the detection efficiency as a function of zenith angle and energy, we calculate the aperture of the Auger Radio Detector and show it as a stacked histogram in Fig. 4 (right). Here, we only account for air showers which have their impact point inside the geometrical area of the Auger SD, i.e., only *contained* events. A factor of $\cos\theta$ arises from the projection of the array area to the plane normal to the incoming direction and explains why the contributions of the most inclined air showers to the aperture become small, even though their detection is fully efficient for the full energy range from $10^{18.6}$ eV to $10^{20.2}$ eV. In contrast, the lower zenith angles provide a larger aperture at the highest energies, but detection becomes inefficient at low energies. Note that, for the highest energies, the aperture approaches the geometrical optimum. These results are in reasonable agreement with the earlier, much more simplified, estimates reported in reference [12].

In a next step, we multiply the aperture with the cosmic-ray flux as published in reference [20] to determine the number of detectable events. For an assumed measurement period of 10 years, the expected event statistics are shown in Fig. 5. The quoted numbers denote the integral event statistics for showers above a given threshold energy. The blue data points refer to the expected event numbers if all detected events are counted; the red points refer to only those bins in zenith angle and energy for which the detection efficiency is at least 97%. For the highest energies of $10^{19}$ eV or higher, we expect between 3,000 and 4,000 cosmic rays in 10 years.

### 4.3 Expected energy resolution

In a next step, we evaluate the resolution of the energy reconstruction achievable with the Auger Radio Detector. It should be noted that radio detectors measure the *electromagnetic energy* of the air shower, i.e., the energy content of the electromagnetic cascade. This can be converted to cosmic-ray energy with knowledge of the *invisible energy*, but strictly speaking is dependent on the mass of the primary particle. In terms of mass composition sensitivity, however, it is anyway preferable to directly use the electromagnetic energy in conjunction with the muon number as measured by the particle detectors, which will be explored in the following subsection.

To reconstruct the electromagnetic energy $E_{\mathrm{em}}$ from radio measurements, one first determines the energy fluence, i.e., the energy deposited in the form of radio emission in the $30-80$ MHz band per unit area, at every antenna position with a detectable signal. After an area integration and the application of corrections for the effects of the angle to the magnetic field and the air density at shower maximum, one can then determine the *corrected radiation energy*, an accurate estimator for the energy in the electromagnetic cascade of an air shower [21–23].

For the Auger Radio Detector, we use a reconstruction procedure tailored specifically to inclined air showers based on a signal model that explicitly separates the geomagnetic and charge-excess emission in inclined air showers and also takes care of geometrical early-late asymmetries in the signal distributions on ground [14]. In the following, we only use air showers with a detectable signal in at least five antennas, and we discard events with zenith angles below $68^\circ$. We estimate an initial impact point of the shower from the SD reconstruction and the arrival direction of the shower from RD signals using a spherical wavefront model. With some loose quality cuts on the sampling of the radio-emission footprint and the quality of the signal-distribution fit, survived by $\sim 95\%$ of the simulated showers, we retrieve the energy reconstruction performance illustrated in Fig. 6 (left). The scatter plot illustrates a 1:1 correlation between reconstructed and true electromagnetic energy, with an improvement of the resolution towards higher energies (and thus signal-to-noise ratios). In the right panel, a more quantitative characterisation of the reconstruction quality is shown in terms of histograms of the ratio of reconstructed versus true electromagnetic energy, both for the simulation set as a whole and for the subsets for particular types of the primary particle. No significant biases are present; the energy resolution is at the level of $\sim 6\%$.

### 4.4 Expected mass composition sensitivity

The goal of the upgraded Pierre Auger Observatory is to exploit separate measurements of the electromagnetic and

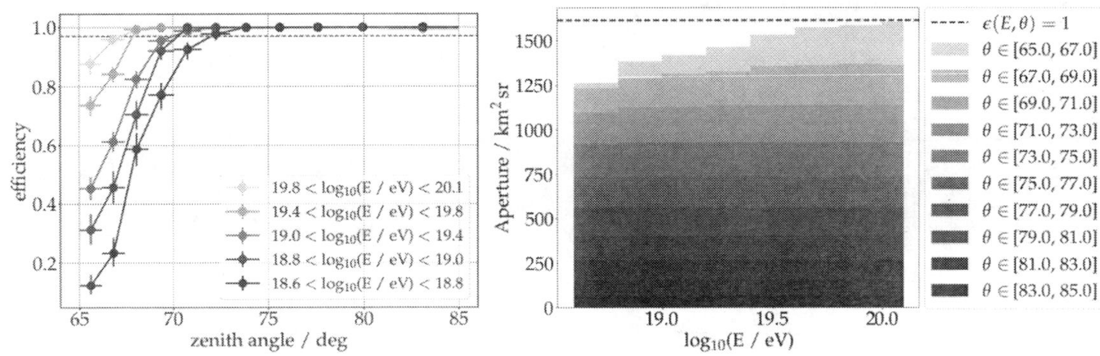


**Figure 4.** Left: Detection efficiency with the Auger Radio Detector as a function of zenith angle an particle energy. The dashed line marks an efficiency of 97%. The error bars denote statistical uncertainties. Right: Stacked histogram of the aperture of the Auger Radio Detector as a function of zenith angle and energy. This applies to *contained* showers which have their impact point in the geometrical 3,000 $km^2$ area of the Pierre Auger Observatory.

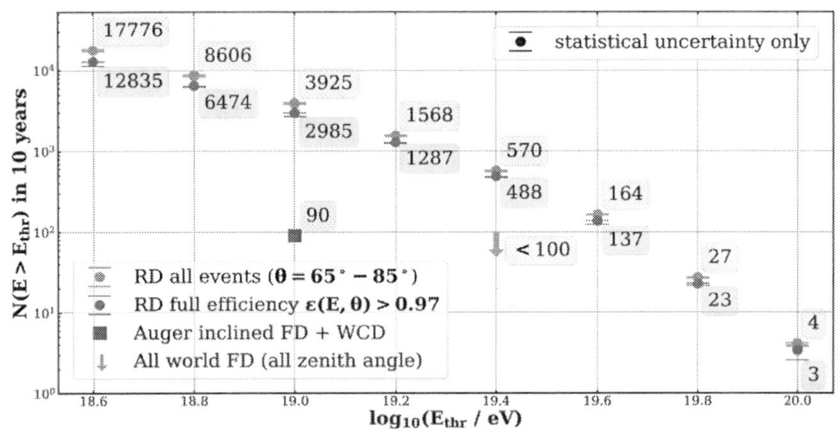


**Figure 5.** Integral number of cosmic rays expected to be detected by the Auger Radio Detector for a measurement period of 10 years. Blue points denote all detected events, red points only those in bins of energy and zenith angle for which the detection efficiency is at least 97%. For comparison, the number of inclined air showers measured with both the Auger Surface and Fluorescence Detectors is shown in green [18], and the total statistics of air showers measured with Fluorescence Detectors worldwide at energies of $10^{19.4}$ eV or higher is shown in yellow [19].

muonic content of air showers for mass composition measurements. (The combined measurement breaks the degeneracy in the muon number scaling with both energy and mass of the primary particle.) While for vertical air showers the separate measurement of the two components is achieved by the combination of the WCD and the SSD, for inclined air showers the combination of the RD and WCD is used. The RD delivers an accurate measurement of the electromagnetic energy of an air shower, while the WCD measures the virtually pure muon content of inclined air showers. The combination thus allows us to study the muon content of extensive air showers, as has previously been performed using the combination of WCD and Fluorescence Detector data [18], yet with orders of magnitude lower statistics than will be possible with the RD, as illustrated already in Fig. 5.

We first investigate an (unrealistic) mixture of 50% proton and 50% iron showers that have been reweighted to follow the actual, steep energy spectrum of cosmic rays at reconstructed electromagnetic energies above $10^{19}$ eV with the statistics expected for ten years of measurement time. With the standard SD reconstruction for inclined air showers, we determine $R_\mu$, a measure for the muon content in air showers relative to the expectation for a simulation template at an energy of $10^{19}$ eV. We remove a non-linear scaling with electromagnetic energy and define $r = R_\mu/(E_{em}/10\,\text{EeV})^{0.91}$. The scatter plot and corresponding profiles for $r$ shown in Fig. 7 (left) illustrate the separation between proton and iron primaries in the normalized relative muon number. The figure of merit $\text{FOM} = |\langle r_p\rangle - \langle r_{Fe}\rangle| / \sqrt{\sigma^2_{r_p} + \sigma^2_{r_{Fe}}} = 1.60 \pm 0.05$, which is

 https://doi.org/10.1051/epjconf/202328306002

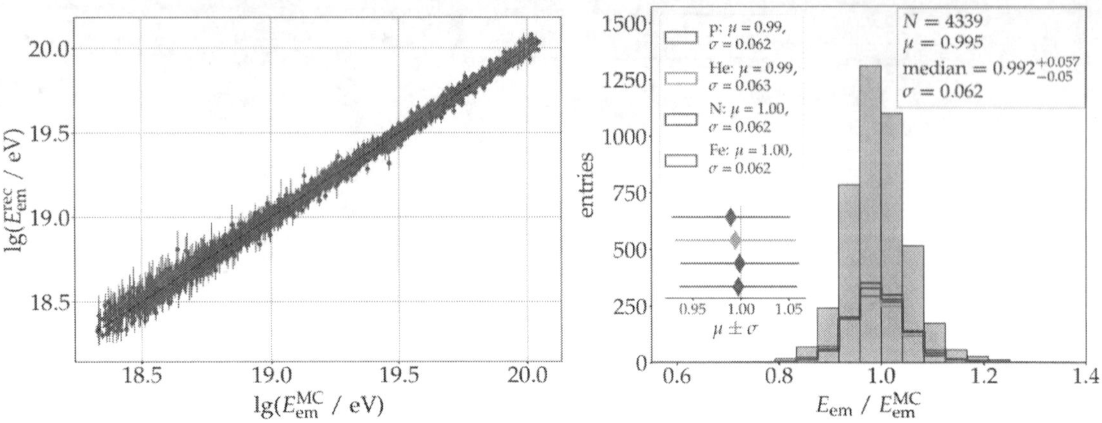


**Figure 6.** Left: Reconstructed electromagnetic energy versus Monte Carlo true electromagnetic energy as expected for the Auger Radio Detector. Right: Histograms illustrating the performance of the energy reconstruction for the whole simulation set as well as the subsets for different primary particle types.

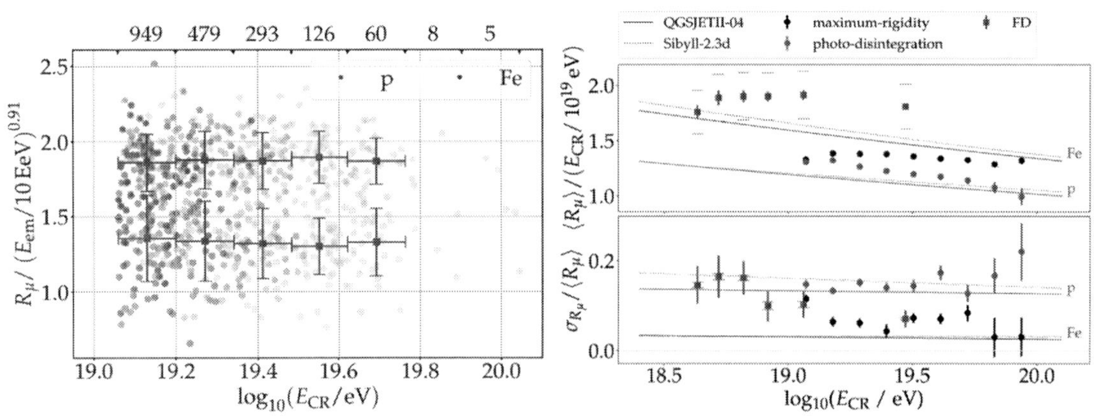


**Figure 7.** Left: Separation power between proton and iron primaries of the relative muon number normalized by the electromagnetic energy as a function of cosmic-ray energy. Right: Discrimination power between to different mass composition scenarios using the mean muon number (top) and the fluctuations in the number of muons (bottom) for combined measurements with the Auger Radio and Water Cherenkov Detectors in 10 years. In green, for comparison, measurements made previously with the Auger Fluorescence and Water Cherenkov Detectors.

an excellent value comparable to that of depth of shower maximum measurements with a near-perfect resolution below $10\,\mathrm{g\,cm^{-2}}$. Key to this success is the excellent energy resolution of $\sim 6\%$ of the Auger Radio detector; the sensitivity would strongly suffer from a worse energy resolution, especially for the steep spectrum of cosmic rays at the highest energies.

Finally, we investigate the mass composition sensitivity for more realistic mixed composition scenarios, namely the ones chosen for the original preliminary design report for the upgrade of the Pierre Auger Observatory [24]. One of the scenarios is a maximum-rigidity scenario whereas the other one represents a mass composition as expected from photo-disintegration. Again, we use the actual, steep cosmic ray spectrum and the expected event statistics for ten years of measurements. The results are shown in Fig. 7. Note that we here normalize on cosmic-ray energy $E_{CR}$ after converting to it from $E_{em}$ with a linear function determined from simulations, $E_{CR} = E_{em} \cdot [1.1426 - 0.0328 \log_{10}(E_{em}/10\,\mathrm{EeV})]$. The upper panel shows the mean number of muons as a function of energy. The error bars show the statistical uncertainties and illustrate the vastly improved statistics with respect to previous measurements with the Fluorescence and Water Cherenkov Detectors [18] shown by the green data points. However, measurements of the mean number of muons suffer from relatively large systematic uncertainties, as illustrated by the error caps of the green data points, arising to a large extent from uncertainties in the absolute energy scale of the measurement. The combined

RD/WCD data are expected to have a systematic uncertainty on a similar scale as shown here for the FD/WCD data. The situation is much more favorable when measuring the *fluctuations* in the number of muons as a function of primary energy, as illustrated in the lower panel. Here, systematic uncertainties due to the absolute energy scale cancel out. Measurements with the Radio and Water Cherenkov Detectors will therefore provide a powerful means to differentiate scenarios for the mass composition of cosmic rays at energies above $10^{19}$ eV.

## 5 Conclusions

We have reported on the design, status, and expected performance of the Auger Radio Detector. This detector will consist of a dual-polarized short aperiodic loaded loop antenna on top of each surface detector, measuring radio signals in the 30 – 80 MHz band. The design of the Radio Detector is final and has been proven to work in an engineering array of seven detectors operating since November 2019. Measurements with the engineering array have successfully confirmed the compatibility of the calibration in the laboratory with an absolute calibration on the Galactic radio emission, and have also successfully demonstrated detection of radio pulses from extensive air showers. Mass production of the RD components is mostly complete, except for the digitizer electronics which suffer from continued delivery chain disruptions. Nevertheless, complete deployment of the Auger Radio Detector is envisaged for the year of 2023.

Meanwhile, the Offline analysis framework of the Pierre Auger Observatory has successfully been enabled for analyzing RD data. The detector has been characterized in detail in the analysis software, allowing us to perform an end-to-end simulation study of the expected performance of the RD on the basis of CoREAS simulations and measured background data. With this, we have shown that the RD will achieve full efficiency for air showers above $\sim 10^{18.6}$ eV for zenith angles above $\sim 70°$. We have quantified the aperture and predicted the expected number of detectable cosmic rays as a function of energy, yielding between 3,000 and 4,000 events in 10 years above $10^{19}$ eV.

Furthermore, an algorithm for the reconstruction of the electromagnetic energy of air showers measured with the RD has been developed and implemented in the Offline analysis framework. Under fully realistic conditions, employing realistic uncertainties and using only information that will also be available in actual measurements, this reconstruction achieves a resolution on the electromagnetic energy of $\sim 6\%$, with no significant bias, not even between different types of primary particles.

This excellent energy resolution is key to exploiting the mass composition sensitivity of combined measurements of the Radio and Water Cherenkov Detectors. We have studied this sensitivity for a realistic, steep spectrum of air showers, showing that the figure of merit for a separation of proton and iron primaries lies at an excellent value of 1.60. For more realistic mass composition scenarios, we have demonstrated that in particular the fluctuations in the number of muons measured by the combination of Radio and Water Cherenkov Detectors will be a powerful means to determine the mass composition of cosmic rays at the highest energies.

## Acknowledgements

We are very grateful to all the agencies and organizations that support the Pierre Auger Observatory (https://www.auger.org/collaboration/funding-agencies). For this work in particular, we acknowledge the support from the European Research Council Advanced Grant #787622 and the Dutch Research Council.

## References

[1] A. Castellina (Pierre Auger), EPJ Web Conf. **210**, 06002 (2019), 1905.04472

[2] C. Berat et al. (Pierre Auger), these proceedings (2023)

[3] T. Huege, A. Haungs, JPS Conference Proceedings **09**, 010018 (2016), proceedings of the UHECR2014 conference, Springdale, USA, arXiv:1507.07769

[4] A. Aab et al. (Pierre Auger), JCAP **10**, 026 (2018), 1806.05386

[5] J.R. Hörandel (Pierre Auger), EPJ Web Conf. **210**, 06005 (2019)

[6] O. Krömer, H. Gemmeke, W.D. Apel, et al., *New Antenna for Radio Detection of UHECR*, in *Proceedings of the 31st ICRC, Lodz, Poland* (2009), 1232

[7] P. Abreu, M. Aglietta, M. Ahlers, et al., JINST **7**, P10011 (2012), 1209.3840

[8] S. Argiró, S.L.C. Barroso, J. Gonzalez, et al., Nucl. Instr. and Meth. A **580**, 1485 (2007)

[9] P. Abreu, M. Aglietta, E.J. Ahn, et al., Nucl. Instr. Meth. A **635**, 92 (2011)

[10] P. Abreu et al. (Pierre Auger), PoS **ICRC2021**, 270 (2021)

[11] M. Büsken, T. Fodran, T. Huege, submitted to Astronomy & Astrophysics (2022), 2211.03086

[12] B.P. for the Pierre Auger Collaboration, PoS(ICRC2019)395 (2019)

[13] T. Huege, M. Ludwig, C.W. James, AIP Conf. Proc. **1535**, 128 (2013), 1301.2132

[14] F. Schlüter, T. Huege, JCAP in press (2022), 2203.04364

[15] D. Heck, J. Knapp, J.N. Capdevielle, G. Schatz, T. Thouw, FZKA Report 6019, Forschungszentrum Karlsruhe (1998)

[16] S. Ostapchenko, Phys. Rev. D **83**, 014018 (2011)

[17] F. Riehn, R. Engel, A. Fedynitch, T.K. Gaisser, T. Stanev, Phys. Rev. D **102**, 063002 (2020), 1912.03300

[18] A. Aab et al. (Pierre Auger), Phys. Rev. Lett. **126**, 152002 (2021), 2102.07797

[19] A. Coleman et al., Astropart. Phys. **147**, 102794 (2023), 2205.05845

[20] A. Aab et al. (Pierre Auger), Phys. Rev. D **102**, 062005 (2020), 2008.06486

[21] A. Aab, P. Abreu, M. Aglietta, et al. (Pierre Auger Collaboration), Phys. Rev. D **93**, 122005 (2016)

[22] A. Aab, P. Abreu, M. Aglietta, et al. (Pierre Auger Collaboration), Phys. Rev. Lett. **116**, 241101 (2016)

[23] C. Glaser, M. Erdmann, J.R. Hörandel, T. Huege, J. Schulz, JCAP **09**, 024 (2016), 1606.01641

[24] A. Aab et al. (Pierre Auger) (2016), 1604.03637

# Current status of the TAx4 surface detectors

*Eiji* Kido[1,*] for the Telescope Array Collaboration[**]

[1]Astrophysical Big Bang Laboratory, RIKEN, Wako, Saitama, Japan

**Abstract.** Telescope Array (TA) is the largest ultrahigh-energy cosmic-ray (UHECR) observatory in the northern hemisphere. It explores the origin of UHECRs using a surface detector (SD) array, covering approximately 700 km$^2$, and a fluorescence detector (FD) stations viewing the skies above the array. TA has found evidence for a cluster of cosmic rays with energies greater than 57 EeV known as a HotSpot. Recently, TA has obtained additional evidence for an excess of cosmic rays with energies above $10^{19.4}$ eV and also found implications of spectrum anisotropy. An expanded SD array and FD stations called TAx4 experiment was designed to increase the data collection rate at the highest energies to confirm the anisotropies with more data. We have constructed and deployed more than half of the SDs for the TAx4 experiment and began stable data taking operations. We present TAx4 SD's current status.

## 1 Introduction

The Telescope Array (TA) is the largest cosmic-ray observatory in the Northern Hemisphere [1]. The aim of TA is to explore the origin and nature of ultrahigh-energy cosmic rays (UHECRs) above $10^{18}$ eV. The TA experiment consists of 507 plastic scintillator surface detectors (SDs) arrayed in a square grid with 1.2 km spacing covering approximately 700 km$^2$. The TA area is overlooked by three fluorescence detector (FD) stations, two of which have 12 telescopes, while the third has 14 [2–4]. The full operation of TA began on May 11, 2008. The duty cycle of the SD array has been 95% throughout the 11-year observation period. The FD duty cycle has been approximately 10%, because the data is taken only on clear moonless nights. The layout of TA is shown in Fig. 1; its central laser facility (CLF) is located at 39.30°N, 112.91°W and 1370 m above sea level in Utah, USA.

In previous work using TA SD data collected over five years, we found that 19 out of 72 observed cosmic rays with energies above 57 EeV were concentrated within a single 20°-radius circle (the "HotSpot") [5]. The result had a maximum local statistical significance of $5.1\sigma$; for the HotSpot to arise by chance anywhere in the field of view would require a $3.4\sigma$ fluctuation, which is the global statistical significance of the result. As reviewed in [6], the global statistical significance does not reach $5\sigma$ even with TA SD data collected over fourteen years. More recently, an additional excess of events with energies greater than $10^{19.4}$ eV with a significance of more than $3\sigma$ in the direction of the Perseus-Pisces supercluster was also found [7].

Several additional indications of cosmic-ray energy-spectrum anisotropy have also recently been observed by the TA experiment [8, 9]. A difference in the spectral cutoffs above and below the declination $\delta = 24.8°$ has been reported in [8]: the high and low declination band cutoffs were found to be $10^{19.85^{+0.03}_{-0.03}}$ eV and $10^{19.59^{+0.05}_{-0.07}}$ eV, respectively. The difference was updated in [10]. The statistical significance of the difference of the breakpoints of the energy spectra was estimated to be $4.3\sigma$. A relative energy distribution test compared events inside oversampled spherical caps of equal exposure to those outside. The center of maximum significance has been found at right ascension $\alpha = 9^h16^m$ and declination $\delta = 45°$, and has an excess of arrival directions for particles with energies above $10^{19.75}$ eV and a deficit for $10^{19.2}$ eV $\leq E < 10^{19.75}$ eV [9]. The post-trial significance was estimated to be $3.7\sigma$. Correlation of arrival directions with energies was also reported in [11]; the post-trial significance of the correlation was estimated to be $4.2\sigma$. These results were obtained using seven to eleven years of observations with the TA SD.

To accelerate the pace of data collection as we further investigate these apparent departures from isotropy, we have developed a quadrupled TA detector. We call this larger detector "TAx4."

The new SD array of the TAx4 experiment is designed expressly to study cosmic rays with energies above 57 EeV [12]. The spacing of the new SD array is 2.08 km, in contrast to the original TA SD's 1.2 km. The area covered by new SDs over the area covered by the TA SDs can be estimated to be $(2.08/1.2)^2 \sim 3$ if there are a large number of SDs. The additional 500 new SDs cover approximately three times more area than the 507 TA SDs; the combined coverage of the TAx4 and TA SDs is approximately 2800 km$^2$. Fig. 1 shows the deployed and planned locations of new detectors for the TAx4 experiment. The first 257 of the new SDs were deployed in 2019. Deployed SDs have been running stably since November 2019. Two new FD stations for the TAx4 experiment were also com-

*e-mail: eiji.kido@riken.jp
**Full author list http://www.telescopearray.org/research/collaborators

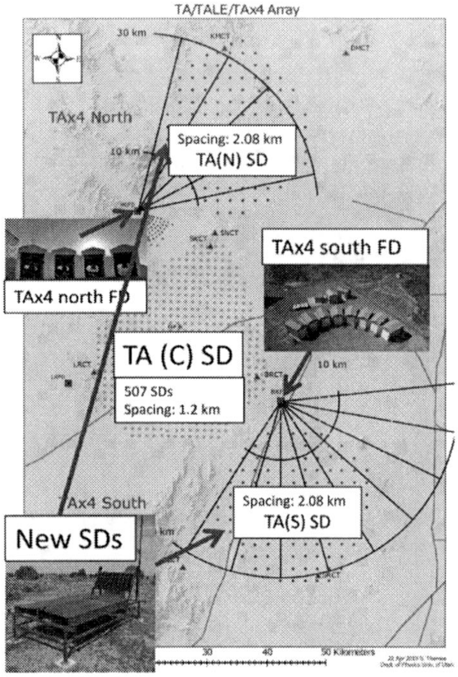


**Figure 1.** Map of the Telescope Array (TA) site. Each red circle in the north (N) and south (S) corresponds to the location of a newly deployed TA surface detector (SD); the spacing of the SDs is 2.08 km. Each yellow circle denotes the future location of a planned TAx4 SD. Each green circle in the west shows the location of a TA SD that is located at the center (C) between north and south array; the spacing of these is 1.2 km. The two fan shapes drawn with black lines are the fields of view of the TAx4 fluorescence detectors (FDs). Four telescopes have been built in the northern Middle Drum FD station, and eight telescopes in the southern Black Rock FD station. The overlap of the locations of the SDs and the fields of view of the FDs enables SD-FD hybrid observation. Pictures of new SDs and FDs are also shown.

pleted. The north FD station runs from June 2018, and the south FD station runs from September 2020. The field of view of each FD covers the sky above the area of the SDs so that FD and SD can observe cosmic-ray events simultaneously; we call such events "hybrids."

The update of the operation of new SDs is shown in Section 2. The status of hybrid observation of new FDs and SDs checks are given in Section 3. Section 4 provides a summary.

## 2 Operation of new SDs

The design and expected performance of new SDs were shown in [12] in detail. The main updates are the wireless communication protocol for the data acquisition and the arrangement of instruments inside the scintillator box. The SDs of the TA experiment used a customized wireless communication protocol, and new SDs use Wi-Fi protocol for new wireless communication devices. Therefore, the required signal strength for stable data transfer required additional study for the TAx4 experiment.

Many collaborators visited the observation site to optimize the wireless communication status for the data acquisition in August, 2022. Amplifiers for the wireless communication were newly installed in many of new SDs, and the signal strength from communication towers was kept above -80 dBm after the maintenance for the stable data transfer.

The performance of deployed SDs was shown in [13]. The statistics of collected UHECRs is still small, but a reasonable agreement between simulations and data was found.

## 3 Hybrid observation of new FDs and SDs

New FDs are running stably except for a few months in 2020 due to COVID-19 pandemic. All geometrical parameters are in reasonable agreement with simulations in the monocular analysis of FDs, as shown in [14].

The trigger efficiency of new SDs is approximately 30% at around 10 EeV (see Fig. 4 in [12]). Therefore, hybrid triggers were implemented to increase the efficiency. FDs send the trigger timing to the communication towers of the SDs within $\pm$ 128 $\mu$sec time window. Three times larger equivalent number of the TA FD and SD hybrid events above 10 EeV is expected from the full TAx4 experiment. Hybrid triggers of new FDs and SDs have been stably operated from June 2020.

## 4 Summary

The TA experiment continues to observe UHECRs from 2008 with the largest detection area in the northern hemisphere. The global statisitical significance of the HotSpot above 57 EeV was estimated to be more than $3\sigma$ with TA SD data collected over five years. A new excess of events with energies above $10^{19.4}$ eV was also found to have a significance greater than $3\sigma$. The global statistical significance of spectrum anisotropy was also estimated to be more than $3\sigma$. However, no global statistical significance of anisotropies exceeds $5\sigma$ even with TA SD data collected over forteen years.

The TAx4 experiment plans to complete the TAx4 upgrade, i.e. totally 500 new SDs with 2.08 km spacing in order to realize approximately 2800 km$^2$ detection area by combining with TA SDs (1.2 km spacing). The first 257 new SDs were deployed in 2019. These SDs are running stably since November, 2019. Two new FD stations have been completed and are operational. New north FD is running stably since June, 2018. New south FD is running stably since September, 2020. Simulations and data of new SDs, FDs, and hybrid events are in reasonable agreement.

## References

[1] H. Kawai, et al., Nuclear Physics B: Proceedings Supplements **175-176**, 221 (2008)
[2] H. Tokuno, et al., Nuclear Instruments and Methods in Physics Research Section A **601**, 364 (2009)
[3] H. Tokuno, et al., Nuclear Instruments and Methods in Physics Research Section A **676**, 54 (2012)
[4] R.U. Abbasi, et al., Phys. Rev. Lett. **92**, 151101 (2004)
[5] R.U. Abbasi, et al., ApJ. **790**, L21 (2014)
[6] J. Kim, *Updates on the Hotspot and the Perseus-Pisces supercluster Excess Observed by the Telescope Array Experiment*, in *6th International Symposium on Ultra High Energy Cosmic Rays* (2022)
[7] R.U. Abbasi, et al. (2021), 2110.014827
[8] R.U. Abbasi, et al. (2018), 1801.07820
[9] R.U. Abbasi, et al., ApJ. **862**, 91 (2018)
[10] D. Ivanov, *Energy Spectrum Measured by the Telescope Array Experiment*, in *36th International Cosmic Ray Conference (ICRC 2019)* (2019), PoS(ICRC2019)298
[11] R.U. Abbasi, et al., ApJ. **899**, 86 (2020)
[12] R.U. Abbasi, et al., Nuclear Instruments and Methods in Physics Research Section A **1019**, 165726 (2021)
[13] K. Fujisue, *TAx4 surface detectors data analysis*, in *6th International Symposium on Ultra High Energy Cosmic Rays* (2022)
[14] M. Potts, *Monocular Energy Spectrum using the TAx4 Fluorescence Detector*, in *6th International Symposium on Ultra High Energy Cosmic Rays* (2022)

# TAx4 surface detectors data analysis

*Kozo* Fujisue[1,*] for the Telescope Array collaboration[**]

[1]Institute for Cosmic Ray Research, Kashiwa-no-ha 5-1-5, Kashiwa city, Japan

**Abstract.** The TAx4 experiment aims to understand UHECR by expanding the observation area of the TA experiment by a factor of 4 and increasing the statistics of UHECR events with energy greater than $10^{19}$ eV. TAx4 consists of newly installed Surface Detectors (SD) and Fluorescence Detectors (FD). The SD was only partially developed before the COVID pandemic and it currently operates with an area 2.5 times TA including the original TA area. The TAx4 SD array has been collecting data since April 2019, and data analysis is underway. In this paper, we will report on the comparison between Monte Carlo simulation and real data acquired by the TAx4 SD array and the preliminary energy spectrum of TAx4 SD array.

## 1 Introduction

Telescope Array (TA) experiment, located in Utah, USA, aims to reveal the origin of Ultra-High-Energy Cosmic Rays (UHECR) by observing extensive air showers using both Surface Detectors (SD) and Fluorescence Detectors (FD) and covering about 700 km$^2$ [1]. TA experiment has observed an indication for a cluster of arrival directions of cosmic rays above 57 EeV within a 20° circle with 5 years data which is known as the Hot Spot. The posterior chance probability to observe the excess anywhere in the field of view with 5 years exposure is 3.4$\sigma$ [2].

TAx4 experiment was proposed to confirm the Hot Spot by increasing the aperture. The detectors have been deployed to the northern side and southern side of TA SDs as shown in Fig 1. Currently, TAx4 experiment consists of 257 TAx4 SDs [3] deployed on a square grid with 2.08 km spacing, which is different from TA SD spacing(1.2 km). TAx4 SDs plus the original TA SDs cover about 2.5 times the area of the original TA SD. TAx4 SDs began the data taking in April 2019.

In this paper, we show current analysis results of TAx4 SD using 1.5 years data for TAx4 SD northern array and 2 years data for TAx4 SD southern array. Monte Carlo (MC) simulation is compared with the observation, and we show they agree within statistical uncertainty for energy greater than 10 EeV. A preliminary energy spectrum observed by TAx4 SD array is also discussed. The TAx4 SD energy spectrum is consistent with the TA SD standard spectrum measured with 11 years of data [4].

## 2 MC simulation

We use the same MC simulation method as TA SD [5]. The method is as follows : (i) showers are generated by the CORSIKA program [6] using QGSJET II-04 [7] for high energy hadronic interaction (for TA SD, QGSJET II-03 was used), FLUKA [8] is used for low energy hadronic interaction, and EGS4 [9] for electromagnetic interaction. The energy range generated goes from $10^{17.5}$ eV up to $10^{20.5}$ eV, and 200 showers are generated for each $\Delta\log_{10}(E/eV) = 0.1$ bin. Zenith angles($\theta$) are simulated following a $\sin\theta\cos\theta$ distribution at the detector level, which means isotropic arrival direction of cosmic rays. Primary particle generated for this study were all proton. (ii) After the shower generation by CORSIKA, all shower particles at ground level are divided spatially into square tiles with 6 m length per side and temporally into 20 ns time bins. Energy deposit to an SD which is virtually located at the center of the square tile for each 20 ns time bin is calculated using a detector response table developed using GEANT4 [10] package. (iii) Energy deposit is converted to ADC counts considering PMT response, electronics response and so on. In this step, real time calibration data is used, so that the simulated air shower events are comparable with observed air shower events. By selecting appropriate tiles and calibration data, we can simulate for various shower core positions, azimuthal angles and event times.

## 3 Data/MC comparison

Comparison between observed real events and simulated events is needed to confirm that the MC simulation accurately reproduces real data. To this aim the distributions of reconstructed parameters of the observed events and those of the MC simulation are here compared.

### 3.1 Event selection

The following 7 conditions are required for events (both in data and simulation) in this analysis : (1) at least 5 SDs are used in reconstruction, (2) the reconstructed core position

*e-mail: fujisue@icrr.u-tokyo.ac.jp
**Full author list: http://www.telescopearray.org/research/collaborators

https://doi.org/10.1051/epjconf/202328306004

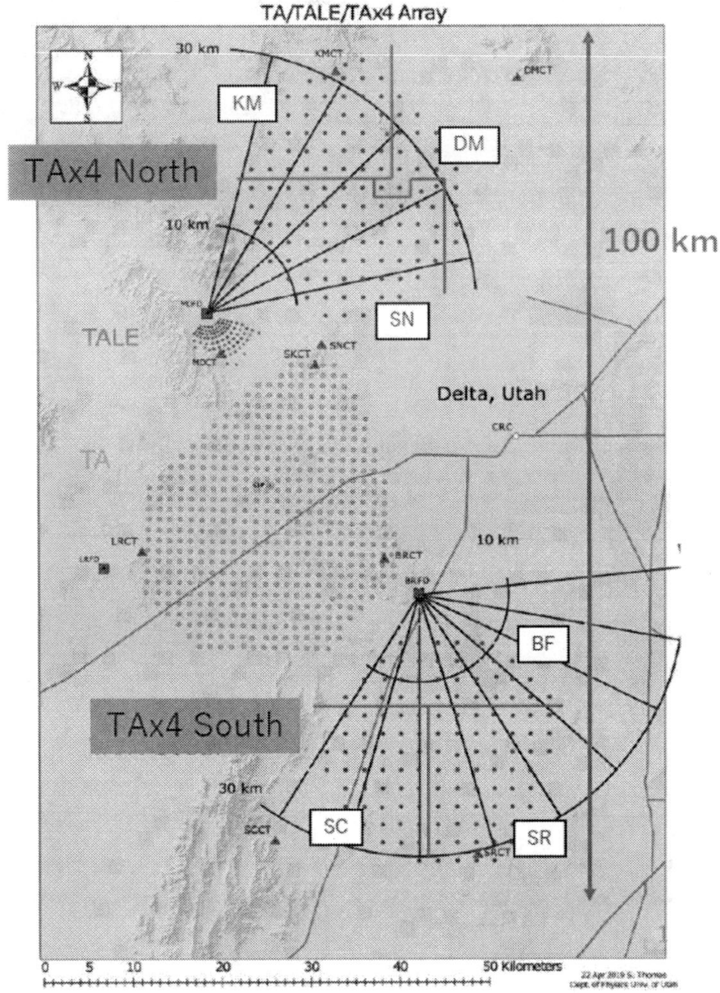


Figure 1: Map of the configuration of TAx4 experiment. The red circles represent the location of the deployed TAx4 SDs. The TAx4 northern array consists of 3 sub-arrays (KM, DM and SN). The TAx4 southern array also consists of 3 sub-arrays (BF, SC and SR). The blue lines represent the borders of TAx4 sub-arrays. The green circles and the blue circles denote TA SDs and TALE SDs respectively.

is at least 400 meters away from the border of the array, (3) the reconstructed zenith angle is less than 55 degrees, (4) the reduced chi square of the reconstruction fit is less than 4, (5) the uncertainty of reconstructed direction is less than 8 degrees, (6) the relative uncertainty of S800 (the density of shower particles at a lateral distance of 800 meters from the shower axis) is less than 0.5 and (7) the reconstructed energy without energy scaling is less than $10^{19}$ eV.

### 3.2 Energy scale

The number of simulated events is large enough to minimize statistical uncertainty of MC simulation, so it is necessary to normalize the simulated events to compare them with observed events. The normalization is often done in terms of the number of observed events, *i.e.* the number of normalized simulated events are the same as that of observed events. In this study, however, we normalize the simulated events in terms the number of expected events, *i.e.* the number of normalized simulated events is the same as that expected to be observed by TA SD in 11 years [4] with the TAx4 SD exposure calculated by the MC simulation.

Because of the power law property of cosmic ray energy spectrum, the normalization in terms of the number of expected events highly depends on an energy scale which is determined experimentally by using those events simultaneously observed by FD and SD. Conversely, we can temporally determine the energy scaling factor of TAx4 SD with QGSJET II-04 by minimizing the difference be-

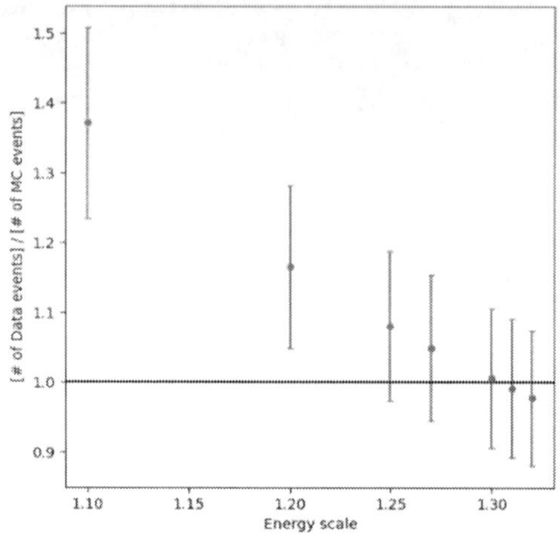


Figure 2: The ratio of the number of observed events to that of normalized MC simulation events for various energy scaling factors. The error bars are statistical uncertainty of observed events. The energy scale which minimizes the difference is 1.3.

tween the number of observed events and that of MC simulation events normalized assuming various energy scaling factors despite the low statistics of TAx4 FD-SD simultaneously observed events. Fig 2 shows the relation between energy scaling factors and the ratio of the number of observed events to that of normalized MC simulation events. From this comparison, we use 1.3 as the temporary energy scaling factor for TAx4 SD with QGSJET II-04 in this study. This is almost the same value as TA SD energy scaling factor with QGSJET II-03 (=1.27) [11]. The actual energy scale will be determined experimentally using simultaneously observed events by TAx4 FD and TAx4 SD.

### 3.3 Comparison of parameters

The comparisons for various parameters between observation and simulation are shown in Fig 3, and the p-values of Kolmogorov–Smirnov test [12] are listed in Table 1. As shown in Fig 3 and Table1, the simulation are statistically consistent with the observation. It shows the accuracy of the MC simulation.

## 4 Energy spectrum

Using 1.5 years TAx4 northern SD array data and 2 years TAx4 southern SD array data, we calculated the energy spectrum. The energy spectrum for each "i"th energy bin (with $\Delta\log_{10}(E/eV) = 0.1$ as a bin width) is calculated by the following equation :

$$\left(\frac{dN}{d\Omega dt dS dE}\right)_i = \frac{\sum_{sub-array} N_{rec}^{Data}(E_i^{rec})}{\sum_{sub-array}[A\Omega T \times Rec.Eff.(E_i^{rec}, E_i^{gen})] \times \Delta E_i^{gen}} \quad (1)$$

where $N_{rec}^{Data}(E_i^{rec})$ is the number of observed events in each reconstructed energy bin, A, $\Omega$ and T are the MC simulation throwing area, throwing solid angle and time duration respectively and $Rec.Eff.(E_i^{rec}, E_i^{gen}) = N_{rec}^{MC}(E_i^{rec})/N_{gen}^{MC}(E_i^{gen})$ is the reconstruction efficiency. The numerator is the number of reconstructed (including the event selection) events in MC simulation for each reconstructed energy bin, while the denominator is the number of thrown events in the MC simulation for each generated energy bin. This term includes bin-to-bin migration effect. $\Delta E_i^{gen}$ is the bin width of the "i"th energy bin.

The effective aperture and the effective exposure of TAx4 SD array calculated by the MC simulation is shown in Fig 4.

The calculated energy spectrum is shown in Fig 5. The red points show the calculated TAx4 energy spectrum. The red solid line and the red dashed line are fits with a broken power law function and single power law function respectively. The gray points and the gray line indicate the TA SD 11 years energy spectrum and the fit result. The fit functions and the fit results are summarized in Table 2. As shown in Fig 5, the TAx4 energy spectrum is consistent with the TA SD energy spectrum and a cut off structure around $E = 10^{19.75}$ eV. The expected number of events above $10^{19.84}$ eV [1] on the assumption that there is no cut off is 27.16, and 10 events have been observed in data. This corresponds to a probability of $1.48 \times 10^{-4}$ (approximately $3.6\sigma$). In addition to this, the deviance of the fits $D/$ndof is 0.62 for the broken power fit and 1.34 for the single power law fit. Both of the fits are acceptable due to the small statistics, however the broken power law fit is preferred.

## 5 Summary

To increase the data acquisition rate of UHECRs, TAx4 experiment is being constructed and has started observation in 2019. In this paper, the initial 1.5 years data of TAx4 northern SD array and 2 years data of TAx4 southern SD array are analyzed. The energy scale of TAx4 SD with QGSJET II-04 is temporally determined in terms of the number of events. The MC simulation is compared with the observation, and it is statistically consistent with the observation. We also analyzed the energy spectrum above $10^{19}$ eV using observed data by TAx4 SD array. It is consistent with the TA SD energy spectrum within statistical uncertainty. The significance of the cut off structure of the TAx4 SD energy spectrum is $3.6\sigma$.

[1] This is the cut off position observed by TA SD with 11 years data

https://doi.org/10.1051/epjconf/202328306004

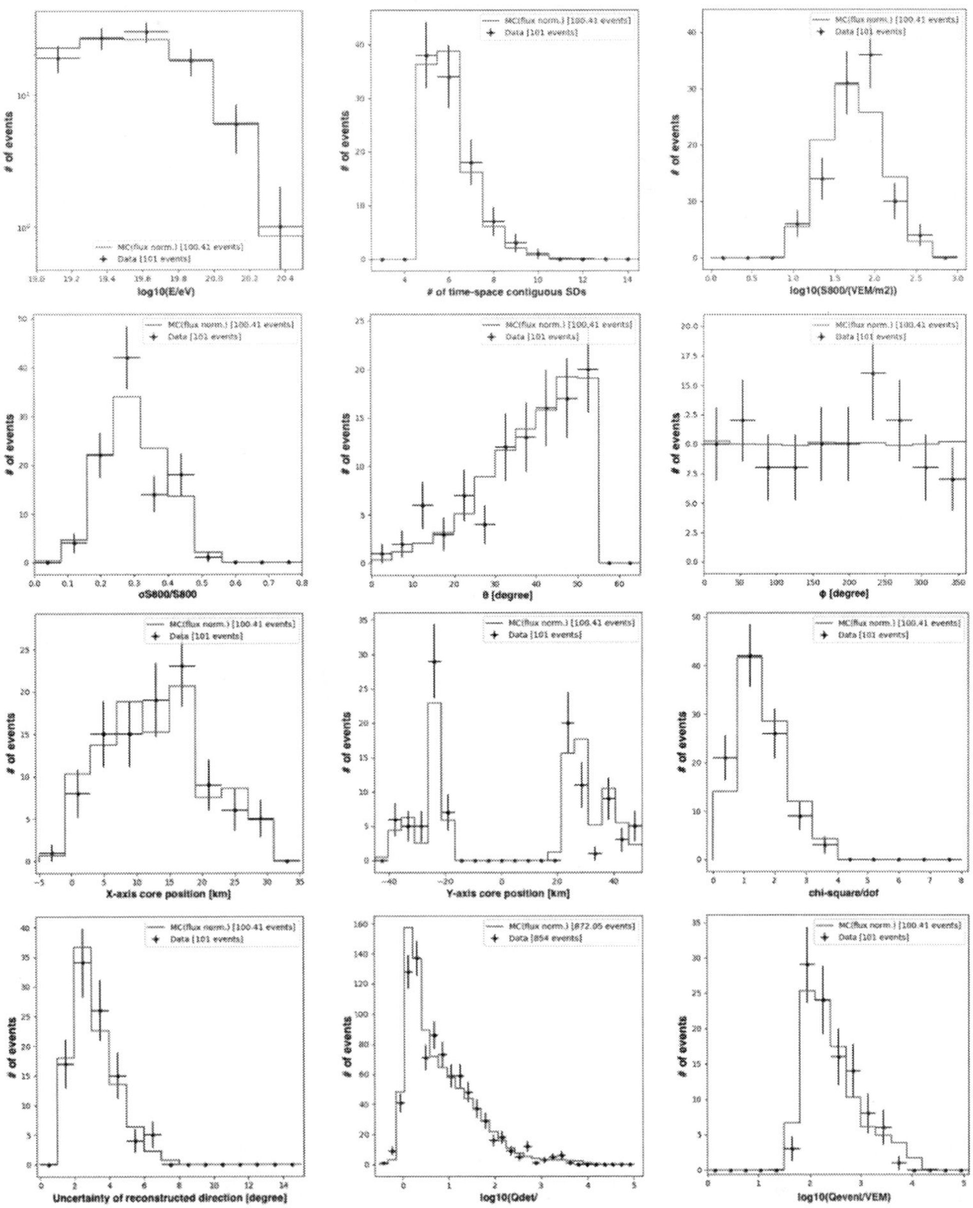


Figure 3: Comparisons between observed data (blue points) and MC simulation (red histogram). MC simulation histogram is normalized in terms of the number of expected events as explained in section 3.2 with 1.3 as an energy scaling factor. KS test p-values of the parameters are listed in Table1. The compared parameters are reconstructed energy, the number of hit SDs, $S_{800}$, $\sigma_{S_{800}}/S_{800}$, zenith angle, azimuthal angle, core positions (west-east direction and south-north direction), $\chi^2$/ndof of fit, uncertainty of reconstructed direction, signal per one SD and signal per one event. The Monte Carlo distributions are in good agreement with the data.

Table 1: Kolmogorov–Smirnov test p-value for parameters.

| parameter | KS test p-value |
|---|---|
| Energy | 0.684 |
| The number of SD | 0.998 |
| $S_{800}$ | 0.356 |
| Relative uncertainty of $S_{800}$ ($\sigma_{S_{800}}/S_{800}$) | 0.337 |
| Zenith angle ($\theta$) | 0.559 |
| Azimuthal angle ($\phi$) | 0.787 |
| Core position (west-east) | 0.559 |
| Core position (south-north) | 0.0598 |
| $\chi^2$/ndof of fit | 0.239 |
| Uncertainty of reconstructed direction | 0.781 |
| Signal/SD | 0.108 |
| Signal/event | 0.845 |

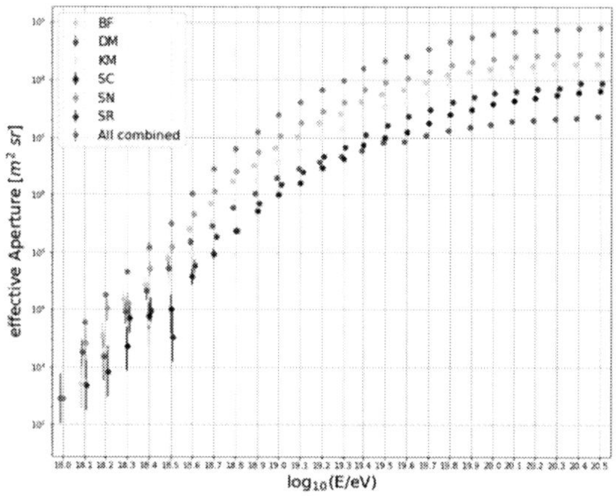


(a) effective aperture of TAx4 SD array.

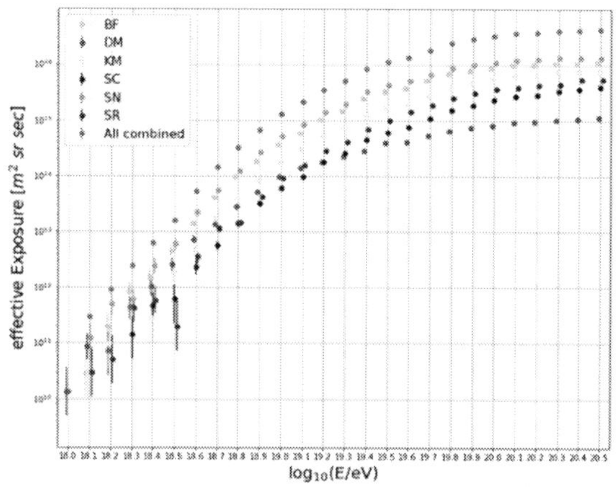


(b) effective exposure of TAx4 SD array

Figure 4: The effective aperture (a) and the effective exposure (b) of TAx4 SDs. The gray circles denote the total TAx4 SD aperture(exposure) which consists of 6 sub-arrays (BF, DM, KM, SC, SN and SR as shown in Fig 1). The energy is MC generated energy without energy scaling.

Table 2: The fit functions and fit results shown in Fig5. The deviance $D$/ndof behaves approximately as $\chi^2$/ndof of chi-square distribution [14]. $\theta(E)$ is a step function.

| | Broken power law | Single power law |
|---|---|---|
| Functional form | $K[\theta(E_{break} - E) \times (E/\mathrm{EeV})^{p_1} + \theta(E - E_{break}) \times (E_{break}/\mathrm{EeV})^{p_1 - p_2} \times (E/\mathrm{EeV})^{p_2}]$ | $K(E/\mathrm{EeV})^{p_1}$ |
| $K$ [$\mathrm{sr}^{-1}\mathrm{s}^{-1}\mathrm{m}^{-2}\mathrm{eV}^{-1}$] | $5.51 \times 10^{-31}$ | $2.28 \times 10^{-30}$ |
| $p_1$ | -2.51 | -2.99 |
| $p_2$ | -4.54 | - |
| $E_{break}$ [eV] | $10^{19.72}$ | - |
| $D$/ndof | 0.62 | 1.34 |

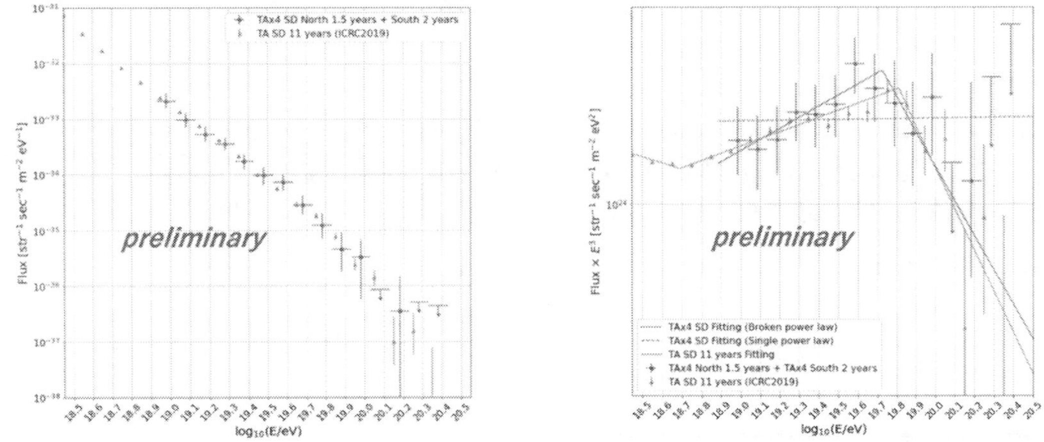


Figure 5: The energy spectrum observed by TAx4 SD array is shown by the red points. The TA SD 11 years energy spectrum [11] is also shown in gray points for comparison. The energy scaling factor between the FD and the SD is assumed to be 1.3. The statistical uncertainty according to [13] of the observed events is considered in the TAx4 SD energy spectrum. For 0 events-observed energy bins, 90% upper limits are shown. The energy spectra, multiplied by $E^3$, are shown in the right figure. The red solid line and the red dashed line in the right figure are the fits with broken power law function and single power law function respectively.

## References

[1] H. Kawai et. al., Nucl. Phys. B Proc. Suppl., **175-176**, 221-226 (2008)

[2] R.U. Abbasi et. al., ApJL, **790**, L21 (2014)

[3] R.U. Abbasi et. al., Nucl. Instrum. Methods in Phys. Res., **1019**, 165726 (2021)

[4] D. Ivanov et al., PoS **ICRC2019**, 298 (2021)

[5] T. Abu-Zayyad et. al., **arXiv:1403.0644** [astro-ph.IM] (2014)

[6] D. Heck, J. Knapp, J. Capdevielle, G. Schatz and T. Thouw, **FZKA-6019** (1998)

[7] S.Ostapchenko, Nucl. Phys. B Proc. Suppl., **151**, 143-146 (2006)

[8] A. Ferrari, P. R. Sala, A. Fasso and J. Ranft, CERN-2005-10 (2005)

[9] W. R. Nelson, H. Hirayama and D.W.O. Rogers, SLAC-265 (1985)

[10] S. Agostinelli et. al., Nucl. Instrum. Methods in Phys. Res., **506**, 250-303 (2003)

[11] T. Abu-Zayyad, et. al., ApJL, **768**, L1 (2013)

[12] F. J. Massey, Journal of the American Statistical Association, **46**, 253, 68–78 (1951)

[13] G. J. Feldman and R. D. Cousins, Phys. Rev. D, **57**, 3873 (1998)

[14] S. Baker and R. D. Cousins, Nucl. Instrum. Methods in Phys. Res., **221**, 437-442 (1984)

# Auger@TA: Deploying an independent Pierre Auger Observatory SD micro-array at the Telescope Array Project

## Auger@TA Working Group Report

*Sonja* Mayotte[1,*], *Jorge* Caraça-Valente[1], *Corbin* Covault[2], *Toshihiro* Fujii[3], *Sungrae* Im[2], *Robin* James[2], *Jeffrey* Johnsen[1], *Karl-Heinz* Kampert[4], *Heiko* Kern[5], *John* Matthews[6], *Eric* Mayotte[1], *Adriel* Bartz Mocellin[1], *Sean* Quinn[2,†], *Julian* Rautenberg[4], *Markus* Roth[5], *Hiroyuki* Sagawa[7], *Takashi* Sako[7], *Frederic* Sarazin[1], *Ricardo* Sato[8], *David* Schmidt[9], *Stan* Thomas[6] and *Günther* Wörner[5], for the Pierre Auger[10,‡] and Telescope Array[11] Collaborations

[1]Department of Physics, Colorado School of Mines, Golden, CO, USA
[2]Case Western Reserve University, Cleveland, OH, USA
[3]Graduate School of Science, Osaka Metropolitan University, Osaka, Japan
[4]Department of Physics, University of Wuppertal, Wuppertal, Germany
[5]Institute for Astroparticle Physics, Karlsruhe Institute of Technology (KIT), Karlsruhe, Germany
[6]Department of Physics and Astronomy, University of Utah, Salt Lake UT, USA
[7]ICRR, University of Tokyo, Kashiwa, Chiba, Japan
[8]Comisión Nacional de Energía Atómica, Malargüe, Argentina
[9]Institute of Experimental Particle Physics, Karlsruhe Institute of Technology (KIT), Karlsruhe, Germany
[10]Observatorio Pierre Auger, Av. San Martín Norte 304, 5613, Malargüe, Argentina.
Full author list: https://www.auger.org/archive/authors_2022_10.html
[11]Telescope Array Project, 201 James Fletcher Bldg, 115 S 1400 East, Salt Lake City, UT 84112–0830, USA.
Full author list: http://www.telescopearray.org/research/collaborators

**Abstract.** The Pierre Auger Observatory (Auger) and the Telescope Array Project (TA) are the two largest ultra-high-energy cosmic ray observatories in the world. They operate in the Southern and Northern hemispheres, respectively, at similar latitudes but with different surface detector (SD) designs. This difference in detector design changes their sensitivity to the various components of extensive air showers. The over-arching goal of the Auger@TA working group is to cross-calibrate the SD arrays of the two observatories in order to identify or rule out systematic causes for the apparent differences in the flux measured at Auger and TA. The project itself is divided into two phases. Phase-I finished in 2020 and consisted of a station-level comparison facilitated by the deployment of two Auger stations, one prototype station with a single central PMT and a standard Auger station, in the middle of the TA SD near the Central Laser Facility, along with a modified TA station to provide external triggers from the TA SD. This provided the opportunity to observe the same extensive air showers with both Auger and TA detectors to directly compare their measurements. Phase-II of Auger@TA is currently underway and aims at building a self-triggering micro-Auger-array inside the TA array. This micro-array consists of eight Auger stations, seven of which use a 1-PMT prototype configuration and form a single hexagon with a traditional 1.5 km Auger spacing. The 8th station is of the standard Auger 3-PMT configuration and is placed at the center of the hexagon, along with a TA station to form a triplet. Each Auger station will also be outfitted with an AugerPrime Surface Scintillator Detector. A custom communication system using readily available components will be used to provide communication between the stations and remote access to each station via a central communications station. The deployment of the micro-array took place at the end of September 2022. A simulation study was carried out to gauge the expected performance of the Auger@TA micro-array and to derive trigger efficiencies and event rates.

## 1 Introduction

Currently, the two largest ultra-high-energy cosmic ray (UHECR) experiments in the world are the Pierre Auger Observatory (Auger) [1] and the Telescope Array Project (TA) [2]. They operate in different hemispheres at similar latitudes, with the Pierre Auger Observatory in the Southern hemisphere and the Telescope Array Project in the Northern hemisphere. In the last 15 years, both experiments have gathered large amounts of data, but have found their results differ. One such difference is apparent in the measured flux of both experiments and is illustrated in Figure 1. There appears to be a discrepancy of about 9 % between the energy scales of the two experiments [3]. This is within the range of the systematic uncertainties for both experiments and could possibly be addressed via

*e-mail: smayotte@mines.edu
†now in industry
‡e-mail: spokespersons@auger.org

re-scaling. However, a large difference in measured flux would remain at the start of the flux suppression region and beyond.

These discrepancies could be due to fundamental differences between the northern and southern UHECR skies or could be due to unresolved discrepancies in the way the two experiments process extensive air shower (EAS) data. As the collaboration between the two experiments grows, it is becoming more and more important to figure out the reason(s) for the differences between the results of both experiments in order to rule out or correct for instrumental/reconstruction biases. This would then allow for high-level joint analyses to be performed using the combined data from both experiments [4–6].

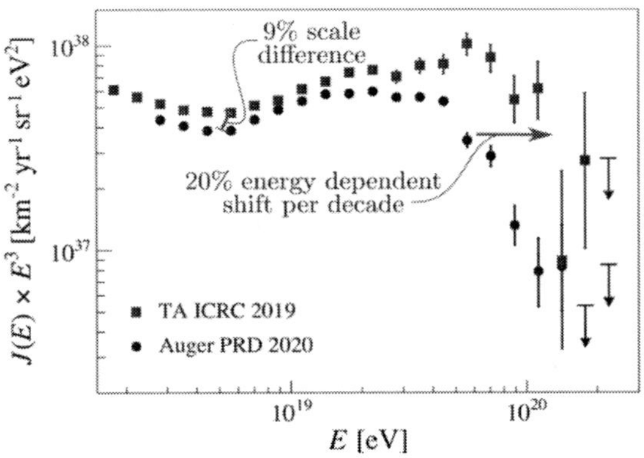


Figure 1: Comparison between the UHECR spectra measured by Auger (black circles) and TA (blue squares). A 9% energy scale difference at the ankle and a growing discrepancy beyond $10^{19}$eV is indicated (modified from [3]).

## 2 The Auger@TA Project

One of the primary similarities between the two experiments is that both use surface detectors (SDs), albeit of very different design, as their main statistics driver. The Auger SD is based on a Water Cherenkov detection (WCD) system, which collects the light produced by charged particles above a certain energy threshold slowing down in the water volume. The baseline unit that all Auger SDs are calibrated to is the energy loss in water of a muon passing vertically through the tank, a Vertical Equivalent Muon (VEM). The WCD is roughly equally sensitive to the electromagnetic and muonic components of a shower. The TA SD instead is based on plastic scintillators that act as particle counters and are predominantly sensitive to the electromagnetic part of the shower, due to electrons being in general more numerous than muons. For scintillators, the baseline calibration unit is the Minimum Ionizing Particle (MIP) energy loss. To directly investigate the impact of these differences, the Auger@TA project was conceived and became an official Auger-TA working group, comprising around a dozen members from both experiments. The central idea behind this project is to perform an in-situ cross-calibration of the SDs of the two experiments by placing Auger detectors at the Telescope Array site and using showers that are measured by both experiments. The central goals of the Auger@TA effort are as follows in order of increasing statistics required.

*Cross-calibration of SDs*

The most integral aspect for the Auger@TA effort is the cross-calibration of the different SD detector types placed in the field in Utah. The primary difference between them is the detection media used, as they are not equally sensitive to the various components of EASs. The cross-calibration will be achieved by making comparisons on a station-by-station level and studying the different responses seen by the different detector types for each shared event.

*Event-by-event reconstruction comparison*

By placing a micro-array of seven Auger-like stations in a hexagon configuration within TA, the Auger reconstruction software can be used to reconstruct measured events in order to compare the different detectors and reconstruction techniques on an event-by-event level as well. This can be used to analyze differences in trigger efficiency and zenith dependence, as well as to study shower component dependent systematic differences.

*Making a fully independent flux measurement*

With the possibility of fully reconstructing events recorded with self-triggering Auger-like detectors in TA, comes the opportunity to potentially make a fully independent flux measurement. This would allow for a direct comparison of the fluxes measured with Auger-like and TA detectors both located in the Northern hemisphere in order to test the nature of the 9 % spectral scale difference.

*Test nature of flux suppression discrepancy*

If high enough statistics are obtained during the lifetime of the Auger@TA project, it would also be possible to extend the studies mentioned above to higher energies in order to possibly shed light on the nature of the differences as flux suppression kicks in.

The Auger@TA project is divided into two phases. Phase I took place between 2018 and 2020 and aimed at performing a station-to-station *in-situ* cross-calibration using three co-located stations (two Auger, one TA) at the site of the TA Central Laser Facility (CLF) [7, 8]. The two Auger stations used consisted of one regular Auger station from Argentina and one station formerly used in an R&D effort for a Northern hemisphere Auger [9]. These *1-PMT prototype stations*, differ in their number of PMTs (one instead of three) as well as their electronics system, the details of which are not relevant here but are described in [10]. The analysis of Phase I data and its interpretation is currently being finalized and will be reviewed by both collaborations prior to publication later this year.

## 3 Auger@TA Phase II: Station-by-station and Event-level Comparisons

Auger@TA Phase II is both a continuation and extension of the efforts of Phase I. There will be an expansion of

https://doi.org/10.1051/epjconf/202328306005

station-level comparisons in order to perform the cross-calibration of detectors, but now it will be possible to also study how the shower reconstructions of Auger and TA perform on the same set of showers and directly compare the results. This study is very much needed as there are significant differences between the way Auger and TA perform their SD reconstructions.

While both Auger and TA rely on a shower size estimator (S(1000) and S(800) respectively) extracted from the lateral distribution function (LDF), there are differences in how the two experiments handle the conversion of this estimator to a quantity normalized against geometric effects, which is eventually calibrated using their respective Fluorescence Detector (FD) energy scale (also known to differ as seen in Figure 1). While Auger uses a Constant Intensity Cut (CIC) method [11, 12] to account for the shower geometry, TA relies on large shower simulation libraries (and a scaling factor) to account for geometric effects [13]. Ideally, such a study should be performed with a large number of stations to push the comparison to the highest energies where the spectrum discrepancy between Auger and TA is the largest. This is unfortunately not realistic at the moment, and comparisons can only be performed using a limited number of stations.

Auger@TA Phase II will do this by making use of all seven remaining 1-PMT prototype stations to build a micro-array. As described below, to lower uncertainties, these stations have been modified to more closely match the regular Auger stations. These stations have been deployed to form a full Auger-like hexagon, with one station in the center, using the same 1.5 km spacing as the southern Auger array. With a full hexagon of stations, the micro-array will provide much higher statistics than were possible in Phase I.

In addition, one regular 3-PMT Auger station and a TA station are also placed at the center of this micro-array to form the triplet illustrated in Figure 2. The operation of the central triplet will provide the high statistics needed to directly study the signal correlations between the Auger 1-PMT and 3-PMT (in VEM), and the TA (in MIP) SD stations, and thus will be used for cross-calibration purposes (thereby allowing for an extension of the Auger@TA Phase I study). Here, the same $\approx 11$ m spacing between stations as used by Auger for doublet and triplet setups is used [14]. These standard Auger hexagon and triplet configurations were chosen to minimize reconstruction biases when using the fine-tuned reconstruction procedure developed for the Observatory.

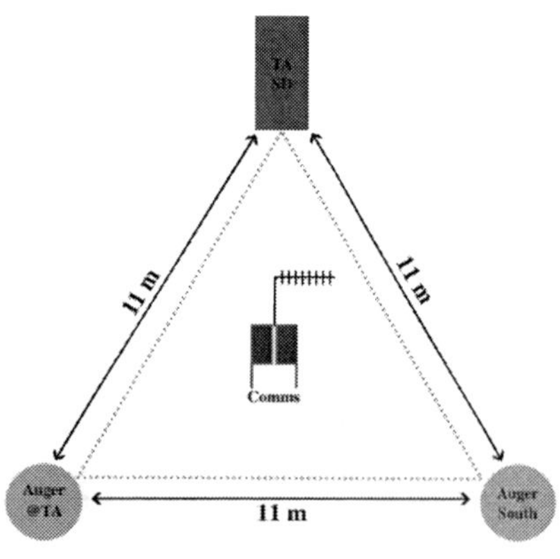


Figure 2: Schematic view of the central triplet. The central communications station is located at the center of micro-array and triplet, while the three stations are spread out evenly around it in an equilateral triangle with a side length of $\approx 11$ m.

### 3.1 The Auger@TA station

The stations used to make up the micro-array hexagon, with the exception of the regular Auger station in the triplet, are prototypes that have been retro-fitted to more closely match a regular Auger station. A schematic overview of these retro-fitted stations, *Auger@TA* stations, can be seen in Figure 3.

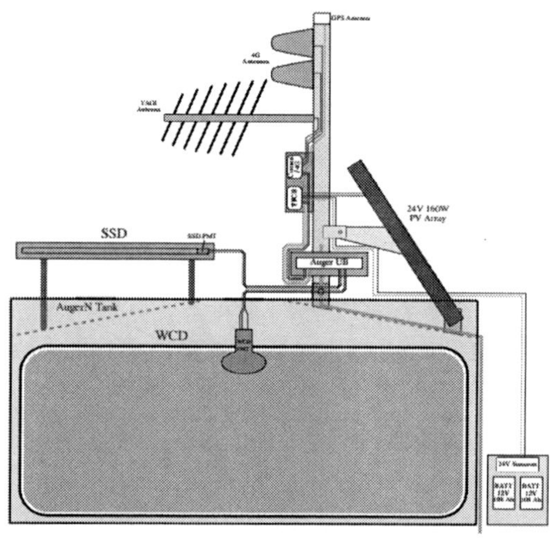


Figure 3: Schematic view of an Auger@TA station.

The Auger@TA station makes use of the 1-PMT prototype shell which has largely the same internal form factor as regular Auger tanks, but uses only one PMT at the center of the station instead of the usual 3-PMT configuration. The bases used with the PMTs have been replaced with regular Auger bases, and are connected to the typical Auger electronics of a Unified Boards (UB) and a Tank Power Control Board (TPCB) [1]. For the regular Auger station, these are placed under the so-called "dome". However, the 1-PMT prototype shells do not have such a structure, as their original electronics were designed to be installed directly inside the tanks. This required procuring of an alternative, which was found in the form of repurposed ammunition boxes hereafter referred to as E-kit boxes. They were chosen as they are water-tight and have a very similar form-factor to the UBs. The E-kit boxes have been painted with white liquid rubber RV roof coating to efficiently reflect sunlight and provide strong heat-protection. The UBs are mounted inside the E-kit boxes on removable drawer slides as this allows quick and easy access to the UB in the field and even makes replacing a board very

 https://doi.org/10.1051/epjconf/202328306005

straightforward. A picture of such an E-kit box with a UB inside can be seen in Figure 4.

For Auger@TA stations, the TPCB box had to be moved as well and is now located in a NEMA enclosure fitted to the communications mast of each station together with each stations' communications electronics. These boxes are also coated in white paint to reflect as much sunlight as possible. An example of such a box can also be seen in Figure 4. The solar power system was upgraded to 24 V/160 W/216 Ah in order to provide sufficient power for the station electronics and the communications system.

Figure 4: Auger@TA station components deployed in the field. Top left: Positioning of the E-kit and the NEMA box on the communications mast. The TPCB and a Raspberry Pi for the communications system can be seen. Top right: Communications mast with GPS (top), 4G (middle), and YAGI (bottom) antennas in place. Bottom left: Close-up of the NEMA enclosure. Bottom right: Inside view of the E-kit box with the UB on the drawer slides.

Additionally, thanks to the efforts of the Karlsruhe Institute of Technology, the Bergische Universität Wuppertal, and the Observatory in Malargüe, each Auger@TA station will also be outfitted with AugerPrime Surface Scintillator Detectors (SSDs) [15, 16] that were assembled with spare material. This includes eight SSDs, SSD support structures, SSD-to-UB connection cables, PMTs, and bases. These SSDs are an addition to the original scope of the micro-array and will provide interesting opportunities for additional cross-calibrations. All the cabling between the two boxes, the PMT hatch cover, the SSD enclosure, and the battery box, are routed through watertight conduits thus providing a completely weather- and dust-proof electronics system.

### 3.2 The Communications System

The communications system for the Auger@TA micro-array is completely custom-made, but uses readily available components and will be used to provide access to the stations directly via the internet.

Communication between stations located at the outside of the hexagonal array and the central triplet are accomplished with YAGI antennas using the Digi Xbee Pro transceiver operating at 900 MHz. An abstraction layer has been implemented that operates over the Xbee native serial line to provide internet access to all nodes at every station, even while science data and commands are relayed between stations via regular Auger communications protocols. This will allow communication with each station directly, for example for debugging purposes. Finally, communication from the central node to the outside world is accomplished via a 4G LTE (mobile) wireless modem. An example from a station in the field using both antenna types can be seen in Figure 4.

## 4 Micro-array Deployment

An area in the south-east corner of the TA array near Black Rock FD was chosen for the site of the micro-array. The chosen site has minimal overlap with land regulated by the US Bureau of Land Management (BLM), thus making a deployment via motorized vehicles (see Figure 5) possible for the Auger@TA effort.

Figure 5: Auger@TA station being put in place at its deployment site with an excavator.

The deployment of the micro-array took place over a span of two weeks at the end of September 2022, with the first week being focused on pre-deployment tasks such as the final assembly of the SSDs, decommissioning of the Phase I stations for re-deployment, liner inflation and inspection, etc. and the second week being used solely to deploy detector stations. The deployment site as well as the deployed triplet can be seen in Figure 6.

Table 1: Overview of deployment status for each station. Check-marks (green) denote accomplished items, while crosses (red) show which tasks are still open. "C" denotes central triplet stations. Unless specified, the station are Auger@TA stations. Station names inspired by [17].

| Site | Station deployed | Components commiss. | Electronics deployed | SSD deployed |
|---|---|---|---|---|
| Sam (C) | ✓ | ✓ | ✓ | × |
| Gollum (3PMT, C) | ✓ | ✓ | ✓ | × |
| Frodo (TA, C) | ✓ | ✓ | × | × |
| Aragorn | ✓ | ✓ | ✓ | × |
| Arwen | ✓ | ✓ | × | × |
| Gimli | ✓ | ✓ | ✓ | × |
| Legolas | ✓ | ✓ | ✓ | × |
| Bilbo | ✓ | ✓ | ✓ | × |
| Galadriel | ✓ | ✓ | ✓ | × |
| Sauron (Comms) | ✓ | ✓ | ✓ | – |

https://doi.org/10.1051/epjconf/202328306005

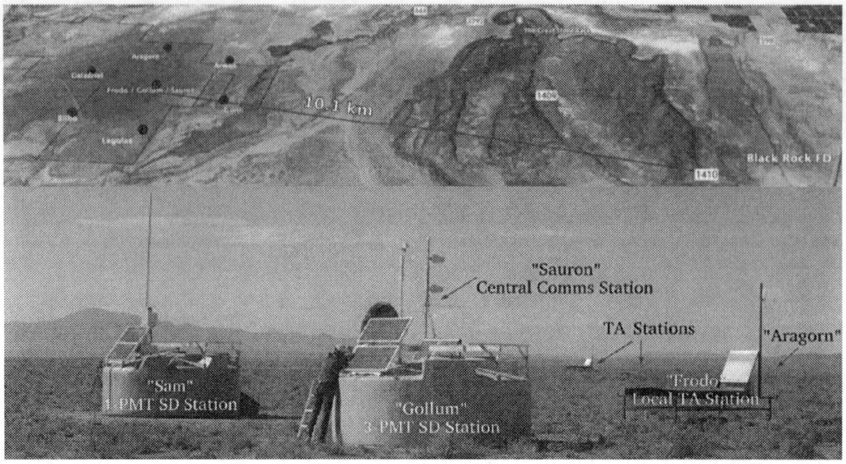


Figure 6: Top: Location of the 9 station micro-array within TA, also showing the distance to the Black Rock FD. Bottom: The center triplet and data acquisition comms station. The "Gollum" station is currently taking data with the 4G-connected "Sauron" central comms receiving event triggers. A communication link is established with "Aragorn" and "Galadriel" (not visible). Note the SSD support structure already installed on "Sam" and "Gollum". Station names inspired by [17].

In the weeks following the deployment, the water delivery to each station took place. As of today, all stations have been deployed in the field, with most stations only requiring PMTs (bases are currently being fabricated), SSDs, and communication systems (components back-ordered) to become operational. A fully instrumented micro-array is expected to be in the field by July 2023. An overview of the deployment status for each individual station is summarized in Table 1.

## 5 Expected Performance

In order to quantify the feasibility of obtaining each of the goals stated in section 1, and to gauge the expected performance of the Auger@TA Phase II micro-array, two sets of event simulations were produced: one set using the full regular Auger array (FA), and one with only a single hexagon (SH) of Auger@TA stations.

These simulations were produced using proton CORSIKA showers thrown with an $E^{-1}$ spectrum in a range $E \in [18.0 - 19.0) \log_{10}(E/\mathrm{eV})$ (see Figure 9). To ensure maximum comparability between the two simulation sets, each CORSIKA shower was thrown at the same geometry, with the same random seeds in both detector configurations, and then matched at the event-by-event level. To avoid edge effects, a fiducial $5 \times 5\,\mathrm{km}^2$ area around the central hexagon was used in both cases which exceeds the triggering range of the SH configuration. The Auger Off<u>line</u> framework was used with small adaptations made to the detector simulations in the case of the SH stations to reflect the changes that come with using Auger@TA stations.

### 5.1 Simulated Station Calibration

To adapt the Auger detector simulations to represent Auger@TA stations the following changes were made:

- Remove all PMTs but one;
- Move PMT to the center of the station;
- Set higher thresholds for single PMT triggers.

Additionally, studies of the station calibration, displayed in Figure 7, have shown that while the noise rate in the calibration histogram is slightly higher, the simulations only show a negligible difference in the relative response to air showers between the 1-PMT and 3-PMT configurations, lowering expected calibration uncertainties for Auger@TA results. Thus, for now the studies shown here use the same calibration constants as for regular Auger stations. However with more statistics and real calibration histograms from the deployed array, this study will be revisited in the future.

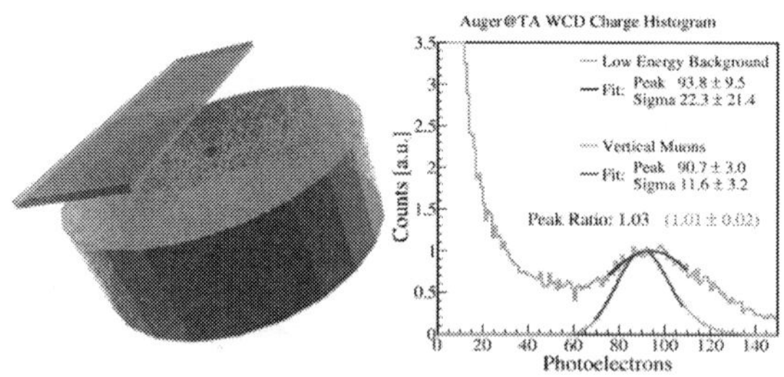


Figure 7: Full implementation of the Auger@TA stations in Geant4-based detector simulations. Left: Geant4 implementation of the 1-PMT Auger@TA detector configuration with SSD. Right: Simulated calibration histograms for the 1-PMT detector configuration.

### 5.2 Array-wide Simulations

A high-quality, high-precision energy reconstruction is a key element for the Auger@TA effort to be successful, but

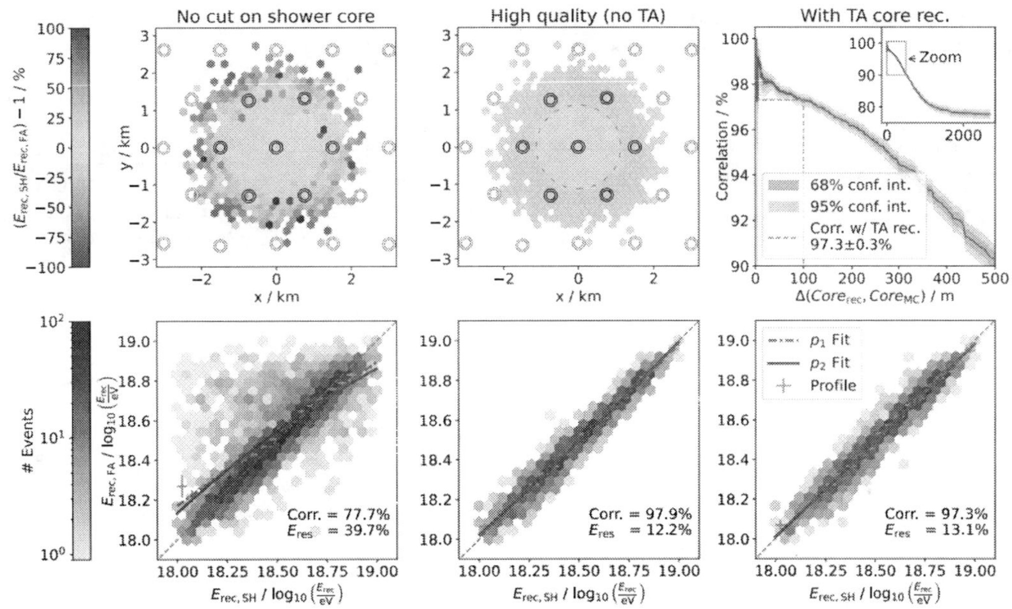


Figure 8: Expected energy reconstruction quality of Auger@TA phase II: Left panels: All reconstructed events from the self-triggered array. Center panels: High-quality reconstructed event sample. Right panels: Estimated reconstruction quality using a TA 100 m core reconstruction uncertainty (see text for details).

is hard to achieve with only one hexagon with 1-PMT stations. The energy reconstruction of the regular Auger array can be used to benchmark the performance of the single hexagon, and their reconstructed energy correlation is shown in Figure 8. When comparing all events that were successfully reconstructed in both simulation sets, the energy resolution of the single hexagon is rather poor with 39.7 % (see Figure 8 top left). This is to be expected since events that have a shower core falling within the single hexagon will have, on average, a better reconstruction. Events where the shower footprint is not contained in the hexagon or the shower core falls outside of the hexagon have a worse reconstruction in direct comparison with the full array simulation set. This can also be verified with the top left plot in Figure 8, which shows the energy reconstruction biases between the single hexagon and the full Auger SD array. A cut on the distance of the reconstructed shower core to the central station ($R_{\mathrm{center}} \leq 1125$ m) allows for the selection of high-quality events and is shown in the middle of Figure 8. As can be seen in the bottom middle of Figure 8, applying this cut to the simulated data significantly improves the energy correlation with the full array.

Reducing the number of usable events in the data set is, of course, not ideal, and such a selection will only be applied for completely independent Auger/TA studies. In most cases, however, the events detected by the Auger micro-array will also be observed by TA, and the TA shower core reconstruction can be used to inform the Auger reconstruction with minimum bias. This is studied in Figure 8 (right panels), by considering how the SH core reconstruction uncertainty affects the energy correlation between the SH and FA data sets. Assuming a TA shower core reconstruction accuracy of $\approx$ 100 m at low energies [13], this indicates that with the TA shower core reconstruction we could reach an energy resolution of 13.1 %, which is very similar to that of the high-quality cut (12.2 %).

### 5.3 Expected Event Rate and Outlook on Flux Comparison

The single hexagon simulation set can also be used to extract the expected trigger efficiency of the micro-array. By folding in the UHECR spectrum [18], an expected yearly event rate can be calculated and is shown in Figure 9.

By comparing the distribution of reconstructed energies (with and without quality cuts) to the thrown Monte Carlo distribution, a region where the reconstructed distributions are flat can be selected. This region ranges from $18.3 - 18.8\ \log_{10}(E/\mathrm{eV})$ and is suitable for making a cosmic ray flux measurement to potentially investigate the nature of the 9 % energy scale difference shown in Figure 1. As illustrated in Figure 8, by using the TA core reconstruction, the majority of reconstructed events are expected to be usable for this measurement. This means an event rate of up to 65 events/yr in the chosen high-energy region can be obtained. At this rate, we expect to achieve an 8.7 % statistical uncertainty on our flux measurement after two years, potentially allowing us to make a 1 $\sigma$ level comparison between Auger and TA flux measurements. A 7-year run-time will be needed to achieve a 2 $\sigma$ level comparison unless lower energy events can eventually be incorporated in the flux measurement. Flux measurements made at lower energies would of course have better statistics, although the quantification of systematic uncertainties is still a work in progress. Refinements to the simulations are planned and likely to optimize the number of events that may be used for the flux comparison study.

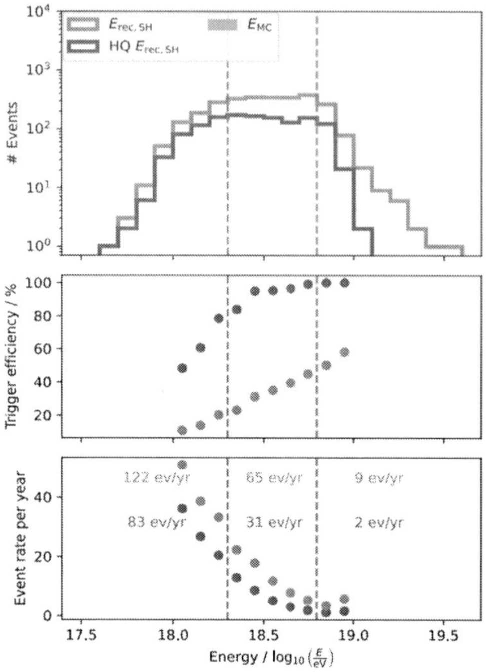


Figure 9: Expected trigger efficiencies and event rates for the Auger@TA micro-array from simulations. All plots: light blue: all reconstructed data, dark blue: high-quality selection (see Figure 8). Top: Reconstructed energy distributions in comparison to the energy distribution of thrown events (filled histogram). The dashed lines represent the energy region that is suitable for making a flux measurement due to trigger efficiencies. Middle: Expected trigger efficiencies. Full efficiency for the high-quality data set is reached at $10^{18.4}$ eV. Bottom: Expected event rate per year with event counts for the low energy, flux measurement, and high energy regions.

## 6 Summary and Outlook

The Auger@TA project is the only ongoing effort that can uncover discrepancies between the Auger and TA SDs using the exact same showers and, as such, is critical in assessing whether the observed discrepancies in the energy scales of the two experiments are due to astrophysical differences between the northern and southern UHECR skies, or due to as-of-yet unresolved discrepancies in how the two experiments analyze and interpret EAS data.

Phase I of the project has already provided promising results that will be published later this year. The second phase of the project is well under way, with the micro-array being deployed, nearing full instrumentation status with the first data likely being taken by the second half of 2023. In the field, one station is already fully up and running for testing purposes and on the communications side lossless transmission during multi-hour trials at a data rate of up to 9600 bps has been verified. These performance results are well within the required specifications to operate the full micro-array.

Currently, the possibility of adding another regular Auger station to form a second doublet of Auger-Auger@TA stations is being evaluated. As shown in subsection 5.3, the statistical budget for making a flux comparison is tight. Because of this, minimizing calibration uncertainties is important to study the compatibility, or lack thereof, of the TA and Auger reconstructions. These uncertainties can be lowered by adding a second 1-PMT/3-PMT doublet within the micro-array to lower systematic uncertainties coming from cross-calibrating two stations that each have their own inherent calibration uncertainties. A second 3-PMT station is already available at the Colorado School of Mines and could be deployed in the micro-array within the next year.

## References


[1] A. Aab et al. (Pierre Auger), Nucl. Instrum. Meth. A **798**, 172 (2015), 1502.01323

[2] T. Abu-Zayyad et al. (Telescope Array), Nucl. Instrum. Meth. A **689**, 87 (2013), 1201.4964

[3] O. Deligny (Pierre Auger, Telescope Array), PoS **ICRC2019**, 234 (2020), 2001.08811

[4] A. Aab et al. (Telescope Array, Pierre Auger), Astrophys. J. **794**, 172 (2014), 1409.3128

[5] P. Tinyakov et al. (Telescope Array, Pierre Auger), PoS **ICRC2021**, 375 (2021), 2111.14593

[6] A. di Matteo et al. (Telescope Array, Pierre Auger), PoS **ICRC2021**, 308 (2021), 2111.12366

[7] S. Quinn et al. (Pierre Auger, Telescope Array), JPS Conf. Proc. **19**, 011033 (2018)

[8] F. Sarazin et al. (Pierre Auger, Telescope Array), EPJ Web Conf. **210**, 05002 (2019)

[9] J. Blumer et al. (Pierre Auger), New J. Phys. **12**, 035001 (2010)

[10] M. Kleifges (Pierre Auger), Astrophys. Space Sci. Trans. **7**, 319 (2011)

[11] J. Hersil, I. Escobar, D. Scott, G. Clark, S. Olbert, Phys. Rev. Lett. **6**, 22 (1961)

[12] J. Abraham et al. (Pierre Auger), Phys. Rev. Lett. **101**, 061101 (2008), 0806.4302

[13] R.U. Abbasi et al. (Telescope Array), Phys. Rev. D **98**, 022002 (2018), 1804.03877

[14] A. Aab et al. (Pierre Auger), JINST **15**, P10021 (2020), 2007.09035

[15] A. Aab et al. (Pierre Auger) (2016), 1604.03637

[16] R. Šmída (Pierre Auger), PoS **ICRC2017**, 390 (2018)

[17] J.R.R. Tolkien, *The Lord of the Rings* (Houghton Mifflin Harcourt, 1954)

[18] A. Aab et al. (Pierre Auger), Phys. Rev. Lett. **125**, 121106 (2020), 2008.06488

# Terzina on board NUSES: a pathfinder for EAS Cherenkov Light Detection from space

*Leonid* Burmistrov[1,*] for the NUSES Collaboration

[1]Département de Physique Nucléaire et Corpusculaire, Faculté de Sciences, Université de Genève, 1205, Switzerland

**Abstract.** In this paper we introduce the Terzina telescope as a part of the NUSES space mission. This telescope aims to detect Ultra High Energy Cosmic Rays (UHECRs) through the Cherenkov light emission from the extensive air showers (EAS) that they create in the Earth's atmosphere. The Cherenkov photons are aligned along the shower axis inside about $\sim 0.2 - 1°$, so that they become detectable by Terzina when it points towards the Earth's limb. A sun-synchronous orbit will allow the telescope to observe only the night side of the Earth's atmosphere. In this contribution, we focus on the description of the telescope detection goals, geometry, optical design and its photon detection camera composed of Silicon Photo-Multipliers (SiPMs). Moreover, we describe the full Monte Carlo simulation chain developed to estimate Terzina's performance for UHECR detection. The estimate of the radiation damage and light background rates, the readout electronics and trigger logic are briefly described. Terzina will be able to study the potential for future physics missions devoted to UHECR detection and to UHE neutrino astronomy. It is a pathfinder for missions like POEMMA or future constellations of similar satellites to NUSES.

## 1 The NUSES mission and its scientific goals

NUSES is a pathfinder satellite project (Fig. 1) for two innovative detectors dedicated to the study of cosmic radiation and the Sun-Earth environment [1]. They are Terzina, devoted to high-energy cosmic rays beyond 100 PeV, and Ziré devoted to cosmic rays below 250 MeV. The NUSES mission will last about three years and the satellite, with the two payloads, will orbit at 550 km altitude at the beginning of life (BoL) and at about 525 km at the end of life (EoL). The orbit inclination will be 97.6° with LTAN 18:00:00.

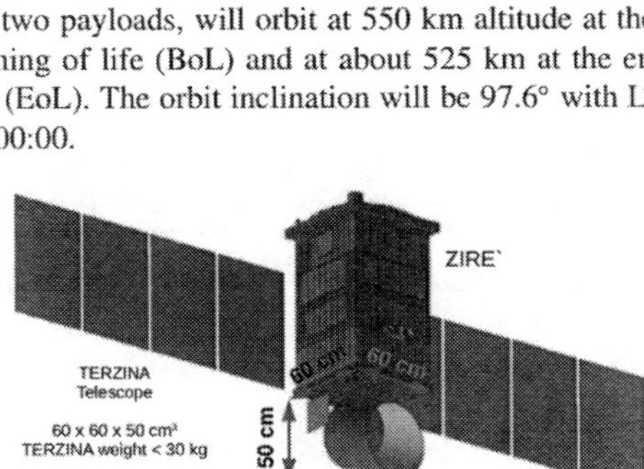


Figure 1: The NUSES satellite containing the Ziré apparatus at the top and the Terzina telescope at the bottom during the flight.

Ziré is made of a scintillating fiber tracker, a stack of plastic scintillator counters and an array of LYSO crystals. An active veto system and a Low Energy Module (LEM) are also part of the payload. SiPM will be used as light sensors. The instrument will perform spectral measurements of electrons, protons and light nuclei from few up to hundreds of MeV, also testing new tools for the detection of 0.1-10 MeV photons [2], and monitoring Van Allen radiation belts and space weather effects.

Terzina is a satellite-based detector designed for Cherenkov light detection from EAS induced by UHECR in the Earth's atmosphere. Terzina will aim to the detection of Earth-skimming ultra-high energy neutrinos [3] and ultra-high energy cosmic rays with energy above 100 PeV [4]. It will prove for the first time the space-based detection of Cherenkov light from EAS. The detection principle is discussed in [4–6].

We expect to see the highest rate of detectable signal from UHECRs when their interactions happen at the atmospheric height of about 30 km for protons of 100 PeV (see Fig. 2)

## 2 The Terzina telescope

The Terzina payload is composed by the following elements: the optical head unit, which is a near-UV-optical telescope, the focal plane assembly (FPA), the thermal control system, and the external harness and electronic units which will be in a separate box.

The optical system of the telescope is based on a dual mirror configuration composed of two parabolic primary and secondary mirrors with a corrector lens in order to cope with aberrations on the photon detection plane, namely the FPA. The dual mirror configuration is chosen to maximize the focal length in the available space which

*email:leonid.burmistrov@unige.ch

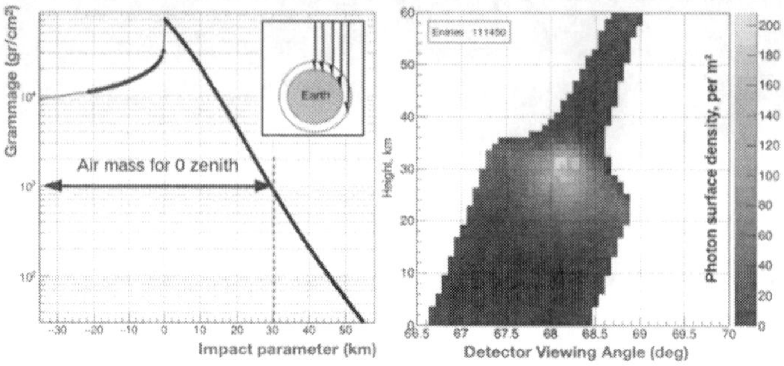


Figure 2: **Left:** Grammage of a cosmic particle as a function of its impact parameter. The insert at the top-right schematically shows incoming particles with different impact parameters. In case the impact parameter is larger than the radius of the Earth, the maximum air mass seen by the particle doubles. **Right :** Photon surface density (z-axis) as a function of the viewing angle and height of a UHECR (primary particle: 100 PeV proton) estimated with EASCherSim [11] (see Sec. 3). The total number of simulated protons is about $1.1 \times 10^5$.

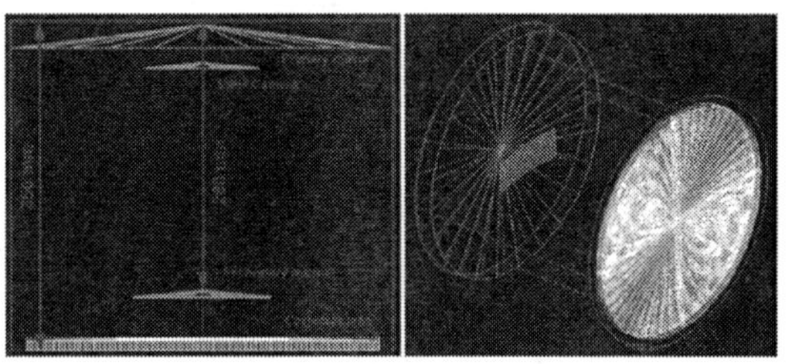

Figure 3: **Left :** View from the top of the payload of Terzina. **Right :** Side view and SiPM camera.

| | RoC* | Distance to primary mirror | Diameter |
|---|---|---|---|
| units | m | mm | mm |
| Big mirror | 0.80 | 0 | 394 |
| Small mirror | 0.36 | 280 | 144 |
| Camera plane | 0.30 | 40 | 121 |
| Corrector | - | 350 | 362 |
| Equivalent focal length : 925 mm | | | |
| *RoC - Radius of Curvature | | | |

Table 1: Parameters of the optical system

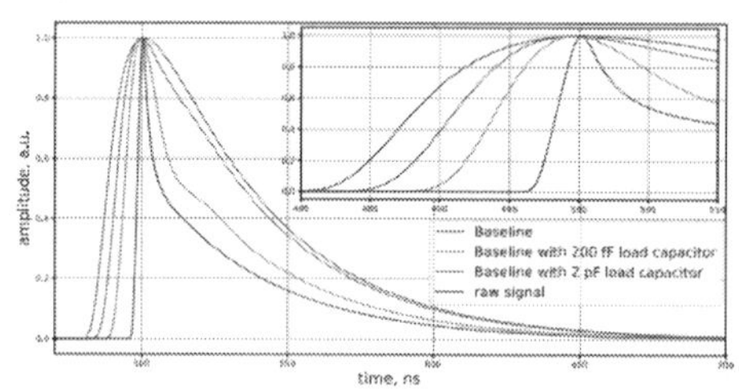


Figure 4: The SiPM signal shape before and after different amplifier configurations for the NUV-HD-MT from FBK with 25 $\mu$m micro-cell size. The amplitudes are normalized to the maximum value.

is a envelope in the shape of a cut-cone with a 394 mm diameter and a 350 mm length. The resulting focal length is about 925 mm. The telescope is inclined by 67.5° with respect to nadir, having an optical axis pointing towards the Earth's limb (see Figures 1 and 3).

The FPA is conceived to detect photons from below and above the limb. It has a rectangular shape with a 2 : 5 aspect ratio. It is composed of 10 SiPM arrays of 8 × 8 pixels each and the point spread function (PSF) of the optical system and the dimension of the limb from the orbit of Terzina are compatible with a $3 \times 3$ mm$^2$ pixel size forming 2 rows of 5 arrays each. The sensors are provided by the Fondazione Bruno Kessler (FBK) and described below. The telescope has a field-of-view of 7.2° horizontally and 2.5°, as each pixel sees 0.18°. It can observe a vast volume of the atmosphere with a cross-section of $140 \times 360$ km$^2$.

The camera frontend electronics is composed of 10 Application Specific Integrated Circuits (ASICs), each reading out one SiPM module with 64 channels. The ASIC has an input amplification stage and only digitizes signals upon validation of trigger conditions described in Sec. 5.1. The currently foreseen amplifier has a bandwidth of 35 MHz forming the amplified signal with ~ 10 ns rising time. In Fig. 4 different scenarios are reported. The red line is the raw signal as provided to the input of the preamplifier. The blue line is the output of the preamplifier with 2 pF output load. After some optimizations to the bias parameters the preamplifier response was improved in terms of speed. The power consumption of the stage was maintained constant, while its noise performance was reduced by a factor 2. Nonetheless the noise increase is tolerable, due to the amplitude of signals to be detected. The orange and green lines represent the preamplifier output with capacitive loads of respectively 200 fF and 2 pF at its output. The 200 fF value is more realistic while the 2 pF is a worst case scenario. The ASIC samples the analog signals at a frequency of 200 MHz. It will digitize at least 3 points on the rising edge of the signal. The digitized signals from the ASICs will be brought out of the telescope system through a hole in the primary mirror partly obscured by the secondary, to a box where the electronic harness is located. The box will include an FPGA collecting the data from all 10 ASICs to form the trigger described in Sec. 5.1. The full Terzina SiPM camera plane with the ASIC system should not consume more than 5 W.

Due to the large NGB and dark current rate (DCR), the sensor technology has to be chosen taking into account many factors. Among the others described in the paragraph below, it is important to take into consideration the power consumption and the radiation tolerance.

Terzina has a technological driver to explore the performance of silicon photomultipliers (SiPMs) in space. Compared to the classic photomultipliers, they are preferred for their typically higher Photon Detection Efficiency (PDE) and their robustness, although they suffer a larger background due to the higher sensitivity, and sensitivity to radiation damage. We developed a parametric waveform simulation based on the knowledge of the sensor's single photoelectron signal shaped by the amplifier and the sensor's noise rate and NGB. This simulation, together with the Geant 4 full simulation, is useful to understand the requirements of the relevant parameters of the sensor and also to define the trigger (see Sec. 5.1). Moreover, the following requirements for the SiPM operating properties are needed: the PDE at peak wavelength should be at least of 50%, the direct and delayed cross-talk CT should be lower than 10% at operation voltage, as well as afterpulse AP and the DCR preferably not higher than 100 kHz/mm$^2$.

The NUV-HD-LowCT SiPM technology [7], developed by FBK, has typical values (for 35 $\mu$m cell-size) of dark count rate (DCR ~ 100 kHz/mm$^2$), afterpulsing (AP ~ 5%;) and optical crosstalk (CT ~ 5%–20%;), and photodetection efficiency (peak PDE ~ 50%–60%.) in the blue region of the light spectrum. The highest values of the PDE are achieved with the largest cell sizes available (e.g. 35 – 40 $\mu$m), as they feature the highest fill factor (i.e. ratio between active area and total area of one microcell / SPAD). On the other hand, the recovery time of the micro-cell also increases with increasing cell size, as it is proportional to the micro-cell capacitance and, as a first approximation, to its area.

To further improve performance, the Collaboration is also evaluating an upgraded version of the NUV-HD technology, the NUV-HD-MT, which employs metal filling of the Deep Trench Isolation (DTI) that separates adjacent micro-cells in the SiPM, to further reduce optical crosstalk probability without sacrificing PDE.

It is to be noted that one of the expected background in space is due to relativistic charged particles which can produce Cherenkov light in the telescope optical/mechanical parts and the SiPM protective resin layer. The potential background created by cosmic ray protons and, to a reduced extent, by relativistic heavier nuclei [1] and electrons is mitigated by reducing these elements close to the sensors and by the telescope chassis with baffle.

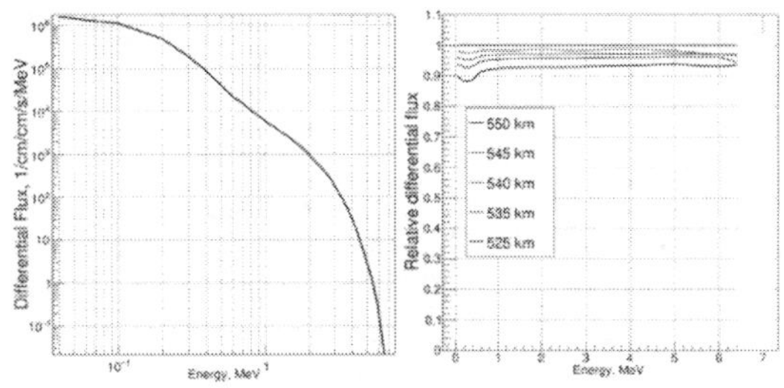


Figure 5: **Left:** Average differential electron flux at 550 km height obtained with SPENVIS [8]. **Right:** Relative variation of the differential electrons flux for different heights.

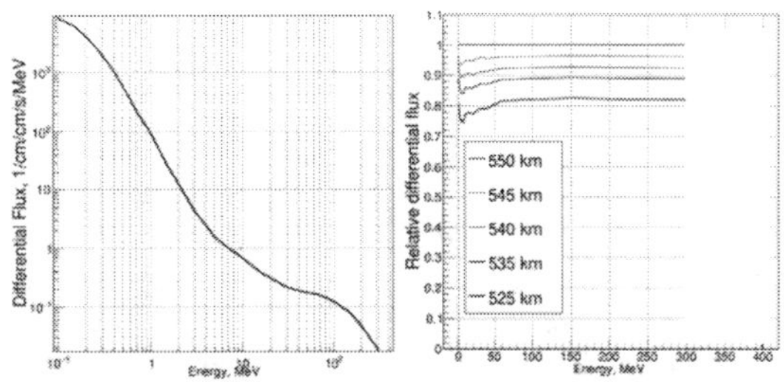


Figure 6: **Left:** Average differential proton flux at 550 km height (SPENVIS [8]). **Right:** Relative variation of the differential proton flux for different heights.

We expect electrons and protons to be the dominant sources of background. Ionizing radiation, like protons and heavier nuclei, induces damage in the silicon of SiPMs. Moreover, neutrons can create dislocations, within the crystal structure, increasing the DCR. Electrons, on the other hand, produce secondary gammas causing nonionising radiation damage. The level of damage depends on the cosmic ray radiation spectra that we obtain with SPENVIS for Terzina's orbit [8] (see Fig. 5 and Fig.6). We use the full simulation of the telescope to estimate the radiation dose integrated by the SiPMs knowing the electron and proton spectra (see sec. 3) .

In Fig. 7 the SiPM camera power consumption at different stages of the lifetime of the mission are shown. The black curve is measured in a laboratory test bench while the others are calculated using the linear extrapolation of the coefficient of the DCR increase with integrated radiation dose. Next, we can estimate NGB + DCR at BoL, after the first and second years of operation, and at EoL, namely for background levels of: 11 MHz, 22 MHz, 33 MHz, 44 MHz respectively. At the end of life, the power consumption of the sensors of the camera, operated at an over-voltage of about 6 V, will reach 0.2 W.

[1] These are mostly He with lower abundance of heavier nuclei up to Fe, particularly dangerous since number of Cherenkov photons is squarely scales with particle charge.

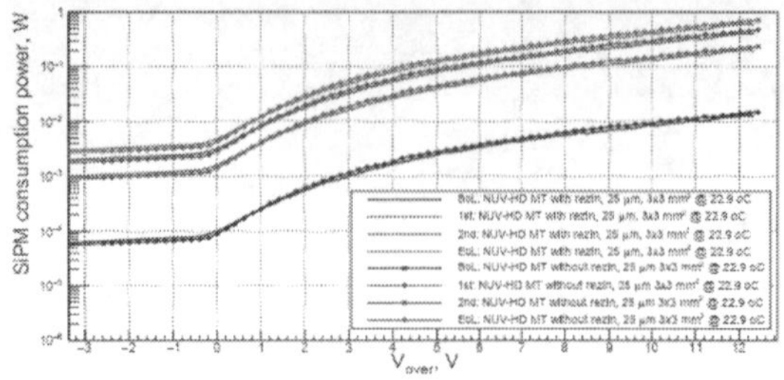


Figure 7: Power consumption by the SiPM camera.

This figure does not include the power consumption of the 10 ASICs, which are expected to consume 3.2 W reading out the 640 channels of the camera, and the trigger system for which the power consumption is estimated to be about 8 W.

Given all considerations and the fact that the NuV-HD-MT tests are not completed yet, our baseline solution is the NUV-HD-lowCT solution of FBK with 25 $\mu$m micro-cell size [7] and the optimal solution is NuV-HD-MT once validated with similar micro-cell size. If tuning of quenching resistor leads to a shorter slow component we can use 35 $\mu$m as well, ensuring larger PDE. Despite we are aware that the NuV-HD-MT can probably be run at higher over-voltage than what we assume in simulation, thanks to its low crosstalk, in Sec. 5.1 we used parameters at 6 V over-voltage of 50% PDE, DCR of 77 kHz/mm$^2$, CR+AP of about 1% + 6.6%, $\tau_s$ = 40 ns.

## 3 The Geant 4 simulation and the performance study

We have produced the preliminary simulation of Terzina based on Geant 4 [9, 10]. While the camera simulation resembles the system we are building, the optical system is preliminary. Recently, a final optimization was conducted in collaboration with a specialized company and this final configuration needs to be implemented in Geant 4 in the next future. Nonetheless, our results should not change substantially as the main optical parameters are similar between the two configurations.

The optical system of the two mirrors, currently implemented as spherical and not parabolic as they will be, is shown in Fig. 3 with the rectangular camera between mirrors and its parameters are provided in Table 1. The profile of the corrector lens is calculated to minimize the PSF for on-axis photons (see Fig. 8).

The optical system simulation takes into account the mirror and corrector lens reflectivity, transparency, the SiPM quantum efficiency and geometry of the photon sensitive camera. The resulting root mean square (RMS) of the light spot on the camera plane is shown in Fig. 9.

The FPA has been simulated using the geometrical shape of arrays and their appropriate fill factors in the micro-cells and in the arrays.

In the simulation we use the Photo-Detection Efficiency (PDE) of the FBK technology NUV-HD, optimized for the UV band shown in Fig. 10, at an over-voltage of $V_{over}$ = 6 V [2]. Also the PDE of a NUV-HD bare sensors and of an RGB sensor (optimized for the optical-IR region) are shown.

To estimate the sensitivity and performance to the expected UHECR signal, a dedicated generator was interfaced to the Geant 4 simulation.

We use the Emission for Extensive Air Showers Cherenkov Simulation (EASCherSim) [11] as physics event generator. EASCherSim provides the average photon density, photon spectral composition (Fig. 11), the photon angle with respect to the shower axis and time (Fig. 12) of photons. We assume a uniform distribution of the azimuth angle with respect to the telescope axis and the absence of the correlation between photon angles, time and wavelength. For the moment it can only simulate proton interactions. The composition of UHECRs is mixed at 100 PeV (see e.g. [13]). Additionally Pierre Auger found out that it is dominated by light elements at 2000 PeV [12] and becomes heavier at higher energies. Hence, in the future we will investigate a mixed composition.

An example of the shower image on the camera plane is shown in Fig. 8 (middle panel) next to the simulated PSF for the same location on the camera plane (right panel).

As an example of results on the Terzina performance from the simulation, we also show the aperture of Terzina. The plot assumes the NGB and DCR of the first year of operation (see Fig. 13). A threshold of 7 p.e. per pixel, which suppresses the background, is assumed.

## 4 The environmental background of Terzina: the night glow.

The Night Glow Background (NGB) has been estimated in two different ways. Firstly, using the formula:

$$\text{Rate per pixel} = S \times d\Omega \times Flux \times Area \times PDE_{eff.} \quad (1)$$

where $Area$ = 0.1 m$^2$; $PDE_{eff}$ = 0.1 is the total optical efficiency calculated from the convolution of the SiPM PDE and the NGB spectrum, as a function of wavelength times an average optical of the 2 mirrors; $\Delta\Omega$ is the pixel solid angle; $\text{Flux}[m^{-2}sr^{-1}ns^{-1}]\Big|_{\lambda=1000\ nm}^{\lambda=300\ nm} = 1.55\times10^4$; $S$ = 6 is a conservative safety factor considering the possible largest variation of the glow, the rate per pixel is ~10 MHz [15].

Secondly, using the full simulation described in Sec. 3, we propagate photons assuming the NGB spectrum and a uniform angular distribution. Increasing the rate by a safety factor of 6 [6], which provide the typical variation with zenith angle of showers, we obtain a pixel background rate of ~5 MHz. Given the discrepancy between the two estimates, we decided to use the more conservative value of 10 MHz per pixel (see Fig. 14).

## 5 Event size and data throughput

The maximum daily data throughput for Terzina is 45 Gbit/day. Assuming that the data throughput will be

[2] $V_{over} = V_{operation}$ (operational voltage) - $V_{bd}$ (break down voltage)

 https://doi.org/10.1051/epjconf/202328306006

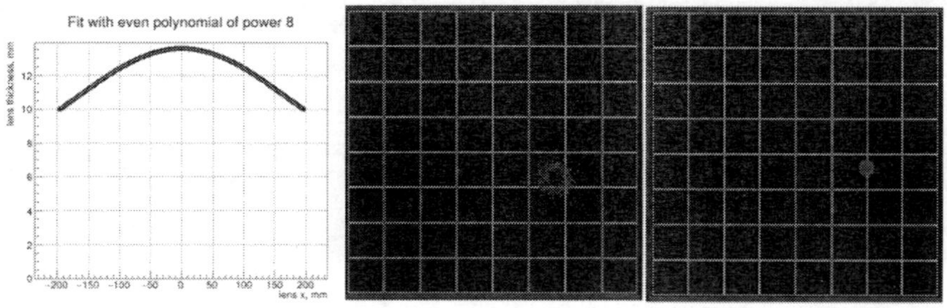


Figure 8: **Left:** The cross-section view of the corrector lens. **Center:** Photons induced by many UHECRs from 100 PeV protons, with equal parameters, superimposed on the camera plane. **Right:** PSF at the same location on the camera plane.

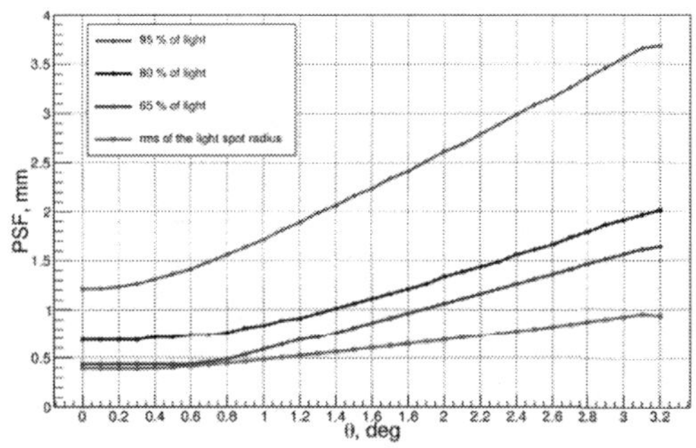


Figure 9: RMS of the radius of the light spot before (lower curve) and after (3 upper curves for different percentage of photon containment) it is processed by the optical elements of the telescope from Geant 4 simulation. For brevity we call this PSF on the y-axis.

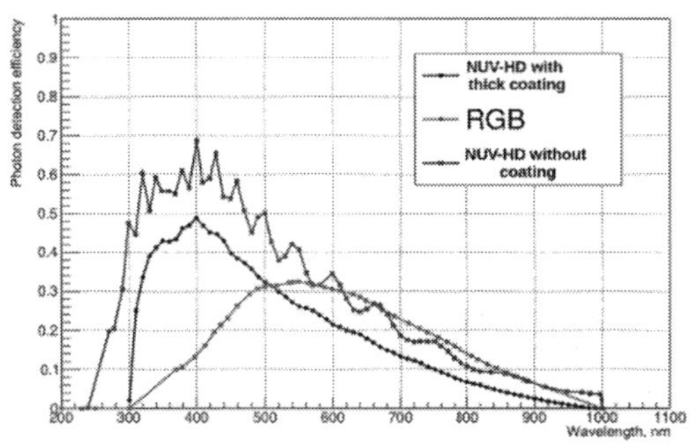


Figure 10: Photon detection efficiency versus photon wavelength for different SiPM types by FBK.

dominated by camera events when compared to telemetry data, we can derive the maximum event rate.

One event will have the size of:

$$size_{event} = size_{event\ header} + N_{samples} \times ADC_{resolution} \times N_{pixels} \quad (2)$$

where $size_{event\ header} = 34$ bits ( includes the bits dedicated to the alignment with the FPGA and and internal address of the readout units), $N_{samples} = 20$, $ADC_{resolution} = 12$ bit and the number of pixels, $N_{pixels} = 9$.

At a trigger rate of 240 Hz, the maximum data throughput for Terzina is reached. From simulations we derive the thresholds to obtain 120 Hz of the events with signal contained in a single pixel and 120 Hz with signal shared between two neighboring pixels.

Within this work, we consider 9 pixels pad to readout. However, at the beginning of the mission, we will read one ASIC or even a full camera for debugging/calibration purposes. Latter as soon we get our system fully understood we will shrink the redout pad, dropping all irrelevant information.

### 5.1 The trigger implementation

The ASIC offers the possibility of two programmable thresholds per channel hereinafter low and high threshold. The trigger will be structured so that, when the high threshold is passed by at least one channel, the waveforms of the 64 channels (one ASIC) are digitized and sent to the FPGA.

The time for the analog to digital conversion of all channels in the given ASIC, assuming the resolution of 12 bits takes $4096 \times 5$ ns, namely $\sim 20.4 \mu s$. The digitised data readout time is 64 (channels) $\times$ 32 (time-cells) $\times$ 12 (resolution) $\times$ 1.25 ns (serialization in DDR) = 30.7 $\mu s$. Thus the total time for the analog to digital conversion and processing in the FPGA amounts to $\sim 51.2 \mu s$.

Considering a power-law increase in the number of UHECR events with a decreasing energy threshold, in order to avoid an obvious increase of the background, the low threshold trigger is defined using the coincidence of 2 adjacent pixels. The coincidence time window is chosen to be twice the rise time of the SiPM signal after amplification, namely $\tau = 20$ ns (however this time window is configurable).

The implementation of the low level threshold trigger is the following: when at least two pixels cross the low threshold in one ASIC or in the 16 pixels along the edges of two neighboring ASICs within the coincidence window $\tau$, a binary hit-map of the corresponding ASIC(s) is sent to the FPGA. The hit-map is analyzed to seek if there are at least two neighboring channels whose threshold is crossed

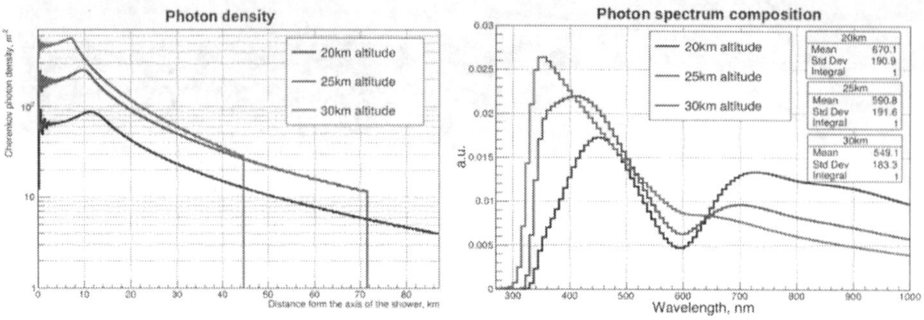


Figure 11: **Left :** Photon density of the Cherenkov ring as seen from the telescope produced by 100 PeV protons for different heights as a function of distance from the shower axis. **Right:** Photon spectral composition for showers produced at different altitudes in the atmosphere. The deeper the shower the more the photons will be scattered in the atmosphere and spectral emission moves toward the red band.

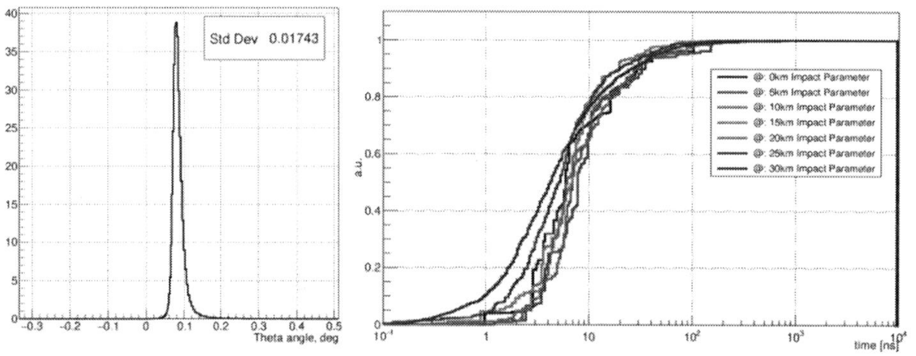


Figure 12: **Left :** Photon angle distribution for a proton of 100 PeV with respect to the shower axis seen by the telescope pointing in the direction of this axis. **Right :** Cumulative distribution of the Cherenkov photons as a function of time from the first one emitted in the direction of the telescope.

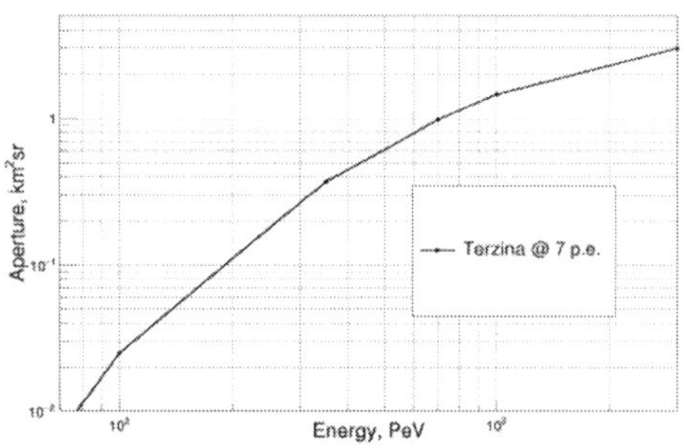


Figure 13: The expected aperture for Terzina versus proton energy during the first year of operation.

(forming a cluster). If a cluster is found, the event is validated and the analog to digital conversion is performed, otherwise the event is discarded.

We decide to assign the same data rate to the two triggering mechanisms, namely 1/2 of the maximum rate each. In particular, considering the aforementioned data transmission limit, the rate per triggering mechanism is set to 120 Hz. We assign the same rate to the two triggering mechanisms because the expected data rate of UHECRs hitting 1 or 2 pixels is about equal. Then, from simulation we evaluate the thresholds corresponding to this rate for both cases. Regardless of whether the event is validated or discarded, the hit-map is used to increment counters to monitor the camera background.

## 5.2 Camera monitor

The camera monitor based on hit-maps for low and high thresholds, aims to monitor the average background rate in the two parts of the camera: the upper row of 5 arrays+ASICs in the camera plane will be sensitive to the Earth region and so to UHE neutrinos, but also to undesired blind occurrences due to storms or city light pollution; the lower part of the camera plane will be sensitive to the limb. Very recent changes in the designing of the read-out system opens the possibility to store on the FPGA the hit-map after the high-threshold trigger is satisfied. The single pixel hit-map is less vulnerable, with respect to dou-

https://doi.org/10.1051/epjconf/202328306006

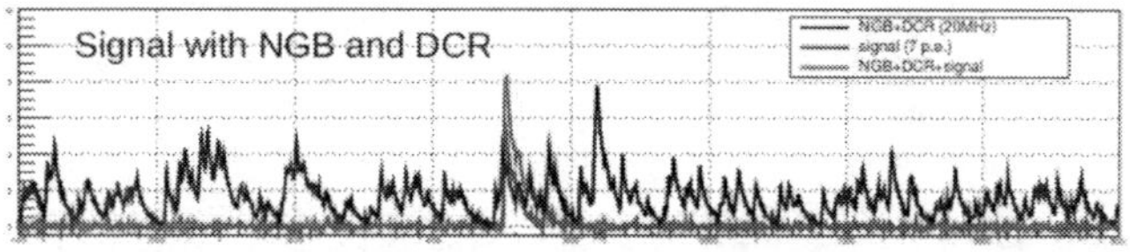


Figure 14: Signal with NGB and DCR

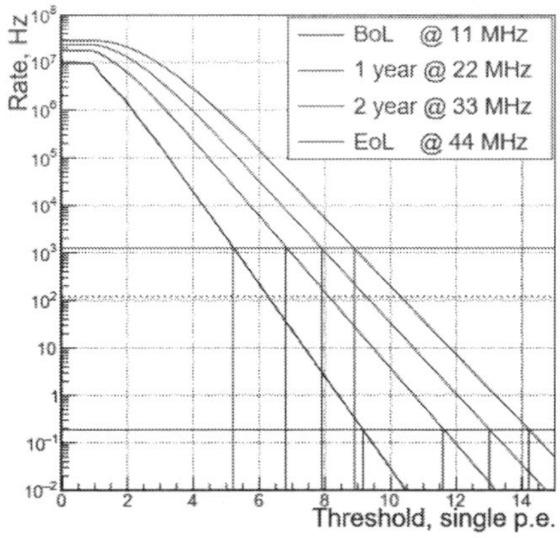


Figure 15: Single pixel trigger rate as a function of the threshold expressed in p.e. for DCR + NGB values estimated at different times during the mission life. The horizontal blue dashed line corresponds to the maximum trigger rate of (120 Hz). Horizontal blue line corresponds to the maximum single pixel trigger rate (120 Hz/640 ~ 0.18 Hz per pixel). The horizontal red line corresponds to 1250 Hz (maximum single pixel rate for configuration with two pixels cluster). The vertical lines shows thresholds for single (blue) and double coincidence (red) trigger logic. Thresholds can be reconfigured during flight.

ble pixel, to the single point failure of a frozen combinatorial logic in the ASIC. This monitor has to operate constantly over mission duration. The monitor will deliver the rate at a given threshold for each pixel. The estimated monitoring data size, running at 1 Hz, is about 1-2% of the event data rate. Hopefully, this method will allow to discriminate the human-induced or meteorological spurious background also through correlations with meteo-maps.

### 5.3 Validation of the trigger through simulation

The parametric waveform simulation was used also to understand the requirements of the sensor and also where to set the trigger thresholds of the two configurations in order to satisfy the total data throughput limit of 45 Gbit/d. We simulate waveforms for a time frame of $10^7$ ns for different NGB+DCR conditions. Counting the numbers of peaks below a given threshold divided by the simulated time frame, we can estimate the trigger rates (see Fig. 15). For the results shown here, we assume the variable NGB during mission lifetime described in Sec. 5 and a cross-talk of sensors of 6.6% and DCR of 77 kHz/mm$^2$, as expected for NUV-HD-MT.

For the single pixel trigger, which requires crossing the threshold in a single pixel, the threshold will have to be increased during the lifetime of the mission due to the increasing radiation damage as discussed in Sec. 2. To satisfy the data throughput requirement, the threshold has to be set at the levels of: 9.2 p.e., 11.7 p.e, 13.1 p.e., 14.3 p.e (blue horizontal line), correspondingly at BoL, after the first year, after the second year, at EoL, respectively.

The number of possible pixel configurations which form a two-pixel cluster (see Fig. 16) in a single ASIC of 64 channels is equal to 210 or 2100 in the full camera. The number of all possible two-pixel configurations in a single ASIC is given by the $n - k$ combination equation:

$$\left.\frac{n!}{k!(n-k)!}\right|_{k=2}^{n=64}, \tag{3}$$

which yields a total of 2016 combinations per ASIC or 20160 in the full camera. Considering the single pixel rate $R_1$ and the already defined concidence window $\tau$, the coincidence rate for clusters in the full camera is computed as follows:

$$2 \times R_1^2 \times \tau \times 2100. \tag{4}$$

On the other hand, the coincidence rate for every possible combination of two pixels firing per ASIC in the full camera is:

$$2 \times R_1^2 \times \tau \times 20160. \tag{5}$$

The maximum single pixel rate for the coincidence configuration that can be afforded, given the data transmission rate, is therefore 1250 Hz (with fixed size of the readout pad - 9 pixels). Hence, from Fig. 15 (red horizontal line) we define the low threshold (in p.e.) for the considered different epochs of the mission as 5.2, 6.8, 7.8, 8.8 p.e. for BoL, after the first year, after the second year, and at EoL, respectively.

Processing two random pixel coincidences will amount to a total of 250 $\mu$s per second, while processing the 240 Hz expected trigger rate from single and two pixels configurations will reach ~7.4 ms which corresponds to a dead time below 1%.

## 6 Conclusions

Terzina is a pathfinder mission to demonstrate Cherenkov detection from a space platform of middle size enabling the possibility to perform neutrino astronomy from space and measuring UHECRs. Its results will be useful information for the planned POEMMA mission [5]. Rapidly evolving progress on the development of the Terzina telescope and its SiPM camera, including the full simulation chain, the amplification stage, and the ASIC, is reported. A preliminary estimation of the trigger rate at a given threshold and readout pad size has been assessed for different mission lifetimes. The radiation dose accumulated by the SiPM and electronics mainly caused by electrons, secondary gammas, and protons has been estimated as well as the SiPM power consumption. An indication of Terzina aperture has been provided.

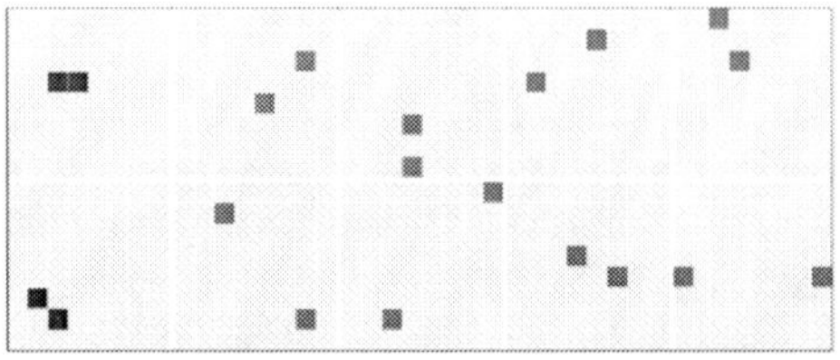

Figure 16: Camera composed of 10 arrays of SIPMs. Each array is readout by one ASIC with $8 \times 8$ channels. On the left, pixels in black forming a clusters can be seen, while the other fired pixels represent examples of background.

## Acknowledgements

NUSES is a joint project of the Gran Sasso Science Institute and Thales Alenia Space Italia, funded by the Italian Government (CIPE n. 20/2019 ), by the Italian Minister of Economic Development and the Abruzzo Region (MISE n. F/130087/00/X38), by the Italian Space Agency (ASI n. 15/2022) and by the Swiss National Foundation (SNF grant n. 178918).

## References

[1] I. De Mitri for the NUSES Collaboration, J. Phys.: Conf. Ser. 2429 (2023) 012007.

[2] A. Di Giovanni and M. di Santo for the NUSES Collaboration, POS 414 (2022) 354, URL: https://pos.sissa.it/414/354.

[3] A. L. Cummings, R. Aloisio and J. F. Krizmanic, "Modeling of the Tau and Muon Neutrino-induced Optical Cherenkov Signals from Upward-moving Extensive Air Showers," Phys. Rev. D **103** (2021) no.4, 043017 [arXiv:2011.09869].

[4] A. L. Cummings, R. Aloisio, J. Eser and J. F. Krizmanic, "Modeling the Optical Cherenkov Signals by Cosmic Ray Extensive Air Showers Directly Observed from Sub-Orbital and Orbital Altitudes," Phys. Rev. D104 (2021) 063029.

[5] A. V. Olinto *et al.* [The POEMMA Collaboration], "The POEMMA (Probe of Extreme Multi-Messenger Astrophysics) observatory," JCAP **06** (2021), 007, [arXiv:2012.07945].

[6] J. Krizmanic, et al. "POEMMA: Probe of extreme multi-messenger astrophysics." EPJ Web of Conferences. Vol. **210**. EDP Sciences, 2019.

[7] A. Gola et al. "NUV-Sensitive Silicon Photomultiplier Technologies Developed at Fondazione Bruno Kessler", Sensor **19** (2019) 308.

[8] https://www.spenvis.oma.be/.

[9] S. Agostinelli et al., "GEANT4: A simulation toolkit", Nucl. Instrum. Meth. A506 (2003) 250–303.

[10] http://geant4.web.cern.ch/geant4.

[11] https://pypi.org/project/easchersim/1.1/, Austin Cummings BSD 3-Clause License.

[12] J. Bellido et al.(Auger Collab.), Proceedings of Science(ICRC2017), 506 (2017).

[13] M. G. Aartsen et al., Phys. Rev. D 100 (2019) 082002.

[14] M. Takahiko et al. "Suppression of the optical crosstalk in a multichannel silicon photomultiplier array" Opt. Express 29 (2021) 16914-16926.

[15] W. Hofmann, et al. "Balloon-borne infrared telescope for absolute surface photometry of the night sky," Appl. Opt. **16** (1977) 3125-3130.

[16] S. Mianowski et al. "Proton Irradiation of SiPM arrays for POLAR-2", Research Square (2022).

# The JEM-EUSO Program for UHECR Studies from Space

*Etienne* Parizot[1,*], *Marco* Casolino[2,3,4], *Piergiorgio* Picozza[2,3], *Toshikazu* Ebisuzaki[4], *Mario Edoardo* Bertaina[5,6], *Christer* Fuglesang[7], *Andreas* Haungs[8], *Fumiyoshi* Kajino[9], *Pavel* Klimov[10], *Angela* Olinto[11], *Marco* Ricci[12], *Hiroyuki* Sagawa[13], *Jacek* Szabelski[14], *Lawrence* Wiencke[15], and *the* JEM-EUSO Collaboration

[1]Université Paris Cité, CNRS, Astroparticule et Cosmologie, F-75013 Paris, France
[2]INFN Section of Rome Tor Vergata, Via della Ricerca Scientifica 1, 00133 Rome, Italy
[3]Department of Physics, University of Rome Tor Vergata, Via della Ricerca Scientifica 1, 00133 Rome, Italy
[4]RIKEN, 2-1 Hirosawa Wako, Saitama 351-0198, Japan
[5]Department of Physics, University of Turin, V. P. Giuria 1, 10125 Turin, Italy
[6]INFN Section of Turin, Via P. Giuria 1, 10125 Turin, Italy
[7]KTH Royal Institute of Technology, Brinellvägen 8, 114 28 Stockholm, Sweden
[8]Karlsruhe Institute of Technology (KIT), Institute for Astroparticle Physics, Karlsruhe, Germany
[9]Department of Physics, Konan University, 8 Chome-9-1 Okamoto, Higashinada Ward Kobe, Hyogo 658-8501, Japan
[10]Skobeltsyn Institute of Nuclear Physics, Lomonosov Moscow State Univ., ul. Kolmogorova 1(2), 119991 Moscow, Russia
[11]Department of Astronomy and Astrophysics, The University of Chicago, 5640 S. Ellis Avenue, Chicago IL 60637, US
[12]INFN National Laboratories of Frascati, Via Enrico Fermi 54, 00044 Frascati, Italy
[13]Institute for Cosmic Ray Research (ICRR), University of Tokyo, Chiba 277-8582, Japan
[14]National Centre for Nuclear Research, ul. Pasteura 7, 02-093 Warszawa, Poland
[15]Department of Physics, Colorado School of Mines, 1523 Illinois St., Golden CO 80401, US

**Abstract.** To take up the challenge of understanding the origin of the ultra-high-energy cosmic rays (UHECRs), new observational means appear necessary. The JEM-EUSO Collaboration has undertaken to open the space road to UHECR studies. For more than a decade, it has been developing a realistic program to measure the UHECRs from space with unprecedented aperture, together with complementary scientific objectives in a broader multidisciplinary context. Several intermediate missions have already been completed (on the ground: EUSO-TA; under stratospheric ballons: EUSO-Balloon and EUSO-SPB1; in space: TUS, and on-board the ISS: MINI-EUSO), and others are in preparation for flight (EUSO-SPB2), under review (K-EUSO: currently on hold), or proposed for the next decade (POEMMA). We report on the general status of the JEM-EUSO program, underlining that its technology has now reached operational maturity, and is ready for actual cosmic-ray shower detection from above.

## 1 Introduction

Despite intense observational efforts and a series of important results in the last two decades, the study of ultra-high-energy cosmic rays (UHECRs) remains one of the most challenging in astronomy, both because their flux is extremely low (one particle per m$^2$ per billion year at the highest known energies) and because their macroscopic energies (tens of Joules) still remain insufficiently large to allow quasi-rectilinear propagation in the Galactic (and extragalactic) magnetic fields. As a consequence, no direct detection of their sources has been possible thus far, and their astrophysical origin as well as their acceleration mechanism remain unknown. A broad and detailed summary of the current status and observational situation in the field, as well as a presentation of the interactions with other fields in high-energy physics and astrophysics can be found in a recent white paper, which involved and was endorsed by a large research community worldwide [1]. It also examines the perspective for progress in the next two decades. One of its conclusions it that new observational means will be necessary, notably to increase the statistics of the UHECR datasets at the highest energies.

As an alternative to building more and more detectors spread over larger and larger areas on the ground (beyond the thousands of km$^2$ currently instrumented), one way to reach significantly larger instantenous apertures with a single instrument has been identified several decades ago [2]: it consists in going to space and looking down to the Earth atmosphere with a large field-of-view telescope, to detect the fluorescence light of UHECR-induced extensive air showers (EAS) over a very large volume. This is of course challenging from the instrumental point of view, given the large distance at which the showers must be observed with sufficient precision to allow a reconstruction that is reliable enough. However, such an achievement appears accessible with current technology, and this is indeed what the JEM-EUSO Collaboration (Joint Experiment Missions for

*e-mail: parizot@apc.in2p3.fr

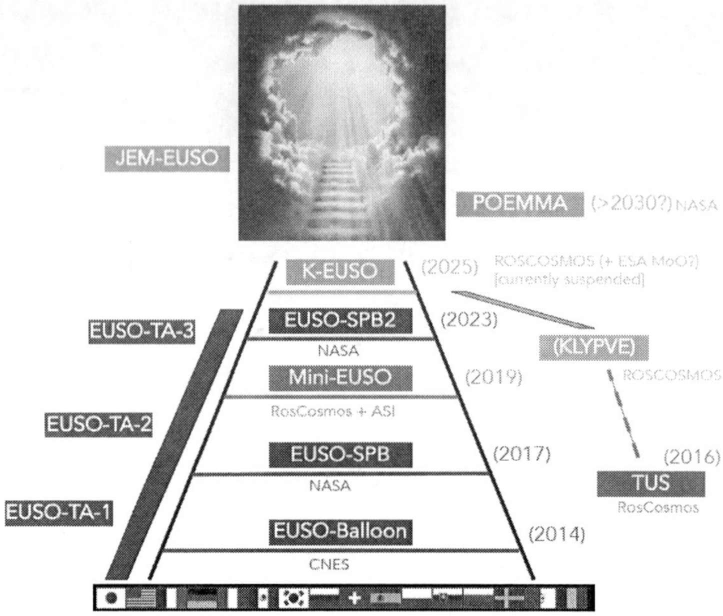


**Figure 1.** Summary of the JEM-EUSO "stairway to Heaven" development program. (Details in the text.)

an Extreme Universe Space Observatory) has undertaken to demonstrate and realise.

Following earlier work in the context of the Orbiting Wide-angle Light-collector (OWL) [3] and Extreme Universe Space Observatory (EUSO) [4] mission projects, the JEM-EUSO Collaboration endeavoured to develop a major space mission to study UHECRs, also allowing the detection of very high energy neutrinos and additional scientific objectives. Its strategy has been to proceed via a series of intermediate missions, each time achieving new technical and scientific milestones, thus climbing step by step what may be loosely referred to as the UHECR "stairway to heaven" (or at least to space!), as summarized in Fig. 1. After a decade of innovative technological developments and successful missions, as briefly summarized below, the JEM-EUSO Collaboration appears ready for a major UHECR mission from space. With the help of concommitant progress in other parts of astrophysics, such as a reliable mapping of the Galactic magnetic field and new progress in the modelling of potential sources, this may indeed lead to a long-awaited breakthrough in the understanding of the origin.

## 2 UHECR state of the art

The current understanding of the UHECR phenomenology is the result of decades of intense activity in the field, which is currently dominated by the two largest observatory ever built: the Pierre Auger Observatory (Auger: 3000 km$^2$; e.g. [5] and Berat, these Proceedings) and the Telescope Array (TA: currently being extended up to 2800 km$^2$; e.g. [6, 7] and Kido, these Proceedings). Very broadly, it may be summarized as follows:

- As expected from simple and solid theoretical considerations, it is clearly established [8, 9] that the spectrum of the UHECRs shows a GZK-like cutoff [10, 11], above a few tens of EeV.
- As generically expected from astrophysical particle acceleration models for which the maximum energy of protons at the source hardly reaches $10^{19}$ EeV, the UHECR composition is getting heavier and heavier above a few EeV [12].
- As expected from simple considerations linking the sources of UHECRs with ordinary matter, the distribution of the UHECR arrival directions is not isotropic. By order of decreasing significance, departure from isotropy has been established through: i) a significant first-order angular modulation coefficient (dipole) above 8 EeV, [13] ii) a correlation with large scale cosmic matter distribution (using catalogs of galaxies as a proxy), [14] and iii) warm spots showing flux excesses on relatively large angular scales, both in the Southern and Northern hemispheres. [15, 16]

In addition, significant signals for a composition anisotropy [17] and a declination-dependent energy spectrum [18] have been reported by the Auger and TA collaborations, respectively.

While these results are important milestones and come as the crowning achievement of years of intense observational efforts, they have not allowed, so far, to discover the origin of the UHECRs. We still do not know what the sources are, nor which physical mechanism is responsible for the acceleration of such high-energy particles. In addition, the experimental situation is still partly confused, primarily by lack of statistics, but also because the two main detectors in operation observe different parts of the sky, with different levels of exposure and limited overlap.

## 3 Forward through space

The main reason why the current generation of detectors were not able to discover the sources of UHECRs, despite their very large aperture and excellent quality data, can be understood from the data themselves. Contrary to what had been (somewhat naively) expected two decades ago, the highest energy particles appear not to be protons, but heavier nuclei, with higher charges and thus larger deflections in the intervening magnetic fields. This makes direct pointing astronomy with cosmic rays essentielly impossible. A better knowledge of the Galactic magnetic field than currently available will be needed to infer the distribution of arrival directions before they enter the Galaxy, which would provide a much clearer view on the location of the sources. In addition, a relatively low rigidity cutoff at the sources (as strongly suggested by the composition [19, 20]) makes the acceleration of the UHECRs less challenging than had been thought, which increases considerably the number of sources potentially contributing to the observed flux. More sources with larger deflections leads to source confusion and overlap, which makes it difficult to exploit the measured anisotropies, especially at the current level of significance, which does not allow to discriminate between competing astrophysical models.

Since we cannot change the phenomenology of UHECRs, we must try and develop strategies to overcome the above problems. This is where a space mission can be of remarkable value, in complement to ground based observatories, by providing much larger statistics at the highest energies as well as near-uniform full sky coverage with one single instrument deployed in Low Earth Orbit (LEO).

Even an intermediate space mission of moderate size, say providing an exposure comparable to that of current ground-based observatory, would be of great value to draw the first full-sky map in UHECR with a single instrument (thus with the uniform systematics), characterise the large scale anisotropies with greater precision, and address the potential discrepancies or differences in the UHECR flux in the Northern and Southern hemispheres. It would also increase by one order of magnitude the instantaneous aperture, and by the same amount the fluorescence aperture, providing a direct measurement of the average of the depth of the maximum shower development ($\langle X_{\max}\rangle$) up to the highest energies. This will help characterising the evolution of the UHECR composition, and its possible non uniformity.

The JEM-EUSO experimental concept has also been enriched by the addition of direct Cherenkov detection capability, which will take effect with the EUSO-SPB2 mission in Spring 2023. This allows to search for high-energy neutrino showers, by looking at the limb of the Earth, either to detect the diffuse cosmogenic flux or by pointing towards targets of opportunities as part of a global multi-messenger strategy [21] (see also Cummings, this conference, and Krizmanic, this conference, and references therein).

In addition, a space-based shower detector allows to target new types of cosmic-ray events, inaccessible from the ground, in particular high-altitude, "atmosphere-skimming" showers, whose entire development takes place in very low density medium [22]. These may provide important information about the physics of the showers, as the balance between the decay of instable particles and the production of new particles, which is the key parameter driving the shower development, directly depends on the ambient density.

Finally, a UHECR space mission will collect a large amount of data that are valuable for other scientific domains, such as planetary science, fundamental physics, atmospheric physics and Earth observation in general. Among the complementary science objectives of current and future missions based on the JEM-EUSO technology, one may cite the study of meteors, transient luminous events (ELVES), putative anitons, ionospheric phenomena, bioluminescence, or the search for nuclearites and strange quark matter.

## 4 The JEM-EUSO program and technology

### 4.1 JEM-EUSO collaboration

The JEM-EUSO Collaboration gathers about 300 physicists and engineers from 12 member countries (Algeria, Czech Republic, France, Germany, Italy, Japan, Mexico, Poland, Russia, Slovakia, Sweden, USA) and 4 associated countries (Republic of Korea, Romania, Spain, Switzerland). It is engaged in an ambitious, consistent experimental project aiming at advancing the understanding of UHECRs through the observation of EAS from space. Taking up this challenge required the establishing of a full program composed of intermediate missions (see Fig. 1), whose progressive development allowed to acquire, optimise and validate the appropriate technology, demonstrate the viability of the experimental and instrumental concepts, and assess the performance that can be realistically expected from a full scale JEM-EUSO mission (cf. Bertaina, this conference).

### 4.2 JEM-EUSO instrumentation

The JEM-EUSO Collaboration has developed technology to operate a *fluorescence telescopes* (FT) from LEO, for the detection of the isotropically emitted fluorescence light and the diffused Cherenkov emission from extensive air showers. The nadir-pointing FT provides the large instantaneous exposure for UHECR detection, which can be further increased by tilting the optical axis to observe larger volumes of atmosphere. In addition, for the forthcoming mission EUSO-SPB2 (see below), a new type of telescope has been added, namely a *Cherenkov telescopes* (CT) for the detection of the direct Cherenkov light emitted (roughly) along the direction of the showers. The CT requires a different pointing configuration, to detect upward going showers from Earth skimming neutrinos (or exotic events) emerging from below the limb of the Earth,

https://doi.org/10.1051/epjconf/202328306007

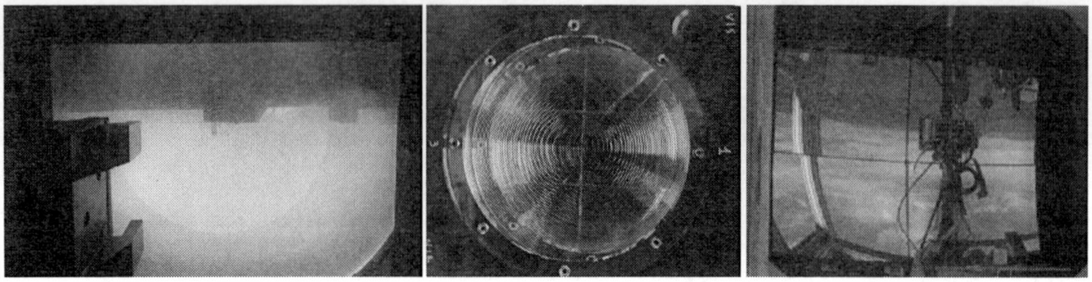

**Figure 2.** JEM-EUSO optics. Left: view through the EUSO-TA refractive Fresnel optics system. Center: front view of the MINI-EUSO Fresnel lens system. Right: side view of the EUSO-SPB2 mirrors.

or horizontal showers developing at high altitude from above the limb.

Operating from space implies specific requirements for the instrumentation. Since the atmospheric showers are observed from a large distance (typically 400 km on the ISS or a bit more for free flyers in LEO), the geometric attenuation of the light intensity is very large. Thus, ensuring an energy threshold compatible with a satisfying overlap with ground-based observatories requires both very sensitive detectors, able to operate in photon counting mode, and very large optics, to collect as much light as possible and allow efficient detection down to a few $10^{19}$ eV.

The JEM-EUSO collaboration has developed two types of large field-of-view optics: refractive optics using large polymethyl methacrylate (PMMA) Fresnel lenses (contributed by Japan) and Schmidt telescope with large segmented spherical mirrors (contributed by Czech Republic). Examples can be seen on Fig. 2, which shows the refractive optics of EUSO-TA (left) and MINI-EUSO (center) and the reflective optics of EUSO-SPB2 (right; NB: a front corrective lens is not visible).

Regarding the photodetectors, the FT uses 64-pixels multi-anode photomultiplier tubes (MAPMT) from Hamamatsu, allowing single photon detection with very high efficiency (~30%). BG3 UV filters are used to narrow the detection window around the $N_2$ fluorescence lines in the UV. A modular focal surface allows to adapt the developed technology to missions of different sizes. The smallest detection unit consists of 4 MAPMTs, gathered into a so-called elementary cell (EC, contributed by France). Three successive generations of ECs have been developed, culminating in the fully integrated EC version 3, used for the K-EUSO prototype and the EUSO-SPB2 mission. It integrates the front-end electronics based on the dedicated ASIC SPACIROC3 (contributed by France), and a specially developed low consumption Cockroft-Walton high-voltage power supply (HVPS) board (contributed by Poland). The EC is then potted to be able to operate in the 3 mbar environment of the JEM-EUSO balloon missions (see Fig. 3). For the CT, the photodetectors are Silicon photomultipliers (SiPM) matrices from Hamamatsu, also assembled into modular elements including the front-end

**Figure 3.** JEM-EUSO FT detection unit. Top left: 64 pixel MAPMTs used by all JEM-EUSO missions. Top right: ASIC boards using SPACIROC 3. Middle left: Cockroft-Walton (CW) high-voltage power supply. Middle right: 4 MAPMTs, 2 ASIC boards and 1 CW board assembled into a potted elementary cell of the third generation for the EUSO-SPB2 mission.

electronics using a Music chip, power supply, amplifier board and microcontroller (contributed by the USA).

Figure 4 shows additional elements of JEM-EUSO's last generation FT electronics, namely the Zynq board ensuring data acquisition and trigger (contributed by Russia), the HVPS DC-DC control board (contributed by Poland) and the Data Processor (contributed by Italy). All these elements, except the latter, are integrated together into a so-called Photo-Detection Module, or PDM, which is the largest FT building block of all JEM-EUSO missions, composed of 9 ECs in a square matrix, i.e. 36 MAPMTs, with a total of 2304 pixels. Its last version, implemented in the EUSO-SPB2 mission, is shown on Fig. 5 with the ECs facing towards the left of the picture.

### 4.3 JEM-EUSO missions

As mentioned above, the JEM-EUSO experimental program has been developed through a series of intermediate balloon and space missions, in parallel with the operation

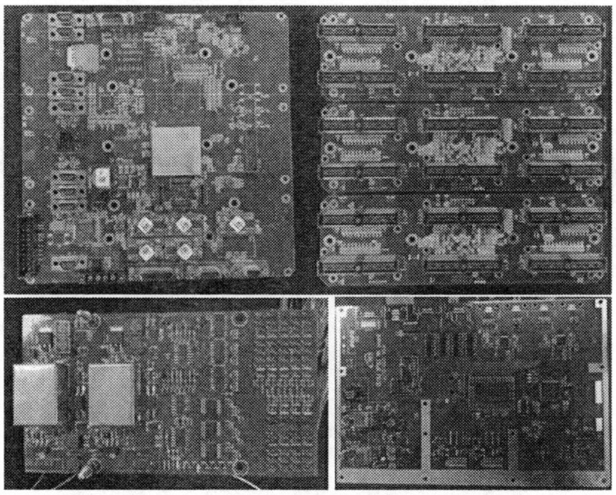

**Figure 4.** JEM-EUSO electronics. Top: Zynq board with FPGA for data acquisition and trigger. Bottom left: HVPS DC-DC control board. Bottom right: data processing and control board.

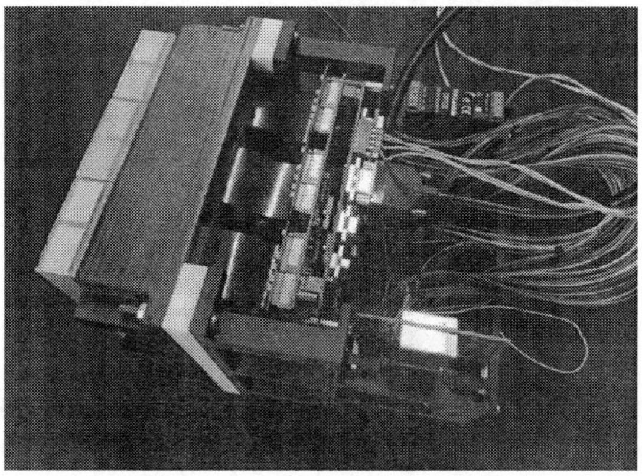

**Figure 5.** JEM-EUSO's last generation PDM.

of the EUSO-TA facility, benefiting from the installation of the TA collaboration at Black Rock Mesa (Delta, Utah, USA). This facility proved very useful for tests, measurements and data taking with the succesive versions of the instrumentation [23].

#### *4.3.1 EUSO-Balloon*

The first JEM-EUSO mission to fly was the EUSO-Balloon mission [24], a one-night mission of the French space agency, CNES, under a stratospheric balloon launched from the joint CNES-CSA facility in Timmins (Canada). The main innovations of the EUSO-Balloon instrument were the Fresnel optics and the first generation PDM using of dedicated ASIC (SPACIROC) for the front-end electronics and a low-consumption HVPS board implementing an automatic switch of the high-voltage scheme in case of a sudden increase of the background light, to protect the MAPMTs from luminous events (such as lightning), as well as efficient data processing.

The mission had a successful flight in August 2014, demonstrating the validity of the experimental concept and the robustness of the instrumentation to operate at 3 mbar. EUSO-Balloon also provided important results for the following steps of the JEM-EUSO program, namely the first imaging measurement of the UV emissivity of the Earth at night, both with and without clouds, which also allowed to demonstrated the anticorrelation between the UV and IR luminosities (using an IR camera contributed by Spain). EUSO-Balloon also demonstrated the serendipitous detection of unexpected flashers on the ground (most probably of artificial origin).

Finally, and most importantly, EUSO-Balloon was able to detect CR-like tracks, produced by laser pulses shot from a helicopter flying below the balloon, crossing the field of view of the telescope. The scattered light from the laser pulse was properly detected for several tens of laser shots, and the reconstruction of the trajectory showed an accuracy of the order of 2 degrees (from the known laser orientation).

#### *4.3.2 EUSO-SPB*

After the success of this first flight, the JEM-EUSO collaboration undertook to develop a second balloon mission with upgraded instrumentation and a much longer flight, so as to be able to detect cosmic-ray showers from above for the first time in fluorescence. This led to the NASA mission EUSO-SPB [25], where SPB stands for Super Pressure Balloon, a closed, pressurised balloon allowing to stay afloat for up to 100 days in principle. The instrument was designed to detect and reconstruct several CR events during such a flight, thanks to an energy threshold around $10^{18}$ eV and a field-of-view similar to that of EUSO-Balloon ($\sim 11° \times 11°$). The focal surface consisted again of 1 PDM, but with several important improvements. Among them a new version of the ASIC, SPACIROC-3, and a new generation of ECs, more compact and stable. These were integrated in the second version of the PDM, shown in Fig. 6 (left picture). The optics performance was also improved compared to EUSO-Balloon, and a new power supply scheme was added, using solar panels, as required for a long duration flight. Finally, the EUSO-SPB instrument had a complete scheme of data acquisition, processing and transmission, implementing an autonomous trigger for showers or other types of events.

Pre-flight tests on the ground at the TA site allowed to assess the performances of the instrument, with a reconstruction of laser shots with $\sim 1°$ accuracy and an energy threshold at 50% detection efficiency at $\sim 3$ EeV, so that one or two events per month could be expected.

EUSO-SPB was successfully launched from the NASA balloon facility in Wanaka (New-Zealand) in April 2017. The instrument proved to be working nominally. Unfortunately, because of a leak in the balloon, the flight was cut short after 12 days, which did not allow to detect an actual cosmic ray shower (nor to recover the instrument). A total of 25.1 hours of data could be downloaded. They contain very useful information about the in-flight

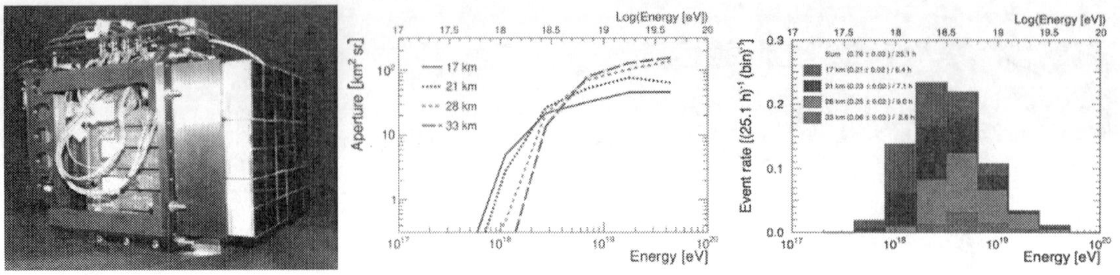


**Figure 6.** Left: JEM-EUSO's second generation PDM, used in the EUSO-SPB mission. Center: EUSO-SPB realised aperture. Right: EUSO-SPB expected event rate. See [25] for details.

performance of the instrument, attesting a photometric stability within ±5%. The UV emissivity of the Earth, which constitutes a background for the EAS detection, could be measured over different types of background: land, ocean, clouds of various types and altitude. Moreover, the effective exposure could be estimated as a function of energy, for different altitudes of the balloon, as shown in Fig. 6 (middle panel). This could in turn be used to derive the expected event rate in different energy bins (right panel). The resulting, integrated number of events expected on average during the entire flight is ~ 0.7 events, which is reduced to 0.4 events when taking clouds into account. Although this is of order unity, in line with the expactations from the mission design, none were actually detected.

Finally, the data collected during the flight allowed to identify various types of events corresponding to direct hits of cosmic rays on the instrument. A classification of such hits (depending on their shape and appearance in the data) was proposed, which will be useful for future missions. Their rate is low enough for the associated dead time to be acceptable, so no veto needs to be implemented to get rid of these undesired events.

### 4.3.3 MINI-EUSO

MINI-EUSO (Multiwavelength Imaging New Instrument for the Extreme Universe Space Observatory, also known as *UV atmosphere* in the Russian Space Program) is the currently active orbital mission of the JEM-EUSO Collaboration [26]. It was precedented in space by the TUS mission [27], launched in 2016, which was subsequently inserted into the JEM-EUSO program both for technology and scientific results and analyses [28]. MINI-EUSO is a joint mission of the Italian (ASI) and Russian (ROSCOSMOS) space agencies, installed inside the International Space Station (ISS) in the Russian Zvezda module. The design of the focal surface is close to that of EUSO-SPB, with 1 PDM using the second generation of ECs. However, the optics is different and consists of two circular Fresnel lenses of 25 cm diameter, to fit in front the UV transparent window of the Zvezda module. The field of view is also much larger, namely ~44°×44°, with a projected size of the individual pixels on the ground of ~6.3×6.3 km$^2$, so that the typical background is similar to that of the other missions, including the future major JEM-EUSO missions, i.e. ~1 photon per pixel per so-called *Gate Time Unit*, or GTU, whose value is 2.5$\mu$s.

**Figure 7.** Top: open MINI-EUSO instrument, showing the circular Fresnel lens, as well as the second generation PDM and its electronics at the focal surface. Bottom: MINI-EUSO locked onto the UV transparent window of the Zvezda module in the ISS.

MINI-EUSO was successfully launched in August 2019, and saw its first light in October 2019. It operates nominally since then, when cosmonauts position it in front of the quartz window and turn it on for 12h sessions, i.e. ~8 orbits. More than 70 sessions have taken place so far, producing data of which a small fraction is downloaded after each session, while the main part if brought back to Earth physically on solid state disks every ~12 months.

Because of the small entrance pupil, the collected light from UHECR showers or other atmospheric phenomena is much reduced, compared to what will be the case for a full-scale mission. As a consequence, the detection threshold of MINI-EUSO for UHECR events is around 3 $10^{21}$ eV, as

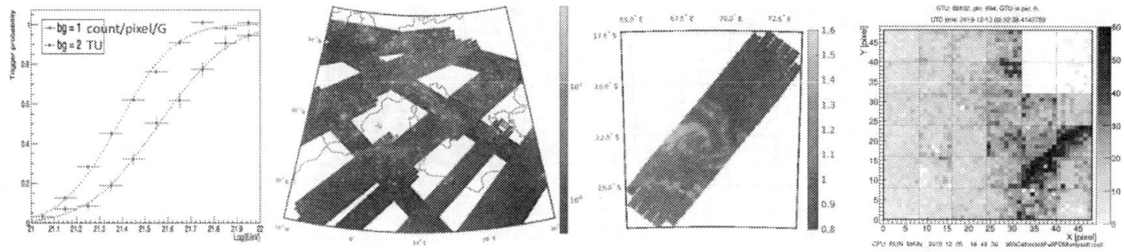


**Figure 8.** Left: trigger efficiency of MINI-EUSO as a function of energy. Center left: UV map of the nightly Earth (the bands correspond to different passages of the ISS with the instrument turned on by the cosmonauts). Center right: UV map of a cyclone. Right: ring of an ELVES (with apparent luminal motion in the field of view).

shown in Fig. 8 (left panel) for two typical levels of background. Given the extremely low flux of particles at such an energy (if there are some at all!), MINI-EUSO is not expected to detect UHECR showers. However, it did already demonstrate its ability to detect such events, by triggering on various signals with similar properties, including artificial UV flashers as well as unidentified events, which are currently under investigation and remain an important target for the mission. Also, the in situ measurements already provided key information allowing to determine the performance of a scaled-up fluorescence telescope for UHECR detection from space, confirming the previous analyses and simulations of the JEM-EUSO collaboration (see Bertaina, this conference).

MINI-EUSO addresses a variety of additional scientific objectives, thanks to a key feature of its electronics and data acquisition system, namely the ability to operate on 3 different timescales in parallel. The first level uses the basic time resolution, namely 1 GTU. This is best suited for rapid transient phenomena, such as man-made flashers, or moving signal in the field of view of the instrument, such as ELVES, potential relativistic grains, and of course UHECR events, above the energy threshold. On the second level, the light integration time is 128 GTU, i.e. 320$\mu$s. This allows to measure emissions from other types of transient luminous events, cities and various anthropogenic signals. Finally, a third level of integration has a time unit of 128×128 GTU, i.e. 40.96 ms, which provides continuous monitoring of clouds, ionospheric waves, lightning, or other natural or non natural emissions.

In addition to the information relevant for future large-scale JEM-EUSO missions, MINI-EUSO provided for the first time UV maps of the Earth at night (Fig. 8). It also detected thousands of meteors down to magnitude 6.5, with full efficiency above magnitude 5. It also allowed unprecedented imaging of ELVES, thanks to its unique sensitivity and spatial resolution, which led to the discovery of new internal feature (paper in preparation).

#### *4.3.4 EUSO-SPB2*

The next mission of the JEM-EUSO Collaboration will be EUSO-SPB2 [29]), a NASA mission under a stratospheric super pressure balloon. Compared to the previous EUSO-SPB mission (now renamed EUSO-SPB1), EUSO-SPB2 has been scaled up with the inclusion of three PDMs instead of one (for a total of 6912 pixels), and the addition of the above-mentioned Cherenkov telescope (CT), whose SiPM-based focal surface is shown in Fig. 9 (see also Cummings, this conference). Moreover, it benefits from several important improvements of the PDM, with new electronics and a third generation of ECs, more compact and more efficient, identical to those developed for the K-EUSO prototype [30]. The length of the GTU has also been reduced compared to previous JEM-EUSO missions, down to 1 $\mu$s instead of 2.5 $\mu$s. This is optimised with respect to the average time it takes a shower to cross the field of view of a pixel, thereby increasing the detection efficiency.

In addition, EUSO-SPB2 will be the first JEM-EUSO mission using reflective optics, with the adoption of a Schmidt camera design for both the FT and the CT. The FT aperture has a diameter of 1 m, and its field of view is ~12°×36°, corresponding to a projected area of ~36 km$^2$ on the ground. The CT also has 1 m diameter aperture, with a field of view of ~6.4°×12.8° covered by 512 SiPM-based pixels. A key feature of its optics is the use of a bi-focal mirror, coupled with a trigger scheme requiring signal on 2 conjugate pixels, to reduce the noise from direct cosmic-ray hit on the focal surface or single pixel fluctuations.

EUSO-SPB2 has been approved for a launch from Wanaka (New-Zealand) in April 2023. Its performances have been tested on the ground, at the TA site in Utah. They confirmed the numerical simulations and validated the different aspects of the instrumental and experimental concept. Based on these results, EUSO-SPB2 is expected to detect and reconstruct several high-energy CR showers, with typical energies around 2 $10^{18}$ eV, thereby completing the development phase of the JEM-EUSO program. The right panel of Fig. 9 shows the expected trigger efficiency as a function of energy, and the resulting number of events per hour of clear sky observation in the different energy bins.

As for the CT, its orientation can be adjusted both in elevation and azimuth. Tilting in elevation allows to chose the position of the limb of the Earth in its ~6.4°×12.8° field of view, to measure the yet unknown UV luminosity both above the limb (as a background for high-altitude,

**Figure 9.** Left: SiPM and electronics modules of the CT camera of EUSO-SPB2. Center: focal surface of the FT of EUSO-SPB2, with 3 PDMs for a total of 6912 pixels. Right: expected trigger efficiency for cosmic-ray events in the FT of EUSO-SPB2 and the corresponding event rate.

horizontal CR showers and other fast atmospheric events) and below the limb (as a background for neutrino shower detection). Complementarily, tilting in azimuth allows to point the CT towards specific targets of opportunity, searching for neutrino counterparts as a response to multi-messenger alerts that may be issued during the flight. While an actual neutrino detection is not likely, the detection of direct Cherenkov light from CRs above the limb is expected to be on the order of ~100 per hour of live time, because of the low energy threshold around $10^{15}$ eV. This should provide a useful benchmark for the Cherenkov method applicable to neutrinos. Additionally, the measurements from the CT towards the limb will be of great important in preparation of a future major neutrino experiment from space, such as POEMMA [31].

## 5 Conclusion

Although the current generation of detectors has collected extremely rich information and high-quality data about the UHECR phenomenology in the last two decades, the current consensus among the interested research community is that a new generation of detectors will be needed to reach a global understanding of the phenomenon and discover the sources of the UHECRs and their acceleration mechanism. A very important step forward will be to significantly increase the statistics and achieve full sky coverage with roughly uniform exposure. This is what a space-based UHECR observatory could naturally provide, and this is the vision that drove the international JEM-EUSO Collaboration in its endeavour to develop suitable and operational technology for a large instrument looking down towards the Earth to detect extensive atmospheric showers from a low Earth orbit.

As summarized above, after one and a half decades of intense instrumental activity, the JEM-EUSO Collaboration has demonstrated its ability to develop efficient technology and meaningful instruments for balloon and space missions of increasing complexity and scientific scope, with the support of many space agencies and research institutes around the globe. All the missions have been experimentally successful, and provided crucial information and results for the development of the future steps.

In this incremental adventure, EUSO-Balloon and EUSO-TA first demonstrated the relevance of the JEM-EUSO technology. The EUSO-SPB1 mission further demonstrated the scalability of the concept with new generations of detectors and electronics, and was able to assess the performance and the robustness of the instruments in flight. With MINI-EUSO, the JEM-EUSO collaboration confirmed the long-term operation of its technology in space, and demonstrated the full potential of complementary and interdisciplinary science accessible with its versatile technology, operating on different timescales in parallel. Finally, EUSO-SPB2 gathered all the past experience in a scaled-up mission. With its improved performance obtained from a third generation of detectors, upgraded electronics and the addition of a Cherenkov telescope targeting high-altitude showers and neutrinos, it will explore the multi-messenger potential of the field.

Most of all, EUSO-SPB2 should be the first mission to detect the fluorescence light of high-energy cosmic-ray showers from above, which will mark the crowning achievement of this preparatory phase, vouching for the maturity of the technology and validating the entire JEM-EUSO program.

From then on, the JEM-EUSO Collaboration can move forward, as it is now ready for the success of a major UHECR space mission, along the lines proposed for the (currently suspended) K-EUSO mission [30] or the POEMMA mission [31].

This paper benefited from the work and contributions of all scientists and engineers who participated in the JEM-EUSO program and its various missions over the years. The JEM-EUSO collaboration acknowledges the institutions and funding agencies involved, in particular the leading space agencies of the developed balloon and space missions, namely the French Space Agency (CNES), the Italian Space Agency (ASI), the Russian Space Agency (ROSCOSMOS), and the US Space Agency (NASA).

## References

[1] A. Coleman et al., Astrop. Phys. **147** 102794 (2023)
[2] Benson, R. & Linsley, J. 1981, International Cosmic Ray Conference, 8, 145
[3] Stecker, F. W., Krizmanic, J. F., Barbier, L. M., et al. 2004, Nuclear Physics B Proceedings Supplements, 136, 433. doi:10.1016/j.nuclphysbps.2004.10.027
[4] Catalano, O. & Petrolini, A. 2005, 29th International Cosmic Ray Conference (ICRC29), Volume 8, 8, 93

[5] Pierre Auger Collaboration 2015, Nuclear Instruments and Methods in Physics Research A, 798, 172. doi:10.1016/j.nima.2015.06.058

[6] Kawai, H., Yoshida, S., Yoshii, H., et al. 2008, Nuclear Physics B Proceedings Supplements, 175, 221. doi:10.1016/j.nuclphysbps.2007.11.002

[7] Kido, E., Telsecope Array Collaboration, EPJ Web Conf. 210 (2019) 06001. doi:10.1051/epjconf/201921006001.

[8] Aab, A., et al., Pierre Auger Collaboration, Phys. Rev. D 102 (6) (2020) 062005, http://dx.doi.org/10.1103/PhysRevD.102.062005, arXiv:2008.06486.

[9] Tsunesada, Y., Array, T., Abbasi, R., et al. 2022, 37th International Cosmic Ray Conference, 337. doi:10.22323/1.395.0337

[10] Greisen, K., Phys. Rev. Lett. 16 (1966) 748–750, http://dx.doi.org/10.1103/PhysRevLett.16.748.

[11] Zatsepin, G.T., Kuzmin, V.A., JETP Lett. 4 (1966) 78–80

[12] Aab, A., et al., Pierre Auger Collaboration, Phys. Rev. D, 90 (2014) 122006, doi: 10.1103/PhysRevD.90.122006

[13] Aab, A., et al., Pierre Auger Collaboration, Science, 357 (2017) 1266, doi: 10.1126/science.aan4338

[14] Aab, A., *et al.*, Pierre Auger Collaboration, Astrophys. J. Lett. **853** (2018) no.2, L29 doi:10.3847/2041-8213/aaa66d

[15] Aab, A., et al., Pierre Auger Collaboration, ApJ 804 (2015) 15 https://doi.org/10.1088/0004-637X/804/1/15

[16] Abbasi, R. U., *et al.*, Telescope Array Collaboration, Astrophys. J. Lett. 790 (2014) L21. doi:10.1088/2041-8205/790/2/L21.

[17] P. Abreu *et al.*, Pierre Auger Collaboration, PoS 37th International Cosmic Ray Conference (2021), 321 doi:10.22323/1.395.0321

[18] D. Ivanov, et al., Telescope Array Collaboration, PoS International Cosmic Ray Conference (2017), 496 doi:10.22323/1.301.0496

[19] D. Allard, N. G. Busca, G. Decerprit, A. V. Olinto and E. Parizot, JCAP **10** (2008), 033 doi:10.1088/1475-7516/2008/10/033 [arXiv:0805.4779 [astro-ph]].

[20] D. Allard, Astropart. Phys. **39-40** (2012), 33-43 doi:10.1016/j.astropartphys.2011.10.011 [arXiv:1111.3290 [astro-ph.HE]].

[21] T. M. Venters, M. H. Reno, J. F. Krizmanic, L. A. Anchordoqui, C. Guépin and A. V. Olinto, Phys. Rev. D **102** (2020), 123013 doi:10.1103/PhysRevD.102.123013

[22] A. Cummings, R. Aloisio, J. Eser and J. Krizmanic, Phys. Rev. D **104** (2021) no.6, 063029 doi:10.1103/PhysRevD.104.063029

[23] Abdellaoui, G., *et al* (JEM-EUSO Coll.), Astrop. Phys. **102** 98-111 (2018)

[24] Adams, J. H. *et al* (JEM-EUSO Coll.), Space Sci. Rev. **218** 3 (2022)

[25] Wiencke, L. and Olinto, A, for the JEM-EUSO Collaboration, PoS(ICRC2017) **301** 1097 (2017) 1097

[26] Bacholle, S. *et al*, ApJS **253** 36 (2021)

[27] P. Klimov *et al* (TUS Collaboration), Space Sci. Rev. **212** 1687-1703 (2017)

[28] D. Barghini, M. Bertaina, A. Cellino, F. Fenu, S. Ferrarese, A. Golzio, O. I. Ruiz-Hernandez, P. Klimov, A. Montanaro and A. Salsi, *et al.* Adv. Space Res. **70** (2022), 2734-2749 doi:10.1016/j.asr.2021.11.044

[29] J. Eser, A.V. Olinto, L. Wiencke for the JEM-EUSO Coll., PoS(ICRC2021) **395** 404 (2021)

[30] P. Klimov *et al* (JEM-EUSO Coll.), Universe **8** 88 (2022)

[31] A. Olinto *et al* (POEMMA Coll.), JCAP **06** 007 (2021)

# Implications of Mini-EUSO measurements for a space-based observation of UHECRs

*Mario Edoardo* Bertaina[1,2,*], *Dario* Barghini[1,2,3], *Matteo* Battisti[2], *Alexander* Belov[4,5], *Marta* Bianciotto[1], *Francesca* Bisconti[6], *Carl* Blaksley[8], *Sylvie* Blin[9], *Karl* Bolmgren[10], *Giorgio* Cambiè[6,7], *Francesca* Capel[11], *Marco* Casolino[6,7,8], *Igor* Churilo[12], *Marino* Crisconio[13], *Christophe* De La Taille[14], *Toshikazu* Ebisuzaki[8], *Johannes* Eser[15], *Francesco* Fenu[16], *George* Filippatos[17], *Massimo Alberto* Franceschi[18], *Christer* Fuglesang[10], *Alessio* Golzio[1,2], *Philippe* Gorodetzky[9], *Fumiyoshi* Kajino[19], *Hiroshi* Kasuga[8], *Pavel* Klimov[5], *Viktoria* Kungel[17], *Vladimir* Kuznetsov[12], *Massimiliano* Manfrin[1,2], *Laura* Marcelli[6], *Gabriele* Mascetti[13], *Włodzimierz* Marszał[20], *Marco* Mignone[2], *Hiroko* Miyamoto[2,21], *Alexey* Murashov[5], *Tommaso* Napolitano[18], *Hitoshi* Ohmori[8], *Angela* Olinto[15], *Etienne* Parizot[9], *Piergiorgio* Picozza[6,7], *Lech Wiktor* Piotrowski[22], *Zbigniew* Plebaniak[1,2,20], *Guillaume* Prévôt[9], *Enzo* Reali[6,7], *Marco* Ricci[18], *Giulia* Romoli[6,7], *Naoto* Sakaki[8], *Sergei* Sharakin[5], *Kenji* Shinozaki[20], *Jacek* Szabelski[20], *Yoshiyuki* Takizawa[8], *Giovanni* Valentini[13], *Michal* Vrabel[20], *Lawrence* Wiencke[17], *Mikhail* Zotov[5], and *the* JEM-EUSO Collaboration

[1]Department of Physics, University of Turin, V. P. Giuria 1, 10125 Turin, Italy
[2]INFN Section of Turin, Via P. Giuria 1, 10125 Turin, Italy
[3]INAF Astrophysics Observatory of Turin, Via Osservatorio 20, 10025 Pino Torinese, Italy
[4]Faculty of Physics, M.V. Lomonosov Moscow State University, ul. Kolmogorova 1(2), 119234 Moscow, Russia
[5]Skobeltsyn Institute of Nuclear Physics, Lomonosov Moscow State Univ., ul. Kolmogorova 1(2), 119991 Moscow, Russia
[6]INFN Section of Rome Tor Vergata, Via della Ricerca Scientifica 1, 00133 Rome, Italy
[7]Department of Physics, University of Rome Tor Vergata, Via della Ricerca Scientifica 1, 00133 Rome, Italy
[8]RIKEN, 2-1 Hirosawa Wako, Saitama 351-0198, Japan
[9]Université Paris Cité, CNRS, Astroparticule et Cosmologie, 10 Rue Alice Domon et Léonie Duquet, 75013 Paris, France
[10]KTH Royal Institute of Technology, Brinellvğen 8, 114 28 Stockholm, Sweden
[11]Technical University of Munich, Arcisstraße 21, 80333 Munich, Germany
[12]S.P. Korolev Rocket and Space Corporation Energia, Lenin str., 4a Korolev, 141070 Moscow area, Russia
[13]ASI, Italian Space Agency, Via del Politecnico, 00133 Rome, Italy
[14]Omega, Ecole Polytechnique, CNRS/IN2P3, Rte de Saclay, 91120 Palaiseau, France
[15]Department of Astronomy and Astrophysics, The University of Chicago, 5640 S. Ellis Avenue, Chicago IL 60637, US
[16]Karlsruhe Institute of Technology, Hermann-von-Helmholtz-Platz 1, 76344 Eggenstein-Leopoldshafen, Germany
[17]Department of Physics, Colorado School of Mines, 1523 Illinois St., Golden CO 80401, US
[18]INFN National Laboratories of Frascati, Via Enrico Fermi 54, 00044 Frascati, Italy
[19]Department of Physics, Konan University, 8 Chome-9-1 Okamoto, Higashinada Ward Kobe, Hyogo 658-8501, Japan
[20]National Centre for Nuclear Research, ul. Pasteura 7, 02-093 Warszawa, Poland
[21]Gran Sasso Science Institute, Viale Francesco Crispi 7, 67100 L'Aquila, Italy
[22]Faculty of Physics, University of Warsaw, Ludwika Pasteura 5, 02-093 Warsaw, Poland

**Abstract.** Mini-EUSO is the first mission of the JEM-EUSO program on board the International Space Station. It was launched in 2019 and it is currently located in the Russian section (Zvezda module) of the station and viewing our planet from a nadir-facing UV-transparent window. The instrument is based on the concept of the original JEM-EUSO mission and consists of an optical system employing two Fresnel lenses and a focal surface composed of 36 Multi-Anode Photomultiplier tubes, 64 channels each, for a total of 2304 channels with single photon counting sensitivity and an overall field of view of 44° × 44°. Mini-EUSO can map the night-time Earth in the near UV range (predominantly between 290 nm and 430 nm), with a spatial resolution of about 6.3 km and different temporal resolutions of 2.5 $\mu$, 320 $\mu$s and 41 ms. Mini-EUSO observations are extremely important to better assess the potential of a space-based detector in studying Ultra-High Energy Cosmic Rays (UHECRs) such as K-EUSO and POEMMA. In this contribution we focus the attention on UV measurements, the observation of clouds and of certain categories of events that Mini-EUSO triggers with the shortest temporal resolution. We place them in the context of UHECR observations from space, namely the estimation of exposure and sensitivity to Extensive Air Showers.

## 1 Introduction

The current main goal in the field of UHECR (Ultra-High Energy Cosmic Ray) science is to identify their as-

*e-mail: bertaina@to.infn.it

https://doi.org/10.1051/epjconf/202328306008

trophysical sources and composition [1]. For this, increased statistics is one of the essential requirements. A space-based detector for UHECR research has the advantage of a very large exposure and a uniform coverage of the celestial sphere. The aim of the JEM-EUSO program [2] is to bring the study of UHECRs to space. The principle of observation is based on the detection of UV light emitted by isotropic uorescence of atmospheric nitrogen excited by Extensive Air Showers (EASs) in Earths atmosphere and forward-beamed Cherenkov radiation reected at the Earths surface or at dense cloud tops. The JEM-EUSO program includes missions on ground (EUSO-TA [3]), on stratospheric balloons (EUSO-Balloon [4], EUSO-SPB1 [5], EUSO-SPB2 [6]), and from space (TUS [7], Mini-EUSO [8]) employing fluorescence detectors to demonstrate the feasibility of the UHECR observation from space and prepare for the large size missions K-EUSO [9] and POEMMA [10]. An updated overview of the JEM-EUSO program can be found in [11]. Mini-EUSO is the first detector of the JEM-EUSO program to observe the Earth from the International Space Station (ISS) and to validate from there the observational principle of a space-based detector for UHECR measurements.

## 2 The Mini-EUSO space telescope

Mini–EUSO (Multiwavelength Imaging New Instrument for the Extreme Universe Space Observatory, known as *UV atmosphere* in the Russian Space Program) [8] is a telescope operating in the near UV range, predominantly between 290 nm and 430 nm, with a square focal surface corresponding to a field of view (FoV) of ~44° × 44°. Its spatial resolution at ground level is approximately $6.3 \times 6.3$ km$^2$, slightly varying with the altitude of the ISS and the pointing direction of the pixel. The detector size is $37 \times 37 \times 62$ cm$^3$, mainly constrained by the size of the nadir-facing UV transparent window in the Russian Zvezda module, where it is attached a couple of times per month during onboard night-time, approximately at 18:30 UTC, with operations lasting about 12 hours. The first observations took place on October 7, 2019. Since then and till October 2022, more than 70 sessions have been performed. Data are stored locally on 512 GB USB pendrives. After each data-taking session samples of data (about 10% of stored data, usually corresponding to the beginning and the end of each session) are copied and transmitted to ground to verify the correct functioning of the instrument and subsequently optimize its working parameters. The pouches containing all stored data are then returned to Earth every ~12 months by the Soyuz spacecraft.

The optics are based on two 25 cm diameter Fresnel lenses in Polymethyl methacrylate (PMMA). The Mini–EUSO focal surface, or Photo Detector Module (PDM), consists of a grid of 36 Multi-Anode Photomultiplier Tubes (MAPMTs, Hamamatsu Photonics R11265-M64). Each MAPMT consists of 8×8 pixels, resulting in a total of 2304 channels. The MAPMTs are grouped in Elementary Cells (ECs) of $2 \times 2$ MAPMTs. MAPMTs are separated by 2 – 3 mm spacing to avoid vibration damage at launch. Consequently, gaps exists between MAPMTs in the collected images. Each EC has an independent high voltage power supply (HVPS) and board connecting the dynodes and anodes of the four photomultipliers. The HVPS system is based on a Cockroft-Walton circuit. The system has an internal safety mechanism which operates either reducing the collection efficiency of the four MAPMTs or reducing the MAPMT gain when particularly bright signals occur (i.e. lightnings or large cities) [8]. The recovery to the nominal mode takes place only few ms after the light level has decreased to a sufficiently low value.

The effective focal length of the system is 300 mm, with a Point Spread Function (PSF) of 1.2 MAPMT pixels. UV bandpass filters (2 mm of BG3 material) with anti-reflective coating are glued in front of the MAPMTs to predominantly select wavelengths between 290 nm and 430 nm.

The system has a single photon-counting capability with a double pulse resolution of ~6 ns. Photon counts are summed in Gate Time Units (GTUs) of 2.5 $\mu$s.

The PDM Data Processor (PDM-DP) stores the 2.5 $\mu$s GTU data stream (D1) in a running buffer on which the trigger code is executed. The algorithm searches for a signal integrated over 8 consecutive GTUs above 16 standard deviations from the average in any pixel of the focal surface. Both the average and standard deviation are calculated in real time to take into account varying illumination conditions. In case of a trigger, the 128 frame buffer (64 frames before the trigger and 64 after it) is stored in memory. The choice to operate on 8 GTUs integration and to act on each pixel independently is based on the time needed by a light signal to cross the FoV of a pixel and the fact that 20 $\mu$s represent a significant portion of an EAS light-track in atmosphere (see Fig. 1).

Independently from the trigger, sums of 128 frames (320 μs, D2) are continuously calculated and stored in another buffer where a similar trigger algorithm, but at this time scale, is running. Similarly, sums of 128 D2 frames (40.96 ms, D3) are calculated in real time and continuously stored. Every 5.24 s, 128 packets of D3 data, up to 4 D2 packets and up to 4 D1 packets (if triggers were present) are sent to the CPU for storage. In this way various classes of phenomena with different durations can be detected with an appropriate time scale (see sec. 3 for details). A more detailed description of the trigger algorithm is reported in [12], which represents an adaptation of the trigger logic conceived for JEM-EUSO [13], while the onboard performance of the trigger system is summarized in [14].

Prior to the launch, the instrument underwent a series of integration and acceptance tests in Rome, Moscow, and Baikonur cosmodrome, where it was placed in the uncrewed Soyuz capsule. A systematic test of the acquisition logic was performed at the TurLab facility [15] of the University of Turin and at the Astrophysical Observatory of Turin (INAF-OATo) [16].

After launch, an end-to-end in-flight calibration of the Mini-EUSO detector has been performed by assembling different UV-flasher systems on ground in Japan, Italy, and

France and by firing them in various observational campaigns. A detailed description of the methodology and of the observational results is reported in [17].

## 3 Mini-EUSO design, expected performance and exposure studies

Mini-EUSO has been designed to detect a photon rate per pixel from diffuse sources (nightglow, clouds, cities, etc.) in the range of values expected from a large mission in space such as the original JEM-EUSO mission [18] or the future detectors K-EUSO or POEMMA. The pixel FoV is, therefore, ~100 times larger in area with respect to the FoV of a JEM-EUSO pixel (~0.5 × 0.5 km$^2$), to compensate for the optical system ~100 times smaller, constrained by the dimension of the UV transparent window (see Fig. 1).

In order to have a precise ratio of the photon rate per pixel from diffuse sources between JEM-EUSO and Mini-EUSO, a full simulation of JEM-EUSO and Mini-EUSO detectors was performed with ESAF simulation software [19]. In case of Mini-EUSO the overall efficiency of the detector was fine-tuned with ESAF, mainly acting at the level of MAPMT response, to match the measured one $\epsilon_{ME} = 0.080 \pm 0.015$ (see the end-to-end in-flight calibration of Mini-EUSO reported in [17]) for a point-like source on ground. A flat diffused UV emission in the range $\lambda$ = 300 - 400 nm was simulated at the detector's aperture either with a range of zenith directions much larger than the FoV of the instrument (±60° for both detectors) or just within the FoV of the detectors (±30° for JEM-EUSO and ±22° for Mini-EUSO). The estimated background ratio ($R(ME/JE)$) between MiniEUSO and JEMEUSO at FS level is $R(ME/JE) = 0.98 - 1.04$ slightly depending on the range of zenith angles. This result confirms that the expected photon rate from diffuse sources is similar in JEM-EUSO and Mini-EUSO instruments.

The energy threshold of Mini-EUSO for point-like sources like UHECRs is roughly 2 orders of magnitude higher than the original JEM-EUSO one. Fig. 2 shows the trigger efficiency of Mini-EUSO for ESAF simulated proton-generated EASs of different energies. Events have been simulated according to a sin(2 · $\theta$) dependence on a larger area than the FoV of the instrument on ground to take into account border effects. The efficiency ($\epsilon$) as a function of the energy ($E$) has then been calculated as:

$$\epsilon(E) = \frac{N_{trig}}{N_{simu}} \cdot \frac{A_{simu}}{A_{FoV}}, \quad (1)$$

where $N_{trig}$ and $N_{simu}$ represent the number of triggered events over the simulated ones for a specific energy bin, while $A_{simu}$ and $A_{FoV}$ indicate the area on which EAS have been injected and the area on ground in the FoV of Mini-EUSO, respectively. The red and blue points assume a fixed nightglow background of 1 and 2 counts/pix/GTU, respectively. The 50% trigger efficiency is located between 3 - 5 × $10^{21}$ eV depending on the UV nightglow level.

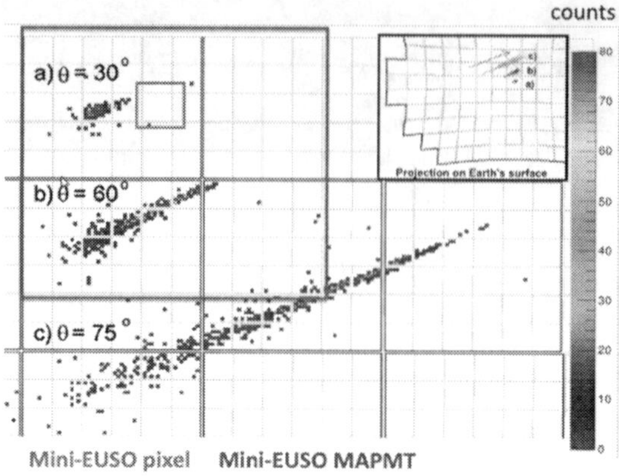


**Figure 1.** Comparison between JEM-EUSO and Mini-EUSO FoVs. The figure shows the detected light track of three simulated EAS with ESAF [19] of energy $10^{20}$ eV with different zenith angles ($\theta$ = 30°, 60°, and 75°) as imaged by JEM-EUSO detector. Each little colored dot represents the pixel FoV of JEM-EUSO projected on ground, each grey square indicates the MAPMT FoV and the large black squares show the FoV of JEM-EUSO PDMs. The green and red squares represent Mini-EUSO pixel's and MAPMT's FoV, respectively. Mini-EUSO pixel's FoV is larger than the FoV of a JEM-EUSO MAPMT and Mini-EUSO MAPMT's FoV is larger than the FoV of one JEM-EUSO PDM. Image adapted from [18].

Mini-EUSO monitors the atmosphere and studies the nature, extension and duration of the almost steady nightglow emissions and rapid transient lights, to investigate the capability of detecting light signals from EASs and minimize the rate of spurious events. Thanks to its large FoV, Mini-EUSO also acts as an atmospheric monitor detector, observing different natural or anthropogenic phenomena ranging from the UV light reflected or diffused by clouds (D3 data), lightning and thunderstorm activity and study of atmospheric electric phenomena termed Transient Luminous Events (TLEs) [20] with D1 and D2 data, in particular elves, up to much slower events (D3 data) like meteors or, possibly, nuclearites [21], with a sensitivity to fainter events beyond the usual capabilities of atmospheric monitor detectors (see [22] for an updated description of Mini-EUSO results).

Fig. 3 shows a comparison between the cloud fraction (%) due to a cyclone in the Indian ocean as foreseen by a post-processing analysis of the weather forecast obtained with the data collected by the the Global Forecast System of NASA [23], and the Mini-EUSO counts per pixel taken in D3 mode but normalized to the D1 GTU. Where the cloud fraction is expected to be higher, Mini-EUSO registers higher counts level. This is obtained in no-moon condition and shows the capability of the UV camera of Mini-EUSO to detect the presence of clouds. More details can be found in [24]. This is very helpful for the calculation of the exposure and for the determination of the atmospheric conditions in case of the detection of an EAS candidate. A dedicated study to assess the capability of Mini-EUSO

https://doi.org/10.1051/epjconf/202328306008

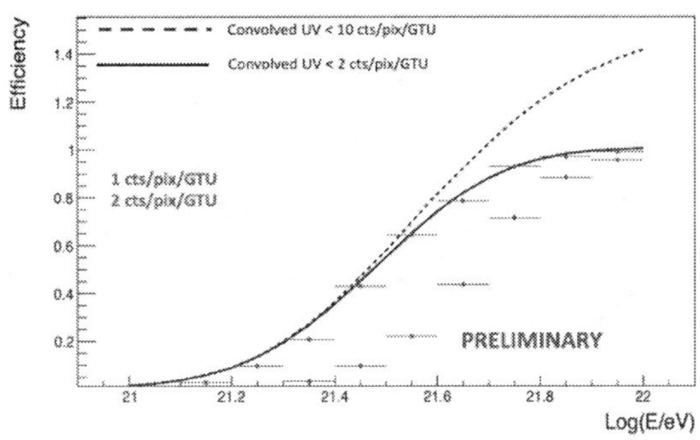


**Figure 2.** Trigger efficiency curves of Mini-EUSO for ESAF simulated proton-generated EASs of different energies on different background levels. The red and blue points assume a fixed nightglow background of 1 and 2 counts/pix/GTU, respectively. The black line represents the convolved trigger efficiency curve in which each background level below 2 counts/pix/GTU is weighted for the relative fraction of time in which it was measured by Mini-EUSO. The dotted line provides the fractional increase in exposure (relative to the black line) if the accepted nightglow background is increased from 2 counts/pix/GTU to 10 counts/pix/GTU. A significant increase in exposure is obtained only at the highest energies. See text for details.

UV camera in recognizing the atmospheric conditions and Mini-EUSO measurements is currently on-going.

The D3 data taken by Mini-EUSO allow a first comparison with the assumed background levels in JEM-EUSO, K-EUSO and POEMMA to verify that the estimated performance is based on justified assumptions. The analysis presented in the following is preliminary as it is based on part of the data collected in 23 sessions between session 05 and 40. A more refined analysis will be performed when all the data collected by Mini-EUSO is available. The weighted moon fraction in those sessions correspond to 0.44 which is well representative of the different moon phases. Fig. 4 shows Mini-EUSO results in terms of UV emission in different conditions: clear and cloudy conditions, sea and land, various lunar phases. A pixel is defined to be in cloudy conditions if the predicted cloud fraction by the GFS model in its FoV is above 1%.

According to Mini-EUSO results in no-moon conditions, in more than 90% of the time of clear sea conditions the count rate belongs to the interval 0.4 - 2.0 counts/pix/GTU, the median being ~0.8 counts/pix/GTU for both clear sea and land conditions. In clear land conditions there is a higher probability of very low counts. These are associated with deserts and forests (see [24] for details). Under the reasonable assumption that conditions above 2 counts/pix/GTU on land are due to the presence of city lights (~20% of the time) and taking into account ~30% land coverage on Earth, this result corresponds to ~6% of the total fraction of time, which is close to the 7% estimation performed for JEM-EUSO in [18]. Moreover, cloudy conditions typically shift curves by a factor 1.5 - 2 towards higher count rates as already measured in JEM-EUSO balloon flights [4]. Taking into account the fact that

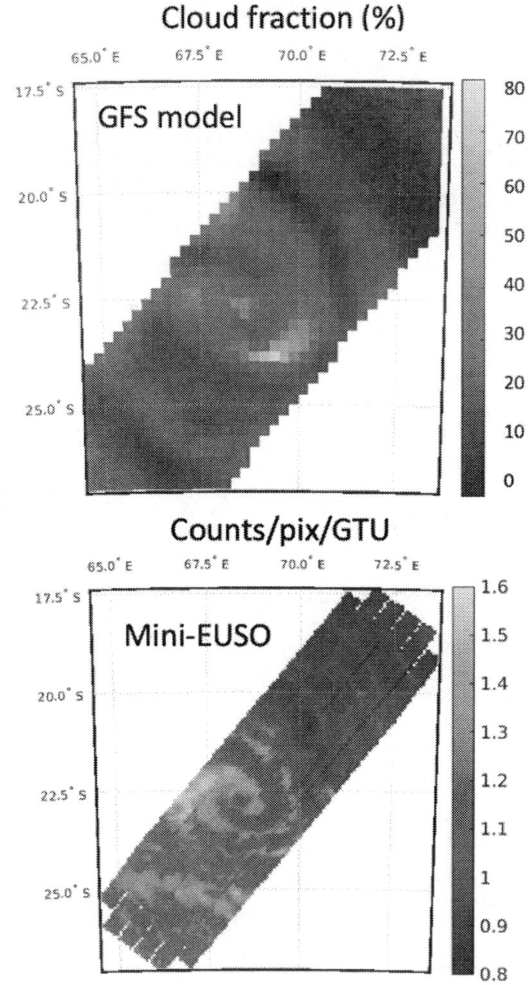


**Figure 3.** The top panel shows the cloud fraction (%) due to a cyclon in the Indian ocean as foreseen by the GFS atmospheric model [23] and the bottom panel shows the Mini-EUSO counts per pixel and D1 GTU. Where the cloud fraction is expected to be higher, Mini-EUSO registers higher counts level. Figure adapted from [24].

JEM-EUSO and Mini-EUSO are expected to measure similar count rates from diffuse sources, these results are consistent with the assumption done in [18] for JEM-EUSO where the exposure was calculated under the nominal condition of ~1.1 count/pix/GTU.

Furthermore, when the moon is visible above the horizon, if the moon fraction is less than 0.5, the increase of UV counts is still moderate and it is possible to accumulate exposure at the highest energies.

In order to have a proper estimation of Mini-EUSO exposure at different energies it is necessary to compute an efficiency curve for the various background levels and convolve each efficiency curve with the fraction of time in which such background level is measured. As it is not

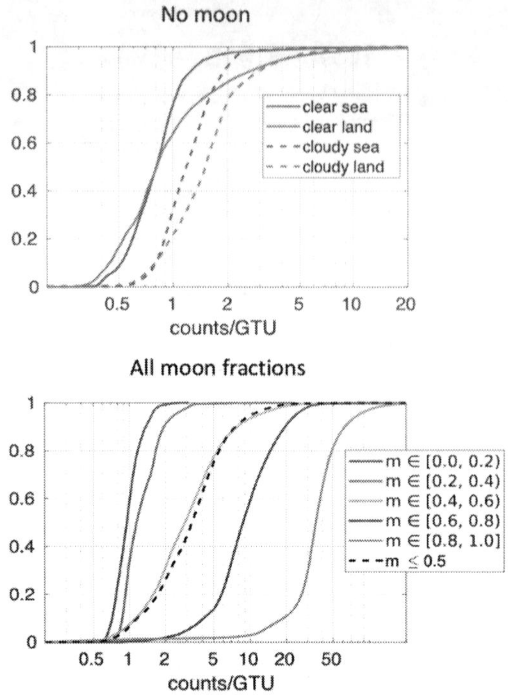


**Figure 4.** Top: cumulative distributions of Mini-EUSO observations over land and sea regions. Solid lines indicate moonless and cloudless conditions. Dashed lines represent moonless and cloudy conditions, i.e. GFS cloud fraction above 1%. Bottom: cumulative distributions of observations with the Moon visible above the horizon, each curve representing a different moon-phase interval. Figure adapted from [24].

possible to simulate the efficiency curves for infinite background levels, as a first step the efficiency curves are estimated by simulated proton-generated EASs of various energies only for a limited set of background levels in the range [0.2;4.0] counts/pix/GTU and the obtained efficiency curves are fitted with a function of the form:

$$\epsilon(E) = 0.5 \cdot [1 + erf(\frac{log_{10}(E/eV) - P_0}{P_1})], \quad (2)$$

where the parameter $P_0$ indicates the energy ($E$) at which the efficiency reaches 50% and $P_1$ measures the slope of the efficiency curve ($\epsilon(E)$). Subsequently, the obtained parameter values at different background levels have been fitted with appropriate functions as shown in Fig. 5 and the dependencies of $P_0$ and $P_1$ as function of the background level have been obtained. Finally, the trigger efficiency curves are modelled at all background levels by equation (2) using as $P_0$ and $P_1$ derived by means of the two functions shown in Fig. 5.

Fig. 6 shows the comparison between the efficiency curve obtained by fitting the simulated data points with equation (2) and the efficiency curve, superimposed on the same data points, derived through the parametrization of the $P_0$ and $P_1$ dependence for the specific background

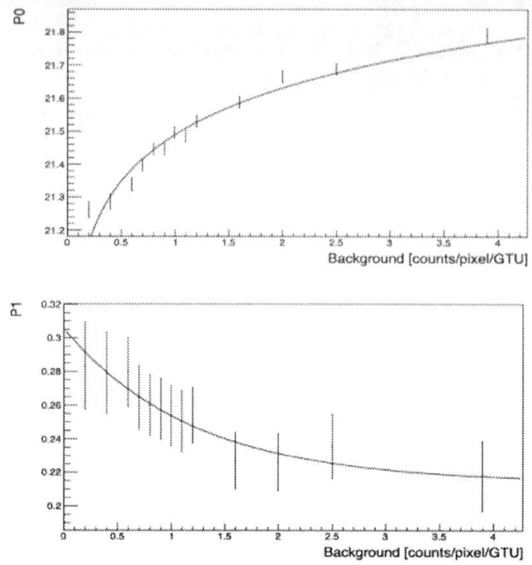


**Figure 5.** The dependence of the $P_0$ and $P_1$ parameters as a function of the background level is displayed. A logarithmic (for $P_0$) and an exponential function (for $P_1$) have been used to parametrize their dependence.

level of 1 count/pix/GTU. The parameterized curve look very similar to the originally fit.

Finally, the convolved exposure curve ($\mathcal{E}(E)$) has been calculated according to equation:

$$\mathcal{E}(E) = \sum_i (\epsilon_i(E) \cdot t_i \cdot A_i) \cdot \Omega \cdot t_{tot} \cdot \eta, \quad (3)$$

where $i$ denotes the UV background level, $t_i$ represents the relative fraction of time in which the measured background is in the interval $i$ (each interval having a width of 0.01 counts/pix/GTU), $A_i$ accounts for the cumulative area on ground with background $i$, $\Omega$ measures the UHECR observational solid angle (which assumes the value of $\pi$), and $t_{tot}$ = 142.5 hours is the total accumulated time. The parameter $\eta$ represents the efficiency factor for detecting the EAS maximum in presence of clouds. This value depends on the type of clouds, their optical depth and height. A detailed analysis was performed at the time of JEM-EUSO and adopted in this calculation ($\eta$ = 60%) and it is applied only to the fraction of time in cloudy conditions. According to Fig. 4 the fraction of time with clear sky and background level below 10 counts/pix/GTU corresponds to ~29% of the total accumulated time, while the fraction in cloudy conditions with similar background conditions corresponds to ~55%. Finally, the exposure curve obtained by integrating all background levels below 2 counts/pix/GTU has been renormalized using the exposure obtained at the plateau level of $10^{22}$ eV. The result, which represents the convolved trigger efficiency curve, is shown as black line in Fig. 2. The black curve is very close to the efficiency curve obtained at the fixed background level of 1 count/pix/GTU, indicating that such a

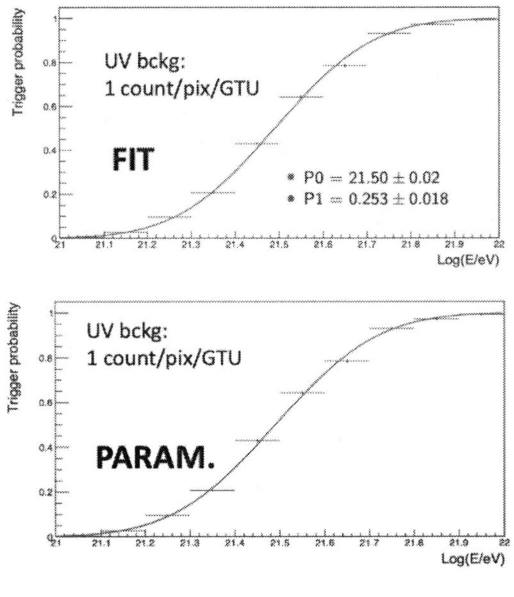


**Figure 6.** Top: efficiency curve obtained by fitting the simulated data points with equation (2) for the UV background level of 1 count/pix/GTU. Bottom: efficiency curve derived through the parametrization of the $P_0$ and $P_1$ dependence as shown in Fig. 5 for the same background level.

curve provides a reasonable approximation of the integrated performance in the various UV background conditions. This result confirms the estimations obtained in the past with JEM-EUSO and the assumptions therein. Similarly, these conclusions can be extended also to the estimation of K-EUSO and POEMMA trigger efficiency curves as they have been derived adopting the very same metodology used for JEM-EUSO.

In addition, if a background level up to 10 counts/pixel/GTU is accepted, at the highest energies the efficiency increases by ~40%. The accumulated exposure of Mini-EUSO in this analysis amounts to 450 Linsley. A very preliminary extrapolation of this result to all good so far recorded data indicates that Mini-EUSO has potentially accumulated till now an exposure equivalent to 2000 Linsley. This value is already about 2% percent of the exposure collected so far by ground-based detectors, roughly 50% in the case of hybrid data [25].

## 4 Level 1 trigger studies and sensitivity to EAS

A detailed analysis of the events collected by the level 1 trigger logic has been performed on a dataset of 34.7 h containing more than $4.7\times10^4$ triggered events. The expected functioning of the logic has been confirmed. The trigger rate on spurious events remains within the requirements in nominal background conditions (~1 Hz), while it saturates in the presence of thunderstorm activity. The dead time related to thunderstorm areas corresponds to locations where an UHECR can not be observed due to the presence of high clouds. In that context such dead time issue does not significantly reduce the observation capabilities of a space-based mission. The trigger logic proves effective in avoiding excessive trigger rates in the presence of static anthropogenic lights such as cities confirming the effectiveness of solutions such as the adaptive thresholds, necessary to prevent static light sources from triggering. Several kinds of events are detected such as elves and anthropogenic flashers. All classes of events have characteristics that make them different from an EAS track. However, this trigger capability demonstrates the possibility of a space-based mission to trigger on events with a time duration and light shape similar to, although still different from, what is expected from UHECRs. An example is shown in Fig. 7. The left panel shows a simulated proton-generated EAS of $5\times10^{19}$ eV with a 60° zenith angle as expected to be seen by JEM-EUSO, while the right panel displays a repetitive flasher signal on which Mini-EUSO has triggered near the Missoula city in Michigan, US. The signal intensity at peak level and the integrated number of counts in a limited portion of the events is comparable. This demonstrates the detectability of an EAS light profile in the energy range of interest with a space-based detector.

This kind of anthropogenic signals can be recognized in Mini-EUSO due to the fact that they are repeating several times with a constant time delay (~10 ms for this specific event). Moreover, a careful inspection of the events show a different combination of the shape of the light curve and of the image on the FS when compared to EASs.

One of the most interesting examples of non repetitive flasher-like signals in the Mini-EUSO data sample is shown in Fig. 8. It has been detected off the coast of Sri Lanka. The trigger was issued by the event in the red circle. The lightcurve presents the characteristic bi-gaussian shape of an EAS, with a faster rise and a slower decay time. The event was compared to different simulated EASs with variable energy and zenith angle. No simulated EAS is compatible with both the image size and the time duration of the light profile. In fact, the light spot is compatible with a nearly vertical event, but the duration is much longer than the time needed by a vertical shower to develop in atmosphere and reach the ground. This event has, therefore, a different nature which is currently under investigation. Moreover, the D3 data of Mini-EUSO indicate that the event occurred in a region potentially at the edge between cloudy and clear atmospheric conditions. This observation is supported also by satellite images and GFS analysis and underlines the importance of D3 data to monitor the actual weather conditions when an interesting event occurs.

Finally, we have searched in Mini-EUSO data for EAS-like events such as those detected by TUS and reported in [26]. Fig. 9 shows the comparison between an event produced by a ground flasher detected by Mini-EUSO (red) and the most promising candidate of an UHECR event detected by TUS above Minnesota on October $3^{rd}$ 2016 (TUS161003 event in blue). The Mini-EUSO event was detected near Traverse City, on the coast of Lake

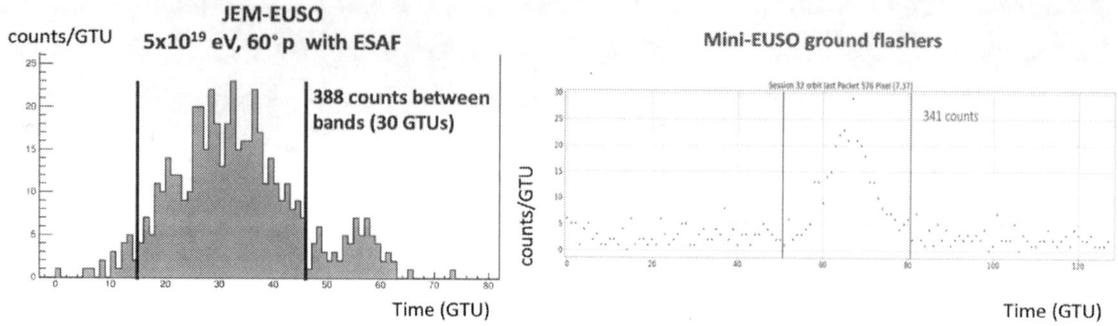


**Figure 7.** Left panel: a light curve of a simulated proton EAS of $5\times10^{19}$ eV with a 60° zenith angle as expected to be seen by JEM-EUSO (image adapted from [15]). Right side: a light curve of a single flash of a repetitive flasher signal detected by Mini-EUSO near the Missoula city in Michigan, US.

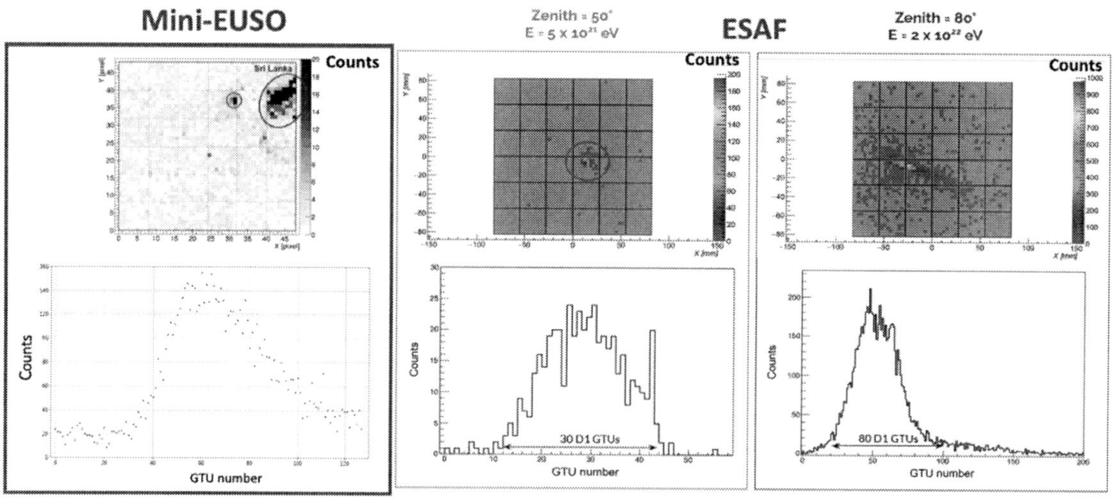


**Figure 8.** Upper left panel: one frame of an event triggered off the coast of Sri Lanka. The blue circle (upper right) encloses a static, bright light source. The trigger was issued by the event in the red circle. Bottom left panel: the lightcurve of a 3×3 pixel box that contains the event. The lightcurve presents the characteristic bi-gaussian shape, with a faster rise and a slower decay time. Central and right panels: Two proton-generated EAS simulated with ESAF at two different energies and zenith angles (top part presents the image of the events while the bottom part the corresponding lightcurves). The central panel shows a simulation at zenith $\theta = 50°$ and energy $E = 5\times10^{21}$ eV. The signal persists in few pixels for ~30 GTUs, much shorter than the one of measured ground source, the sharp cutoff is given by the shower reaching the ground. The right panel is a simulation with $\theta = 80°$ and primary energy $E = 2\times10^{22}$ eV. The signal is much longer in time but the footprint on the focal plane is much more elongated. Image adapted from [14].

Michigan. The Mini-EUSO event has an anthropogenic nature, because it was triggered four times with a repetition time of ~33 ms, and was found in an area near three small airports. The similar structure of the lightcurve indicated that ground sources can generate light profiles similar to what was detected by TUS, however, we underline that in the TUS161003 event the signal appears to be moving among pixels with a relativistic speed, while in Mini-EUSO data all the signals with a duration compatible with EAS showers appear stationary. More information on the onboard performance of the Mini-EUSO first level trigger and search for EAS-like events can be found in [14].

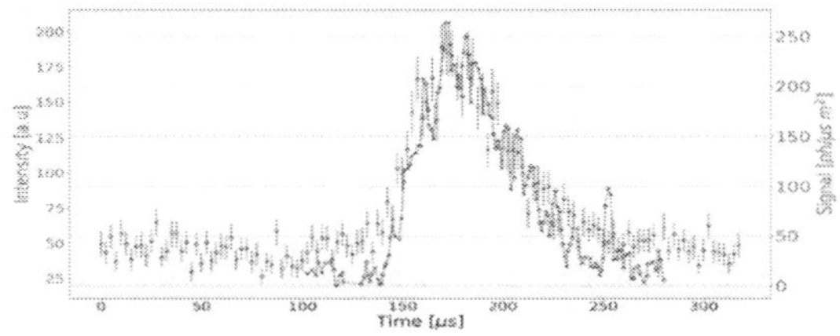

**Figure 9.** Blue: The lightcurve of the TUS161003 event as seen by TUS [26]. Even though the event has probably an anthropogenic origin, it presents all the features of an EAS signal. Red: An event detected by Mini-EUSO near lake Michigan. It presents the same time profile of the event seen by TUS, even though it is ~10 times brighter. It is produced by a ground source in an area near three small airports. Image adapted from [14].

## 5 Conclusions

Mini-EUSO is operating on the ISS since more than three years and observes events of different nature demonstrating the multi-disclipinarity of an UHECR detector in space. It shows that it is possible, with larger detectors, to perform UHECR observation from space. Preliminary results indicate that measurements are in agreement with predictions from simulations performed for JEM-EUSO and background issues are under control. These results can be extrapolated directly to K-EUSO and POEMMA and support the outcomes on the instrument performance obtained on those studies.

## 6 Acknowledgements

This work was supported by the Italian Space Agency through the agreement n. 2020-26-Hh.0, by the French space agency CNES, and by the National Science Centre in Poland grants 2017/27/B/ST9/02162 and 2020/37/B/ST9/01821. This research has been supported by the Interdisciplinary Scientific and Educational School of Moscow University "Fundamental and Applied Space Research" and by Russian State Space Corporation Roscosmos. The article has been prepared based on research materials collected in the space experiment "UV atmosphere". We thank the Altea-Lidal collaboration for providing the orbital data of the ISS.

## References

[1] A. Coleman et al., Astrop. Phys. **147** 102794 (2023)
[2] M. Ricci (JEM-EUSO Coll.), J.Phys.Conf.Ser. **718 n.5** 052034 (2016)
[3] G. Abdellaoui *et al* (JEM-EUSO Coll.), Astrop. Phys. **102** 98-111 (2018)
[4] J.H. Adams Jr. *et al* (JEM-EUSO Coll.), Space Sci. Rev. **218** 3 (2022)
[5] L. Wiencke and A. Olinto for the JEM-EUSO Coll., PoS(ICRC2017) **301** 1097 (2017) 1097
[6] J. Eser, A.V. Olinto, L. Wiencke for the JEM-EUSO Coll., PoS(ICRC2021) **395** 404 (2021)
[7] P. Klimov *et al* (TUS Coll.), Space Sci. Rev. **212** 1687-1703 (2017)
[8] S. Bacholle *et al*, ApJS **253** 36 (2021)
[9] P. Klimov *et al* (JEM-EUSO Coll.), Universe **8** 88 (2022)
[10] A. Olinto *et al* (POEMMA Coll.), JCAP **06** 007 (2021)
[11] E. Parizot et al. (JEMEUSO Coll.), This Conference Proceedings (2023)
[12] A. Belov *et al.* (JEM-EUSO Coll.), ASR **62/10** 2966-2976 (2018)
[13] G. Abdellauoi *et al.* (JEM-EUSO Coll.), NIMA **866** 150-163 (2017)
[14] M. Battisti *al.* (JEM-EUSO Coll.), ASR **70/9** 2750-2766 (2022)
[15] P. Barrillon *et al.* (JEM-EUSO Coll.), Exp. Astr. **https://doi.org/10.1007/s10686-022-09871-8** (2022)
[16] F. Bisconti *et al.* (JEM-EUSO Coll.), Exp. Astr. **53/1** 133-158 (2021)
[17] H. Miyamoto et al. (JEMEUSO Coll.), This Conference Proceedings (2023)
[18] J.H. Adams Jr. *et al.* (JEM-EUSO Coll.), Astrop. Phys. **44** 76-90 (2013)
[19] F. Fenu et al. (JEMEUSO Coll.), PoS(ICRC2019) **1472** 252 (2019)
[20] V.P. Pasko et al., Space Sci. Rev. **168** 475-516 (2012)
[21] J.H. Adams *et al.* (JEM-EUSO Coll.) Exp. Astr. **40** 253-279 (2015)
[22] M. Casolino et al. (JEMEUSO Coll.), This Conference Proceedings (2023)
[23] National Centers for Environmental Prediction, National Weather Service, NOAA, U.S. Department of Commerce, 2007, http://dx.doi.org/10.5065/D65Q4TSG.
[24] M. Casolino *et al.* (JEM-EUSO Coll.), Remote Sens. Environ. **284** 113336 (2023)
[25] V. Novotny for the Pierre Auger Coll., PoS(ICRC2021) **395** 324 (2021); H. Shin for the Telescope Array Coll., PoS(ICRC2021) **395** 305 (2021)
[26] B. Khrenov et al. (JEM-EUSO Coll.), JCAP **03** 033 (2020)

# EUSO-SPB2: A balloon experiment for UHECR and VHE neutrino observation

*Austin* Cummings[1], *Johannes* Eser[2,*], *George* Filippatos[3], *Angela* Olinto[2], *Tonia* Venters[4], *Lawrence* Wiencke[3], and *The JEM-EUSO* Collaboration

[1]Departments of Physics and Astronomy & Astrophysics, Institute for Gravitation and the Cosmos, Pennsylvania State University, University Park, PA 16802, USA
[2]Department of Astrophysics & Astronomy, The University of Chicago, Chicago, IL 60637, USA
[3]Department of Physcis, Colorado School of Mines, Golden, CO 80401, USA
[4]NASA Goddard Space Flight Center, Greenbelt, MD 20771, USA

**Abstract.** The Extreme Universe Space Observatory on a Super Pressure Balloon 2 (EUSO-SPB2) experiment will make new measurements from suborbital space as a precursor for future space missions that will address the challenge of the extremely low fluxes of ultra-high energy cosmic rays (UHECR) and very high energy (VHE) neutrinos.

The EUSO-SPB2 detector is comprised of two 1m diameter aperture telescopes. The Fluorescence Telescope (FT) will point in nadir and will record fluorescence light from cosmic ray EAS with energies above 1 EeV in its field of view of 36 by 12 degrees. The Cherenkov Telescope (CT) features a silicon photomultiplier focal surface with a field of view of 12 by 6 degrees. The CT will switch between two observation modes: one which points the CT above the limb to measure the Cherenkov emission of cosmic ray EAS with energies above 1 PeV and one which points the CT below the limb to record the Cherenkov emission produced by PeV scale EAS initiated by neutrino-sourced tau decay. As it is the first time such an instrument has been flown, one of the priorities of the CT will be the study of the optical background for observing neutrinos in this way.

EUSO-SPB2 is undergoing the final integration steps for launch on a NASA super pressure balloon payload in the spring of 2023 from Wanaka NZ. The CT was field-tested in March 2022 and the FT in August - September 2022.

The data collected during the EUSO-SPB2 mission will be essential to advancing the development of a space-based multi-messenger observatory such as the Probe of Extreme Multi-Messenger Astrophysics (POEMMA). In this contribution, we discuss the EUSO-SPB2 science goals, the instruments, the expected performance and the current status.

## 1 Introduction

Ultra-High Energy Cosmic Rays (UHECR) and very high energy (E>1 PeV) neutrinos provide a unique window into the most extreme astrophysical events in the universe, delivering complimentary information that may be used to identify the production and acceleration mechanisms within sources, the cosmological evolution of sources, and the propagation behavior of the resultant particles. As cosmic rays pass through Earth's atmosphere, they may interact, resulting in km-scale particle cascades, called extensive air showers (EAS). The secondary particles within the EAS produce observable emission in multiple channels, including radio, air fluorescence, and optical Cherenkov, which are frequently used in conjunction with direct secondary observation by ground-based observatories to detect cosmic rays. The flux of cosmic rays and neutrinos drops rapidly with increasing energy, and at the highest energies, requires massive instrumental volumes to reach sufficient statistics for proper analysis. For neutrinos, which have minuscule interaction probabilities even at the highest energies, this issue is compounded further. A detailed discussion of the current status and the future of UHECR including the role of multi-messenger Astrophysics is given in [1].

One cost-effective method to increase experimental acceptance to both cosmic rays and neutrinos is to observe from high altitudes and use the entire Earth and its atmosphere as the corresponding target volume for particle interaction, allowing for gargantuan increases in geometric acceptance. Neutrinos can be observed from a high altitude detector via the Earth-skimming technique, whereby a neutrino propagates through the Earth and interacts near the Earth's surface to produce a charged lepton, which then either interacts or decays in the atmosphere to produce an upward moving EAS.

The Extreme Universe Space Observatory on a Super Pressure Balloon 2 (EUSO-SPB2) is a pathfinder mission for a space based mission, i.e. POEMMA [2], designed to identify and quantify the various backgrounds relevant for the space-based detection of cosmic rays and neutri-

*e-mail: jeser@uchicago.edu

https://doi.org/10.1051/epjconf/202328306009

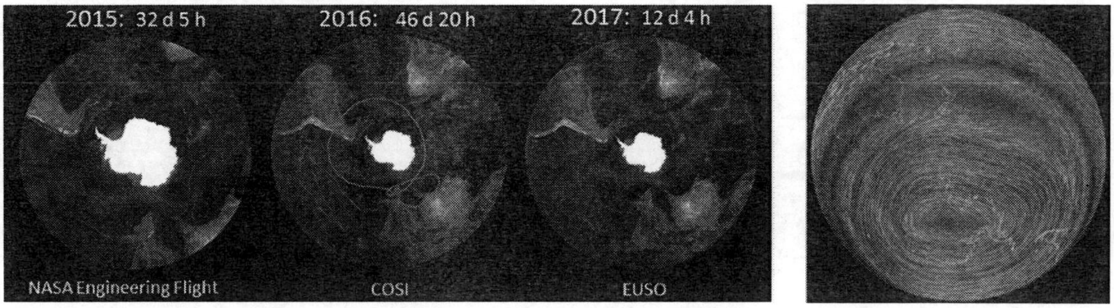


Figure 1: The previous flight trajectories of the NASA super pressure balloons launched from Wanaka, NZ in 2015 (32 d), 2016 (40 d) and 2017 (12 d). The very right panel displays the circular air pattern from 2017/04/30 (see https://earth.nullschool.net/#2017/04/30/2300Z/wind/isobaric/10hPa/orthographic=-177.79,-57.63,40/loc=171.065,-43.450.

nos via the air fluorescence and optical Cherenkov emission channels, and to verify the overall detection strategy. It is also the third generation balloon mission undertaken by the JEM-EUSO collaboration. The first was EUSO-Balloon [3] showing the proof of concept in an overnight flight in 2014 and followed by the first attempt of a long duration flight in 2017 of EUSO-SPB1 [4].which had to be terminated early due to a leaking balloon after only 12 days at float. To address the science objectives mentioned above, EUSO-SPB2 is comprised of two telescopes: a fluorescence telescope (FT), which points directly downward and is designed for the measurement of cosmic rays, and a Cherenkov telescope (CT), which points near the Earth's limb and is designed for the measurement of Earth-skimming neutrinos and direct (above-the-limb) cosmic rays. The two telescopes of EUSO-SPB2 will fly onboard a super pressure balloon, which reaches sub-orbital altitudes of ~ 33 km and will launch from Wanaka, NZ in Spring of 2023, with the hope of achieving flight of up to 100 d. While at float, the balloon will follow a stratospheric air current that develops twice a year at these latitudes allowing for circulation around the globe. The previous trajectories are shown in Fig. 1.

The payload design is shown in the left panel of fig. 2, highlighting all the required provisions for such a long duration flight. The right panel of fig. 2 shows the fully assembled payload during hang testing at CSBF in Palestine, TX.

## 2 The Fluorescence Telescope (FT)

### 2.1 The Instrument

The fluorescence telescope, designed to observe ultraviolet light emitted via fluorescence of UHECR induced EAS, consists of three photo-detection modules (PDMs). Four Hamamatsu Photonics R11265-M64multi-anode photo-multiplier tubes (MAPMTs Hamamatsu Photonics R11265-M64) controlled by a custom application specific integrated circuit [5], are potted together in a gelatinous gel to form an elementary cell (EC). Nine of these ECs are connected to xilinx FPGA to form a PDM with 36 MAPMTs and 2,304 total pixels. Each PDM has a square field of view of roughly $12^o$x$12^o$ and can trigger internally [6]. The three PDMs (shown in top panel of fig. 3) are connected via a central data processor which issues triggers received from each PDM to the other two for a unified readout. Additionally, the data processor [7] combines the data from the PDMs and onboard differential GPS, and prioritizes data for download. The three PDMs, behind a BG3 filter[1] and field flattener (bootom panel fig. 3), are laid out in a rectangle at the focus of a Schmidt telescope consisting of a one meter diameter entrance pupil and six spherical mirror segments. An end-to-end measurement with calibrated LEDs confirm the expected overall efficiency of the detector of 19%. The telescope as assembled for field testing in August of 2022 is shown in Fig. 4 including the infrared camera on the bottom right for cloud detection.

### 2.2 Expected Cosmic Ray Detection Performance of the FT

The expected performance of the FT has been extensively benchmarked via simulations [8]. These simulations have been carried out utilizing the JEM-EUSO OffLine framework [9] which provides an end-to-end simulation of the response. Starting from an EAS simulated in Conex, the light production and atmospheric absorption are simulated, followed by the detectors response including the trigger algorithm. By using a Monte Carlo approach, an isotropic flux of UHECRs can be simulated. Showers are thrown over an over-sized disk on ground beneath the detector of 100 km radius. Unlike ground based experiments, there is no area on ground the EAS must land in order to be detected. As such, the over-sized sampling area allows for all possible observing geometries to be simulated. By combining the energy dependent trigger efficiency, simulated aperture, and the UHECR energy spectrum measured by the Pierre Auger Observatory [10], an expected event rate can be estimated.

Based on these simulations, the peak energy sensitivity of the FT is expected to be around 2.5 EeV. Across all

[1]https://www.schott.com/shop/advanced-optics/en/Matt-Filter-Plates/BG3/c/glass-BG3

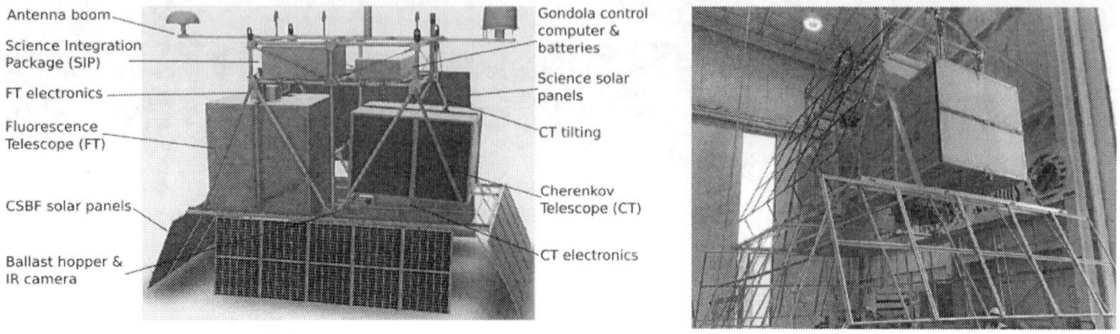


Figure 2: The EUSO-SPB2 payload. The left panel shows a labeled design drawing of EUSO-SPB2 including all the long duration balloon flight equipment such as solar panels and telecommunication antennas next to the two telescopes and the IR camera for cloud monitoring. The right panel shows the fully assembled payload during hang testing at the CSBF facility in Palestine, TX in November 2022.

Figure 3: The FT camera: Top panel shows the 3 PDMs with a total of 6,912 pixels, the bottom panel shows the PDMs mounted in the telescope with the BG3 filter and field flattener in front of the MAPMTs.

energies, the expected event rate is one event roughly every five to eight hours of clear observations (see Fig. 5).

Figure 4: The Fluorescence Telescope during the field Test in August 2022. Yoke structure used for transportation and ground observations allows for full azimuthal rotation and tilting.

## 3 The Cherenkov Telescope (CT)

### 3.1 The Instrument

To record the Cherenkov emission from air showers sourced by neutrinos or cosmic rays, a Silicon Photomultipler (SiPM) camera was built using the S14521-6050AN-04 from Hamamatsu with a spectral range between 400 nm and 800 nm. The 512 pixel camera is shown in Fig. 6. A more technical description of the camera including its

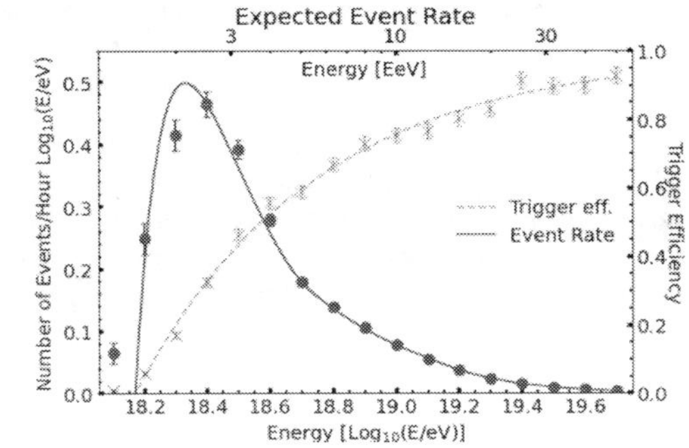


Figure 5: Expected event rate of the FT.

electronics which provide a 10 ns integration time (which matches the typical duration of a Cherenkov pulse sourced from an EAS) is given in [11]. The optical system is a modified Schmidt system with a 1 m diameter aperture providing a vertical field of view of 6.4° and 12.8° in azimuth. To be able to accommodate both major science objectives as outlined in section 3.2 with such a field of view, the pointing direction can be changed during the flight to any point in azimuth and from 3 deg above horizontal to $\sim$10° below the Earth's limb (15° below horizontal). The 4 segments of the mirror are aligned in a bi-focal manner, meaning parallel light from outside the instrument is focused into two distinct spots separated by 12 mm. This way, the background rate given by direct hits of charged particles can be reduced as such events produce only one spot in the camera, and coincident direct hits within the 10 ns integration time are rare. EUSO-SPB2 is the first mission equipped with such an instrument as a payload on a stratospheric balloon flight following an initial prototype flown on EUSO-SPB1 [12].

Figure 6: The 512 SiPM camera of the CT.

### 3.2 The Science of the CT

#### *3.2.1 Background Measurements*

In addition to the desired science goals of measuring cosmic rays and neutrinos, the CT of EUSO-SPB2 will quantify, for the first time, the night-sky airglow background from sub-orbital space. Current estimations of the airglow background are based on ground measurements, and may not be accurate at such high altitudes. Precise measurements of the airglow background are critical to the successful design of future space-based observatories, and will be provided by the CT across wide viewing ranges (both above and below the limb) and observation conditions.

The CT will also measure and quantify other sources of backgrounds relevant for future space-based missions, such as the rate of direct hits from cosmic rays (which will be well identified due to the bifocal optics) and man-made or natural phenomena. The latter category is particularly interesting, providing measurements to explore secondary science goals, such as the measurement of transient luminous events (TLEs).

#### *3.2.2 Above-the-Limb Cosmic Rays*

Direct Cherenkov emission is expected from above the limb due to EAS initiated by atmosphere-skimming cosmic rays. With potential near-limb arrival directions, such events represent a relevant background for neutrino searches, and need to be well understood.

EUSO-SPB2's CT sensitivity to above-the-limb cosmic rays has been investigated using a full Monte Carlo analysis where the `EASCherSim`[2] code has been used to generate the optical Cherenkov signal of showers, taking into account geometric and atmospheric effects [13]. The estimates provided by this study have since been updated to consider realistic assumptions of the trigger threshold of the CT obtained during field testing (see section 4). Fig. 7 shows the estimated cumulative event rate of above-the-limb cosmic rays for the CT as a function of primary energy, while fig. 8 shows the angular distribution of the accepted cosmic rays for a few primary energies.

Fig. 7 shows a CT threshold of $\sim$ 1 PeV to EAS induced by above-the-limb cosmic rays, resulting in a total event rate of $\sim$ 100 events per hour of live time. Fig. 8 shows that while the majority of PeV-scale events are observed a few degrees above the limb, signals generated by more energetic primaries compete with the heavy atmospheric attenuation present for trajectories which pass closer to the Earth.

The simulated Cherenkov signals generated by above-the-limb cosmic ray EAS show similar properties to those generated by neutrino-sourced showers, having comparable intensities, wavelength spectra, and timing and angular distributions of arriving photons. In this way, the frequently expected low energy cosmic rays from relatively high above the limb represent a reliable benchmark on which to calibrate the CT and verify the Cherenkov detection method towards the observation of potential neutrino

[2]https://c4341.gitlab.io/easchersim/

https://doi.org/10.1051/epjconf/202328306009

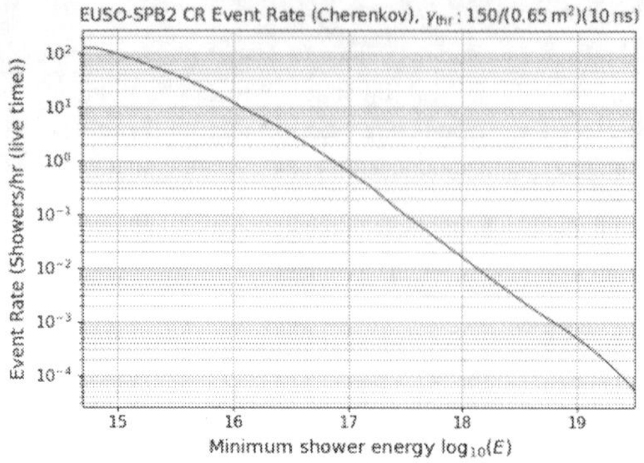


Figure 7: Cumulative event rate of above-the-limb cosmic rays as a function of primary energy for EUSO-SPB2's CT, calculated using `EASCherSim` and trigger thresholds determined from field testing.

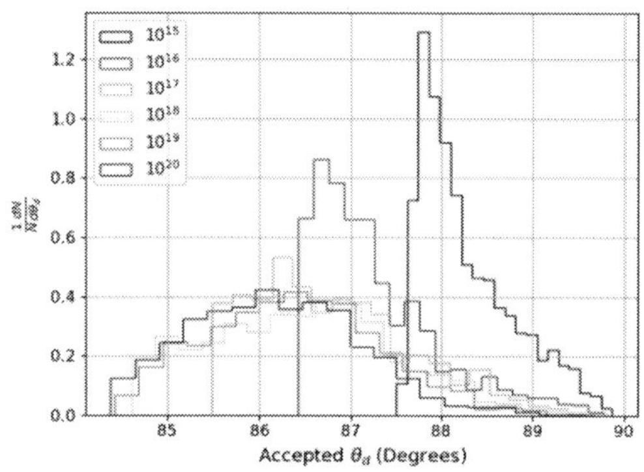


Figure 8: Normalized distributions of (arrival) zenith angles of accepted above-the-limb cosmic rays for various primary cosmic ray energies. For the 33 km altitude of EUSO-SPB2, Earth's limb appears at 84.18°.

events. Consequently, the more rarely occurring, high energy cosmic rays from near the limb constitute a background for neutrino signals, particularly considering the expected atmospheric refraction of the generated optical emission.

### 3.2.3 *Target of Opportunity*

A variety of astrophysical transient events are suspected of harboring the conditions for high-energy and very-high energy neutrino production [e.g., 14–18]. Neutrino oscillations over large distances ensure a flux of tau neutrinos that would be detectable by EUSO-SPB2's CT through their upward-going EASs resulting from their interactions in the Earth.

The acceptance of EUSO-SPB2's CT to tau neutrinos and its transient sensitivity were studied in [19]. These calculations made use of detailed simulations of tau neutrino propagation through the Earth, the development of upward-going EASs, and the transmission of Cherenkov signals through the atmosphere [20].

Fig. 9 and 10 show separate calculations of EUSO-SPB2's sensitivity to long-duration (lasting $\gtrsim$ 1 day) and short-duration (lasting $\lesssim 10^3$ s) events, respectively. For both scenarios, the sensitivity is presented as a band to account for variations in sensitivity that arise due to the positions of the sources behind the Earth as viewed by EUSO-SPB2 as a function of time. For long-duration events, the sensitivity presented in Fig. 9 is the average over the assumed flight duration (30 days or 100 days) in order to account for the day-to-day variation in the observation period due to the Sun and the Moon. For short-duration events, the sensitivity presented in Fig. 10 is integrated over $10^3$ s and assumes the "best-case" scenario in which the source is in the optimal position for neutrino detection (*i.e.*, when source just dips below the Earth's limb) and there is no interference from the Sun or the Moon.

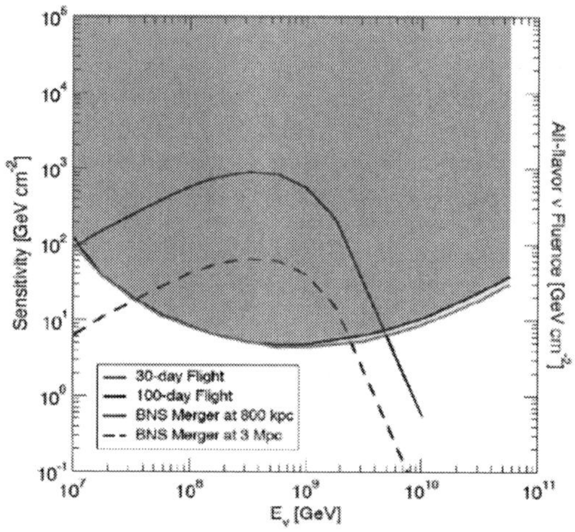


Figure 9: All-flavor long-duration transient sensitivity band for EUSO-SPB2 assuming a 30-day (red) or a 100-day (black) flight. For comparison, solid and dashed blue lines present the all-flavor neutrino fluences from the binary neutron star merger model of [16], placing the source at various distances.

Fig. 11 provides a sky plot of the time-averaged tau neutrino acceptance at $10^{8.5}$ GeV. The plot demonstrates EUSO-SPB2's sky coverage, averaged over 100 days, assuming launch on April 1, 2023 and accounting for the reduction in the observation period due to the Sun and the Moon. For reference, select nearby astrophysics sources and reported hotspots from the Telescope Array experiment can be found in [21].

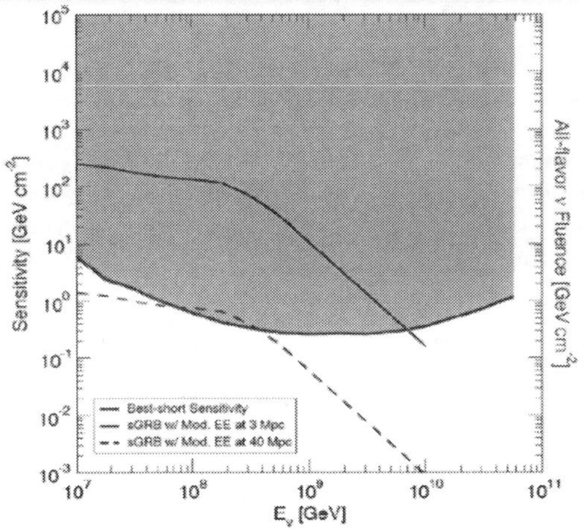


Figure 10: All-flavor "best-case" short-duration transient sensitivity band for EUSO-SPB2. For comparison, solid and dashed blue lines present the all-flavor neutrino fluences from the short gamma-ray burst with extended emission model of [18], placing the source at various distances.

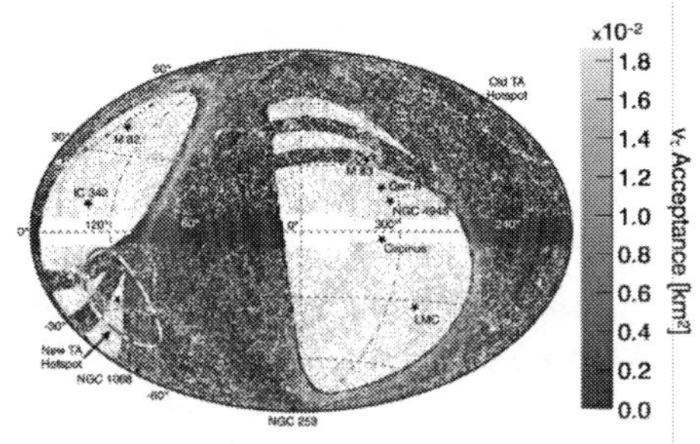


Figure 11: Sky plot of the tau neutrino acceptance at $10^{8.5}$ GeV averaged over 100 days, assuming launch on April 1, 2023.

## 4 Field Test of the EUSO-SPB2 Telescopes

The full characterization of both telescopes using realistic signals prior to flight is essential to providing high quality data during the mission. In addition, the inherent risk of balloon flight makes the data taken in the field the only guaranteed data. Such full-scale end-to-end tests were performed in the field at the TA site in Delta, UT, U.S.A. for both telescopes. These tests were especially important for the CT as it was the first time such an instrument has been built. A UV laser system [22] was fired at a distance across the field of view to mimic the signals of extensive air showers in order to validate the telescopes. The same laser system has been successfully used to characterize the EUSO-SPB1 instrument in a similar setup [23], and has reliably provided test beams for several other telescopes.

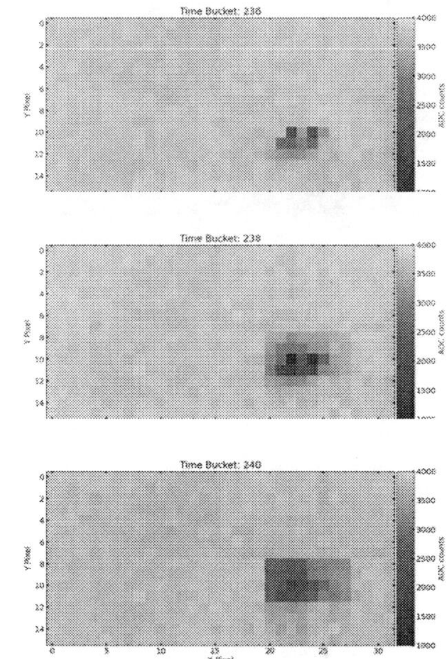


Figure 12: Progression of downward going cosmic ray as observed by the CT of EUSO-SPB2 during field tests. The three panels cover a 30 ns time window (Each time bucket is an integration of 10 ns), and clearly show the bifocal focusing of the light.

The CT was tested in February of 2022 for a period of 9 days. During these tests, the CT recorded 10000s of laser shots and UV LED flashes at various atmospheric and environmental conditions as well as downward going cosmic ray candidate signals. An example of a recorded signal for the CT is shown in fig. 12. These cosmic ray candidates were measured with a frequency of roughly $\sim 10$ Hz, with the rate dependent on the elevation angle of the CT, which varied between 90° and 60°. A full monte carlo analysis using the `EASCherSim` code was performed to estimate the event rate of downward going cosmic rays and compare with what was measured, with the aim to confirm the validity of these candidates as cosmic ray events. The simulated event rate as a function of energy is shown in figure 13 for a CT zenith pointing angle of 30°, and results in a total rate of 2.3 Hz. While this estimate fails to reach the measured 10 Hz observed in the field, it should be noted that the simulation results do not consider multiple pixel triggering, which can increase the event rate by allowing for triggering on less energetic events. While a full analysis is required, it is reasonable to state that the measured rate of cosmic ray candidates is a good estimate of the rate of downward going showers, therefore providing one measure of their validity.

The fluorescence telescope was tested over a period of 12 days at the end August of 2022 at the same location using the same artificial light sources as for the

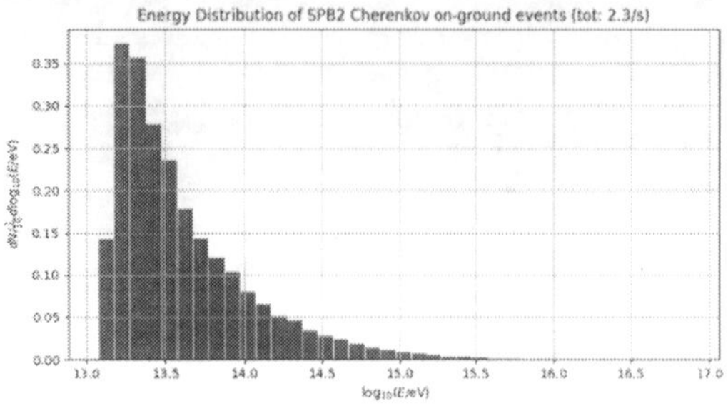


Figure 13: Simulated event rate of downward going cosmic rays for the CT of EUSO-SPB. In this configuration, the CT points 60° in elevation. During field testing, the CT's orientation was varied between perfectly vertical (90°) and 60°.

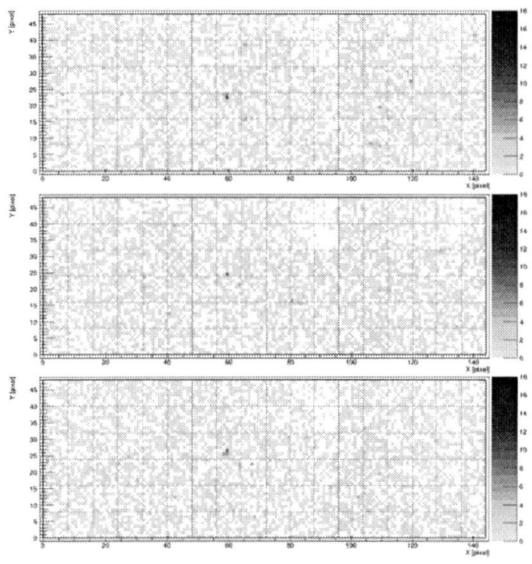

Figure 14: Example frames, each 1 $\mu$s, of a laser track crossing the camera of the FT on 31/08/2022 at 06:27:10. The color bar shows the photoelectron count on the focal surface. The laser energy was set to 1.8 mJ and the laser at a distance of 24 km was pointed straight upwards while the telescope was pointed upwards by 7° from horizontal.

CT, recording 10000s of laser shots and LED events at various monitoring conditions. Three example frames for a recorded laser track is shown in fig. 14. The tests performed included a photometric end-to-end calibration, a field of view measurement and a energy threshold study among others.

The recorded field-test data plays a special role for the FT: besides its use for an instrument characterization, the data is also the basis for the training of the neural network run on board the instrument as a download prioritizer. This method will mitigate the very limited download bandwidth of a balloon payload, making this data mission critical.

Preliminary comparisons between the recorded data and simulation show that both instruments behave as expected and validate the expected performance studies mentioned in section 3.2 and 2.2. A detailed analysis of all the collected data and a final characterization of the instruments based on these data and combined with laboratory measurements is still ongoing for both telescopes.

## 5 Summary

EUSO-SPB2 is the next major stepping stone to a space based multi-messenger detector. As such, a successful launch will raise the technological readiness for such a mission significantly and provide the proof of concept of the measurement techniques from suborbital space for the first time. The instrument is en route and launch is expected in April 2023 from Wanaka, NZ.

With an expected flight duration of at least a couple of weeks, it is expected that multiple extensive air showers with the FT will be detected, following extensive simulation studies which predict an overall event rate of 0.12 events per hour. The assumptions used in these studies were tentatively confirmed by the preliminary field test data. In addition to looking for UHECR tracks, the FT will also look for event candidates with conditions matching those of the ANITA anomalous event candidates (two classes: i) steeply upgoing, high-energy events and ii) near limb, upgoing high energy events) [24–26].

Studies of the novel SiPM based CT of EUSO-SPB2 show that it does not have sensitivity to the diffuse neutrino flux competitive with existing ground based experiments. However, with its ability to be pointed in zenith and azimuth on short time scales, it is possible to track potential sources across the sky following multimessenger alerts, significantly increasing the chance of neutrino observation, provided the event is within our cosmic neighbourhood. To validate and improve detection and reconstruction techniques for future missions, the CT will be pointed above the limb during flight and will record more than 100 direct cosmic rays per hour of observation. These events are guaranteed to exist, and the properties of their recorded Cherenkov emission strongly mimic those resulting from neutrino-sourced showers. Being a novel instrument, another important goal is the measurement and quantification of all relevant backgrounds for future space-based observation. The preliminary analysis of the field test data has shown that the CT behaves as expected and provides more confidence in the performed simulation studies.

**Acknowledgements**

This work was partially supported by NASA awards 1 11-APRA-0058, 16-APROBES16-0023, 17-APRA17-0066, NNX17AJ82G, NNX13AH54G, 80NSSC18K0246, 80NSSC18K0473, 80NSSC19K0626, 80NSSC18K0464, and 80NSSC22K1488 in the USA. This research used resources of the National Energy Research Scientific Computing Center (NERSC), a U.S. Department of Energy Office of Science User Facility operated under Contract No. DE-AC02-05CH11231. We acknowledge the NASA Balloon Program Office and the Columbia Scientific Balloon Facility and staff for extensive support, the Telescope Array Collaboration for the use of their facilities in Utah. We also acknowledge the invaluable contributions of the administrative and technical staffs at our home institutions.

## References

[1] A. Coleman et al., Astropart. Phys. **147**, 102794 (2023)
[2] A. Olinto et al., J. Cosmol. Astropart. Phys. **2021**, 007 (2021)
[3] P. von Ballmoos et al., Adv. Space Res. **53**, 1544 (2014)
[4] L. Wiencke, A. Olinto et al., PoS **ICRC2017**, 1097 (2017)
[5] S. Blin, P. Barrillon, C. de La Taille et al., NIM-A **912**, 363 (2018)
[6] G. Filippatos, M. Battisti, A. Belov, M. Bertaina, F. Bisconti, J. Eser, M. Mignone, F. Sarazin, L. Wiencke, Adv. Space Res. **70**, 2794 (2022)
[7] G. Osteria et al., PoS **ICRC2019**, 420 (2019)
[8] G. Filippatos, M. Battisti, M. Bertaina et al., PoS **ICRC2021**, 405 (2021)
[9] T. Paul et al., PoS **ICRC2017** (2017)
[10] A. Aab, P. Abreu, M. Aglietta et al., Phys. Rev. **D102**, 062005 (2020)
[11] M. Bagheri et al., PoS **ICRC2021**, 1091 (2021)
[12] W. Painter, A. Haungs, M.E. Bertaina, A. Bortone, A. Menshikov, M. Renschler, PoS **ICRC2017**, 442 (2017)
[13] A. Cummings, R. Aloisio, J. Eser, J. Krizmanic, Phys. Rev. **D104** (2021)
[14] C. Lunardini, W. Winter, Phys. Rev. **D95**, 123001 (2017), 1612.03160
[15] M.G. Aartsen et al., Science **361**, eaat1378 (2018), 1807.08816
[16] K. Fang, B.D. Metzger, Astrophys. J. **849**, 153 (2017), 1707.04263
[17] E. Waxman, J.N. Bahcall, Phys. Rev. Lett. **78**, 2292 (1997), astro-ph/9701231
[18] S.S. Kimura, K. Murase, P. Mészáros, K. Kiuchi, Astrophys. J. **848**, L4 (2017), 1708.07075
[19] T.M. Venters, M.H. Reno, J.F. Krizmanic (POEMMA, JEM-EUSO), PoS **ICRC2021**, 977 (2021), 2206.02730
[20] M.H. Reno, J.F. Krizmanic, T.M. Venters, Phys. Rev. **D100**, 063010 (2019), 1902.11287
[21] J. Kim, D. Ivanov, K. Kawata, H. Sagawa, G. Thomson (Telescope Array), PoS **ICRC2021**, 328 (2021)
[22] P. Hunt et al., PoS **ICRC2015** (2015)
[23] J.H. Adams et al., Exp. Astron. **52**, 125 (2021)
[24] P. Gorham et al., Phys. Rev. Lett. **117** (2016)
[25] P. Gorham et al., Phys. Rev. Lett. **121** (2018)
[26] P. Gorham et al., Phys. Rev. Lett. **126** (2021)

# Recent results from prototypes of the Fluorescence detector Array of Single-pixel Telescopes (FAST) in both hemispheres

*Toshihiro* Fujii[1,*], *Justin* Albury[2], *Fraser* Bradfield[1], *Jose* A. Bellido[2], *Ladislav* Chytka[3], *John* Farmer[4], *Petr* Hamal[3], *Pavel* Horvath[5], *Miroslav* Hrabovsky[3,5], *Hiromu* Iwasaki[6], *Vlastimil* Jilek[5], *Jakub* Kmec[5], *Jiri* Kvita[5], *Max* Malacari[4], *Dusan* Mandat[3,5], *Massimo* Mastrodicasa[7], *John* N. Matthews[8], *Stanislav* Michal[3], *Hiromu* Nagasawa[6], *Hiroki* Namba[6], *Xiaochen* Ni[4], *Libor* Nozka[5], *Tomohiko* Oka[5], *Miroslav* Palatka[5], *Miroslav* Pech[5], *Paolo* Privitera[4], *Petr* Schovanek[5], *Francesco* Salamida[7], *Radomir* Smida[4], *Zuzana* Svozilikova[5], *Stan* B. Thomas[8], *Akimichi* Taketa[9], *Kenta* Terauchi[6], *Petr* Travnicek[3,5], and *Martin* Vacula[5] (The FAST Collaboration)

[1]Graduate School of Science, Osaka Metropolitan University, Sumiyoshi-ku, Osaka, Japan
[2]Department of Physics, University of Adelaide, Adelaide, S.A., Australia
[3]Institute of Physics of the Academy of Sciences of the Czech Republic, Prague, Czech Republic
[4]Kavli Institute for Cosmological Physics, University of Chicago, Chicago, IL, USA
[5]Joint Laboratory of Optics of PU and IF of CAS, Palacky University, Olomouc, Czech Republic
[6]Graduate School of Science, Kyoto University, Sakyo-ku, Kyoto, Japan
[7]Department of Physical and Chemical Sciences, University of L'Aquila and INFN LNGS
[8]High Energy Astrophysics Institute and Department of Physics and Astronomy, University of Utah, Salt Lake City, UT, USA
[9]Earthquake Research Institute, University of Tokyo, Bunkyo-ku, Tokyo, Japan

**Abstract.** The origin and nature of ultrahigh-energy cosmic rays (UHECRs) are of uppermost importance in astroparticle physics. Motivated by the need for an unprecedented aperture for further advancements, the Fluorescence detector Array of Single-pixel Telescopes (FAST) is a prospective next-generation, ground-based UHECR observatory that aims to cover an enormous area by deploying a large array of low-cost fluorescence telescopes. The full-scale FAST prototype consists of four 20 cm photomultiplier tubes at the focus of a segmented mirror 1.6 m in diameter. Three FAST prototypes have been installed at the Telescope Array Experiment in Utah, USA, and two prototypes at the Pierre Auger Observatory in Mendoza, Argentina, commencing remote observation of UHECRs in both hemispheres. We report on recent results of the full-scale FAST prototypes operated in both hemispheres, including telescope calibrations, atmospheric monitoring, ongoing electronics upgrades, development of sophisticated reconstruction methods and UHECR detections.

## 1 Ultra-high-energy cosmic rays

The origins and acceleration mechanisms of ultra-high energy cosmic rays (UHECRs) are still largely unknown [1, 2] after the first detection of the cosmic ray with a energy of 100 EeV 60 years ago [3]. They are one of the most intriguing mysteries in a field of the particle astrophysics.

Immediately after the discovery of the 100 EeV cosmic ray, K. Greisen, G. T. Zatsepin and V. A. Kuzmin predicted the UHECR energy spectrum to be suppressed above 60 EeV with the 3 K cosmic microwave background radiation via pion production, known as GZK cutoff [4, 5]. If the GZK cutoff exists, the origin of UHECRs is significantly restricted to nearby sources distributed non-uniformly within 50-100 Mpc. Additionally as UHECRs are deflected less strongly by magnetic fields due to their enormous kinetic energies, their arrival directions are more significantly correlated with their sources. Charged-particle astronomy with UHECRs is hence a potentially viable probe of extremely energetic phenomena in the nearby universe. However, due to their limited flux of UHECRs, less than one particle per century per square kilometer at the highest energies, a very large area must be instrumented to collect significant statistics. The energy, arrival direction, and mass composition of UHECRs can be inferred from studies of the cascades of secondary particles (Extensive Air Shower, EAS) produced by their interaction with the atmosphere of the Earth.

Two well-established methods are used for UHECR detection: arrays of detectors such as plastic scintillators or water-Cherenkov stations that sample EAS particles at the ground level, and large-field-of-view telescopes that directly measure atmospheric shower development by observing ultra-violet nitrogen fluorescence. The two largest UHECR observatories are hybrid detectors that combine both techniques, employing arrays of ground detectors overlooked by fluorescence detectors (FDs). These are the Pierre Auger Observatory (Auger) in Mendoza, Argentina [7], and the Telescope Array Experiment (TA) in Utah, USA [8, 9].

*e-mail: toshi@omu.ac.jp

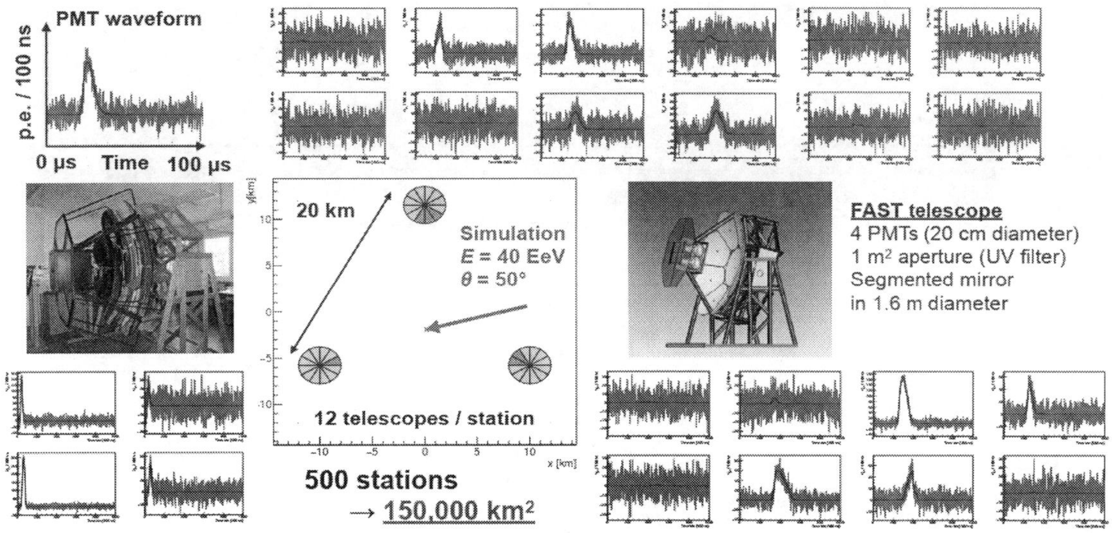


**Figure 1.** The Fluorescence detector Array of Single-pixel Telescopes: a possible solution for a future giant ground array [6]. The traces show simulated signals emitted from a UHECR with an energy of 40 EeV and a zenith of 50°. The black curves in the waveforms show the simulated traces without contributions of the expected night sky backgrounds

Although the statistics of UHECRs was significantly improved, their reported results at the highest energies are still limited by the statistics due to sharp cutoff of the energy spectrum. Future ground arrays will require an unprecedented aperture exceeding current experiments by an order of magnitude with the mass composition sensitivity above 100 EeV. Future detectors should hence be low-cost and easy to deploy, operate and maintain. A worldwide collaboration is necessary to construct such an array.

## 2 Fluorescence detector Array of Single-pixel Telescopes (FAST)

A possible solution to fulfill these requirements is a ground-based fluorescence detector array. The Fluorescence detector Array of Single-pixel Telescopes (FAST)[1] features compact FD telescopes with a smaller light-collecting area and fewer pixels than current-generation FD designs, leading to a significant reduction in cost that allows for the production of more FD units [10]

In the FAST design, a 30° × 30° field-of-view is covered by four 20 cm photomultiplier-tubes (PMTs)at the focal plane of a compact segmented mirror of 1.6 m diameter [11]. Its smaller light-collecting optics, smaller telescope housing, and fewer number of PMTs significantly reduces its cost to be ~35 kUSD per telescope. Each FAST station would consist of 12 such telescopes, covering 360° in azimuth and 30° in elevation. These stations would be deployed in a triangular array with a 20 km spacing, suggested by simulations.

Figure 1 shows the simulated waveforms from a UHECR shower detected in 3-fold coincidence by such an array. The arrival direction and shower profile of the primary particle are included in the shapes of the recorded waveforms. The most sensitive parameter of the mass composition, $X_{\max}$ at which the number of EAS particles reaches at maximum, is reconstructed from the observed shower profile. To achieve the unprecedented exposure exceeding current experiments by an order of magnitude, 500 stations covering 150,000 km$^2$ are required, after accounting for the standard FD duty-cycle and additional moon-night operation.

## 3 Recent progress of developments on the FAST prototypes

Motivated by UHECR detections with a single 20 cm PMT at the focus of a 1 m$^2$ Fresnel lens in 2014 [6], we installed three full-scale FAST prototypes at the TA site in Figure 2(a)-left [12]. We assembled the telescope frames on-site, mounted the PMTs in their camera boxes and installed ultra-violet band-pass filters at their apertures. We then astrometrically aligned the telescopes using a camera mounted to their frames' exteriors [11]. Following this, we began observation via remote connection, using external triggers from the adjacent TA fluorescence detector. We used an automated all-sky monitoring camera to record cloud coverage and atmospheric transparency [13].

As shown in Figure 2(a)-right, two identical FAST prototypes were installed at the Auger site. These identical FAST prototypes installed at TA and Auger will allow for a cross-calibration of the energy and $X_{\max}$ scales between TA and Auger, as well as a comparison between the atmospheric transparency at both sites, an important source of systematic uncertainty in the fluorescence technique.

[1] https://www.fast-project.org

(a) FAST prototypes installed at TA (left) and Auger (right) observatories

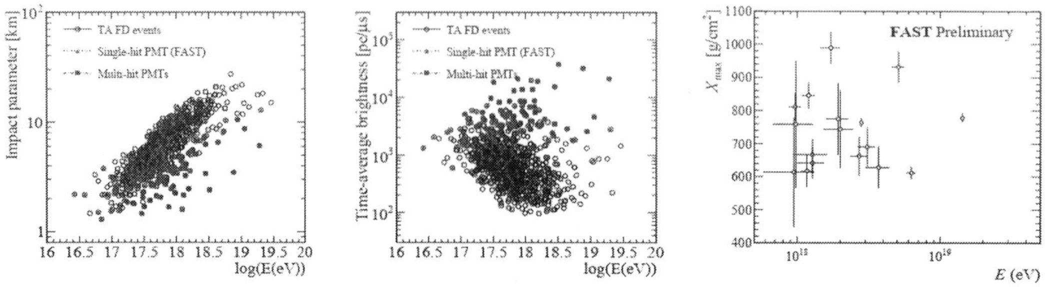


(b) Impact parameter for coincidences between TA FD and FAST (c) Time-average brightness for coincidences between TA FD and FAST (d) Energy and $X_{max}$ reconstructed by FAST top-down reconstruction

**Figure 2.** (a) The three FAST prototypes installed at the Black Rock Mesa site of the Telescope Array Experiment and the one prototype installed at Los Leones site of the Pierre Auger Observatory. (b) Impact parameter and (c) time-average brightness for the coincidence search between TA FD and FAST. (d) Preliminary result of top-down Energy and $X_{max}$ reconstructions for multi-hit events above 1 EeV.

Analyzing 224 hours of data measured by the FAST prototypes at the TA site from March 2018 to October 2019, we found 964 showers with corresponding monocular reconstructions from the TA FD [14]. We searched for significant signals (defined as a $\geq 6\sigma$ signal-to-noise ratio over $\geq$ 500 nanoseconds) in time coincidence with these FD events and found 179 significant FAST events out of the 964 TA EASs, with 59 events producing significant signals in more than one PMT. Figure 2(b) and (c) show the impact parameter and time-average brightness of the detected EASs as a function of energy, split by single-PMT and multi-PMT events. These parameters are reconstructed by the TA FD.

A "top-down" reconstruction algorithm has been implemented that determines the best-fit shower parameters by comparing our measured traces to the simulated ones [12]. Because FAST features only four pixels, rather than use the entry and exit times for each pixel as traditional reconstruction methods do, we extract timing information from each individual bin of the traces. Figure 2(d) shows preliminary $X_{max}$ and energy values reconstructed by this method for multi-hit events above 1 EeV using only FAST prototypes.

### 3.1 Neural network first-guess estimation

The top-down reconstruction requires a reasonable first-guess geometry to reduce computational time. This is provided by a neural network first-guess estimation [17]. The total signal, centroid time, and pulse height of each PMT with a significant signal are used as inputs. The outputs are six parameters: $X_{max}$, energy, zenith, azimuth and west-east/south-north core positions. The model uses the Keras/Tensorflow library with two hidden fully-connected layers.

The resolution and detection bias on $X_{max}$ are evaluated by only applying this first-guess estimation for EASs of four primaries (proton, helium, nitrogen, and iron) with three hadronic interaction models (EPOS-LHC, QGSJetII-04 and Sibyll 2.3c) [18]. The EASs are generated with uniformly-distributed arrival directions and core positions randomly generated in the triangular array's inner circle. The typical resolutions are 4.2 degrees in arrival direction, 465 m in core position, 8% in energy, and 30 g/cm$^2$ on $X_{max}$ at 40 EeV for the 3-fold coincidence events.

Figure 3 shows a preliminary detection bias on $\langle X_{max} \rangle$ and $\sigma(X_{max})$, and also reconstructed $X_{max}$ distributions in each energy bin. Although the reconstructed elongation rate of $X_{max}$ is biased below 10 EeV due to the limitation of field-of-view, it is reproduced above 10 EeV. As seen in the reconstructed $X_{max}$ distribution, the FAST with the neural network estimation is possible to distinguish primary species using a statistical method. Note that this performance is evaluated by only the neural network first-guess estimation. The full-chain performance of both the top-down reconstruction and the neural network first-guess estimation is being investigated.

With this simulation study, the trigger efficiency for 3-fold detections is evaluated as shown in Figure 4(a). The FAST array has a 100% efficiency above 20 EeV.

 https://doi.org/10.1051/epjconf/202328306010

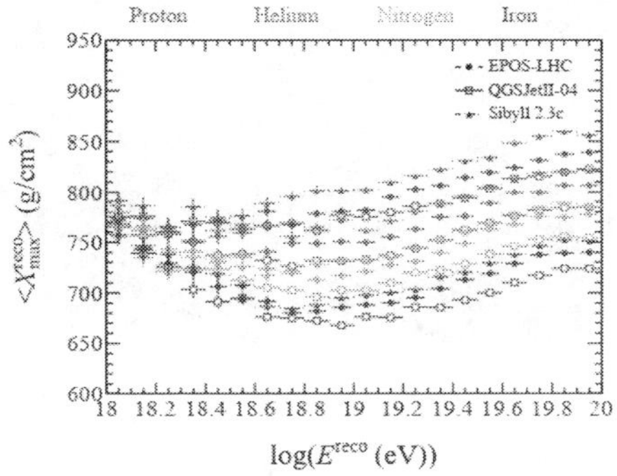


(a) Reconstructed $\langle X_{max} \rangle$

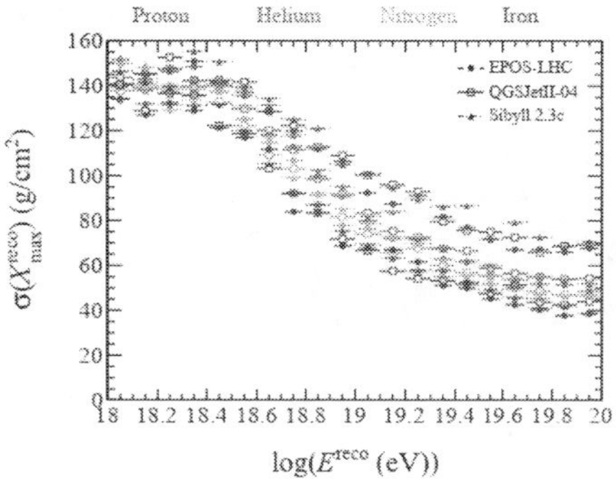


(b) Reconstructed $\sigma(X_{max})$

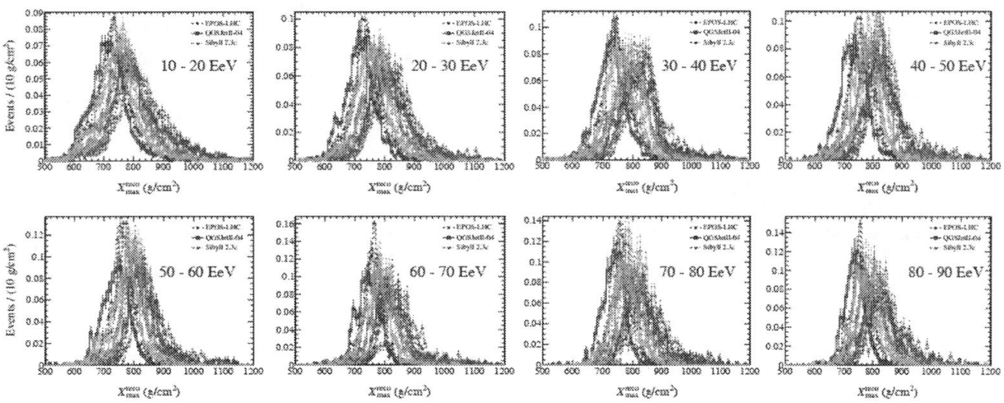


(c) Reconstructed $X_{max}$ distributions

**Figure 3.** Reconstruction bias on (a) $\langle X_{max} \rangle$ and (b) $\sigma(X_{max})$ of four primaries (proton, helium, nitrogen and iron) with three interaction models (EPOS-LHC, QGSJetII-04, Sibyll 2.3c) reconstructed by the neural network first-guess estimation. (c) Their reconstructed $X_{max}$ distributions in each energy bin.

The energy threshold is related to the bias on the average $\langle X_{max} \rangle$ and $\sigma(X_{max})$ as shown in Figure 3. Figure 4(b) is the expected sensitivity on the energy spectrum with a full-size FAST array. Assuming an effective exposure of 90,000 km$^2$ sr per year, the detectable flux under the 95% confidence level is evaluated with FAST, compared to the latest measurements reported from TA and Auger. A full-sized FAST array will extend UHECR measurements beyond 300 EeV.

### 3.2 Developments for stand-alone observation of FAST array

Since these tests, several advances have been made: improvements in our telescope design and the robust hut for telescope storage, improvements in PMT calibration systems, updates of electronics with low-power consumption as shown in Figure 5. The compact and robust hut is constructed for a long-term operation more than 20 years. The shutter of the hut will be operated by the solar panel and batteries.

The PMTs are calibrated to measure the absolute gain using single photo-electron measurement and the gain curve as a function of the high voltage. The non-uniformity of gain on the PMT surface is also investigated by the commercial robotic arm with the position sensitivity of 0.2 mm. The new PMT (R14866) is recently developed to improve the non-uniformity on the PMT surface, showing more homogeneous sensitivity compared to the current one (R5912).

The improved electronics is particularly important as previous tests have capitalized on the infrastructure of existing FD detectors. These new electronics will allow for the first deployment of an independent, solar-powered FAST station, as well as permit stand-alone operation with the FAST array, an important step in validating our design and testing our expected resolution. The potential infield calibration will be performed using an extended uniform light source such as the integrating sphere [19].

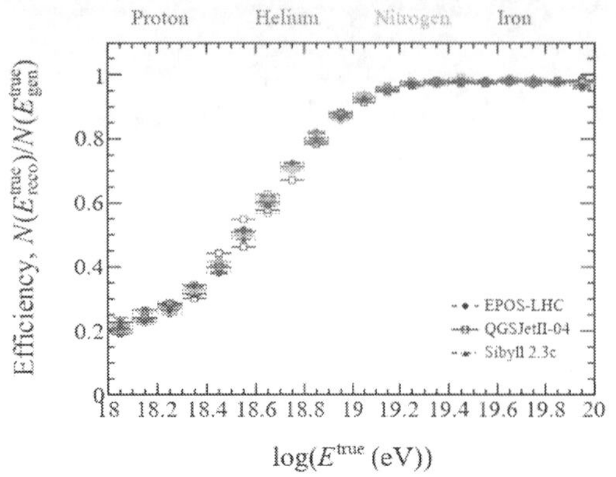


(a) 3-fold Trigger efficiency

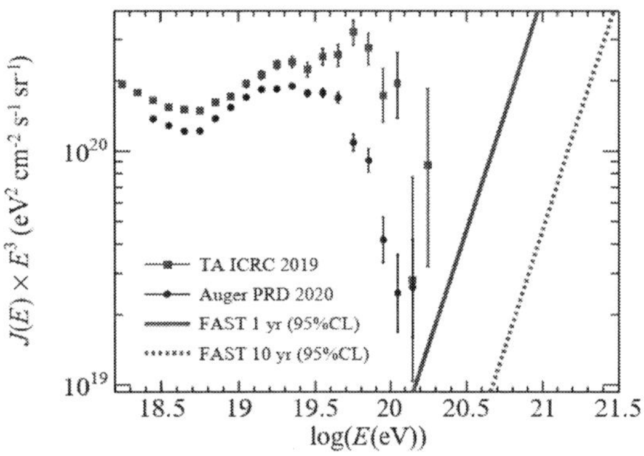


(b) Expected sensitivity of full FAST array

**Figure 4.** (a) Trigger efficiency for 3-fold detections with a hypothetical FAST array. (b) Expected 95% confidence level detectable sensitivities of the energy spectrum with the full FAST array of 500 stations compared to the spectra reported from TA [15] and Auger [16].

## 4 Summary

We have developed a low-cost, easily-deployed fluorescence detector optimized for detection of the highest energy cosmic rays in anticipation of a future ground array with an unprecedented exposure exceeding current experiments by an order of magnitude with the mass composition sensitivity. Three FAST prototypes have been installed at the Telescope Array Experiment, and two prototype has been installed at the Pierre Auger Observatory. We have begun observations in both hemispheres and have demonstrated the viability of the reconstruction methods such as top-down reconstruction and neural network first-guess reconstruction. Furthermore, measurements of UHECRs with the identical FAST prototypes at both observatories is important to perform a cross-calibration of the TA and Auger energy and $X_{max}$ scales. We will continue the steady operation of these FAST prototypes and developments for the stand-alone operation toward the full-sized FAST array.

**Figure 5.** Developments for future stand-alone operations: the robust hut being constructed, PMT calibration system at laboratory and the new electronics under development.

## Acknowledgements

This work was supported by JSPS KAKENHI Grant Number 21H04470, 20H05852, 18KK0381, 18H01225, 15H05443, and a research fund from the Hakubi Center for Advanced Research, Kyoto University. This work was partially carried out by the joint research program of the Institute for Cosmic Ray Research (ICRR) at the University of Tokyo. This work was supported in part by NSF grant PHY-1713764, PHY-1412261 and by the Kavli Institute for Cosmological Physics at the University of Chicago through grant NSF PHY-1125897 and an endowment from the Kavli Foundation and its founder Fred Kavli. The Czech authors gratefully acknowledge the support of the Ministry of Education, Youth and Sports of the Czech Republic project No. LTAUSA17078, CZ.02.1.01/0.0/17_049/0008422, LTT 18004. The Czech authors gratefully acknowledge the support of the Czech Academy of Sciences and Japan Society for the Promotion of Science within the bilateral joint research project with Kyoto University and Osaka Metropolitan University (Mobility Plus project JSPS 21-10). The Czech authors gratefully acknowledge the support of the project IGA_PrF_2023_005 of Palacky University. The Australian authors acknowledge the support of the Australian Research Council, through Discovery Project DP150101622. The authors thank the Pierre Auger and Telescope Array Collaborations for providing logistic support, part of the instrumentation to perform the FAST prototype measurement, and productive discussions.

## References

[1] A. Coleman et al., Astropart. Phys. **147**, 102794 (2023), 2205.05845
[2] R. Alves Batista et al., Front. Astron. Space Sci. **6**, 23 (2019), 1903.06714
[3] J. Linsley, Phys. Rev. Lett. **10**, 146 (1963)
[4] K. Greisen, Phys. Rev. Lett. **16**, 748 (1966)
[5] G. Zatsepin, V. Kuzmin, JETP Lett. **4**, 78 (1966)
[6] T. Fujii et al. (FAST), Astropart. Phys. **74**, 64 (2016), 1504.00692
[7] A. Aab et al. (Pierre Auger), Nucl. Instrum. Meth. **A798**, 172 (2015), 1502.01323
[8] H. Tokuno, Y. Tameda, M. Takeda, K. Kadota, D. Ikeda et al., Nucl. Instrum. Meth. **A676**, 54 (2012), 1201.0002
[9] T. Abu-Zayyad et al. (Telescope Array), Nucl. Instrum. Meth. **A689**, 87 (2012), 1201.4964
[10] T. Fujii et al. (FAST), PoS **ICRC2021**, 402 (2021), 2107.02949
[11] D. Mandat et al. (FAST), JINST **12**, T07001 (2017)
[12] M. Malacari et al. (FAST), Astropart. Phys. **119**, 102430 (2020), 1911.05285
[13] L. Chytka et al. (FAST), JINST **15**, T10009 (2020)
[14] R.U. Abbasi et al. (Telescope Array), Astropart. Phys. **80**, 131 (2016), 1511.07510
[15] D. Ivanov (Telescope Array), PoS **ICRC2019**, 298 (2020)
[16] A. Aab et al. (Pierre Auger), Phys. Rev. Lett. **125**, 121106 (2020), 2008.06488
[17] J. Albury, Ph.D. thesis p. University of Adelaide (2021)
[18] D. Heck, G. Schatz, T. Thouw, J. Knapp, J. Capdevielle, Forschungszentrum Karlsruhe Report FZKA p. 6019 (1998)
[19] M. Vacula et al., Optik **242**, 167169 (2021)

https://doi.org/10.1051/epjconf/202328306011

# Progress and future prospect of the CRAFFT project for the next generation UHECR observatory

*Yuichiro* Tameda[1,*], *Takayuki* Tomida[2], *Daisuke* Ikeda[3], *Katsuya* Yamazaki[4], *Jihyun* Kim[5], *Yuya* Nakamura[2], *Yuto* Kubota[2], *Norimichi* Shibata[1], *Eiji* Nishio[1], *Miyato* Murakami[1], *Yasuki* Ishimoto[1], *Tomoki* Katayama[1], and *Yuga* Kobayashi[1] for the CRAFFT collaboration

[1]Osaka Electro Communication University, Neyagawa, Osaka, Japan
[2]Shinshu University, Nagano, Nagano, Japan
[3]Kanagawa University, Yokohama, Kanagawa, Japan
[4]Chubu University, Kasugai, Aichi, Japan
[5]University of Utah, Salt Lake City, Utah, USA

**Abstract.** The next generation of ultra-high energy cosmic ray observations will require large detector arrays to achieve large statistics. In order to realize next-generation large-scale detector arrays, the Cosmic Ray Air Fluorescence Fresnel lens Telescope (CRAFFT) project is developing a low-cost simple fluorescence detector (FD). The simple structure of the CRAFFT detector will reduce the cost to about 1/10 of the current FD. We also aim to realize a fully automated observation system. A prototype of the CRAFFT detector has been successfully used to detect cosmic ray air showers. Since the spatial resolution of the simple FD is rougher than that of the current FD, we are developing a new air shower reconstruction method using the waveform fitting method. In this presentation, we report the performance of the CRAFFT detector, detector optimization, and future prospect.

## 1 Introduction

It is important to clarify the origin of ultra-high energy cosmic rays (UHECRs) above $10^{18}$ eV for astrophysics. During cosmic rays are propagation, they are diffused in the galactic or extra galactic space due to their electric charge. That make difficult to trace back the origin of the cosmic rays. However, assuming that cosmic rays are light nuclei such as protons, they are expected to travel straight through the galactic or extra-galactic magnetic field at the energy of $10^{20}$ eV.

The UHECR has been observed by various experiments. Recently, the Telescope Array (TA) experiment reported a "hotspot" in the direction of arrival of cosmic rays above 57 EeV[1]. Pierre Auger Observatory (Auger) reported the existence of a dipole structure in the direction of arrival of cosmic rays above 8 EeV[2]. As described above, various anisotropies in the direction of arrival of cosmic rays have been reported, and such a fact enriches our understanding of the origin of the cosmic rays.

How can we identify the origin of UHECR? First of all, it is necessary to measure the arrival direction distribution with large statistics. Since the flux of UHECR is quit low, a larger scale observation facility than the current one will be required. Since the cost will inevitably increase with a larger scale, a strategy to reduce the cost will be necessary. All-sky observation is also important for the measurement of the arrival direction distribution. Therefore, global deployment of detectors will be necessary. The mass composition measurement is also essential because the propagation mechanism is also an important point.

The TAx4 and AugerPrime experiments are the upgrade of TA and Auger experiments. What should we do as the next step? The concept of Global Cosmic Ray Observatory (GCOS) has been proposed as the next generation UHECR experiment. Candidate detectors for this experiment include the rich SD and Fluorescence detector (FD) developed at AugerPrime, which are reliable. In addition, there are challenging detectors such as POEMA for space-based observations and GRAND for radio observations. The CRAFFT (Cosmic Ray Air Fluorescence Fresnel lens Telescope) or FAST (Fluorescence detector Array of Single-pixel Telescopes) project are also developing simple FDs for the next generation experiment [3].

## 2 CRAFFT project

The CRAFFT project is developing detectors for the next generation of UHECR observations [4]. In order to clarify the origin of UHECRs, it is necessary to make large-statistics, all-sky observations with a detector that can measure the mass composition. Large-scale observations are inevitable for large-statistics observations, and cost reduction is also an important point. CRAFFT is a research and development project aiming to overcome these problems.

*e-mail: tameda@osakac.ac.jp

**Table 1.** Roadmap of the CRAFFT project

| | |
|---|---|
| Phase 1 | Confirmation of the concept of detectors |
| Phase 1.5 | Optimization of the detector design |
| Phase 2 | Confirmation of the concept of observation |
| Phase 3 | Large scale deployment |

The flux of UHECRs is very low, so a larger detection area is required to achieve high statistics. The increase in cost is inevitable for a larger detection area, which could make difficult the realization of next-generation observations. We propose the following strategies to reduce the cost. We can reduce the cost of detector fabrication by simplifying the detector configuration. We can also reduce the maintenance cost by that. Additionally, we want to reduce operating costs by constructing a fully automatic observation system.

From the viewpoint to clarify the origin of UHECRs, it is important to measure the mass composition, which affects the propagation mechanism of cosmic rays, and the distribution of arrival direction in the entire sky. Therefore, detectors are required to be sensitive to the mass composition. FD is one of the most reliable detector in the field, even if surface detector array is also being studied for mass composition analysis. Radio detection using antennas also has a potential in the future. In order to observe the entire sky, it is necessary to deploy detectors both in the northern and southern hemisphere. In order to minimize systematic errors, it is useful to observe the entire sky with the same type detector, and it is also necessary to consider easy transportation of the detector for deployment. Therefore, a simple structure may be more preferred than the detectors now in use.

Another viewpoint is the reduction of damage to the environment, which has become increasingly important in recent years. The larger the detection area, the greater the damage to the environment. Therefore, it is desirable to make the detector density sparsely.

With above considerations, the CRAFFT project is developing an FD with a simple structure. The detector can be a mass composition sensitive by measuring longitudinal development of extensive air showers. The simple structure will allow for low cost and easy transport. The FDs will be installed at intervals of several tens of kilometers, which means that the detector density is sparse. This reduces the number of installation sites and facilitates maintenance. We are currently developing a simple FD according to the roadmap shown in 1. We are currently in the process of moving from phase 1.5 to 2. The development of the simplified FD and the large-scale deployment of 360-degree stations will realize the next generation of UHECR observations.

### 2.1 Detector

The CRAFFT detector has a simple structure to reduce costs. The original concept is proposed in UHECR 2012 [5]. The main components are a Fresnel lens, a photomultiplier tube (PMT), a UV transmission filter, and electronic circuits such as a FADC and a high-voltage power supply. The Fresnel lens (NTKJ, CF1200-B) is mounted at the aperture and has a size of $1.4\,\mathrm{m}^2$. The transmission of the lens in the UV region is approximately 90%. The lens is supported by two beams to prevent deflection. An 8-inch PMT (Hamamatsu R5912) is mounted at the focal point of the lens. A UV transmission filter (Hoya U330) is placed just in front of the PMT, and the transparency in the UV region is approximately 90%. In the prototype, the UV transmission filter has a spatial filter attached to it, which limits the field of view from $12° \times 12°$ to $8° \times 8°$ degrees. This is to avoid the indefinite sensitivity at the edge of the spherical PMT. Signals are recorded with a 12-bit, 80-MHz FADC (TokushuDensiKairo, Cosmo-Z) through an amplifier. Compared to conventional FDs, this system has fewer channels and is more compact in size, The cost of production can be reduced.

**Figure 1.** The exterior of CRAFFT prototype detector deployed at TA site in Utah, USA.

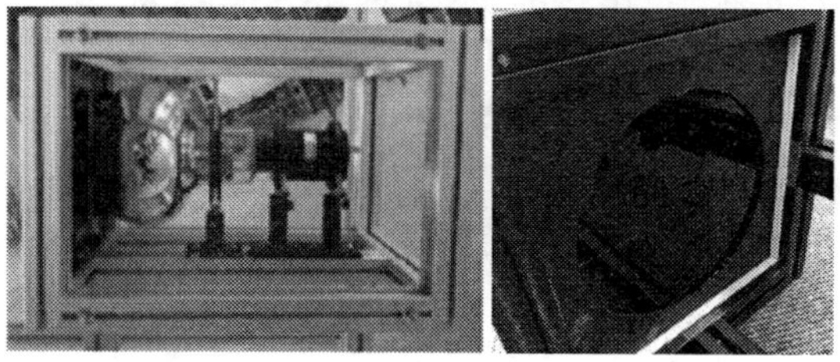

**Figure 2.** 8 in. PMT with UV transmitting filter. For the prototype, the appeature

The CRAFFT detector is an FD, and it is a sensitive detector for measuring mass composition. The detector is a lens-based telescope, which means that there is no obstruction between the aperture and the focal point, and thus it is characterized by high light collection. The CRAFFT detector consists of an aluminum frame and a galvalume steel plate covering it. All components can be installed inside the detector. A curtain just behind the lens protects the interior during the day time. Because of this packaging, the telescope does not require a building for installation and

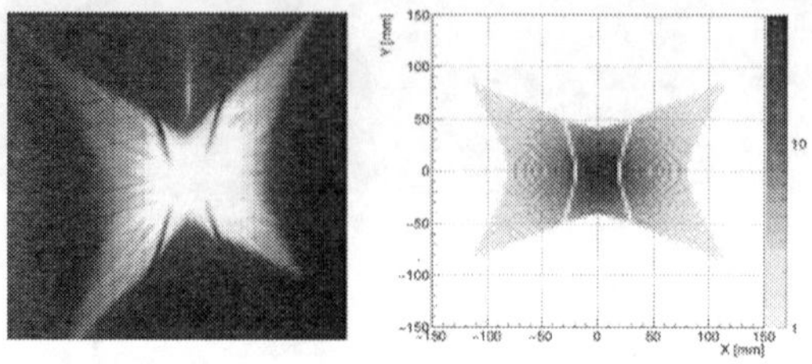

**Figure 3.** Left: Shape of spot around the focal plane. Right: Reproduced spot shape by raytracing simulation.

**Figure 4.** Prototype CRAFFT detectors were deployed next to the TA FD building as red circle. The center of field of view is pointed to the CLF.

can be placed directly on the ground. The detector can be easily transported to multiple experimental sites.

An FD with a Fresnel lens is a unique detector. Therefore, it is necessary to understand the details of the detector performance. The optical performance of the detector was evaluated by ray-tracing simulations. The ray-tracing simulation was performed using ROBAST [6]. Figure 3 shows the actual spot and the shape of the spot obtained by the raytracing simulation, which is well reproduced. The spot size at the focal plane is 44 mm defined in the 95% interval. In the detector simulation, expected waveform has already been simulated, and it is estimated that even a $10^{20}$ eV cosmic ray air shower can be detected within 30 km with a high signal-to-noise (S/N) ratio.

Test observations were performed using a prototype detector. The four prototypes were installed next to the Black Rock Mesa FD station at the TA experimental site as shown in Fig. 4. The center of the field of view was pointed toward the direction where the Central Laser Facility (CLF) was installed. The elevation angle of three of the detectors was set to 24°–32° and that of the other one to 16°–24°. The test observation took place in November 2017 and observation time was 63.5 hours (10 nights). There were at least 10 obvious air shower events during this observation period. The expected number of events at energies above $10^{17}$ eV is around 8, which is consistent with the number of events observed. In order to synchronize with TA FD, data acquisition was performed by trigger pulses from TA FD.

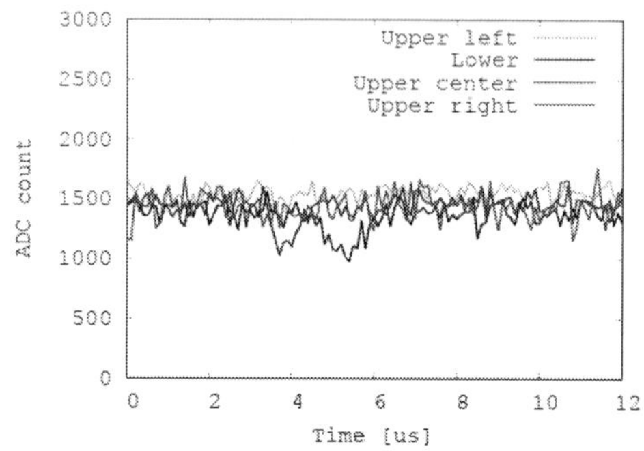


**Figure 5.** Example of actual waveform observed by CRAFFT detector. Vertical and horizontal axes are FADC count and time slot, respectively.

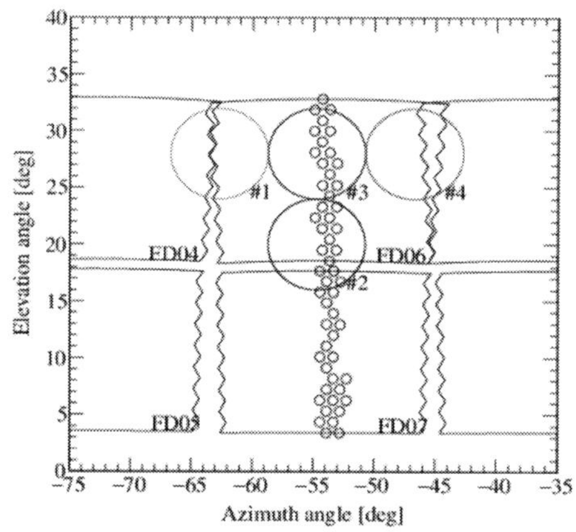


**Figure 6.** Event display of the air shower event same as Fig. 5 observed by TA FD. Field of view of CRAFFT is superposed. The circle color of CRAFFT field of view correspond to the waveform in Fig. 5

Figure 5 and 6 show an example of an actual observed event. This event was also observed by TA FD, and Fig. 6 shows the field of view of the CRAFFT detector superimposed on the TA FD event display for this event. Tracks of the cosmic ray air shower can be seen in the CRAFFT fields of view numbered 2 and 3. Figure 6 is the waveform of this event. Significant signals are seen in detectors 2 and 3 against night sky background. When this event is analyzed by TA FD, the distance from the detector to the air shower is 2.6 km and the energy is $10^{17.7}$ eV. Thus, in the CRAFFT experiment, the establishment of the detector concept of Phase 1 was completed.

### 2.2 Air shower reconstruction method

As shown in Figure 6, the spatial resolution of CRAFFT is lower than that of conventional telescopes and a shower detector plane cannot be obtained. Therefore, we have developed a reconstruction method by waveform fitting. The waveform generated by the simulation as shown in Fig. 7 is compared with the waveform of the real data, and when the simulation agrees with the true value,

$$\chi^2 = \sum_{i=1}^{n} \left( \frac{x_{data,i} - x_{sim,i}}{\sigma_i} \right)^2 \quad (1)$$

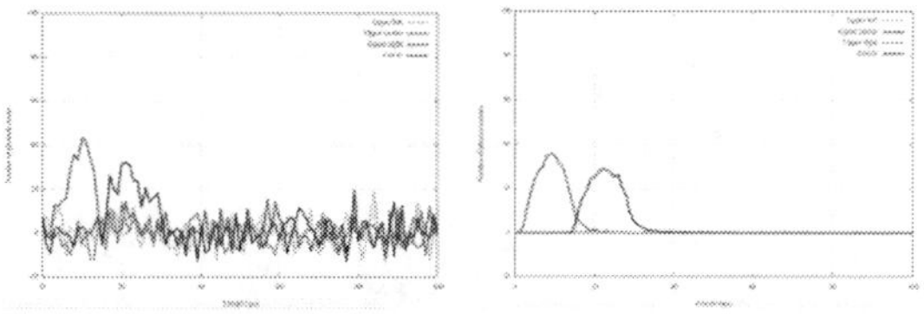

**Figure 7.** Observed waveform by CRAFFT detector (Left) and simulated waveform (Right) to be compared to observed one.

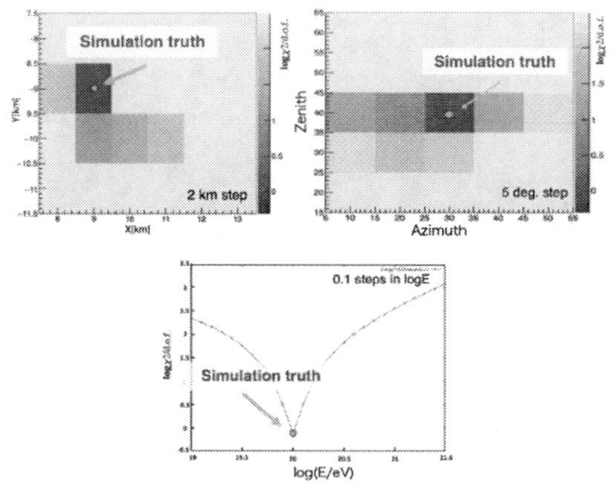


**Figure 8.** Example of a distribution of $\chi^2$ of a waveform fitting with 5 parameters.

is in principle the minimum. $x_{data,i}$ and $x_{sim,i}$ are waveforms obtained from real data and simulated, respectively, where $\sigma_i$ is noise due to night sky background, and $n$ is the number of time slots. Figure 8 shows the convergence of $\chi^2$ for a grid search with five parameters. The step of core position, direction and $\log(E/\mathrm{eV})$ are 2 km, 5°, and 0.1 for grid searching of waveform fitting, respectively. This is a monocular analysis under the same conditions as those of the previous test observation. As can be seen, $\chi^2$ converges around the true value. However, $X_{\max}$ is fixed here. In a Monte Carlo study at $10^{20}$ eV, the accuracy of geometry determination by four-parameter fitting is estimated to be about $\pm 3$ degree in the direction of arrival and $\pm 200$ m m in the core position in the monocular analysis. Currently, we are developing a procedure for analysis based on six-parameters fitting by adding $X_{\max}$ and energy. The six-parameters fitting however requires a more efficient algorithm for reaching the convergence. Stereo and triplet observations will be very effective in constraining the parameters.

## 3 Detector optimization and future plan

### 3.1 Optimization of detector configuration

Since we have successfully detected cosmic rays with the CRAFFT prototype, we are trying to optimize the configuration while maintaining the cost reduction[7]. One guideline for optimization is to increase the field of view per detector, from the viewpoint of cost reduction. To increase the field of view per telescope, the detection area of the light detection area composed by PMTs at the focal point should be increased. This is easier than with conventional FDs, since a refracting telescope does not screen the incident light even if the focal plane is enlarged. Another guideline is to improve the accuracy and S/N ratio of the reconstruction. To improve the accuracy of reconstruction and S/N ratio, the field of view per channel should be narrower. From this point of view, we performed Monte Carlo studies on various configurations of the PMT size and arrangement, and the light guide used to guide the light incident in the gap to the PMT, to estimate the most appropriate arrangement of the PMT. As shown in Fig 9, a four by four array of 5-inch PMTs seems to be a good configuration. For this configuration, the accuracy of geometry determination by four-parameter fitting was estimated to be 2.3° in the arrival direction and 160 km in the core position. The field of view per PMT is about 6.5° , and the total field of view of the telescope is about 26°. The new configuration of the CRAFFT detector can cover the field of view of 2.6 TA FDs.

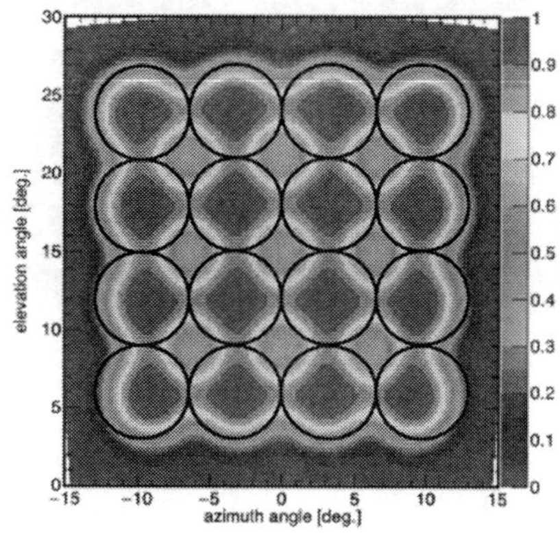


**Figure 9.** Sensitivity map as a function of arrival direction of incident light at the focal plane with 16 PMTs with 5 in. diameter. The arrangement of PMTs of new configuration is four by four matrix.

### 3.2 Preparation for the confirmation of observation concept

Based on the optimized new configuration of the detector, each new component should be evaluated. The PMT selected is the Hamamatsu R877-100. The diameter of the incident window of this PMT is 5 inches, the surface is flat, and the material of photocathode is super bialkali. Because of the large incident window of the PMT, two-dimensional non-uniformity of sensitivity is important. We have developed a system to measure a two-dimensional non-uniformity of PMT gain consisting a UV LED and an XY state, and we are measuring the sensitiv-

 https://doi.org/10.1051/epjconf/202328306011

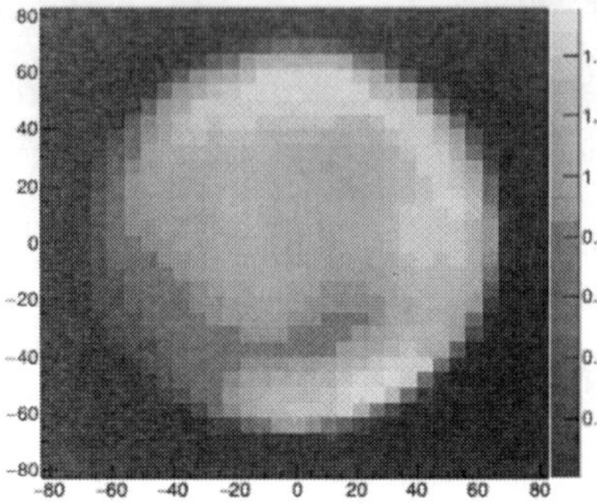

**Figure 10.** Example of a measurement of non-uniformity of PMT gain to be used for new configuration.

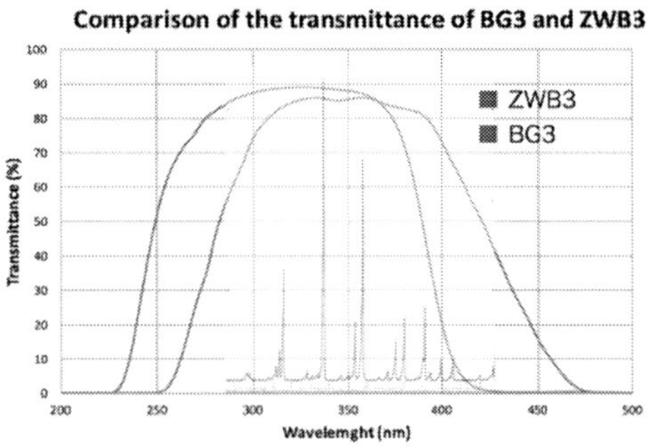


**Figure 11.** Transmittance of UV transmitting filter. ZWB3 is planed to be used for CRAFFT and BG3 is used for TA FD. Fluorescence spectrum is superposed.

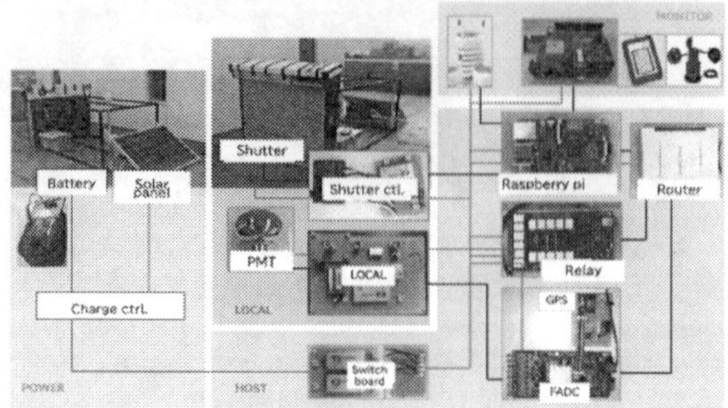


**Figure 12.** Block diagram of the automation system of CRAFFT for long-term tests at Shinshu University.

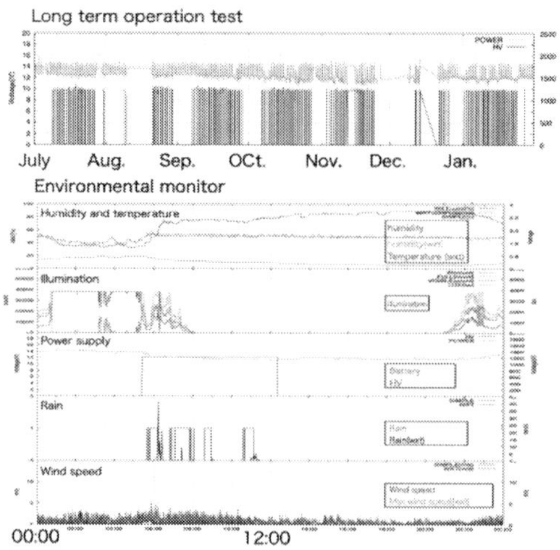


**Figure 13.** Example of the monitor plot of the automation system in long-term tests.

ity of incident window of all the PMTs to be used for the test observations scheduled for next summer. The upper of Fig. 10 is an example of the measured sensitivity relative to the center. These results should be used for simulation and shower reconstruction.

ZWB3 (Haian Subei Optical Glass Factory) will be used as the UV transmission filter. Figure 11 shows the measured transmittance at each wavelength of ZWB3 and BG3 used for FD in the TA experiment and the fluorescence spectra are overlaid . The transmittance of ZWB3 is about 90% at the main peaks of fluorescence, which is sufficient.

In cost reduction, not only production costs but also operating costs must be considered. To reduce operating costs, it is important to have maintenance-free and automatic observations. We have conducted long-term tests for automatic observation. The automatic observation system consists of an environmental monitor and a control unit shown in Fig. 12. The environmental monitor measures the weather conditions such as wind speed, humidity and rainfall, and brightness, and determines whether the observation is possible or not before starting the data acquisition. Figure 12 shows examples of monitors for humidity, illumination, rainfall, wind speed and so on. Cloud monitors using CCD monitors are also useful [8]. When it is determined that the observation is feasible, a high-voltage power supply is automatically applied before the start of the observation, and the electric shutter is opened, then the data taking is started.

Long-term tests were conducted to demonstrate the robustness of the automation system. Detectors for cosmic ray observations, such as CRAFFT, are supposed to operate without a commercial power supply. Therefore, all the power for these systems is provided by a solar power system. The site is the rooftop of the Nagano campus of Shinshu University. Figure 13 shows the results of the seven-month test. This figure shows that the power supply system was stable and the high-voltage power is applied during the observable hours on the days when it is determined that the system is observable. We are planning to conduct test observations using such a system to establish the observation concept as phase 2.

Test observation at TA BR site to confirm the observation concept by simple FDs is planed. We will use four telescope which can cover the field of view of one TA FD station approximately. In this test, we want to realize stable observation with automation system.

## 4 Summary

Various experiments have observed UHECRs and it is expected to clarify their origin. Large-scale experiments with large statistics and sensitivity to the mass composition will be required, and large scale experiments cannot be avoided that will make the cost high. The CRAFFT project aims to develop a simple FD to realize the next generation of UHECR observations with the strategy of cost-reduced FD. We have demonstrated the principle of the simple FD using a Fresnel lens through test observations. We are now planning to establish the concept of stereo or triplet observations. In the future, we would like to install the simple FD globally to realize all-sky observation. Since simple SDs are easy to transport, simple FDs can be distributed to any location. Simple FD stations with $\sim 20\,\mathrm{km}$ spacing are to be set up to provide a global view of the entire perimeter. If the total area of $400,000\,\mathrm{km}^2$ is realized, sufficient statistics will be obtained even considering the FD duty cycle of about 15%. The development of such simple FD can be a powerful strategy for future projects such as GCOS.

## Acknowledgements

This work was supported by the JSPS KAKENHI Grant Numbers 25610051 and JP16K17710. This work was partially carried out by the joint research program of the Institute for Cosmic Ray Research (ICRR), The University of Tokyo. This study was also supported by the Earthquake Research Institute The University of Tokyo Joint Usage/Research Program. This work was partially supported by the collaborative project of Fundamental Electronics Research Institute of Osaka Electro-Communication University. The Telescope Array Collaboration supported CRAFFT as an associated experiment and allowed us to use TA equipments and FD event displays. We wish to thank the staffs at the University of Utah, especially Prof. J.N. Matthews.

## References

[1] R. U. Abbasi *et al.*, ApJL **790** L21 (2014)
[2] The Pierre Auger Collaboration, Science **357** 1266–1270 (2017)
[3] T. Fujii *et al.*, Astroparticle Physics **74**, 64-72 (2016)
[4] Y. Tameda *et al.*, Prog. Theor. Exp. Phys. **2019**, 043F01 (2019)
[5] P. Privitera *et al.*, Internal Symposium on Future Directions in UHECR Physics (2012)
[6] A. Okumura, K. Noda, C. Rulten, Astroparticle Physics **76**, 38-47 (2016)
[7] Y. Kubota *et al.*, PoS **ICRC2021**, 216 (2022)
[8] K. Yamazaki *et al.*, PoS **ICRC2019**, 474 (2021)

# Muon counting with the Underground Muon Detector of The Pierre Auger Observatory

*Marina* Scornavacche[1,2,*] for the Pierre Auger Collaboration[3,**]

[1]Instituto de Tecnologías en Detección y Astropartículas, Av. Gral Paz 1499, 1650 Buenos Aires, Argentina
[2]Institut für Astroteilchenphysik, Hermann-von-Helmholtz-Platz 1, 76131 Karlsruhe, Germany
[3]Observatorio Pierre Auger, Av. San Martín Norte 304, 5613 Malargüe, Argentina

**Abstract.** The Underground Muon Detector (UMD) aims to extend the detection range of the Pierre Auger Observatory to observe cosmic rays of energies greater than $10^{16.6}$ eV and performs a direct measurement of the muon component (sensitive observable to the mass composition) of the air showers. In this work, we will present an overview of the final design of the UMD and its reconstruction techniques alongside the observations obtained during the engineering array phase. First results suggest that the current hadronic interaction models have a muon deficit at energies between $2\times10^{17}$ eV and $2\times10^{18}$ eV.

## 1 Introduction

The Pierre Auger Observatory was designed to answer the key questions about the origin and composition of ultra-high energy cosmic rays. The Observatory is currently undergoing an upgrade dubbed *AugerPrime* [1]. The purpose of *AugerPrime* is to improve the sensitivity to mass composition observables, such as the muon content of the air showers. In order to probe the characteristics of cosmic rays above $10^{17}$ eV, *AugerPrime* will provide a cleaner separation between the electromagnetic and muonic components of the showers. The Underground Muon Detector (UMD) plays a key role in *AugerPrime* since it is designed to perform a direct measurement of the muon component of the air showers.

In the framework of the Heitler-Matthews model of a cascade [2], the number of muons $N_\mu$ is a composition sensitive observable since it scales with energy $E$ and mass number $A$ as $N_\mu \propto \frac{A\,E^\beta}{(A\,\xi_c)^\beta}$, where $\xi_c$ is the critical energy at which charged pions are likely to decay into muons ($\xi_c \sim 20-30$ GeV, $\beta \sim 0.9$). Assuming the validity of the superposition model, the number of muons can be written as $N_\mu^A = A^{1-\beta}\,N_\mu^p$ indicating that more muons are produced in an iron shower than in a proton shower at the same total primary energy.

## 2 The underground muon detector: Final design and current production status

The Pierre Auger Observatory consists of a surface detector of about 1660 water-Cherenkov stations distributed in a triangular grid with 1500 m spacing (SD-1500) covering a total area of 3000 km$^2$. To extend the energy threshold of the SD to lower energies, two denser arrays with spacings of 750 m (SD-750) and 433 m (SD-433) were deployed (over areas of 23.5 km$^2$ and 1.9 km$^2$, respectively). The two arrays SD-750 and SD-433 will be equipped with UMDs. Currently, 26 positions (37% of the full UMD array) are in acquisition (see Figure 1). The full UMD array is expected to be completed by end-2023.

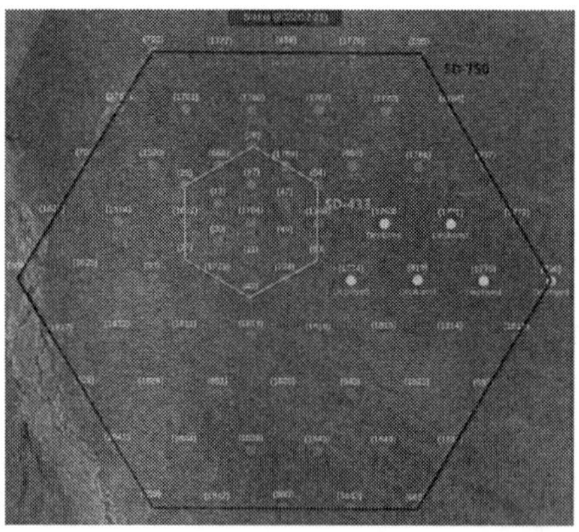


**Figure 1.** Schematics of the water-Cherenkov surface detectors (circles) where UMD positions were deployed (yellow) and are in acquisition (green) in the SD-750 (blue grid) and SD-433 (light blue grid).

In the final design, three modules of 10 m$^2$ are buried at 2.3 m next to a water-Cherenkov detector of the SD-750 and the SD-433 (see Figure 2). The soil shielding ensures the exclusive detection of the muon content of the air showers. Each module consists of 64 plastic scintillator strips containing wavelength shifting optical fiber con-

*e-mail: marina.scornavacche@iteda.cnea.gov.ar
**e-mail: spokespersons@auger.org, full author list available at https://www.auger.org/archive/authors_2022_10.html

 https://doi.org/10.1051/epjconf/202328306012

nected to an array of 64 silicon photomultipliers (SiPMs). When a muon impinges the scintillator, the photons produced are collected and propagated along the fibers to the photodetector. In Figure 3, an image of a complete deployed station is shown, along with a module under construction.

For the full UMD array, the SiPM option was chosen because of their better photodetection efficiency, reduced power consumption and lower cost [3]. Each SiPM is an array of 1584 avalanche photo-diodes (also referred to as cells) operated in Geiger mode, where the number of triggered cells is referred to as photon equivalents (PE). For a detailed review of the design, characterization and implementation of the SiPMs for the data acquisition see references [4][5].

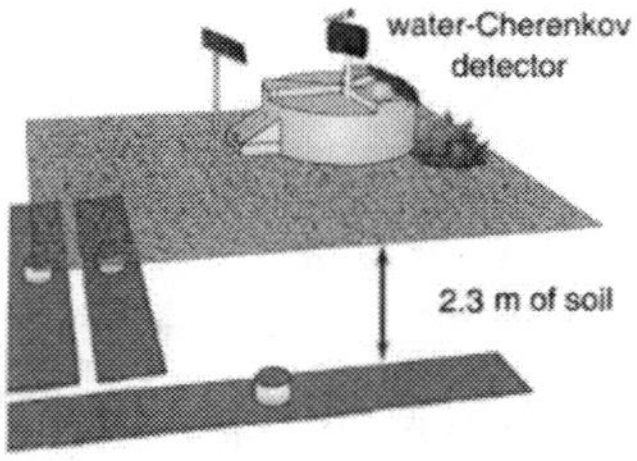


**Figure 2.** Schematics of the final UMD design with three modules of 10 $m^2$.

**Figure 3.** (Left) Three modules of 10 $m^2$ deployed in the field. (Right) Scintillator module constructed in the assembly building.

In the final design, the number of muons can be estimated in two complementary ways: *counting* and *integrated signal*. The first relies on the amplitude of the signals, the latter on its charge. The two modes work in a complementary way to extend the dynamic range of the detector: *counting* for low particle densities (far from the shower core) and *integrated signal* for high particle densities (close to the shower core).

In the *counting* mode the 64 SiPMs signals are handled independently through a pre-amplifier, fast-shaper, and a discriminator. A "1-bit" is produced if the signal goes above the discriminator threshold and "0-bit" otherwise. An example of a simulated single-muon signal [6] is shown in Figure 4 (top). This discriminator threshold is set at 2.5 photon equivalents (PE) to filter most of the SiPM dark current (due to the crosstalk between the SiPM cells). The calibration procedure [7] ensures an equalized response in the whole SiPM array.

The single-muon signal in the *counting* mode was characterized employing laboratory measurements. Muons deposit wider signals than the background in the binary trace, so muons can be identified as sequences of '1111x'. Muons are then identified by applying a time window starting from the first identified match, during which the muon search is inhibited. The *counting* strategy was proposed as a compromise between the under-counting due to the strip efficiency and over-counting due to both the detector noise and wide muon signals. An optimal inhibition window of 12 samples from the first match was selected, since it can completely contain approximately 99% of the muon signals.

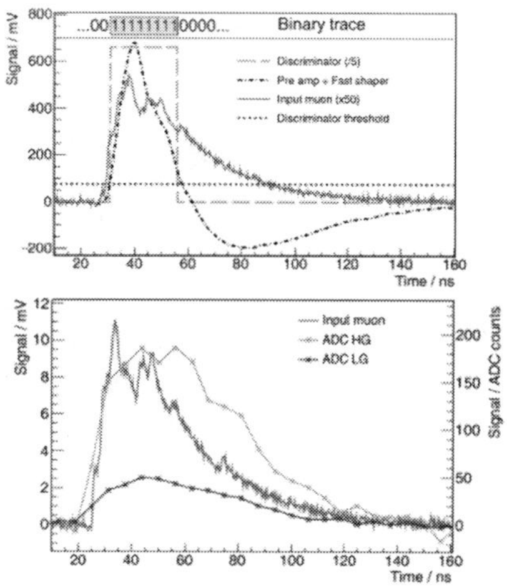


**Figure 4.** Two simulated single-muon signal at 2 m for the *Counting* mode (top) and for the *Integrated signal* mode (bottom). In the first case, the signal from the SiPM and the discriminator pulse are re-scaled for illustration.

In the *integrated signal* mode, the 64 SiPMs signals are summed [4], and the output is the total signal charge. The number of muons can be obtained by dividing the signal charge by the mean charge deposit of a single muon. Two independent channels are employed: a high gain (HG) and a low gain (LG) (see Figure 4 (bottom)). An algorithm is implemented in the electronics to extract the calibration histograms [8]. The calibration histograms are built for each channel selecting individual muons with the counting mode. An example is shown in Figure 5. The performance and calibration of the detectors are constantly being monitored to ensure high quality data.

The results obtained with the *counting* mode strategy from the UMD engineering array [9] will be shown in section 3. Currently, improvements have been made in the low level reconstruction of the *integrated signal* mode. Regarding the *counting* mode, a new *counting* strategy has been developed [10] that generates a smaller standard deviation of the estimated number of muons at high input number of muons and that reconstructs the time structure of the muon signal to a single time-bin resolution, opening the door to new studies of the temporal structure. UMDs

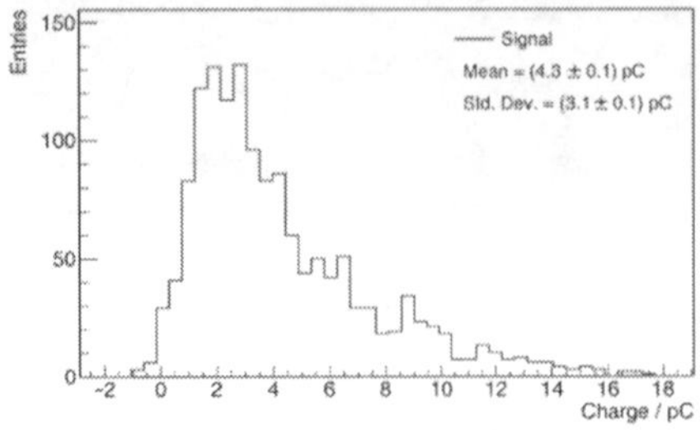


**Figure 5.** Example of a single muon charge histogram used to calibrate the *integrated signal* mode.

will significantly increase the amount of high quality data. These improvements will ensure high-level physics analyses with the data.

## 3 Results from engineering array

A data set of 1742 events measured by the UMD engineering array [9] was selected, with energies of $[2{\times}10^{17}, 2{\times}10^{18}]$ eV and $\theta \leq 45°$. These cuts allow us to minimize uncertainties and to work in a regime of full efficiency. The data used correspond to a 1-year period.

The deposited muon density $\rho(r)$ is modeled with a muon lateral distribution function (MLDF) at a perpendicular distance $r$ to the shower axis. The optimal distance to minimize the uncertainties of the signal is 450 m. An average MLDF is determined from Monte-Carlo simulations of extensive air showers and the signal deposited by an event is adjusted to this average by scaling its normalization in a fitting procedure. On an event-by-event basis, the value of $\rho(450)$ is estimated with a maximum likelihood. For a detailed description of the likelihood method, see reference [9]. An example of the MDLF for an event is shown in Figure 6.

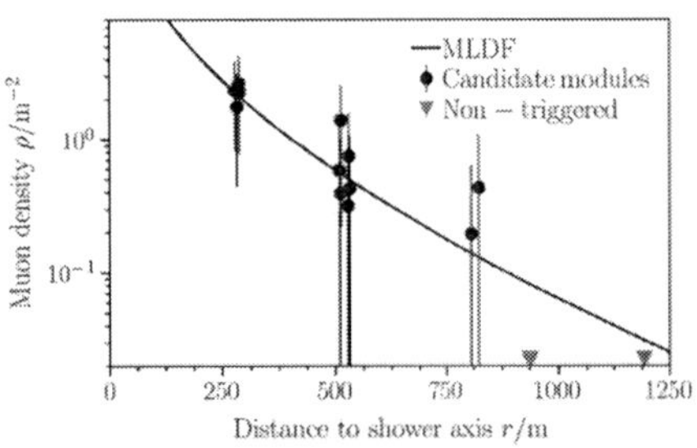


**Figure 6.** Example of a MLDF fit to a real event.

The muon density estimator is corrected for attenuation $\rho_{35} = \rho(450)/f_{\rm att}(\theta)$ (for the soil and the longer path in the atmosphere) based on the constant intensity cut method [11]. The reconstructed attenuation-free muon density $\rho_{35}$ is shown in Figure 7 as a function of the energy estimated by the SD. In this energy range, a single power-law dependence of $\rho_{35}$ with energy is expected from Monte-Carlo simulations. Hence, a fit of such a kind was performed and is also shown in Figure 7.

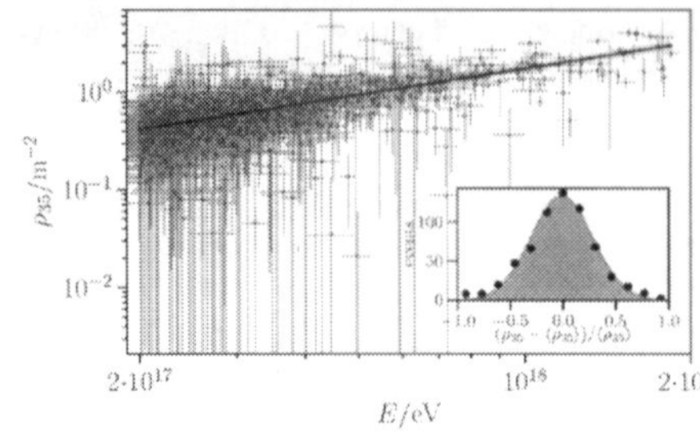


**Figure 7.** Experimental muon density estimator $\rho_{35}$ (dots) as a function of the reconstructed energy, and a single power-law fit (line). (Inset) The distribution of the normalized residuals.

The evolution of the muon content in data is compared to simulations of proton (red) and iron (blue) primaries in Figure 8. The muon densities are normalized by the energy. The number of events in each energy bin is stated at the top of the figure. The obtained fit curve is shown as the black solid line with a shaded band corresponding to the statistical uncertainties. The systematic uncertainties are indicated by square brackets, corresponding 14% to the SD energy estimate. The muon densities are larger in data than in the models. However, a shift of the data points within the systematic uncertainties (mainly inherited from the energy scale) is enough to bring them in the region of the iron primaries.

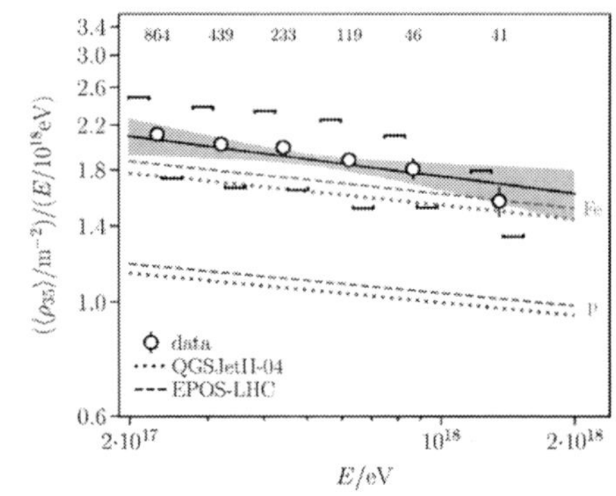


**Figure 8.** Energy-normalized muon densities compared to the expectations from simulations using QGSJetII-04 (dotted) and EPOS-LHC (dashed).

The depth of the shower maximum $X_{\rm max}$ can be directly observed with the fluorescence detectors (FD). In the generalised Heitler model, $< \ln \rho_{35} >$ and $< X_{\rm max} >$ are linear with the mean logarithmic mass $< \ln A >$. With the $< X_{\rm max} >$ data extracted from [12] we obtained Figure 9. For the EPOS-LHC model to explain the data, an increase of 38% in the muon density is required at $10^{17.5}$ eV and $10^{18}$ eV, while for QGSJetII-04, an increment of 50% and 53% is needed.

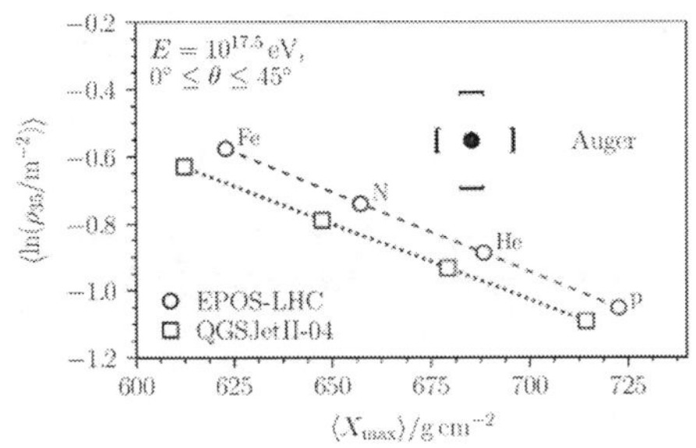


**(a)** $E = 10^{17.5}$eV

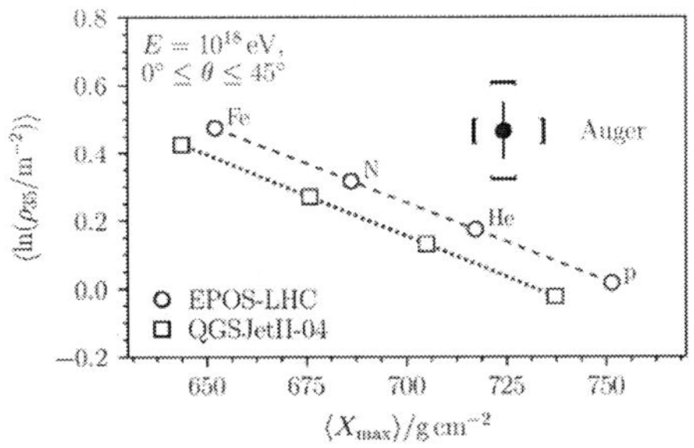


**(b)** $E = 10^{18}$eV

**Figure 9.** Mean logarithmic muon density $< \ln \rho_{35} >$ as a function of the mean depth of the shower maximum $< X_{\max} >$ for simulations with primary energies of $10^{17.5}$ eV (a) and $10^{18}$ eV (b) compared to Auger data.

The $z$-scale factor is defined by the following equation

$$z = \frac{< \ln x > - < \ln x >_{\mathrm{p}}}{< \ln x >_{\mathrm{Fe}} - < \ln x >_{\mathrm{p}}}, \tag{1}$$

where $x$ is the muon-density estimator and the symbols $< . >_{\mathrm{p}}$ and $< . >_{\mathrm{Fe}}$ represent the expected muon densities for proton and iron showers, simulated with a given model and accounting for detector effects.

We can extend previous results in inclined air showers at higher energies [13] with the $z$-scale factor, where $x$ is the muon-density estimator $\rho_{35}$ (in this work) and $R_\mu$ (in ref. [13]). The measurements of $< X_{\max} >$ by the FD can be converted into the mean logarithmic mass $< \ln A >$ assuming the validity of the superposition model and finally into the $z$-scale factor $z = \frac{<\ln A>}{\ln 56}$.

The hadronic interaction models fail to reproduce the $z$-scale factor consistently, as shown in Figure 10. This tension is higher in the lower energies (this work) due to the smaller uncertainties. However, the results are sensitive to a change in energies. A positive shift of our energy scale within its systematic uncertainties would reduce the tensions on the muon numbers.

## 4 Conclusions

In this work, we present the first direct measurement by the Pierre Auger Collaboration of the muon content of air showers at energies between $2{\times}10^{17}$ eV and $2{\times}10^{18}$ eV. We have shown that the hadronic interaction models indicate fewer muons are produced in simulated showers than those observed in extensive air showers. These results have extended previous results in inclined air showers at higher energies [13].

We have presented an overview of the final UMD design. Currently 37% of positions of the full UMD array have been installed and are in acquisition. The full UMD array will be completed by end-2023, increasing significantly the amount of UMD data available for performing high-level physics analyses.

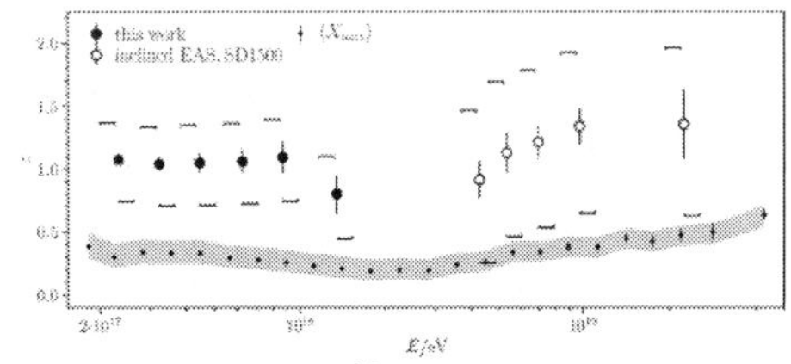


**(a)** EPOS-LHC

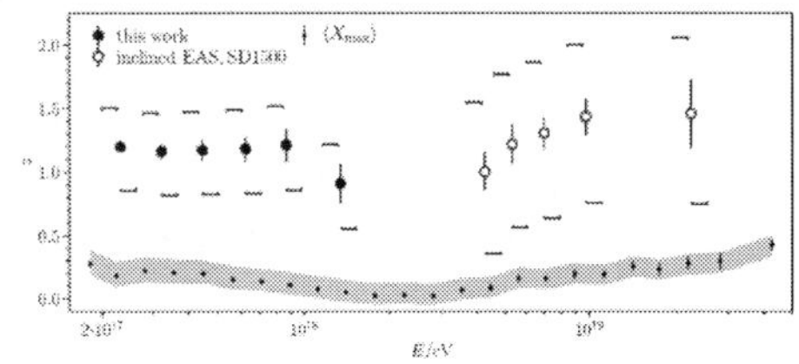


**(b)** QGSJetII-04

**Figure 10.** $z$-scale factor from this work (filled circles), from measurements of inclined air showers at higher energies (open circles) and from $< X_{\max} >$ measurements with the FD.

## References

[1] A. Aab et al. [Pierre Auger Coll.], Science Reviews **1**, 8 (2020)
[2] J. Matthews, Astropart. Phys. **22**, 387 (2005)
[3] A. Aab et al. [Pierre Auger Coll.], JINST **12**, P03002 (2017)
[4] A. Aab et al. [Pierre Auger Coll.], JINST **16**, P01026 (2021)
[5] A. Aab et al. [Pierre Auger Coll.], JINST **16**, T07008 (2021)
[6] A. Botti et al., JINST **16**, P07059 (2021)
[7] A. Botti, Proc. Int. Cosmic Ray Conf. **37**, 233 (2021)
[8] A. Aab et al. [Pierre Auger Coll.], JINST **16**, P04003 (2021)
[9] A. Aab et al. [Pierre Auger Coll.], Eur. Phys. J. C **80**, 751 (2020)
[10] F. Gesualdi, A. D. Supanitsky, Eur. Phys. J. C **82**, 925 (2022)
[11] J. Hersil et al., Phys. Rev. Lett. **6** , 22 (1961)
[12] J. Bellido, Proc. Int. Cosmic Ray Conf. **35**, 506 (2017)
[13] A. Aab et al. [Pierre Auger Coll.], Phys. Rev. D **91**, 032003 (2015)

# The XY-Scanner for Absolute End-to-End Calibration of Fluorescence Detectors

*Martin* Vacula[1,*] for the Pierre Auger Collaboration[2,**]

[1]Palacký University, Faculty of Science, Joint Laboratory of Optics of Palacky University and Institute of Physics of CAS, 17. listopadu 12, 771 46, Olomouc, Czech Republic
[2]Observatorio Pierre Auger, Av. San Martín Norte 304, 5613 Malargüe, Argentina
Full author list: https://www.auger.org/archive/authors_2022_10.html

**Abstract.** The precise determination of the energy scale is a key part of experiments in astroparticle physics. At the Pierre Auger Observatory, the energy scale is set by the calorimetric measurement of extensive air showers with fluorescence detectors. Thus, the absolute end-to-end calibration of the fluorescence detectors is of utmost importance. In the past, this calibration was performed by illuminating the whole optical system of a fluorescence telescope with a large-scale extended uniform light source of the same diameter as the telescope aperture. However, handling difficulties, excessive manpower requirements, and degradation of such a source led to the need for a different approach for the absolute end-to-end calibration. The fundamental idea of the novel approach is to significantly reduce the geometrical size of the calibration light source, which is a near-UV LED source implemented in a portable integrating sphere with specifically designed interior. This light source is moved over the aperture by a rail mechanism with two independent linear stages named the XY-Scanner. Calibration data are evaluated from a series of light source positions instead of illuminating the entire aperture at once. The absolute photometric determination of the light source emission intensity is performed in a dedicated laboratory setup with a measurement uncertainty of 3.5 %. The XY-Scanner mechanics installed at the aperture gives also the opportunity to install other, devices for instance a narrow, collimated beam source to investigate local impurities of the telescopes. This contribution gives an overview of this novel XY-Scanner calibration method and presents preliminary results and discusses plans for the future.

## 1 Introduction

The Pierre Auger Observatory [1] is located in Argentina near the city of Malargüe. It covers the area of ~3000 km$^2$ deployed with water-Cherenkov detectors (WCDs) forming the surface detector (SD). There are ~1660 WCDs evenly distributed among the area (the exception is the North-West region, where the layout is denser). The whole SD array is overlooked by the 27 telescopes. Six telescopes at Los Leones, Los Morados, and Loma Amarilla sites, and nine telescopes at Coihueco site. Those three additional telescopes at Coihueco are called HEAT (High Elevation Auger Telescopes). Altogether, they form the fluorescence detector (FD), see the layout of the observatory in Fig. 1. Both SD and FD utilize different approaches to detect cosmic-ray-induced air showers. With this hybrid approach, Pierre Auger Observatory is a precise and highly sensitive observing instrument.

However, every instrument has to be calibrated to provide reliable data. This contribution summarizes the current calibration method as well as describes a novel method for the absolute end-to-end calibration of Auger FD. The main goal of the calibration of FD is to verify/set the energy scale at the Pierre Auger Observatory.

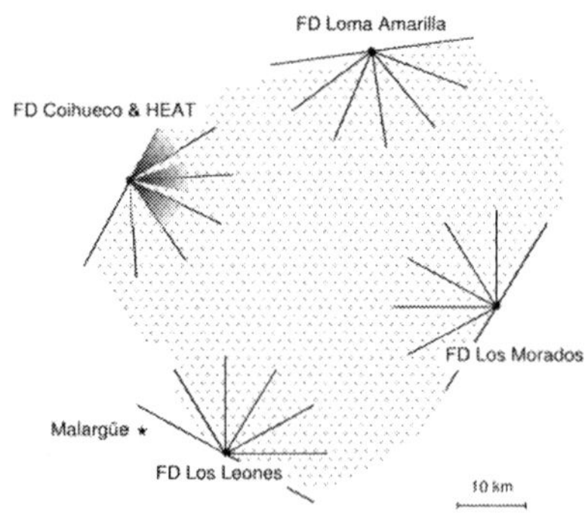


**Figure 1.** The layout of the Pierre Auger Observatory near the city of Malargüe. The SD array consisting of ~1660 WCDs is overlooked by the 27 FD telescopes at four sites.

## 2 The calibration of fluorescence detector

To perform the calibration task, all 27 FD telescopes need to be calibrated. The FD telescope [2] consists of a large spherical segmented mirror with a radius of 3400 mm. In the focal plane of the mirror, there is a camera with 440 PMT pixels. The whole aperture window of the telescope is 2200 mm in diameter. At the outer aperture perimeter, a

*e-mail: martin.vacula@upol.cz
**e-mail: spokespersons@auger.org

 https://doi.org/10.1051/epjconf/202328306013

corrector ring that provides a partial correction for spherical aberration of the mirror is installed. The aperture is filled with segments of near-UV filter glass transparent for wavelengths ranging from 290 nm to 410 nm.

To obtain the calibration constant for each individual PMT, the entire aperture has to be illuminated (in the ideal case) with a light pulse from an extended Lambertian source of known emission intensity (photon flux) located outside of the aperture. During the pulse, all PMT pixels of the camera are illuminated as well as FD telescope components. Thus, all of the components such as mirror reflectance, filter transmittance or quantum efficiency of the PMT, etc. contribute to the resulting signal detected on PMTs. The signal on individual PMTs is read out simultaneously at the time of the light pulse and the calibration constant is set according to the photon flux of the source.

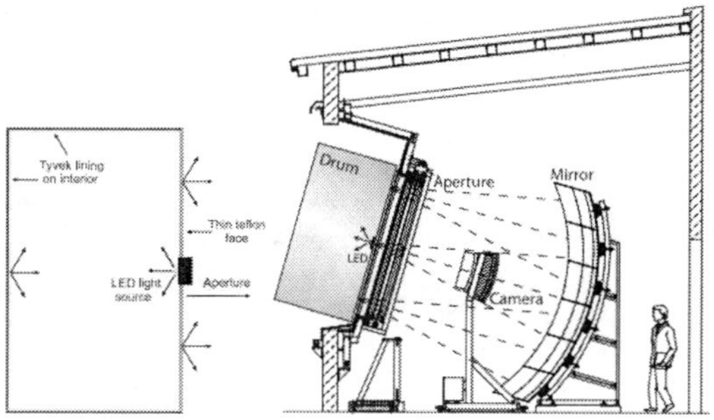


**Figure 2.** *Left:* The schematics of the Drum interior with the calibration LED source. *Right:* Schematics of the Drum absolute end-to-end calibration process [4].

### 2.1 Current calibration method

The current calibration method to determine the energy scale (performed in the past) is based on a unique design solution of an extended uniform light source (EULS) developed for calibrating Auger FD telescopes. According to its shape, it was called the Drum. In brief, the Drum was a hollow diffuse light source of a cylindrical shape with an outer diameter of 2800 mm, depth of 1500 mm, and the output aperture of 2500 mm. The sides and back surfaces of the Drum interior were lined with Tyvek, a material diffusely reflective in the near-UV spectral region. The front face of the Drum was made of a 0.38 mm thick Teflon sheet, which transmitted the light diffusely [2]. The primary point light source consisted of a near-UV LED with a nominal wavelength of 365 nm and a Teflon cylinder in front of the LED. Thus, uniformity was achieved by three diffuse components in the Drum. See Fig. 2 left. In addition, the Drum had an option enabling the calibration at multiple wavelengths within the region of 300-400 nm by means of a Xenon lamp with a series of 5 narrow-band optical filters.

During the calibration, the Drum was placed on the input aperture of the FD telescope and illuminated all 440 PMT pixels of the camera simultaneously, Fig. 2 right. Moreover, all individual components of the entire optomechanical structure of the telescope contributed to the calibration process. Therefore, this method of calibration is known as an absolute end-to-end calibration [3, 4].

Unfortunately, time has shown that the size and weight of the drum were very limiting in terms of its manipulation and maintenance. The absolute calibration of the Drum was rather cumbersome providing results with higher and higher uncertainties. In addition, the logistics of the Drum caused more uncertainties even within a single measurement campaign. In time, the calibration with the Drum has proved unsustainable and the need for replacement has become urgent.

### 2.2 Novel calibration method

The fundamental idea of the novel approach to absolute end-to-end calibration is based on the significant reduction of the size of the calibration light source, which reduces issues of the manipulation, weight, and absolute calibration of the new light source. However, with a smaller EULS, the aperture of the FD telescope will be illuminated partially instead of being illuminated at once. This is solved by summing every partial illumination, which in the end corresponds to the single flash of the Drum. The scanning of the FD aperture is done by moving the smaller calibration light source with a known photon flux (integrating sphere with an LED) on a motorized rail mechanism named XY-Scanner shown in Fig. 3. XY-Scanner utilizes two vertical stages (Y-axis) and one horizontal stage (X-axis). Vertical stages and their motor are permanently fixed at the aperture of every FD telescope. The horizontal stage with its motor is removable, portable, and easily mountable on a desired FD telescope aperture for calibration. On the horizontal stage, a universal holder to the calibration light source of a maximum weight of 2.5 kg can be mounted. Both motors and the calibration light source are connected via cables to the electronics box. The whole system is controlled by scripted software.

**Figure 3.** Photo of an FD telescope aperture with fully installed XY-Scanner. The Y axes and Y motor are permanently mounted at the aperture while the X axis with X motor, light source, and electronics box are portable.

Before each calibration measurement, a *Calibration A* reference measurement (consisting of an LED aiming directly at the PMT camera from the center of the FD telescope mirror) is performed [2]. During the calibration

 https://doi.org/10.1051/epjconf/202328306013

itself, a large number of scanning positions of triangular grid evenly distributed over the FD telescope aperture are measured, see Fig. 4. The calibration light source is moved across the FD telescope aperture following the black dots of the triangular grid with the sub-millimeter relative precision. The red circle is drawn at the border of the FD telescope aperture while the black-dashed circle shows the inner edge of the corrector ring. At each position, the calibration light source emits a 5 µs long light pulse of fixed intensity with the frequency of 1 Hz. This flashing frequency is limited by the read-out speed of the whole FD telescope camera. The calibration constant is obtained by integrating over all flashes from all measurement positions as long as the overall coverage of the FD aperture is sufficient.

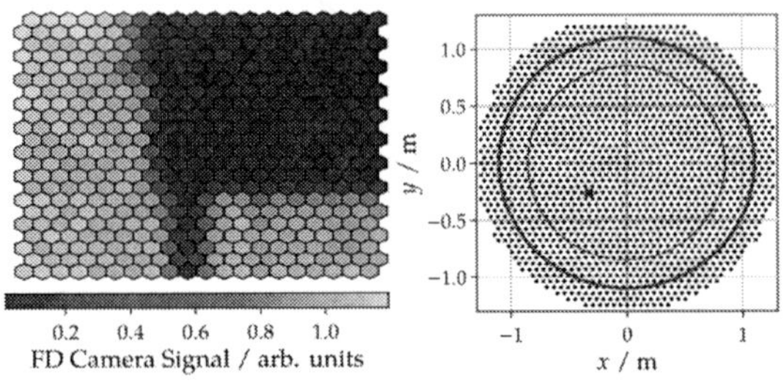


**Figure 4.** *Left:* an example of a signal detected at each PMT pixel for a single position of the calibration light source. The shadow of the camera and supporting structure are clearly seen. The position of the light source is marked as a blue spot on the right side of the figure. *Right:* a triangular grid (black dots) of all positions of the calibration light source at the FD telescope aperture.

The spacing of 6 cm between scanning positions results in ~1700 iterations that are performed within an hour of measuring time. With a calibration light source of 5.08 cm diameter, the total coverage of the FD telescope aperture is ~65 %, which is sufficient at the moment according to the preliminary measurements and simulations. The coverage could be possibly lowered in order to decrease the calibration measurement time. After the measurement run, another *Calibration A* reference is performed to determine the overall stability of PMT pixels over the calibration process. If they are stable, the calibration measurement is considered valid. Otherwise, it has to be discarded. Based on the previous results, an average of ~1 % change in PMT signal is observed between the two measurements.

The up-to-date status by the end of 2022 after the autumn campaign is 19 fully installed XY-Scanner systems. The installation of the remaining 8 XY-Scanners (4 at the Los Leones site and 4 at the Loma Amarilla site) is planned to be completed in early 2023.

# 3 Light sources compatible with XY-Scanner

The versatile nature of the XY-Scanner allows the use of various light sources for different purposes. At the moment, the most important is the portable EULS that serves as a tool for the end-to-end calibration of the FD telescopes allowing them to set the calibration constant for their photomultiplier tubes (PMTs). While the requirement on the calibration source is the highest spatial and angular radiance uniformity possible, there is also another light source that utilizes a collimated beam and can be used with the XY-Scanner as well. Its purpose is to investigate the long-term stability of the local properties of the FD telescope.

## 3.1 XY-Scanner calibration light source

The calibration light source consists of a modified portable general-purpose integrating sphere 3P-GPS-053-SL with a diameter of 13.5 cm [5] and of a temperature-stabilized LED head holding a light source. The issue of integrating spheres as EULSs is summarized in [6]. The light source is a near-UV LED with a peak wavelength of 365 nm covered with an opal glass diffuser to approach the Lambertian emission profile. To achieve the uniform emission profile of the whole light source (integrating sphere with an LED), an exit port is reduced to a diameter of 5.08 cm (see Fig. 5). Moreover, a UV transparent plexiglass cover is mounted at the exit port to prevent any mechanical damage or dust pollution of the integrating sphere interior due to its use in the outdoor environment. The whole light source is mounted on the XY-Scanner using a universal holder enabling the rotation of the source along its optical axis with 15° stepping.

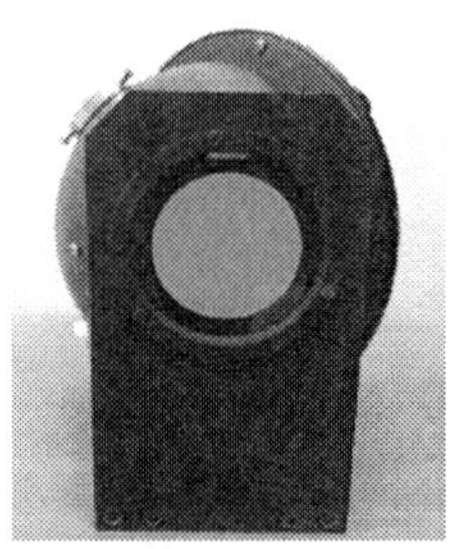

**Figure 5.** Front view on the integrating sphere exit port covered with the plexiglass. The integrating sphere is mounted in a universal holder.

To determine the far-field emission profile of the calibration light source a laboratory setup is developed. The light pulse of the calibration light source is detected by a 3.8 cm PMT located at 5 m distance from the source (the distance corresponds to the FD telescope dimensions), which is rotated using two motorized rotation stages. The setup is built in a dark room. Preliminary results of the measurement are shown in Fig. 6. These results are compared with the cosine emission profile of the ideal Lambertian source. The angular area of interest corresponding to the field of view of the Auger FD telescope is from −16° to 16°. The angle $\beta$ represents the rotation of the source around the optical axis.

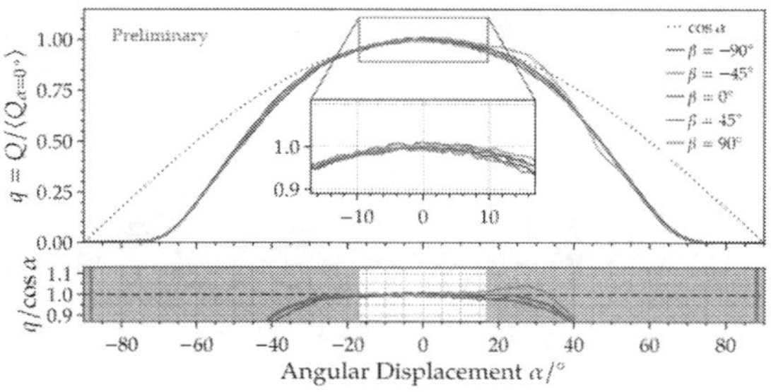


**Figure 6.** The emission profile of the calibration light source. Angular displacement area of interest corresponding to the field of view of Auger FD telescope is from $-16°$ to $16°$.

### 3.2 Optical collimator

The optical collimator is a device developed to further study the optical properties of the FD telescope such as local impurities or degradation of the FD telescope components. The impurities (for example the level of dust pollution) can be investigated at the UV filter, corrector ring, and mirror segments. The ideal spot of the emitting light beam has to fit in a single PMT after passing through transmissive components and after reflecting on the mirror. Any other detection on the neighboring PMTs indicates the higher degradation resulting in higher scattering. Thus, the emitting light beam has to be narrow with minimal scatter component.

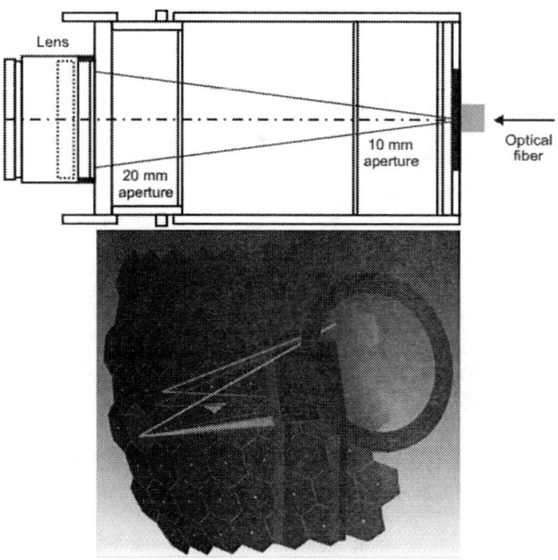


**Figure 7.** *Top:* the schematics of the optical collimator. *Bottom:* the layout of the optical simulation showing several light beams of the optical collimator reflecting on the FD telescope mirror.

The Fig. 7 (top) shows the schematics of the designed optical collimator that consists of an air-spaced doublet lens (focal distance of 99.9 mm and numerical aperture of 0.14) located at the front of a non-reflecting cylindrical tube with a diameter of 5.08 cm. Inside the tube, there are two apertures with diameters of 20 mm and 10 mm, respectively. The light is guided to the very back of the tube by an optical fiber with a core diameter of 0.1 mm and a numerical aperture of 0.1 resulting in a low divergence beam. The light source is a fiber-coupled LED with a peak wavelength of 365 nm.

Fig. 7 (bottom) shows the layout of the optical simulation with a collimated light beam reflecting on the FD telescope mirror and impacting the PMTs at the focal plane of the FD telescope. The optical collimator can be further used to study the point spread function (PSF) of the telescope affected by the FD telescope components. Either locally or for the whole aperture using the same principle of the end-to-end calibration utilizing the XY-Scanner.

## 4 Conclusion

The contribution summarizes the current status of the calibration of Auger FD telescopes and presents the XY-Scanner as a novel method for absolute end-to-end calibration of FD telescopes. The novel method utilizes the motorized rail system of two vertical axes and one horizontal axis, where a portable light source can be mounted. Furthermore, the contribution describes two of the possible sources designed for different purposes. One for the calibration itself and the second for the study of local properties of the FD telescopes based on PSF analysis.

At the moment, there are 19 XY-Scanner systems fully installed at Pierre Auger Observatory. The aim is to finish the installation for all 27 (4 remaining at the Los Leones site and 4 at the Loma Amarilla site) FD telescopes during early 2023. Regarding the calibration itself, the recent measurements show promising results that will be verified during the next campaign and discussed in 2023.

## Acknowledgement

The Czech authors gratefully acknowledge the support from the project IGA_PrF_2023_005 of Palacký University, and from the project of Ministry of Education, Youth and Sports of the Czech Republic no. LTT 18004. Czech authors also gratefully acknowledge the support of the Czech Academy of Sciences and Japan Society for the Promotion of Science within the bilateral joint research project with Kyoto University (Mobility Plus project JSPS 21-10).

## References

[1] A.Aab et al. [Pierre Auger Coll.], Nucl. Instrum. Meth. A **798**, 172 (2015)

[2] A.Aab et al. [Pierre Auger Coll.], Nucl. Instrum. Meth. A **620**, 227 (2010)

[3] J. T. Brack et al., Astropart Phys. **20**, 653 (2004)

[4] J. T. Brack et al., JINST **8**, P05014 (2013)

[5] Labsphere Inc., 231 Shaker St., North Sutton, NH 03260, USA, (2008), www.labsphere.com

[6] M. Vacula et al., Optik **242**, 18 (2021)

# The Pierre Auger Observatory: studying atmospheric electricity with cosmic-ray detectors

*Roberta* Colalillo[1,2,*] *for the Pierre Auger Collaboration*[3,**]

[1]Dipartimento di Fisica "Ettore Pancini", Università degli Studi di Napoli Federico II, via Cintia 80126 Napoli, Italy
[2]INFN, Sezione di Napoli, via Cintia 80126 Napoli, Italy
[3]Full author list: https://www.auger.org/archive/authors_2022_10.html

**Abstract.** The Pierre Auger Observatory, designed to detect extensive air showers (EAS) generated by ultrahigh-energy cosmic rays, proved to be a unique instrument to study phenomena associated with atmospheric electricity: Terrestrial Gamma-ray Flashes (TGFs) and ELVES. TGFs are the most explosive events emitted by thunderstorms and have usually been observed by spacecraft, but the Auger surface detector has collected events with a morphology very different from that of EAS, that are likely downward TGFs. Ground observations are a crucial addition to detection in space, as detectors are closer to TGF sources, and a single event can be observed with several complementary instruments. In this talk, I will discuss the comparison of the already collected data with simulations as well as the ongoing work to adapt the data-taking system to detect TGFs. I will also discuss the status of the study of ELVES, upper atmospheric optical phenomena associated with thunderstorms, which we have detected since 2013 with the fluorescence detectors. We have recorded with unprecedented time and space resolution both single and double ELVES, the latter being classified using the radial variation of the time gap and the photon flux ratio between flashes. The detection and classification algorithms were improved by detecting ELVES from closer lightning thanks to the three High Elevation Auger Telescopes, which observe the night sky with an enhanced time resolution at elevation angles between 30 and 60 degrees.

## 1 Introduction

Thunderstorms are the most energetic natural particle accelerators on Earth [1] producing very bright events in coincidence with lightning. The Pierre Auger Observatory [2], despite its main goal being the detection of ultrahigh-energy cosmic rays, proved to be a unique instrument to study these phenomena, in particular ELVES and downward TGFs.

The Observatory, located in the Argentinian Pampa, is a hybrid system of a giant surface array of particle detectors (SD) coupled with air fluorescence telescopes (FD). When a cosmic ray crosses the atmosphere, it produces an extensive air shower interacting with atmospheric molecules. The surface detector, which consists of 1660 water-Cherenkov detectors (WCDs), samples the footprint of the air showers as they hit the ground, while the FD, with its 27 telescopes distributed over four sites on the perimeter of the SD, records the evolution of the showers in the atmosphere.

In 2005, the FD captured by chance the first Emission of Light and Very low-frequency perturbations due to Electromagnetic pulse Sources (ELVES), and now the Observatory has a dedicated trigger to study ELVES with unprecedented time and space resolution.

The SD also observed, serendipitously, peculiar events during thunderstorms. Their morphology, as well as the signals observed in the detectors, are totally different from what is observed when an extensive air shower strikes the array [5]. Their correlation with lightning activity, their time evolution, and their huge deposited energy at the ground suggested that they are most likely downward TGFs [7].

In the following sections, the status of the two analyses will be described.

## 2 Downward Terrestrial Gamma-ray Flashes

TGFs are millisecond pulses of gamma rays usually observed from space [3]. A "standard" upward TGF produces about $10^{17}$ gamma rays of energy $> 1$ MeV in the primary beam. We expect a similar beam for downward TGFs, but only a handful have ever been detected on the ground. Recently, the knowledge of downward TGFs has been increasing thanks to an increase in observations, and the Pierre Auger Observatory is contributing to this effort.

A typical footprint of a TGF-like event observed at the Auger Observatory is shown in panel "a" of figure 1. It covers about 200 $km^2$. Most triggered stations are the so-called "long-signal stations" (upper plot in panel "b" of figure 1). The typical signal of TGF-like events lasts more

*e-mail: colalillo@na.infn.it
**e-mail: spokespersons@auger.org

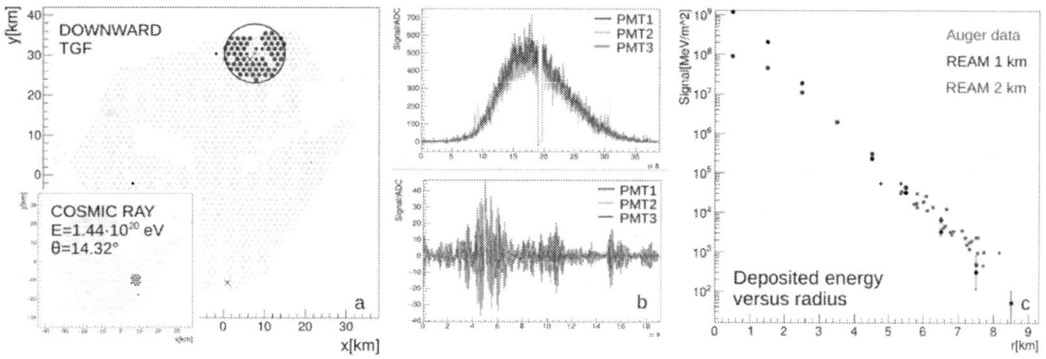


**Figure 1.** **Panel "a"**: the large footprint of a TGF-like event (blue dots) compared to the footprint of an extensive air shower (black dots). **Panel "b"**: the typical long signal of a TGF-like event is shown in the upper plot. This specific trace is the combination of the two long signals collected in two consecutive events. The time evolution of the TGF-like events is longer than that of a cosmic-ray shower. The lower plot shows the so-called lightning signal, whose presence in each TGF-like event suggested their correlation with lightning. **Panel "c"**: deposited energy at the ground as a function of the distance from the center of the footprint compared with the expectation for a standard downward TGF.

than 10 $\mu$s, at least one order of magnitude longer than the duration of the signal produced by cosmic-ray showers. The WCDs have high efficiency for gamma-ray detection, and Auger is observing the brightest events ($\sim$ 1000 MeVm$^2$) at the ground connected with lightning activity. We are probably underestimating the deposited energy at the ground because the signal is saturated in several cases, the time acquisition window is not long enough to detect the whole trace and, in particular, because there is a lack of signal in the central part of the footprint. It was verified that this signature, as well as the low rate of TGF-like events (2 events/year), has not a physical origin [7], but is due to the Auger trigger and acquisition chain optimized for cosmic-ray events. An algorithm to tag long signals has been recently developed using differences of integrals performed over predefined parts of the trace and it is now installed in the whole array [8]. We expect to increase the detected rate modifying the read-out logic of the Central Data Acquisition System to prioritise events containing long signals. At least 30 events/year are expected according to the upward TGF frequency. With a detector closer to the source, a population of slightly fainter TGFs could be discovered. Systematic measurements of downward TGFs are necessary to understand this phenomenon better.

It is known that TGFs are produced by the Relativistic Runaway Electron Avalanche (RREA) process. When a "seed electron", travelling in a thundercloud region that contains a strong electric field, gains more energy than it loses predominately due to ionization, we say that it "runs away" and activates the electron avalanche, but it is not yet clear what provides the "seed electron". The two main models, "Lightning Leader" and "Relativistic Feedback", can be constrained by studying the rising and falling edge of the long signals and also the footprint at the ground because they could produce a different angular distribution. A first comparison between data and simulations was performed. In panel "c" of figure 1, the deposited energy per m$^2$ of a TGF-like event as a function of the distance from the center of the footprint is shown together with the expectation for a standard downward TGF with an isotropic emission into the lower hemisphere and the source at 1 and 2 km above the ground [6]. This height is compatible with the source height obtained fitting the signal arrival time in the Auger detectors assuming a spherical propagation [5]. Simulations were produced using REAM [4], a Monte Carlo simulation of the avalanche region. Nothing can be said about the source models because they predict the same spectrum, but we can conclude that the deposited energy at the ground of the Auger TGF-like events is compatible with the one of a standard TGFs. More information about the source altitude will be available if the central signal is recovered thanks to the new algorithm for long signals.

## 3 Emission of Light and Very Low Frequency perturbations due to Electromagnetic Pulse Sources

ELVES are transient luminous events occurring at the base of the ionosphere triggered by lightning in the troposphere [9]. They appear as rapidly expanding rings. More than 95% of the observed ELVES at the Auger Observatory are 250–1000 km away, where the FoV of a telescope crosses the ionosphere, and direct light from lightning is blocked by the limb of the Earth. Thanks to the Auger FD time resolution (100 ns), the temporal structure in the light emission of ELVES can be observed very finely, and data can be sorted into two categories, single-peaked and multipeaked ELVES [11], studying the phototraces collected by the PMTs of the FD. During the period 2014–20, about 27% of the events detected at the Auger Observatory were multi-ELVES, in particular with double and triple peaks. It is known that ELVES are produced by the ElectroMagnetic Pulse (EMP) emitted during the lightning re-

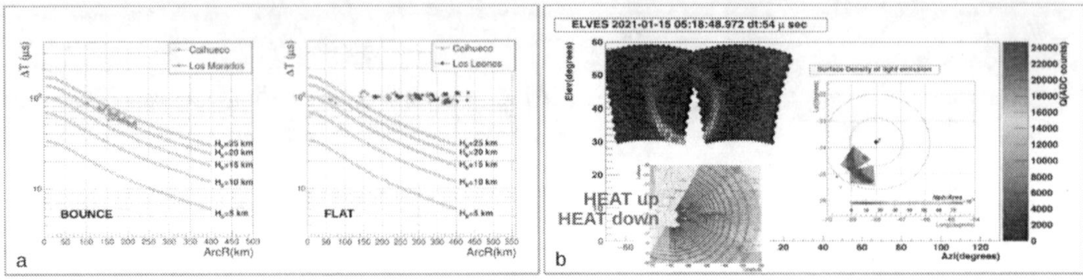


**Figure 2. Panel "a"**: classification of double ELVES according to the time gap between the two pulses vs arc distance, which is the distance on the emission layer between the projection of the position of the lightning triggering the ELVES and the region where the EMP is converted into the UV light which we observe in a given pixel of our camera. **Panel "b"**: bottom left - projection of each camera pixel at the base of the ionosphere for HEAT telescopes in their configuration "up" and "down". In the configuration "up", it is possible to observe ELVES closer to the Observatory; top left - an ELVES ring shape in the HEAT camera; right - ELVES light density at the emission layer.

turn stroke or in the initial breakdown stages of the discharge, but the production mechanism for two or more consecutive ELVES rings in less than 1 ms is not clear yet. By studying the time gap between pulses in each pixel, we found that only a small fraction of double ELVES can be explained as ground reflection [10] ("bounce" in figure 2) of the EMP. The study of the ratio between the intensity of the two pulses shows a variety of patterns, suggesting that multiple ELVES are due to several causes [13].

The ELVES trigger was also implemented in the three High Elevation Auger Telescopes (HEAT). As these three detectors can be tilted to an elevation angle of 45°, the light of closer lightning does not reach their camera and ELVES triggered by lightning closer to the Observatory can be studied [12], improving the knowledge of the region of maximum emission, which has a radius of about 80 km around the vertical of the lightning. Moreover, HEAT electronics is faster than the one of standard FD telescopes, and a time resolution of 50 ns can be reached. In the right panel of figure 2, an ELVES observed by HEAT telescopes is shown.

Another important information about ELVES origin was obtained by analyzing the ELVES correlated with lightning located by GLD360, which provides current and polarity for each strike. Double ELVES are more frequently generated in negative lightning strikes [12]. The comparison of ELVES light curves with the signals from ENTLN antennas [14], deployed in Argentina in 2018, will allow us to better investigate this aspect. ENTLN provides high bandwidth data, and so an accurate discrimination between cloud-to-ground and intracloud lightning.

## References

[1] J.R. Dwyer, M.A. Uman, Physics Reports **534** 147-241 (2014)

[2] Pierre Auger Coll., Nucl. Instrum. Meth. A **798** 172–213 (2015)

[3] J.R. Dwyer, D.M. Smith, and S.A. Cummer, Space Sci. Rev. **173** 133–196 (2012)

[4] J. R. Dwyer, Geophys. Res. Lett. **30(20)** 2055 (2003) - J.R. Dwyer and D.M. Smith, Geophys. Res. Lett. **32** L08811 (2005)

[5] R. Colalillo [for the Pierre Auger Coll.], PoS(ICRC2017)**314**

[6] R. Colalillo [for the Pierre Auger Coll.] and J.R. Dwyer, EGU22-12119, https://doi.org/10.5194/egusphere-egu22-12119 (2022)

[7] R. Colalillo [for the Pierre Auger Coll.], PoS(ICRC2021)**395**

[8] M. Schimassek [for the Pierre Auger Coll.], J. Phys.: Conf. Ser. **2398** 012003 (2022)

[9] U.S. Inan et al., Geophys.Res.Lett. **18** 705 (1991)

[10] N. Liu et al., J. Geophys. Res. **122** 10563 (2017)

[11] Pierre Auger Coll., Earth Space Sci. **7** e2019EA000582 (2020)

[12] A. Vásquez-Ramírez [for the Pierre Auger Coll.], PoS(ICRC2021)**327**

[13] R. Mussa [for the Pierre Auger Coll.], J. Phys.: Conf. Ser. **2398** 012006 (2022)

[14] https://www.earthnetworks.com

EPJ Web of Conferences **283**, 06015 (2023)
*UHECR 2022* https://doi.org/10.1051/epjconf/202328306015

# Performance of the TALE infill experiment as a TA-TALE extension down to the PeV region

*Aoi* Iwasaki[1,*], *Keitaro* Fujita, *Shoichi* Ogio[2,**], *Toshihiro* Fujii[1,***], and *Yoshiki* Tsunesada[1,****] for the Telescope Array Collaboration

[1]Osaka Metropolitan University
[2]University of Tokyo

**Abstract.** The TALE infill experiment is a further extension of TA-TALE detectors to observe low-energy cosmic rays down to the PeV region. TALE infill utilizes the existing TALE-FD detectors, and newly developed "infill" surface detectors with 100 m and 200 m spacing. The new detectors will be deployed at the TALE site in October-November 2022. We present the design and performance of the TALE infill array in the hybrid mode, in terms of the resolutions and biases of arrival direction, energy, and $X_{\max}$.

## 1 TALE infill : Science goals

This is the first experiment in the world to observe low-energy cosmic rays using both a Surface Detector (SD) and Fluorescence Detector (FD). By using TALE infill hybrid, we want to accurately measure mass composition in the low energy range, where results vary from experiment to experiment(Fig.1), and unravel the knee structure which is thought to be the energy range where cosmic rays transition from lighter nuclei to heavier nuclei. Therefore, it is necessary to distinguish at least protons and iron in the PeV region.

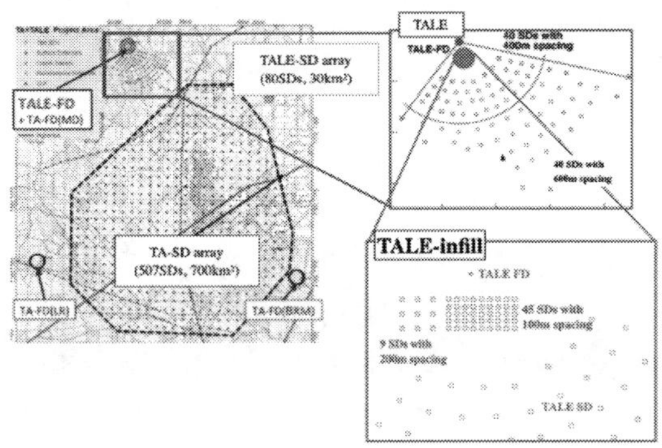


Figure 2: The layout of the TA detector, the TALE detector, and the TALE infill detector. TALE infill SDs, represented by purple square, are located between TALE-FD and TALE-SDs.

## 2 Detectors

### 2.1 TALE infill SD array

The TALE-SDs are located northwest of the TA-SDs. To observe cosmic rays with energies lower than the TALE experiment, the TALE infill SDs are deployed closer to the TALE-FD and denser than the TALE-SDs. TALE infill SDs have 100 m spacing area and 200 m spacing area. By creating a high density area, it's possible to observe low energy CRs. By creating a low density area, it's possible to eliminate energy gap between the TALE hybrid experiment.

*e-mail: m21sa004@st.osaka-cu.ac.jp
**e-mail: sogio@icrr.u-tokyo.ac.jp
***e-mail: o22412a@omu.ac.jp
****e-mail: tsunesada@omu.ac.jp

### 2.2 Hybrid observation

FD observes the air fluorescence and cherenkov photons to determine the longitudinal development of air showers. Since September 2013, TALE-FD has been operational with 10 telescopes. TALE-FD station has a field of view from 30° to 57° in the Elevation direction and 114° in the azimuth direction. On the other hand, SD measures the lateral distribution of air shower charged particles. TALE-infill SD was created in October 2021. We will deploy that in October-November 2022. By using both information on longitudinal and ground, the hybrid observation has better resolution than other observation methods.

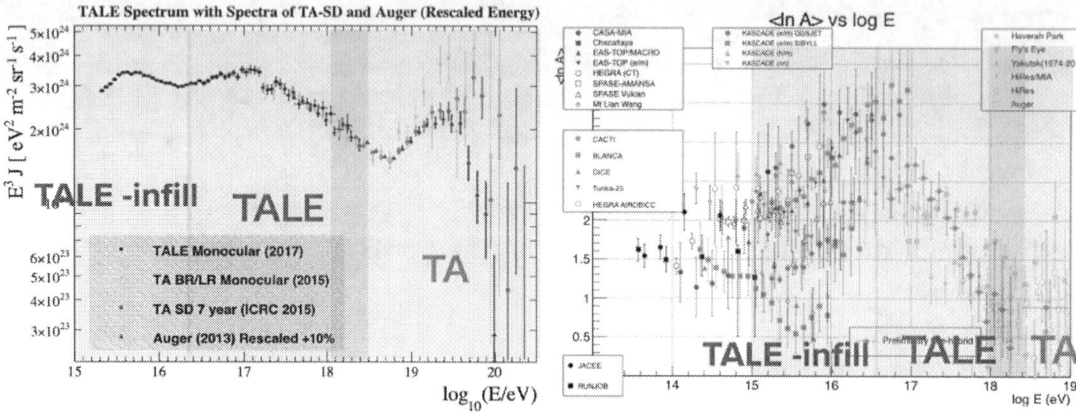


Figure 1: Energy spectrum (left) and composition (right) from the knee region onwards. Measurements of $\langle X_{\max}\rangle$ with direct observation(JACEE [1], RUNJOB [2]) are shown in black. Experiments using chamber light obesrvation(CACTI [3], BALANCA [4], DICE [5], Tunka-25 [6], HEGRA AIROBICC [7]) are shown in orange. Purple represents $\langle X_{\max}\rangle$ by the experiments observing around the knee range(CASA_MIA [8], Chacaltaya [9], EAS_TOPMACRO [10], EAS_TOP(em) [11], HEGRA(CT) [12], SPASE_AMANSA [13], SPASE Vulcan [14], Mt Lian Wang [15]), and light blue represents $\langle X_{\max}\rangle$ by the KASCADE experiment group(KASCADE(em) QGSJET [16], KASCADE (em) SIBYLL [16], KASCADE(hm) [17], KASCADE(nn) [18]), green represents $\langle X_{\max}\rangle$ by the experiments observing ultra high energy range(Haverah Park [19], Fly's Eye [20], Yakutsk [21], HiResMIA [22], HiRes [23], Auger [24]), and pink represents $\langle X_{\max}\rangle$ by the TA hybrid. The blue, red and green regions represent the observed energies of the TA, TALE, and TALE infill experiments respectively.

(a) The Fluorescence Detector (FD).

(b) The Surface Detector (SD).

Figure 3: (a) : The Middle Drum station for TA experiment is on the left, and the TALE-FD station on the right. (b) : The scintillator under the brown iron observes the charged particles.

## 3 Simulation

We simulated air showers by the CORSIKA for proton and iron primaries. The energy is $10^{15.2}$ eV, $10^{15.4}$ eV, $10^{15.6}$ eV, $10^{15.8}$ eV, $10^{16.0}$ eV, $10^{16.2}$ eV and $10^{16.5}$ eV (fixed for each energy). Zenith angle and Azimuthal angle ranges are 0° to 60° and 0° to 360° with uniformly random distributions, respectively. The core position is uniformly random distributed within a semi-circle of 1.4 km radius shown in Fig.4. All of the calibration factors with time dependence are applied to SD and FD detector simulations.

We select events that satisfy the quality cuts conditions in Table.1 from those that pass reconstruction, in order to remove poorly reconstructed events and ensure good detector resolution. We define fluorescence events as fractional contribution to the total signal of Fluorescence Light (FL) > 0.75, and Cherenkov events as fractional contribution to the total signal of FL $\leq$ 0.75.

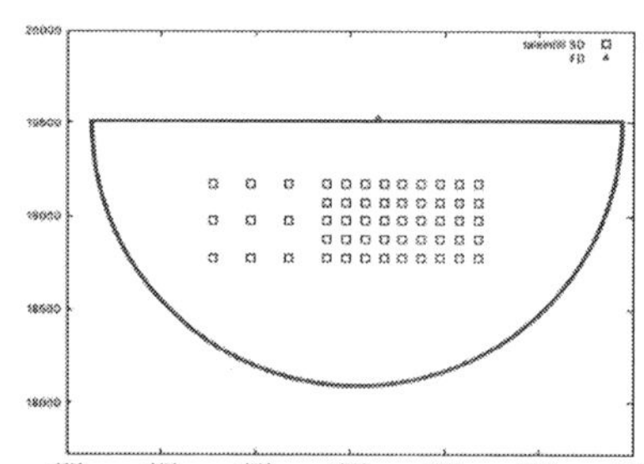

Figure 4: Core location distribution of Monte Carlo showers.

## 4 Event distribution @$10^{16.0}$ eV

An example of event distribution that passed reconstruction are shown in Fig.5. The reason for the small number of events with azimuths of 150° is that the shutters of the FD stations viewing this angle were closed at the time of the simulation.

 https://doi.org/10.1051/epjconf/202328306015

| Variable | CL | FL |
|---|---|---|
| No saturated PMTs in FD | applied | |
| SD detected ≥ 3 MIPs | applied | |
| $X_{max}$ in TALE-FD viewing angle | applied | |
| Angular track-length [deg] | > 8° | - |
| Event duration [$\mu$s] | > 0.1 $\mu$s | - |
| # of PMTs | > 10 | - |
| # of Photo-electrons / # of PMTs | > 50 | - |
| # of Photo-electrons | - | > 2000 |

Table 1: quality cuts applied in this study.

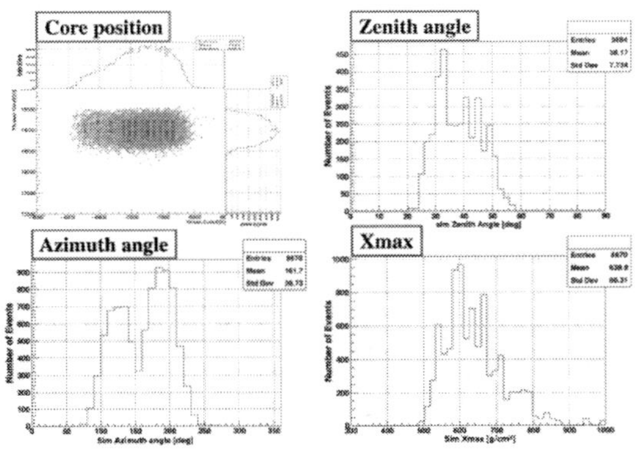


Figure 5: The event distribution of core position (upper left), zenith angle (upper right), azimuth angle (lower left), and $X_{max}$ (lower right).

## 5 Parameter resolution

The resolution of $X_{max}$ and Energy are shown in Fig.6.

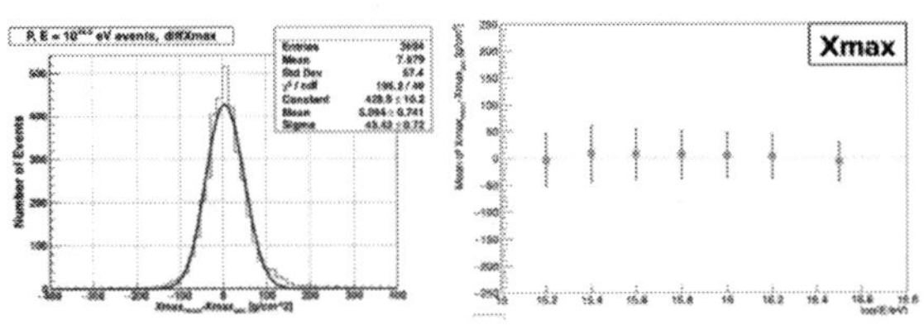


(a) The resolution of $X_{max}$(5.1 ± 43.4 g/cm$^2$ at $10^{16.0}$ eV).

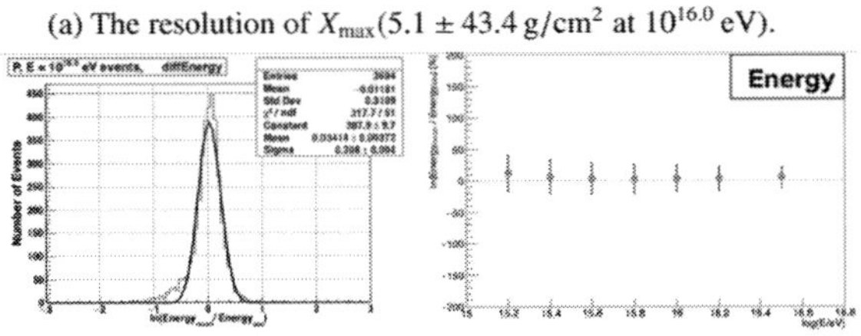


(b) The resolution of Energy(3.4 ± 20.8 % at $10^{16.0}$ eV).

Figure 6: The resolution of parameter. The plots and error bars on the right are the mean and sigma obtained by figging the histogram on the left with the Gaussian function.

## 6 Mean $X_{max}$ bias

Fig.7 shows thrown $\langle X_{max}^{true} \rangle$ without any reconstructions, triggered $\langle X_{max}^{true} \rangle$, reconstructed $\langle X_{max}^{true} \rangle$, and reconstructed $\left\langle X_{max}^{reconstructed} \right\rangle$ with quality cuts using TALE infill hybrid. The $\langle X_{max} \rangle$ bias is estimated from difference between thrown $\langle X_{max}^{true} \rangle$ and reconstructed $\left\langle X_{max}^{reconstructed} \right\rangle$ with quality cuts.The maximum value of the bias os 50 g/cm$^2$. The mass composition measurement of CRs at the knee region has been validated from this simulation studies.

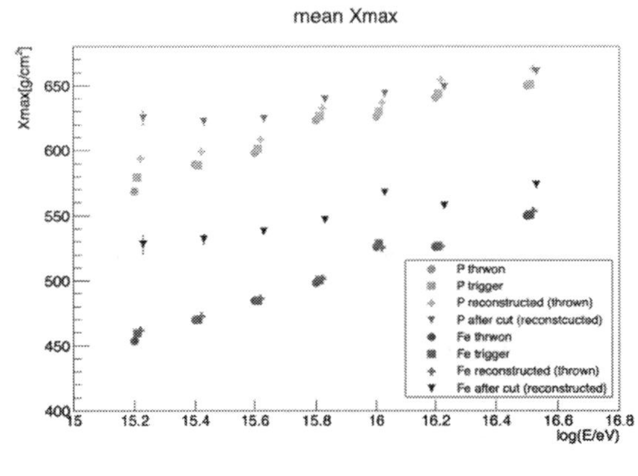


Figure 7: The $\langle X_{max} \rangle$ bias for proton (orange) and iron(blue). These points represent thrown $\langle X_{max}^{true} \rangle$ (circle), triggered $\langle X_{max}^{true} \rangle$ (square), reconstructed $\langle X_{max}^{true} \rangle$ (plus), and reconstructed $\left\langle X_{max}^{reconstructed} \right\rangle$ with quality cuts (inverse triangle).

## 7 Future work

Considering the bias of the parameter resolution and mean $X_{max}$, the $X_{max}$ resolution is better than 40 g/cm$^2$ by optimizing the quality cuts.

## acknowledgements

The TALE SD production and the TALE hybrid operations are supported by the Japan Society for the Promotion of Science(JSPS) through Grants-in-Aid for Scientific Research (S) 15H05741 and 19H05607; by the joint research program of the Institute for Cosmic Ray Research (ICRR), The University of Tokyo. The experimental site became available through the cooperation of the Utah School and Institutional Trust Lands Administration (SITLA), U.S. Bureau of Land Management (BLM), and the U.S. Air Force. We appreciate the assistance of the State of Utah and Fillmore offices of the BLM in crafting the Plan of Development for the site. The people and the officials of Millard County, Utah have been a source of steadfast and warm support for our work which we greatly appreciate. We gratefully acknowledge the contribution from the technical staffs of our home institutions.

 https://doi.org/10.1051/epjconf/202328306015

## References


[1] Y. Takahashi, Nuclear Physics B - Proceedings Supplements **60**, 83 (1998)

[2] V.A. Derbina et al. (RUNJOB), Astrophys. J. Lett. **628**, L41 (2005)

[3] S. Paling et al., ,, in *25th International Cosmic Ray Conference* (1997), p. 253

[4] J.W. Fowler, L.F. Fortson, C.C.H. Jui, D.B. Kieda, R.A. Ong, C.L. Pryke, P. Sommers, Astropart. Phys. **15**, 49 (2001), astro-ph/0003190

[5] S.P. Swordy, D.B. Kieda, Astropart. Phys. **13**, 137 (2000), astro-ph/9909381

[6] D. Chernov et al., Int. J. Mod. Phys. A **20**, 6799 (2006), astro-ph/0411139

[7] F. Arqueros et al. (HEGRA), Astron. Astrophys. **359**, 682 (2000), astro-ph/9908202

[8] M.A.K. Glasmacher et al., *"The cosmic ray energy spectrum from* $10^{14}$ eV *to* $10^{16}$ eV,", in *26th International Cosmic Ray Conference* (1999)

[9] C. Aguirre et al., Phys. Rev. D **62**, 032003 (2000)

[10] C. Aguirre, H. Aoki, K. Hashimoto, K. Honda, N. Inoue, N. Kawasumi, Y. Maeda, N. Martinic, T. Matano, N. Ohmori et al., Phys. Rev. D **62**, 032003 (2000)

[11] M. Aglietta et al. (EAS-TOP), Astropart. Phys. **21**, 583 (2004)

[12] K. Bernlohr, W. Hofmann, G. Leffers, V. Matheis, M. Panter, R. Zink, Astropart. Phys. **8**, 253 (1998), astro-ph/9801042

[13] K. Rawlins (SPASE, AMANDA), *"Measurement of the cosmic ray composition at the knee with the SPASE-2/AMANDA-B10 detectors,"*, in *28th International Cosmic Ray Conference* (2003), pp. 173–176

[14] J.E. Dickinson et al., *"Studies of the mass composition of cosmic rays with the SPASE-2/VULCAN instrument at the South Pole,"*, in *26th International Cosmic Ray Conference* (1999)

[15] M. Cha et al., ,, in *27th International Cosmic Ray Conference* (2001), p. 132

[16] T. Antoni et al. (KASCADE), Astropart. Phys. **24**, 1 (2005), astro-ph/0505413

[17] J. Horandel et al., in *16th European Cosmic Ray Symposium* (1998), p. 579

[18] T. Antoni et al. (KASCADE), Astropart. Phys. **16**, 245 (2002), astro-ph/0102443

[19] M. Ave, L. Cazon, J.A. Hinton, J. Knapp, J. LLoyd-Evans, A.A. Watson, Astropart. Phys. **19**, 61 (2003), astro-ph/0203150

[20] D.J. Bird et al. (HiRes), Astrophys. J. **424**, 491 (1994)

[21] S.P. Knurenko, A.A. Ivanov, A.V. Sabourov, I.Y. Sleptsov, *"Average Mass Composition of Primary Cosmic Rays in the Superhigh Energy Region by the Yakutsk Complex EAS Array Data,"*, in *30th International Cosmic Ray Conference* (2007), Vol. 4, pp. 167–170, 0711.2348

[22] T. Abu-Zayyad et al. (HiRes-MIA), Astrophys. J. **557**, 686 (2001), astro-ph/0010652

[23] R.U. Abbasi et al. (HiRes), Astrophys. J. **622**, 910 (2005), astro-ph/0407622

[24] M. Unger (Pierre Auger), Astron. Nachr. **328**, 614 (2007), 0706.1495

# Absolute calibration of the JEM-EUSO photodetection modules

*Daniil* Trofimov[1,2,3,*], *Alexander* Belov[2,3], *Sylvie* Blin-Bondil[1], *Alexandre* Creusot[1], *Pavel* Klimov[2], *Etienne* Parizot[1], *Guillaume* Prévôt[1], and *the JEM-EUSO* Collaboration

[1]Université Paris Cité, CNRS, Astroparticule et Cosmologie, F-75013, Paris, France
[2]Skobeltsyn Institute of Nuclear Physics, Lomonosov Moscow State University, 1(2), Leninskiye Gory, Moscow 119991, Russia
[3]Faculty of Physics, Lomonosov Moscow State University, 1(2), Leninskiye Gory, Moscow 119991, Russia

**Abstract.** The JEM-EUSO (Joint Experiment Missions for Extreme Universe Space Observatory) collaboration constructs a series of balloon and orbital telescopes to detect fluorescent UV emission from the Earth atmosphere, with the primary aim to study ultra-high-energy cosmic rays (UHECRs) from space.
The detectors have wide field-of-view (more than 20 degrees FOV), high temporal resolution (1-2.5 $\mu s$) and high sensitivity provided by a large aperture. Currently, one of these detectors is operating onboard the International Space Station (Mini-EUSO). The next one is planned to be launched in the spring of 2023 (EUSO-SPB2) and another one is in preparation stage (K-EUSO). These projects use the same photo-detection modules (PDMs) composed of 36 multi-anode photomultiplier tubes (MAPMTs), each with 64 pixels, for a total of 2304 pixels. Mini-EUSO uses one PDM, EUSO-SPB2 uses three and the future full-scale missions will use several tens of PDMs.
In the process of preparing and testing of PDMs, a new technique was developed to characterize their performance and provide absolute calibration of the MAPMTs used in the different JEM-EUSO missions. The method provides the efficiency of each pixel (including the sub-pixel structures), as well as the actual area occupied by the different pixels on the photocathode of the MAPMT.
The method and its application to EUSO-SPB2 PDMs at different high voltages and in different modes of operation are presented.

## 1 Introduction

Ultra-High-Energy Cosmic Rays (UHECRs) are the most energetic particles in the Universe with energies above $10^{18}$ eV. Although ground-based installations have observed these energetic particles for decades, their source and acceleration mechanisms remain largely unknown, and their study is observationally challenging because of their extremely low flux at Earth's surface [1].

The goal of the JEM-EUSO (Joint Experiment Missions for Extreme Universe Space Observatory) international collaboration is to extend the study of UHECRs by operating for the first time an observatory from space, which allows us to observe a very large volume of atmosphere at once with one single instrument. A first attempt to measure UHECRs from orbit was carried out in the TUS experiment [2–4], now integrated in the global JEM-EUSO program. For a review and update on this program, see Parizot et al. [5](this conference).

The Extreme Universe Space Observatory on a Super Pressure Balloon 2 (EUSO-SPB2) [6] is the third and most advanced balloon mission undertaken by the JEM-EUSO collaboration. EUSO-SPB2 is built on the experience of previous stratosphere missions: EUSO-Balloon [7] and EUSO-SPB [8], and of the Mini-EUSO space mission currently active onboard the International Space Station (ISS) [9]. It is particularly relevant in the context of future large-scale orbital missions, such as the proposed K-EUSO [10] and POEMMA [11], of which EUSO-SPB2 can be considered a pathfinder.

While the EUSO-SPB2 payload is composed of two complementary instruments, namely a Fluorescence Telescope (FT) and a Cherenkov Telescope (CT), this paper focuses on the calibration of the former, which is expected to allow the detection of extensive air showers from ultra-high-energy cosmic rays through fluorescence light for the first time. (See also Cummings et al. [12], this conference, for more details.)

## 2 Structure of the Fluorescence Telescope

The telescope is built according to the Schmidt scheme and has 6 mirror segments with a radius of curvature of 1659.8 mm and an effective focal length of 860 mm. The aperture is 1 m diameter, the field of view is 37.4°×11.4°, for one pixel 0.2°×0.2° and the direction of the FOV is nadir [13]. The FT camera consists of 3 photodetection modules (PDMs), each PDM contains 9 elementary cells (EC – units). A PDM and an EC-unit are shown in Figure 1. The main parts of each EC are 4 multi-anode photomultiplier tubes (MAPMTs) Hamamatsu type R11265 with 64 channels of registration. Thus each PDM has 64×4×9 = 2304 channels of registration. The MAPMT signals are

*e-mail: trofimov@apc.in2p3.fr

https://doi.org/10.1051/epjconf/202328306016

read by SPACIROC3 specialized chips designed to count single photoelectron pulses for a certain time [14]. The basis of the chip is an amplitude discriminator.

An important task of the pre-flight preparation of the equipment is the pixel-by-pixel calibration of such a multichannel detector.

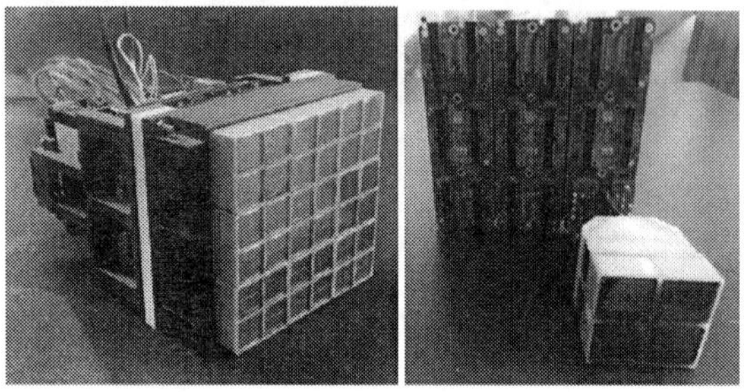

**Figure 1.** Assembled photodetection module (PDM) of EUSO-SPB2's FT (left) and Elementary Cell (EC) with 64×4=256 detection channels (right).

## 3 PDM calibration

### 3.1 The method of pre-flight PDM calibration

The calibration uses a controlled light source, either in full illumination mode (all pixels at once) or in single pixel illumination mode (with a collimator). The first method provides pixel efficiency and allows us to determine the optimal value of the electronics threshold used to measure a single photoelectron pulse for each pixel (see Sect. 3.2), and the second is used in conjunction with a mechanical scanning procedure to determine response of the photosensitive area to local illumination, also allowing to estimate the actual borders of each pixel and its internal structure. The principle is shown in Figure 2: the PDM is placed in a black box, with negligible background. The PDM is connected to the computer, which also has connections to a power-meter for precise light intensity control and to an $XY$-movement system providing precise positioning of the light source in front of the PDM. The light control system uses an integrating sphere with three windows: one is used to received light from LED (input), another one to illuminate the PDM pixels (output), and the third one to measure the light intensity inside the sphere with a NIST photodiode (control). The output window is connected to a collimator, which allows us to illuminate a surface even smaller than one pixel. The intensity of the output light is in a constant ratio with the intensity of light on the control NIST. Thus, it can be precisely calibrated at high intensity, and then used at low intensity for detailed calibration in single photon counting mode.

### 3.2 S-curve measurements

When a photon hits the photocathode of the MAPMT, it generates an internal electron cascade which results in

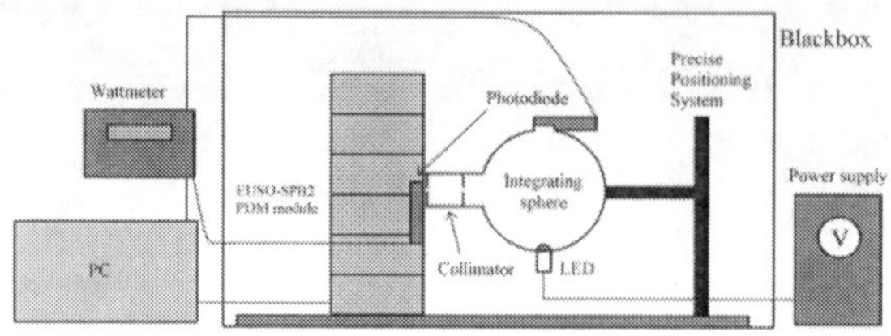


**Figure 2.** Scheme of the calibration process, with a collimator attached to the integrating sphere for single pixel illumination.

a signal on the anode. This signal is read by an ASIC (SPACIROC3) which counts one pulse if the signal amplitude is above a given threshold, adjustable through a DAC value. The so-called "S-curve" is the curve giving the number of counts as a function of that threshold or DAC. S-curves were studied in details for the Mini-EUSO project previously [15]. At low threshold (high DAC), the ASIC essentially counts electronic noise (pedestal). At higher thresholds (lower DAC), it counts only real photoelectrons, with an efficiency given by the number of counts per incoming photons. If the threshold increases, the number of counts decreases (i.e. some photoelectrons are missed). The ASIC can adjust its threshold through a combination of two DAC values: a "DAC-10" (10 bits, from 0 to 1023), set at the level of the MAPMT (for all pixels) and a "DAC-7" (7 bits, from 0 to 127), set at the level of individual pixels (before the pedestal). Figure 3 (left) demonstrates the S-curve for one pixel in DAC-10 mode. The dashed line is the S-curve derivative, i.e. amplitude distribution of single photoelectron pulses (at DAC-10 $\approx$ 370). In the right Figure 3 S-curve plots for all 2304 pixels in DAC-10 mode with individual DAC-7 thresholds adjusted so that all pedestals appear around the same DAC-10 value.

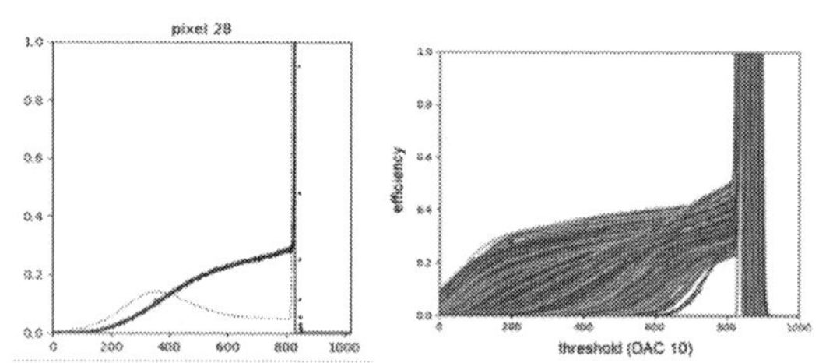


**Figure 3.** Left: S-curve for one pixel in DAC-10 mode. Vertical axis: efficiency, horizontal axis: DAC-10 threshold. S-curve derivative (the dashed line) demonstrates one photoelectron peak. Right: S-curve plot for all (2304) pixels in DAC-10 mode with the adjustment of all pedestals thanks to proper DAC-7 choices.

### 3.3 PDM scans

After setting individual thresholds, the full scans for each PDM were obtained. Each full scan has a resolution

200 × 200 points, it allows us to see details in the internal structure of the pixel (Figure 4). Full scans were made for all PDMs using 2 LEDs with 375 nm or 405 nm wavelengths, and with 2 different values of cathode high voltage: with a normal one around 1 kV and a reduced one for bright events.

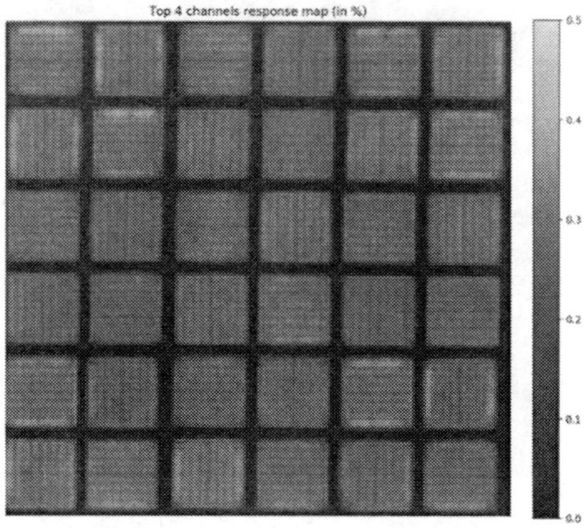


**Figure 4.** Full scan of PDM 3, at a wavelength of 375 nm

Precise scans were obtained for several MAPMTs with a resolution of 100 × 100 points revealing additional details on the internal structure of the photocathode, down to sub-pixel resolution. The left panel of Figure 5 shows a response map of a MAPMT where, for each position, the sum photon count of the 4 brightest pixels were used to derive the total detection efficiency at that position ("top-4 channel response map"). In the right panel of Figure 5 there is the "maximum response map" (photon count rate of the brightest pixel only). This figure reveals that border pixels are larger than the central ones – causing "border effects" in full illumination mode (Figure 8).

It is interesting to note that the structure of the pixel's efficiency depends on the direction of scanning (vertically or horizontally). It is demonstrated in Figure 6. On the right panel (horizontal scan) the shape of the curve for each pixel is smooth with one maximum. On the left figure (vertical scan) there is a dip in the efficiency in the center of each pixel.

### 3.4 Pixel sizes accounting

The size of a pixel is an important parameter of the photodetector and should be investigated and taken into account because of the possible nonuniformity of the MAPMT. Area of the pixel is included in the formula for calculating the number of photons during calibrations:

$$N_{ph} = \frac{c_{cross} \cdot \lambda \cdot t_{GTU}}{hc} \cdot \frac{S_{pix}}{S_{pd}} \cdot W \qquad (1)$$

where $N_{ph}$ is the average number of photon received and counted in one Gate Time Unit = 1 $\mu s$ (GTU), $c_{cross}$ is the cross calibration coefficient (ratio of two NIST photodiodes measurements), $\lambda$ - wavelength of light source, $t_{GTU}$ - Gate Time Unit, $S_{pix}$ - pixel area, $S_{pd}$ - photodiode area, $W$ - the light power from the photodiode inside the integrating sphere, and $h$ and $c$ - the Planck constant and the speed of light, respectively. The detection efficiency is $\varepsilon = N_{p.e.}/N_{ph}$. Therefore, the pixel area is needed to determine $\varepsilon$. The scanning procedure allows us to determine the position of pixel borders and in turn provides a measurement of the area of each pixel individually (Figure 7). The pixel borders are determined by moving the center of the light spot until the photon count of the pixel under investigation is half of the maximum photon count, obtained when the light spot is entirely contained within that pixel.

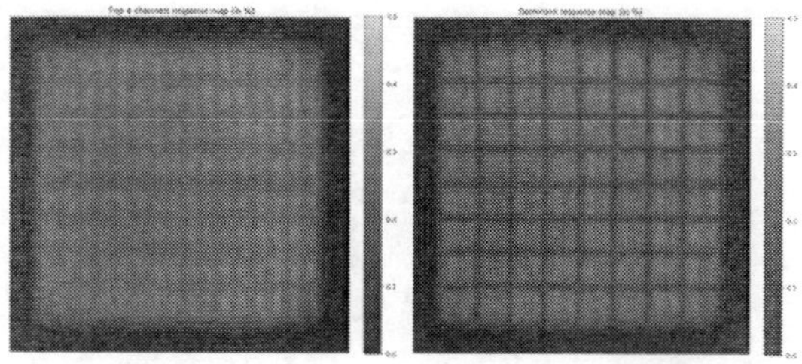

**Figure 5.** Left: "top-4" channel response map. Right: "maximum response map". The Figure reveals that border pixels are larger than central ones – causing "border effects" in full illumination mode. PMT22 of PDM2, $\lambda$ = 405 nm, 100×100 points resolution.

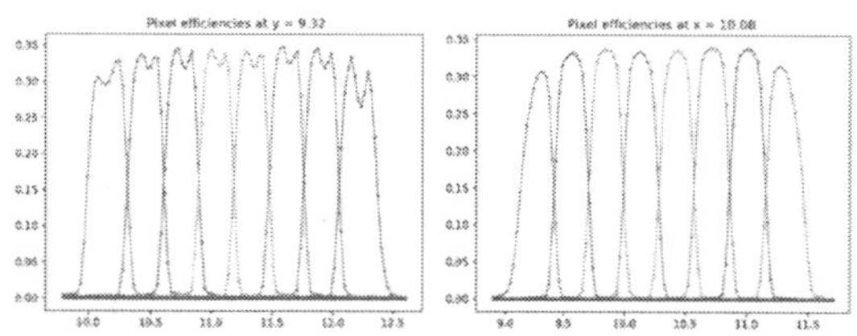

**Figure 6.** Vertial (left) and horizontal (right) scans of PMT22 from PDM2, $\lambda$ = 405 nm.

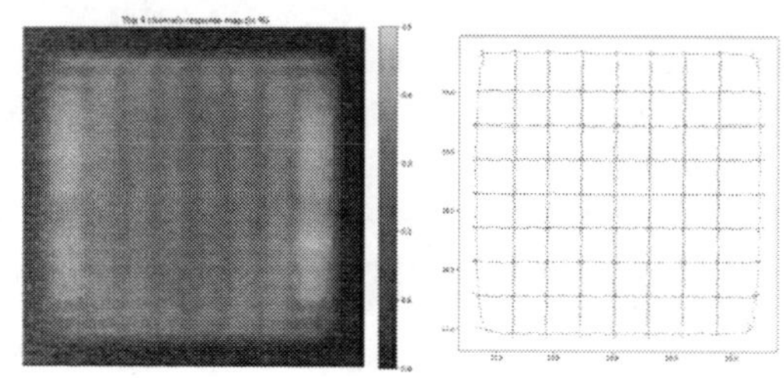

**Figure 7.** Left: "top-4 response map", right: scheme of pixel borders obtained during scan procedure.

The individual pixel areas allow us to renormalize the efficiency obtained in full illumination mode. In Figure 8 the PDM is uniformly illuminated. On the left panel, the efficiency is calculated assuming the average pixel size for all pixels according to Hamamatsu datasheet (equal for each pixel), on the right panel, renormalized efficiency, taking into account the individual pixel areas is demostrated. The efficiency map appears more uniform.

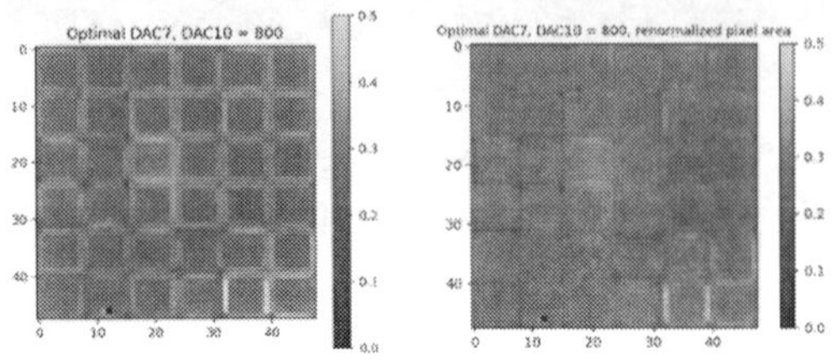


**Figure 8.** Full PDM illumination. Color means detection efficiency. Left: efficiency assuming the average pixel size equal for all pixels. Right: renormalized efficiency, taking into account the individual pixel areas.

### 3.5 Pile-up effect

For high light intensity the superimposition of pulses begins to play a significant role (so called "pile-up" effect) because the time interval between consecutive photoelectrons is smaller than the time resolution of the ASIC (a few ns). The number of counts thus saturates and eventually decreases as the photon rate increases (more and more photons being missed). Figure 9 shows saturation and consequtive decrease of number of counts/GTU as a function of incoming photons/GTU. On the left panel the results of simulations are shown. The expected average number of counts per GTU is defined by the following expression:

$$N_{p.e.} = \varepsilon N_{ph} \times \exp(-\varepsilon N_{ph}\delta t / t_{GTU}) \qquad (2)$$

where $N_{ph}$ is the average number of photon received and counted in one GTU, $\varepsilon$ is the detection efficiency, $\delta t$ is the double pulse resolution (dead time) and $t_{GTU}$ is GTU. The source of photons is assumed as poissonian. The right panel shows a measured pile-up curve in the relatively low incoming photon flux, which corresponds to the left part of the simulated curve.

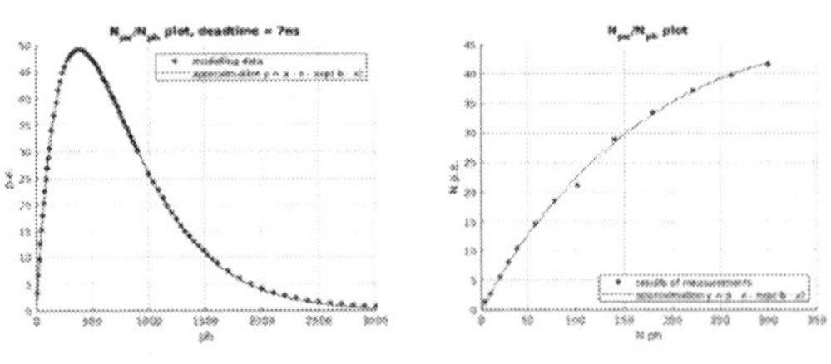

**Figure 9.** Left: simulated pile-up curve with $\delta t$ =7.5 ns. Right: Example of a pile-up curve, measured for one pixel and fitted with $\varepsilon$ = 0.25 and $\delta t$ = 8.3 ns.

## 4 Summary

All three PDMs of the EUSO-SPB2 fluorescence telescope have been calibrated with high precision in different modes. Detailed information about the size and efficiency of each pixel will be used in the analyses of extensive air showers and other events detected during the mission, which is expected to be launched in spring 2023 from Wanaka (New-Zealand). The calibration methods presented here will also be useful for the future missions of the JEM-EUSO Collaboration and can be applied more generally to any instrument using MAPMTs.

D.T. acknowledges the support by the French space agency CNES, the Vernadski program (BGF scolarship) and the Interdisciplinary Scientific and Educational School of Moscow University "Fundamental and Applied Space Research"

## References

[1] A. Coleman et al., Ultra-High-Energy Cosmic Rays: The Intersection of the Cosmic and Energy Frontiers, arXiv:2205.05845v3 (2022)

[2] D. Barghini et al., UV telescope TUS on board Lomonosov satellite: Selected results of the mission. Advances in Space Research, 70(9), 2734-2749 (2022)

[3] P.A. Klimov et al., The TUS detector of extreme energy cosmic rays on board the Lomonosov satellite. Space science reviews, **212**, 1687-1703 (2017).

[4] B.A. Khrenov et al., First results from the TUS orbital detector in the extensive air shower mode. Journal of Cosmology and Astroparticle Physics, 09, 006 (2017).

[5] E. Parizot et al. (JEM-EUSO Coll.), The JEM-EUSO program for UHECR studies from space, This Conference Proceedings (2023)

[6] J. Eser et al. (JEM-EUSO Coll.), Science and mission status of EUSO-SPB2, arXiv:2112.08509v1 (2021)

[7] J.H. Adams et al. (JEM-EUSO Coll.), A Review of the EUSO-Balloon Pathfinder for the JEM-EUSO Program. Space Sci Rev 218, 3 (2022).

[8] J. Eser et al. (JEM-EUSO Coll.), Results of the EUSO-SPB1 flight, Volume **358** (ICRC2019)

[9] S. Bacholle *et al*, Mini-EUSO Mission to Study Earth UV Emissions on board the ISS ApJS **253** 36 (2021)

[10] P. Klimov et al. (JEM-EUSO Coll.), Status of the K-EUSO Orbital Detector of Ultra-High Energy Cosmic Rays. Universe, 8, 88 (2022)

[11] A.V. Olinto et al. (JEM-EUSO Coll.), The POEMMA (Probe of Extreme Multi-Messenger Astrophysics) observatory, JCAP 2021, V. **06** (2021)

[12] A. Cummings et al. (JEM-EUSO Coll.), EUSO-SPB2: A balloon experiment for UHECR and VHE neutrino observation, This Conf. Proceedings (2023)

[13] V. Kungel et al. (JEM-EUSO Coll.), EUSO-SPB2 Telescope Optics and Testing, PoS (ICRC2021) **412**

[14] S. Blin et al. (JEM-EUSO Coll.), SPACIROC3: 100 MHz photon counting ASIC for EUSO-SPB, Nuclear Instruments and Methods in Physics Research ,Volume **912**, 363-367 (2018)

[15] C. J. Blaksley, *Photodetection Aspects of JEM-EUSO and Studies of the UHECR Sky* (2013) , 331, https://arxiv.org/pdf/1406.5818.pdf

# An end-to-end in-flight calibration of Mini-EUSO detector

*Hiroko* Miyamoto[1,2*], *Matteo* Battisti[2**], *Dario* Barghini[2,3,4], *Alexander* Belov[5,6], *Mario* Bertaina[2,3], *Marta* Bianciotto[3], *Francesca* Bisconti[7], *Carl* Blaksley[8], *Sylvie* Blin[9], *Karl* Bolmgren[10], *Giorgio* Cambiè[7,11], *Francesca* Capel[12], *Marco* Casolino[7,8,11], *Igor* Churilo[13], *Marino* Crisconio[14], *Christophe* De La Taille[15], *Toshikazu* Ebisuzaki[8], *Johannes* Eser[16], *Francesco* Fenu[17], *George* Filippatos[18], *Massimo Alberto* Franceschi[19], *Christer* Fuglesang[10], *Alessio* Golzio[2,3], *Philippe* Gorodetzky[9], *Fumiyoshi* Kajino[20], *Hiroshi* Kasuga[8], *Pavel* Klimov[6], *Viktoria* Kungel[18], *Vladimir* Kuznetsov[13], *Massimiliano* Manfrin[2,3], *Laura* Marcelli[7], *Gabriele* Mascetti[14], *Włodzimierz* Marszał[21], *Marco* Mignone[2], *Alexey* Murashov[6], *Tommaso* Napolitano[19], *Hitoshi* Ohmori[8], *Angela* Olinto[16], *Etienne* Parizot[9], *Piergiorgio* Picozza[7,11], *Lech Wiktor* Piotrowski[22], *Zbigniew* Plebaniak[2,3,21], *Guillaume* Prévôt[9], *Enzo* Reali[7,11], *Marco* Ricci[19], *Giulia* Romoli[7,11], *Naoto* Sakaki[8], *Sergei* Sharakin[6], *Kenji* Shinozaki[21], *Jacek* Szabelski[21], *Yoshiyuki* Takizawa[8], *Giovanni* Valentini[14], *Michal* Vrabel[21], *Lawrence* Wincke[18], and *Mikhail* Zotov[6] *for the JEM-EUSO Collaboration*

[1]Gran Sasso Science Institute, Viale Francesco Crispi 7, 67100 l'Aquila, Italy
[2]INFN Section of Turin, Via P. Giuria 1, 10125 Turin, Italy
[3]Department of Physics, University of Turin, Via P. Giuria 1, 10125 Turin, Italy
[4]INAF Astrophysics Observatory of Turin, Via Osservatorio 20, 10025 Pino Torinese, Italy
[5]Faculty of Phyiscs, M.V.Lomonosov Moscow State University, ul. Kolmogorova 1(2), 119234 Moscow, Russia
[6]Skobeltsyn Institute of Nuclear Physics, Lomonosov Moscow State Univ., ul. Kolmogorova 1(2), 119234 Moscow, Russia
[7]INFN Section of Rome Tor Vergata, Via della Ricerca Scientifica 1, 00133 Rome, Italy
[8]RIKEN, 2-1 Hirosawa Wako, Saitama 351-0198, Japan
[9]Université Paris Cité, CNRS, Astroparticule et Cosmologie, 10 Rue Alice Domon et Lèonie Duquet, 75013 Paris, France
[10]KTH Royal Institute of Technology, Brinellvägen 8, 114 28 Stockholm, Sweden
[11]Department of Physics, University of Rome Tor Vergata, Via della Ricerca Scientifica 1, 00133 Rome, Italy
[12]Technical University of Munich, Arcisstraße 21, 80333 Munich, Germany
[13]S.P. Korolev Rocket and Space Corporation Energia, Lenin str., 4a Korolev, 141070 Moscow area, Russia
[14]ASI, Italian Space Agency, Via del Politecnico, 00133 Rome, Italy
[15]Omega, Ecole Polytechnique, CNRS/IN2P3, Rte de Saclay, 91120 Palaiseau, France
[16]Department of Astronomy and Astrophysics, The University of Chicago, 5640 S. Ellis Avenue, Chicago IL 60637, US
[17]Karlsruhe Institute of Technology, Hermann-von-Helmholtz-Platz 1, 76344 Eggenstein-Leopoldshafen, Germany
[18]Department of Physics, Colorado School of Mines, 1523 Illinois St., Golden CO 80401, US
[19]INFN National Laboratories of Frascati, Via Enrico Fermi 54, 00044 Frascati, Italy
[20]Department of Physics, Konan University, 8 Chome-9-1 Okamoto, Higashinada Ward Kobe, Hyogo 658-8501, Japan
[21]National Centre for Nuclear Research, 28 Pułku Strzelców Kaniowskich 69, 90-558 Łódź, Poland
[22]Faculty of Physics, University of Warsaw, Ludwika Pasteura 5, 02-093 Warsaw, Poland

**Abstract.** Mini-EUSO is a wide Field-of-View (FoV, 44°) telescope currently in operation from a nadir-facing UV-transparent window in the Russian Zvezda module on the International Space Station (ISS). It is the first detector of the JEM-EUSO program deployed on the ISS, launched in August 2019. The main goal of Mini-EUSO is to measure the UV emissions from the ground and atmosphere, using an orbital platform. Mini-EUSO is mainly sensitive in the 290–430 nm bandwidth. Light is focused by a system of two Fresnel lenses of 25 cm diameter each on the Photo-Detector-Module (PDM), which consists of an array of 36 Multi-Anode Photomultiplier Tubes (MAPMTs), arranged in blocks of $2 \times 2$ called Elementary Cells (ECs), for a total of 2304 pixels working in photon counting mode, in three different time resolutions of 2.5 $\mu$s (defined as 1 Gate Time Unit, GTU), 320 $\mu$s and 40.96 ms operating in parallel. In the longest time scale, the data is continuously acquired to monitor the UV emission of the Earth. It is best suited for the observation of ground sources and therefore has been used for the observational campaigns of the ground-based UV flasher in order to perform an end-to-end calibration of Mini-EUSO. In this contribution, the assembled UV flasher, the operation of the field campaign and the analysis of the obtained data are presented. The result is compared with the overall efficiency computed from the expectations which takes into account the atmospheric attenuation and the parametrisation of different effects such as the optics efficiency, the MAPMT detection efficiency, BG3 filter transmittance and the transparency of the ISS window.

*email: miyamoto@to.infn.it
**email: matteo.battisti@to.infn.it

https://doi.org/10.1051/epjconf/202328306017

## 1 Introduction

As part of the JEM-EUSO Program [1], the main goal of Mini-EUSO is to measure the UV emissions from the ground and atmosphere and to assess the feasibility and performance of the measurement of ultra-high energy cosmic rays by means of a space-based detector. Fig. 1 shows schematic views of the full Mini-EUSO telescope (top) and the PDM (bottom). A more detailed explanation of the Mini-EUSO detector and data acquisition chain is reported in [2].

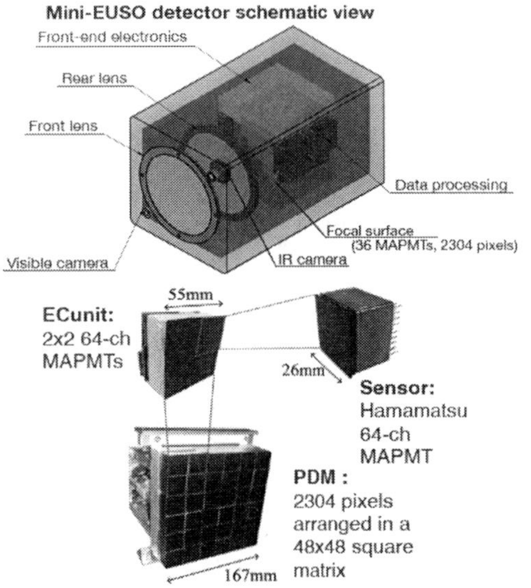


Figure 1: Schematic view of Mini-EUSO telescope (top) and the PDM (bottom).

Several kinds of ground-based flashers have been developed in different groups of the JEM-EUSO collaboration such as in Japan, Italy and France. One of such flashers has been developed in Turin ("Torino UV flasher") and calibrated with the Torino EC telescope (see Fig. 2). The flasher consists of an array of 9 100W COB-UV LEDs, batteries and an Arduino circuit (see the left panel of Fig. 3). LEDs were programmed to pulse 6 times in 12 s with a pulse duration of 1600 ms on and 400 ms off each, followed by 12 pulses in 9.6 s with 400 ms on and 400 ms off each. The durations were decided taking into account that a Mini-EUSO pixel changes completely its FoV every ~800 ms as it corresponds to ~6.3 km on ground and the ISS speed is ~7.5 km/s. In this way it was both guaranteed to have a light signal lasting the entire transit of the flasher in a pixel FoV also considering the mismatch between the flasher firing and the GTU start (1600 ms on), and the possibility to measure at the same time in a pixel the flasher light and the background (400 ms on).

In May 2021, an observational campaign was performed at Piana di Castelluccio in central Italy at the altitude of ~ 1550 m a.s.l. The place was chosen based on the very low light pollution in an area of several km radius. In this campaign, photons from the Torino UV flasher were detected and we preliminarily estimated the overall efficiency of Mini-EUSO telescope. Unfortunately, the flasher signal was detected by Mini-EUSO at the edge of the FoV of an MAPMT. Therefore, some light was missed making the calibration effort more uncertain.

Prior to the measurement, the flasher was tested with the Torino EC telescope at the TurLab facility [3]. As shown in the right panel of Fig. 3, this facility allows one to place the telescope at 40 m distance from a light source in a dark environment. Fig. 4 shows examples of the Torino flasher data taken by the Torino EC telescope (top) and by the Mini-EUSO detector in space (bottom), where one can see the image of one frame (integration of 40.96 ms) on the left and the light curves of the UV flasher signal (right) after the background subtraction.

The overall efficiency of Mini-EUSO telescope can be estimated as:

$$Eff = N_{\rm det}/N_{\rm w}, \qquad (1)$$

where $N_{\rm det}$ is the photon counts detected by Mini-EUSO telescope and $N_{\rm w}$ is the number of photons arriving at the UV transparent window on the ISS. For the analysis, we selected two pixels, pixel(31,5) and pixel(31,13), which are located at 18° and 12° from nadir, respectively, where the flasher is flashing in the center of a pixel according to the following methods for each:

- method 1: Search for a pixel with flasher signal when the neighbouring top, bottom and left pixels have the same counts (the right pixel belongs to a different MAPMT and gaps exist between MAPMTs).

- method 2: Estimate the geographical position from the orbital information (UTC time, distance and the angle)

As a result, $N_{\rm det}$ are 9.7 cts/GTU (pix(31,5)) and 8.3 cts/GTU (pix(31,13)), respectively. In the following we describe how $N_{\rm w}$ has been estimated.

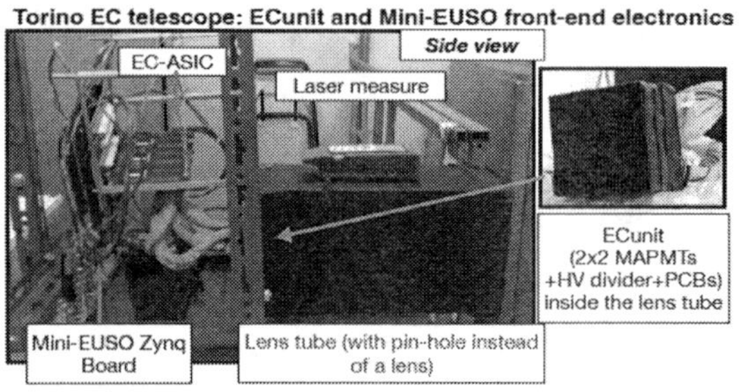


Figure 2: The Torino EC Telescope consists of an EC unit, EC-ASIC, Zynq Board, lens tube, 1" plano-convex lens, CPU (PC), external LV and HV power supplies. For the flasher measurements in the lab, a pin-hole of 0.1 mm in diameter, instead of the lens, is used to reduce light from the flasher LEDs.

## 2 Calibration of the Flasher

An EC based telescope, so-called "Torino EC telescope" (Fig. 2), has been originally built in Turin for fundamental functionality tests, the study and development of the trigger for the Mini-EUSO telescope at the TurLab facility and open-sky conditions [4]. It consists of a 30 cm lens

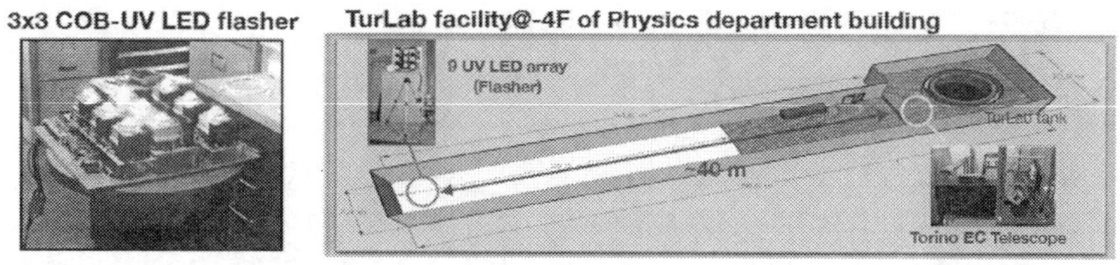


Figure 3: Left: 3 × 3 COB-UV LED flasher built in Turin. Right: schematic view of TurLab facility where the positions of the flasher and Torino EC telescope are indicated.

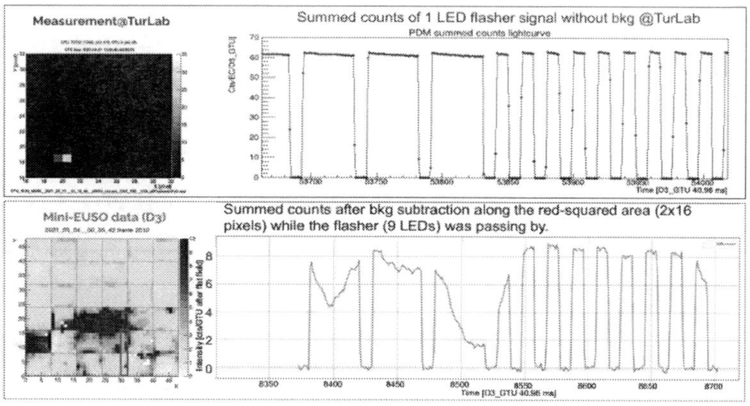


Figure 4: Image of a frame (left) and the lightcurve (right) of the flasher LED photons detected by the Torino EC at 40 m distance (top) and by Mini-EUSO on the ISS (bottom).

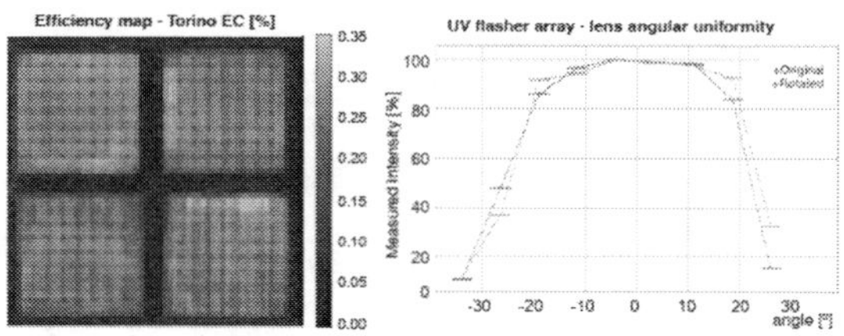


Figure 5: Left: PDE map of Torino EC. Right: Results of the flasher angular measurement.

tube with a 1-inch plano-convex lens, an ECunit, front-end electronics based on SPACIROC3 ASICs, and the Zynq board connected to a PC via ethernet cable, where dedicated software for the Mini-EUSO data processing system is installed. MAPMTs and electronics boards are powered by external High, and Low Voltage DC Power Supply (HVPS, LVPS), respectively. For the end-to-end calibration, it is possible to compare the flasher signals in the same Focal Surface (FS) detector, data acquisition system and electronics on the ground and in space, taking into account the factors of distance, incident angles, atmospheric attenuation, the transmission and transmittance of the optics and the ISS window. The Torino ECunit together with EC_ASIC front-end board were absolutely calibrated in France (APC/Univ. Paris Cité) using the method and equipment for the standard JEM-EUSO FS detector calibration [5]. The left plot of Fig. 5 shows the resulted absolute Photo Detection Efficiency (PDE) map of the Torino EC. The telescope is set at 40 m distance from the flasher LEDs to estimate the light intensity of the UV flasher in the large dark room of the TurLab facility located at the fourth basement of the Physics department building of University of Turin. A high precision 0.1 mm pin-hole is attached to the lens tube instead of the lens to reduce light from the flasher. We repeated the measurement at different times to verify that no significant difference in the measurement exists depending on the time or setup conditions. The number of emitted photons is estimated from the detected photon counts by the Torino EC telescope as following:

1. Apply pile-up correction (pixel by pixel).
2. Apply the absolute efficiency of the EC taking into account of a correction factor of 2% as it is calibrated at 395 nm in wavelength, while the flasher is of ~ 395–400 nm in wavelength.
3. Sum all the counts of the pixels with dominant signal.
4. Average peak counts.

As a result, the total number of photons obtained from the measurements performed in Apr 2022 and Jun 2022 are 3564±304 photons/GTU and 3400±296 photons/GTU, respectively.

The same measurement was repeated at different emission angles between the flasher and the telescope to study the angular response of the flasher. In this measurement, we used a single UV LED which is of the same type of the ones employed in the UV flasher. The right plot of Fig. 5 shows the light intensity at each angle comparing

Table 1: List of the obtained parameters of the flasher and the quadratic sum of the errors

| | $N_{\mathrm{det}}$ | $N_{\mathrm{ph}}$ | PDE | Lens uniformity | Atmospheric Attenuation | Angle ($\cos^3\theta$) |
|---|---|---|---|---|---|---|
| Value | 9.7–8.3 | 3482 | / | 0.9–0.95 | 0.22 | 18.6–12.3 |
| Error [%] | 10 | 6 | 10 | 5 | 10 | 2 |

Table 2: Parameters for the arrival photons at ISS window.

| | $pix(31,5)$ | $pix(31,13)$ |
|---|---|---|
| $\theta$ [°] | 18 | 12 |
| $Eff_\theta$ [%] | 0.9 | 0.95 |
| $D_{\mathrm{EC}}/D_{\mathrm{ME}}$ | $9.26\times10^{-9}\cos^2\theta$ | |
| $\cos^3\theta$ | 0.86 | 0.94 |

the obtained photon counts with "on-axis" value (red dotted curve). The measurement was then repeated after rotating the lens of flasher LED to check the dependency on the uniformity of the lens (blue dotted curve).

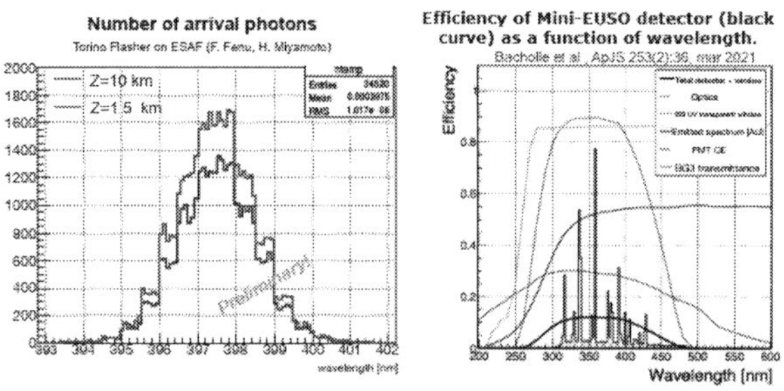


Figure 6: Left: The number of arrival photons at the ISS window as a function of the emitted wavelength as simulated with ESAF for two different heights of the flasher. Right: Efficiency of the Mini-EUSO detector (black curve) as a function of wavelength, resulted from the Optics transmittance (purple), transmission of the BG3 bandpass filter (red), PMT detector efficiency (blue) and ISS UV transparent window (green). The histogram shows the fluorescence light from extensive air showers.

For the attenuation by the atmosphere, we refer to the simulation result. The left panel of Fig. 6 shows the UV flasher simulated by the EUSO Simulation and Analysis Framework (ESAF) setting the flasher position at the altitude of 1.5 km (blue) as for the campaign, and 10 km (red), where no attenuation by the atmosphere occurs [6]. This result indicates that the number of arrival photons from the flasher is reduced to ~78% due to the attenuation. The number of arrival photons at the ISS window is:

$$N_{\mathrm{w}} = N_{\mathrm{ph}}\times Eff_\theta\times(1-Attn)\times A_{\mathrm{ME}}/A_{\mathrm{EC}}\times(D_{\mathrm{EC}}/D_{\mathrm{ME}})^2\times\cos\theta,$$

where $N_{\mathrm{ph}}$ is the number of photons emitted by the flasher estimated by the TurLab measurement, $\theta$ and $Eff_\theta$ are the incident angle and the Mini-EUSO lens efficiency of the corresponding angle, $Attn$ is the atmospheric attenuation, $A_{\mathrm{ME}}$ and $A_{\mathrm{EC}}$ are the lens and pin-hole areas of Mini-EUSO and Torino EC respectively, $D_{\mathrm{EC}}$ and $D_{\mathrm{ME}}$ are the distances between the flasher and each detectors. Applying parameters shown in Table 2, $N_{\mathrm{w}}$ for pix(31,5) and pix(31,13) are 120.4 cts/GTU and 139.3 cts/GTU. As it is indicated in Eq. (1), the estimated efficiencies for those pixels are 8.1 ± 1.5% and 6.0 ± 1.5%, respectively. The total error (19%) comes from the quadratic sum of the errors listed in Table 1. Note that there is no value of "PDE" in the table as the error is independent from it.

## 3 Comparison with theoretical value

The right panel of Fig. 6 shows the theoretical overall efficiency of the Mini-EUSO detector (black curve) as a function of wavelength. It is the result of the optics transmittance (purple curve), the transmission of the BG3 bandpass filter (red curve), the PMT detector efficiency of the photocathodes (blue curve) and of the UV transparent window of the ISS (green curve). The system has been designed to optimize observations of the fluorescence light emitted by nitrogen atoms excited by the extensive air shower of cosmic rays (grey histogram). The theoretical efficiency for perpendicular light emitted at 397.5 nm is 11%. Applying the optical efficiency of 96% and 98% at 18° and 12°, the theoretical efficiency for those angles are 10.6% and 10.8%, respectively.

## 4 Acknowledgements

This work was supported by the Italian Space Agency through the agreement n. 2020-26-Hh.0, by the French space agency CNES, and by the National Science Centre in Poland grants 2017/27/B/ST9/2162 and 202/37/B/ST9/01821. This research has been supported by the Interdisciplinary Scientific and Educational School of Moscow University "Fundamental and Applied Space Research" and by Russian State Space Corporation Roscosmos. The article has been prepared based on research materials collected in the space experiment "UV atmosphere". We thank the Altea-Lidal collaboration for providing the orbital data of the ISS.

## References

[1] M. Ricci (JEM-EUSO Coll.), J. Phys. Conf. Ser. **718** n.5 052034 (2016)

[2] S. Bacholle et al., ApJ Supplement Series 253(2):**36** (2021).

[3] P. Barrillon et al. (JEM-EUSO Coll.), Exp. Astr. **368** https://doi.org/10.1007/s10686-022-09871-8 (2022)

[4] F. Bisconti et al. (JEM-EUSO Coll.), Exp. Astr. **53** 133-158 (2021)

[5] D. Trofimov et al. (JEM-EUSO Coll.), This Conference Proceedings (2023)

[6] H. Miyamoto et al. (JEM-EUSO Coll.), PoS (ICRC2021) **334**

# Modeling Neutrino and Background Signals for the Payload for Ultrahigh Energy Observations (PUEO) Experiment

*Austin* Cummings[1,*], *John* Krizmanic[2], and *Stephanie* Wissel[1]

[1]Departments of Physics and Astronomy Astrophysics, Institute for Gravitation and the Cosmos, Pennsylvania State University, University Park, PA 16802, USA
[2]NASA Goddard Space Flight Center, Greenbelt, MD 20771, USA

**Abstract.** In the current age of multi-messenger astrophysics, Very High Energy (VHE) cosmic neutrinos (E>1 PeV) represent a unique observation window into the most extreme astrophysical events in the universe. Measurements of the neutrino flux in this high energy regime provide information regarding the distribution and composition of Ultra-High Energy Cosmic Rays (UHECR), details of source acceleration mechanics, and definitive tests of physics beyond the standard model. Observations in multiple different detection channels (e.g. optical, radio, direct particle counting) are being explored to extend the sensitivity to neutrinos above PeV energies.

The Payload for Ultrahigh Energy Observations (PUEO) is a long duration balloon experiment that builds on the successes of the Antarctic Impulsive Transient Antenna (ANITA) experiment to probe the cosmic neutrino flux above EeV energies. PUEO, like ANITA, seeks to measure coherent radio emission in the form of Askaryan emission from in-ice neutrino interactions and both geomagnetic and Askaryan emission from Extensive Air Showers (EAS) produced by the decays of Earth-emergent $\tau$-leptons sourced from $\tau$ neutrinos.

To evaluate the sensitivity of PUEO (and other detectors) to a given flux of neutrinos, it is necessary to accurately model i) the propagation of neutrinos through the Earth ii) the emission generated from the neutrino-sourced particle shower iii) the detector performance to a generated signal and iv) relevant backgrounds. Numerous software frameworks exist to model these different mechanisms independently, but often require significant work to use together.

$\nu$SpaceSim is an open source, end-to-end neutrino simulation package that models these described mechanisms for an arbitrary detector geometry in both the optical and radio emission channels. For a given experimental design, $\nu$SpaceSim is designed to model the sensitivity to both the cosmogenic neutrino flux and to astrophysical neutrino transient events. The highly modular design of $\nu$SpaceSim allows for comparison and exploration of different physics models within the same code base.

In this contribution, we detail the state of the $\nu$SpaceSim code as it pertains to modeling the sensitivity of the PUEO experiment, and highlight the mechanisms necessary to perform this calculation. We also describe the near-term improvements to the radio emission calculation in $\nu$SpaceSim and the modeling of the most relevant backgrounds for PUEO: the direct (above-the-limb) cosmic rays, and the indirect (ice-reflected) cosmic ray signals.

## 1 Introduction

The Payload for Ultrahigh Energy Observations (PUEO) is a long-duration, high-altitude (33 km) balloon experiment that aims to probe the cosmic neutrino flux above EeV energies [1]. During flight, PUEO is expected to observe coherent radio emission (via the geomagnetic and Askaryan production mechanisms) from four different classes of events: i) Direct (above the horizon) cosmic rays ii) Indirect cosmic rays, with emission reflected off the Antarctic ice iii) Upgoing extensive air showers (EAS) produced through the decay of neutrino-sourced $\tau$-leptons iv) In-ice neutrino interactions. A diagram of PUEO's observation strategy and the major event classifications is shown in Figure 1.

PUEO is built on the heritage of the Antarctic Impulsive Transient Antenna (ANITA) experiment, which has completed 4 long duration flights between 2006 and 2016 [2–4]. During these flights, ANITA collectively observed 64 reflected cosmic rays candidates, 7 direct cosmic ray candidates, and several candidates with non-inverted polarity from below the horizon, consistent with the phenomenology of upgoing EAS. This latter category of anomalous events can be divided into two sub-categories based on the angle with which they were observed: i) steeply upgoing, which both ANITA-I and ANITA-III recorded with elevation angles of $-27°$ and $-35°$ and energies of $\sim$ 0.5 EeV and ii) near-horizon, which were observed four times during the flight of ANITA-IV. While

*e-mail: alc6658@psu.edu

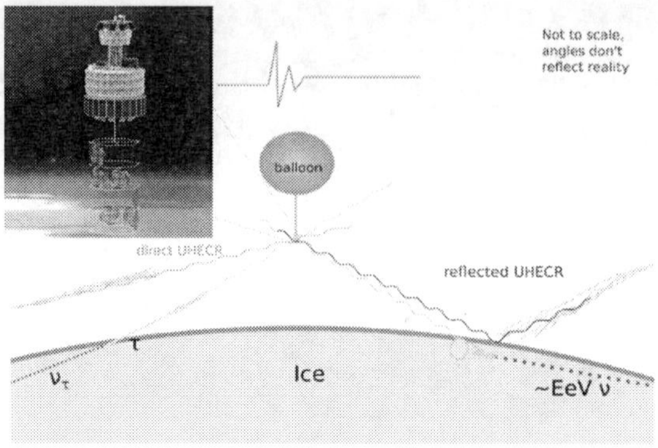


**Figure 1.** Diagram of PUEO's observation scheme of cosmic rays and neutrinos, modified from original image from Cosmic Deaconu.

each of these measurements is in tension with existing limits set on the diffuse neutrino flux by existing ground based observatories, the steeply upgoing candidates in particular cannot be explained as neutrino-sourced EAS under standard model assumptions of neutrino propagation inside the Earth. An example signal from each type of these candidate events is shown in figure 2, superimposed and offset in time and electric field values for easier comprehension.

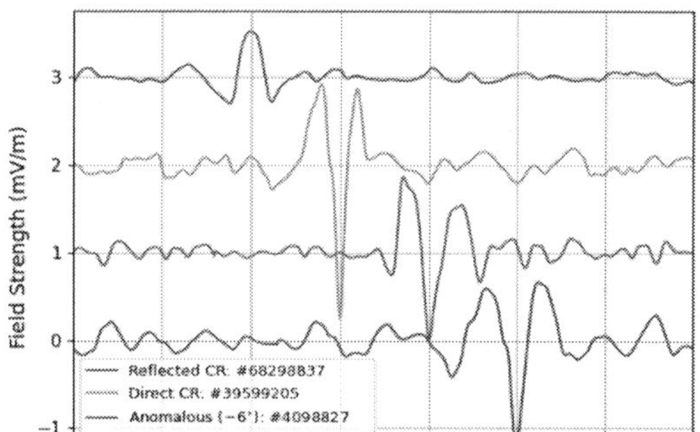


**Figure 2.** Four candidate H-Pol signals from different event classifications observed during the flights of ANITA, superimposed on the same plot, shifted in time and field strength for clarity. The four event classifications shown are: i) reflected cosmic rays, ii) direct (above-the-horizon) cosmic rays iii) slightly below horizon upgoing EAS iv) steeply upgoing EAS [2–4].

PUEO will further resolve the measurements made by ANITA by lowering energy thresholds to record more event candidates and improving detection strategies to optimize observations and improve event quality of these candidates. During the course of a single 30 d flight, PUEO will either set stringent limits on the diffuse neutrino flux above EeV energies or measure the highest energy neutrino to date. In this work, we describe the PUEO experiment, it's improvements over ANITA, and the most current attempts to model its performance regarding neutrino signals and potential backgrounds.

## 2 PUEO

### 2.1 Main Instrument

The main instrument of PUEO conists of 108 dual-polarization quad-ridge horn antennas, improving significantly on ANITA's 48 antenna layout. This is made possible by increasing the low frequency cutoff of the antenna from ANITA's 200 MHz to cover a frequency range of 300 MHz to 1200 MHz. PUEO's antennas are arranged in four 24-antenna rings, with each column representing one "phi-sector", and a 12-antenna drop-down ring, downward-canted at 40° to increase sensitivity to upgoing EAS.

PUEO's biggest improvement over ANITA comes from the inclusion of an interferometric phased array (or beamforming) trigger. By coherently summing signals received in a number of antennas $N_{\text{ant}}$, the effective signal to noise ratio (SNR) for an event increases as $\sqrt{N_{\text{ant}}}$, resulting in lower energy thresholds, better chance of off-axis detection, and improved event quality. A given signal in PUEO's main instrument is expected to be received in 4 phi-sectors, or 16 antennas, resulting in a typical SNR increase of ~ 4. A plot of PUEO's single-antenna trigger efficiency as a function of SNR is shown in Figure 3, compared to those of ANITA-IV and the Askaryan Radio Array (ARA).

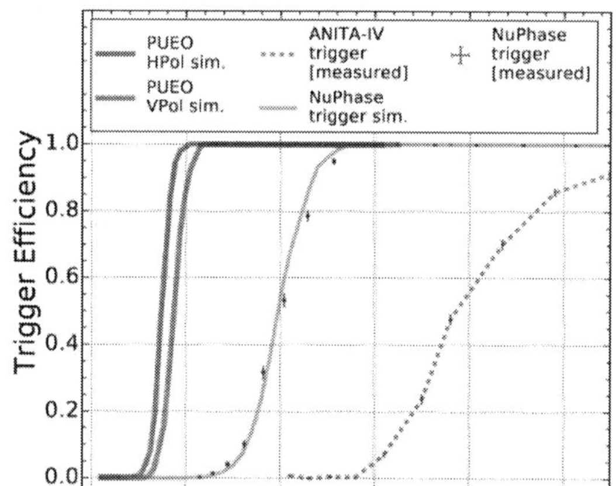


**Figure 3.** Simulated trigger efficiency of PUEO using a 16-antenna beamforming trigger compared to the trigger efficiency of ANITA-IV and ARA (Askaryan Radio Array) [1]

PUEO will also use a Radio-Frequency System-on Chip (RFSoC) as a combined readout and interferometric trigger system. The RFSoC will allow for improved filtering of anthropogenic noise via digital notch filtering and will replace the analog filters that were used during the flight of ANITA-IV, thereby improving overall event quality.

### 2.2 Low Frequency (LF) Instrument

PUEO will include a drop-down array of 8 dual-polarization sinuous antennas below the main array that operate in the frequency range of 50 MHz to 300 MHz (see Figure 1). A separate beamforming trigger will be included for the LF instrument, and has been shown to provide an expected SNR increase of ~ 2.2 via extensive simulation studies. The LF instrument operates independently of the main instrument, and will both help improve event quality and lower energy thresholds by providing an additional trigger.

The primary goal of the LF is to increase PUEO's sensitivity to air showers by measuring events further away from the shower axis, where the generated radio emission is only coherent for lower frequencies. Figure 4 shows the frequency integrated SNR as a function of viewing angle for a sample reflected cosmic ray for the main instrument and LF.

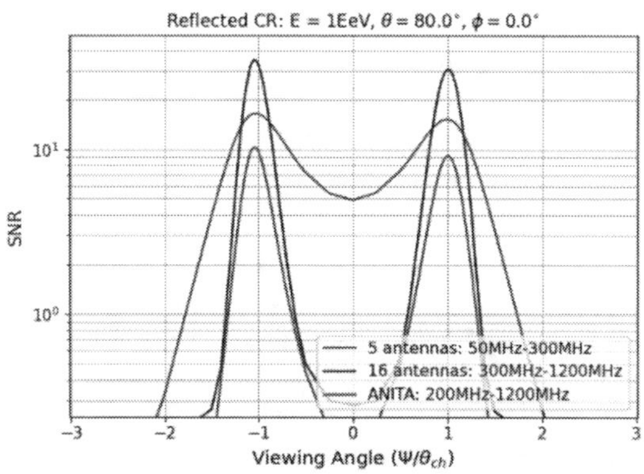


**Figure 4.** Frequency integrated SNR of a reflected cosmic ray event as a function of viewing angle for the LF instrument compared against PUEO's main instrument. The same curve is also plotted for the ANITA-IV antenna for comparison.

Figure 4 shows, for an example air shower event, that although the peak SNR is lower for the LF than the main instrument (primarily due to increased noise temperatures at lower frequencies), the signal is significantly broader, and has amplified SNR further outside the Cherenkov cone. This behavior ultimately manifests in a higher overall energy threshold, but amplified acceptances at higher energies.

## 3 Modeling

### 3.1 PUEOSim

`PUEOSim`[1] is an end-to-end Monte Carlo simulation package developed to accurately model full time domain signals and backgrounds for PUEO and calculate its neutrino sensitivity under various configurations, observation strategies, and source characteristics, taking into account the complete properties of the instrument. `PUEOSim` is based on the `ANITASim`[2] package, which has been used to calculate ANITA's sensitivity to in-ice neutrino interactions via the Askaryan effect (which is itself modeled by the `nicemc`[3] [5] simulation code). Figure 5 shows PUEO's estimated single event sensitivity to neutrinos via the Askaryan effect, calculated with `ANITASim` using reasonably scaled thresholds based on the design requirements listed in [1]. The $\tau$-lepton air shower sensitivity was generated using an independent code, and provides additional sensitivity below energies of ~ 1 EeV.

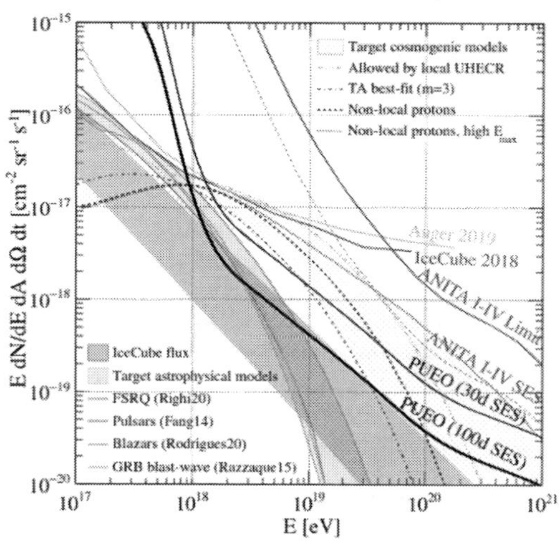


**Figure 5.** PUEO's single-event sensitivity (SES) to diffuse neutrinos for 30 d and 100 d flights. Also shown are existing limits set by various experiments and some cosmogenic and astrophysical models [1].

Both the new antennas of PUEO's main instrument and those of the LF instrument are fully modeled in `PUEOSim`, as are the beamforming triggers for both instruments. As a novel feature to `PUEOSim`, time domain signals from both ice-reflected cosmic ray and upgoing $\tau$-lepton initiated EAS are being included via tables generated by the ZHAireS air shower simulation code [6]. In the near future, signals from direct above-the-horizon cosmic rays will also be included using a modified form of ZHAireS. In addition, the model of Askaryan emission used in `nicemc`

[1] https://github.com/PUEOCollaboration/pueoSim
[2] https://github.com/anitaNeutrino/anitaSim
[3] https://github.com/PUEOCollaboration/nicemc

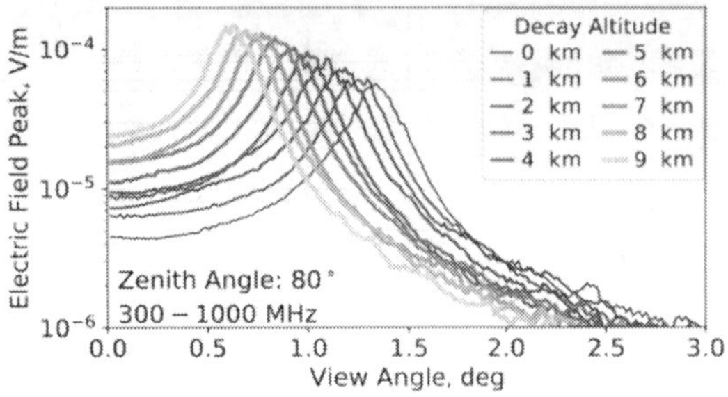


**Figure 6.** Peak electric field value for $\tau$-lepton induced air showers with an 80° zenith angle and varied decay altitudes generated by ZHAireS for use in the $\nu$SpaceSim framework [8].

has recently been updated, improving PUEO's sensitivity to neutrinos across nearly all energies, outperforming the expected curves shown in Figure 5. A full update of PUEO's science capability using `PUEOSim` is in progress.

### 3.2 $\nu$SpaceSim

$\nu$SpaceSim[4] is an open-source, modular, end-to-end simulation which models the signals produced by upwards moving extensive air showers (EAS) sourced by cosmic neutrinos [7]. $\nu$SpaceSim is designed to evaluate the sensitivity of an experiment with an arbitrary geometry and measuring technique (both optical Cherenkov and radio) to diffuse and transient neutrino fluxes.

$\nu$SpaceSim models radio emission generated by upgoing $\tau$-lepton initiated EAS via lookup tables generated by ZHAireS, taking into account geometric scaling. The radio emission lookup tables are integrated in the frequency domain to yield peak electric field values, which are used during runtime to provide simple thresholding-based acceptance arguments. Peak electric field values for an example upgoing EAS with different starting altitudes are shown in Figure 6 as a function of shower viewing angle, integrated withing the frequency range 300 MHz to 1000 MHz. Using this approach, $\nu$SpaceSim has been able to reproduce the ANITA neutrino sensitivity (via the $\tau$-lepton upgoing EAS observation channel).

To further improve upon sensitivity estimates for PUEO and other proposed neutrino telescopes utilizing the radio technique, $\nu$SpaceSim will be updated in the near future to consider full time domain waveforms of radio signals from EAS sourced by neutrinos and cosmic rays (both direct and ice-reflected). This improvement will allow for implementation of realistic detector properties, signal chains, and trigger mechanisms beyond simple thresholding arguments, resulting in more accurate estimations of neutrino sensitivity and more realistic comparisons between $\nu$SpaceSim and other computation schemes. For the calculation of transient neutrino sensitivity, $\nu$SpaceSim necessarily requires time based positioning of a given instrument, a feature which is currently under development. One potential future development to consider in $\nu$SpaceSim involves the inclusion of Askaryan emission from in-ice neutrino interactions when measuring over large ice sheets in Antarctica or Greenland.

[4] https://github.com/$\nu$SpaceSim/$\nu$SpaceSim

## 4 Summary

PUEO builds on the framework of the ANITA missions to target the flux of very high energy neutrinos through radio emission produced by in-ice neutrino interactions and upgoing EAS initiated by neutrino-sourced $\tau$-leptons. PUEO significantly improves upon the performance of ANITA by including i) more than double the number of antennas with respect to ANITA ii) an interferometric phased array trigger iii) a dedicated low-frequency instrument iv) a dedicated drop-down antenna ring dedicated to air shower measurements and v) better filtering of man-made noise at the trigger level. Over the course of a single 30 d flight, PUEO will either measure the first neutrino with energy exceeding 1 EeV or set strict limits on the diffuse flux, which helps to constrain models of cosmic ray composition and source evolution.

Comprehensive simulation efforts have been made to model PUEO's sensitivity to diffuse and transient fluxes of cosmic neutrinos. In this work, we have discussed two potential simulation packages that can be used to approach this problem: `PUEOSim`, the dedicated simulation for PUEO, based on the framework of the successful `ANITASim` package, and $\nu$SpaceSim, a generalized simulation package for both optical Cherenkov and radio experiments. `PUEOSim` uses the `nicemc` simulation package to model Askaryan emission from in-ice neutrino interactions, which forms the dominant observation channel for PUEO, while $\nu$SpaceSim uses the ZHAireS air shower simulation to model the geomagnetic emission from upgoing $\tau$-lepton initiated EAS. Both packages are making progress in including all candidate signal classes shown in Figure 2 and have overlap between one another, providing two robust methodologies for evaluating the impact of signals and backgrounds for the PUEO experiment.

## References

[1] The PUEO collaboration, Journal of Instrumentation **16**, P08035 (2021)
[2] P. Gorham et al., Physical Review Letters **117** (2016)
[3] P. Gorham et al., Physical Review Letters **121** (2018)
[4] P. Gorham et al., Physical Review Letters **126** (2021)
[5] L. Cremonesi et al., Journal of Instrumentation **14**, P08011 (2019)
[6] J. Alvarez-Muñiz, W.R. Carvalho, E. Zas, Astroparticle Physics **35**, 325 (2012)
[7] J. Krizmanic et al. (NuSpaceSim), PoS **ICRC2019**, 936 (2019)
[8] A. Romero-Wolf (NuSpaceSim), PoS **ICRC2021**, 1031 (2021)

# AUTHOR INDEX

# AUTHOR INDEX

# AUTHOR INDEX

**EDP Sciences – Web of Conferences**
17, Avenue du Hoggar
Parc d'Activité de Courtabouf
BP 112
F-91944 Les Ulis Cedex A
France

ISBN 978-1-7138-7337-2

CURRAN ASSOCIATES INC.
proceedings.com

9 781713 873372

# Fifteenth International Conference on Systems (ICONS 2020)

Held at NexComm 2020

Lisbon, Portugal
23 - 27 February 2020

IARIA